D. CARMICHAEL

PRINCIPAL SYMBOLS

j	unit vector (Y-axis) instantaneous current density (6, 7)		neutron (7) number of interest periods (8)
K	modulus of volume elasticity (2) equilibrium constant (4)	P	power (2, 4, 6) total load (2) present amount (8)
k	unit vector (Z-axis) radius of gyration (2) surface roughness (3) compressibility (4) thermal conductivity (4, 5) incremental interest rate (8)	p	pressure stress (3, 4, 5) number of poles (6) dipole electric moment (6) minimum acceptable return (8)
L	length inductance (6) thermal diffusion length (7) life of project (8)	Q	heat (2, 3, 4, 5) electric charge (6)
ℓ	equivalent length (2) mixing length (3) length of path (6)	q	volumetric flow rate (3) rate of heat flow (5) unit charge (6) project interest rate (8)
M	moment of force (2) molecular weight (3, 4, 5) mutual inductance (6)	R	resultant force (2) gas constant (4) thermal resistance (5) electric resistance (6) net receipt (8)
m	mass magnetic moment (6)		
N	normal force (2) number (3) number of turns (6) number density (7)	r	radius radial coordinate (3) roentgen (7) continuous interest rate (8)
n	frequency (2) concentration (3) number of moles (4)	$\mathscr{R}$	relucta...

THE ENGINEER'S COMPANION

THE ENGINEER'S COMPANION:

A CONCISE HANDBOOK OF ENGINEERING FUNDAMENTALS

MOTT SOUDERS

FORMERLY DIRECTOR OF OIL DEVELOPMENT
SHELL DEVELOPMENT COMPANY,
EMERYVILLE, CALIFORNIA

JOHN WILEY AND SONS, INC.
NEW YORK · LONDON · SYDNEY

5 6 7 8 9 10

Copyright © 1966 by John Wiley & Sons, Inc.

All Rights Reserved
This book or any part thereof
must not be reproduced in any form
without the written permission of the publisher.
ISBN 471 81395 8

Library of Congress Catalog Card Number: 65-26851
Printed in the United States of America

PREFACE

"We need more to be reminded than instructed"—Samuel Johnson

Engineers must deal quantitatively with a wide scope of scientific information, much of which is imperfectly recollected. Hence there is need for a handbook of engineering fundamentals that is comprehensive, concise, and portable—a book that can be held in one hand or carried in a briefcase, an engineer's companion. This book is intended to be a ready reference useful both to students in science and engineering and to practicing engineers whose formal education may be some years behind them. As such it is expected to remind the user of relationships already known to him but imperfectly recalled, rather than to serve as a textbook. Derivations of equations are therefore kept concise, and techniques and examples of application are omitted. For explicit derivations or deeper understanding the user is urged to consult the supplementary references given at the end of each section.

It is not expected that everything in this book will be accessible to every engineer. Nor is it likely that the specialist in any of the fields treated here will find the depth and scope adequate. Finally, this book is not intended to be a shortcut to design. The art of engineering includes much more than its scientific and mathematical bases. Science makes it known but engineering makes it work, in spite of imperfect knowledge.

Each section has been read by two or more competent reviewers, many of whose suggestions have been incorporated. Especially helpful reviews were provided by E. H. Heiberg, Glenn Murphy, Otto Redlich, W. M. Rohsenow, R. F. Shea, H. H. Skilling, C. V. Sternling, and R. M. Wilson. Specific credit is due R. W. Kunstman for help with transistors, J. F. McGarry for aid with tables, and Betty Souders for organization of the manuscript.

For Mathematics, Electricity and Magnetism, and Mathematical Tables extensive use was made of material from other handbooks published by John Wiley and Sons. Special credit is due J. L. and M. S. Barnes for the use of material adapted from their chapter on mathematics in O. W. Eshbach, *Handbook of Engineering Fundamentals*, second edition, John Wiley and Sons, New York, 1952.

MOTT SOUDERS

Piedmont, California
September 1965

CONTENTS

Section 1 Mathematics — 1

- Dimensions and Units — 1
- Mensuration — 6
- Analytic Geometry — 13
- Algebra — 28
- Trigonometry — 46
- Calculus — 58
- Differential Equations — 76
- Laplace Transformation — 82
- Complex Variable — 89
- Vector Analysis — 92
- Machine Computation — 97

Section 2 Mechanics — 99

- Basic Concepts — 100
- Statics — 100
- Kinematics — 111
- Dynamics — 114
- Mechanics of Materials — 119

Section 3 Fluid Mechanics — 145

- Definitions — 145
- Fluid Statics — 146
- Fluid Dynamics — 148

Section 4 Thermodynamics — 165

- Definitions — 165
- General Equations — 167
- Relations for Simple Systems — 169
- Thermodynamic Engines — 179
- Phase Equilibria — 185

CONTENTS

    Chemical Reactions   186
    Generalized Thermodynamic Properties   190

Section 5 Heat Transfer   194

    Conduction   195
    Convection   197
    Radiation   200

Section 6 Electricity and Magnetism   209

    General Relations   209
    Electrostatics   212
    Electromagnetism   216
    Direct-Current Circuits   224
    Sinusoidal Currents   227
    Transients   228
    Electrical Machines   230
    Electronics   236

Section 7 Nuclear Physics   253

    General Relations   253
    Nuclear Reactions   255
    Nuclear Reactors   260

Section 8 Engineering Economy   268

    Time Value of Money   268
    Economic Choice   271
    Interest Tables   274

Section 9 Mathematical and Physical Tables   286

    Tables   287
    Conversion Factors   399

    INDEX   415

THE ENGINEER'S COMPANION

SECTION 1

MATHEMATICS

DIMENSIONS AND UNITS

The statement of a physical quantity Q involves the two concepts of a number of standard units and the general class of the defined quantity that identifies the unit. Thus

$$Q = nU = n[U]$$

where n is the number of units of a standard size U, and $[U]$ is the *dimension* of the generalized physical quantity. Thus, all units of length (centimeters, feet, or light years) have the general dimension $[L]$. Each kind of physical quantity has a dimension, and there can be as many dimensions as there are kinds of physical quantities. For convenience in practice, however, only a few basic quantities are assigned dimensions, and all other quantities are expressed in dimensions derived from the basic ones. In this book the basic dimensions are mass $[m]$, length $[L]$, time $[t]$, temperature $[T]$, and electric charge $[Q]$. Hence all physical quantities are expressed dimensionally in terms of the five basic quantities,

$$[U] = [L]^a [m]^b [t]^c [T]^d [Q]^e$$

where the exponents are small positive or negative integers, or zero when the basic dimension is not involved in the unit.

The choice of basic quantities is a matter of convention and convenience. Any other set of basic quantities would be equally valid since none of the physical quantities are *a priori* fundamental in theory.

Dimensionally homogeneous equations are those in which all terms have the same dimensions. When such an equation is divided by one of its terms, all the terms of the resulting equation are dimensionless. Such equations can be used with any set of consistent units if the corresponding proportionality constants are included. All equations in this book are dimensionally homogeneous unless the units are specifically stated.

2 MATHEMATICS

Certain dimensionless groups of quantities occur time after time in technical equations. Since such groups yield the same number regardless of the system of consistent units used, they are usually called *dimensionless numbers*. Many of the widely used groups with the name and symbols associated with them are listed on page 194.

Most of the fundamental equations of physics are statements of proportionalities and involve a *proportionality constant*. Thus, in Newton's definition of force as a function of mass and acceleration,

$$Fg_c = ma$$

the proportionality constant, $1/g_c$, is a function of the units specified for the three physical quantities. Physicists like to choose units that make the proportionality constant equal to unity and have proposed many ingenious systems of units designed to suppress one or more proportionality constants. Engineers, however, prefer not to be so constrained since they are required to use, with equal facility, both scientific and practical units and to convert quantities from one system to another. The use of g_c in engineering equations removes confusion

Table 1-1 Dimensional Constant in Newton's Law, g_c

Force	Mass	Length	Time	g_c
poundal	pound	foot	second	1
pound	slug	foot	second	1
pound	pound	foot	second	32.17
pound	pound	mile	minute	21.94
dyne	gram	centimeter	second	1
gram	gram	centimeter	second	980.7
newton	kilogram	meter	second	1
kilogram	kilogram	meter	second	9.807
kilogram	kilogram	kilometer	minute	35.30

among systems of units and avoids the use of awkward units of mass such as the slug and the gram-seven. In general, g_c is likely to be needed in every engineering equation that contains both force and mass terms. Table 1-1 gives numerical values of g_c for the various units of force, mass, length, and time.

In the dimensional system chosen here, the proportionality constant $1/g_c$ is dimensionless; many proportionality constants, however, have dimensions including such fundamental physical constants as the gravitational constant k_g, Planck's constant h, the gas constant R_o, the velocity of light c_o, the absolute permeability of free space μ_o, and the absolute dielectric constant of free space ϵ_o.

Table 1-2 gives the principal physical quantities, their symbols and dimensions, with the systematic units in the English and Metric practical systems and the International MKSC system. The two groups of practical units (foot, pound, second, pound-force and meter, kilogram, second, kilogram-force) provide consistent systems of units for mechanics and thermodynamics (when the corresponding value of g_c is used), but not for electricity and magnetism. The

International MKSC system of units is the only recognized comprehensive system that provides a single set of consistent units for mechanics, thermodynamics, electrostatics, and electromagnetics. The MKSC system is based on the five units for the fundamental dimensions, the newton of force, and the "absolute" practical electrical units adopted January 1, 1948 by the International Committee on Weights and Measures. The MKSC system still contains an ambiguity; it permits two different definitions of the absolute permeability of free space which differ by a factor of 4π. Table 1-2 uses the so-called rationalized definition: $\mu_o = 4\pi \times 10^{-7}$ henry per meter so that the factor 4π appears only in the force equations and where required by spherical geometry.

Conversion of Units. A physical quantity Q is the product of a number n and the standard unit U in which the quantity is measured. For example, a viscosity (Q) equals $10^{-5}(n)$ lb/ft-sec (U). The quantity Q must be independent of the size of the units used; hence

$$Q = n_1 U_1 = n_2 U_2 = n_1(n_c U_2)$$

where n_c is the unit conversion factor, or the number of units of U_2 which are equal to one unit of U_1. In the example above the unit of viscosity is a compound unit so that the overall conversion factor is the product of the separate unit conversion factors, i.e.,

$$n_c = (n_{\text{mass}})(n_{\text{length}})^{-1}(n_{\text{time}})^{-1}$$

For example, to convert viscosity from centipoises to lb/ft-sec:

1 centipoise = 10^{-2} gm/cm-sec
$n_{\text{mass}} = 2.205 \times 10^{-3}$; $n_{\text{length}} = 3.281 \times 10^{-2}$; $n_{\text{time}} = 1$
$n_c = \dfrac{2.205 \times 10^{-3}}{3.281 \times 10^{-2}} = 6.72 \times 10^{-2}$
$n_1(n_c U_2) = 10^{-2}(6.72 \times 10^{-2})U_2$
or 1 centipoise = 6.72×10^{-4} lb/ft-sec

4 MATHEMATICS

Table 1-2 Quantities, Dimensions, and Units

Symbol	Quantity	Dimensions	English Practical	Systematic Units — Metric Practical	International MKSC
Fundamental Units:					
L	Length	(L)	foot (ft)	meter (m)	meter
m	Mass	(m)	pound (lb)	kilogram (kg)	kilogram
t	Time	(t)	second (sec)	second	second
T	Temperature	(T)	°F or °R	°C or °K	°C or °K
Q	Electric charge	(Q)	coulomb	coulomb	coulomb
Derived Units:					
F	Force	$(m)(L)(t)^{-2}$	pound force (lbf)	kilogram force (kgf)	newton
E	Energy	$(m)(L)^2(t)^{-2}$	ft lbf	m kgf	joule
P	Power	$(m)(L)^2(t)^{-3}$	ft lbf/sec	m kgf/sec	watt
p	Pressure, stress	$(m)(L)^{-1}(t)^{-2}$	lbf/ft²	kgf/m²	newton/m²
I	Moment of inertia	$(m)(L)^2$	lb ft²	kg m²	kg m²
μ	Viscosity	$(m)(L)^{-1}(t)^{-1}$	lb/ft-sec	kg/m sec	kg/m sec
S	Entropy	$(m)(L)^2(t)^{-2}(T)^{-1}$	ft lbf/°F	m kgf/°C	joule/°C
c	Heat capacity	$(L)^2(t)^{-2}(T)^{-1}$	ft lbf/lb °F	m kgf/kg °C	joule/kg °C
k	Thermal conductivity	$(m)(L)(t)^{-3}(T)^{-1}$	lbf/sec °F	kgf/sec °C	watt/m °C
q	Heat transfer rate	$(m)(L)^2(t)^{-3}$	ft lbf/sec	m kgf/sec	joule/sec
U	Heat transfer coef	$(m)(t)^{-3}(T)^{-1}$	lbf/ft-sec °F	kgf/m sec °C	watt/m² °C

DIMENSIONS AND UNITS 5

Table 1-2 (*Continued*)

Symbol	Quantity	Dimensions	English Practical	Metric Practical	International MKSC
I	Current	$(Q)(t)^{-1}$	ampere	ampere	ampere
R	Resistance	$(m)(Q)^{-2}(L)^2(t)^{-1}$	ohm	ohm	ohm
V	Potential, Voltage	$(m)(Q)^{-1}(L)^2(t)^{-2}$	volt	volt	volt
C	Capacitance	$(m)^{-1}(Q)^2(L)^{-2}(t)^2$			farad
L	Inductance	$(m)(Q)^{-2}(L)^2$			henry
ϕ	Magnetic flux	$(m)(Q)^{-1}(L)^2(t)^{-1}$			weber
ϵ	Dielectric const	$(m)^{-1}(Q)^2(L)^{-3}(t)^2$			farad/m
μ	Permeability	$(m)(Q)^{-2}(L)$			henry/m
E	Electric field strength	$(m)(Q)^{-1}(L)(t)^{-2}$			volt/m
B	Induction	$(m)(Q)^{-1}(t)^{-1}$			weber/m²
H	Magnetic intensity	$Q(L)^{-1}(t)^{-1}$			amp/m
Dimensional Constants:					
g_c	Factor, Newton's Law	none	32.17	9.807	1.0
k_g	Gravitational constant	$(m)^{-1}(L)^3(t)^{-2}$	1.0671×10^{-9} ft³/lb(sec)²	6.664×10^{-11} m³/kg(sec)²	6.664×10^{-11} m³/kg(sec)²
R_0	Gas constant (mole)	$(m)(L)^2(t)^{-2}(T)^{-1}$	1545 ft lbf/°F	.8478 m kgf/°K	8.314 joule/°K
c_0	Velocity of light	$(L)(t)^{-1}$	9.84×10^8 ft/sec	3×10^8 m/sec	3×10^8 m/sec
ϵ_0	Dielectric constant of free space	$(m)^{-1}(Q)^2(L)^{-3}(t)^2$			8.85×10^{-12} farad/m
μ_0	Permeability of free space	$(m)(Q)^{-2}(L)$			$4\pi \times 10^{-7}$ henry/m
h	Planck constant	$(m)(L)^2(t)^{-1}$			6.624×10^{-34} joule-sec

Systematic Units

6 MATHEMATICS

Notation. Lines, $a, b, c, \cdots$; angles, $\alpha, \beta, \gamma, \cdots$; altitude (perpendicular height), h; side, l; diagonals, $d, d_1, \cdots$; perimeter, p; radius of inscribed circle, r; radius of circumscribed circle, R; area, A.

1. Right Triangle

(One angle 90°)
$p = a + b + c;\ c^2 = a^2 + b^2;$
$$A = \frac{ab}{2} = \frac{a^2}{2} \tan \beta = \frac{c^2}{4} \sin 2\beta = \frac{c^2}{4} \sin 2\alpha.$$
For additional formulas, see *General Triangle* below, and also trigonometry.

2. General Triangle (and Equilateral Triangle)

For *General Triangle*:
$p = a + b + c.$ Let $s = 1/2\,(a + b + c).$
$$r = \frac{\sqrt{s(s-a)(s-b)(s-c)}}{s}\ ;\ R = \frac{a}{2 \sin \alpha} = \frac{abc}{4rs}\ ;$$
$$A = \frac{ah}{2} = \frac{ab}{2} \sin \gamma = \frac{b^2 \sin \gamma \sin \alpha}{2 \sin \beta} = rs = \frac{abc}{4R}.$$
Length of median to side $c = 1/2 \sqrt{2(a^2 + b^2) - c^2}.$
Length of bisector of angle $\gamma = \dfrac{\sqrt{ab[(a+b)^2 - c^2]}}{a+b}.$

For *Equilateral Triangle* ($a = b = c = l$ and $\alpha = \beta = \gamma = 60°$):
(Equal sides and equal angles)
$$p = 3l,\ r = \frac{l}{2\sqrt{3}}\ ;\ R = \frac{l}{\sqrt{3}} = 2r;$$
$$h = \frac{l\sqrt{3}}{2}\ ;\ l = \frac{2h}{\sqrt{3}}\ ;\ A = \frac{l^2\sqrt{3}}{4}.$$
For additional formulas, see trigonometry.

3. Rectangle (and Square)

For *Rectangle*:
$p = 2(a+b);\ d = \sqrt{a^2 + b^2};\ A = ab.$
For *Square* ($a = b = l$):
$$p = 4l;\ d = l\sqrt{2};\ l = \frac{d}{\sqrt{2}}\ ;\ A = l^2 = \frac{d^2}{2}.$$

4. General Parallelogram (and Rhombus)

For *General Parallelogram* (*Rhomboid*):
(Opposite sides parallel)
$p = 2(a+b);\ d_1 = \sqrt{a^2 + b^2 - 2ab \cos \gamma};$
$d_2 = \sqrt{a^2 + b^2 + 2ab \cos \gamma};\ d_1{}^2 + d_2{}^2 = 2(a^2 + b^2);$
$A = ah = ab \sin \gamma.$
For *Rhombus* ($a = b = l$):
(Opposite sides parallel and all sides equal)
$p = 4l;\ d_1 = 2l \sin \dfrac{\gamma}{2}\ ;\ d_2 = 2l \cos \dfrac{\gamma}{2}\ ;\ d_1{}^2 + d_2{}^2 = 4l^2;$
$d_1 d_2 = 2l^2 \sin \gamma;\ A = lh = l^2 \sin \gamma = \dfrac{d_1 d_2}{2}.$

5. General Trapezoid (and Isosceles Trapezoid)

Let mid-line bisecting non-parallel sides $= m$. Then $m = \dfrac{a+b}{2}.$

For *General Trapezoid*:
(Only one pair of opposite sides parallel)
$p = a + b + c + d;\ A = \dfrac{(a+b)h}{2} = mh.$

For *Isosceles Trapezoid* ($d = c$):
(Non-parallel sides equal)
$$A = \frac{(a+b)h}{2} = mh = \frac{(a+b)c \sin \gamma}{2}$$
$$= (a - c \cos \gamma)c \sin \gamma = (b + c \cos \gamma)c \sin \gamma.$$

MENSURATION

6. General Quadrilateral (Trapezium)

(No sides parallel)

$p = a + b + c + d$

$A = 1/2 \, d_1 d_2 \sin \alpha$ = sum of areas of the two triangles formed by either diagonal and the four sides.

7. Quadrilateral Inscribed in Circle

(Sum of opposite angles = 180°)

$ac + bd = d_1 d_2$.

Let $s = 1/2 \, (a + b + c + d) = \dfrac{p}{2}$ and α = angle between sides a and b.

$A = \sqrt{(s-a)(s-b)(s-c)(s-d)} = 1/2 \, (ab + cd) \sin \alpha$.

8. Regular Polygon (and General Polygon)

For *Regular Polygon:*

(Equal sides and equal angles)

Let n = number of sides.

Central angle $= 2\alpha = \dfrac{2\pi}{n}$ radians;

Vertex angle $= \beta = \dfrac{(n-2)}{n} \pi$ radians.

$p = ns; \quad s = 2r \tan \alpha = 2R \sin \alpha;$

$r = \dfrac{s}{2} \cot \alpha; \quad R = \dfrac{s}{2} \csc \alpha;$

$A = \dfrac{nsr}{2} = nr^2 \tan \alpha = \dfrac{nR^2}{2} \sin 2\alpha = \dfrac{ns^2}{4} \cot \alpha$ = sum of areas of the n equal triangles such as OAB.

For *General Polygon:*

A = sum of areas of constituent triangles into which it can be divided.

1b. Plane Curvilinear Figures

Notation. Lines, $a, b, \cdots$; radius, r; diameter, d; perimeter, p; circumference, c; central angle n radians, θ; arc, s; chord of arc s, l; chord of half arc $s/2, l'$; rise, h; area, A.

9. Circle (and Circular Arc)

For *Circle:*

$d = 2r; \quad c = 2\pi r = \pi d; \quad A = \pi r^2 = \dfrac{\pi d^2}{4} = \dfrac{c^2}{4\pi}$.

For *Circular Arc:*

Let arc $PAQ = s$; and chord $PA = l'$. Then, $s = r\theta = \dfrac{d\theta}{2}$; $s = \dfrac{8l' - l}{3}$. (The latter equation is Huygens' approximate formula. For θ small; error is very small; for $\theta = 120°$, error is about 0.25%; for $\theta = 180°$, error is less than 1.25%.)

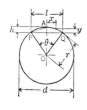

$l = 2r \sin \dfrac{\theta}{2}; \quad l = 2\sqrt{2hr - h^2}$ (approximate formula)

$r = \dfrac{s}{\theta} = \dfrac{l}{2 \sin \dfrac{\theta}{2}}; \quad r = \dfrac{4h^2 + l^2}{8h}$ (approximate formula)

$h = r \mp \sqrt{r^2 - \dfrac{l^2}{4}}$ ($-$ if $\theta \leq 180°$; $+$ if $\theta \geq 180°$) $= r \left(1 - \cos \dfrac{\theta}{2}\right)$

$= r \operatorname{versin} \dfrac{\theta}{2} = 2r \sin^2 \dfrac{\theta}{4} = \dfrac{l}{2} \tan \dfrac{\theta}{4} = r + y - \sqrt{r^2 - x^2}$.

Side ordinate $y = h - r + \sqrt{r^2 - x^2}$.

8 MATHEMATICS

10. Circular Sector (and Semicircle)

For *Circular Sector*:
$$A = \frac{\theta r^2}{2} = \frac{sr}{2}.$$

For *Semicircle*:
$$A = \frac{\pi r^2}{2}.$$

11. Circular Segment

$$A = \frac{r^2}{2}(\theta - \sin\theta)$$
$$= 1/2 \,[sr \mp l(r - h)](-\text{ if } h \leq r; +\text{ if } h \geq r).$$
$A = \frac{2lh}{3}$ or $\frac{h}{15}(8l' + 6l)$. (Approximate formulas. For h small compared with r, error is very small; for $h = \frac{r}{4}$, first formula errs about 3.5% and second less than 1.0%.)

12. Annulus

(Region between two concentric circles)
$$A = \pi(r_1{}^2 - r_2{}^2) = \pi(r_1 + r_2)(r_1 - r_2);$$
$$A \text{ of sector } ABCD = \frac{\theta}{2}(r_1{}^2 - r_2{}^2) = \frac{\theta}{2}(r_1 + r_2)(r_1 - r_2)$$
$$= \frac{t}{2}(s_1 + s_2).$$

13. Ellipse

$$p = \pi(a+b)\left(1 + \frac{R^2}{4} + \frac{R^4}{64} + \frac{R^4}{256} + \cdots\right) \text{ where } R = \frac{a-b}{a+b}.$$
$$p = \pi(a+b)\frac{64 - 3R^4}{64 - 16R^2} \text{ (approximate formula)}.$$
$A = \pi ab$; A of quadrant $AOB = \frac{\pi ab}{4}$;
A of sector $AOP = \frac{ab}{2}\cos^{-1}\frac{x}{a}$; A of sector $POB = \frac{ab}{2}\sin^{-1}\frac{x}{a}$;
A of section $BPP'B' = xy + ab\sin^{-1}\frac{x}{a}$;
A of segment $PAP'P = -xy + ab\cos^{-1}\frac{x}{a}$.
For additional formulas, see analytic geometry.

14. Parabola

Arc $BOC = s = 1/2\sqrt{l^2 + 16h^2} + \frac{l^2}{8h}\log_e\frac{4h + \sqrt{l^2 + 16h^2}}{l}$.

Let $R = \frac{h}{l}$. Then,
$$s = l\left(1 + \frac{8R^2}{3} - \frac{32R^4}{5} + \cdots\right) \text{ (approximate formula)}.$$
$d = \frac{h}{l^2}(l^2 - l_1{}^2); l_1 = l\sqrt{\frac{h-d}{h}}; h = \frac{dl^2}{l^2 - l_1{}^2};$
A of segment $BOC = \frac{2hl}{3}$;
A of section $ABCD = \frac{2}{3}d\left(\frac{l^3 - l_1{}^3}{l^2 - l_1{}^2}\right).$
For additional formulas, see analytic geometry.

15. Hyperbola

A of figure $OPAP'O = ab\log_e\left(\frac{x}{a} + \frac{y}{b}\right) = ab\cosh^{-1}\frac{x}{a}$;

A of segment $PAP' = xy - ab\log_e\left(\frac{x}{a} + \frac{y}{b}\right) = xy - ab\cosh^{-1}\frac{x}{a}.$
For additional formulas, see analytic geometry.

MENSURATION

16. Cycloid

Arc $OP = s = 4r\left(1 - \cos\dfrac{\phi}{2}\right)$; arc $OMN = 8r$;

A under curve $OMN = 3\pi r^2$.

For additional formulas, see analytic geometry.

17. Epicycloid

Arc $MP = s = \dfrac{4r}{R}(R + r)\left(1 - \cos\dfrac{R\phi}{2r}\right)$;

Area $MOP = A = \dfrac{r}{2R}(R + r)(R + 2r)\left(\dfrac{R\phi}{r} - \sin\dfrac{R\phi}{r}\right)$.

For additional formulas, see analytic geometry.

18. Hypocycloid

Arc $MP = s = \dfrac{4r}{R}(R - r)\left(1 - \cos\dfrac{R\phi}{2r}\right)$;

Area $MOP = A = \dfrac{r}{2R}(R - r)(R - 2r)\left(\dfrac{R\phi}{r} - \sin\dfrac{R\phi}{r}\right)$.

For additional formulas, see analytic geometry.

19. Catenary

If d is small compared with l:

Arc $MPN = s = l\left[1 + \dfrac{2}{3}\left(\dfrac{2d}{l}\right)^2\right]$ (approximately).

For additional formulas, see analytic geometry.

20. Helix (a skew curve)

Let length of helix = s; radius of coil (= radius of cylinder in figure) = r; distance advanced in one revolution = pitch = h; and number of revolutions = n. Then,

$s = n\sqrt{(2\pi r)^2 + h^2}$.

21. Spiral of Archimedes

Let $a = \dfrac{r}{\phi}$. Then,

Arc $OP = s = \dfrac{a}{2}[\phi\sqrt{1 + \phi^2} + \log_e(\phi + \sqrt{1 + \phi^2})]$.

For additional formulas, see analytic geometry.

22. Irregular Figure

Divide the figure into an *even* number, n, of strips by means of $(n + 1)$ ordinates, y_i, spaced equal distances, w. The area can then be determined approximately by any of the following formulas, which are presented in the order of usual increasing approach to accuracy. In any of the first three cases, the greater the number of strips used, the more nearly accurate will be the result.

(Approximate Formulas)

Trapezoidal Rule.......... $A = w\left[\dfrac{y_0 + y_n}{2} + y_1 + y_2 + \cdots + y_{n-1}\right]$;

Durand's Rule............ $A = w[0.4(y_0 + y_n) + 1.1(y_1 + y_{n-1}) + y_2 + y_3 + \cdots + y_{n-2}]$;

Simpson's Rule.......... $A = \dfrac{w}{3}[(y_0 + y_n) + 4(y_1 + y_3 + \cdots + y_{n-1}) +$
(n *must* be even)

$2(y_2 + y_4 + \cdots + y_{n-2})]$;

Weddle's Rule............ $A = \dfrac{3w}{10}[5(y_1 + y_5) + 6y_3 + y_0 + y_2 + y_4 + y_6]$.
(for 6 strips only)

Areas of irregular regions can often be determined more quickly by such methods as plotting on squared paper and counting the squares; graphical coordinate representation (see analytic geometry); or use of a planimeter.

Notation. Lines, $a, b, c, \cdots$; altitude (perpendicular height), h; slant height, s; perimeter of base, p_b or p_B; perimeter of a right section, p_r; area of base, A_b or A_B; area of a right section, A_r; total area of lateral surfaces, A_l; total area of all surfaces, A_t; volume, V.

23. Wedge (and Right Triangular Prism)

For Wedge:
(Narrow-side rectangular); $V = \dfrac{ab}{6}(2l_1 + l_2)$.

For Right Triangular Prism (or wedge having parallel triangular bases perpendicular to sides) $l_2 = l_1 = l$:
$$V = \dfrac{abl}{2}.$$

24. Rectangular Prism (or Rectangular Parallelepiped) (and Cube)

For Rectangular Prism or Rectangular Parallelepiped:
$A_l = 2c(a + b)$; $A_t = 2(de + ac + bc)$;
$V = A_r c = abc$.

For Cube (letting $b = c = a$):
$A_t = 6a^2$; $V = a^3$; Diagonal $= a\sqrt{3}$.

25. General Prism

$A_l = hp_b = sp_r = s(a + b + \cdots + n)$;
$V = hA_b = sA_r$.

26. General Truncated Prism (and Truncated Triangular Prism)

For General Truncated Prism:
$V = A_r \cdot$ (length of line BC joining centers of gravity of bases).

For Truncated Triangular Prism:
$$V = \dfrac{A_r}{3}(a + b + c).$$

27. Prismatoid

Let area of mid-section $= A_m$.
$$V = \dfrac{h}{6}(A_B + A_b + 4A_m).$$

28. Right Regular Pyramid (and Frustum of Right Regular Pyramid)

For Right Regular Pyramid:
$A_l = \dfrac{sp_B}{2}$; $V = \dfrac{hA_B}{3}$.

For Frustum of Right Regular Pyramid:
$A_l = \dfrac{s}{2}(p_B + p_b)$; $V = \dfrac{h}{3}(A_B + A_b + \sqrt{A_B A_b})$.

29. General Pyramid (and Frustum of Pyramid)

For General Pyramid:
$$V = \dfrac{hA_B}{3}.$$

For Frustum of General Pyramid:
$$V = \dfrac{h}{3}(A_B + A_b + \sqrt{A_B A_b}).$$

30. Regular Polyhedrons

Tetrahedron Cube Octahedron

Dodecahedron Icosahedron

Let edge $= a$, and radius of inscribed sphere $= r$. Then,
$r = \dfrac{3V}{A_t}$, and:

Number of Faces	Form of Faces	Total Area A_t	Volume V
4	Equilateral triangle	$1.7321a^2$	$0.1179a^3$
6	Square	$6.0000a^2$	$1.0000a^3$
8	Equilateral triangle	$3.4641a^2$	$0.4714a^3$
12	Regular pentagon	$20.6457a^2$	$7.6631a^3$
20	Equilateral triangle	$8.6603a^2$	$2.1817a^3$

(Factors shown only to four decimal places.)

MENSURATION 11

Notation. Lines, $a, b, c, \cdots$; altitude (perpendicular height), $h, h_1, \cdots$; slant height, s; radius, r; perimeter of base, p_b; perimeter of a right section, p_r; angle in radians, ϕ; arc, s; chord of segment, l; rise, h; area of base, A_b or A_B; area of a right section, A_r; total area of convex surface, A_l; total area of all surfaces, A_t; volume, V.

31. Right Circular Cylinder (and Truncated Right Circular Cylinder)	For Right Circular Cylinder: $A_l = 2\pi rh; \; A_t = 2\pi r(r + h);$ $V = \pi r^2 h.$ For Truncated Right Circular Cylinder: $A_l = \pi r(h_1 + h_2); \; A_t = \pi r \left[h_1 + h_2 + r + \sqrt{r^2 + \left(\frac{h_1 - h_2}{2}\right)^2} \right];$ $V = \frac{\pi r^2}{2} (h_1 + h_2).$
32. Ungula (Wedge) of Right Circular Cylinder	$A_l = \frac{2rh}{b} [a + (b - r)\phi];$ $V = \frac{h}{3b} [a(3r^2 - a^2) + 3r^2(b - r)\phi]$ $= \frac{hr^3}{b} \left[\sin \phi - \frac{\sin^3 \phi}{3} - \phi \cos \phi \right].$ For Semicircular Base (letting $a = b = r$): $A_l = 2rh; \; V = \frac{2r^2 h}{3}.$
33. General Cylinder	$A_l = p_b h = p_r s;$ $V = A_b h = A_r s.$
34. Right Circular Cone (and Frustum of Right Circular Cone)	For Right Circular Cone: $A_l = \pi r_B s = \pi r_B \sqrt{r_B^2 + h^2}; \; A_t = \pi r_B (r_B + s);$ $V = \frac{\pi r_B^2 h}{3}.$ For Frustum of Right Circular Cone: $s = \sqrt{h_1^2 + (r_B - r_b)^2}; \; A_l = \pi s(r_B + r_b);$ $V = \frac{\pi h_1}{3} (r_B^2 + r_b^2 + r_B r_b).$
35. General Cone (and Frustum of General Cone)	For General Cone: $V = \frac{A_B h}{3}.$ For Frustum of General Cone: $V = \frac{h_1}{3} (A_B + A_b + \sqrt{A_B A_b}).$
36. Sphere	Let diameter $= d.$ $A_t = 4\pi r^2 = \pi d^2;$ $V = \frac{4\pi r^3}{3} = \frac{\pi d^3}{6}.$
37. Spherical Sector (and Hemisphere)	For Spherical Sector: $A_t = \frac{\pi r}{2} (4h + l); \; V = \frac{2\pi r^2 h}{3}.$ For Hemisphere $\left(\text{letting } h = \frac{l}{2} = r\right):$ $A_t = 3\pi r^2; \; V = \frac{2\pi r^3}{3}.$

38. Spherical Zone (and Spherical Segment)

For Spherical Zone Bounded by Two Planes:
$$A_l = 2\pi r h; \quad A_t = \frac{\pi}{4}(8rh + a^2 + b^2).$$
For Spherical Zone Bounded by One Plane (b = 0):
$$A_l = 2\pi r h = \frac{\pi}{4}(4h^2 + a^2);$$
$$A_t = \frac{\pi}{4}(8rh + a^2) = \frac{\pi}{2}(2h^2 + a^2).$$
For Spherical Segment with Two Bases:
$$V = \frac{\pi h}{24}(3a^2 + 3b^2 + 4h^2).$$
For Spherical Segment with One Base (b = 0):
$$V = \frac{\pi h}{24}(3a^2 + 4h^2) = \pi h^2\left(r - \frac{h}{3}\right).$$

39. Spherical Polygon (and Spherical Triangle)

For Spherical Polygon:
Let sum of angles in radians $= \theta$ and number of sides $= n$.
$$A = [\theta - (n-2)\pi]r^2$$
(The quantity $[\theta - (n-2)\pi]$ is called "spherical excess.")
For Spherical Triangle (n = 3):
$$A = (\theta - \pi)r^2$$
For additional formulas, see trigonometry.

40. Torus

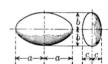

$$A_l = 4\pi^2 Rr;$$
$$V = 2\pi^2 Rr^2.$$

41. Ellipsoid (and Spheroids)

For Ellipsoid:
$$V = \frac{4}{3}\pi abc.$$
For Prolate Spheroid:
Let $c = b$ and $\dfrac{\sqrt{a^2 - b^2}}{a} = e.$
$$A_t = 2\pi b^2 + 2\pi ab\frac{\sin^{-1} e}{e}; \quad V = \frac{4}{3}\pi ab^2.$$
For Oblate Spheroid:
Let $c = a$ and $\dfrac{\sqrt{a^2 - b^2}}{a} = e.$
$$A_t = 2\pi a^2 + \frac{\pi b^2}{e}\ln\left(\frac{1+e}{1-e}\right); \quad V = \frac{4}{3}\pi a^2 b.$$

42. Paraboloid of Revolution

A_l of segment $DOC = \dfrac{2\pi l}{3h^2}\left[\left(\dfrac{l^2}{16} + h^2\right)^{3/2} - \left(\dfrac{l}{4}\right)^3\right].$
For Paraboloidal Segment with Two Bases:
$$V \text{ of } ABCD = \frac{\pi d}{8}(l^2 + l_1^2).$$
For Paraboloidal Segment with One Base ($l_1 = 0$ and $d = h$):
$$V \text{ of } DOC = \frac{\pi h l^2}{8}.$$

43. Hyperboloid of Revolution

$$V \text{ of segment } AOB = \frac{\pi h}{24}(l^2 + 4l_1^2).$$

ANALYTIC GEOMETRY

Equations of a Straight Line

General form	$Ax + By + C = 0$
Slope form	$y = mx + b$
Point-slope form	$y - y_1 = m(x - x_1)$
Intercept form	$\dfrac{x}{a} + \dfrac{y}{b} = 1$
Two-point form	$\dfrac{y - y_1}{x - x_1} = \dfrac{y_2 - y_1}{x_2 - x_1}$
Normal form (Fig. 1-1)	$x \cos \alpha + y \sin \alpha - p = 0$
Polar form	$r \sin \theta = p$

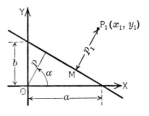

Fig. 1-1

Points and Lines

Distance p_1 between point P_1 and line (Fig. 1-1).

$$p_1 = \frac{Ax_1 + By_1 + C}{\pm(A^2 + B^2)^{1/2}}$$

Parallel lines	$\dfrac{A_1}{A_2} = \dfrac{B_1}{B_2}$ or $m_1 = m_2$
Perpendicular lines	$A_1 A_2 = B_1 B_2$ or $m_1 = -1/m_2$
Intersecting lines	$x_1 = \dfrac{B_2 C_1 - B_1 C_2}{A_2 B_1 - A_1 B_2}$, $y_1 = \dfrac{A_1 C_2 - A_2 C_1}{A_2 B_1 - A_1 B_2}$
Angle θ between lines	$\tan \theta = \dfrac{A_1 B_2 - A_2 B_1}{A_1 A_2 + B_1 B_2} = \dfrac{m_2 - m_1}{1 + m_1 m_2}$

Transformation of Coordinates

Change of origin 0 to $0'$ (Fig. 1-2):

$$x = x' + h \quad \text{and} \quad y = y' + k$$

Rotation of axes about the origin (Fig. 1-3):

$$x = x' \cos \theta - y' \sin \theta \quad \text{and} \quad y = x' \sin \theta + y' \cos \theta$$

Axes both translated and rotated:

$$x = x' \cos \theta - y' \sin \theta + h \quad \text{and} \quad y = x' \sin \theta + y' \cos \theta + k$$

Relations between rectangular and polar coordinates:

$$x = r \cos \theta, \quad y = r \sin \theta, \quad r = (x^2 + y^2)^{1/2}, \quad \theta = \tan^{-1} y/x$$

Conic Sections

A conic section is the locus of a point P moving so that the ratio (*eccentricity*) of its distances from a fixed point (*focus*) and from a fixed line (*directrix*) is constant. In Figs. 1-4, 1-5, and 1-6 the eccentricity is PF/PM. Table 1-3 gives some properties and equations of conics.

Table 1-3 Properties of Conics Having Principal Axis Parallel to OX

Curve	Circle	Ellipse Fig. 1-4	Hyperbola Fig. 1-5	Parabola Fig. 1-6
General equation with center (x_0, y_0)	$(x - x_0)^2 + (y - y_0)^2 = r^2$	$\dfrac{(x - x_0)^2}{a^2} + \dfrac{(y - y_0)^2}{b^2} = 1$	$\dfrac{(x - x_0)^2}{a^2} - \dfrac{(y - y_0)^2}{b^2} = 1$	$(y - y_0)^2 = 4a(x - x_0)$
Polar equation focus as pole	$r = a$	$r = \dfrac{a(1 - e^2)}{1 - e\cos\theta}$	$r = \dfrac{a(e^2 - 1)}{1 - e\cos\theta}$	$r = \dfrac{2a}{1 - \cos\theta}$
Parametric equation	$x = a\cos\theta$ $y = a\sin\theta$	$x = a\cos\theta$ $y = b\sin\theta$	$x = a\cosh\phi$ $y = a\sinh\phi$	$x = at^2$ $y = 2at$
Standard form origin at	Center	Center	Center	Vertex
Eccentricity, e	0	$\left(1 - \dfrac{b^2}{a^2}\right)^{1/2} < 1$	$\left(1 + \dfrac{b^2}{a^2}\right)^{1/2} > 1$	1
Standard equation	$x^2 + y^2 = r^2$	$\dfrac{x^2}{a^2} + \dfrac{y^2}{b^2} = 1$	$\dfrac{x^2}{a^2} - \dfrac{y^2}{b^2} = 1$	$y^2 = 4ax$
Foci	(0, 0)	$(ae, 0)$ $(-ae, 0)$	$(ae, 0)$ $(-ae, 0)$	$(a, 0)$
Equation of directrices	$x = 0$	$x = \dfrac{a}{e}$ $x = \dfrac{-a}{e}$	$x = \dfrac{a}{e}$ $x = \dfrac{-a}{e}$	$x = -a$
Length of latus rectum	$2r$	$\dfrac{2b^2}{a}$	$\dfrac{2b^2}{a}$	$4a$
Focal radius of a point (x_1, y) on the curve	r	$r_1 = a + x_1 e$ $r_2 = a - x_1 e$	$r_1 = a + x_1 e$ $r_2 = a + x_1 e$	$r = x_1 + a$

ANALYTIC GEOMETRY 15

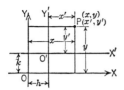

Fig. 1-2

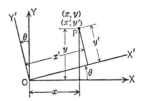

Fig. 1-3

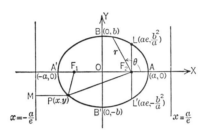

Fig. 1-4

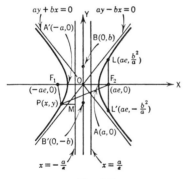

Fig. 1-5

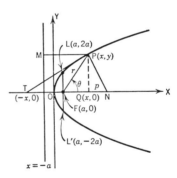

Fig. 1-6

Higher Plane Curves

Plane Curves. The point (x, y) describes a plane curve if x and y are continuous functions of a variable t (parameter), as $x = x(t)$, $y = y(t)$. The elimination of t from the two equations gives $F(x, y) = 0$, or in explicit form $y = f(x)$. The angle τ which a tangent to the curve makes with OX can be found from

$$\sin \tau = \frac{dy}{ds}, \quad \cos \tau = \frac{dx}{ds}, \quad \tan \tau = \frac{dy}{dx} = y'$$

where ds is the element of arc length:

$$ds = \sqrt{dx^2 + dy^2} = \sqrt{1 + y'^2}\, dx$$

In polar coordinates,

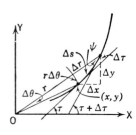

Fig. 1-7

$$ds = \sqrt{dr^2 + r^2\, d\theta^2} = \left[\left(\frac{dr}{d\theta}\right)^2 d\theta + r^2\right]^{1/2}$$

From Fig. 1-7, it may be seen that

$$\sin \psi = \frac{r\, d\theta}{ds}, \quad \cos \psi = \frac{dr}{ds}, \quad \tan \psi = \frac{r\, d\theta}{dr}$$

The equation of the *tangent* to the curve $y = f(x)$ at the point (x_1, y_1) is

$$y - y_1 = \left(\frac{dy}{dx}\right)_{x=x_1} (x - x_1)$$

The equation of the *normal* to the curve $y = f(x)$ at the point (x_1, y_1) is

$$y - y_1 = -\frac{1}{\left(\dfrac{dy}{dx}\right)_{x=x_1}} (x - x_1)$$

The *radius of curvature* of the curve at the point (x, y) is

$$\rho = \frac{ds}{d\tau} = \frac{\left[1 + \left(\dfrac{dy}{dx}\right)^2\right]^{3/2}}{\dfrac{d^2y}{dx^2}} = \frac{[1 + y'^2]^{3/2}}{y''}$$

The reciprocal $1/\rho$ is called the *curvature of the curve* at (x, y).

The coordinates (x_0, y_0) of the center of curvature for the point (x, y) on the curve (the center of the circle of curvature tangent to the curve at (x, y) and of radius ρ) are

$$\left.\begin{aligned}x_0 &= x - \rho \frac{dy}{ds} = x - y'\frac{(1 + y'^2)}{y''} \\ y_0 &= y + \rho \frac{dx}{ds} = y + \frac{(1 + y'^2)}{y''}\end{aligned}\right\}$$

A curve has a *singular point* if simultaneously

$$F(x, y) = 0, \quad \frac{\partial F}{\partial x} = 0, \quad \frac{\partial F}{\partial y} = 0$$

Let

$$D = \left(\frac{\partial^2 F}{\partial x\, \partial y}\right)^2 - \frac{\partial^2 F}{\partial x^2}\frac{\partial^2 F}{\partial y^2}$$

Then for $D > 0$, the curve has a *double point* with two real different tangents. For $D = 0$, the curve has a *cusp* with two coincident tangents.

For $D < 0$, the curve has an *isolated point* with no real tangent.

Semicubic, or Neil's, Parabola   **Logarithmic Curve**   **Exponential Curve**

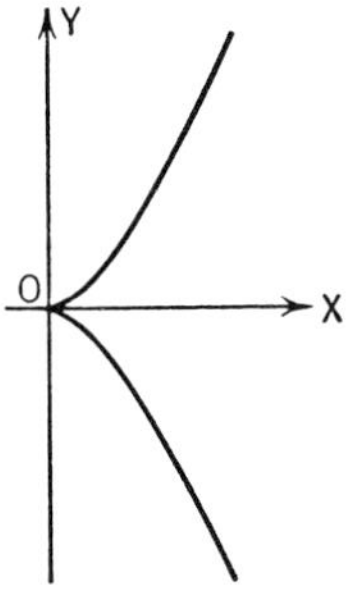

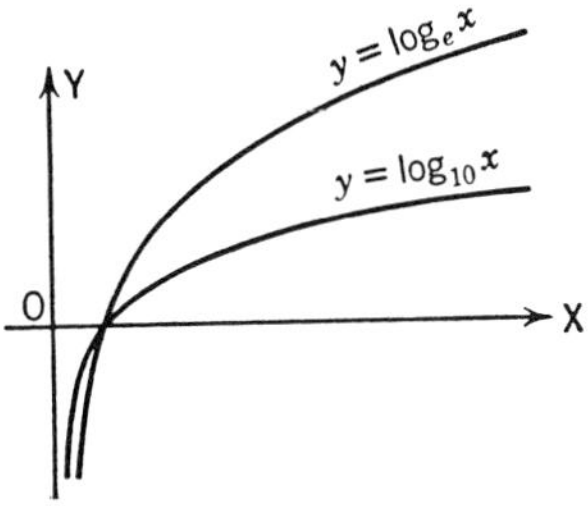

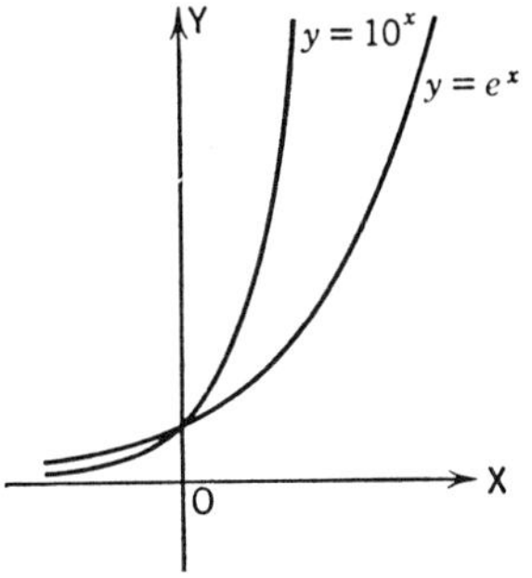

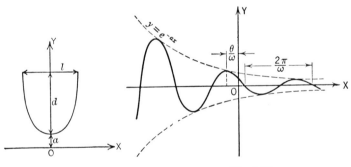

Fig. 1-8   Fig. 1-9   Fig. 1-10
$y^2 = ax^3$ $y = \log_b x$ $y = b^x$

Catenary   **Damped Wave**

Fig. 1-11   Fig. 1-12

$$y = \frac{a}{2}(e^{x/a} + e^{-x/a}) = a \cosh \frac{x}{a} \qquad y = e^{-ax} \cos(\omega x + \theta)$$

For l large compared with d,

$$s \approx l\left[1 + \frac{2}{3}\left(\frac{2d}{l}\right)^2\right]$$

Trochoid. A curve traced by a point at a distance b from the center of a circle of radius a as the circle rolls on a straight line.

$$x = a\phi - b\sin\phi, \quad y = a - b\cos\phi$$

Cycloid

$$a = b$$

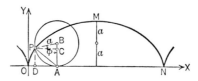

Fig. 1-13

$$x = a(\phi - \sin \phi), \quad y = a(1 - \cos \phi)$$

$$x = a \cos^{-1} \frac{a - y}{a} \pm \sqrt{(2a - y)y}$$

For one arch, arc length $= 8a$, area $= 3\pi a^2$

Prolate Cycloid   **Curtate Cycloid**
$a < b$ $a > b$

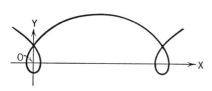

Fig. 1-14

Fig. 1-15

Hypotrochoid. A curve traced by a point at a distance b from the center of a circle of radius a as the circle rolls on the inside of a fixed circle of radius R.

$$x = (R - a) \cos \phi + b \cos \frac{R - a}{a} \phi, \quad y = (R - a) \sin \phi - b \sin \frac{R - a}{a} \phi$$

For a Hypocycloid, $b = a$.

Hypocycloid of Four Cusps, or Astroid

$$b = a = \tfrac{1}{4} R$$

$$x = R \cos^3 \phi, \quad y = R \sin^3 \phi$$

$$x^{2/3} + y^{2/3} = R^{2/3}$$

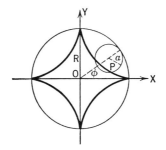

Fig. 1-16

Epitrochoid. A curve traced by a point at a distance b from the center of a circle of radius a as the circle rolls on the outside of a fixed circle of radius R.

$$x = (R + a)\cos\phi - b\cos\frac{R+a}{a}\phi, \quad y = (R + a)\sin\phi - b\sin\frac{R+a}{a}\phi$$

Epicycloid

$$b = a$$

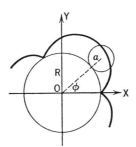

Fig. 1-17

Limaçon of Pascal

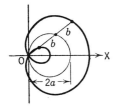   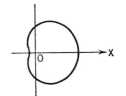

Fig. 1-18   Fig. 1-19

$b < 2a$ $b > 2a$

$$r = b + 2a\cos\theta$$

Other forms of the right-hand side of the equation, $b + 2a \sin \theta$, $b - 2a \cos \theta$, $b - 2a \sin \theta$, give curves rotated through 1, 2, 3 right angles, respectively.

Cardioid

Limaçon in which $b = 2a$

Epicycloid in which $R = a$

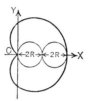

Fig. 1-20

$$r = 2a(1 + \cos \theta)$$

$$(x^2 + y^2 - 2ax)^2 = 4a^2(x^2 + y^2)$$

Involute of a Circle	Spiral of Archimedes
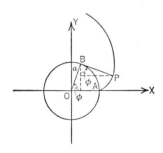	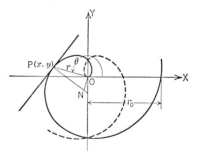
Fig. 1-21	Fig. 1-22
$x = a(\cos \phi + \phi \sin \phi)$	$r = a\theta$
$y = a(\sin \phi - \phi \cos \phi)$	Polar subnormal $ON = a$
$\theta = \sqrt{r^2/a^2 - 1} - \tan^{-1} \sqrt{r^2/a^2 - 1}$	Length of arc $OP = s = \frac{1}{2}a(\theta \sqrt{1 + \theta^2} + \sinh^{-1} \theta)$
Spiral traced by the end of a taut string unwinding from a circle.	For many turns, $s \approx \frac{1}{2}a\theta^2$

Hyperbolic, or Reciprocal, Spiral

Logarithmic, or Equiangular, Spiral

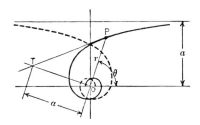

Fig. 1-23

$r\theta = a$

Polar subtangent $OT = -a$

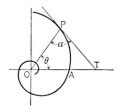

Fig. 1-24

$r = ae^{m\theta}, \quad m > 0$

or

$$\ln \frac{r}{a} = m\theta$$

As $\theta \to \infty, r \to 0$. The curve winds an indefinite number of times around the origin.

As $\theta \to 0, r \to \infty$. The curve has an asymptote parallel to the polar axis at a distance a.

The tangent to the curve at any point makes a constant angle α ($= \cot^{-1} m$) with the radius vector.

As $\theta \to -\infty, r \to 0$. The curve winds an indefinite number of times around the origin.

Lemniscate of Bernoulli

Three-Leaved Roses

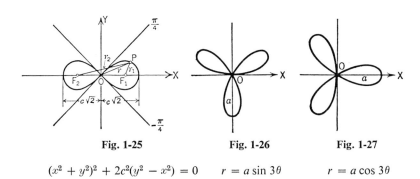

Fig. 1-25

$(x^2 + y^2)^2 + 2c^2(y^2 - x^2) = 0$

$r^2 = 2c^2 \cos 2\theta$

Fig. 1-26

$r = a \sin 3\theta$

Fig. 1-27

$r = a \cos 3\theta$

Locus of a point P, the product of whose distances from two fixed points F_1 and F_2 is equal to the square of half the distance between them, $r_1 \cdot r_2 = c^2$.

Four-Leaved Roses

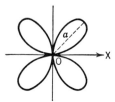

Fig. 1-28

$r = a \sin 2\theta$

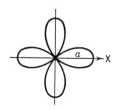

Fig. 1-29

$r = a \cos 2\theta$

The roses, $r = a \sin n\theta$ and $r = a \cos n\theta$, have, for n even, $2n$ leaves; for n odd, n leaves.

Cissoid of Diocles

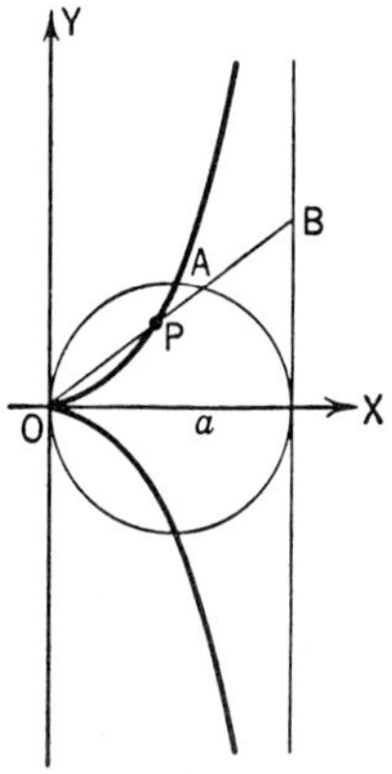

Strophoid

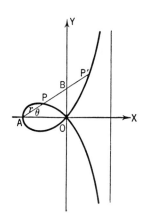

Fig. 1-30

Fig. 1-31

$$y^2 = \frac{x^3}{a - x}$$

$$y^2 = \frac{x^2(a + x)}{a - x}$$

$r = a(\sec \theta - \cos \theta)$

$r = a(\sec \theta - \tan \theta)$

Locus of point P such that $OP = AB$.

If the line AB rotates about A, intersecting the y axis at B, and if $PB = BP' = OB$, the locus of P and P' is the strophoid.

Conchoid of Nicomedes

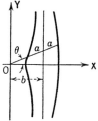

Fig. 1-32

$(x^2 + y^2)(x - b)^2 = a^2 x^2$

$r = b \sec \theta - a$

Witch of Agnesi

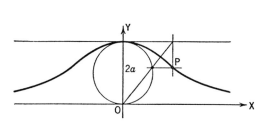

Fig. 1-33

$y = \dfrac{8a^3}{x^2 + 4a^2}$

$x = 2a \tan \phi$

$y = 2a \cos^2 \phi$

Folium of Descartes

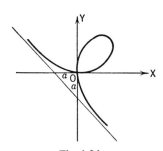

Fig. 1-34

$x^3 + x^3 - 3axy = 0$

$r = \dfrac{3a \sin \theta \cos \theta}{\sin^3 \theta + \cos^3 \theta}$

Tractrix

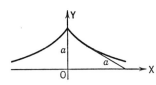

Fig. 1-35

$x = a \cosh^{-1} \dfrac{a}{y} - \sqrt{a^2 - y^2}$

Locus of one end P of tangent line of length a as the other end Q is moved along the x axis.

Circles in Polar Coordinates

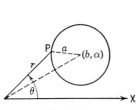

Fig. 1-36

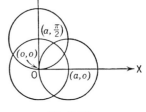

Fig. 1-37

$r^2 + b^2 - 2rb \cos(\theta - \alpha) = a^2$

Center at (b, α), radius a

Center $(0, 0)$ $r = a$

Center $(a, 0)$ $r = 2a \cos \theta$

Center $\left(a, \dfrac{\pi}{2}\right)$ $r = 2a \sin \theta$

Frequency-Modulated Wave

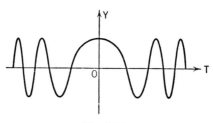

Fig. 1-38

$y = k \cos [\phi(t)]$, instantaneous frequency $= \Omega(t) = \dfrac{d\phi}{dt}$

In Fig. 1-38, $y = \cos \pi/2\ t^2$, $\Omega(t) = \pi t$.

Solid Analytic Geometry

Coordinate Systems

Right-Hand Rectangular (Fig. 1-39). The position of a point $P(x, y, z)$ is fixed by its distances x, y, z from the mutually perpendicular planes yz, xz, and xy, respectively.

Spherical, or Polar (Fig. 1-40). The position of a point $P(r, \theta, \phi)$ is fixed by its distance from a given point O, the origin, and its direction from O, determined by the angles θ and ϕ.

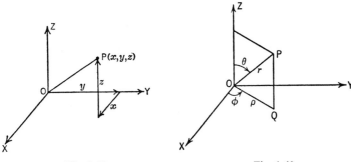

Fig. 1-39   Fig. 1-40

Cylindrical (Fig. 1-40). The position of a point $P(\rho, \phi, z)$ is fixed by its distance z from a given plane and the polar coordinates (ρ, ϕ) of the projection Q of P on the given plane.

Relations among coordinates of the three systems.

$$x = r \sin \theta \cos \phi = \rho \cos \phi$$
$$y = r \sin \theta \sin \phi = \rho \sin \phi$$
$$z = r \cos \theta$$
$$\rho = \sqrt{x^2 + y^2} = r \sin \theta$$
$$\phi = \tan^{-1} \frac{y}{x}$$
$$r = \sqrt{x^2 + y^2 + z^2} = \sqrt{\rho^2 + z^2}$$
$$\theta = \tan^{-1} \frac{\sqrt{x^2 + y^2}}{z} = \tan^{-1} \frac{\rho}{z}$$

Point, Line, and Plane

The Distance between Two Points $P_1(x_1, y_1, z_1)$ and $P_2(x_2, y_2, z_2)$ is

$$s = \sqrt{(x_2 - x_1)^2 + (y_2 - y_1)^2 + (z_2 - z_1)^2}$$

The Angles α, β, γ which the line P_1P_2 makes with the coordinate directions x, y, z, respectively, are the *direction angles* of P_1P_2. The cosines

$$\cos \alpha = \frac{x_2 - x_1}{s}, \quad \cos \beta = \frac{y_2 - y_1}{s}, \quad \cos \gamma = \frac{z_2 - z_1}{s}$$

are the *direction cosines* of P_1P_2, and

$$\cos^2 \alpha + \cos^2 \beta + \cos^2 \gamma = 1$$

If $l:m:n = \cos \alpha : \cos \beta : \cos \gamma$, then

$$\cos \alpha = \frac{l}{\sqrt{l^2 + m^2 + n^2}} \quad \cos \beta = \frac{m}{\sqrt{l^2 + m^2 + n^2}} \quad \cos \gamma = \frac{n}{\sqrt{l^2 + m^2 + n^2}}$$

The Angle θ between Two Lines in terms of their direction angles $\alpha_1, \beta_1, \gamma_1$ and $\alpha_2, \beta_2, \gamma_2$ is obtained from

$$\cos \theta = \cos \alpha_1 \cos \alpha_2 + \cos \beta_1 \cos \beta_2 + \cos \gamma_1 \cos \gamma_2$$

If $\cos \theta = 0$, the lines are perpendicular to each other.

Equation of a Plane
General form $Ax + By + Cz + D = 0$
Normal form $x \cos \alpha + y \cos \beta + z \cos \gamma - p = 0$
Intercept form $\dfrac{x}{a} + \dfrac{y}{b} + \dfrac{z}{c} = 1$
Parallel planes $A_1:B_1:C_1 = A_2:B_2:C_2$

Distance between Point P and Plane

$$= \frac{Ax_1 + By_1 + Cz_1 + D}{\pm(A^2 + B^2 + C^2)^{1/2}}$$

where the sign of the denominator is opposite to that of D.

The Angle θ between Two Planes is given by

$$\cos \theta = \frac{A_1A_2 + B_1B_2 + C_1C_2}{\pm \sqrt{(A_1^2 + B_1^2 + C_1^2)(A_2^2 + B_2^2C_2^2)}}$$

The two planes are perpendicular if $A_1A_2 + B_1B_2 + C_1C_2 = 0$

Transformation of Coordinates

Changing the Origin. Let the coordinates of a point P with respect to the original axes be x, y, z and with respect to the new axes x', y', z'. For a parallel displacement of the axes with x_0, y_0, z_0 the coordinates of the new origin

$$x = x_0 + x', \quad y = y_0 + y', \quad z = z_0 + z'$$

Rotation of the Axes about the Orgin. Let the cosines of the angles of the new axes x', y', z', with the x axis be λ_1, μ_1, ν_1, with the y axis be λ_2, μ_2, ν_2, with the z axis be λ_3, μ_3, ν_3. Then

$$\begin{aligned} x &= \lambda_1 x' + \mu_1 y' + \nu_1 z' \\ y &= \lambda_2 x' + \mu_2 y' + \nu_2 z' \\ z &= \lambda_3 x' + \mu_3 y' + \nu_3 z' \end{aligned} \qquad \left. \begin{aligned} x' &= \lambda_1 x + \lambda_2 y + \lambda_3 z \\ y' &= \mu_1 x + \mu_2 y + \mu_3 z \\ z' &= \nu_1 x + \nu_2 y + \nu_3 z \end{aligned} \right\}$$

ANALYTIC GEOMETRY

The following relations exist:

(1) $\lambda_1^2 + \mu_1^2 + \nu_1^2 = 1$
$\lambda_2^2 + \mu_2^2 + \nu_2^2 = 1$
$\lambda_3^2 + \mu_3^2 + \nu_3^2 = 1$

(2) $\lambda_1^2 + \lambda_2^2 + \lambda_3^2 = 1$
$\mu_1^2 + \mu_2^2 + \mu_3^2 = 1$
$\nu_1^2 + \nu_2^2 + \nu_3^2 = 1$

(3) $\lambda_1 \lambda_2 + \mu_1 \mu_2 + \nu_1 \nu_2 = 0$
$\lambda_2 \lambda_3 + \mu_2 \mu_3 + \nu_2 \nu_3 = 0$
$\lambda_3 \lambda_1 + \mu_3 \mu_1 + \nu_3 \nu_1 = 0$

(4) $\lambda_1 \mu_1 + \lambda_2 \mu_2 + \lambda_3 \mu_3 = 0$
$\mu_1 \nu_1 + \mu_2 \nu_2 + \mu_3 \nu_3 = 0$
$\nu_1 \lambda_1 + \nu_2 \lambda_2 + \nu_3 \lambda_3 = 0$

(5) $\lambda_1 = \mu_2 \nu_3 - \nu_2 \mu_3$
$\mu_1 = \nu_2 \lambda_3 - \lambda_2 \nu_3$
$\nu_1 = \lambda_2 \mu_3 - \mu_2 \lambda_3$

(6) $\lambda_2 = \nu_1 \mu_3 - \mu_1 \nu_3$
$\mu_2 = \lambda_1 \nu_3 - \nu_1 \lambda_3$
$\nu_2 = \mu_1 \lambda_3 - \lambda_1 \mu_3$

(7) $\lambda_3 = \mu_1 \nu_2 - \nu_1 \mu_2$
$\mu_3 = \nu_1 \lambda_2 - \lambda_1 \nu_2$
$\nu_3 = \lambda_1 \mu_2 - \mu_1 \lambda_2$

(8) $\begin{vmatrix} \lambda_1 & \mu_1 & \nu_1 \\ \lambda_2 & \mu_2 & \nu_2 \\ \lambda_3 & \mu_3 & \nu_3 \end{vmatrix} = 1$

For a combination of displacement and rotation, apply the corresponding equations simultaneously.

Quadric Surfaces

The surface defined by an equation of the second degree in x, y, and z is a quadric surface.

Sphere. The equation

$$x^2 + y^2 + z^2 + zx + by + cz + d = 0$$

represents a sphere with radius $r = \frac{1}{2}(a^2 + b^2 + c^2 - 4d)^{\frac{1}{2}}$ and center $x_0 = -a/2$, $y_0 = -b/2$, $z_0 = -c/2$. If the origin is at the center, $x^2 + y^2 + z^2 = r^2$.

Cylinder. The equation of a cylinder normal to one of the coordinate planes is the same as the equation of a section of the cylinder in the same plane. See p. 11.

Cone with Vertex at the Origin

Right circular $\quad x^2 + y^2 - c^2 z^0 = 0$

Elliptic $\quad \dfrac{x^2}{a^2} + \dfrac{y^2}{b^2} - \dfrac{z^2}{c^2} = 0$

Ellipsoids (Fig. 1-41)

$$\frac{x^2}{a^2} + \frac{y^2}{b^2} + \frac{z^2}{c^2} = 1$$

When $b = c$ the surface is an oblate spheroid if $a < b$ and a prolate spheroid if $a > b$.

Hyperboloids

One-sheet (Fig. 1-42) $\quad \dfrac{x^2}{a^2} + \dfrac{y^2}{b^2} - \dfrac{z^2}{c^2} = 1$

Two-sheet (Fig. 1-43) $\quad \dfrac{x^2}{a^2} + \dfrac{y^2}{b^2} - \dfrac{z^2}{c^2} = -1$

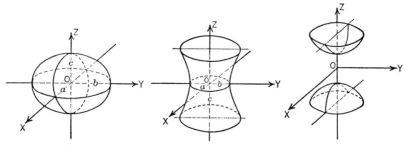

Fig. 1-41   Fig. 1-42   Fig. 1-43

Paraboloids

Elliptic (Fig. 1-44) $\dfrac{x^2}{a^2} + \dfrac{y^2}{b^2} = 2cz$

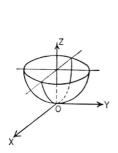

Fig. 1-44

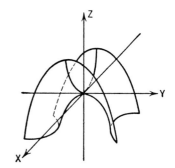

Fig. 1-45

Hyperbolic (Fig. 1-45) $\dfrac{x^2}{a^2} - \dfrac{y^2}{b^2} = 2cz$

Paraboloid of revolution $x^2 + y^2 = 2cz$

ALGEBRA

Logarithms

If $N = b^x$, then x is the *logarithm* of the number N to the *base b*; or $\log_b N = x$

$$\log_b b = 1; \quad \log_b 1 = 0; \quad \log_b 0 = -\infty \quad (1 < b < \infty)$$

$$\log_b b^N = N; \quad b^{\log_b N} = N; \quad \log_b N = \frac{\log_a N}{\log_a b}$$

$$\log xy = \log x + \log y; \quad \log \frac{x}{y} = \log x - \log y$$

$$\log x^n = n \log x; \quad \log (x^n)^{1/p} = \frac{n}{p} \log x$$

Common Logarithms

Common or *Briggsian* logarithms (abbreviated $\log_{10}$ or log) have the base 10. The integral part of a common logarithm is the *characteristic*, and the decimal part is the *mantissa*. Any finite positive number N may be expressed as $N = M(10^n)$ where M lies between 1 and 10. The characteristic, positive or negative, of $\log_{10} N$ is the exponent n and the mantissa, always positive, is the $\log_{10} M$ found in tables.

Natural Logarithms

Natural or *Naperian* logarithms (abbreviated $\log_e$ or ln) have the base $e = 2.71828$. Since $N = M(10^n)$, $\ln N = \ln M + n \ln 10$; $\ln 10 = 2.302585$.

$$\ln N = \frac{\log_{10} N}{\log_{10} e} = 2.302585 \log_{10} N$$

$$\log_{10} N = 0.434294 \ln N$$

Theory of the Slide Rule

The slide rule is simply a mechanical device for adding or subtracting lengths. When the lengths are scaled in proportion to the logarithms of numbers it is the logarithms that are added and subtracted. Thus,

$$\text{length } a - \text{length } c + \text{length } b = \log a - \log c + \log b = \log \frac{ab}{c}$$

Common slide rule scales are:

A and B, proportioned to $\frac{1}{2} \log N$
C and D, proportioned to $\log N$
K, proportioned to $\frac{1}{3} \log N$
S, proportioned to $\log \sin$
T, proportioned to $\log \tan$
LL, proportioned to $\log \log N$

Identities

Powers

$$a^{1/n} = \sqrt[n]{a}$$
$$a^m a^n = a^{m+n}$$
$$a^m a^{-n} = a^{m-n}$$

$$1/a^n = a^{-n}$$
$$(ab)^n = a^n b^n$$
$$(a^m)^n = a^{mn}$$

Fractions and Proportions

$$\frac{a}{c} \pm \frac{b}{d} = \frac{ad \pm bc}{cd},$$

$$\frac{a}{c} \pm \frac{a}{d} = \frac{a(d \pm c)}{cd}$$

If $\dfrac{a}{b} = \dfrac{c}{d}$, then $\dfrac{c+d}{d} = \dfrac{a+b}{b}$ and $\dfrac{a-b}{a+b} = \dfrac{c-d}{c+d}$

Factors

$a^2 - b^2 = (a - b)(a + b)$, $a^2 + b^2 = (a + b\sqrt{-1})(a - b\sqrt{-1})$
$a^3 + b^3 = (a \mp b)(a^2 \pm ab + b^2)$
$a^4 + b^4 = (a^2 + ab\sqrt{2} + b^2)(a^2 - ab\sqrt{2} + b^2)$
$a^n - b^n = (a - b)(a^{n-1} + a^{n-2}b + \cdots + b^{n-1})$
$a^n + b^n = (a + b)(a^{n-1} - a^{n-2}b + \cdots + b^{n-1})$ n is odd

Factorial

$n! = 1 \cdot 2 \cdot 3 \cdots n = e^{-n} n^n (2\pi n)^{1/2}$ approximately

Expansions

$(a + b + c)^2 = a^2 + b^2 + c^2 + 2ab + 2ac + 2bc$
$(a + b + c)^3 = a^3 + b^3 + c^3 + 3a^2(b + c) + 3b^2(a + c)$
$\hspace{5cm} + 3c^2(a + b) + 60bc$
$(a + b + c + d + \cdots)^2 = a^2 + b^2 + c^2 + d^2 + \cdots + 2a(b + c + d + \cdots)$
$\hspace{3cm} + 2b(c + d + \cdots) + 2c(d + \cdots) + \cdots$

Permutations and Combinations

The number P of arrangements or *permutations* of n objects taken t at a time is

$$P = n(n - 1)(n - 2) \cdots (n - t + 1) = \frac{n!}{(n - t)!}$$

If $t = 1$, $P = n$ and   if $t = n$, $P = n!$
The number C of unarranged selections or *combinations* of n objects taken t at a time is

$$C = \frac{P}{t!} = \frac{n!}{t!(n - t)!}$$

If $t = 1$, $C = n$ and if $t = n$, $C = 1$
Combinations taken 1 to n at a time, $C = 2^n - 1$

Probability

If, in a set M of m events which are mutually exclusive and equally likely, one event will occur, and if in the set M there is a subset N of n events ($n \leq m$), then the *a priori probability* p that the event which will occur is one of the subset N is n/m. The probability q that the event which will occur does not belong to N is $1 - n/m$.

Example. If the probability of drawing one of the 4 aces from a deck of 52 cards is to be found, then $m = 52$, $n = 4$, and $p = \frac{4}{52} = \frac{1}{13}$. The probability of drawing a card that is not an ace is $q = 1 - \frac{1}{13} = \frac{12}{13}$.

If, out of a large number r of observations in which a given event might or might not occur, the event has occurred s times, then a useful approximate value of the *experimental*, or *a posteriori, probability* of the occurrence of the event under the same conditions is s/r.

ALGEBRA   31

Example. From the American Experience Mortality Table, out of 100,000 persons living at age 10 years, 749 died within a year. Here $r = 100,000$, $s = 749$, and the probability that a person of age 10 will die within a year is $749/100,000$.

If p is the probability of receiving an amount A, then the *expectation* is pA.

Addition Rule (either or). The probability that any one of several mutually exclusive events will occur is the sum of their separate probabilities.

Example. The probability of drawing an ace from a deck of cards is $\frac{1}{13}$, and the probability of drawing a king is the same. Then the probability of drawing either an ace or a king is $\frac{1}{13} + \frac{1}{13} = \frac{2}{13}$.

Multiplication Rule (both and). (*a*) The probability that two (or more) independent events will both (or all) occur is the product of their separate probabilities.

(*b*) If p_1 is the probability that an event will occur, and if, after it has occurred, p_2 is the probability that another event will occur, then the probability that both will occur in the given order is $p_1 p_2$. This rule can be extended to more than 2 events.

Example. (*a*) The probability of drawing an ace from a deck of cards is $\frac{1}{13}$, and the probability of drawing a king from another deck is $\frac{1}{13}$. Then the probability that an ace will be drawn from the first deck and a king from the second is $\frac{1}{13} \cdot \frac{1}{13} = \frac{1}{169}$. (*b*) After an ace has been drawn from a deck of cards, the probability of drawing a king is $\frac{4}{51}$. If two cards are drawn in succession without the first being replaced, the probability that the first is an ace and the second a king is $\frac{1}{13} \cdot \frac{4}{51} = \frac{4}{663}$.

Repeated Trials. If p is the probability that an event will occur in a single trial, then the probability that it will occur exactly s times in r trials is the *binomial* or *Bernoulli* distribution function

$$_rC_s p^s (1-p)^{r-s} \quad \text{where} \quad _rC_s = \frac{r!}{s!(r-s)!}$$

The probability that it will occur at least s times is

$$p^r + {}_rC_{r-1} p^{r-1}(1-p) + {}_rC_{r-2} p^{r-2}(1-p)^2 + \cdots + {}_rC_s p^s (1-p)^{r-s}$$

Example. If five cards are drawn, one from each of five decks, the probability that exactly three will be aces is ${}_5C_3(\frac{1}{13})^3(\frac{12}{13})^2$. The probability that at least three will be aces is $(\frac{1}{13})^5 + {}_5C_4(\frac{1}{13})^4(\frac{12}{13}) + {}_5C_3(\frac{1}{13})^3(\frac{12}{13})^2$.

Equations

Quadratic Equation

$$f(x) = ax^2 + bx + c = 0$$

The two roots, both real or both complex are given by,

$$x_1, x_2 = \frac{-b \pm \sqrt{b^2 - 4ac}}{2a} = \frac{2c}{-b \mp \sqrt{b^2 - 4ac}}$$

Cubic Equation

$$f(x) = x^3 + ax^2 + bx + c = 0$$

By substituting $(y - a/3)$ for x, $f(x)$ reduces to

$$y^3 + dy + e = 0$$

where $d = \tfrac{1}{3}(3b - a^2)$ and $e = \tfrac{1}{27}(2a^3 - 9ab + 27c)$.

Let $A = \left(-\dfrac{e}{2} + \sqrt{\dfrac{e^2}{4} + \dfrac{d^3}{27}}\right)^{\!1/3}$ and $B = \left(-\dfrac{e}{2} - \sqrt{\dfrac{e^2}{4} + \dfrac{d^3}{27}}\right)^{\!1/3}$

Then the three values of y are

$$A + B, \quad -\dfrac{A+B}{2} + \dfrac{(A-B)(-3)^{1/2}}{2} \quad \text{and} \quad -\dfrac{A+B}{2} - \dfrac{(A-B)(-3)^{1/2}}{2}$$

When $(e^2/4 + d^2/27) < 0$ the trigonometric solution avoids finding the cube roots of complex quantities. Evaluate the angle θ from

$$\cos\theta = -\dfrac{e}{2} \bigg/ \left(-\dfrac{d^3}{27}\right)^{\!1/2}$$

Then the values of y will be

$$2\left(-\dfrac{d}{3}\right)^{\!1/2}\cos\dfrac{\theta}{3}, \quad 2\left(-\dfrac{d}{3}\right)^{\!1/2}\cos\left(\dfrac{\theta}{3} + 120°\right) \quad \text{and} \quad 2\left(-\dfrac{d}{3}\right)^{\!1/2}\cos\left(\dfrac{\theta}{3} + 240°\right)$$

Equations of Higher Degree

$$f(x) = a_0 x^n + a_1 x^{n-1} + a_2 x^{n-2} + \cdots a_{n-1} x + a_n = 0$$

For all values of x when n is an integer and $a_0 = 0$

$$a_0 x^n + a_1 x^{n-1} + \cdots a_{n-1} x + a_n = a_0(x - r_1)(x - r_2) \cdots (x - r_n) = 0$$

and the n roots, some of which may be identical, are the r_i. If n is odd there is at least one real root.

Relations between Roots and Coefficients

$$r_1 + r_2 + \cdots + r_n = -\dfrac{a_1}{a_0}$$

$$r_1 r_2 + r_1 r_3 + \cdots + r_{n-1} r_n = \dfrac{a_2}{a_0}$$

$$\cdots\cdots\cdots\cdots\cdots\cdots$$

$$r_1 r_2 r_3 \cdots r_n = (-1)^n \dfrac{a_n}{a_0}$$

If $f(x)$ is divided by $(x - r)$ until a numerical remainder is obtained, this remainder is the value of $f(r) = f(x)$ when $x = r$. If $(x - r)$ is a factor of $f(x)$, then $f(r) = 0$ and $x = r$ is a root of $f(x)$.

ALGEBRA

Solution of $f(x) = 0$ may be by successive approximation, synthetic division, or the methods of Horner, Newton, and Graeffe (cf. MacDuffee, *Theory of Equations* John Wiley and Sons, 1954.)

Series

Progressions

a = first term, f = last term, d = common difference,
r = common ratio, n = number of terms

Arithmetic

$$f = a + (n-1)d, \quad \sum(n) = \frac{n(a+f)}{2} = \frac{n}{2}\{2a + (n-1)d\}$$

Arithmetic mean is $\dfrac{a+b}{2}$

Geometric

$$f = ar^{n-1}, \quad \sum(n) = \frac{a(r^n - 1)}{r - 1} = \frac{fr - a}{r - 1}$$

When n is infinity and $r^2 < 1$, $\sum(n) = \dfrac{a}{1 - r}$

Geometric mean is $(ab)^{1/2}$

Exponential

$$\sum(n^2) = 1^2 + 2^2 + \cdots + n^2 = \frac{n(n+1)(2n+1)}{6}$$

$$\sum(n^3) = 1^3 + 2^3 + \cdots + n^3 = \frac{n^2(n+1)^2}{4}$$

$$e = 1 + \frac{1}{1} + \frac{1}{2!} + \frac{1}{3!} + \cdots + \frac{1}{(n-1)!}$$

Harmonic

When the reciprocals of the terms form an arithmetic progression, the progression is harmonic.

Harmonic mean is $\dfrac{2ab}{a+b}$.

The relation among arithmetic, geometric, and harmonic means is $G^2 = AH$.

Binomial Series

$$(a \pm b)^n = a^n \pm na^{n-1}b + \frac{n(n-1)}{2!} a^{n-2}b^2$$
$$\pm \frac{n(n-1)(n-2)}{3!} a^{n-3}b^3 + (\pm 1)^r \frac{n(n-1)\cdots(n-r+1)}{r!} a^{n-r}b^r + \cdots$$

in which the last term shown is the $(r + 1)$th.

If n is a positive integer, the series is finite and the last term is b^n. If n is less than unity, the series is infinite and converges only when $b < a$.
For any value of n, and convergent when x^2 is less than unity,

$$(1 \pm x)^n = 1 \pm nx + \frac{n(n-1)}{2!}x^2 \pm \frac{n(n-1)(n-2)}{3!}x^3$$
$$+ \frac{n(n-1)(n-2)(n-3)x^4}{4!} \pm \cdots$$

$$(1 \pm x)^{-n} = 1 \mp nx + \frac{n(n-1)}{2!} \mp \frac{n(n-1)(n+2)}{3!}x^3 + \cdots$$

$$(1 \pm x)^{-1} = 1 \pm x + x^2 \mp x^3 + x^4 \mp x^5 + \cdots$$

$$(1 \pm x)^{\frac{1}{2}} = 1 \pm \tfrac{1}{2} \times - \frac{1}{2 \cdot 4}x^2 \pm \frac{1 \cdot 3}{2 \cdot 4 \cdot 6}x^3 - \cdots$$

$$(1 \pm x)^{-\frac{1}{2}} = 1 \mp \tfrac{1}{2} \times + \frac{1 \cdot 3}{2 \cdot 4}x^2 \pm \frac{1 \cdot 3 \cdot 5}{2 \cdot 4 \cdot 6}x^3 + \cdots$$

Infinite Series

Test for Convergence

According to Cauchy's convergence principle, a sequence of numbers $u_1, u_2 \cdots$ converges if, and only if, for every positive number ϵ there exists an integer N such that

$$|u_n - u_n| < \epsilon \quad \text{when } m > N, \quad n > N$$

In other words, somewhere in the series are located two terms u_m, u_n that are separated by less than any ϵ however small.

Ratio Test. In a series $u, u_2 \cdots u_n$, let

$$L = \lim \left| \frac{u_{n+1}}{u_n} \right| \quad \text{where } u_n \neq 0$$

The series converges absolutely if $L < 1$, diverges if $L > 1$ or L does not exist. If $L = 1$ the test fails.

Comparison Test for Series of All Positive Terms. If one such series is known to be convergent and is compared to another series where after some term all the following terms become as small or smaller than the remaining terms of the first series, then the second series is convergent. In brief, if $u_n \leqq w_n$, the series u converges if the series w converges.

Root Test. In a series $u_n^{1/n}, u_{n+1}^{1/(n+1)} \cdots$ if $L = \lim_{n \to \infty} |u_n|^{1/n}$ the series converges if $L < 1$ and diverges if $L > 1$. If $L = 1$, the test fails.

ALGEBRA

Taylor's Series. If $f(x)$ has continuous derivatives in the neighborhood of a point $x = a$, then

$$f(x) = f(a) + \frac{f'(a)}{1!}(x - a) + \frac{f''(a)}{2!}(x - a)^2 + \cdots$$
$$+ \frac{f^{(n-1)}(a)}{(n - 1)!}(x - a)^{n-1} + \cdots$$

with the remainder after n terms

$$R_n = \frac{f^{(n)}(\xi)}{n!}(x - a)^n, \quad \xi = a + \theta(x - a), \quad 0 < \theta < 1$$

Another form of Taylor's series is

$$f(x + h) = f(x) + \frac{h}{1!}f'(x) + \frac{h^2}{2!}f''(x) + \cdots + \frac{h^{n-1}}{(n - 1)!}f^{(n-1)}(x) + \cdots$$

with the remainder after n terms

$$R_n = \frac{h^n}{n!}f^{(n)}(\xi), \quad \xi = x + \theta h, \quad 0 < \theta < 1$$

Maclaurin's Series. If $a = 0$ in equation (41),

$$f(x) = f(0) + \frac{f'(0)}{1!}x + \frac{f''(0)}{2!}x^2 + \cdots + \frac{f^{(n-1)}(0)}{(n - 1)!}x^{n-1} + \cdots$$

with the remainder after n terms

$$R_n = \frac{f^{(n)}(\xi)}{n!}x^n, \quad \xi = \theta x, \quad 0 < \theta < 1$$

A Taylor's or Maclaurin's series represents a function in an interval if and only if $R_n \to 0$ as $n \to \infty$.

Example. Expand e^{ax} in powers of x

$$f(x) = e^{ax}, f'(x) = ae^{ax}, f''(x) = a^2 e^{ax}, f'''(x) = a^3 e^{ax}, \ldots$$
$$f(0) = 1, f'(0) = a, f''(0) = a^2, f'''(0) = a^3, \ldots$$
$$f(x) = e^{ax} = 1 + \frac{a}{1!}x + \frac{a^2}{2!}x^2 + \frac{a^3}{3!}x^3 + \cdots$$

Since $\lim\limits_{n \to \infty} \frac{a^{n-1}/(n - 1)!}{a^n/n!} = \lim\limits_{n \to \infty} \frac{n}{a} = \infty$, the series converges for all values of x.

Taylor's Series for Two Variables

$$f(x + h, y + k) = f(x, y) + \frac{1}{1!}\left(h\frac{\partial}{\partial x} + k\frac{\partial}{\partial y}\right)f(x, y)$$

$$+ \frac{1}{2!}\left(h\frac{\partial}{\partial x} + k\frac{\partial}{\partial y}\right)^2 f(x, y) + \cdots + \frac{1}{(n-1)!}\left(h\frac{\partial}{\partial x} + k\frac{\partial}{\partial y}\right)^{n-1} f(x, y) + \cdots$$

with the remainder

$$R_n = \frac{1}{n!}\left(h\frac{\partial}{\partial x} + k\frac{\partial}{\partial y}\right)^n f(x + \theta h, y + \theta k), \qquad 0 < \theta < 1$$

Fourier Series. If $f(x)$ is of bounded variation over an interval of length $2l$, that is, if it can be expressed as the difference of two non-decreasing or non-increasing bounded functions, then

$$f(x) = \frac{a_0}{2} + \sum_{n=1}^{\infty}\left(a_n \cos\frac{n\pi x}{l} + b_n \sin\frac{n\pi x}{l}\right)$$

$$= \frac{a_0}{2} + a_1 \cos\frac{\pi x}{l} + a_2 \cos\frac{2\pi x}{l} + \cdots + b_1 \sin\frac{\pi x}{l} + b_2 \sin\frac{2\pi x}{l} + \cdots$$

in which

$$a_n = \frac{1}{l}\int_k^{k+2l} f(x) \cos\frac{n\pi x}{l}\, dx, \quad b_n = \frac{1}{l}\int_k^{k+2l} f(x) \sin\frac{n\pi x}{l}\, dx, \qquad n = 0, 1, 2, \ldots$$

In exponential form

$$f(x) = \sum_{n=-\infty}^{\infty} c_n e^{in\pi x/l}, \quad c_n = \frac{1}{2l}\int_k^{k+2l} f(x) e^{-in\pi x/l}\, dx, \quad n = \ldots, -2, -1, 0, 1, 2, \ldots$$

At a point of discontinuity a Fourier series gives the value at the midpoint of the jump.

Fourier Series for Even or Odd Functions. If $f(-x) = f(x)$, it is an *even* function. Then

$$a_n = \frac{2}{l}\int_0^l f(x) \cos\frac{n\pi x}{l}\, dx, \qquad n = 0, 1, 2, \ldots$$

$$b_n = 0$$

If $f(-x) = -f(x)$, it is an *odd* function. Then

$$a_n = 0$$

$$b_n = \frac{2}{l}\int_0^l f(x) \sin\frac{n\pi x}{l}\, dx, \qquad n = 0, 1, 2, \ldots$$

Table 1-4 Functions Expanded in Series
($\log = \log_e$)

$(a + x)^n = a^n + na^{n-1}x + \dfrac{n(n-1)}{2!}a^{n-2}x^2 + \dfrac{n(n-1)(n-2)}{3!}a^{n-3}x^3 + \cdots$ \hfill $(x^2 < a^2)$

$e^x = 1 + x + \dfrac{x^2}{2!} + \dfrac{x^3}{3!} + \dfrac{x^4}{4!} + \cdots$ \hfill $(-\infty < x < \infty)$

$a^x = 1 + x \log a + \dfrac{(x \log a)^2}{2!} + \dfrac{(x \log a)^3}{3!} + \cdots$ \hfill $(-\infty < x < \infty)$

$e^{-x^2} = 1 - x^2 + \dfrac{x^4}{2!} - \dfrac{x^6}{3!} + \dfrac{x^8}{4!} - \cdots$ \hfill $(-\infty < x < \infty)$

$e^{\sin x} = 1 + x + \dfrac{x^2}{2!} - \dfrac{3x^4}{4!} - \dfrac{8x^5}{5!} - \dfrac{3x^6}{6!} + \dfrac{56x^7}{7!} + \cdots$ \hfill $(-\infty < x < \infty)$

$e^{\cos x} = e\left(1 - \dfrac{x^2}{2!} + \dfrac{4x^4}{4!} - \dfrac{31x^6}{6!} + \cdots\right)$ \hfill $(-\infty < x < \infty)$

$e^{\tan x} = 1 + x + \dfrac{x^2}{2!} + \dfrac{3x^3}{3!} + \dfrac{9x^4}{4!} + \dfrac{37x^5}{5!} + \cdots$ \hfill $\left(-\dfrac{\pi}{2} < x < \dfrac{\pi}{2}\right)$

$\log x = \dfrac{x-1}{x} + \dfrac{1}{2}\left(\dfrac{x-1}{x}\right)^2 + \dfrac{1}{3}\left(\dfrac{x-1}{x}\right)^3 + \cdots$ \hfill $\left(x > \dfrac{1}{2}\right)$

$\log x = 2\left[\dfrac{x-1}{x+1} + \dfrac{1}{3}\left(\dfrac{x-1}{x+1}\right)^3 + \dfrac{1}{5}\left(\dfrac{x-1}{x+1}\right)^5 + \cdots\right]$ \hfill $(x > 0)$

$\log(1 + x) = x - \dfrac{x^2}{2} + \dfrac{x^3}{3} - \dfrac{x^4}{4} + \cdots$ \hfill $(-1 < x < 1)$

$\log \dfrac{1+x}{1-x} = 2\left(x + \dfrac{x^3}{3} + \dfrac{x^5}{5} + \dfrac{x^7}{7} + \cdots\right)$ \hfill $(-1 < x < 1)$

$\log \dfrac{x+1}{x-1} = 2\left(\dfrac{1}{x} + \dfrac{1}{3x^3} + \dfrac{1}{5x^5} + \cdots\right)$ \hfill $(x^2 > 1)$

$\log \sin x = \log x - \dfrac{x^2}{6} - \dfrac{x^4}{180} - \dfrac{x^6}{2835} - \cdots$ \hfill $(-\pi < x < \pi)$

$\log \cos x = -\dfrac{x^2}{2} - \dfrac{x^4}{12} - \dfrac{x^6}{45} - \dfrac{17x^8}{2520} - \cdots$ \hfill $\left(-\dfrac{\pi}{2} < x < \dfrac{\pi}{2}\right)$

$\log \tan x = \log x + \dfrac{x^2}{3} + \dfrac{7x^4}{90} + \dfrac{62x^6}{2835} + \cdots$ \hfill $\left(-\dfrac{\pi}{2} < x < \dfrac{\pi}{2}\right)$

$\sin x = x - \dfrac{x^3}{3!} + \dfrac{x^5}{5!} - \dfrac{x^7}{7!} + \cdots$ \hfill $(-\infty < x < \infty)$

$\cos x = 1 - \dfrac{x^2}{2!} + \dfrac{x^4}{4!} - \dfrac{x^6}{6!} + \cdots$ \hfill $(-\infty < x < \infty)$

$\tan x = x + \dfrac{x^3}{3} + \dfrac{2x^5}{15} + \dfrac{17x^7}{315} + \dfrac{62x^9}{2835} + \cdots$ \hfill $\left(-\dfrac{\pi}{2} < x < \dfrac{\pi}{2}\right)$

$\cot x = \dfrac{1}{x} - \dfrac{x}{3} - \dfrac{x^3}{45} - \dfrac{2x^5}{945} - \dfrac{x^7}{4725} - \cdots$ \hfill $(-\pi < x < \pi)$

$\sec x = 1 + \dfrac{x^2}{2!} + \dfrac{5x^4}{4!} + \dfrac{61x^6}{6!} + \cdots$ \hfill $\left(-\dfrac{\pi}{2} < x < \dfrac{\pi}{2}\right)$

Table 1-4 (*Continued*)

$$\csc x = \frac{1}{x} + \frac{x}{3!} + \frac{7x^3}{3\cdot 5!} + \frac{31x^5}{3\cdot 7!} + \cdots \qquad (-\pi < x < \pi)$$

$$\sin^{-1} x = x + \frac{x^3}{2\cdot 3} + \frac{3x^5}{2\cdot 4\cdot 5} + \frac{3\cdot 5 x^7}{2\cdot 4\cdot 6\cdot 7} + \cdots \qquad (-1 \leqq x \leqq 1)$$

$$\cos^{-1} x = \frac{\pi}{2} - \sin^{-1} x$$

$$\tan^{-1} x = \frac{\pi}{2} - \frac{1}{x} + \frac{1}{3x^3} - \frac{1}{5x^5} + \cdots \qquad (x^2 \geqq 1)$$

$$= x - \frac{x^3}{3} + \frac{x^5}{5} - \frac{x^7}{7} + \cdots \qquad (-1 \leqq x \leqq 1)$$

$$\cot^{-1} x = \frac{\pi}{2} - \tan^{-1} x$$

$$\sec^{-1} x = \frac{\pi}{2} - \frac{1}{x} - \frac{1}{2\cdot 3x^3} - \frac{3}{2\cdot 4\cdot 5x^5} - \frac{3\cdot 5}{2\cdot 4\cdot 6\cdot 7x^7} - \cdots \qquad (x^2 > 1)$$

$$\csc^{-1} x = \frac{\pi}{2} - \sec^{-1} x$$

$$\sinh x = x + \frac{x^3}{3!} + \frac{x^5}{5!} + \frac{x^7}{7!} + \cdots \qquad (-\infty < x < \infty)$$

$$\cosh x = 1 + \frac{x^2}{2!} + \frac{x^4}{4!} + \frac{x^6}{6!} + \frac{x^8}{8!} + \cdots \qquad (-\infty < x < \infty)$$

$$\tanh x = x - \frac{x^3}{3} + \frac{2x^5}{15} - \frac{17x^7}{315} + \cdots \qquad \left(-\frac{\pi}{2} < x < \frac{\pi}{2}\right)$$

$$\coth x = \frac{1}{x} + \frac{x}{3} - \frac{x^3}{45} + \frac{2x^5}{945} - \frac{x^7}{4725} + \cdots \qquad (-\pi < x < \pi)$$

$$\operatorname{sech} x = 1 - \frac{x^2}{2!} + \frac{5x^4}{4!} - \frac{61x^6}{6!} + \frac{1385 x^8}{8!} - \cdots \qquad \left(-\frac{\pi}{2} < x < \frac{\pi}{2}\right)$$

$$\operatorname{csch} x = \frac{1}{x} - \frac{x}{6} + \frac{7x^3}{360} - \frac{31 x^5}{15{,}120} + \cdots \qquad (-\pi < x < \pi)$$

$$\sinh^{-1} x = x - \frac{x^3}{2\cdot 3} + \frac{3x^5}{2\cdot 4\cdot 5} - \frac{3\cdot 5 x^7}{2\cdot 4\cdot 6\cdot 7} + \cdots \qquad (-1 < x < 1)$$

$$\sinh^{-1} x = \log 2x + \frac{1}{2\cdot 2x^2} - \frac{3}{2\cdot 4\cdot 4x^4} + \frac{3\cdot 5}{2\cdot 4\cdot 6\cdot 6\cdot x^6} + \cdots \qquad (x^2 > 1)$$

$$\cosh^{-1} x = \pm \left(\log 2x - \frac{1}{2\cdot 2x^2} - \frac{1\cdot 3}{2\cdot 4\cdot 4x^4} - \frac{1\cdot 3\cdot 5}{2\cdot 4\cdot 6\cdot 6x^6} - \cdots \right) \qquad (x > 1)$$

$$\tanh^{-1} x = x + \frac{x^3}{3} + \frac{x^5}{5} + \frac{x^7}{7} + \cdots \qquad (-1 < x < 1)$$

$$\coth^{-1} x = \frac{1}{x} + \frac{1}{3x^3} + \frac{1}{5x^5} + \frac{1}{7x^7} + \cdots \qquad (x^2 > 1)$$

$$\operatorname{sech}^{-1} x = \pm \left(\log \frac{2}{x} - \frac{1}{2\cdot 2} x^2 - \frac{1\cdot 3}{2\cdot 4\cdot 4} x^4 - \frac{1\cdot 3\cdot 5}{2\cdot 4\cdot 6\cdot 6} x^6 - \cdots \right) \qquad (0 < x < 1)$$

$$\operatorname{csch}^{-1} x = \frac{1}{x} - \frac{1}{2\cdot 3x^3} + \frac{3}{2\cdot 4\cdot 5x^5} - \frac{3\cdot 5}{2\cdot 4\cdot 6\cdot 7x^7} + \cdots \qquad (x^2 > 1)$$

Matrices and Determinants

Definitions

1. A *matrix* is a system of *mn* quantities, called *elements*, arranged in a rectangular array of m rows and n columns.

$$A = \begin{pmatrix} a_{11} & a_{12} & \cdots & a_{1n} \\ a_{21} & a_{22} & \cdots & a_{2n} \\ \vdots & & & \vdots \\ a_{m1} & a_{m2} & \cdots & a_{mn} \end{pmatrix} = \begin{Vmatrix} a_{11} & a_{12} & \cdots & a_{1n} \\ a_{21} & a_{22} & \cdots & a_{2n} \\ \vdots & & & \vdots \\ a_{m1} & a_{m2} & \cdots & a_{mn} \end{Vmatrix}$$

$$= (a_{ij}) = \|a_{ij}\|, \qquad (i = 1, \ldots, m; \ j = 1, \ldots, n)$$

2. If $m = n$, then A is a *square* matrix of *order n*.
3. Two matrices are *equal* if and *only if* they have the same number of rows and of columns, and corresponding elements are equal.
4. Two matrices are *transposes* (sometimes called *conjugates*) of each other, if either is obtained from the other by interchanging rows and columns.
5. The *complex conjugate* of a matrix (a_{ij}) with complex elements is the matrix $(\bar{a}_{ij})$.
6. A matrix is *symmetric* if it is equal to its transpose, that is, if $a_{ij} = a_{ji}$; $i, j = 1, \ldots, n$.
7. A matrix is *skew-symmetric*, or *anti-symmetric*, if $a_{ij} = -a_{ji}$; $i, j = 1, \ldots, n$. The diagonal elements $a_{ii} = 0$.
8. A matrix all of whose elements are zero is a *zero matrix*.
9. If the nondiagonal elements a_{ij}, $i \neq j$, of a square matrix A are all zero, then A is a *diagonal matrix*. If, furthermore, the diagonal elements are all equal, the matrix is a *scalar matrix;* if they are all 1, it is an *identity* or *unit matrix*, denoted by I.
10. The *determinant* $|A|$ of a square matrix (a_{ij}); $i, j = 1, \ldots, n$, is the sum of the $n!$ products $a_{1r_1} a_{2r_2} \cdots a_{nr_n}$, in which $r_1, r_2, \ldots, r_n$ is a permutation of $1, 2, \ldots, n$, and the sign of each product is $+$ or $-$ according as the permutation is obtained from $1, 2, \ldots, n$ by an even or an odd number of interchanges of two numbers.

Symbols used are

$$|A| = \begin{vmatrix} a_{11} & a_{12} & \cdots & a_{1n} \\ a_{21} & a_{22} & \cdots & a_{2n} \\ \vdots & & & \vdots \\ a_{n1} & a_{n2} & \cdots & a_{nn} \end{vmatrix} = |a_{ij}|, \qquad i, j = 1, \ldots, n$$

11. A square matrix (a_{ij}) is *singular* if its determinant $|a_{ij}|$ is zero.
12. The determinants of the square submatrices of any matrix A, obtained by striking out certain rows or columns, or both, are called the *determinants* or *minors* of A. A matrix is of *rank r* if it has at least one *r*-rowed determinant which is not zero, while all its determinants of order higher than r are zero.

The *nullity d* of a square matrix of order n is $d = n - r$. The zero matrix is of rank 0.

13. The *minor* D_{ij} of the element a_{ij} of a square matrix is the determinant of the submatrix obtained by striking out the row and column in which a_{ij} lies. The *cofactor* A_{ij} of the element a_{ij} is $(-1)^{i+j}D_{ij}$. A *principal minor* is the minor obtained by striking out the same rows as columns.

14. The *inverse* of the square matrix A is

$$A^{-1} = \begin{pmatrix} \dfrac{A_{11}}{|A|} & \cdots & \dfrac{A_{n1}}{|A|} \\ \cdots & & \cdots \\ \dfrac{A_{1n}}{|A|} & \cdots & \dfrac{A_{nn}}{|A|} \end{pmatrix}$$

$$AA^{-1} = A^{-1}A = I$$

15. The *adjoint* of A is

$$\text{Adj } A = \begin{pmatrix} A_{11} & \cdots & A_{n1} \\ \cdots & & \cdots \\ A_{1n} & \cdots & A_{nn} \end{pmatrix}$$

16. *Elementary transformations* of a matrix are
    (1) The interchange of two rows or of two columns.
    (2) The addition to the elements of a row (or column) of any constant multiple of the corresponding elements of another row (or column).
    (3) The multiplication of each element of a row (or column) by any non-zero constant.

17. Two $m \times n$ matrices A and B are *equivalent* if it is possible to pass from one to the other by a finite number of elementary transformations.
    (1) The matrices A and B are equivalent if and only if there exist two non-singular square matrices E and F, having m and n rows respectively, such that $EAF = B$.
    (2) The matrices A and B are equivalent if and only if they have the same rank.

Matrix Operations

1. *Addition and Subtraction.* The sum or difference of two matrices (a_{ij}) and (b_{ij}) is the matrix $(a_{ij} \pm b_{ij})$, $i = 1, \ldots, m$; $j = 1, \ldots, n$.

2. *Scalar Multiplication.* The product of the scalar k and the matrix (a_{ij}) is the matrix (ka_{ij}).

3. *Matrix Multiplication.* The product (p_{ik}), $i = 1, \ldots, m$; $k = 1, \ldots, q$, of two matrices (a_{ij}), $i = 1, \ldots, m$; $j = 1, \ldots, n$, and (b_{jk}), $j = 1, \ldots, n$; $k = 1, \ldots, q$, is the matrix whose elements are

$$p_{ik} = \sum_{j=1}^{n} a_{ij}b_{jk} = a_{i1}b_{1k} + a_{i2}b_{2k} + \cdots + a_{in}b_{nk}$$

ALGEBRA

The element in the ith row and kth column of the product is the sum of the n products of the n elements of the ith row of (a_{ij}) by the corresponding n elements of the kth column of (b_{jk}).

Example.

$$\begin{pmatrix} a_{11} & a_{12} \\ a_{21} & a_{22} \end{pmatrix} \begin{pmatrix} b_{11} & b_{12} & b_{13} \\ b_{21} & b_{22} & b_{23} \end{pmatrix} = \begin{pmatrix} a_{11}b_{11} + a_{12}b_{21} & a_{11}b_{12} + a_{12}b_{22} & a_{11}b_{13} + a_{12}b_{23} \\ a_{21}b_{11} + a_{22}b_{21} & a_{21}b_{12} + a_{22}b_{22} & a_{21}b_{13} + a_{22}b_{23} \end{pmatrix}$$

All the laws of ordinary algebra hold for the addition and subtraction of matrices and for scalar multiplication.

Multiplication of matrices is not in general commutative, but it is associative and distributive.

If the product of two or more matrices is zero, it does not follow that one of the factors is zero. The factors are *divisors of zero*.

Example.

$$\begin{pmatrix} a & 0 \\ b & 0 \end{pmatrix} \begin{pmatrix} 0 & 0 \\ c & d \end{pmatrix} = \begin{pmatrix} 0 & 0 \\ 0 & 0 \end{pmatrix}$$

Linear Dependence

1. The quantities $l_1, l_2, \ldots, l_n$ are *linearly dependent* if there exist constants $c_1, c_2, \ldots, c_n$, not all zero, such that

$$c_1 l_1 + c_2 l_2 + \cdots + c_n l_n = 0$$

If no such constants exist, the quantities are *linearly independent*.

2. The linear functions

$$l_i = a_{i1} x_1 + a_{i2} x_2 + \cdots + a_{in} x_n, \qquad i = 1, 2, \ldots, m$$

are *linearly dependent* if and only if the matrix of the coefficients is of rank $r < m$. Exactly r of the l_i form a linearly independent set.

3. For $m > n$, any set of m linear functions are linearly dependent.

Consistency of Equations

1. The system of homogeneous linear equations

$$a_{i1} x_1 + a_{i2} x_2 + \cdots + a_{in} x_n = 0, \qquad i = 1, 2, \ldots, m$$

has solutions not all zero if the rank r of the matrix (a_{ij}) is less than n.

If $m < n$, there always exist solutions not all zero. If $m = n$, there exist solutions not all zero if $|a_{ij}| = 0$.

If r of the equations are so selected that their matrix is of rank r, they determine uniquely r of the variables as homogeneous linear functions of the remaining $n - r$ variables. A solution of the system is obtained by assigning arbitrary values to the $n - r$ variables and finding the corresponding values of the r variables.

2. The system of linear equations

$$a_{i1} x_1 + a_{i2} x_2 + \cdots + a_{in} x_n = k_i, \qquad i = 1, 2, \ldots, m$$

is consistent if and only if the *augmented* matrix derived from (a_{ij}) by annexing the column $k_1, \ldots, k_m$ has the same rank r as (a_{ij}).

As in the case of a system of homogeneous linear equations, r of the variables can be expressed in terms of the remaining $n - r$ variables.

Linear Transformations

1. If a linear transformation

$$x'_i = a_{i1}x_1 + a_{i2}x_2 + \cdots + a_{in}x_n, \quad i = 1, 2, \ldots, n$$

with matrix (a_{ij}) transforms the variables x_i into the variables x'_i, and a linear transformation

$$x''_i = b_{i1}x'_1 + b_{i2}x'_2 + \cdots + b_{in}x'_n, \quad i = 1, 2, \ldots, n$$

with matrix (b_{ij}) transforms the variables x'_i into the variables x''_i, then the linear transformation with matrix $(b_{ij})(a_{ij})$ transforms the variables x_i into the variables x''_i directly.

2. A real *orthogonal* transformation is a linear transformation of the variables x_i into the variables x'_i such that

$$\sum_{i=1}^{n} x_i^2 = \sum_{i=1}^{n} x'_i{}^2$$

A transformation is orthogonal if and only if the transpose of its matrix is the inverse of its matrix.

3. A *unitary* transformation is a linear transformation of the variables x_i into the variables x'_i such that

$$\sum_{i=1}^{n} x_i \bar{x}_i = \sum_{i=1}^{n} x'_i \bar{x}'_i$$

A transformation is unitary if and only if the transpose of the conjugate of its matrix is the inverse of its matrix.

Quadratic Forms

A *quadratic form* in n variables is

$$\sum_{i,j=1}^{n} a_{ij}x_i x_j = a_{11}x_1^2 + a_{12}x_1 x_2 + \cdots + a_{1n}x_1 x_n$$
$$+ a_{21}x_2 x_1 + a_{22}x_2^2 + \cdots + a_{2n}x_2 x_n$$
$$\cdots\cdots\cdots\cdots\cdots\cdots\cdots\cdots\cdots\cdots$$
$$+ a_{n1}x_n x_1 + a_{n2}x_n x_2 + \cdots + a_{nn}x_n^2$$

in which $a_{ji} = a_{ij}$. The symmetric matrix (a_{ij}) of the coefficients is the *matrix* of the quadratic form and the rank of (a_{ij}) is the *rank* of the quadratic form.

A real quadratic form of rank r can be reduced by a real non-singular linear transformation to the *normal form*

$$x_1^2 + \cdots + x_p^2 - x_{p+1}^2 - \cdots - x_r^2$$

in which the *index p* is uniquely determined.

If $p = r$, a quadratic form is *positive*, and, if $p = 0$, it is *negative*. If, furthermore, $r = n$, both are *definite*. A quadratic form is positive definite if and only if the determinant and all the principal minors of its matrix are positive.

Hermitian Forms

A *Hermitian form* in n variables is

$$\sum_{i,j=1}^{n} a_{ij} x_i \bar{x}_j, \quad a_{ji} = \bar{a}_{ij}$$

The matrix (a_{ij}) is a *Hermitian matrix*. Its transpose is equal to its conjugate. The rank of (a_{ij}) is the *rank* of the Hermitian form.

A Hermitian form of rank r can be reduced by a nonsingular linear transformation to the *normal form*

$$x_1 \bar{x}_1 + \cdots + x_p \bar{x}_p - x_{p+1} \bar{x}_{p+1} - \cdots - x_r \bar{x}_r$$

in which the *index* p is uniquely determined.

If $p = r$, the Hermitian form is *positive*, and, if $p = 0$, it is *negative*. If, furthermore, $r = n$, both are *definite*.

Determinants

Second- and third-order determinants are formed from their square symbols by taking diagonal products, down from left to right being positive and up negative.

$$\begin{vmatrix} a_{11} & a_{12} \\ a_{21} & a_{22} \end{vmatrix} = a_{11}a_{22} - a_{21}a_{12}$$

$$\begin{vmatrix} a_{11} & a_{12} & a_{13} \\ a_{21} & a_{22} & a_{23} \\ a_{31} & a_{32} & a_{33} \end{vmatrix} = \begin{matrix} a_{11}a_{22}a_{33} + a_{12}a_{23}a_{31} + a_{13}a_{32}a_{21} \\ - a_{31}a_{22}a_{13} - a_{32}a_{23}a_{11} - a_{33}a_{12}a_{21} \end{matrix}$$

Third and higher order determinants are formed by selecting any row or column and taking the sum of the products of each element and its cofactor. This process is continued until second- or third-order cofactors are reached.

$$\begin{vmatrix} a_{11} & a_{12} & a_{13} \\ a_{21} & a_{22} & a_{23} \\ a_{31} & a_{32} & a_{33} \end{vmatrix} = a_{11} \begin{vmatrix} a_{22} & a_{23} \\ a_{32} & a_{33} \end{vmatrix} - a_{21} \begin{vmatrix} a_{12} & a_{13} \\ a_{32} & a_{33} \end{vmatrix} + a_{31} \begin{vmatrix} a_{12} & a_{13} \\ a_{22} & a_{23} \end{vmatrix}$$

The determinant of a matrix A is

(1) Zero, if two rows or two columns of A have proportional elements.
(2) Unchanged, if
  (*a*) The rows and columns of A are interchanged.
  (*b*) To each element of a row or column of A is added a constant multiple of the corresponding element of another row or column.

(3) Changed in sign, if two rows or two columns of A are interchanged.

(4) Multiplied by c, if each element of any row or column of A is multiplied by c.

(5) The sum of the determinants of two matrices B and C, if A, B, and C have all the same elements, except that in one row or column each element of A is the sum of the corresponding elements of B and C.

Example.

$$\begin{vmatrix} 2 & 9 & 9 & 4 \\ 2 & -3 & 12 & 8 \\ 4 & 8 & 3 & -5 \\ 1 & 2 & 6 & 4 \end{vmatrix} = \begin{vmatrix} 2 & 5 & 9 & 4 \\ 2 & -7 & 12 & 8 \\ 4 & 0 & 3 & -5 \\ 1 & 0 & 6 & 4 \end{vmatrix} = 3 \begin{vmatrix} 2 & 5 & 3 & 4 \\ 2 & -7 & 4 & 8 \\ 4 & 0 & 1 & -5 \\ 1 & 0 & 2 & 4 \end{vmatrix}$$

        Multiply 1st column   Factor 3 out of
        by -2 and add to 2nd.   the 3rd column.

$$= 3 \times (-5) \begin{vmatrix} 2 & 4 & 8 \\ 4 & 1 & -5 \\ 1 & 2 & 4 \end{vmatrix} + 3 \times (-7) \begin{vmatrix} 2 & 3 & 4 \\ 4 & 1 & -5 \\ 1 & 2 & 4 \end{vmatrix}$$

Expand according to second column.

$$= 0 \qquad\qquad -21 \begin{vmatrix} 1 & 1 & 0 \\ 4 & 1 & -5 \\ 1 & 2 & 4 \end{vmatrix}$$

      First and third   Subtract third
      rows proportional.   row from first.

$$= -21 \begin{vmatrix} 1 & -5 \\ 2 & 4 \end{vmatrix} - (-21) \begin{vmatrix} 4 & -5 \\ 1 & 4 \end{vmatrix} = -21[(4 + 10) - (16 + 5)] = +147$$

Expand according to first row.

Boolean Algebras

Algebra of Sets

A *set* is a collection of objects each of which is an *element* of the set. Thus a as an element of set A is denoted by $a \in A$. The *null set*, one which has no elements, is denoted by ϕ. Two sets are equal ($A = B$) if the elements in each set are the same.

If every element of A is also an element of B, A is a *subset* of B, denoted by $A \subset B$ or $B \supset A$. When B contains at least one element which is not an element of A, then A is a proper *subset* of B. By convention ϕ is a subset of every set.

If all the elements considered are elements of a universal set I, then all sets are subsets of I. The *complementary set* of A relative to I, A', is the set which contains all the elements of I that are not elements of A.

The *union of sets* A and B, $A \cup B$, is the set of all elements which are elements of A or B or both.

The *intersection of sets* A and B, $A \cap B$, is the set of all elements which are elements of both A and B.

Properties of sets A, B, C in a universal set I.

Commutative laws:
$$A \cup B = B \cup A \quad \text{and} \quad A \cap B = B \cap A$$

Associative laws:
$$(A \cup B) \cup C = A \cup (B \cup C)$$
$$(A \cap B) \cap C = A \cap (B \cap C)$$

Distributive laws:
$$A \cup (B \cap C) = (A \cup B) \cap (A \cup C)$$
$$A \cap (B \cup C) = (A \cap B) \cup (A \cap C)$$

Idempotent laws:
$$A \cup A = A \quad \text{and} \quad A \cap A = A$$

Properties of I and ϕ:
$$A \cap I = A \quad \text{and} \quad A \cup I = I$$
$$A \cap \phi = \phi \quad \text{and} \quad A \cup \phi = A$$

Properties of $\subset$:
$$A \subset (A \cup B) \quad \text{and} \quad (A \cap B) \subset A$$
$$A \subset I \quad \text{and} \quad \phi \subset A$$
$$\text{If } A \subset B, \text{ then } A \cup B = B \text{ and } A \cap B = A$$

Properties of complement:
$$A \cup A' = I \quad \text{and} \quad A \cap A' = \phi$$
$$(A \cup B)' = A' \cap B' \quad \text{and} \quad (A \cap B)' = A' \cap B'$$

$\cup$ and $\cap$, ϕ and I, or $\subset$ and $\supset$ may be interchanged in any correct formula to obtain another correct formula.

Application to Logic and Circuits

	Operational Concept		
Symbol	Algebra of Sets	Symbolic Logic	Switching Circuits
$\cup$	union	or	parallel
$\cap$	intersection	and	series
/	complement	negation	complement
$\leq$	order	if a then b	
		a if and only if b	
a, b	elements	true, false	on, off

46 MATHEMATICS

The elements are usually represented by the binary numbers 0 and 1 to denote "off" and "on" in switching circuits and "false" and "true" in symbolic logic. The following rules then apply:

$$a \cup a = a, \quad a \cap a = a$$
$$a \cup a' = 1, \quad a \cap a' = 0$$

Truth tables in logic are used to show the relationship of variables. In the example below, the first four columns apply to switching circuits as well.

a	b	$a \cup b$	$a \cap b$	a'	$\leqq$	$\subseteq$
1	1	1	1	0	1	1
1	0	1	0	0	0	0
0	1	1	0	1	1	0
0	0	0	0	1	1	1

TRIGONOMETRY

Circular Functions of Plane Angles

Definitions and Values

$$\text{Sine } \alpha = \frac{y}{r} = \sin \alpha$$

$$\text{Cosine } \alpha = \frac{x}{r} = \cos \alpha$$

$$\text{Tangent } \alpha = \frac{y}{x} = \tan \alpha$$

$$\text{Cotangent } \alpha = \frac{x}{y} = \cot \alpha$$

$$\text{Secant } \alpha = \frac{r}{x} = \sec \alpha$$

$$\text{Cosecant } \alpha = \frac{r}{y} = \csc \alpha$$

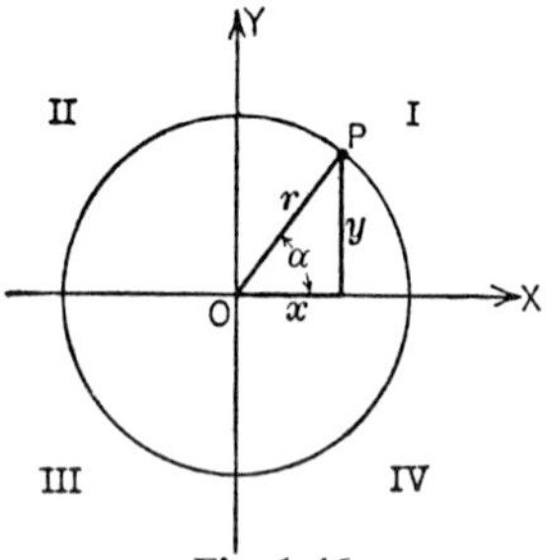

$$\text{Versine } \alpha = \frac{r - x}{r} = \text{vers } \alpha = 1 - \cos \alpha$$

$$\text{Coversine } \alpha = \frac{r - y}{r} = \text{covers } \alpha = 1 - \sin \alpha$$

$$\text{Haversine } \alpha = \frac{r - x}{2r} = \text{hav } \alpha = \frac{1}{2} \text{vers } \alpha$$

Fig. 1-46

TRIGONOMETRY   47

Positive and Negative Values. An angle α (Fig. 1-46), if measured in a *counterclockwise* direction, is said to be *positive*; if measured *clockwise*, *negative*. Following the convention that x is positive if measured along OX to the right of the OY axis and negative if measured to the left, and similarly, y is positive if measured along OY above the OX axis and negative if measured below, the signs of the trigonometric functions are different for angles in the quadrants I, II, III, and IV.

Table 1-5   Signs of Trigonometric Functions

Quadrant	sin	cos	tan	cot	sec	csc
I	+	+	+	+	+	+
II	+	−	−	−	−	+
III	−	−	+	+	−	−
IV	−	+	−	−	+	−

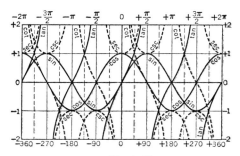

Fig. 1-47

Values of Trigonometric Functions are periodic, the period of the sin, cos, sec, csc being 2π radians, and that of the tan and cot, π radians. For example, in Fig. 1-47 (n an integer)

$$\sin(\alpha + 2\pi n) = \sin \alpha$$
$$\tan(\alpha + \pi n) = \tan \alpha$$

Table 1-6   Functions of Angles in Any Quadrant in Terms of Angles in the First Quadrant

	$-\alpha$	$90° \pm \alpha$	$180° \pm \alpha$	$270° \pm \alpha$	$360° \pm \alpha$
sin	− sin α	+ cos α	∓ sin α	− cos α	± sin α
cos	+ cos α	∓ sin α	− cos α	± sin α	+ cos α
tan	− tan α	∓ cot α	± tan α	∓ cot α	± tan α
cot	− cot α	∓ tan α	± cot α	∓ tan α	± cot α
sec	+ sec α	∓ csc α	− sec α	± csc α	+ sec α
csc	− csc α	+ sec α	∓ csc α	− sec α	± csc α

Table 1-7 Functions of Certain Angles

	0°	30°	45°	60°	90°	180°	270°	360°
sin	0	$\frac{1}{2}$	$\frac{1}{2}\sqrt{2}$	$\frac{1}{2}\sqrt{3}$	1	0	-1	0
cos	1	$\frac{1}{2}\sqrt{3}$	$\frac{1}{2}\sqrt{3}$	$\frac{1}{2}$	0	-1	0	1
tan	0	$\frac{1}{3}\sqrt{3}$	1	$\sqrt{3}$	∞	0	∞	0
cot	∞	$\sqrt{3}$	1	$\frac{1}{3}\sqrt{3}$	0	∞	0	∞
sec	1	$\frac{2}{3}\sqrt{3}$	$\sqrt{2}$	2	∞	-1	∞	1
csc	∞	2	$\sqrt{2}$	$\frac{2}{3}\sqrt{3}$	1	∞	-1	∞

Inverse, or Anti-functions. The symbol $\sin^{-1} x$ means the angle whose sine is x, and is read inverse sine of x, anti-sine of x, or arc sine x. Similarly for $\cos^{-1} x$, $\tan^{-1} x$, $\cot^{-1} x$, $\sec^{-1} x$, $\csc^{-1} x$, $\text{vers}^{-1} x$, the last meaning an angle α such that $(1 - \cos \alpha) = x$. While the direct functions (sine, etc.) are single valued, the indirect are many valued; thus $\sin 30° = 0.5$, but $\sin^{-1} 0.5 = 30°, 150°, \cdots$.

Table 1-8 Functions of an Angle in Terms of Each of the Others

	$\sin \alpha = a$	$\cos \alpha = a$	$\tan \alpha = a$	$\cot \alpha = a$	$\sec \alpha = a$	$\csc \alpha = a$
sin	a	$\sqrt{1-a^2}$	$\dfrac{a}{\sqrt{1+a^2}}$	$\dfrac{1}{\sqrt{1+a^2}}$	$\dfrac{\sqrt{a^2-1}}{a}$	$\dfrac{1}{a}$
cos	$\sqrt{1-a^2}$	a	$\dfrac{1}{\sqrt{1+a^2}}$	$\dfrac{a}{\sqrt{1+a^2}}$	$\dfrac{1}{a}$	$\dfrac{\sqrt{a^2-1}}{a}$
tan	$\dfrac{a}{\sqrt{1-a^2}}$	$\dfrac{\sqrt{1-a^2}}{a}$	a	$\dfrac{1}{a}$	$\sqrt{a^2-1}$	$\dfrac{1}{\sqrt{a^2-1}}$
cot	$\dfrac{\sqrt{1-a^2}}{a}$	$\dfrac{a}{\sqrt{1-a^2}}$	$\dfrac{1}{a}$	a	$\dfrac{1}{\sqrt{a^2-1}}$	$\sqrt{a^2-1}$
sec	$\dfrac{1}{\sqrt{1-a^2}}$	$\dfrac{1}{a}$	$\sqrt{1+a^2}$	$\dfrac{\sqrt{1+a^2}}{a}$	a	$\dfrac{a}{\sqrt{a^2-1}}$
csc	$\dfrac{1}{a}$	$\dfrac{1}{\sqrt{1-a^2}}$	$\dfrac{\sqrt{1+a^2}}{a}$	$\sqrt{1+a^2}$	$\dfrac{a}{\sqrt{a^2-1}}$	a

Note. The sign of the radical is to be determined by the quadrant.

Functions of the Sum and Difference of Two Angles

$$\sin(\alpha \pm \beta) = \sin \alpha \cos \beta \pm \cos \alpha \sin \beta$$
$$\cos(\alpha \pm \beta) = \cos \alpha \cos \beta \mp \sin \alpha \sin \beta$$
$$\tan(\alpha \pm \beta) = (\tan \alpha \pm \tan \beta)/(1 \mp \tan \alpha \tan \beta)$$
$$\cot(\alpha \pm \beta) = (\cot \beta \cot \alpha \mp 1)/(\cot \beta \pm \cot \alpha)$$

TRIGONOMETRY

If x is small, say 3° or 4°, then the following are close approximations, in which the quantity x is to be expressed in radians (1° = 0.01745 radian).

$$\sin \alpha \approx \alpha, \quad \cos \alpha \approx 1, \quad \tan \alpha \approx \alpha$$
$$\sin(\alpha \pm x) \approx \sin \alpha \pm x \cos \alpha, \quad \cos(\alpha \pm x) \approx \cos \alpha \mp x \sin \alpha$$

Functions of Half Angles

$$\sin \tfrac{1}{2}\alpha = \sqrt{\tfrac{1}{2}(1 - \cos \alpha)} = \tfrac{1}{2}\sqrt{1 + \sin \alpha} - \tfrac{1}{2}\sqrt{1 - \sin \alpha}$$
$$\cos \tfrac{1}{2}\alpha = \sqrt{\tfrac{1}{2}(1 + \cos \alpha)} = \tfrac{1}{2}\sqrt{2 + \sin \alpha} + \tfrac{1}{2}\sqrt{1 - \sin \alpha}$$
$$\tan \tfrac{1}{2}\alpha = \sqrt{(1 - \cos \alpha)/(1 + \cos \alpha)} = (1 - \cos \alpha)/\sin \alpha = \sin \alpha/(1 + \cos \alpha)$$
$$\cot \tfrac{1}{2}\alpha = \sqrt{(1 + \cos \alpha)/(1 - \cos \alpha)} = (1 + \cos \alpha)/\sin \alpha = \sin \alpha/(1 - \cos \alpha)$$

Functions of Multiples of Angles

$$\sin 2\alpha = 2 \sin \alpha \cos \alpha$$
$$\tan 2\alpha = 2 \tan\alpha/(1 - \tan^2 \alpha)$$
$$\cos 2\alpha = \cos^2 \alpha - \sin^2 \alpha = 2\cos^2 \alpha - 1 = 1 - 2\sin^2 \alpha$$
$$\cot 2\alpha = (\cot^2 \alpha - 1)/2 \cot \alpha$$
$$\sin 3\alpha = 3 \sin \alpha - 4 \sin^3 \alpha$$
$$\cos 3\alpha = 4 \cos^3 \alpha - 3 \cos \alpha$$
$$\sin 4\alpha = 8 \cos^3 \alpha \sin \alpha - 4 \cos \alpha \sin \alpha$$
$$\cos 4\alpha = 8 \cos^4 \alpha - 8 \cos^2 \alpha + 1$$
$$\sin n\alpha = 2 \sin(n-1)\alpha \cos \alpha - \sin(n-2)\alpha$$
$$= n \sin \alpha \cos^{n-1} \alpha - {}_nC_3 \sin^3 \alpha \cos^{n-3} \alpha + {}_nC_5 \sin^5 \alpha \cos^{n-5} \alpha - \cdots$$
$$\cos n\alpha = 2 \cos(n-1)\alpha \cos \alpha - \cos(n-2)\alpha$$
$$= \cos^n \alpha - {}_nC_2 \sin^2 \alpha \cos^{n-2} \alpha + {}_nC_4 \sin^4 \alpha \cos^{n-4} \alpha - \cdots$$

(For ${}_nC_t$, see p. 30.)

Products and Powers of Functions

$$\sin \alpha \sin \beta = \tfrac{1}{2}\cos(\alpha - \beta) - \tfrac{1}{2}\cos(\alpha + \beta)$$
$$\cos \alpha \cos \beta = \tfrac{1}{2}\cos(\alpha - \beta) + \tfrac{1}{2}\cos(\alpha + \beta)$$
$$\sin \alpha \cos \beta = \tfrac{1}{2}\sin(\alpha - \beta) + \tfrac{1}{2}\sin(\alpha + \beta)$$
$$\tan \alpha \cot \alpha = \sin \alpha \csc \alpha = \cos \alpha \sec \alpha = 1$$

$$\sin^2 \alpha = \tfrac{1}{2}(1 - \cos 2\alpha); \quad \cos^2 \alpha = \tfrac{1}{2}(1 + \cos 2\alpha)$$
$$\sin^3 \alpha = \tfrac{1}{4}(3 \sin \alpha - \sin 3\alpha); \quad \cos^3 \alpha = \tfrac{1}{4}(3 \cos \alpha + \cos 3\alpha)$$
$$\sin^4 \alpha = \tfrac{1}{8}(3 - 4 \cos 2\alpha + \cos 4\alpha); \quad \cos^4 \alpha = \tfrac{1}{8}(3 + 4 \cos 2\alpha + \cos 4\alpha)$$
$$\sin^5 \alpha = \tfrac{1}{16}(10 \sin \alpha - 5 \sin 3\alpha + \sin 5\alpha)$$
$$\sin^6 \alpha = \tfrac{1}{32}(10 - 15 \cos 2\alpha + 6 \cos 4\alpha - \cos 6\alpha)$$
$$\cos^5 \alpha = \tfrac{1}{16}(10 \cos \alpha + 5 \cos 3\alpha + \cos 5\alpha)$$
$$\cos^6 \alpha = \tfrac{1}{32}(10 + 15 \cos 2\alpha + 6 \cos 4\alpha + \cos 6\alpha)$$

Sums and Differences of Functions

$$\sin \alpha + \sin \beta = 2 \sin \tfrac{1}{2}(\alpha + \beta) \cos \tfrac{1}{2}(\alpha - \beta)$$
$$\sin \alpha - \sin \beta = 2 \cos \tfrac{1}{2}(\alpha + \beta) \sin \tfrac{1}{2}(\alpha - \beta)$$
$$\cos \alpha + \cos \beta = 2 \cos \tfrac{1}{2}(\alpha + \beta) \cos \tfrac{1}{2}(\alpha - \beta)$$
$$\cos \alpha - \cos \beta = -2 \sin \tfrac{1}{2}(\alpha + \beta) \sin \tfrac{1}{2}(\alpha - \beta)$$
$$\tan \alpha + \tan \beta = \frac{\sin(\alpha + \beta)}{\cos \alpha \cos \beta}; \quad \cot \alpha + \cot \beta = \frac{\sin(\alpha + \beta)}{\sin \alpha \sin \beta}$$
$$\tan \alpha - \tan \beta = \frac{\sin(\alpha - \beta)}{\cos \alpha \cos \beta}; \quad \cot \alpha - \cot \beta = -\frac{\sin(\alpha - \beta)}{\sin \alpha \sin \beta}$$
$$\sin^2 \alpha - \sin^2 \beta = \sin(\alpha + \beta) \sin(\alpha - \beta)$$
$$\cos^2 \alpha - \cos^2 \beta = -\sin(\alpha + \beta) \sin(\alpha - \beta)$$
$$\cos^2 \alpha - \sin^2 \beta = \cos(\alpha + \beta) \cos(\alpha - \beta)$$

Anti-Trigonometric or Inverse Functional Relations. In the following formulas the periodic constant is omitted.

$$\sin^{-1} x = -\sin^{-1}(-x) = \frac{\pi}{2} - \cos^{-1} x = \cos^{-1} \sqrt{1 - x^2} = \tan^{-1} \frac{x}{\sqrt{1 - x^2}}$$
$$= \cot^{-1} \frac{\sqrt{1 - x^2}}{x} = \csc^{-1} \frac{1}{x} = \sec^{-1} \frac{1}{\sqrt{1 - x^2}}$$

$$\cos^{-1} x = \pi - \cos^{-1}(-x) = \frac{\pi}{2} - \sin^{-1} x = \tfrac{1}{2} \cos^{-1}(2x^2 - 1) = \sin^{-1} \sqrt{1 - x^2}$$
$$= \tan^{-1} \frac{\sqrt{1 - x^2}}{x} = \cot^{-1} \frac{x}{\sqrt{1 - x^2}} = \sec^{-1} \frac{1}{x} = \csc^{-1} \frac{1}{\sqrt{1 - x^2}}$$

$$\tan^{-1} x = -\tan^{-1}(-x) = \frac{\pi}{2} - \cot^{-1} x = \sin^{-1} \frac{x}{\sqrt{1 + x^2}} = \cos^{-1} \frac{1}{\sqrt{1 + x^2}}$$
$$= \cot^{-1} \frac{1}{x} = \sec^{-1} \sqrt{1 + x^2} = \csc^{-1} \frac{\sqrt{1 + x^2}}{x}$$

$$\cot^{-1} x = \tan^{-1} \frac{1}{x}; \quad \sec^{-1} x = \cos^{-1} \frac{1}{x}; \quad \csc^{-1} x = \sin^{-1} \frac{1}{x}$$

$$\sin^{-1} x \pm \sin^{-1} y = \sin^{-1} y = \sin^{-1} \{x \sqrt{1 - y^2} \pm y \sqrt{1 - x^2}\}$$
$$\cos^{-1} x \pm \cos^{-1} y = \cos^{-1} \{xy \pm \sqrt{(1 - x^2)(1 - y^2)}\}$$
$$\sin^{-1} x \pm \cos^{-1} y = \sin^{-1} \{xy \pm \sqrt{(1 - x^2)(1 - y^2)}\} = \cos^{-1} \{y \sqrt{1 - x^2} \pm x \sqrt{1 - y^2}\}$$

$$\tan^{-1} x \pm \tan^{-1} y = \tan^{-1} \frac{x \pm y}{1 \mp xy}$$

$$\tan^{-1} x \pm \cot^{-1} y = \tan^{-1} \frac{xy \pm 1}{y \mp x} = \cot^{-1} \frac{y \mp x}{xy \pm 1}$$

Solution of Triangles

Relations between Angles and Sides of Plane Triangles. Let a, b, c = sides of triangle; α, β, γ = angles opposite, a, b, c, respectively; A = area of triangle; $s = \frac{1}{2}(a + b + c)$; r = radius of inscribed circle (Fig. 1-48).

$$\frac{a}{\sin \alpha} = \frac{b}{\sin \beta} = \frac{c}{\sin \gamma} \quad \text{(Law of Sines)}$$

$$a^2 = b^2 + c^2 - 2bc \cos \alpha \quad \text{(Law of Cosines)}$$

$$\frac{a - b}{a + b} = \frac{\tan \frac{1}{2}(\alpha - \beta)}{\tan \frac{1}{2}(\alpha + \beta)} \quad \text{(Law of Tangents)}$$

$$\alpha + \beta + \gamma = 180°$$

$a = b \cos \gamma + c \cos \beta; \quad b = c \cos \alpha + a \cos \gamma; \quad c = a \cos \beta + b \cos \alpha$

$A = \sqrt{s(s - a)(s - b)(s - c)}$

$\sin \alpha = \dfrac{2}{bc} A; \quad \sin \beta = \dfrac{2}{ca} A; \quad \sin \gamma = \dfrac{2}{ab} A$

$\sin \dfrac{\alpha}{2} = \left[\dfrac{(s - b)(s - c)}{bc}\right]^{\frac{1}{2}};$

$\sin \dfrac{\beta}{2} = \left[\dfrac{(s - c)(s - a)}{ca}\right]^{\frac{1}{2}};$

$\sin \dfrac{\gamma}{2} = \left[\dfrac{(s - a)(s - b)}{ab}\right]^{\frac{1}{2}}$

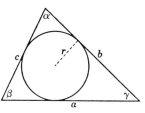

Fig. 1-48

$\cos \dfrac{\alpha}{2} = \left[\dfrac{s(s - a)}{bc}\right]^{\frac{1}{2}}; \quad \cos \dfrac{\beta}{2} = \left[\dfrac{s(s - b)}{ca}\right]^{\frac{1}{2}}; \quad \cos \dfrac{\gamma}{2} = \left[\dfrac{s(s - c)}{ab}\right]$

$\tan \dfrac{\alpha}{2} = \left[\dfrac{(s - b)(s - c)}{s(s - a)}\right]^{\frac{1}{2}};$

$\tan \dfrac{\beta}{2} = \left[\dfrac{(s - c)(s - a)}{s(s - b)}\right]^{\frac{1}{2}};$

$\tan \dfrac{\gamma}{2} = \left[\dfrac{(s - a)(s - b)}{s(s - c)}\right]^{\frac{1}{2}}$

Solution of Plane Oblique Triangles

Given a, b, c. (If logarithms are to be used, use 1.)

(1) $r = \left[\dfrac{(s - a)(s - b)(s - c)}{s}\right]^{\frac{1}{2}}; \quad A = \sqrt{s(s - a)(s - b)(s - c)} = rs;$

$\tan \dfrac{\alpha}{2} = \dfrac{r}{s - a}, \quad \tan \dfrac{\beta}{2} = \dfrac{r}{s - b}, \quad \tan \dfrac{\gamma}{2} = \dfrac{r}{s - c}$

(2) $\cos \alpha = \dfrac{b^2 + c^2 - a^2}{2bc}, \quad \cos \beta = \dfrac{a^2 + c^2 - b^2}{2ac};$

$\cos \gamma = \dfrac{a^2 + b^2 - c^2}{2ab}, \quad \text{or} \quad \gamma = 180° - (\alpha + \beta)$

Given a, b, α.

$\sin \beta = \dfrac{b \sin \alpha}{a}$ (if $a > b$, $\beta < \dfrac{\pi}{2}$ and has only one value, if $b > a$, β has two values,

β_1 and $\beta_2 = 180° - \beta_1$), $\gamma = 180° - (\alpha + \beta)$, $c = \dfrac{a \sin \gamma}{\sin \alpha}$, $A = \tfrac{1}{2}ab \sin \gamma$

Given a, α, β.

$$b = \dfrac{a \sin \beta}{\sin \alpha}, \gamma = 180° - (\alpha + \beta), c = \dfrac{a \sin \gamma}{\sin \alpha}, A = \tfrac{1}{2}ab \sin \gamma$$

Given a, b, γ. (If logarithms are to be used, use 1.)

(1) $\tan \tfrac{1}{2}(\alpha - \beta) = \dfrac{a - b}{a + b} \cot \tfrac{1}{2}\gamma$, $\tfrac{1}{2}(\alpha + \beta) = 90° - \tfrac{1}{2}\gamma$, $c = \dfrac{a \sin \gamma}{\sin \alpha}$,

$A = \tfrac{1}{2}ab \sin \gamma$

(2) $c = \sqrt{a^2 + b^2 - 2ab \cos \gamma}$, $\sin \alpha = \dfrac{a \sin \gamma}{c}$, $\beta = 180° - (\alpha + \gamma)$

(3) $\tan \alpha = \dfrac{a \sin \gamma}{b - a \cos \gamma}$, $\beta = 180° - (\alpha + \gamma)$, $c = \dfrac{a \sin \gamma}{\sin \alpha}$

Mollweide's Check Formulas

(1) $\dfrac{a - b}{c} = \dfrac{\sin \tfrac{1}{2}(\alpha - \beta)}{\cos \tfrac{1}{2}\gamma}$

(2) $\dfrac{a + b}{c} = \dfrac{\cos \tfrac{1}{2}(\alpha - \beta)}{\sin \tfrac{1}{2}\gamma}$

Solution of Plane Right Triangles. Let $\gamma = 90°$ and c be the hypotenuse. Given any two sides or one side and an acute angle α.

$a = \sqrt{c^2 - b^2} = \sqrt{(c + b)(c - b)} = b \tan \alpha = c \sin \alpha$

$b = \sqrt{c^2 - a^2} = \sqrt{(c + a)(c - a)} = \dfrac{a}{\tan \alpha} = c \cos \alpha$

$c = \sqrt{a^2 + b^2} = \dfrac{a}{\sin \alpha} = \dfrac{b}{\cos \alpha}$

$\alpha = \sin^{-1} \dfrac{a}{c} = \cos^{-1} \dfrac{b}{c} = \tan^{-1} \dfrac{a}{b}$; $\beta = 90° - \alpha$

$A = \dfrac{ab}{2} = \dfrac{a^2}{2 \tan \alpha} = \dfrac{b^2 \tan \alpha}{2} = \dfrac{c^2 \sin 2\alpha}{4}$

Spherical Trigonometry

Spherical Trigonometry. Let O be the center of the sphere and a, b, c the sides of a triangle on the surface with opposite angles α, β, γ, respectively, the sides being measured by the angle subtended at the center of the sphere. Let $s = \tfrac{1}{2}(a + b + c)$, $\sigma = \tfrac{1}{2}(\alpha + \beta + \gamma)$, $E = \alpha + \beta + \gamma - 180°$, the spherical excess.

TRIGONOMETRY   53

The following formulas are valid usually only for triangles of which the sides and angles are all between 0° and 180°. To each such triangle there is a polar triangle, whose sides are $180° - \alpha$, $180° - \beta$, $180° - \gamma$, and whose angles are $180° - a$, $180° - b$, $180° - c$.

General Formulas

$\dfrac{\sin a}{\sin \alpha} = \dfrac{\sin b}{\sin \beta} = \dfrac{\sin c}{\sin \gamma}$ (Law of Sines)

$\cos a = \cos b \cos c + \sin b \sin c \cos \alpha$ (Law of Cosines)

$\cos \alpha = -\cos \beta \cos \gamma + \sin \beta \sin \gamma \cos a$ (Law of Cosines)

$\cos a \sin b = \sin a \cos b \cos \gamma + \sin c \cos \alpha$

$\cot a \sin b = \sin \gamma \cot \alpha + \cos \gamma \cos b$

$\cos \alpha \sin \beta = \sin \gamma \cos a - \sin \alpha \cos \beta \cos c$

$\cot \alpha \sin \beta = \sin c \cot a - \cos c \cos \beta$

$\sin \dfrac{a}{2} = \left[\dfrac{-\cos \sigma \cos (\sigma - \alpha)}{\sin \beta \sin \gamma} \right]^{\frac{1}{2}}; \quad \sin \dfrac{\alpha}{2} = \left[\dfrac{\sin (s - b) \sin (s - c)}{\sin b \sin c} \right]^{\frac{1}{2}}$

$\cos \dfrac{a}{2} = \left[\dfrac{\cos (\sigma - \beta) \cos (\sigma - \gamma)}{\sin \beta \sin \gamma} \right]^{\frac{1}{2}}, \quad \cos \dfrac{\alpha}{2} = \left[\dfrac{\sin s \sin (s - a)}{\sin b \sin c} \right]^{\frac{1}{2}}$

$\tan \dfrac{a}{2} = \left[\dfrac{-\cos \sigma \cos (\sigma - \alpha)}{\cos (\sigma - \beta) \cos (\sigma - \gamma)} \right]^{\frac{1}{2}}, \quad \tan \dfrac{\alpha}{2} = \left[\dfrac{\sin (s - b) \sin (s - c)}{\sin s \sin (s - a)} \right]^{\frac{1}{2}}$

$\tan \dfrac{E}{4} = \left[\tan \dfrac{s}{2} \tan \dfrac{s-a}{2} \tan \dfrac{s-b}{2} \tan \dfrac{s-c}{2} \right]^{\frac{1}{2}},$

$\cot \dfrac{E}{2} = \dfrac{\cot \dfrac{a}{2} \cot \dfrac{b}{2} + \cos \gamma}{\sin \gamma}$

$\tan \dfrac{a+b}{2} = \dfrac{\cos \dfrac{\alpha - \beta}{2}}{\cos \dfrac{\alpha + \beta}{2}} \tan \dfrac{c}{2}, \quad \tan \dfrac{a-b}{2} = \dfrac{\sin \dfrac{\alpha - \beta}{2}}{\sin \dfrac{\alpha + \beta}{2}} \tan \dfrac{c}{2}$

$\tan \dfrac{\alpha + \beta}{2} = \dfrac{\cos \dfrac{a-b}{2}}{\cos \dfrac{a+b}{2}} \cot \dfrac{\gamma}{2}, \quad \tan \dfrac{\alpha - \beta}{2} = \dfrac{\sin \dfrac{a-b}{2}}{\sin \dfrac{a+b}{2}} \cot \dfrac{\gamma}{2}$

$\cos \dfrac{\alpha + \beta}{2} \cos \dfrac{c}{2} = \cos \dfrac{a+b}{2} \sin \dfrac{\gamma}{2}, \quad \sin \dfrac{\alpha + \beta}{2} \cos \dfrac{c}{2} = \cos \dfrac{a-b}{2} \cos \dfrac{\gamma}{2}$

$\cos \dfrac{\alpha - \beta}{2} \sin \dfrac{c}{2} = \sin \dfrac{a+b}{2} \sin \dfrac{\gamma}{2}, \quad \sin \dfrac{\alpha - \beta}{2} \sin \dfrac{c}{2} = \sin \dfrac{a-b}{2} \cos \dfrac{\gamma}{2}$

54 MATHEMATICS

The Right Spherical Triangle. Let $\gamma = 90°$ and c be the hypotenuse.

$$\cos c = \cos a \cos b = \cot \alpha \cot \beta, \quad \cos a = \frac{\cos \alpha}{\sin \beta}, \quad \cos b = \frac{\cos \beta}{\sin \alpha},$$

$$\sin \alpha = \frac{\sin a}{\sin c}, \quad \cos \alpha = \frac{\tan b}{\tan c}, \quad \tan \alpha = \frac{\tan a}{\sin b}$$

Hyperbolic Trigonometry

Hyperbolic Angles are defined in a manner similar to circular angles but with reference to an *equilateral hyperbola*. The comparative relations are shown in Figs. 1-49 and 1-50. A *circular angle* is a central angle measured in radians by the ratio s/r or the ratio $2A/r^2$, where A is the area of the sector included by the angle α and the arc s (Fig. 1-49). For the *hyperbola* the radius ρ is not constant and only the value of the *differential hyperbolic angle* $d\theta$ is defined by the ratio ds/ρ. Thus, $\theta = \int ds/\rho = 2A/a^2$, where A represents the shaded area in Fig. 1-50. If both s and ρ are measured in the same units the angle is expressed in *hyperbolic radians*.

Hyperbolic Functions are defined by ratios similar to those defining functions of circular angles and also named similarly. Their names and abbreviations are:

$$\text{Hyperbolic sine } \theta = \frac{y}{a} = \sinh \theta$$

$$\text{Hyperbolic cosine } \theta = \frac{x}{a} = \cosh \theta$$

$$\text{Hyperbolic tangent } \theta = \frac{y}{x} = \tanh \theta$$

$$\text{Hyperbolic cotangent } \theta = \frac{x}{y} = \coth \theta$$

$$\text{Hyperbolic secant } \theta = \frac{a}{x} = \text{sech } \theta$$

$$\text{Hyperbolic cosecant } \theta = \frac{a}{y} = \text{csch } \theta$$

Values and Exponential Equivalents. The values of hyperbolic functions may be computed from their exponential equivalents. The graphs are shown in Fig. 1-51. Values for increments of 0.01 radian are given in Section 9, p. 310.

$$\sinh \theta = \frac{e^\theta - e^{-\theta}}{2}; \quad \cosh \theta = \frac{e^\theta + e^{-\theta}}{2}; \quad \tanh \theta = \frac{e^\theta - e^{-\theta}}{e^\theta + e^{-\theta}}$$

If θ is extremely small, $\sinh \theta \approx \theta$, $\cosh \theta \approx 1$, and $\tanh \theta \approx \theta$. For large values of θ, $\sinh \theta \approx \cosh \theta$, and $\tanh \theta \approx \coth \theta \approx 1$.

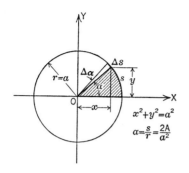
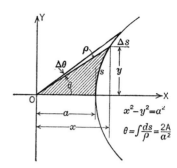

Fig. 1-49   Fig. 1-50

Fundamental Identities

$$\operatorname{csch} \theta = \frac{1}{\sinh \theta}; \quad \operatorname{sech} \theta = \frac{1}{\cosh \theta}, \quad \coth \theta = \frac{1}{\tanh \theta}$$

$\cosh^2 \theta - \sinh^2 \theta = 1, \quad \operatorname{sech}^2 \theta = 1 - \tanh^2 \theta, \quad \operatorname{csch}^2 \theta = \coth^2 \theta - 1$
$\cosh \theta + \sinh \theta = e^\theta, \quad \cosh \theta - \sinh = e^{-\theta}$
$\sinh(-\theta) = -\sinh \theta, \quad \cosh(-\theta) = \cosh \theta$
$\tanh(-\theta) = -\tanh \theta, \quad \coth(-\theta) = -\coth \theta$
$\sinh(\theta_1 \pm \theta_2) = \sinh \theta_1 \cosh \theta_2 \pm \cosh \theta_1 \sinh \theta_2$
$\cosh(\theta_1 \pm \theta_2) = \cosh \theta_1 \cosh \theta_2 \pm \sinh \theta_1 \sinh \theta_2$

$$\tanh(\theta_1 \pm \theta_2) = \frac{\tanh \theta_1 \pm \tanh \theta_2}{1 \pm \tanh \theta_1 \tanh \theta_2}, \quad \coth(\theta_1 \pm \theta_2) = \frac{1 \pm \coth \theta_1 \coth \theta_2}{\coth \theta_1 \pm \coth \theta_2}$$

$$\sinh 2\theta = 2 \sinh \theta \cosh \theta = \frac{2 \tanh \theta}{1 - \tanh^2 \theta}$$

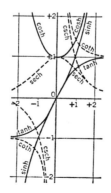

Fig. 1-51

MATHEMATICS

$$\cosh 2\theta = \sinh^2 \theta + \cosh^2 \theta = 1 + 2\sinh^2 \theta = 2\cosh^2 \theta - 1 = \frac{1 + \tanh^2 \theta}{1 - \tanh^2 \theta}$$

$$\tanh 2\theta = \frac{2 \tanh \theta}{1 + \tanh^2 \theta}; \quad \coth 2\theta = \frac{1 + \coth^2 \theta}{2 \coth \theta}$$

$$\sinh \theta/2 = \sqrt{(\cosh \theta - 1)/2}; \quad \cosh \theta/2 = \sqrt{(\cosh \theta + 1)/2}$$

$$\tanh \theta/2 = \sqrt{\frac{\cosh \theta - 1}{\cosh \theta + 1}} = \frac{\sinh \theta}{\cosh \theta + 1} = \frac{\cosh \theta - 1}{\sinh \theta}$$

$$\sinh \theta_1 \pm \sinh \theta_2 = 2 \sinh \frac{(\theta_1 \pm \theta_2)}{2} \cosh \frac{(\theta_1 \mp \theta_2)}{2}$$

$$\cosh \theta_1 + \cosh \theta_2 = 2 \cosh \frac{(\theta_1 + \theta_2)}{2} \cosh \frac{(\theta_1 - \theta_2)}{2}$$

$$\cosh \theta_1 - \cosh \theta_2 = 2 \sinh \frac{(\theta_1 + \theta_2)}{2} \sinh \frac{(\theta_1 - \theta_2)}{2}$$

$$\tanh \theta_1 \pm \tanh \theta_2 = \frac{\sinh(\theta_1 \pm \theta_2)}{\cosh \theta_1 \cosh \theta_2}$$

$$(\cosh \theta \pm \sinh \theta)^n = \cosh n\theta \pm \sinh n\theta$$

Anti-Hyperbolic or Inverse Functions. The inverse hyperbolic sine of u is written: $\sinh^{-1} u$. Values of the inverse functions may be computed from their logarithmic equivalents.

$$\sinh^{-1} u = \log_e (u + \sqrt{u^2 + 1}); \quad \cosh^{-1} u = \log_e (u + \sqrt{u^2 - 1})$$

$$\tanh^{-1} u = \tfrac{1}{2} \log_e \frac{1 + u}{1 - u}; \quad \coth^{-1} u = \tfrac{1}{2} \log_e \frac{u + 1}{u - 1}$$

Functions of Imaginary and Complex Angles

Relations of Hyperbolic to Circular Functions. By comparison of the exponential equivalents of hyperbolic and circular functions the following identities are established ($i = \sqrt{-1}$):

$$\sin \alpha = -i \sinh i\alpha \qquad \sinh \beta = -i \sin i\beta$$
$$\cos \alpha = \cosh i\alpha \qquad \cosh \beta = \cos i\beta$$
$$\tan \alpha = -i \tanh i\alpha \qquad \tanh \beta = -i \tan i\beta$$
$$\cot \alpha = i \coth i\alpha \qquad \coth \beta = i \cot i\beta$$
$$\sec \alpha = \operatorname{sech} i\alpha \qquad \operatorname{sech} \beta = \sec i\beta$$
$$\csc \alpha = i \operatorname{csch} i\alpha \qquad \operatorname{csch} \beta = i \csc i\beta$$

Relations between Inverse Functions.

$$\sin^{-1} A = -i \sinh^{-1} iA \qquad \sinh^{-1} B = -i \sin^{-1} iB$$
$$\cos^{-1} A = -i \cosh^{-1} A \qquad \cosh^{-1} B = i \cos^{-1} B$$
$$\tan^{-1} A = -i \tanh^{-1} iA \qquad \tanh^{-1} B = -i \tan^{-1} iB$$
$$\cot^{-1} A = i \coth^{-1} iA \qquad \coth^{-1} B = i \cot^{-1} iB$$
$$\sec^{-1} A = -i \operatorname{sech}^{-1} A \qquad \operatorname{sech}^{-1} B = i \sec^{-1} B$$
$$\csc^{-1} A = i \operatorname{csch}^{-1} iA \qquad \operatorname{csch}^{-1} B = i \csc^{-1} iB$$

TRIGONOMETRY

Functions of a Complex Angle. In complex notation

$$c = a + ib = |c|(\cos\theta + i\sin\theta) = |c|e^{i\theta}$$

where $|c| = \sqrt{a^2 + b^2}$, $i = \sqrt{-1}$, and $\theta = \tan^{-1}(b/a)$.
$\log_e |c| e^{i\theta} = \log |c| + i(\theta + 2k\pi)$ and is infinitely many valued. By its principal part will be understood $\log_e |c| + i\theta$. Some convenient identities are:

$$\log_e 1 = 0, \quad \log_e(-1) = i\pi, \quad \log_e i = i\frac{\pi}{2}, \quad \log_e(-i) = i\frac{3\pi}{2}$$

$$(\cos\theta \pm i\sin\theta)^n = \cos n\theta \pm i\sin n\theta,$$

$$\sqrt[n]{\cos\theta \pm i\sin\theta} = \cos\frac{\theta + 2\pi k}{n} \pm i\sin\frac{\theta + 2\pi k}{n}$$

The use of complex angles occurs frequently in electric-circuit problems where it is often necessary to express the functions of them as a complex number.

$$\sin(\alpha \pm i\beta) = \sin\alpha\cosh\beta \pm i\cos\alpha\sinh\beta = \sqrt{\cosh^2\beta - \cos^2\alpha}\, e^{\pm i\theta}$$

where $\theta = \tan^{-1}\cot\alpha\tanh\beta$.

$$\cos(\alpha \pm i\beta) = \cos\alpha\cosh\beta \mp i\sin\alpha\sinh\beta = \sqrt{\cosh^2\beta - \sin^2\alpha}\, e^{\pm iv}$$

where $\theta = \tan^{-1}\tan\alpha\tanh\beta$.

$$\sinh(\alpha \pm i\beta) = \sinh\alpha\cos\beta \pm i\cosh\alpha\sin\beta$$
$$= \sqrt{\sinh^2\alpha + \sin^2\beta}\, e^{\pm i\theta} = \sqrt{\cosh^2\alpha - \cos^2\beta}\, e^{\pm i\theta}$$

where $\theta = \tan^{-1}\coth\alpha\tan\beta$.

$$\cosh(\alpha \pm i\beta) = \cosh\alpha\cos\beta \pm i\sinh\alpha\sin\beta$$
$$= \sqrt{\sinh^2\alpha + \cos^2\beta}\, e^{\pm i\theta} = \sqrt{\cosh^2\alpha - \sin^2\beta}\, e^{\pm i\theta}$$

where $\theta = \tan^{-1}\tanh\alpha\tan\beta$.

$$\tan(\alpha \pm i\beta) = \frac{\sin 2\alpha \pm i\sinh 2\beta}{\cos 2\alpha + \cosh 2\beta}, \quad \tanh(\alpha \pm i\beta) = \frac{\sinh 2\alpha \pm i\sin 2\beta}{\cosh 2\alpha + \cos 2\beta}$$

The hyperbolic sine and cosine have the period $2\pi i$; the hyperbolic tangent has the period πi.

$$\sinh(\alpha + 2k\pi i) = \sinh\alpha, \quad \cosh(\alpha + 2k\pi i) = \cosh\alpha$$
$$\tanh(\alpha + k\pi i) = \tanh\alpha, \quad \coth(\alpha + k\pi i) = \coth\alpha$$

Inverse Functions of Complex Numbers

$$\sin^{-1}(A \pm iB) = \sin^{-1}\left[\frac{\sqrt{B^2 + (1+A)^2} - \sqrt{B^2 + (1-A)^2}}{2}\right]$$

$$\pm i\cosh^{-1}\left[\frac{\sqrt{B^2 + (1+A)^2} + \sqrt{B^2 + (1-A)^2}}{2}\right]$$

$$\cos^{-1}(A \pm iB) = \cos^{-1}\left[\frac{\sqrt{B^2 + (1+A)^2} - \sqrt{B^2 + (1-A)^2}}{2}\right]$$

$$\mp i\cosh^{-1}\left[\frac{\sqrt{B^2 + (1+A)^2} + \sqrt{B^2 + (1-A)^2}}{2}\right]$$

$$\tan^{-1}(A \pm iB) = \left[\frac{\pi - \tan^{-1}\frac{A}{\pm B - 1} + \tan^{-1}\frac{A}{\pm B + 1}}{2}\right]$$

$$\pm i\tfrac{1}{4}\log_e \frac{A^2 + (1 \pm B)^2}{A^2 + (1 \mp B)^2}$$

$$\sinh^{-1}(A \pm iB) = \cosh^{-1}\left[\frac{\sqrt{A^2 + (1+B)^2} + \sqrt{A^2 + (1-B)^2}}{2}\right]$$

$$\pm i\sin^{-1}\left[\frac{\sqrt{A^2 + (1+B)^2} - \sqrt{A^2 + (1-B)^2}}{2}\right]$$

$$\cosh^{-1}(A \pm iB) = \cosh^{-1}\left[\frac{\sqrt{B^2 + (1+A)^2} + \sqrt{B^2 + (1-A)^2}}{2}\right]$$

$$\pm i\cos^{-1}\left[\frac{\sqrt{B^2 + (1+A)^2} - \sqrt{B^2 + (1-A)^2}}{2}\right]$$

$$\tanh^{-1}(A \pm iB) = \tfrac{1}{2}\tanh^{-1}\frac{2A}{1 + A^2 + B^2} + i\tfrac{1}{2}\tan^{-1}\frac{\pm 2B}{1 - A^2 - B^2}$$

CALCULUS

Differential Calculus

Definitions and Relations

$$\frac{dy}{dx} = \lim_{\Delta x \to 0}\frac{\Delta y}{\Delta x} = \lim_{\Delta x \to 0}\frac{f(x + \Delta x) - f(x)}{\Delta x} = \frac{d}{dx}f(x) = f'(x) = y'$$

$$\frac{d^2y}{dx^2} = \frac{d}{dx}\left(\frac{dy}{dx}\right) = \frac{d}{dx}f'(x) = f''(x)$$

$$\frac{d^n y}{dx^n} = \frac{d}{dx}\left(\frac{d^{n-1}y}{dx^{n-1}}\right) = \frac{d}{dx}f^{n-1}(x) = f^n(x)$$

$f'(x) = \dfrac{dy}{dx} = \tan\theta$ represents the slope of a curve at point $P(xy)$ where θ is the angle between the tangent line at P and the X axis.

$$f(a) = f(x) \quad \text{when } x = a$$

Table 1-9 Differentiation Formulas

Let $u, v, w, \cdots$ be functions of x; a and n be constants; and e be the base of the natural or Napierian logarithms. Then $e = 2.7183^-$.

$$\frac{d}{dx} a = 0$$

$$\frac{d}{dx}(u + v + w + \cdots) = \frac{du}{dx} + \frac{dv}{dx} + \frac{dw}{dx} + \cdots$$

$$\frac{d}{dx} au = a \frac{du}{dx}$$

$$\frac{d}{dx} uv = u \frac{dv}{dx} + v \frac{du}{dx}$$

$$\frac{d}{dx}(uvw\cdots) = \left(\frac{1}{u}\frac{du}{dx} + \frac{1}{v}\frac{dv}{dx} + \frac{1}{w}\frac{dw}{dx} + \cdots\right)(uvw\cdots)$$

$$\frac{d}{dx}\left(\frac{u}{v}\right) = \frac{v\frac{du}{dx} - u\frac{dv}{dx}}{v^2}$$

$$\frac{d}{dx} u^n = nu^{n-1}\frac{du}{dx}$$

$$\frac{d}{dx} \log_e u = \frac{1}{u}\frac{du}{dx}$$

$$\frac{d}{dx} \log_{10} u = \frac{1}{u}\frac{du}{dx} \log_{10} e = (0.4343)\frac{1}{n}\frac{du}{dx}$$

$$\frac{d}{dx} e^u = e^u \frac{du}{dx}$$

$$\frac{d}{dx} u^v = vu^{v-1}\frac{du}{dx} + u^v \frac{dv}{dx} \log_e u$$

$$\frac{d}{dx} f(u) = \frac{df(u)}{du} \cdot \frac{du}{dx}$$

$$\frac{d^2 f(u)}{dx^2} = \frac{df(u)}{du} \cdot \frac{d^2 u}{dx^2} + \frac{d^2 f(u)}{du^2}\left(\frac{du}{dx}\right)^2$$

$$\frac{d}{dx} \sin u = \cos u \frac{du}{dx}$$

$$\frac{d}{dx} \cos u = -\sin u \frac{du}{dx}$$

$$\frac{d}{dx} \tan u = \sec^2 u \frac{du}{dx}$$

$$\frac{d}{dx} \cot u = -\csc^2 u \frac{du}{dx}$$

$$\frac{d}{dx} \sec u = \sec u \tan u \frac{du}{dx}$$

$$\frac{d}{dx} \csc u = -\csc u \cot u \frac{du}{dx}$$

$$\frac{d}{dx} \sin^{-1} u = \frac{1}{\sqrt{1-u^2}}\frac{du}{dx} \quad \left(-\frac{\pi}{2} \leqq \sin^{-1} u \leqq \frac{\pi}{2}\right)$$

$$\frac{d}{dx} \cos^{-1} u = -\frac{1}{\sqrt{1-u^2}}\frac{du}{dx} \quad (0 \leqq \cos^{-1} u \leqq \pi)$$

$$\frac{d}{dx} \tan^{-1} u = \frac{1}{1+u^2}\frac{du}{dx}$$

$$\frac{d}{dx} \cot^{-1} u = -\frac{1}{1+u^2}\frac{du}{dx}$$

$$\frac{d}{dx} \sec^{-1} u = \frac{1}{u\sqrt{u^2-1}}\frac{du}{dx} \ *$$

$$\frac{d}{dx} \csc^{-1} u = -\frac{1}{u\sqrt{u^2-1}}\frac{du}{dx} \ *$$

$$\frac{d}{dx} \sinh u = \cosh u \frac{du}{dx}$$

$$\frac{d}{dx} \cosh u = \sinh u \frac{du}{dx}$$

$$\frac{d}{dx} \tanh u = \text{sech}^2 u \frac{du}{dx}$$

$$\frac{d}{dx} \coth u = -\text{csch}^2 u \frac{du}{dx}$$

$$\frac{d}{dx} \text{sech}\, u = -\text{sech}\, u \tanh u \frac{du}{dx}$$

$$\frac{d}{dx} \text{csch}\, u = -\text{csch}\, u \coth u \frac{du}{dx}$$

$$\frac{d}{dx} \sinh^{-1} u = \frac{1}{\sqrt{u^2+1}}\frac{du}{dx}$$

$$\frac{d}{dx} \cosh^{-1} u = \frac{1}{\sqrt{u^2-1}}\frac{du}{dx}$$

$$\frac{d}{dx} \tanh^{-1} u = \frac{1}{1-u^2}\frac{du}{dx}$$

$$\frac{d}{dx} \coth^{-1} u = \frac{1}{1-u^2}\frac{du}{dx}$$

$$\frac{d}{dx} \text{sech}^{-1} u = -\frac{1}{u\sqrt{1-u^2}}\frac{du}{dx}$$

$$\frac{d}{dx} \text{csch}^{-1} u = -\frac{1}{u\sqrt{u^2+1}}\frac{du}{dx}$$

* For angles in the first and third quadrants. Use the opposite sign in the second and fourth quadrants.

$x = x(t)$ represents that x is a function of t, i.e., $x = f(t)$

$$\frac{dy}{dx} = 1 \bigg/ \frac{dx}{dy} \quad \text{when } x = f(y)$$

$$\frac{dy}{dx} = \frac{dy}{dt} \bigg/ \frac{dx}{dt} \quad \text{when } x = x(t) \text{ and } y = y(t)$$

$$\frac{dy}{dx} = \frac{dy}{du} \cdot \frac{du}{dx} \quad \text{when } y = y(u) \text{ and } u = u(x)$$

$f(x + h) = f(x) + hf'(x + \theta h)$, where $0 < \theta < 1$ and $f(x)$ is single valued and continuous over the interval chosen.

Maximum at $x = x_1$: $f'(x_1) = 0$ or ∞, and $f''(x_1) < 0$
Minimum at $x = x_1$: $f'(x_1) = 0$ or ∞, and $f''(x_1) > 0$

Table 1-10 Derivatives of Common Functions

$f(x)$	$f'(x)$	$f^n(x)$
x^a	ax^{a-1}	$a(a-1)(a-2) \cdots (a-n+1)x^{a-n}$
x^x	$x^x(1 + \log_e x)$	
e^{ax}	ae^{ax}	$a^n e^{ax}$
a^x	$a^x \log_e a$	$a^x(\log_e a)^n$
$\log_e x$	$1/x$	$(-1)^{n-1}(n-1)! \dfrac{1}{x} n$
$\log_a x$	$\dfrac{1}{x} \log_a e$	$(-1)^{n-1}(n-1)! \dfrac{1}{x} n \log_a e$
$\sin x$	$\cos x$	$\sin\left(x + \dfrac{\pi n}{2}\right)$
$\cos x$	$-\sin x$	$\cos\left(x + \dfrac{\pi n}{2}\right)$

Partial Derivatives

Functions of Two Variables. If three variables $f(x, y)$, x, y are so related that to each pair of values of x and y in a given range there corresponds a value of $f(x, y)$, then $f(x, y)$ is a function of x and y in that range. If x is considered as the only variable while y is taken as constant, then the derivative of $f(x, y)$ with respect to x is called the *partial derivative* of f with respect to x and is denoted by

$$\frac{\partial f}{\partial x} = f_x = \lim_{\Delta x \to 0} \frac{f(x + \Delta x, y) - f(x, y)}{\Delta x}$$

Likewise, the partial derivative of f with respect to y is obtained by considering x to be constant while y varies:

$$\frac{\partial f}{\partial y} = f_y = \lim_{\Delta y \to 0} \frac{f(x, y + \Delta y) - f(x, y)}{\Delta y}$$

If $\partial f / \partial x$ and $\partial f / \partial y$ are again differentiable, the partial derivatives of the second order may be found.

$$\frac{\partial}{\partial x}\left(\frac{\partial f}{\partial x}\right) = \frac{\partial^2 f}{\partial x^2} = f_{xx} \qquad \frac{\partial}{\partial y}\left(\frac{\partial f}{\partial y}\right) = \frac{\partial^2 f}{\partial y^2} = f_{yy}$$

$$\frac{\partial}{\partial x}\left(\frac{\partial f}{\partial y}\right) = \frac{\partial^2 f}{\partial x\,\partial y} = f_{yx} \qquad \frac{\partial}{\partial y}\left(\frac{\partial f}{\partial x}\right) = \frac{\partial^2 f}{\partial y\,\partial x} = f_{xy}$$

If the derivatives in question are continuous, the order of differentiation is immaterial, that is,

$$\frac{\partial^2 f}{\partial y\,\partial x} = \frac{\partial^2 f}{\partial x\,\partial y}$$

Similarly, the third and higher partial derivatives of $f(x, y)$ may be found. The third partial derivatives, if continuous, are four in number:

$$\frac{\partial}{\partial x}\left(\frac{\partial^2 f}{\partial x^2}\right) = \frac{\partial^3 f}{\partial x^3} \qquad \frac{\partial}{\partial x}\left(\frac{\partial^2 f}{\partial y^2}\right) = \frac{\partial}{\partial y}\left(\frac{\partial^2 f}{\partial x\,\partial y}\right) = \frac{\partial^2}{\partial y^2}\left(\frac{\partial f}{\partial x}\right) = \frac{\partial^3 f}{\partial x\,\partial y^2}$$

$$\frac{\partial}{\partial y}\left(\frac{\partial^2 f}{\partial y^2}\right) = \frac{\partial^3 f}{\partial y^3} \qquad \frac{\partial}{\partial y}\left(\frac{\partial^2 f}{\partial x^2}\right) = \frac{\partial}{\partial x}\left(\frac{\partial^2 f}{\partial x\,\partial y}\right) = \frac{\partial^2}{\partial x^2}\left(\frac{\partial f}{\partial y}\right) = \frac{\partial^3 f}{\partial x^2\,\partial y}$$

Functions of N Variables. The formulas above may be generalized to the case where f is a function of more than two variables, that is, there corresponds a value of $f(x, y, z, \ldots)$ to every set of values of $x, y, z, \ldots$.

In general,

$$d^n f = \left(\frac{\partial}{\partial x}dx + \frac{\partial}{\partial y}dy + \frac{\partial}{\partial z}dz + \cdots\right)^n f(x, y, z, \ldots)$$

Exact Differential. In order for the expression $P(x, y)\,dx + Q(x, y)\,dy$ to be the *exact* or *complete* differential of a function of two variables, it is necessary and sufficient that

$$\frac{\partial Q}{\partial x} = \frac{\partial P}{\partial y} \quad \text{(integrability condition)}$$

For three variables, $P\,dx + Q\,dy + R\,dz$, the corresponding conditions are

$$\frac{\partial Q}{\partial z} = \frac{\partial R}{\partial y}, \quad \frac{\partial R}{\partial x} = \frac{\partial P}{\partial z}, \quad \frac{\partial P}{\partial y} = \frac{\partial Q}{\partial x}$$

Differentiation of Composite Functions. If $u = f(x, y, z, \cdots w)$, and $x, y, z, \cdots w$ are functions of a single variable t, then

$$\frac{du}{dt} = \frac{\partial u}{\partial x}\frac{dx}{dt} + \frac{\partial u}{\partial y}\frac{dy}{dt} + \cdots + \frac{\partial u}{\partial w}\frac{dw}{dt}$$

which is the total derivative of u with respect to t.

Integral Calculus

$F(x)$ is an *indefinite integral* of $f(x)$ if $dF(x) = f(x)\,dx$. If $f(x)$ is single valued and integrable over the range (a, b),

$$\int_a^x f(u)\,du = F(x)\Big]_a^x = F(x) - F(a), \qquad a \le x \le b$$

In general,

$$\int f(x)\,dx = F(x) + C$$

The constant C is usually omitted in tables of integrals.

Integrals containing fractional powers of polynomials above the first degree usually cannot be integrated in terms of the elementary integral forms except for the square root of polynomials of the second degree.

Table 1-11   Table of Integrals

Elementary Indefinite Integrals

1. $\int a\,dx = ax$

2. $\int (u + v + w + \cdots)\,dx = \int u\,dx + \int v\,dx + \int w\,dx + \cdots$

3. $\int u\,dv = uv - \int v\,du$, integration by parts

4. $\int \log_e x\,dx = x \log_e x - x$

5. $\int x^n\,dx = \dfrac{x^{n+1}}{n+1}, (n \ne -1)$

6. $\int \dfrac{dx}{x} = \log_e x + c = \log_e c_1 x, [\log_e x = \log_e (-x) + (2k+1)\pi i]$

7. $\int e^{ax}\,dx = \dfrac{1}{a} e^{ax}$

8. $\int a^x\,dx = \dfrac{a^x}{\log_e a}$

9. $\int a^x \log_e a\,dx = a^x$

10. $\int \sin ax\,dx = -\dfrac{1}{a} \cos ax$

11. $\displaystyle\int \cos ax\, dx = \frac{1}{a} \sin ax$

12. $\displaystyle\int \tan ax\, dx = -\frac{1}{a} \log_e \cos ax = \frac{1}{a} \log_e \sec ax$

13. $\displaystyle\int \cot ax\, dx = \frac{1}{a} \log_e \sin ax = -\frac{1}{a} \log_e \csc ax$

14. $\displaystyle\int \sec ax\, dx = \frac{1}{a} \log_e (\sec ax + \tan ax) = \frac{1}{a} \log_e \tan \left(\frac{ax}{2} + \frac{\pi}{4}\right)$

15. $\displaystyle\int \csc ax\, dx = \frac{1}{a} \log_e (\csc ax - \cot ax) = \frac{1}{a} \log_e \tan \frac{ax}{2}$

16. $\displaystyle\int \frac{dx}{\sqrt{a^2 - x^2}} = \sin^{-1} \frac{x}{a} = -\cos^{-1} \frac{x}{a}\ (x^2 < a^2)$

17. $\displaystyle\int \frac{dx}{a^2 + x^2} = \frac{1}{a} \tan^{-1} \frac{x}{a} = -\frac{1}{a} \cot^{-1} \frac{x}{a}$

18. $\displaystyle\int \sinh ax\, dx = \frac{1}{a} \cosh ax$

19. $\displaystyle\int \cosh ax\, dx = \frac{1}{a} \sinh ax$

20. $\displaystyle\int \tanh ax\, dx = \frac{1}{a} \log_e (\cosh ax)$

21. $\displaystyle\int \coth ax\, dx = \frac{1}{a} \log_e (\sinh ax)$

22. $\displaystyle\int \operatorname{sech} ax\, dx = \frac{1}{a} \sin^{-1} (\tanh ax) = \frac{1}{a} \tan^{-1} (\sinh ax)$

23. $\displaystyle\int \operatorname{csch} ax\, dx = \frac{1}{a} \log_e \left(\tanh \frac{ax}{2}\right)$

24. $\displaystyle\int \sin^2 ax\, dx = \frac{1}{2} x - \frac{1}{2a} \sin ax \cos ax = \frac{1}{2} x - \frac{1}{4a} \sin 2ax$

25. $\displaystyle\int \cos^2 ax\, dx = \frac{1}{2} x + \frac{1}{2a} \sin ax \cos ax = \frac{1}{2} x + \frac{1}{4a} \sin 2ax$

26. $\displaystyle\int \tan^2 ax\, dx = \frac{1}{a} \tan ax - x$

27. $\displaystyle\int \cot^2 ax\, dx = -\frac{1}{a} \cot ax - x$

28. $\int \sec^2 ax\, dx = \frac{1}{a} \tan ax$

29. $\int \csc^2 ax\, dx = -\frac{1}{a} \cot ax$

30. $\int \sin^{-1} ax\, dx = x \sin^{-1} ax + \frac{1}{a}\sqrt{1 - a^2x^2}$

31. $\int \cos^{-1} ax\, dx = x \cos^{-1} ax - \frac{1}{a}\sqrt{1 - a^2x^2}$

32. $\int \tan^{-1} ax\, dx = x \tan^{-1} ax - \frac{1}{2a} \log_e (1 + a^2x^2)$

33. $\int \cot^{-1} ax\, dx = x \cot^{-1} ax + \frac{1}{2a} \log_e (1 + a^2x^2)$

34. $\int \sec^{-1} ax\, dx = x \sec^{-1} ax - \frac{1}{a} \log_e (ax + \sqrt{a^2x^2 - 1})$

35. $\int \csc^{-1} ax\, dx = x \csc^{-1} ax + \frac{1}{a} \log_e (ax + \sqrt{a^2x^2 - 1})$

Integrals Involving $(ax + b)$

36. $\int (ax + b)^n\, dx = \frac{1}{a(n + 1)} (ax + b)^{n+1} (n \neq -1)$

37. $\int \frac{dx}{ax + b} = \frac{1}{a} \log_e (ax + b)$

38. $\int x(ax + b)^n\, dx = \frac{1}{a^2(n + 2)} (ax + n)^{n+2} - \frac{b}{a^2(n + 1)} (ax + b)^{n+1} (n \neq -1, -2)$

39. $\int \frac{x\, dx}{ax + b} = \frac{x}{a} - \frac{b}{a^2} \log_e (ax + b)$

40. $\int \frac{x\, dx}{(ax + b)^2} = \frac{b}{a^2(ax + b)} + \frac{1}{a^2} \log_e (ax + b)$

41. $\int \frac{x^2\, dx}{ax + b} = \frac{1}{a^3} \left[\frac{1}{2}(ax + b)^2 - 2b(ax + b) + b^2 \log_e (ax + b) \right]$

42. $\int \frac{x^2\, dx}{(ax + b)^2} = \frac{1}{a^3} \left[(ax + b) - 2b \log_e (ax + b) - \frac{b^2}{ax + b} \right]$

43. $\int \frac{x^2\, dx}{(ax + b)^3} = \frac{1}{a^3} \left[\log_e (ax + b) + \frac{2b}{ax + b} - \frac{b^2}{2(ax + b)^2} \right]$

44. $\int \frac{dx}{x(ax + b)} = \frac{1}{b} \log_e \frac{x}{ax + b}$

CALCULUS 65

45. $\int \dfrac{dx}{x^2(ax+b)} = -\dfrac{1}{bx} + \dfrac{a}{b^2}\log_e \dfrac{ax+b}{x}$

46. $\int \dfrac{dx}{x(ax+b)^2} = \dfrac{1}{b(ax+b)} - \dfrac{1}{b^2}\log_e \dfrac{ax+b}{x}$

47. $\int \dfrac{dx}{x^2(ax+b)^2} = -\dfrac{b+2ax}{b^2x(ax+b)} + \dfrac{2a}{b^3}\log_e \dfrac{ax+b}{x}$

48. $\int \dfrac{dx}{x\sqrt{ax+b}} = \dfrac{1}{\sqrt{b}}\log_e \dfrac{\sqrt{ax+b} - \sqrt{b}}{\sqrt{ax+b} + \sqrt{b}}$ (b positive)

49. $\int \dfrac{dx}{x\sqrt{ax+b}} = \dfrac{2}{\sqrt{-b}}\tan^{-1}\left(\dfrac{ax+b}{-b}\right)^{1/2}$ (b negative)

50. $\int \dfrac{\sqrt{ax+b}}{x}\,dx = 2\sqrt{ax+b} + \sqrt{b}\log_e \dfrac{\sqrt{ax+b} - \sqrt{b}}{\sqrt{ax+b} + \sqrt{b}}$ (b positive)

51. $\int \dfrac{\sqrt{ax+b}}{x}\,dx = 2\sqrt{ax+b} - 2\sqrt{-b}\tan^{-1}\left(\dfrac{ax+b}{-b}\right)^{1/2}$ (b negative)

52. $\int \dfrac{dx}{x^2\sqrt{ax+b}} = -\dfrac{\sqrt{ax+b}}{bx} - \dfrac{a}{2b\sqrt{b}}\log_e \dfrac{\sqrt{ax+b} - \sqrt{b}}{\sqrt{ax+b} + \sqrt{b}}$ (b positive)

53. $\int \dfrac{dx}{x^2\sqrt{ax+b}} = -\dfrac{\sqrt{ax+b}}{bx} - \dfrac{a}{b\sqrt{-b}}\tan^{-1}\left(\dfrac{ax+b}{-b}\right)^{1/2}$ (b negative)

54. $\int \dfrac{ax+b}{fx+g}\,dx = \dfrac{ax}{f} + \dfrac{bf-ag}{f^2}\log_e (fx+g)$

55. $\int \dfrac{dx}{(ax+b)(fx+g)} = \dfrac{1}{bf-ag}\log_e \left(\dfrac{fx+g}{ax+b}\right)$ ($ag \neq bf$)

56. $\int \dfrac{x\,dx}{(ax+b)(fx+g)} = \dfrac{1}{bf-ag}\left[\dfrac{b}{a}\log_e (ax+b) - \dfrac{g}{f}\log_e (fx+g)\right]$ ($ag \neq bf$)

57. $\int \dfrac{dx}{(ax+b)^2(fx+g)} = \dfrac{1}{bf-ag}\left(\dfrac{1}{ax+b} + \dfrac{f}{bf-ag}\log_e \dfrac{fx+g}{ax+b}\right)$ ($ag \neq bf$)

Integrals Involving $(ax^n + b)$

58. $\int (ax^2+b)^n x\,dx = \dfrac{1}{2a}\dfrac{(ax^2+b)^{n+1}}{n+1}$ ($n \neq -1$)

59. $\int \dfrac{dx}{ax^2+b} = \dfrac{1}{\sqrt{ab}}\tan^{-1}(x\sqrt{a/b})$ (a and b positive)

60. $\int \dfrac{dx}{ax^2+b} = \dfrac{1}{2\sqrt{-ab}}\log_e \dfrac{x\sqrt{a} - \sqrt{-b}}{x\sqrt{a} + \sqrt{-b}}$ (a positive, b negative)

$\phantom{60.\int \dfrac{dx}{ax^2+b}} = \dfrac{1}{2\sqrt{-ab}}\log_e \dfrac{\sqrt{b} - x\sqrt{-a}}{\sqrt{b} - x\sqrt{-a}}$ (a negative, b positive)

61. $\int \dfrac{dx}{x(ax^2 + b)} = \dfrac{1}{2b} \log_e \dfrac{x^2}{ax^2 + b}$

62. $\int \dfrac{dx}{(ax^2 + b)^n} = \dfrac{1}{2(n-1)b} \dfrac{x}{(ax^2 + b)^{n-1}} + \dfrac{2n-3}{2(n-1)b} \int \dfrac{dx}{(ax^2 + b)^{n-1}}$

$(n \text{ integer} > 1)$

63. $\int \dfrac{x^2 \, dx}{ax^2 + b} = \dfrac{x}{a} - \dfrac{b}{a} \int \dfrac{dx}{ax^2 + b}$

64. $\int \dfrac{x^2 \, dx}{(ax^2 + b)^n} = -\dfrac{1}{2(n-1)a} \dfrac{x}{(ax^2 + b)^{n-1}} + \dfrac{1}{2(n-1)a} \int \dfrac{dx}{(ax^2 + b)^{n-1}}$

$(n \text{ integer} > 1)$

65. $\int \dfrac{dx}{x^2(ax^2 + b)^n} = \dfrac{1}{b} \int \dfrac{dx}{x^2(ax^2 + b)^{n-1}} - \dfrac{a}{b} \int \dfrac{dx}{(ax^2 + b)^n}$ $(n = \text{positive integer})$

66. $\int \sqrt{ax^2 + b} \, dx = \dfrac{x}{2} \sqrt{ax^2 + b} + \dfrac{b}{2\sqrt{a}} \log_e \dfrac{x\sqrt{a} + \sqrt{ax^2 + b}}{\sqrt{b}}$ $(a \text{ positive})$

67. $\int \sqrt{ax^2 + b} \, dx = \dfrac{x}{2} \sqrt{ax^2 + b} + \dfrac{b}{2\sqrt{-a}} \sin^{-1}(x\sqrt{-a/b})$ $(a \text{ negative})$

68. $\int \dfrac{dx}{\sqrt{ax^2 + b}} = \dfrac{1}{\sqrt{a}} \log_e (x\sqrt{a} + \sqrt{ax^2 + b})$ $(a \text{ positive})$

69. $\int \dfrac{dx}{\sqrt{ax^2 + b}} = \dfrac{1}{\sqrt{-a}} \sin^{-1}(x\sqrt{-a/b})$ $(a \text{ negative})$

70. $\int \dfrac{x \, dx}{\sqrt{ax^2 + b}} = \dfrac{1}{a} \sqrt{ax^2 + b}$

71. $\int \dfrac{\sqrt{ax^2 + b}}{x} \, dx = \sqrt{ax^2 + b} + \sqrt{b} \log_e \dfrac{\sqrt{ax^2 + b} - \sqrt{b}}{x}$ $(b \text{ positive})$

72. $\int \dfrac{\sqrt{ax^2 + b}}{x} \, dx = \sqrt{ax^2 + b} - \sqrt{-b} \tan^{-1} \dfrac{\sqrt{ax^2 + b}}{\sqrt{-b}}$ $(b \text{ negative})$

73. $\int x\sqrt{ax^2 + b} \, dx = \dfrac{1}{3a} (ax^2 + b)^{3/2}$

74. $\int x^2 \sqrt{ax^2 + b} \, dx = \dfrac{x}{4a} (ax^2 + b)^{3/2} - \dfrac{bx}{8a} \sqrt{ax^2 + b} - \dfrac{b^2}{8a\sqrt{a}} \log_e (x\sqrt{a} + \sqrt{ax^2 + b})$

$(a \text{ positive})$

75. $\int x^2 \sqrt{ax^2 + b} \, dx = \dfrac{x}{4a} (ax^2 + b)^{3/2} - \dfrac{bx}{8a} \sqrt{ax^2 + b} - \dfrac{b^2}{8a\sqrt{-a}} \sin^{-1}(x\sqrt{-a/b})$

$(a \text{ negative})$

76. $\int \dfrac{dx}{x\sqrt{ax^2 + b}} = \dfrac{1}{\sqrt{b}} \log_e \dfrac{\sqrt{ax^2 + b} - \sqrt{b}}{x}$ $(b \text{ positive})$

77. $\displaystyle\int \frac{dx}{x\sqrt{ax^2+b}} = \frac{1}{\sqrt{-b}} \sec^{-1}(x\sqrt{-a/b})$ (b negative)

78. $\displaystyle\int \frac{x^2\,dx}{\sqrt{ax^2+b}} = \frac{x}{2a}\sqrt{ax^2+b} - \frac{b}{2a\sqrt{a}}\log_e(x\sqrt{a}+\sqrt{ax^2+b})$ (a positive)

79. $\displaystyle\int \frac{x^2\,dx}{\sqrt{ax^2+b}} = \frac{x}{2a}\sqrt{ax^2+b} - \frac{b}{2a\sqrt{-a}}\sin^{-1}(x\sqrt{-a/b})$ (a negative)

80. $\displaystyle\int \frac{\sqrt{ax^2+b}}{x^2}\,dx = -\frac{\sqrt{ax^2+b}}{x} + \sqrt{a}\,\log_e(x\sqrt{a}+\sqrt{ax^2+b})$ (a positive)

81. $\displaystyle\int \frac{\sqrt{ax^2+b}}{x^2}\,dx = -\frac{\sqrt{ax^2+b}}{x} - \sqrt{-a}\,\sin^{-1}(x\sqrt{-a/b})$ (a negative)

82. $\displaystyle\int \frac{dx}{x(ax^n+b)} = \frac{1}{bn}\log_e \frac{x^n}{ax^n+b}$

83. $\displaystyle\int \frac{dx}{x\sqrt{ax^n+b}} = \frac{1}{n\sqrt{b}}\log_e \frac{\sqrt{ax^n+b}-\sqrt{b}}{\sqrt{ax^n+b}+\sqrt{b}}$ (b positive)

84. $\displaystyle\int \frac{dx}{x\sqrt{ax^n+b}} = \frac{2}{n\sqrt{-b}}\sec^{-1}\sqrt{-ax^n/b}$ (b negative)

Integrals Involving (ax^2+bx+d)

85. $\displaystyle\int \frac{dx}{ax^2+bx+d} = \frac{1}{\sqrt{b^2-4ad}}\log_e \frac{2ax+b-\sqrt{b^2-4ad}}{2ax+b+\sqrt{b^2-4ad}}$ ($b^2 > 4ad$)

86. $\displaystyle\int \frac{dx}{ax^2+bx+d} = \frac{2}{\sqrt{4ad-b^2}}\tan^{-1}\frac{2ax+b}{\sqrt{4ad-b^2}}$ ($b^2 < 4ad$)

87. $\displaystyle\int \frac{dx}{ax^2+bx+d} = -\frac{2}{2ax+b}$ ($b^2 = 4ad$)

88. $\displaystyle\int \frac{dx}{\sqrt{ax^2+bx+d}} = \frac{1}{\sqrt{a}}\log_e(2ax+b+2\sqrt{a(ax^2+bx+d)})$ (a positive)

89. $\displaystyle\int \frac{dx}{\sqrt{ax^2+bx+d}} = \frac{1}{\sqrt{-a}}\sin^{-1}\frac{-2ax-b}{\sqrt{b^2-4ad}}$ (a negative)

90. $\displaystyle\int \frac{x\,dx}{ax^2+bx+d} = \frac{1}{2a}\log_e(ax^2+bx+d) - \frac{b}{2a}\int \frac{dx}{ax^2+bx+d}$

91. $\displaystyle\int \frac{x\,dx}{\sqrt{ax^2+bx+d}} = \frac{\sqrt{ax^2+bx+d}}{a} - \frac{b}{2a}\int \frac{dx}{\sqrt{ax^2+bx+d}}$

92. $\displaystyle\int \frac{dx}{x\sqrt{ax^2+bx+d}} = -\frac{1}{\sqrt{d}}\log_e\left(\frac{\sqrt{ax^2+bx+d}+\sqrt{d}}{x} + \frac{b}{2\sqrt{d}}\right)$ (d positive)

93. $\int \frac{dx}{x\sqrt{ax^2 + bx + d}} = \frac{1}{\sqrt{-d}} \sin^{-1} \frac{bx + 2d}{x\sqrt{b^2 - 4ad}}$ (d negative)

94. $\int \frac{dx}{x\sqrt{ax^2 + bx}} = -\frac{2}{bx} \sqrt{ax^2 + bx}$

95. $\int \sqrt{ax^2 + bx + d} \, dx = \frac{2ax + d}{4a} \sqrt{ax^2 + bx + d} + \frac{4ad - b}{8a} \int \frac{dx}{\sqrt{ax^2 + bx + d}}$

96. $\int x\sqrt{ax^2 + bx + d} \, dx = \frac{(ax^2 + bx + d)^{3/2}}{3a} - \frac{b}{2a} \int \sqrt{ax^2 + bx + d} \, dx$

Integrals Involving $\sin^n ax$

97. $\int \sin^3 ax \, dx = -\frac{1}{a} \cos ax + \frac{1}{3a} \cos^3 ax$

98. $\int \sin^4 ax \, dx = \frac{3}{8} x - \frac{1}{4a} \sin 2ax + \frac{1}{32a} \sin 4ax$

99. $\int \sin^n ax \, dx = -\frac{\sin^{n-1} ax \cos ax}{na} + \frac{n-1}{n} \int \sin^{n-2} ax \, dx$ (n = positive integer)

100. $\int x \sin ax \, dx = \frac{\sin ax}{a^2} - \frac{x \cos ax}{a}$

101. $\int x^2 \sin ax \, dx = \frac{2x}{a^2} \sin ax - \left(\frac{x^2}{a} - \frac{2}{a^3}\right) \cos ax$

102. $\int x^3 \sin ax \, dx = \left(\frac{3x^2}{a^2} - \frac{6}{a^4}\right) \sin ax - \left(\frac{x^3}{a} - \frac{6x}{a^3}\right) \cos ax$

103. $\int x^n \sin ax \, dx = -\frac{x^n}{a} \cos ax + \frac{n}{a} \int x^{n-1} \cos ax \, dx$ ($n > 0$)

104. $\int \frac{\sin ax}{x^n} dx = -\frac{1}{n-1} \frac{\sin ax}{x^{n-1}} + \frac{a}{n-1} \int \frac{\cos ax}{x^{n-1}} dx$

105. $\int \frac{dx}{\sin^n ax} = -\frac{1}{a(n-1)} \frac{\cos ax}{\sin^{n-1} ax} + \frac{n-2}{n-1} \int \frac{dx}{\sin^{n-2} ax}$ (n integer > 1)

106. $\int \frac{x \, dx}{\sin^2 ax} = -\frac{x}{a} \cot ax + \frac{1}{a^2} \log_e \sin ax$

107. $\int \frac{dx}{1 + \sin ax} = -\frac{1}{a} \tan \left(\frac{\pi}{4} - \frac{ax}{2}\right)$

108. $\int \frac{dx}{1 - \sin ax} = \frac{1}{a} \cot \left(\frac{\pi}{4} - \frac{ax}{2}\right)$

CALCULUS   69

109. $\int \dfrac{x\,dx}{1+\sin ax} = -\dfrac{x}{a}\tan\left(\dfrac{\pi}{4}-\dfrac{ax}{2}\right)+\dfrac{2}{a^2}\log_e\cos\left(\dfrac{\pi}{4}-\dfrac{ax}{2}\right)$

110. $\int \dfrac{x\,dx}{1-\sin ax} = \dfrac{x}{a}\cot\left(\dfrac{\pi}{4}-\dfrac{ax}{2}\right)+\dfrac{2}{a^2}\log_e\sin\left(\dfrac{\pi}{4}-\dfrac{ax}{2}\right)$

111. $\int \dfrac{dx}{b+d\sin ax} = \dfrac{-2}{a\sqrt{b^2-d^2}}\tan^{-1}\left[\left(\dfrac{b-d}{b+d}\right)^{1/2}\tan\left(\dfrac{\pi}{4}-\dfrac{ax}{2}\right)\right]$ $(b^2 > d^2)$

112. $\int \dfrac{dx}{b+d\sin ax} = \dfrac{-1}{a\sqrt{d^2-b^2}}\log_e\dfrac{d+b\sin ax+\sqrt{d^2-b^2}\cos ax}{b+d\sin ax}$ $(d^2 > b^2)$

113. $\int \sin ax \sin bx\,dx = \dfrac{\sin(a-b)x}{2(a-b)} - \dfrac{\sin(a+b)x}{2(a+b)}$ $(a^2 \ne b^2)$

Integrals Involving $\cos^n ax$

114. $\int \cos^3 ax\,dx = \dfrac{1}{a}\sin ax - \dfrac{1}{3a}\sin^3 ax$

115. $\int \cos^4 ax\,dx = \dfrac{3}{8}x + \dfrac{1}{4a}\sin 2ax + \dfrac{1}{32a}\sin 4ax$

116. $\int \cos^n ax\,dx = \dfrac{\cos^{n-1}ax\,\sin ax}{na} + \dfrac{n-1}{n}\int \cos^{n-2}ax\,dx$ (n = positive integer)

117. $\int x\cos ax\,dx = \dfrac{\cos ax}{a^2} + \dfrac{x\sin ax}{a}$

118. $\int x^2\cos ax\,dx = \dfrac{2x}{a^2}\cos ax + \left(\dfrac{x^2}{a}-\dfrac{2}{a^3}\right)\sin ax$

119. $\int x^3\cos ax\,dx = \left(\dfrac{3x^2}{a^2}-\dfrac{6}{a^4}\right)\cos ax + \left(\dfrac{x^3}{a}-\dfrac{6x}{a^3}\right)\sin ax$

120. $\int x^n\cos ax\,dx = \dfrac{x^n\sin ax}{a} - \dfrac{n}{a}\int x^{n-1}\sin ax\,dx$ $(n > 0)$

121. $\int \dfrac{\cos ax}{x^n}\,dx = -\dfrac{1}{n-1}\dfrac{\cos ax}{x^{n-1}} - \dfrac{a}{n-1}\int \dfrac{\sin ax}{x^{n-1}}\,dx$

122. $\int \dfrac{dx}{\cos^n ax} = \dfrac{1}{a(n-1)}\dfrac{\sin ax}{\cos^{n-1}ax} + \dfrac{n-2}{n-1}\int \dfrac{dx}{\cos^{n-2}ax}$ (n integer > 1)

123. $\int \dfrac{x\,dx}{\cos^2 ax} = \dfrac{x}{a}\tan ax + \dfrac{1}{a^2}\log_e\cos ax$

124. $\int \dfrac{dx}{1+\cos ax} = \dfrac{1}{a}\tan\dfrac{ax}{2}$

125. $\int \dfrac{dx}{1-\cos ax} = -\dfrac{1}{a}\cot\dfrac{ax}{2}$

126. $\int \dfrac{x\,dx}{1+\cos ax} = \dfrac{x}{a}\tan\dfrac{ax}{2} + \dfrac{2}{a^2}\log_e \cos\dfrac{ax}{2}$

127. $\int \dfrac{x\,dx}{1-\cos ax} = -\dfrac{x}{a}\cot\dfrac{ax}{2} + \dfrac{2}{a^2}\log_e \sin\dfrac{ax}{2}$

128. $\int \dfrac{dx}{b+d\cos ax} = \dfrac{2}{a\sqrt{b^2-d^2}}\tan^{-1}\left(\left(\dfrac{b-d}{b+d}\right)\tan\dfrac{ax}{2}\right)\quad (b^2 > d^2)$

129. $\int \dfrac{dx}{b+d\cos ax} = \dfrac{1}{a\sqrt{d^2-b^2}}\log_e \dfrac{d+b\cos ax + \sqrt{d^2-b^2}\sin ax}{b+d\cos ax}\quad (d^2 > b^2)$

130. $\int \cos ax \cos bx\, dx = \dfrac{\sin(a-b)x}{2(a-b)} + \dfrac{\sin(a+b)x}{2(a+b)}\quad (a^2 \neq b^2)$

Integrals Involving $\sin^n ax$, $\cos^n ax$

131. $\int \sin ax \cos bx\, dx = -\dfrac{1}{2}\left[\dfrac{\cos(a-b)x}{a-b} + \dfrac{\cos(a+b)x}{a+b}\right]\quad (a^2 \neq b^2)$

132. $\int \sin^2 ax \cos^2 ax\, dx = \dfrac{x}{8} - \dfrac{\sin 4ax}{32a}$

133. $\int \sin^n ax \cos ax\, dx = \dfrac{1}{a(n+1)}\sin^{n+1} ax\quad (n \neq -1)$

134. $\int \sin ax \cos^n ax\, dx = -\dfrac{1}{a(n+1)}\cos^{n+1} ax\quad (n \neq -1)$

135. $\int \sin^n ax \cos^m ax\, dx = -\dfrac{\sin^{n-1} ax \cos^{m+1} ax}{a(n+m)} + \dfrac{n-1}{n+m}\int \sin^{n-2} ax \cos^m ax\, dx$

$(m, n$ pos$)$

136. $\int \dfrac{\sin^n ax}{\cos^m ax}\, dx = \dfrac{\sin^{n+1} ax}{a(m-1)\cos^{m-1} ax} - \dfrac{n-m+2}{m-1}\int \dfrac{\sin^n ax}{\cos^{m-2} ax}\, dx$

$(m, n$ pos, $m \neq 1)$

137. $\int \dfrac{\cos^m ax}{\sin^n ax}\, dx = \dfrac{-\cos^{m+1} ax}{a(n-1)\sin^{n-1} ax} + \dfrac{n-m-2}{(n-1)}\int \dfrac{\cos^m ax}{\sin^{n-2} ax}\, dx$

$(m, n$ pos, $n \neq 1)$

138. $\int \dfrac{dx}{\sin ax \cos ax} = \dfrac{1}{a}\log_e \tan ax$

139. $\int \dfrac{dx}{b\sin ax + d\cos ax} = \dfrac{1}{a\sqrt{b^2+d^2}}\log_e \tan \tfrac{1}{2}\left(ax + \tan^{-1}\dfrac{d}{b}\right)$

140. $\int \dfrac{\sin ax}{b+d\cos ax}\, dx = -\dfrac{1}{ad}\log_e(b+d\cos ax)$

141. $\int \dfrac{\cos ax}{b+d\sin ax}\, dx = \dfrac{1}{ad}\log_e(b+d\sin ax)$

Integrals Involving $\tan^n ax$, $\cot^n ax$, $\sec^n ax$, $\csc^n ax$

142. $\int \tan^n ax \, dx = \dfrac{1}{a(n-1)} \tan^{n-1} ax - \int \tan^{n-2} ax \, dx$ (n integer > 1)

143. $\int \cot^n ax \, dx = -\dfrac{1}{a(n-1)} \cot^{n-1} ax - \int \cot^{n-2} ax \, dx$ (n integer > 1)

144. $\int \sec^n ax \, dx = \dfrac{1}{a(n-1)} \dfrac{\sin ax}{\cos^{n-1} ax} + \dfrac{n-2}{n-1} \int \sec^{n-2} ax \, dx$ (n integer > 1)

145. $\int \csc^n ax \, dx = -\dfrac{1}{a(n-1)} \dfrac{\cos ax}{\sin^{n-1} ax} + \dfrac{n-2}{n-1} \int \csc^{n-2} ax \, dx$ (n integer > 1)

146. $\int \dfrac{dx}{b + d \tan ax} = \dfrac{1}{b^2 + d^2} \left[bx + \dfrac{d}{a} \log_e (b \cos ax + d \sin ax) \right]$

147. $\int \dfrac{dx}{\sqrt{b + d \tan^2 ax}} = \dfrac{1}{a\sqrt{b-d}} \sin^{-1} \left[\left(\dfrac{b-d}{b} \right)^{1/2} \sin ax \right]$ (b pos, $b^2 > d^2$)

148. $\int \tan ax \sec ax \, dx = \dfrac{1}{a} \sec ax$

149. $\int \tan^n ax \sec^2 ax \, dx = \dfrac{1}{a(n+1)} \tan^{n+1} ax$ ($n \neq -1$)

150. $\int \dfrac{\sec^2 ax \, dx}{\tan ax} = \dfrac{1}{a} \log_e \tan ax$

151. $\int \cot ax \csc ax \, dx = -\dfrac{1}{a} \csc ax$

152. $\int \cot^n ax \csc^2 ax \, dx = -\dfrac{1}{a(n+1)} \cot^{n+1} ax$ ($n \neq -1$)

153. $\int \dfrac{\csc^2 ax}{\cot ax} dx = -\dfrac{1}{a} \log_e \cot ax$

Integrals Involving b^{ax}, e^{ax}, $\sin bx$, $\cos bx$

154. $\int xb^{ax} \, dx = \dfrac{xb^{ax}}{a \log_e b} - \dfrac{b^{ax}}{a^2 (\log_e b)^2}$

155. $\int xe^{ax} \, dx = \dfrac{e^{ax}}{a^2} (ax - 1)$

156. $\int x^n b^{ax} \, dx = \dfrac{x^n b^{ax}}{a \log_e b} - \dfrac{n}{a \log_e b} \int x^{n-1} b^{ax} \, dx$ (n positive)

157. $\int x^n e^{ax} \, dx = \dfrac{1}{a} x^n e^{ax} - \dfrac{n}{a} \int x^{n-1} e^{ax} \, dx$ (n positive)

158. $\int \dfrac{dx}{b + de^{ax}} = \dfrac{1}{ab}[ax - \log_e (b + de^{ax})]$

159. $\int \dfrac{e^{ax}\, dx}{b + de^{ax}} = \dfrac{1}{ad} \log_e (b + de^{ax})$

160. $\int \dfrac{dx}{be^{ax} + de^{-ax}} = \dfrac{1}{a\sqrt{bd}} \tan^{-1}(e^{ax}\sqrt{b/d})$ (b and d positive)

161. $\int \dfrac{e^{ax}}{x}\, dx = \log_e x + ax + \dfrac{(ax)^2}{2 \cdot 2!} + \dfrac{(ax)^3}{3 \cdot 3!} + \cdots$

162. $\int \dfrac{e^{ax}}{x^n}\, dx = \dfrac{1}{n-1}\left(-\dfrac{e^{ax}}{x^{n-1}} + a\int \dfrac{e^{ax}}{x^{n-1}}\, dx\right)$ (n integer > 1)

163. $\int e^{ax} \sin bx\, dx = \dfrac{e^{ax}}{a^2 + b^2}(a \sin bx - b \cos bx)$

164. $\int e^{ax} \cos bx\, dx = \dfrac{e^{ax}}{a^2 + b^2}(a \cos bx + b \sin bx)$

165. $\int xe^{ax} \sin bx\, dx = \dfrac{xe^{ax}}{a^2 + b^2}(a \sin bx - b \cos bx)$

$\qquad\qquad\qquad\qquad\qquad - \dfrac{e^{ax}}{(a^2 + b^2)^2}[(a^2 - b^2) \sin bx - 2ab \cos bx]$

166. $\int xe^{ax} \cos bx\, dx = \dfrac{xe^{ax}}{a^2 + b^2}(a \cos bx + b \sin bx)$

$\qquad\qquad\qquad\qquad\qquad - \dfrac{e^{ax}}{(a^2 + b^2)^2}[(a^2 - b^2) \cos bx + 2ab \sin bx]$

Integrals Involving $\log_e ax$

167. $\int \log_e ax\, dx = x \log_e ax - x$

168. $\int (\log_e ax)^n\, dx = x(\log_e ax)^n - n\int (\log_e ax)^{n-1}\, dx$ (n pos.)

169. $\int x^n \log_e ax\, dx = x^{n+1}\left[\dfrac{\log_e ax}{n+1} - \dfrac{1}{(n+1)^2}\right]$, $n \neq -1$

170. $\int \dfrac{(\log_e ax)^n}{x}\, dx = \dfrac{(\log_e ax)^{n+1}}{n+1}$, $n \neq -1$

171. $\int \dfrac{dx}{x \log_e ax} = \log_e \log_e ax$

172. $\int \dfrac{dx}{\log_e ax} = \dfrac{1}{a}\left[\log_e (\log_e ax) + \log_e ax + \dfrac{(\log_e ax)^2}{2 \cdot 2!} + \cdots\right]$

173. $\int x^m (\log_e ax)^n \, dx = \dfrac{x^{m+1}(\log_e ax)^n}{m+1} - \dfrac{n}{m+1} x^m (\log_e ax)^{n-1} \, dx, \quad m, n \neq 1$

174. $\int \dfrac{x^m \, dx}{(\log_e ax)^n} = -\dfrac{x^{m+1}}{(n-1)(\log_e ax)^{n-1}} + \dfrac{m+1}{n-1} \dfrac{x^m \, dx}{(\log_e ax)^{n-1}}$

Some Definite Integrals

1. $\displaystyle\int_0^a \sqrt{a^2 - x^2} \, dx = \dfrac{\pi a^2}{4}$

2. $\displaystyle\int_0^a \sqrt{2ax - x^2} \, dx = \dfrac{\pi a^2}{4}$

3. $\displaystyle\int_0^\infty \dfrac{dx}{a + bx^2} = \dfrac{\pi}{2\sqrt{ab}} \quad (a \text{ and } b \text{ positive})$

4. $\displaystyle\int_0^{\sqrt{a/b}} \dfrac{dx}{a + bx^2} \, dx = \int_{\sqrt{a/b}}^\infty \dfrac{dx}{a + bx^2} = \dfrac{\pi}{4\sqrt{ab}} \quad (a \text{ and } b \text{ positive})$

5. $\displaystyle\int_0^{\sqrt{a/b}} \dfrac{dx}{\sqrt{a - bx^2}} = \dfrac{\pi}{2\sqrt{b}} \quad (a \text{ and } b \text{ positive})$

6. $\displaystyle\int_0^\infty \dfrac{\sin bx}{x} \, dx = \dfrac{\pi}{2} \quad (b > 0)$
   $= 0 \quad (b = 0)$
   $= -\dfrac{\pi}{2} \quad (b < 0)$

7. $\displaystyle\int_0^\infty \dfrac{\tan x}{x} \, dx = \dfrac{\pi}{2}$

8. $\displaystyle\int_0^{\pi/2} \sin^{2n+1} x \, dx = \int_0^{\pi/2} \cos^{2n+1} x \, dx = \dfrac{2 \cdot 4 \cdot 6 \cdot \ldots \cdot 2n}{3 \cdot 5 \cdot 7 \cdot \ldots \cdot (2n+1)} \quad (n > 0)$

9. $\displaystyle\int_0^{\pi/2} \sin^{2n} x \, dx = \int_0^{\pi/2} \cos^{2n} x \, dx = \dfrac{1 \cdot 3 \cdot 5 \cdot \ldots \cdot (2n-1)}{2 \cdot 4 \cdot 6 \cdot \ldots \cdot 2n} \cdot \dfrac{\pi}{2} \quad (n > 0)$

10. $\displaystyle\int_0^\pi \sin ax \sin bx \, dx = \int_0^\pi \cos ax \cos bx \, dx = 0 \quad (a \neq b)$

11. $\displaystyle\int_0^\pi \sin^2 ax \, dx = \int_0^\pi \cos^2 ax \, dx = \dfrac{\pi}{2}$

12. $\displaystyle\int_0^{\pi/2} \log_e \cos x \, dx = \int_0^{\pi/2} \log_e \sin x \, dx = -\dfrac{\pi}{2} \log_e 2$

13. $\displaystyle\int_0^\infty e^{-ax^2} \, dx = \tfrac{1}{2}\sqrt{\pi/a}$

14. $\int_0^\infty x^n e^{-ax}\, dx = \dfrac{n!}{a^{n+1}} \quad (a > 0,\, n = 1, 2, 3, \ldots)$

15. $\int_0^1 \dfrac{\log_e x}{1 - x}\, dx = -\dfrac{\pi^2}{6}$

16. $\int_0^1 \dfrac{\log_e x}{1 + x}\, dx = -\dfrac{\pi^2}{12}$

17. $\int_0^1 \dfrac{\log_e x}{1 - x^2}\, dx = \dfrac{\pi^2}{8}$

Properties of Definite Integrals

$$\int_a^b f(x)\, dx = -\int_b^a f(x)\, dx,$$

$$\int_a^b f(x)\, dx = \int_a^c f(x)\, dx + \int_c^b f(x)\, dx$$

$$\int_a^b f(x)\, dx = \int_{u(a)}^{u(b)} f[x(u)]\, \dfrac{dx}{du}\, du = \int_{u(a)}^{u(b)} f[x(u)] \left(\dfrac{du}{dx}\right)^{-1} du$$

$$\dfrac{\partial}{\partial t} \int_{u(t)}^{u(t)} f(x, t)\, dx = \int_{u(t)}^{u(t)} \dfrac{\partial}{\partial t} f(x, t)\, dx + f(v, t)\dfrac{\partial v}{\partial t} - f((u, t)\dfrac{\partial u}{\partial t}$$

$f(x) \leq g(x)$ implies $\int_a^b f(x)\, dx \leq \int_a^b g(x)\, dx$

$$\int_a^b f(x)\, dx = (b - a) f(X), \qquad a \leq X \leq b$$

$\lim_{x \to 0}(1 + x)^{1/x} = e, \quad \lim_{x \to \infty}\left(1 + \dfrac{1}{n}\right)^n = e$

$\lim_{x \to 0} \dfrac{c^x - 1}{x} = \log_e c, \quad \lim_{x \to 0} x^x = 1$

$\lim_{x \to 0} \dfrac{\sin x}{x} = \lim_{x \to 0} \dfrac{\tan x}{x} = \lim_{x \to 0} \dfrac{\sinh x}{x} = \lim_{x \to 0} \dfrac{\tanh x}{x} = 1$

$\lim_{x \to 0} x^a \log_e x = \lim_{x \to \infty} x^a e^{-x} = 0, \qquad a > 0$

Applications of Integration

Area of a Surface. The area of the surface of a solid of revolution generated by revolving the curve $y = f(x)$ between $x = a$ and $x = b$:

about the x axis is $\qquad 2\pi \int_a^b y \left[1 + \left(\dfrac{dy}{dx}\right)^2\right]^{1/2} dx$

about the y axis is $\qquad 2\pi \int_c^d x \left[1 + \left(\dfrac{dx}{dy}\right)^2\right]^{1/2} dy$

where $c = f(a)$ and $d = f(b)$.

If the equation of the surface is written as $x = u, y = v, z = f(u, v) = f(x, y)$,

the arc length $\quad s = \int \left[(1 + p^2)\left(\frac{dx}{dt}\right)^2 + 2pq\frac{dx}{dt}\frac{dy}{dt} + (1 + q^2)\left(\frac{dy}{dt}\right)^2 \right]^{1/2} dt$

the area $\quad S = \iint \sqrt{1 + p^2 + q^2}\, dx\, dy, \quad$ where $p = \dfrac{\partial z}{\partial x}, q = \dfrac{\partial z}{\partial y}$

(the limits of integration to be supplied)

Volume. By triple integration:

Rectangular coordinates $\quad V = \iiint dx\, dy\, dz$

Spherical coordinates $\quad V = \iiint r^2 \sin\theta\, d\theta\, d\phi\, dr$

Cylindrical coordinates $\quad V = \iiint \rho\, d\rho\, d\phi\, dz$

(the limits of integration to be supplied)

Moment. The moment of a mass m

about the yz plane, $\qquad M_{yz} = \int x\, dm;$

about the xz plane, $\qquad M_{xz} = \int y\, dm;$

about the xy plane, $\qquad M_{xy} = \int z\, dm$

(the limits of integration to be supplied)

Center of Gravity. The coordinates of the center of gravity of a mass m are

$$x = \frac{\int x\, dm}{\int dm}, \quad y = \frac{\int y\, dm}{\int dm}, \quad z = \frac{\int z\, dm}{\int dm}$$

(the limits of integration to be supplied)

Moment of Inertia. The moments of inertia I are

for a plane curve about the x axis, $\qquad I_x = \int y^2\, ds$

for a plane curve about the y axis, $\qquad I_y = \int x^2\, ds$

for a plane curve about the origin, $\qquad I_0 = \int (x^2 + y^2)\, ds$

for a plane area about the x axis, $\quad I_z = \int y^2 \, dA$

for a plane area about the y axis, $\quad I_y = \int x^2 \, dA$

for a plane area about the origin, $\quad I_0 = \int (x^2 + y^2) \, dA$

for a solid of mass m about the yz plane, $\quad I_{yz} = \int x^2 \, dm$

for a solid of mass m about the xz plane, $\quad I_{xz} = \int y^2 \, dm$

for a solid of mass m about the xy plane, $\quad I_{xy} = \int z^2 \, dm$

for a solid of mass m about the x axis, $\quad I_x = I_{xz} + I_{xy}$, etc.

(the limits of integration to be supplied)

DIFFERENTIAL EQUATIONS

First-Order Equations

Separation of Variables. A differential equation of the first order

$$f\left(x, y, \frac{dy}{dx}\right) = 0$$

can be brought into the form

$$P(x, y) \, dx + Q(x, y) \, dy = 0$$

For the special case where P is a function of x only and Q a function of y only,

$$P(x) \, dx + Q(y) \, dy = 0$$

the variables are separated. The solution is

$$\int P(x) \, dx + \int Q(y) \, dy = c$$

Homogeneous Equations. A function $f(x, y)$ is homogeneous of the nth degree in x and y, if $f(kx, ky) = k^n f(x, y)$. An equation

$$P(x, y) \, dx + Q(x, y) \, dy = 0$$

is homogeneous if the functions $P(x, y)$ and $Q(x, y)$ are homogeneous in x and y. By substituting $y = vx$, the variables can be separated.

Linear Differential Equation. The differential equation

$$\frac{dy}{dx} + P(x)y = Q(x)$$

in which y and dy/dx appear only in the first degree, and P and Q are functions of x, is a *linear equation of the first order*. This has the general solution

$$y = e^{-\int P(x)\,dx}\left[\int Q(x)e^{\int P(x)\,dx}\,dx + c\right]$$

The Bernoulli Equation is

$$\frac{dy}{dx} + P(x)y = Q(x)y^n$$

in which $n \neq 1$. By making the substitution $z = y^{1-n}$, a linear equation is obtained and the general solution is

$$y = e^{-\int P(x)\,dx}\left[(1-n)\int e^{(1-n)\int P(x)\,dx}Q(x)\,dx + c\right]^{1/(1-n)}$$

Exact Differential Equation. The equation

$$P(x, y)\,dx + Q(x, y)\,dy = 0$$

is an *exact differential equation* if its left side is an exact differential

$$du = P\,dx + Q\,dy$$

that is, if $\partial P/\partial y = \partial Q/\partial x$. Then,

$$\int P\,dx + \int\left[Q - \frac{\partial \int P\,dx}{\partial y}\right]dy = c$$

is a solution.

Integrating Factor. If the left member of the differential equation $P(x, y)\,dx + Q(x, y)\,dy = 0$ is not an exact differential, look for a factor $v(x, y)$ such that $du = v(P\,dx + Q\,dy)$ is an exact differential. Such an *integrating factor* satisfies the equation

$$Q\frac{\partial v}{\partial x} - P\frac{\partial v}{\partial y} + \left(\frac{\partial Q}{\partial x} - \frac{\partial P}{\partial y}\right)v = 0$$

Riccati's Equation is

$$\frac{dy}{dx} + P(x)y^2 + Q(x)y + R(x) = 0$$

If a particular integral y_1 is known, place $y = y_1 + \dfrac{1}{z}$ and obtain a linear equation in z.

Second-Order Equations

The differential equation

$$F\left(x, y, \frac{dy}{dx}, \frac{d^2y}{dx^2}\right) = 0$$

is of the *second order*. If some of these variables are missing there is a straightforward method of solution.

Case 1. With y and dy/dx missing.

$$\frac{d^2y}{dx^2} = f(x)$$

This has the solution

$$y = \int dx \int f(x)\, dx + cx + c_1$$

Case 2. With x and dy/dx missing.

$$\frac{d^2y}{dx^2} = f(y)$$

Multiply both sides by $2(dy/dx)$ and obtain

$$x = \int \frac{dy}{\sqrt{c + 2\int f(y)\, dy}} + c_1$$

as a solution.

Case 3. With x and y missing.

$$\frac{d^2y}{dx^2} = f\left(\frac{dy}{dx}\right)$$

Place

$$\frac{dy}{dx} = p, \quad \frac{d^2y}{dx^2} = \frac{dp}{dx}$$

Then

$$x = \int \frac{dp}{f(p)} + c$$

Solve for p, replace p by dy/dx, and solve the resulting first-order equation.

Case 4. With y missing.

$$\frac{d^2y}{dx^2} = f\left(\frac{dy}{dx}, x\right)$$

Place $dy/dx = p$ and obtain the first-order equation $dp/dx = f(p, x)$. If this can be solved for p, then

$$y = \int p(x)\, dx + c$$

Case 5. With x missing.

$$\frac{d^2y}{dx^2} = f\left(\frac{dy}{dx}, y\right)$$

Place $dy/dx = p$ and obtain the first-order equation $p(dp/dy) = f(p, y)$. If this can be solved for p, then

$$x = \int \frac{dy}{p(y)} + c$$

Linear Equations

General Theorem. The differential equation

$$\frac{d^n y}{dx^n} + P_1(x)\frac{d^{n-1}y}{dx^{n-1}} + \cdots + P_{n-1}(x)\frac{dy}{dx} + P_n(x)y = F(x)$$

is called the general nth order linear differential equation. If $F(x) = 0$, the equation is *homogeneous*; otherwise it is *nonhomogeneous*. If $\phi(x)$ is a solution of the nonhomogeneous equation and $y_1, y_2, \ldots, y_n$ are linearly independent solutions of the homogeneous equation, then the *general solution* is

$$y = c_1 y_1 + c_2 y_2 + \cdots + c_n y_n + \phi(x)$$

The part $\phi(x)$ is called the *particular integral*, and the part $c_1 y_1 + \cdots + c_n y_n$ is the *complementary function*.

Homogeneous Differential Equation with Constant Coefficients.

$$\frac{d^n y}{dx^n} + a_1 \frac{d^{n-1}y}{dx^{n-1}} + \cdots + a_{n-1}\frac{dy}{dx} + a_n y = 0$$

A solution of this equation is

$$y_k = c e^{r_k x}$$

if r_k is a root of the algebraic equation

$$r^n + a_1 r^{n-1} + \cdots + a_{n-1} r + a_n = 0$$

If all the n roots $r_1, r_2, \ldots, r_n$ are different, then

$$y = c_1 e^{r_1 x} + c_2 e^{r_2 x} + \cdots + c_n e^{r_n x}$$

is a general solution. If k of the roots are equal, $r_1 = r_2 = \cdots = r_k$ while $r_{k+1}, \ldots, r_n$ are different, then

$$y = (c_1 + c_2 x + \cdots + c_k x^{k-1})e^{r_1 x} + c_{k+1} e^{r_{k+1} x} + \cdots + c_n e^{r_n x}$$

is a general solution. If $r_1 = p + iq$, $r_2 = p - iq$ are conjugate complex roots, then

$$c_1 e^{r_1 x} + c_2 e^{r_2 x} = e^{px}(C_1 \cos qx + C_2 \sin qx)$$

Nonhomogeneous Differential Equation with Constant Coefficients.

$$\frac{d^n y}{dx^n} + a_1 \frac{d^{n-1}y}{dx^{n-1}} + \cdots + a_{n-1}\frac{dy}{dx} + a_n y = F(x)$$

The complementary function is found as above. To find the particular integral, replace

$$\frac{dy}{dx} \text{ by } D, \quad \frac{d^2 y}{dx^2} \text{ by } D^2, \ldots, \frac{d^n y}{dx^n} \text{ by } D^n$$

$$P(D)y = (D^n + a_1 D^{n-1} + \cdots + a_{n-1} D + a_n)y = F(x)$$

Euler's Homogeneous Equation.

$$x^n \frac{d^n y}{dx^n} + ax^{n-1} \frac{d^{n-1} y}{dx^{n-1}} + \cdots + a_{n-1} x \frac{dy}{dx} + a_n y = 0$$

Place $x = e^t$, and since

$$x \frac{dy}{dx} = \frac{dy}{dt}, \quad x^2 \frac{d^2 y}{dx^2} = \left[\frac{d}{dt}\left(\frac{d}{dt} - 1\right)\right] y, \quad x^3 \frac{d^3 y}{dx^3} = \left[\frac{d}{dt}\left(\frac{d}{dt} - 1\right)\left(\frac{d}{dt} - 2\right)\right] y, \ldots$$

Euler's equation is transformed into a linear homogeneous differential equation with constant coefficients.

Depression of Order. If a particular integral of a linear homogeneous differential equation is known, the order of the equation can be lowered. If y_1 is a particular integral of

$$\frac{d^n y}{dx^n} + P_1(x) \frac{d^{n-1} y}{dx^{n-1}} + \cdots + P_{n-1}(x) \frac{dy}{dx} + P_n(x) = 0$$

substitute $y = y_1 z$. The coefficient of z will be zero, and then by placing $dz/dx = u$, the equation is reduced to the $(n-1)$st order.

Systems of Linear Differential Equations with Constant Coefficients. For a system of n linear equations with constant coefficients in n dependent variables and one independent variable t, the symbolic algebraic method of solution may be used. If $n = 2$,

$$(D^n + a_1 D^{n-1} + \cdots + a_n)x + (D^m + b_1 D^{m-1} + \cdots + b_m)y = R(t)$$
$$(D^p + c_1 D^{p-1} + \cdots + c_p)x + (D^q + d_1 D^{q-1} + \cdots + d_q)y = S(t)$$

where $D = d/dt$. The equations may be written as

$$P_1(D)x + Q_1(D)y = R, \quad P_2(D)x + Q_2(D)y = S$$

Treating these as algebraic equations, eliminate either x or y and solve the equation thus obtained.

Partial Differential Equations

First Order

Definition. If $x_1, x_2, \ldots, x_n$ are n independent variables, $z = z(x_1, x_2, \ldots, x_n)$ the dependent variable, and if

$$\frac{\partial z}{\partial x_1} = p_1, \ldots, \frac{\partial z}{\partial x_n} = p_n$$

then

$$F(x_1, x_2, \ldots, x_n, z, p_1, p_2, \ldots, p_n) = 0$$

is a partial differential equation of the first order. An equation

$$f(x_1, x_2, \ldots, x_n, z, c_1, \ldots, c_n) = 0$$

Linear Differential Equations.

$$P(x, y, z)p + Q(x, y, z)q = R(x, y, z), \quad \text{in which } p = \frac{\partial z}{\partial x}, \quad q = \frac{\partial z}{\partial y}$$

is a linear partial differential equation. From the system of ordinary equations

$$\frac{dx}{P} = \frac{dy}{Q} = \frac{dz}{R},$$

the two independent solutions $u(x, y, z) = c_1$, $v(x, y, z) = c_2$ are obtained. Then $\Phi(u, v) = 0$, where Φ is an arbitrary function, is the *general solution* of

$$Pp + Qq = R.$$

General Method of Solution. Given $F(x, y, z, p, q) = 0$, the partial differential equation to be solved. Since z is a function of x and y, it follows that $dz = p\,dx + q\,dy$. If another relation can be found among x, y, z, p, q, such as $f(x, y, z, p, q) = 0$, then p and q can be eliminated. The solution of the ordinary differential equation thus formed, involving x, y, z, will satisfy the given equation, $F(x, y, z, p, q) = 0$. The unknown function f must satisfy the following linear partial differential equation:

$$\frac{\partial F}{\partial p}\frac{\partial f}{\partial x} + \frac{\partial F}{\partial q}\frac{df}{\partial y} + \left(p\frac{\partial F}{\partial p} + q\frac{\partial F}{\partial q}\right)\frac{\partial f}{\partial z} - \left(\frac{\partial F}{\partial x} + p\frac{\partial F}{\partial z}\right)\frac{\partial f}{\partial p} - \left(\frac{\partial F}{\partial y} + q\frac{\partial F}{\partial z}\right)\frac{\partial f}{\partial q} = 0$$

which is satisfied by any of the solutions of the system

$$\frac{\partial x}{\frac{\partial F}{\partial p}} = \frac{\partial y}{\frac{\partial F}{\partial q}} = \frac{dz}{p\frac{\partial F}{\partial p} + q\frac{\partial F}{\partial q}} = \frac{-dp}{\frac{\partial F}{\partial x} + p\frac{\partial F}{\partial z}} = \frac{-dq}{\frac{\partial F}{\partial y} + q\frac{\partial F}{\partial z}}$$

Second Order

Definition. A linear partial differential equation of the second order with two independent variables is of the form

$$L = Ar + 2Bs + Ct + Dp + Eq + Fz = f(x, y)$$

where

$$r = \frac{\partial^2 z}{\partial x^2}, \quad s = \frac{\partial^2 z}{\partial x\,\partial y}, \quad t = \frac{\partial^2 z}{\partial y^2}, \quad p = \frac{\partial z}{\partial x}, \quad q = \frac{\partial z}{\partial y}$$

The coefficients $A, \ldots, F$ are real continuous functions of the real variables x and y.

Method of Separation of Variables. As an example of this method, the solution will be given to Laplace's equation

$$\nabla^2 u = \frac{\partial^2 u}{\partial x^2} + \frac{\partial^2 u}{\partial y^2} = 0$$

Assume that
$$u = X(x) \cdot Y(y)$$
where X is a function of x only, and Y a function of y only. By substitution and dividing by $X \cdot Y$,
$$\frac{1}{X}\frac{d^2X}{dx^2} = -\frac{1}{Y}\frac{d^2Y}{dy^2}$$

Since the left side does not contain y, the right side does not contain x, and the two sides are equal, they must equal a constant, say $-k^2$.
$$\frac{1}{X}\frac{d^2X}{dx^2} = -k^2, \quad \frac{1}{Y}\frac{d^2Y}{dy^2} = k^2$$

The solutions of these homogeneous linear differential equations with constant coefficients are
$$X = c_1 \cos kx + c_2 \sin kx, \quad Y = c_3 e^{ky} + c_4 e^{-ky}$$
Hence,
$$u = (c_1 \cos kx + c_2 \sin kx)(c_3 e^{ky} + c_4 e^{-ky})$$
$$= e^{ky}(k_1 \cos kx + k_2 \sin kx) + e^{-ky}(k_3 \cos kx + k_4 \sin kx)$$

The sum of any number of solutions is again a solution. An infinite number of solutions may be taken provided the series converges and may be differentiated term by term. Then
$$u = \sum_{n=0}^{\infty} [e^{ky}(A_n \cos kx + B_n \sin kx) + e^{-ky}(D_n \cos kx + E_n \sin kx)]$$
is a solution. The coefficients are determined by using the series as a Fourier series to fit the boundary conditions.

LAPLACE TRANSFORMATION

Transformation Principles

The Laplace and Fourier transformation methods and the Heaviside operational calculus are in essence different aspects of the same method. This method simplifies the solving of linear constant-coefficient integrodifferential equations and convolution type integral equations. For brevity the conditions under which the steps of the method may be validly applied will be omitted. Hence the correctness of a final result should be checked in each case by showing that the formal solution satisfies the given equation and conditions.

1. *Direct Laplace Transformation.* Let t be a real variable, s a complex variable, $f(t)$ a real function of t which equals zero for $t < 0$, $F(s)$ a function of s, and e the base of the natural logarithms. If the Lebesgue integral
$$\int_0^{\infty} e^{-st} f(t)\, dt = F(s)$$

then $F(s)$ is the *direct Laplace transform* of $f(t)$; in simpler notation

$$\mathscr{L}[f(t)] = F(s)$$

2. Inverse Laplace Transformation. Under certain conditions the direct transformation can be inverted, giving as one explicit representation

$$\frac{1}{2\pi i} \int_{c-i\infty}^{c+i\infty} e^{ts} F(s)\, ds\, (=) f(t)$$

in which c is a real constant chosen so that the path of integration lies to the right of all the singularities of $F(s)$, and ($=$) means equals, except possibly for a set of values of t of measure zero. If this relation holds, then $f(t)$ is the *inverse Laplace transform* of $F(s)$. In simpler notation the transformation is written

$$\mathscr{L}^{-1}[F(s)] = f(t)$$

3. Transformation of nth Derivative. If $\mathscr{L}[f(t)] = F(s)$, then

$$\mathscr{L}\left[\frac{d^n f(t)}{dt^n}\right] = s^n F(s) - \sum_{k=0}^{n-1} f^{(k)}(0+) \cdot s^{n-1-k}$$

where $f^{(2)}(0+)$ means $d^2 f(t)/dt^2$ evaluated for $t \to 0$, and $f^{(0)}(0+)$ means $f(0+)$, and $n = 1, 2, 3, \ldots$.

4. Transformation of nth Integral. If $\mathscr{L}[f(t)] = F(s)$, then

$$\mathscr{L}\left[\iint \cdots \int f(t)\, dt\right] = s^{-n} F(s) + \sum_{k=-1}^{-n} f^{(k)}(0+) \cdot s^{-n-1-k}$$

where $f^{(-2)}(0+)$ means $\iint f(t)\, dt\, dt$ evaluated for $t \to 0$, and $n = 1, 2, 3, \ldots$

5. Inverse Transformation of Product. If

$$\mathscr{L}^{-1}[F_1(s)] = f_1(t), \quad \mathscr{L}^{-1}[F_2(s)] = f_2(t)$$

then

$$\mathscr{L}^{-1}[F_1(s) \cdot F_2(s)] = \int_0^t f_1(t-\lambda) \cdot f_2(\lambda)\, d\lambda$$

6. Linear Transformations $\mathscr{L}$ and $\mathscr{L}^{-1}$. Let k_1, k_2 be real constants. Then

$$\mathscr{L}[k_1 f_1(t) + k_2 f_2(t)] = k_1 \mathscr{L}[f_1(t)] + k_2 \mathscr{L}[f_2(t)]$$

and

$$\mathscr{L}^{-1}[k_1 F_1(s) + k_2 F_2(s)] = k_1 \mathscr{L}^{-1}[F_1(s)] + k_2 \mathscr{L}^{-1}[F_2(s)]$$

Procedure

To illustrate the application of the rules of procedure the following simple initial-value problem will be solved. Given the equation

$$k_1 \frac{dy(t)}{dt} + k_2 y(t) + k_3 \int y(t)\, dt = u(t)$$

and initial values $y(0)$, $y^{(-1)}(0)$ where $u(t) = 0$ for $t < 0$, and 1 for $0 < t$, and k_1, k_2, k_3 are real constants. Assume that $y(t)$ has a Laplace transform $Y(s)$, that is, $\mathscr{L}[y(t)] = Y(s)$.

Step A. Find the Laplace transform of the equation to be solved and express it in terms of the transform of the unknown function.
Thus,

$$\mathscr{L}\left[k_1 \frac{dy(t)}{dt} + k_2 y(t) + k_3 \int y(t)\, dt\right] = \mathscr{L}[u(t)]$$

By 6 this becomes

$$k_1 \mathscr{L}\left[\frac{dy(t)}{dt}\right] + k_2 \mathscr{L}[y(t)] + k_3 \mathscr{L}\left[\int y(t)\, dt\right] = \mathscr{L}[u(t)]$$

By 3 and 4 and the given initial conditions of the problem the equation becomes

$$k_1[sY(s) - y(0)] + k_2 Y(s) + k_3[s^{-1}Y(s) + y^{(-1)}(0) \cdot s^{-1}] = \mathscr{L}[u(t)]$$

Step B. Solve the resulting equation for the transform of the unknown function. Thus,

$$Y(s) = \frac{\mathscr{L}[u(t)] + k_1 y(0) - y^{(-1)}(0) \cdot s^{-1}}{k_1 s + k_2 + k_3 s^{-1}}$$

Step C. Evaluate the direct transform of the given function (right member) in the original equation.
Since

$$\mathscr{L}[u(t)] = \frac{1}{s}$$

$$Y(s) = \frac{k_1 y(0) \cdot s - y^{(-1)}(0) + 1}{k_1 s^2 + k_2 s + k_3}$$

Step D. Obtain the solution of the problem by evaluating the inverse Laplace transform of the function obtained by the preceding steps.
One way to carry out Step D is to find the inverse transform from a table such as Table 1-13. To use the table, the denominator of the fraction should be factored.

$$y(t) = \mathscr{L}^{-1}[Y(s)] = \mathscr{L}^{-1}\left[\frac{k_1 y(0) \cdot s - y^{(-1)}(0) + 1}{k_1 s^2 + k_2 s + k_3}\right]$$

$$= \mathscr{L}^{-1}\left[\frac{k_1 y(0) \cdot s - y^{(-1)}(0) + 1}{k_1 (s + K_1)(s + K_2)}\right]$$

in which

$$K_1 \equiv \frac{k_2}{2k_1} - \frac{1}{2k_1}(k_2^2 - 4k_1 k_3)^{1/2}$$

$$K_2 \equiv \frac{k_2}{2k_1} + \frac{1}{2k_1}(k_2^2 - 4k_1 k_3)^{1/2}$$

To find the result it is necessary to distinguish between two cases.

Case 1. If $K_1 \neq K_2$,

$$y(t) = \{[k_1 y(0) K_1 + y^{(-1)}(0) - 1]e^{-K_1 t}$$
$$- [k_1 y(0) K_2 + y^{(-1)}(0) - 1]e^{-K_2 t}\}/[k_1(K_1 - K_2)]$$

for $0 < t$, and $= 0$ for $t < 0$.

Case 2. If $K_1 = K_2 = K$, then $K = \dfrac{k_2}{2k_1}$, and

$$y(t) = \mathscr{L}^{-1}\left[\frac{k_1 y(0) \cdot s - y^{(-1)}(0) + 1}{k_1(s + K)^2}\right]$$

From the table,

$$y(t) = \{k_1 y(0) e^{-Kt} - [y^{(-1)}(0) - 1 + k_1 y(0) K] t e^{-Kt}\}/k_1$$

for $0 < t$, and $= 0$ for $t < 0$.

The solutions can be shown to satisfy the original equation and initial conditions.

The use of Step C can be avoided by using Steps E, F, and G in place of Steps C and D in the following way.

Step E. Factor the transform of the unknown function obtained by Step B, and evaluate the inverse Laplace transform of each factor.

Note. The inverse transform of a rational fraction can be found only if it is a proper fraction.

Thus,

$$Y(s) = \frac{k_1 y(0) \cdot s - y^{(-1)}(0)}{k_1 s^2 + k_2 s + k_3} + \frac{s \mathscr{L}[u(t)]}{k_1 s^2 + k_2 s + k_3}$$

Let

$$y_1(t) \equiv \mathscr{L}^{-1}\left[\frac{k_1 y(0) \cdot s - y^{(-1)}(0)}{k_1(s + K_1)(s + K_2)}\right] = \{[k_1(y)(0) K_1 + y^{(-1)}(0)]e^{-K_1 t}$$
$$- [k_1 y(0) K_2 + y^{(-1)}(0)]e^{-K_2 t}\}/[k_1(K_1 - K_2)]$$

for $0 < t$, and $= 0$ for $t < 0$. Also

$$\mathscr{L}^{-1}\left[\frac{s}{k_1(s + K_1)(s + K_2)}\right] = (K_1 e^{-K_1 t} - K_2 e^{-K_2 t})/[k_1(K_1 - K_2)]$$

for $0 < t$, and $= 0$ for $t < 0$. Finally, $\mathscr{L}^{-1}\{\mathscr{L}[u(t)]\} = u(t)$.

Step F. Use 5 to find the inverse transform of the product.

Thus, by 6 and Step F,

$$y(t) = y_1(t) + [k_1(K_1 - K_2)]^{-1} \int_0^t [K_1 e^{-K_1(t-\tau)} - K_2 e^{-K_2(t-\tau)}] u(\tau)\, d\tau$$

Step G. Evaluate the (convolution) integral arising from Step F. Thus,

$$y(t) = y_1(t) + [k_1(K_1 - K_2)]^{-1}(e^{-K_2 t} - e^{-K_1 t})$$

for $0 < t$, and $= 0$ for $t < 0$.

For the particular problem treated above it is much simpler to use Steps C and D than Steps E, F, and G. However, for a more complicated right member of the original equation it could happen that Step G would be easier to carry out than Step C, in which case the second method (A, B, E, F, G) should be used rather than the first (A, B, C, D).

Table 1-12 Laplace Transformation Theorems and Corresponding Functions

t Function	s Function
1. $u(t) =$ unit function	$1/s$
2. $af_1(t) + bf_2(t)$	$aF_1(s) + bF_2(s)$
3. $f'(t)$	$sF(s) - f(+0)$†
4. $f^n(t)$	$s^n F(s) - s^{n-1}f(+0) - s^{n-2}f'(+0) - \cdots f^{n-1}(+0)$
5. $\int_0^t f(t)\, dt$	$\dfrac{1}{s} F(s)$
6. $e^{-at}f(t)$	$F(s + a)$
7. $f(t - a) = f_a(t)$ ‡	$e^{-as}F(s)$
8. $\dfrac{t^{n-1}}{(n-1)!}$	$\dfrac{1}{s^n}$
9. e^{-at}	$\dfrac{1}{s + a}$
10. $\sin at$	$\dfrac{a}{s^2 + a^2}$
11. $\cos at$	$\dfrac{s}{s^2 + a^2}$
12. $\int_0^t f_1(t - \lambda)f_2(\lambda)\, d\lambda = f_1 * f_2$ §	$F_1(s)F_2(s)$
13. $\sum_{n=1}^{n=k} \dfrac{p(a_n)}{q'(a_n)} e^{a_n t}$	$\dfrac{p(s)}{q(s)}$, where $q(s) = (s - a_1)(s - a_2) \cdots (s - a_k)$
14. $e^{at} \sum_{n=1}^{n=r} \dfrac{\gamma^{(r-n)}(a)}{(r-n)!} \dfrac{t^{n-1}}{(n-1)!} + \cdots$	$\dfrac{p(s)}{q(s)} = \dfrac{\gamma(s)}{(s-a)^r}$

† $f(+0)$ is the value of $f(t)$ as t approaches 0 through positive values.
‡ $f(t - a) = f_a(t)$ is a function which is zero for all times prior to $t = a$.
§ $f_1 * f_2$ is read the convolution of f_1 and f_2.

Table 1-13 Some Laplace Transforms

	$F(s) = \mathscr{L}\{f(t)\}$	$f(t)$
1	$1/s$	1
2	$1/s^2$	t
3	$1/s^n$ $(n = 1, 2, \ldots)$	$t^{n-1}/(n-1)!$
4	$1/\sqrt{s}$	$1/\sqrt{\pi t}$
5	$1/s^{3/2}$	$2\sqrt{t/\pi}$
6	$1/s^a$ $(a > 0)$	$t^{a-1}/\Gamma(a)$
7	$\dfrac{1}{s-a}$	e^{at}
8	$\dfrac{1}{(s-a)^2}$	te^{at}
9	$\dfrac{1}{(s-a)^n}$ $(n = 1, 2, \ldots)$	$\dfrac{1}{(n-1)!} t^{n-1} e^{at}$
10	$\dfrac{1}{(s-a)^k}$ $(k > 0)$	$\dfrac{1}{\Gamma(k)} t^{k-1} e^{at}$
11	$\dfrac{1}{(s-a)(s-b)}$ $(a \neq b)$	$\dfrac{1}{(a-b)}(e^{at} - e^{bt})$
12	$\dfrac{s}{(s-a)(s-b)}$ $(a \neq b)$	$\dfrac{1}{(a-b)}(ae^{at} - be^{bt})$
13	$\dfrac{1}{s^2 + \omega^2}$	$\dfrac{1}{\omega} \sin \omega t$
14	$\dfrac{s}{s^2 + \omega^2}$	$\cos \omega t$
15	$\dfrac{1}{s^2 - a^2}$	$\dfrac{1}{a} \sinh at$
16	$\dfrac{s}{s^2 - a^2}$	$\cosh at$
17	$\dfrac{1}{(s-a)^2 + \omega^2}$	$\dfrac{1}{\omega} e^{at} \sin \omega t$
18	$\dfrac{s-a}{(s-a)^2 + \omega^2}$	$e^{at} \cos \omega t$

Table 1-13 (*Continued*)

	$F(s) = \mathscr{L}\{f(t)\}$	$f(t)$
19	$\dfrac{1}{s(s^2 + \omega^2)}$	$\dfrac{1}{\omega^2}(1 - \cos \omega t)$
20	$\dfrac{1}{s^2(s^2 + \omega^2)}$	$\dfrac{1}{\omega^3}(\omega t - \sin \omega t)$
21	$\dfrac{1}{(s^2 + \omega^2)^2}$	$\dfrac{1}{2\omega^3}(\sin \omega t - \omega t \cos \omega t)$
22	$\dfrac{s}{(s^2 + \omega^2)^2}$	$\dfrac{t}{2\omega} \sin \omega t$
23	$\dfrac{s^2}{(s^2 + \omega^2)^2}$	$\dfrac{1}{2\omega}(\sin \omega t + \omega t \cos \omega t)$
24	$\dfrac{s}{(s^2 + a^2)(s^2 + b^2)}$ $(a^2 \neq b^2)$	$\dfrac{1}{b^2 - a^2}(\cos at - \cos bt)$
25	$\dfrac{1}{s^4 + 4a^4}$	$\dfrac{1}{4a^3}(\sin at \cosh at - \cos at \sinh at)$
26	$\dfrac{s}{s^4 + 4a^4}$	$\dfrac{1}{2a^2} \sin at \sinh at$
27	$\dfrac{1}{s^4 - a^4}$	$\dfrac{1}{2a^3}(\sinh at - \sin at)$
28	$\dfrac{s}{s^4 - a^4}$	$\dfrac{1}{2a^2}(\cosh at - \cos at)$
29	$\sqrt{s - a} - \sqrt{s - b}$	$\dfrac{1}{2\sqrt{\pi t^3}}(e^{bt} - e^{at})$
30	$\dfrac{1}{\sqrt{s + a}\sqrt{s + b}}$	$e^{-(a+b)t/2} I_0\left(\dfrac{a - b}{2} t\right)$
31	$\dfrac{1}{\sqrt{s^2 + a^2}}$	$J_0(at)$
32	$\dfrac{s}{(s - a)^{3/2}}$	$\dfrac{1}{\sqrt{\pi t}} e^{at}(1 + 2at)$
33	$\dfrac{1}{(s^2 - a^2)^k}$ $(k > 0)$	$\dfrac{\sqrt{\pi}}{\Gamma(k)}\left(\dfrac{t}{2a}\right)^{k-1/2} I_{k-1/2}(at)$

FUNCTIONS OF A COMPLEX VARIABLE

Table 1-13 (Continued)

	$F(s) = \mathscr{L}^s\{f(t)\}$	$f(t)$
34	$\dfrac{1}{s} e^{-k/s}$	$J_0(2\sqrt{kt})$
35	$\dfrac{1}{\sqrt{s}} e^{-k/s}$	$\dfrac{1}{\sqrt{\pi t}} \cos 2\sqrt{kt}$
36	$\dfrac{1}{s^{3/2}} e^{k/s}$	$\dfrac{1}{\sqrt{\pi k}} \sinh 2\sqrt{kt}$
37	$\dfrac{1}{s} \ln s$	$-\ln t - \gamma \;(\gamma \approx 0.5772)$
38	$\ln \dfrac{s-a}{s-b}$	$\dfrac{1}{t}(e^{bt} - e^{at})$
39	$\ln \dfrac{s^2 + \omega^2}{s^2}$	$\dfrac{2}{t}(1 - \cos \omega t)$
40	$\ln \dfrac{s^2 - a^2}{s^2}$	$\dfrac{2}{t}(1 - \cosh at)$
41	$\arctan \dfrac{\omega}{s}$	$\dfrac{1}{t} \sin \omega t$
42	$\dfrac{1}{s} \operatorname{arc\,cot} s$	$Si(t)$

FUNCTIONS OF A COMPLEX VARIABLE

Complex Numbers

A *complex number* A is a combination of two real numbers a_1, a_2 in the ordered pair $(a_1, a_2) = A = a_1 + ia_2$, where $i = (-1)^{1/2}$. Real and imaginary numbers are special cases of complex numbers obtained by placing $(a_1, 0) = a_1$, $(0, a_2) = ia_2$.

1. If $a_1 + ia_2 = 0$, then $a_1 = 0$, $a_2 = 0$.
2. If $a_1 + ia_2 = b_1 + ib_2$, then $a_1 = b_1$, $a_2 = b_2$.
3. $a_1 + ia_2$ and $a_1 - ia_2$ are *conjugate* complex numbers. The complex conjugate of A is $\bar{A}$ or A^*.

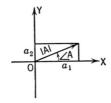

Fig. 1-52

4. $A + B = (a_1 + ia_2) + (b_1 + ib_2) = (a_1 + b_1) + i(a_2 + b_2)$
5. $a_1 + ia_2 = |A|(\cos \angle A + i \sin \angle A) = |A| e^{i\angle A}$
   $a_1 - ia_2 = |A|(\cos \angle A - i \sin \angle A) = |A| e^{-i\angle A}$

where $|A| = \sqrt{a_1^2 + a_2^2}$, $\sin \angle A = \dfrac{a_2}{|A|}$, $\cos \angle A = \dfrac{a_1}{|A|}$, $|A|$ is the *absolute value* (*modulus*), and $\angle A$ is the *angle* of A.

6. $AB = (a_1 + ia_2)(b_1 + ib_2) = (a_1 b_1 - a_2 b_2) + i(a_2 b_1 + a_1 b_2)$
   $= |A||B| e^{i(\angle A + \angle B)}$
7. $A\bar{A} = (a_1 + ia_2)(a_1 - ia_2) = a_1^2 + a_2^2 = |A|^2$
8. $\dfrac{A}{B} = \dfrac{a_1 + ia_2}{b_1 + ib_2} = \dfrac{(a_1 + ia_2)(b_1 - ib_2)}{(b_1 + ib_2)(b_1 - ib_2)} = \dfrac{a_1 b_1 + a_2 b_2}{b_1^2 + b_2^2} + i\dfrac{a_2 b_1 - a_1 b_2}{b_1^2 + b_2^2}$
   $= \dfrac{|A|}{|B|} e^{i(\angle A - \angle B)}$
9. $A^n = (a_1 + ia_2)^n = [|A|(\cos \angle A + i \sin \angle A)]^n = |A|^n e^{in\angle A}$
   $= |A|^n (\cos n \angle A + i \sin n \angle A)$
   $\bar{A}^n = (a_1 - ia_2)^n = [|A|(\cos \angle A - i \sin \angle A)]^n = |A|^n e^{-in\angle A}$
   $= |A|^n (\cos n \angle A - i \sin n \angle A)$
10. $\sqrt[n]{A} = \sqrt[n]{a_1 + ia_2} = \sqrt[n]{|A|} \left(\cos \dfrac{\angle A + 2k\pi}{n} + i \sin \dfrac{\angle A + 2k\pi}{n} \right)$
    $= \sqrt[n]{|A|} e^{i \frac{\angle A + 2k\pi}{n}}$

where k is an integer. For $k = 0, 1, 2, \ldots, n - 1$, all of the n roots are obtained.

Complex Variables

Analytic Functions of a Complex Variable. A function $w = f(z)$, $z = x + iy$, which has a derivative

$$\frac{df}{dz} = f'(z) = \lim_{h \to 0} \frac{f(z + h) - f(z)}{h}$$

at a point z independent of the manner of approach of $z + h$ to z is *analytic* at z and may be expanded in a convergent power series there. A function which is analytic at every point of a region is *analytic in the region*. If $f(z) = w = u(x, y) + iv(x, y)$ and $f(z)$ is analytic at z, then the Cauchy-Riemann differential equations

$$\frac{\partial u}{\partial x} = \frac{\partial v}{\partial y}, \quad \frac{\partial u}{\partial y} = -\frac{\partial v}{\partial x}$$

hold at z. If in a neighborhood of a point z these four partial derivatives exist and are continuous and if the Cauchy-Riemann equations hold, then $f(z)$ is analytic at z. The functions u and v satisfy Laplace's equation

$$\frac{\partial^2 \phi}{\partial x^2} + \frac{\partial^2 \phi}{\partial y^2} = 0$$

Examples of analytic functions are z, $1/z$, e^z, and $\sin z$. An example of a non-analytic function is $w = x - iy$.

Conformal Mapping. The function $w = f(z)$, analytic in a region R_z of the z plane, *conformally* maps each point in R_z on a point of the w plane in the region R_w if $f'(z) \neq 0$ at all points of R_z. This mapping is also *isogonal*, that is, the angle between two curves starting at z_0 is equal to the angle between their mapped curves starting at w_0.

Examples. 1. $w = z + b$, b complex, is a *translation* of magnitude $|b|$ in the direction $\angle b$.
 2. $w = az$, a complex, is a *rotation* through $\angle a$ and a *magnification* by $|a|$.
 3. $w = az + b$, a, b complex, the *integral linear transformation*, is a combination of 1 and 2.
 4. $w = 1/z$, the *inversion transformation*, carries the origin of the z plane into the point at infinity in the *enlarged w* plane.
 5. $w = \dfrac{az + b}{cz + d}$, $ad - bc \neq 0$, the *general linear* or *bilinear transformation*, can be resolved into two linear integral and one inversion transformations.

Integrals of Analytic Functions. If $dF(z)/dz = f(z)$ in a simply connected region R_z, then $F(z) = \int f(z)\, dz$ is analytic throughout R_z. If $f_1(z)$, $f_2(z)$ are analytic in R_z and the path of integration is in R_z, then

1. $\displaystyle\int_{z_0}^{z_0} f_1(z)\, dz = 0$

2. $\displaystyle\int_{z_0}^{z_1} [k_1 f_1(z) + k_2 f_2(z)]\, dz = k_1 \int_{z_0}^{z_1} f_1(z)\, dz + k_2 \int_{z_0}^{z_1} f_2(z)\, dz$

3. $\displaystyle\int_{z_1}^{z_0} f_1(z)\, dz = -\int_{z_0}^{z_1} f_1(z)\, dz$

4. $\displaystyle\int_{z_0}^{z_1} f_1(z)\, dz + \int_{z_1}^{z_2} f_1(z)\, dz = \int_{z_0}^{z_2} f_1(z)\, dz$

Cauchy's Integral Theorem. If $f(z)$ is analytic and single-valued on and within a simple closed contour C, then $\int_C f(z)\, dz = 0$.

A *contour* is a continuous curve made up of a finite number of elementary arcs.

If $f(z)$ is continuous on a simple closed contour C and analytic in the region bounded by C, then *Cauchy's integral formula*

$$f(z) = \frac{1}{2\pi i} \int_C \frac{f(\zeta)}{\zeta - z}\, d\zeta$$

holds; also

$$f^{(n)}(z) = \frac{n!}{2\pi i} \int_C \frac{f(\zeta)}{(\zeta - z)^{n+1}}\, d\zeta$$

Laurent Series. A function $f(z)$ has a *zero of order n* at z_1 if it can be put in the form $f(z) = (z - z_1)^n f_1(z)$, n a positive integer, $f_1(z_1) \neq 0$. A function $f(z)$ has a *pole of order n* at z_1 if it can be put in the form $f(z) = f_2(z)/(z - z_1)^n$, n a positive integer, $f_2(z_1) \neq 0$.

If $f(z)$ is analytic in a ring between and on two concentric circles C_1 and C_2 with radii R_1 and R_2, $R_1 < R_2$, and center z_1, then the *Laurent series*

$$f(z) = \sum_{n=-\infty}^{\infty} c_n(z - z_1)^n, \quad R_1 < |z - z_1| < R_2$$

is convergent everywhere in the ring, and

$$c_n = \frac{1}{2\pi i} \int_C \frac{f(z)}{(z - z_1)^{n+1}} dz$$

C is circle $|z - z_1| = r$, $R_1 < r < R_2$.

If a single-valued analytic function $f(z)$ is expanded in a Laurent series in the neighborhood of an isolated singularity z_1, then the *residue* of $f(z)$ at z_1 is

$$c_{-1} = \frac{1}{2\pi i} \int_C f(z) \, dz$$

C is any circle with center at z_1 which excludes all other singularities of $f(z)$.

A function is *meromorphic* in a region if it is analytic in the region except for a finite number of poles. If $f(z)$ is analytic on an inside a contour C, except for a finite number of poles, and has no zeros on C, then

$$\frac{1}{2\pi i} \int_C \frac{f'(z)}{f(z)} dz = N - P$$

N the total order of the zeros and P the total order of the poles within the contour.

VECTOR ANALYSIS

Vector Algebra

A *scalar* is a quantity which has magnitude, such as mass, density, and temperature. A *vector* is a quantity which has magnitude and direction, such as force, velocity, acceleration. A vector may be represented geometrically by an oriented line segment.

Two vectors **A** and **B** are equal if they have the same magnitude and direction. A free vector may be displaced parallel to itself provided it retains the same magnitude and direction. A vector having the same magnitude but direction opposite to that of **A** is the negative of **A** and is written $-\mathbf{A}$. If **A** is a vector of magnitude, or length, a, then $|\mathbf{A}| = a$. A vector parallel to **A** but with magnitude equal to the reciprocal of the magnitude of **A** is written $\mathbf{A}^{-1} = 1/\mathbf{A}$. A unit vector $\frac{\mathbf{A}}{|\mathbf{A}|}$ ($\mathbf{A} \neq 0$) has the direction of **A** and magnitude 1.

The sum of two vectors **A** and **B** is $\mathbf{A} + \mathbf{B}$ (Fig. 1-53). Similarly, the sum of three or more vectors can be found by adding them end to end.

The sum of **A** and $-\mathbf{B}$ is $\mathbf{A} - \mathbf{B}$ (Fig. 1-53), the *difference* of two vectors. Let **A**, **B**, **C** be vectors and p, q scalars.

$p\mathbf{A} = \mathbf{A}p$, a vector p times as long as $\mathbf{A}$ with the same direction as $\mathbf{A}$ if p is positive and opposite if p is negative.

$$(p + q)\mathbf{A} = p\mathbf{A} + q\mathbf{A}; \quad p(\mathbf{A} + \mathbf{B}) = p\mathbf{A} + p\mathbf{B}$$
$$\mathbf{A} + \mathbf{B} = \mathbf{B} + \mathbf{A}$$
$$\mathbf{A} + (\mathbf{B} + \mathbf{C}) = (\mathbf{A} + \mathbf{B}) + \mathbf{C}$$

$|\mathbf{A} + \mathbf{B}| \leq |\mathbf{A}| + |\mathbf{B}|$, where the equality sign holds only for $\mathbf{A}$ parallel to $\mathbf{B}$.

Rectangular Coordinates. Figure 1-54 shows a right-hand coordinate system. Let $\mathbf{i}, \mathbf{j}, \mathbf{k}$ be unit vectors with the directions OX, OY, OZ, respectively. The vector $\mathbf{R}$ with initial point O and end point $P(x, y, z)$ can be expressed as the sum of its components

$$\mathbf{R} = \mathbf{i}x + \mathbf{j}y + \mathbf{k}z$$

If $\mathbf{A} = \mathbf{i}a_1 + \mathbf{j}a_2 + \mathbf{k}a_3$ and $\mathbf{B} = \mathbf{i}b_1 + \mathbf{j}b_2 + \mathbf{k}b_3$, then

$$\mathbf{A} + \mathbf{B} = \mathbf{i}(a_1 + b_1) + \mathbf{j}(a_2 + b_2) + \mathbf{k}(a_3 + b_3)$$

The Scalar, Inner, or Dot Product of two vectors $\mathbf{A}$ and $\mathbf{B}$ is $\mathbf{A} \cdot \mathbf{B} = |\mathbf{A}| |\mathbf{B}| \cos \theta$ (Fig. 1-55).

$$\mathbf{A} \cdot \mathbf{B} = \mathbf{B} \cdot \mathbf{A}$$
$$\mathbf{A} \cdot (\mathbf{B} + \mathbf{C}) = \mathbf{A} \cdot \mathbf{B} + \mathbf{A} \cdot \mathbf{C}$$
$$\mathbf{A} \cdot \mathbf{A} = \mathbf{A}^2 = |\mathbf{A}|^2$$
$$\mathbf{i} \cdot \mathbf{i} = \mathbf{j} \cdot \mathbf{j} = \mathbf{k} \cdot \mathbf{k} = 1$$
$$\mathbf{i} \cdot \mathbf{j} = \mathbf{j} \cdot \mathbf{k} = \mathbf{k} \cdot \mathbf{i} = 0$$

If $\mathbf{A} \cdot \mathbf{B} = 0$, then either $\mathbf{A} = 0$, $\mathbf{B} = 0$, or $\mathbf{A}$ is perpendicular to $\mathbf{B}$.

If $\mathbf{A} = \mathbf{i}a_1 + \mathbf{j}a_2 + \mathbf{k}a_3$ and $\mathbf{B} = \mathbf{i}b_1 + \mathbf{j}b_2 + \mathbf{k}b_3$, then $\mathbf{A} \cdot \mathbf{B} = a_1b_1 + a_2b_2 + a_3b_3$.

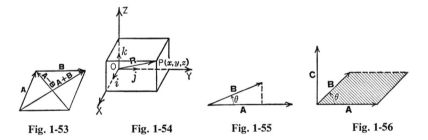

Fig. 1-53   Fig. 1-54   Fig. 1-55   Fig. 1-56

The Vector, Restricted Outer, or Cross Product of two vectors $\mathbf{A}$ and $\mathbf{B}$ is $\mathbf{A} \times \mathbf{B} = \mathbf{C}$, where $\mathbf{C}$ is perpendicular to the plane of $\mathbf{A}$ and $\mathbf{B}$ with the magnitude $|\mathbf{C}| = |\mathbf{A}| |\mathbf{B}| \sin \theta$ (the area of the parallelogram made by $\mathbf{A}$ and $\mathbf{B}$, Fig. 1-56) and so directed that a right-hand rotation of less than $180°$ carries $\mathbf{A}$ into $\mathbf{B}$.

$$\mathbf{A} \times \mathbf{B} = -\mathbf{B} \times \mathbf{A}$$
$$\mathbf{A} \times (\mathbf{B} + \mathbf{C}) = \mathbf{A} \times \mathbf{B} + \mathbf{A} \times \mathbf{C}$$
$$(\mathbf{B} + \mathbf{C}) \times \mathbf{A} = \mathbf{B} \times \mathbf{A} + \mathbf{C} \times \mathbf{A}$$

$$\mathbf{i} \times \mathbf{i} = \mathbf{j} \times \mathbf{j} = \mathbf{k} \times \mathbf{k} = 0$$
$$\mathbf{i} \times \mathbf{j} = \mathbf{k} = -\mathbf{j} \times \mathbf{i}$$
$$\mathbf{j} \times \mathbf{k} = \mathbf{i} = -\mathbf{k} \times \mathbf{j}$$
$$\mathbf{k} \times \mathbf{i} = \mathbf{j} = -\mathbf{i} \times \mathbf{k}$$

If $\mathbf{A} \times \mathbf{B} = 0$, then either $\mathbf{A} = 0$, $\mathbf{B} = 0$, or $\mathbf{A}$ is parallel to $\mathbf{B}$.

If $\mathbf{A} = \mathbf{i}a_1 + \mathbf{j}a_2 + \mathbf{k}a_3$ and $\mathbf{B} = \mathbf{i}b_1 + \mathbf{j}b_2 + \mathbf{k}b_3$, then $\mathbf{A} \times \mathbf{B} = \begin{vmatrix} \mathbf{i} & \mathbf{j} & \mathbf{k} \\ a_1 & a_2 & a_3 \\ b_1 & b_2 & b_3 \end{vmatrix}$

If $\mathbf{A} = \mathbf{i}a_1 + \mathbf{j}a_2 + \mathbf{k}a_3$, $\mathbf{B} = \mathbf{i}b_1 + \mathbf{j}b_2 + \mathbf{k}b_3$, $\mathbf{C} = \mathbf{i}c_1 + \mathbf{j}c_2 + \mathbf{k}c_3$, then the scalar triple product

$$\mathbf{A} \cdot (\mathbf{B} \times \mathbf{C}) = (\mathbf{A} \times \mathbf{B}) \cdot \mathbf{C} = \mathbf{B} \cdot (\mathbf{C} \times \mathbf{A}) = (ABC) = \begin{vmatrix} a_1 & a_2 & a_3 \\ b_1 & b_2 & b_3 \\ c_1 & c_2 & c_3 \end{vmatrix}$$

and is equal to the volume of a parallelepiped whose three determining edges are $\mathbf{A}$, $\mathbf{B}$, $\mathbf{C}$.

$$(\mathbf{A} \times \mathbf{B}) \times \mathbf{C} = (\mathbf{A} \cdot \mathbf{C})\mathbf{B} - (\mathbf{B} \cdot \mathbf{C})\mathbf{A} = -\mathbf{C} \times (\mathbf{A} \times \mathbf{B})$$
$$(\mathbf{A} \times \mathbf{B}) \cdot (\mathbf{C} \times \mathbf{D}) = (\mathbf{A} \cdot \mathbf{C})(\mathbf{B} \cdot \mathbf{D}) - (\mathbf{A} \cdot \mathbf{D})(\mathbf{B} \cdot \mathbf{C})$$

Differentiation and Integration of Vectors

Differentiation. A vector function of one or more scalar variables is called a *variable vector* or *field vector*. The derivative is

$$\frac{d\mathbf{F}}{dt} = \mathbf{F}'(t) = \lim_{\Delta t \to 0} \frac{\mathbf{F}(t + \Delta t) - \mathbf{F}(t)}{\Delta t} = \lim_{\Delta t \to 0} \frac{\Delta \mathbf{F}}{\Delta t} \quad \text{(Fig. 1-57)}$$

Fig. 1-57

If the length of $\mathbf{F}$ remains unaltered, then $\mathbf{F} \cdot d\mathbf{F} = 0$. If the direction of $\mathbf{F}$ remains unaltered, then $\mathbf{F} \times d\mathbf{F} = 0$.

$$d(\mathbf{A} + \mathbf{B}) = d\mathbf{A} + d\mathbf{B}$$
$$d(\mathbf{A} \cdot \mathbf{B}) = \mathbf{A} \cdot d\mathbf{B} + \mathbf{B} \cdot d\mathbf{A}$$
$$d(\mathbf{A} \times \mathbf{B}) = d\mathbf{A} \times \mathbf{B} + \mathbf{A} \times d\mathbf{B} = \mathbf{A} \times d\mathbf{B} - \mathbf{B} \times d\mathbf{A}$$
$$d(\mathbf{A} \cdot \mathbf{B} \cdot \mathbf{C}) = \mathbf{A} \cdot \mathbf{B} \cdot d\mathbf{C} + \mathbf{B} \cdot \mathbf{C} \cdot d\mathbf{A} + \mathbf{C} \cdot \mathbf{A} \cdot d\mathbf{B}$$

The Differential Operators.

$$\nabla = \text{del} = \mathbf{i}\frac{\partial}{\partial x} + \mathbf{j}\frac{\partial}{\partial y} + \mathbf{k}\frac{\partial}{\partial z}$$

$$\nabla^2 = \text{Laplacian} = \frac{\partial^2}{\partial x^2} + \frac{\partial^2}{\partial y^2} + \frac{\partial^2}{\partial z^2}$$

If V is a scalar function, then

$$\nabla V = \text{grad } V = \mathbf{i}\frac{\partial V}{\partial x} + \mathbf{j}\frac{\partial V}{\partial y} + \mathbf{k}\frac{\partial V}{\partial z}$$

If $\mathbf{A}$ is a vector function with components $\mathbf{A}_x$, $\mathbf{A}_y$, $\mathbf{A}_z$, then

$$\nabla \cdot \mathbf{A} = \text{div } \mathbf{A} = \frac{\partial \mathbf{A}_x}{\partial x} + \frac{\partial \mathbf{A}_y}{\partial y} + \frac{\partial \mathbf{A}_z}{\partial z}$$

$$\nabla \times \mathbf{A} = \text{curl } \mathbf{A} = \text{rot } \mathbf{A} = \begin{vmatrix} \mathbf{i} & \mathbf{j} & \mathbf{k} \\ \frac{\partial}{\partial x} & \frac{\partial}{\partial y} & \frac{\partial}{\partial z} \\ \mathbf{A}_x & \mathbf{A}_y & \mathbf{A}_z \end{vmatrix}$$

Formulas for Differentiation. Let U and V be scalar functions and $\mathbf{A}$ and $\mathbf{B}$ be vector functions of x, y, z. Then:

$$\nabla(U + V) = \nabla U + \nabla V, \quad \nabla \cdot (\mathbf{A} + \mathbf{B}) = \nabla \cdot \mathbf{A} + \nabla \cdot \mathbf{B},$$
$$\nabla \times (\mathbf{A} + \mathbf{B}) = \nabla \times \mathbf{A} + \nabla \times \mathbf{B}$$
$$\nabla(UV) = V\nabla U + U\nabla V, \quad \nabla \cdot (U\mathbf{A}) = U\nabla \cdot \mathbf{A} + \mathbf{A} \cdot \nabla U$$
$$\nabla \times (U\mathbf{A}) = \nabla U \times \mathbf{A} + U\nabla \times \mathbf{A}$$
$$\nabla \cdot (\mathbf{A} \times \mathbf{B}) = \mathbf{B} \cdot \nabla \times \mathbf{A} - \mathbf{A} \cdot \nabla \times \mathbf{B}$$
$$\nabla(\mathbf{A} \cdot \mathbf{B}) = \mathbf{A} \cdot \nabla \mathbf{B} + \mathbf{B} \cdot \nabla \mathbf{A} + \mathbf{A} \times (\nabla \times \mathbf{B}) + \mathbf{B} \times (\nabla \times \mathbf{A})$$
$$\nabla \times (\mathbf{A} \times \mathbf{B}) = \mathbf{B} \cdot \nabla \mathbf{A} - \mathbf{A} \cdot \nabla \mathbf{B} + \mathbf{A}(\nabla \cdot \mathbf{B}) - \mathbf{B}(\nabla \cdot \mathbf{A})$$
$$\nabla \times (\nabla \times \mathbf{A}) = \nabla(\nabla \cdot \mathbf{A}) - \nabla^2 \mathbf{A}$$
$$\nabla \cdot (\nabla \times \mathbf{A}) = 0$$
$$\nabla \times (\nabla U) = 0$$

If $\mathbf{R} = \mathbf{i}x + \mathbf{j}y + \mathbf{k}z$ (Fig. 1.54), then

$$\nabla \cdot \mathbf{R} = 3, \quad \nabla \times \mathbf{R} = 0, \quad \mathbf{A} \cdot \nabla |\mathbf{R}| = |\mathbf{A}|, \quad \nabla \cdot \frac{1}{|\mathbf{R}|} = -\frac{\mathbf{R}}{|\mathbf{R}|^3}, \quad \nabla^2 \frac{1}{|\mathbf{R}|} = 0$$

Integration. The line integral of a vector $\mathbf{F}$ along a curve AB denotes the integral of the tangential component of the vector along the curve; thus

$$\int_A^B \mathbf{F} \cdot d\mathbf{R} = \int_A^B |\mathbf{F}_c| \, ds \quad \text{(Fig. 1-58)}$$

where $d\mathbf{R} = \mathbf{i}\, dx + \mathbf{j}\, dy + \mathbf{k}\, dz$.

Fig. 1-58

If $\mathbf{F} = \nabla U$ is the gradient of a single-valued continuous function $U(x, y, z)$ the line integral of $\mathbf{F}$ depends only on the end points. Conversely, if $\mathbf{F}(x, y, z)$ is continuous and $\int_C \mathbf{F} \cdot d\mathbf{R} = 0$ for any closed path C in a three-dimensional region, there is a function $U(x, y, z)$ such that $\mathbf{F} = \nabla U$.

Theorems and Formulas

Let $\mathbf{n}$ be the vector of unit length perpendicular to a surface at a point P and extending on the positive side (the outward normal); dS, the element of surface, and dv, the element of volume.

The Divergence (Gauss) Theorem. If a field vector $\mathbf{F}$ and its first derivatives are continuous at all points in a region of volume v bounded by a closed elementary surface S, then

$$\iint_S \mathbf{F} \cdot \mathbf{n} \, dS = \iiint_v \nabla \cdot \mathbf{F} \, dv$$

Stokes's Theorem. If a field vector $\mathbf{F}$ and its first derivatives are continuous at all points in a region of area S bounded by a closed curve C, then

$$\iint_S \nabla \times \mathbf{F} \cdot \mathbf{n} \, dS = \int_C \mathbf{F} \cdot d\mathbf{R}$$

Green's Theorem. Under the conditions of the divergence theorem,

$$\iint_S \mathbf{n} \cdot U \nabla V \, dS = \iiint_v U \nabla^2 V \, dv + \iiint_v (\nabla U \cdot \nabla V) \, dv$$

$$\iint_S \mathbf{n} \cdot (U \nabla V - V \nabla U) \, dS = \iiint_v (U \nabla^2 V - V \nabla^2 U) \, dv$$

Cylindrical Coordinates.

$$x = r \cos \theta, \quad y = r \sin \theta, \quad z = z$$

The element of volume $dv = r \, dr \, d\theta \, dz$. The unit vectors $\mathbf{u}_r$, $\mathbf{u}_\theta$, $\mathbf{u}_z$ are perpendicular to each other.

$$\text{grad } V = \nabla V = \frac{\partial V}{\partial r} \mathbf{u}_r + \frac{1}{r} \frac{\partial V}{\partial \theta} \mathbf{u}_\theta + \frac{\partial V}{\partial z} \mathbf{u}_z$$

$$\text{div } \mathbf{F} = \nabla \cdot \mathbf{F} = \frac{1}{r} \frac{\partial}{\partial r} (r F_r) + \frac{1}{r} \frac{\partial}{\partial \theta} (F_\theta) + \frac{\partial}{\partial z} (F_z)$$

$$\text{curl } \mathbf{F} = \nabla \times \mathbf{F} = \begin{vmatrix} \dfrac{\mathbf{u}_r}{r} & \mathbf{u}_\theta & \dfrac{\mathbf{u}_z}{r} \\ \dfrac{\partial}{\partial r} & \dfrac{\partial}{\partial \theta} & \dfrac{\partial}{\partial z} \\ F_r & r F_\theta & F_z \end{vmatrix}$$

$$\nabla^2 V = \frac{1}{r} \frac{\partial V}{\partial r} + \frac{\partial^2 V}{\partial r^2} + \frac{1}{r^2} \frac{\partial^2 V}{\partial \theta^2} + \frac{\partial^2 V}{\partial z^2}$$

Spherical Coordinates.

$$x = r \cos \phi \sin \theta, \quad y = r \sin \phi \sin \theta, \quad z = r \cos \theta$$

The unit vectors $\mathbf{u}_r$, $\mathbf{u}_\phi$, $\mathbf{u}_\theta$ are perpendicular to each other.

$$\operatorname{grad} V = \nabla V = \frac{\partial V}{\partial r} \mathbf{u}_r + \frac{1}{r \sin \theta} \frac{\partial V}{\partial \phi} \mathbf{u}_\phi + \frac{1}{r} \frac{\partial V}{\partial \theta} \mathbf{u}_\theta$$

$$\operatorname{div} \mathbf{F} = \nabla \cdot \mathbf{F} = \frac{1}{r^2} \frac{\partial}{\partial r}(r^2 \mathbf{F}_r) + \frac{1}{r \sin \theta} \frac{\partial \mathbf{F}_\phi}{\partial \phi} + \frac{1}{r \sin \theta} \frac{\partial}{\partial \phi}(\sin \theta \mathbf{F}_\theta)$$

$$\operatorname{curl} \mathbf{F} = \nabla \times \mathbf{F} = \begin{vmatrix} \dfrac{\mathbf{u}_r}{r^2 \sin \theta} & \dfrac{\mathbf{u}_\theta}{r \sin \theta} & \dfrac{\mathbf{u}^\phi}{r} \\ \dfrac{\partial}{\partial r} & \dfrac{\partial}{\partial \theta} & \dfrac{\partial}{\partial \phi} \\ \mathbf{F}_r & r \mathbf{F}_\theta & r \sin \theta \mathbf{F}_\phi \end{vmatrix}$$

$$\nabla^2 V = \frac{1}{r^2} \frac{\partial}{\partial r}\left(r^2 \frac{\partial V}{\partial r}\right) + \frac{1}{r^2 \sin^2 \theta} \frac{\partial^2 V}{\partial \phi^2} + \frac{1}{r^2 \sin \theta} \frac{\partial}{\partial \theta}\left(\sin \theta \frac{\partial V}{\partial \theta}\right)$$

MACHINE COMPUTATION

The development of automatic digital computers has had a profound influence on numerical analysis. The high-speed computer makes it possible to solve problems otherwise not attempted. Also the rigor of machine instructions and the computational detail employed have forced a re-examination of the classical methods of analysis and the development of new ones.

Programming of a problem for a digital computer is a process of translating instructions from human language to computer language while using an economical sequence of orders. For instruction in techniques of programming see the manuals by D. D. McCracken published by John Wiley and Sons, *Digital Computer Programming, A Guide to Fortran Programming, A Guide to Algol Programming*, and *A Guide to IBM 1401 Programming*.

For an index of available computer programs see *Catalog of Programs for IBM Data Processing Systems*, IBM Corp., White Plains, New York.

Examples of available programs for mathematical computation are listed below.

Bessel functions, various
Binomial coefficients
Boolean algebra minimizer
Definite integral evaluation
Determinants, evaluation
Differential equations, solution of N simultaneous
Differentiation and partial differentiation
Elliptical integrals

98 MATHEMATICS

Exponential integrals
Fourier series, partial derivatives
Least square curve fitting
Logarithms to base 10 and base e
Matrix inversion
Nth root calculation
Newton's method for roots of polynomials
Nonlinear simultaneous equations, solution
Ordinary differential equations, solution
Polar to cartesian coordinates
Roots of polynomials with real coefficients
Second-order differential equations, solution

REFERENCES

Churchill, R. V., *Operational Mathematics*, McGraw-Hill, New York, 1958.
Hoel, P. G., *Introduction to Mathematical Statistics*, John Wiley, New York, third edition, 1962.
Korn, G. A. and T. M. Korn, editors, *Mathematical Handbook for Scientists and Engineers*, McGraw-Hill, New York, 1961.
Kreyszig, Erwin, *Advanced Engineering Mathematics*, John Wiley, New York, 1962.
MacDuffee, C. C., *Theory of Equations*, John Wiley, New York, 1954.
Nering, E. D., *Linear Algebra and Matrix Theory*, John Wiley, New York, 1963.
Petit Bois, G., *Tables of Indefinite Integrals*, Dover Publications, New York, 1961.
Ralston, A. and H. S. Wilf, editors, *Mathematical Methods for Digital Computers*, John Wiley, New York, 1960.
Rose, I. H., *A Modern Introduction to College Mathematics*, John Wiley, New York, 1959.
Steen, F. H., *Differential Equations*, Ginn and Co., Boston, Mass., 1955.

SECTION 2

MECHANICS

Principal Symbols

Boldface in the text indicates vector; for units see Table 1-2.

a	linear acceleration		r	radius
A	area		R	resultant force
b	breadth		s	distance
c	distance between neutral plane and remotest fiber		S	section modulus
			t	time, thickness
C	a constant		T	torque, temperature
d	depth of beam		U	internal energy
D	diameter		v	linear velocity
e	normal deformation		V	volume, shearing force
E	modulus of elasticity		w	weight
f	coefficient of friction		W	work
F	force			
g	acceleration of gravity		α	angular acceleration, coef. linear thermal expansion
g_c	dimensional constant			
G	shear modulus		β	coefficient volumetric thermal expansion
h	height			
I	moment of inertia		γ	unit shear strain
J	polar moment of inertia		δ	lateral deformation in shear
k	radius of gyration		Δ	difference
K	bulk modulus		ϵ	unit normal strain
l	equivalent length		θ	angle, radians
L	length		μ	coefficient of viscosity
m	mass		ν	Poisson's ratio
M	moment		σ	unit normal stress
P	total load, power		τ	unit shear stress
Q	heat evolved		ω	angular velocity

BASIC CONCEPTS

Statics deals with the equilibrium of rigid bodies subjected to applied forces.

Kinematics deals with the motion of particles and rigid bodies without regard to the forces which cause the motion. Kinematics is concerned only with geometry and time.

Dynamics deals with the effect of unbalanced external forces in modifying the motion of particles and rigid bodies.

Mechanics of Materials considers, in addition to external forces, the internal stresses and strains of a body.

Force and Mass are usually related in terms of Newton's second law, i.e., the time rate of change of momentum is proportional to the force which produces the change:

$$F \propto \frac{d(mv)}{dt} \quad (2\text{-}1)$$

$$F = C(ma) = \frac{m}{g_c} a \quad (2\text{-}2)$$

The proportionality constant, the coefficient g_c, is the conversion factor for the units used. For example, if the units are pound-force, pound-mass, foot, and second, the value of g_c is 32.2. The factor g_c will usually be implied in equations where both force and mass appear. For all equations in Section 2, Mechanics, g_c is omitted; hence either the units used must be chosen to make g_c equal to unity or the value of g_c must be introduced. Values of g_c for various units are given in Table 1-1, page 2.

STATICS

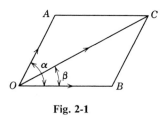

Fig. 2-1

A **force** is completely defined by its magnitude, direction, and point of application as shown in Fig. 2-1, where the magnitude is represented by the line OA, the direction by the arrow, and the point of application by O. The *sense of a force* refers to one of the two directions along the line of action of the force. According to the principle of the transmissibility of force the effect of a force on a rigid body is the same no matter where along the line of action the force is applied. The projection of a force on any axis in its plane of action is the product of the magnitude of the force and the cosine of the angle between the line of action and the selected axis, i.e., $F_x = F \cos \alpha$ and $F_y = F \sin \alpha$.

Concurrent Forces

If several forces are applied to a body we have a *system of forces* which may be classified as *coplanar* or *noncoplanar*. The general problem of statics is the reduction of a given system of forces to the simplest equivalent or the *resultant*.

The conditions under which this resultant vanishes are the *conditions of equilibrium* of the system.

Parallelogram Law. The resultant of the forces OA and OB of Fig. 2-1 is represented by the diagonal OC of the parallelogram $OACB$. The resultant can also be represented analytically by

$$OC = [(OA)^2 + (OB)^2 + 2OA \cdot OB \cos \alpha]^{1/2} \tag{2-3}$$

and

$$\sin \beta = \frac{OA}{OC} \sin \alpha \tag{2-4}$$

where α is the angle between OA and OB and β the angle between OB and OC.

Triangle Law. Geometric addition of the vectors represented by OA and BC gives the resultant as shown in Fig. 2-1.

Composition of Several Coplanar Forces. The resultant of several forces lying in the same plane may be obtained by geometric addition of the force vectors to construct a *force polygon* in which the closing side is the resultant force. See Fig. 2-2 where the magnitude of the resultant AE is 114 lb and the angle θ is $8°20'$.

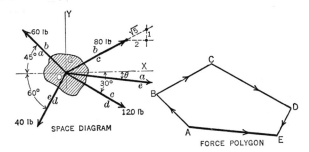

Fig. 2-2

The resultant may also be obtained algebraically from the x and y components of the forces

$$R = [(\sum F_x)^2 + (\sum F_y)^2]^{1/2} \tag{2-5}$$

$$\tan \theta = \frac{\sum F_y}{\sum F_x} \tag{2-6}$$

Composition of Noncoplanar Forces. The resultant of three forces lying in different planes may be obtained by constructing to scale a parallelepiped of forces as in Fig. 2-3 where the diagonal represents the resultant R of the forces P, Q, and S.

For any number of concurrent forces the resultant may be obtained analytically from the x, y, z components of the forces

$$R = [(\sum F_x)^2 + (\sum F_y)^2 + (\sum F_z)^2]^{1/2} \qquad (2\text{-}7)$$

$$\cos \alpha = \frac{\sum F_x}{R}, \quad \cos \beta = \frac{\sum F_y}{R}, \quad \cos \gamma = \frac{\sum F_z}{R} \qquad (2\text{-}8)$$

Moments and Couples

Moment or torque M of a force about a point or center of moments is the product of the force magnitude and the arm or perpendicular distance s from the point to the action line of the force,

$$M = Fs \qquad (2\text{-}9)$$

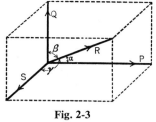

Fig. 2-3

The unit of moment is usually called the pound-foot, kilogram-meter, newton-meter, etc., to distinguish it from the unit of energy.

From the *principle of moments* the moment of the resultant of forces about a point in their plane (coplanar forces) or about a line (any forces) is the algebraic sum of the moments of those forces about the point or line.

The Force Couple. Two equal, opposite, and parallel forces acting on a body are called a couple, of which the *arm* is the perpendicular distance between the lines of action, and the *moment* is the product of one of the force magnitudes and the arm. The sum of the moments of the two forces is the same about any moment center in their plane of action. The couple tends to produce rotation.

The resultant of any number of couples is another couple. A couple may be represented by a vector with length equal to the magnitude of the moment and direction perpendicular to the plane of the couple, pointed the way a right-hand screw would advance if turned by the couple. A number of couple vectors may be added to obtain the vector that defines the resultant couple. Thus

$$\mathbf{M} = (\mathbf{M}_x^2 + \mathbf{M}_y^2 + \mathbf{M}_z^2)^{1/2} \qquad (2\text{-}10)$$

$$\cos \theta_x = \frac{M_x}{M}, \quad \cos \theta_y = \frac{M_y}{M}, \quad \cos \theta_z = \frac{M_z}{M} \qquad (2\text{-}11)$$

Centroid and Center of Gravity

The centroid of a system of parallel forces with fixed points of application is the point through which their resultant will always pass however the forces may be turned while being kept parallel.

The center of gravity of a body, or system of bodies, is the centroid of the forces of gravitation or the point of application of the resultant of the gravity forces (for practical purposes considered as parallel) acting upon the various elements. The force of gravity on a body is, of course, equal to the weight w of the body. The coordinates of the center of gravity of a group of bodies are:

$$\bar{x} = \frac{\sum w_i \bar{x}_i}{\sum w_i}, \quad \bar{y} = \frac{\sum w_i \bar{y}_i}{\sum w_i}, \quad \bar{z} = \frac{\sum w_i \bar{z}_i}{\sum w_i} \quad (2\text{-}12)$$

where w_i is the weight of one element and $\bar{x}_i$, $\bar{y}_i$, $\bar{z}_i$ are the coordinates of the center of gravity of this element.

A system of bodies supported at the center of gravity will remain stable in any position if the only forces acting are gravitational.

Moment of Inertia

The moment of inertia I of a body is the moment of the inertia forces caused by a unit angular acceleration α, about an axis, of all the elements of mass dm of the body.

$$\text{Moment} = \alpha I = \alpha \int x\, dmx \quad (2\text{-}13)$$

or

$$I = \int x^2\, dm$$

in which x is the moment arm or distance from axis to centroid of the element of mass.

General equations for *statical moment, moment of inertia, radius of gyration,* and *product of inertia* are given in Table 2-1. Properties of specific figures and bodies are given in Table 2-2. Properties of sections of standard structural shapes are given in Tables 2-13 through 2-23.

Table 2-1 Equations for Moment, Inertia, and Gyration

	Area	Mass
Statical Moment M	$M = Ax$	$M = mx$
Moment of Inertia I		
Y axis	$I_y = \int x^2\, dA$	$I_y = \int x^2\, dm$
x axis	$I_x = \int y^2\, dA$	$I_x = \int y^2\, dm$
z axis	$I_z = \int x^2\, dA$	$I_z = \int x^2\, dm$
Product of Inertia I_{nm}		
YZ and XZ planes	$I_{xy} = \int xy\, dA$	$I_{xy} = \int xy\, dm$
XZ and XY planes	$I_{yz} = \int yz\, dA$	$I_{yz} = \int yz\, dm$
XY and YZ planes	$I_{zx} = \int zx\, dA$	$I_{zx} = \int zx\, dm$
Radius of Gyration k	$k = (I/A)^{1/2}$	$k = (I/m)^{1/2}$

Table 2-2 Properties of Lines, Surfaces, and Bodies

I_i rectangular moment of inertia
R_i corresponding radius of gyration
J_o polar moment of inertia about axis through O normal to plane
R_o corresponding radius of gyration

Lines

Figure	Centroid Location
1. Any Plane Curve	C.G. is at point having coordinates $\bar{x}, \bar{y}$, where $$\bar{x} = \frac{\int x\, ds}{\text{Length}} \text{ where } ds = \sqrt{1 + \left(\frac{dy}{dx}\right)^2}\, dx$$ $$\bar{y} = \frac{\int y\, ds}{\text{Length}} \text{ where } ds = \sqrt{1 + \left(\frac{dx}{dy}\right)^2}\, dy$$
2. Circular Arc	C.G. is on axis of symmetry at $$\bar{x} = \frac{r \sin \alpha}{\alpha} = \frac{rc}{s}.$$ If α is small, distance from C.G. to chord = approx. $\frac{2h}{3}$. (Error is small even for $\alpha = 45°$) For semi-circle: $\bar{x} = \frac{2r}{\pi}$ For quadrant: $\bar{x} = \frac{2r\sqrt{2}}{\pi}$, and distance from radius drawn to either end of arc $= \frac{2r}{\pi}$.

Plane Surfaces

Figure	Centroid Location; Moments of Inertia; Radii of Gyration
3. Any Plane Surface	C.G. is at point having coordinates $\bar{x}, \bar{y}$, where $$\bar{x} = \frac{\iint x\, dx dy}{\text{Area}} = \frac{\iint \rho^2 \cos\theta\, d\rho d\theta}{\text{Area}};$$ $$\bar{y} = \frac{\iint y\, dx dy}{\text{Area}} = \frac{\iint \rho^2 \sin\theta\, d\rho d\theta}{\text{Area}}$$ $$I_x = \iint y^2\, dx dy;\quad I_y = \iint x^2\, dx dy;\quad J_o = \iint \rho^3\, d\rho d\theta = I_x + I_y$$ $$k_x = \sqrt{\frac{I_x}{\text{Area}}};\quad k_y = \sqrt{\frac{I_y}{\text{Area}}};\quad k_o = \sqrt{\frac{J_o}{\text{Area}}} = \sqrt{\frac{I_x + I_y}{\text{Area}}}$$

Table 2-2 *(Continued)*
Plane Surfaces *(Continued)*

Figure	Centroid Location; Moments of Inertia; Radii of Gyration
4. Triangle 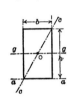	C.G. is at O = intersection of medians. Perpendicular distance from $a-a = \dfrac{h}{3}$. $I_g = \dfrac{bh^3}{36}; \quad I_a = \dfrac{bh^3}{12}; \quad I_c = \dfrac{bh^3}{4}$ $k_g = \dfrac{h}{3\sqrt{2}}; \quad k_a = \dfrac{h}{\sqrt{6}}; \quad k_c = \dfrac{h}{\sqrt{2}}$
5. Solid Rectangle (or Square) 	C.G. is at O = intersection of diagonals. *For Rectangle:* $I_g = \dfrac{bh^3}{12}; \quad I_a = \dfrac{bh^3}{3}; \quad I_c = \dfrac{b^3h^3}{6(b^2+h^2)}; \quad J_o = \dfrac{bh(b^2+h^2)}{12}$ $k_g = \dfrac{h}{2\sqrt{3}}; \quad k_a = \dfrac{h}{\sqrt{3}}; \quad k_c = \dfrac{bh}{\sqrt{6(b^2+h^2)}}; \quad k_o = \sqrt{\dfrac{b^2+h^2}{12}}$ *For Square* (letting $b = h = s$): $I_g = \dfrac{s^4}{12}; \quad I_a = \dfrac{s^4}{3}; \quad I_c = \dfrac{s^4}{12}; \quad J_o = \dfrac{s^4}{6}$ $k_g = \dfrac{s}{2\sqrt{3}}; \quad k_a = \dfrac{s}{\sqrt{3}}; \quad k_c = \dfrac{s}{2\sqrt{3}}; \quad k_o = \dfrac{s}{\sqrt{6}}$
6. Hollow Rectangle (or Square) 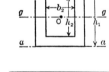	C.G. is at O = intersection of diagonals. *For Hollow Rectangle:* $I_g = \dfrac{b_1 h_1^3 - b_2 h_2^3}{12}; \quad I_a = \dfrac{b_1 h_1^3}{3} - \dfrac{b_2 h_2(3h_1^2 + h_2^2)}{12}$ $k_g = \sqrt{\dfrac{b_1 h_1^3 - b_2 h_2^3}{12(b_1 h_1 - b_2 h_2)}}; \quad J_o = \dfrac{b_1 h_1 (b_1^2 + h_1^2) - b_2 h_2 (b_2^2 + h_2^2)}{12}$ *For Hollow Square* (letting $b_1 = h_1 = s_1$ and $b_2 = h_2 = s_2$): $I_g = \dfrac{s_1^4 - s_2^4}{12}; \quad I_a = \dfrac{s_1^4}{3} - \dfrac{s_2^2(3s_1^2 + s_2^2)}{12}$ $k_g = \sqrt{\dfrac{s_1^2 + s_2^2}{12}}; \quad J_o = \dfrac{s_1^4 - s_2^4}{6}$ (Note: For a diagonal $c-c$, $I_c = I_g$ and $k_c = k_g$)
7. Trapezoid	C.G. is at O, located as shown. $I_g = \dfrac{h^3(B^2 + 4Bb + b^2)}{36(B+b)}; \quad I_a = \dfrac{h^3(B + 3b)}{12}$ $k_g = \dfrac{h\sqrt{2(B^2 + 4Bb + b^2)}}{6(B+b)}; \quad k_a = \dfrac{h}{\sqrt{6}}\sqrt{\dfrac{B + 3b}{B + b}}$
8. Quadrilateral	C.G. is at O, located as follows: Divide the sides into thirds and construct the parallelogram with sides passing through the third-points as shown. The intersection of the diagonals of this parallelogram is the desired centroid.
9. Regular Polygon	C.G. is at O = geometrical center. Let $g-g$ be any axis through O and in plane of polygon. Then $I_g = \dfrac{\text{Area} \cdot (6R^2 - a^2)}{24} = \dfrac{\text{Area} \cdot (12r^2 + a^2)}{48};$ $J_o = \dfrac{\text{Area} \cdot (6R^2 - a^2)}{12} = \dfrac{\text{Area} \cdot (12r^2 + a^2)}{24}$ $k_g = \sqrt{\dfrac{6R^2 - a^2}{24}} = \sqrt{\dfrac{12r^2 + a^2}{48}};$ $k_o = \sqrt{\dfrac{6R^2 - a^2}{12}} = \sqrt{\dfrac{12r^2 + a^2}{24}}$

Table 2-2 (*Continued*)
Plane Surfaces (*Continued*)

Figure	Centroid Location; Moments of Inertia; Radii of Gyration
10. Circle	C.G. is at O = geometrical center. $I_g = \dfrac{\pi r^4}{4} = \dfrac{\pi d^4}{64}$; $J_o = \dfrac{\pi r^4}{2} = \dfrac{\pi d^4}{32}$ $k_g = \dfrac{r}{2} = \dfrac{d}{4}$; $k_o = \dfrac{r}{\sqrt{2}} = \dfrac{d}{\sqrt{8}}$
11. Circular Sector	C.G. is on axis of symmetry at O. Distance from $a\text{-}a = \dfrac{2r \sin \alpha}{3\alpha} = \dfrac{2rc}{3s}$. A = area = $r^2 \alpha$ $I_g = \dfrac{Ar^2}{4}\left(1 - \dfrac{\sin \alpha \cos \alpha}{\alpha}\right)$; $I_a = \dfrac{Ar^2}{4}\left(1 + \dfrac{\sin \alpha \cos \alpha}{\alpha}\right)$ $k_g = \dfrac{r}{2}\sqrt{1 - \dfrac{\sin \alpha \cos \alpha}{\alpha}}$; $k_a = \dfrac{r}{2}\sqrt{1 + \dfrac{\sin \alpha \cos \alpha}{\alpha}}$
12. Semi-circle	C.G. is on axis of symmetry at O. Distance from $a\text{-}a = \dfrac{4r}{3\pi} = 0.424r$ $I_g = \dfrac{d^4(9\pi^2 - 64)}{1152\pi} = \dfrac{r^4(9\pi^2 - 64)}{72\pi} = 0.1098r^4$; $I_a = I_b = \dfrac{\pi d^4}{128} = \dfrac{\pi r^4}{8}$; $J_o = r^4\left(\dfrac{\pi}{4} - \dfrac{8}{9\pi}\right) = 0.5025r^4$ $k_g = \dfrac{d\sqrt{9\pi^2 - 64}}{12\pi} = \dfrac{r\sqrt{9\pi^2 - 64}}{6\pi} = 0.264r$; $k_a = k_b = \dfrac{d}{4} = \dfrac{r}{2}$; $k_o = r\sqrt{\dfrac{1}{2} - \dfrac{16}{9\pi^2}} = 0.566r$
13. Circular Segment	C.G. is on axis of symmetry at O. Distance from $a\text{-}a = \dfrac{2r^3 \sin^3 \alpha}{3A} = \dfrac{c^3}{12A}$ where A = area = $\dfrac{r^2(2\alpha - \sin 2\alpha)}{2}$. $I_g = \dfrac{Ar^2}{4}\left(1 - \dfrac{2 \sin^3 \alpha \cos \alpha}{3(\alpha - \sin \alpha \cos \alpha)}\right)$; $I_a = \dfrac{Ar^2}{4}\left(1 + \dfrac{2 \sin^3 \alpha \cos \alpha}{(\alpha - \sin \alpha \cos \alpha)}\right)$ $k_g = \dfrac{r}{2}\sqrt{1 - \dfrac{2 \sin^3 \alpha \cos \alpha}{3(\alpha - \sin \alpha \cos \alpha)}}$; $k_a = \dfrac{r}{2}\sqrt{1 + \dfrac{2 \sin^3 \alpha \cos \alpha}{(\alpha - \sin \alpha \cos \alpha)}}$
14. Annulus	C.G. is at O = geometrical center. $I_g = \dfrac{\pi(d_1^4 - d_2^4)}{64} = \dfrac{\pi(r_1^4 - r_2^4)}{4}$; $J_o = \dfrac{\pi(d_1^4 - d_2^4)}{32} = \dfrac{\pi(r_1^4 - r_2^4)}{2}$ $k_g = \dfrac{\sqrt{d_1^2 + d_2^2}}{4} = \dfrac{\sqrt{r_1^2 + r_2^2}}{2}$; $k_o = \sqrt{\dfrac{d_1^2 + d_2^2}{8}} = \sqrt{\dfrac{r_1^2 + r_2^2}{2}}$
15. Ellipse	C.G. is at O = geometrical center. For semi-ellipse ABB', C.G. is on OA at distance to right of $c\text{-}c = \dfrac{4a}{3\pi}$. For quarter-ellipse ABO, C.G. is at distance to right of $c\text{-}c = \dfrac{4a}{3\pi}$ and at distance above $g\text{-}g = \dfrac{4b}{3\pi}$. A = area = πab $I_g = \dfrac{\pi ab^3}{4} = \dfrac{Ab^2}{4}$; $I_c = \dfrac{\pi a^3 b}{4} = \dfrac{Aa^2}{4}$; $J_o = \dfrac{A(a^2 + b^2)}{4}$ $k_g = \dfrac{b}{2}$; $k_c = \dfrac{a}{2}$; $k_o = \dfrac{\sqrt{a^2 + b^2}}{2}$
16. Parabolic Segment	C.G. is on axis of symmetry at O. Distance from $c\text{-}c = \dfrac{3a}{5}$. $I_g = \dfrac{4ab^3}{15}$; $I_c = \dfrac{4a^3 b}{7}$ $k_g = \dfrac{b}{\sqrt{5}} = 0.447b$; $k_c = a\sqrt{\dfrac{3}{7}} = 0.654a$
17. Structural Shapes	

Table 2-2 (*Continued*)

Homogeneous Bodies
(Including Nonplanar Surfaces)
("Body" is to be understood unless "Surface" is indicated.)

Figure	Centroid Location; Moments of Inertia; Radii of Gyration
18. Any Surface or Body of Revolution	Let axis of revolution be X axis. Then generating curve is $y = f(x)$. C.G. is at point having coordinates $\bar{x}, \bar{y}, \bar{z}$. *For Surface:* $$\bar{x} = \frac{\int 2\pi xy\,ds}{\int 2\pi y\,ds} = \frac{\int xy\sqrt{1+\left(\frac{dy}{dx}\right)^2}\,dx}{\int y\sqrt{1+\left(\frac{dy}{dx}\right)^2}\,dx}; \quad \bar{y}=0; \quad \bar{z}=0$$ *For Body* $\left(\text{letting } \delta = \text{density} = \dfrac{m}{\text{volume}}\right):$ $$\bar{x} = \frac{\int \pi xy^2\,dx}{\int \pi y^2\,dx}; \quad \bar{y}=0; \quad \bar{z}=0.$$ $$I_x = \frac{\pi\delta}{2}\int y^4\,dx; \quad I_y = I_z = \pi\delta\int\left(\frac{y^4}{4}+x^2y^2\right)dx$$ $$k_x = \sqrt{\frac{I_x}{m}}; \quad k_y = k_z = \sqrt{\frac{I_y}{m}} = \sqrt{\frac{I_z}{m}}$$ *For Thin Shell* having mass: C.G. coordinates are same as for surface. $$I_x = 2\pi\delta\int y^3\,ds = 2\pi\delta\int y^3\sqrt{1+\left(\frac{dy}{dx}\right)^2}\,dx$$ $$k_x = \sqrt{\frac{I_x}{m}}$$
19. Thin Straight Rod	C.G. is at O = geometrical center. $$I_g = \frac{ml^2}{12}; \quad I_b = \frac{ml^2}{3}; \quad I_c = \frac{ml^2\sin^2\alpha}{12}; \quad I_d = \frac{ml^2\sin^2\alpha}{3}$$ $$k_g = \frac{l}{\sqrt{12}}; \quad k_b = \frac{l}{\sqrt{3}}; \quad k_c = \frac{l\sin\alpha}{\sqrt{12}}; \quad k_d = \frac{l\sin\alpha}{\sqrt{3}}$$
20. Thin Rod Bent into Circular Arc	C.G. is on axis of symmetry at $\bar{x} = \dfrac{r\sin\alpha}{\alpha}$. $$I_x = \frac{mr^2}{2}\left(1 - \frac{\sin\alpha\cos\alpha}{\alpha}\right); \quad I_y = \frac{mr^2}{2}\left(1 + \frac{\sin\alpha\cos\alpha}{\alpha}\right); \quad I_z = mr^2$$ $$k_x = r\sqrt{\frac{1}{2} - \frac{\sin\alpha\cos\alpha}{2\alpha}}; \quad k_y = r\sqrt{\frac{1}{2} + \frac{\sin\alpha\cos\alpha}{2\alpha}}; \quad k_z = r$$
21. Rectangular Parallelepiped (or Cube)	C.G. is at O = geometrical center. *For Parallelepiped:* $$I_g = \frac{m(b^2+c^2)}{12}; \quad I_d = \frac{m(a^2+b^2)}{12}; \quad I_e = \frac{m(4a^2+b^2)}{12}$$ $$k_g = \sqrt{\frac{b^2+c^2}{12}}; \quad k_d = \sqrt{\frac{a^2+b^2}{12}}; \quad k_e = \sqrt{\frac{4a^2+b^2}{12}}$$ *For Cube* (letting $a=b=c=s$): $$I_g = I_d = \frac{ms^2}{6}; \quad I_e = \frac{5ms^2}{12}$$ $$k_g = k_d = \frac{s}{\sqrt{6}}; \quad k_e = s\sqrt{\frac{5}{12}}$$
22. Right Rectangular Pyramid	C.G. is on axis of symmetry at O. Distance from base $= \dfrac{h}{4}$. Drawing g–g axis through O parallel to side a: $$I_g = \frac{m}{20}\left(b^2 + \frac{3h^2}{4}\right); \quad I_c = \frac{m}{20}(a^2+b^2)$$ $$k_g = \sqrt{\frac{4b^2+3h^2}{80}}; \quad k_c = \sqrt{\frac{a^2+b^2}{20}}$$

Table 2-2 (*Continued*)

Homogeneous Bodies (*Continued*)

Figure	Centroid Location; Moments of Inertia; Radii of Gyration
23. Pyramid (or Frustum of Pyramid) 	*For Surface of Any Pyramid*: C.G. of surface (base excluded) is on line joining apex with centroid of perimeter of base, at a distance two-thirds its length from the apex. *For Body of Any Pyramid*: C.G. of body is on line joining apex with centroid of base, at a distance three-fourths its length from the apex. *For Surface of Frustum of Pyramid having Regular Bases*: Letting R and r be the lengths of sides of the larger and smaller bases respectively, and h the altitude: C.G. of surface (bases excluded) is at distance from larger base = $\dfrac{h(R + 2r)}{3(R + r)}$. *For Body of Frustum of Any Pyramid*: Letting A and a be the areas of the larger and smaller bases, respectively, and h the altitude: C.G. of body is at distance from larger base = $\dfrac{h(A + 2\sqrt{Aa} + 3a)}{4(A + \sqrt{Aa} + a)}$
24. Right Elliptical Cylinder (or Circular Cylinder) 	C.G. is at O = geometrical center. *For Right Elliptical Cylinder*: $I_g = \dfrac{m}{12}(3b^2 + h^2)$; $I_c = \dfrac{m}{4}(a^2 + b^2)$; $I_e = \dfrac{m}{12}(3r^2 + 4h^2)$ $k_g = \sqrt{\dfrac{3b^2 + h^2}{12}}$; $k_c = \dfrac{\sqrt{a^2 + b^2}}{2}$; $k_e = \sqrt{\dfrac{3r^2 + 4h^2}{12}}$ *For Right Circular Cylinder* (letting $a = b = r$): $I_g = \dfrac{m}{12}(3r^2 + h^2)$; $I_c = \dfrac{mr^2}{2}$ $k_g = \sqrt{\dfrac{3r^2 + h^2}{12}}$; $k_c = \dfrac{r}{\sqrt{2}}$
25. Hollow Right Circular Cylinder	C.G. is at O = geometrical center. $I_g = \dfrac{m}{12}(3R^2 + 3r^2 + h^2)$; $I_c = \dfrac{m}{2}(R^2 + r^2)$; $I_e = \dfrac{m}{12}(3R^2 + 3r^2 + 4h^2)$ $k_g = \sqrt{\dfrac{3R^2 + 3r^2 + h^2}{12}}$; $k_c = \sqrt{\dfrac{R^2 + r^2}{2}}$; $k_e = \sqrt{\dfrac{3R^2 + 3r^2 + 4h^2}{12}}$ *For Thin Shell* (radius R): $I_g = \dfrac{m}{12}(6R^2 + h^2)$; $I_c = mR^2$; $I_e = \dfrac{m}{6}(3R^2 + 2h^2)$ $k_g = \sqrt{\dfrac{6R^2 + h^2}{12}}$; $k_c = R$; $k_e = \sqrt{\dfrac{3R^2 + 2h^2}{6}}$
26. Right Circular Cone	C.G. is on axis of symmetry at O. Distance from base = $\dfrac{h}{4}$. Drawing g-g axis through O and d-d axis through apex, both parallel to base: $I_g = \dfrac{3m}{20}\left(r^2 + \dfrac{h^2}{4}\right)$; $I_c = \dfrac{3mr^2}{10}$; $I_d = \dfrac{3m}{20}(r^2 + 4h^2)$ $k_g = \sqrt{\dfrac{3}{80}(4r^2 + h^2)}$; $k_c = \dfrac{3r}{\sqrt{30}}$; $k_d = \sqrt{\dfrac{3}{20}(r^2 + 4h^2)}$
27. Frustum of Right Circular Cone 	C.G. is on axis of symmetry at O. Distance from base = $\dfrac{h(R^2 + 2Rr + 3r^2)}{4(R^2 + Rr + r^2)}$. $I_c = \dfrac{3m(R^5 - r^5)}{10(R^3 - r^3)}$; $k_c = \sqrt{\dfrac{3(R^5 - r^5)}{10(R^3 - r^3)}}$

Table 2-2 *(Continued)*
Homogeneous Bodies *(Continued)*

Figure	Centroid Location; Moments of Inertia; Radii of Gyration
28. Cone (or Frustum of Cone)	*For Surface of Any Cone:* C.G. of surface (base excluded) is on line joining apex with centroid of perimeter of base, at a distance two-thirds its length from the apex. *For Body of Any Cone:* C.G. of body is on line joining apex with centroid of base, at a distance three-fourths its length from the apex. *For Surface of Frustum of a Circular Cone:* Letting R and r be the radii of the larger and smaller bases, respectively, and h the altitude: C.G. of surface (bases excluded) is at distance from larger base = $\dfrac{h(R + 2r)}{3(R + r)}$. *For Body of Frustum of a Circular Cone:* Letting R and r be the radii of the larger and smaller bases, respectively, and h the altitude: C.G. of body is at distance from larger base = $\dfrac{h(R^2 + 2Rr + 3r^2)}{4(R^2 + Rr + r^2)}$
29. Thin Circular Lamina	C.G. is at O = geometrical center. $I_g = \dfrac{mr^2}{4}$; $I_c = \dfrac{mr^2}{2}$ (where c–c axis is perpendicular to the plane). $k_g = \dfrac{r}{2}$; $k_c = \dfrac{r}{\sqrt{2}}$
30. Sphere	C.G. is at O = geometrical center. $I_g = \dfrac{2mr^2}{5}$ $k_g = \dfrac{2r}{\sqrt{10}}$
31. Hollow Sphere	C.G. is at O = geometrical center. $I_g = \dfrac{2m}{5}\left(\dfrac{R^5 - r^5}{R^3 - r^3}\right)$; $k_g = \sqrt{\dfrac{2}{5}\left(\dfrac{R^5 - r^5}{R^3 - r^3}\right)}$ *For Thin Shell* (radius R): $I_g = \dfrac{2mR^2}{3}$; $k_g = \dfrac{2R}{\sqrt{6}}$
32. Spherical Sector	C.G. is on axis of symmetry at O. Distance from center of sphere = $\dfrac{3(2r - h)}{8}$. $I_g = \dfrac{m}{5}(3rh - h^2)$ $k_g = \sqrt{\dfrac{3rh - h^2}{5}}$
33. Hemisphere	*For Surface:* C.G. is on axis of symmetry at distance from center of sphere = $\dfrac{r}{2}$. *For Body:* C.G. is on axis of symmetry at distance from center of sphere = $\dfrac{3r}{8}$. $I_g = \dfrac{2mr^2}{5}$; $k_g = \dfrac{2r}{\sqrt{10}}$
34. Spherical Segment	C.G. is on axis of symmetry at distance from center of sphere = $\dfrac{3(2r - h)^2}{4(3r - h)}$. $I_g = m\left(r^2 - \dfrac{3rh}{4} + \dfrac{3h^2}{20}\right)\dfrac{2h}{(3r - h)}$ $k_g = \sqrt{\left(r^2 - \dfrac{3rh}{4} + \dfrac{3h^2}{20}\right)\dfrac{2h}{3r - h}}$

Table 2-2 (*Continued*)
Homogeneous Bodies (*Continued*)

Figure	Centroid Location; Moments of Inertia; Radii of Gyration
35. Torus	C.G. is at O = geometrical center. $$I_g = \frac{m(4R^2 + 5r^2)}{8}; \quad I_c = \frac{m(4R^2 + 3r^2)}{4}$$ $$k_g = \sqrt{\frac{4R^2 + 5r^2}{8}}; \quad k_c = \frac{\sqrt{4R^2 + 3r^2}}{2}$$
36. Ellipsoid	C.G. is at O = geometrical center. C.G. of *one octant* is at point having coordinates: $$\bar{x} = \frac{3a}{8}; \quad \bar{y} = \frac{3b}{8}; \quad \bar{z} = \frac{3c}{8}$$ For Complete Ellipsoid: $$I_x = \frac{m}{5}(b^2 + c^2); \quad I_y = \frac{m}{5}(a^2 + c^2); \quad I_z = \frac{m}{5}(a^2 + b^2)$$ $$k_x = \sqrt{\frac{b^2 + c^2}{5}}; \quad k_y = \sqrt{\frac{a^2 + c^2}{5}}; \quad k_z = \sqrt{\frac{a^2 + b^2}{5}}$$
37. Paraboloid	C.G. is on axis of symmetry at O. Distance from base $= \frac{h}{3}$. $$I_g = \frac{mr^2}{3}; \quad I_c = \frac{m}{18}(3r^2 + h^2)$$ $$k_g = \frac{r}{\sqrt{3}}; \quad k_c = \sqrt{\frac{3r^2 + h^2}{18}}$$

Rectangular and Polar Moments of Inertia. The moment of inertia of a plane area is rectangular ($I = \int x^2 \, dA$) if the axis is in the plane of the area, and polar ($J = \int r^2 \, dA$) if the axis is perpendicular to the plane of the area. If I_x, I_y, and J_z are the moments of inertia about the X, Y, and Z axes for a plane figure in the XY plane, and k_x, k_y, and k_z are the corresponding radii of gyration, then

$$J_z = I_x + I_y \quad \text{and} \quad k_z^2 + k_x^2 = k_y^2 \tag{2-14}$$

Principal Axes of Inertia

For a plane surface the principal axes are the two axes through a point in the plane about which the moments of inertia are a maximum and a minimum, and the corresponding moments of inertia are called the *principal moments of inertia of the plane.* The principal axes are always at right angles to each other, and the product of inertia with respect to them is zero.

For a body there are three principal axes through one point: the two axes at right angles for which the moments of inertia are a maximum and a minimum and the axis at right angles to the plane of these two. The corresponding moments of inertia of the body are the *principal moments of inertia of the body* at the point. If the point is the center of gravity of the body, the properties are called *central principal axes* and *moments of inertia.* For the planes determined by a set of principal axes, the three products of inertia are zero.

Transformation Formulas

Parallel Axes. If the moment of inertia about any axis is I and about a parallel axis through the centroid is I_0, k and k_0 are the corresponding radii of gyration, and s is the distance between axes, then

$$\text{for a plane} \quad I = I_0 + As^2, \quad k^2 = k_0^2 + s^2 \tag{2-15}$$

$$\text{for a body} \quad I = I_0 + ms^2, \quad k^2 = k_0^2 + s^2 \tag{2-16}$$

Rotated Axes for a Plane Figure. If X and Y, U and V are two sets of rectangular coordinate axes with common origin and I_x, I_y, I_u, and I_v are the moments of inertia about these areas, I_{xy} and I_{uv} are the products of inertia with respect to the sets of axes, and α is the angle of rotation (positive if counterclockwise), then

$$I_u + I_v = I_x + I_y \tag{2-17}$$

$$I_u = I_x \cos^2 \alpha + I_y \sin^2 \alpha - I_{xy} \sin 2\alpha \tag{2-18}$$

$$I_{uv} = \tfrac{1}{2}(I_x - I_y) \sin 2\alpha + I_{xy} \cos 2\alpha \tag{2-19}$$

Rotated Axes for a Body. If I_x, I_y, and I_z are the moments of inertia of a body about the X, Y, and Z axes and I_{xy} (YZ plane), I_{yz} (ZX plane), I_{zx} (XY plane) are the products of inertia, and I is the moment of inertia about a line through the origin of coordinates having direction-angles α, β, γ, then

$$I = I_x \cos^2 \alpha + I_y \cos^2 \beta + I_z \cos^2 \gamma - 2I_{yz} \cos \beta \cos \gamma - 2I_{zx} \cos \gamma \cos \alpha \tag{2-20}$$

KINEMATICS

Displacement s of a particle is the vector change of position. ***Velocity*** $\mathbf{v}$ is the time rate of displacement and is a vector quantity. ***Speed*** is the magnitude of the instantaneous velocity, the absolute value of velocity, and is a scalar quantity. ***Acceleration*** a is the time rate of change of velocity and is a vector quantity.

Motion of a Particle

Instantaneous velocity

$$\mathbf{v} = \lim_{\Delta t \to 0} \frac{\Delta \mathbf{s}}{\Delta t} = \frac{d\mathbf{s}}{dt} \tag{2-21}$$

is a vector with direction along the line tangent to the curve of position and with magnitude equal to the speed of the particle, ds/dt. The tangential component of velocity v_t is ds/dt, and the normal component v_n is v^2/r, where r is the radius of curvature. The axial components of velocity are:

$$\mathbf{v}_x = \frac{dx}{dt}, \quad \mathbf{v}_y = \frac{dy}{dt}, \quad \mathbf{v}_z = \frac{dz}{dt} \tag{2-22}$$

The resultant velocity and its direction cosines are:

$$\mathbf{v} = (\mathbf{v}_x^2 + \mathbf{v}_y^2 + \mathbf{v}_z^2)^{1/2} \tag{2-23}$$

$$\cos \theta_x = \frac{v_x}{v}, \quad \cos \theta_y = \frac{v_y}{v}, \quad \cos \theta_z = \frac{v_z}{v} \tag{2-24}$$

112   MECHANICS

Instantaneous acceleration:

$$\mathbf{a} = \lim_{\Delta t \to 0} \frac{\Delta \mathbf{v}}{\Delta t} = \frac{d\mathbf{v}}{dt} \tag{2-25}$$

The tangential component of acceleration a_t is $dv/dt = d^2s/dt^2$, and the normal component a_n is v^2/r, where r is the radius of curvature. The axial components acceleration are:

$$a_x = \frac{d^2x}{dt^2}, \quad a_y = \frac{d^2y}{dt^2}, \quad a_z = \frac{d^2z}{dt^2} \tag{2-26}$$

Table 2-3   Equations for Translation and Rotation

	Rectilinear or Plane Motion	Rotation Fixed Axis
Velocity	$v = \dfrac{ds}{dt}$ $v = v_x + v_y$	$\omega = \dfrac{d\theta}{dt}$
Acceleration	$a = \dfrac{dv}{dt} = \dfrac{d^2s}{dt^2}$	$\alpha = \dfrac{d\omega}{dt} = \dfrac{d^2\theta}{dt^2}$
Displacement General*	$s_2 - s_1 = \displaystyle\int_{t_1}^{t_2} v\, dt$	$\theta_2 - \theta_1 = \displaystyle\int_{t_1}^{t_2} \omega\, dt$
Uniform accel.	$s = \tfrac{1}{2}at^2 + v_0 t$	$\theta = \tfrac{1}{2}\alpha t^2 + \omega_0 t$
Velocity acquired General*	$v_2 - v_1 = \displaystyle\int_{t_1}^{t_2} a\, dt$ $v_2 - v_1 = 2\displaystyle\int_{s_1}^{s_2} a\, ds$	$\omega_2 - \omega_1 = \displaystyle\int_{t_1}^{t_2} \alpha\, dt$ $\omega_2 - \omega_1 = 2\displaystyle\int_{\theta_1}^{\theta_2} \alpha\, d\theta$
Uniform accel.	$v_t = at + v_0$ $v_t = (2as + v_0^2)^{1/2}$	$\omega_t = \alpha t + \omega_0$ $\omega_t = (2\alpha\theta + \omega_0^2)^{1/2}$
Time for displacement*	$t_2 - t_1 = \displaystyle\int_{s_1}^{s_2} \dfrac{ds}{v} = \displaystyle\int_{v_1}^{v_2} \dfrac{dv}{a}$	$t_2 - t_1 = \displaystyle\int_{\theta_1}^{\theta_2} \dfrac{d\theta}{dt} = \displaystyle\int_{\omega_1}^{\omega_2} \dfrac{d\omega}{a}$
Uniform accel.	$t = \dfrac{(2as + v_0^2)^{1/2} - v_0}{a}$	$t = \dfrac{(2\alpha\theta + \omega_0^2)^{1/2} - \omega_0}{a}$
Uniform accel. required for displacement in time t	$a = \dfrac{2(s - v_0 t)}{t^2}$	$\alpha = \dfrac{2(\theta - \omega_0 t)}{t^2}$

* Rectilinear motion only.

The resultant acceleration and its direction cosines are:

$$\mathbf{a} = (\mathbf{a}_x^2 + \mathbf{a}_y^2 + \mathbf{a}_z^2)^{1/2} \tag{2-27}$$

$$\cos \theta_x = \frac{a_x}{a}, \quad \cos \theta_y = \frac{a_y}{a}, \quad \cos \theta_z = \frac{a_z}{a} \tag{2-28}$$

Relations between Rectilinear and Rotational Motions. If the axis of rotation is fixed, these relations are between scalar quantities, with displacement in radians and distance r from the axis of rotation:

$$v = r\omega, \quad a_t = r\alpha, \quad a_n = r\omega^2, \quad a = r(\alpha^2 + \omega^4)^{1/2} \tag{2-29}$$

If the axis of rotation is not fixed, the relations are in terms of vector products:

$$\mathbf{v} = \boldsymbol{\omega} \times \mathbf{r}, \quad \mathbf{a}_t = \boldsymbol{\alpha} \times \mathbf{r}, \quad \mathbf{a}_n = \boldsymbol{\omega} \times (\boldsymbol{\omega} \times \mathbf{r}) \tag{2-30}$$

Table 2-3 gives equations for velocity, acceleration, displacement, and time for rectilinear and rotational motions of a particle.

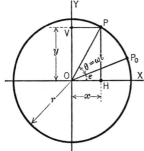
Fig. 2-4 Simple harmonic motion.

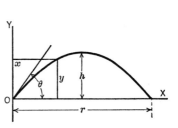
Fig. 2-5 Path of projectile.

Simple Harmonic Motion

Simple harmonic motion is any rectilinear motion in which the acceleration is directed toward a fixed point in the path and is proportional to the distance between that point and the moving point.

The motion of the projection on a diameter of a point moving with constant speed around a circle is an example of simple harmonic motion. (See Fig. 2-4.) The distance of the projection from the center is the *displacement*, and the radius of the circle is the *amplitude*. The angle ϵ of lead or lag is the *phase constant*.

Displacement,
$$\left. \begin{array}{l} x = r \cos(\omega t + \epsilon) \\ y = r \sin(\omega t + \epsilon) \end{array} \right\} \tag{2-31a}$$

Velocity,
$$\left. \begin{array}{l} v_x = -r\omega \sin(\omega t + \epsilon) = -\omega y \\ v_y = r\omega \cos(\omega t + \epsilon) = \omega x \end{array} \right\} \tag{2-31b}$$

114   MECHANICS

Acceleration,

$$a_x = -r\omega^2 \cos(\omega t + \epsilon) = -\omega^2 x \\ a_y = -r\omega^2 \sin(\omega t + \epsilon) = -\omega^2 y \quad (2\text{-}31c)$$

Motion of a Projectile

The trajectory or path of a particle, neglecting air resistance, is a parabola with coordinates x and y at any time t. The initial velocity of projection is v_0, total time of flight is T, and downward acceleration is g. See Fig. 2-5 for other symbols. Then

$$x = (v_0 \cos \theta_0)t, \quad y = (v_0 \sin \theta_0)t - \frac{gt^2}{2} \quad (2\text{-}32a)$$

$$v = (v_0^2 - 2gy)^{1/2}, \quad v_x = v_0 \cos \theta, \quad v_y = v_0 \sin \theta - gt \quad (2\text{-}32b)$$

$$h = \frac{v_0^2}{2g} \sin 2\theta, \quad r = \frac{v_0^2}{g} \sin 2\theta, \quad T = \frac{2v_0}{g} \sin \theta \quad (2\text{-}32c)$$

If projection is horizontal, i.e., a bomb released from airplane in level flight, $\theta = 0$, $x = v_0 t$, and $y = -gt^2/2$.

Pendulum

For a *simple pendulum* with the mass concentrated at a point L distance from the axis, the period of oscillation

$$t = 2\pi (L/g)^{1/2} \quad (2\text{-}33)$$

For a *conical pendulum* where h is the height of the cone of revolution, the period of revolution

$$t = 2\pi (h/g)^{1/2} \quad (2\text{-}34)$$

and the angular velocity

$$\omega = (g/h)^{1/2} \quad (2\text{-}35)$$

For a *compound pendulum* where k is the radius of gyration about the axis and s is the distance between the axis and the center of gravity, the length of the equivalent simple pendulum

$$L = k^2/s \quad (2\text{-}36)$$

DYNAMICS

Newton's Laws of Motion

First Law. Every body persists in a state of rest or of uniform linear motion until acted upon by some force.

$$mv = \text{const.} \quad (2\text{-}37)$$

Second Law. The time rate of change of momentum is proportional to the force producing it and occurs in the direction of the resultant of the forces acting.

$$Fg_c = \frac{d(mv)}{dt} \quad (2\text{-}38)$$

Third Law. The mutual actions of any two bodies are always equal and oppositely directed; to every action there is an equal and opposite reaction:

$$F = -F_R \qquad (2\text{-}39)$$

Equations of Motion

With fixed coordinates and g_c equal to unity (see Table 1-1),

$$F = \frac{d(mv)}{dt} = m\frac{d^2s}{dt^2}$$

$$F_x = m\frac{d^2x}{dt^2}, \quad F_y = m\frac{d^2y}{dt^2}, \quad F_z = m\frac{d^2z}{dt^2} \qquad (2\text{-}40)$$

which may be replaced by the single vector equation

$$\mathbf{F} = m\mathbf{a} \qquad (2\text{-}41)$$

With moving coordinates where **d** is the displacement with respect to the moving coordinates whose origin has translational velocity $d\mathbf{R}/dt$ and angular velocity $\boldsymbol{\omega}$, the vector equations are

$$\frac{d\mathbf{d}}{dt} \frac{d\mathbf{d}_r}{dt} + \boldsymbol{\omega}\mathbf{d} \qquad (2\text{-}42)$$

where $\mathbf{d}_r/dt$ is the velocity relative to the moving coordinates.

$$\mathbf{F} = m\frac{d^2\mathbf{R}}{dt^2} + m\boldsymbol{\omega} \times \boldsymbol{\omega} \times \mathbf{d} + m\frac{d\boldsymbol{\omega}}{dt} \times \mathbf{d} + m\frac{d^2\mathbf{d}_r}{dt^2} + 2m\boldsymbol{\omega} \times \frac{d\mathbf{d}_r}{dt} \qquad (2\text{-}43)$$

Euler's equations for the motion of a rigid body refer to centroidal axes x, y, and z which are fixed in the body, i.e., rotate with it so that the moments of inertia become constants, invariant with time, if x, y, z are principal axes.

$$M_x = I_{xx}\alpha_x + (I_{zz} - I_{yy})\omega_y\omega_z \qquad (2\text{-}44a)$$

$$M_y = I_{yy}\alpha_y + (I_{xx} - I_{zz})\omega_z\omega_x \qquad (2\text{-}44b)$$

$$M_z = I_{zz}\alpha_z + (I_{yy} - I_{xx})\omega_x\omega_y \qquad (2\text{-}44c)$$

where M_x, etc., are moments of force and I_{xx}, etc., are moments of inertia.

Work, Power, and Energy

Work is the integral product of force and the displacement over which the force acts.

$$\text{Work of a force,} \quad W = \int_{s_1}^{s_2} F \cos \alpha \, ds = \int_{s_1}^{s_2} F_t \, ds \qquad (2\text{-}45)$$

where ds is the elementary length of path, α is the angle between force and path, and F_t is the tangential component of force.

$$\text{Work of a torque,} \quad W = \int_{\theta_1}^{\theta_2} T \, d\theta = \int_{\theta_1}^{\theta_2} F_t r \, d\theta \qquad (2\text{-}46)$$

where $d\theta$ is the elementary angular displacement and r is the torque radius.

Table 2-4 Kinetic Equations
Constant F, v, a, T, ω, α

	Translation	Rotation
Force, torque	$F = ma$	$T = I\alpha = mk^2\alpha$
Impulse	$\text{Imp} = Ft$ $\text{Imp} = m(v_2 - v_1)$	$\text{Imp} = Tt$ $\text{Imp} = I(\omega_2 - \omega_1)$
Momentum change	$m(v_2 - v_1) = Ft$	$I(\omega_2 - \omega_1) = Tt$
Work	$W = Fs$	$W = T\theta$
Power	$P = Fv$	$P = T\omega$
Kinetic energy	$E_K = \tfrac{1}{2}mv^2$	$E_K = \tfrac{1}{2}I\omega^2 = \tfrac{1}{2}mk^2\omega^2$
Kinetic energy change	$\tfrac{1}{2}m(v_2^2 - v_1^2) = F(s_2 - s_1)$	$\tfrac{1}{2}I(\omega_2^2 - \omega_1^2) = T(\theta_2 - \theta_1)$

Power is the time rate of doing work.

$$\text{Power of a force,} \quad P = \frac{dW}{dt} = F_t \frac{ds}{dt} = F_t v. \tag{2-47}$$

where v is the instantaneous velocity of the point of application of the force.

$$\text{Power of a torque,} \quad P = \frac{dW}{dt} = T\frac{d\theta}{dt} = T\omega \tag{2-48}$$

where ω is the instantaneous angular velocity.

Energy of a system is its potential for doing work. *Kinetic energy* ($mv^2/2$ or $I\omega^2/2$) is the energy of motion. V is the velocity of the mass center; I is the moment of inertia of mass about an axis through the mass center perpendicular to the plane of motion. *Potential energy* (mgh) is the energy of position. The *total energy* of a mechanical system is the sum of kinetic energy, potential energy, and internal energy U.

$$E = \frac{mv^2}{2} + \frac{I\omega^2}{2} + mgh + U \tag{2-49}$$

According to the *principle of conservation of energy*, the total energy of a system remains constant although one form of energy may be converted to another. This principle leads to the mechanical-energy balance,

$$\frac{mv_1^2}{2} + \frac{I\omega_1^2}{2} + mgh_1 + U_1 = \frac{mv_2^2}{2} + \frac{I\omega_2^2}{2} + mgh_2 + U_2 + W_s + Q \tag{2-50}$$

where W_s is the work done by the system, and Q is the heat evolved by the system (heat of friction, for example).

If only transformations of potential energy and translational kinetic energy are involved,

$$\Delta E_p = -\Delta E_K, \quad mg(h_2 - h_1) = \tfrac{1}{2}m(v_1^2 - v_2^2) \tag{2-51}$$

If only work and translational kinetic energy are involved,

$$W_s = -\Delta E_K = \tfrac{1}{2}m(v_1^2 - v_2^2) \tag{2-52}$$

Impulse, Momentum, and Impact

Integration of the equation of motion,

$$\int_{t_1}^{t_2} \mathbf{F}\, dt = \int_{t_1}^{t_2} m \frac{d\mathbf{v}}{dt}\, dt \tag{2-53a}$$

gives

$$\int_{t_1}^{t_2} \mathbf{F}\, dt = m\mathbf{v}_2 - m\mathbf{v}_1 \tag{2-53b}$$

The term on the left is the impulse of the force $\mathbf{F}$ and the term $m\mathbf{v}$ is the momentum of the particle. Thus the impulse equals the change in momentum. Since linear impulse is a vector quantity,

$$\mathbf{Imp} = [(F_x t)^2 + (F_y t)^2 + (F_z t)^2]^{1/2} \tag{2-54}$$

Angular impulse of a force about a line is

$$\mathrm{Imp} = \int_{t_1}^{t_2} T\, dt \tag{2-55}$$

Impact occurs when two bodies collide. In the absence of external forces, the impulses exerted by the two bodies are equal and opposite, hence the total momentum is conserved. On *direct central impact* where motion is normal to the striking surfaces and along the line of their mass centers,

$$m_1 v_1 + m_2 v_2 = m_1 v'_1 + m_2 v'_2 \tag{2-56}$$

where the subscripts refer to the two bodies and v and v' denote velocity before and after collision.

A *coefficient of restitution e* is introduced to account for imperfect elasticity. If e is known from experiment or analysis of stress wave propagation, the rebound velocities and kinetic energy loss Q may be computed by

$$v'_1 - v'_2 = -e(v_1 - v_2) \tag{2-57a}$$

and

$$Q = \frac{(v_1 - v_2)^2 (1 - e^2) m_1 m_2}{2(m_1 + m_2)} \tag{2-57b}$$

Friction

The general equation for the frictional resistance F to sliding between two dry surfaces pressed together by the normal force N is

$$F = fN \tag{2-58}$$

If F is the force required to start sliding, f is the coefficient of static friction. If F is the somewhat smaller force required to maintain sliding once started, f is the coefficient of kinetic friction. The values of f vary widely for different materials and surface conditions, and tend to decrease as relative velocity increases. See Table 2-5 for approximate values of f.

Table 2-5 Approximate Coefficients of Friction

Materials	Condition	Sliding Friction θ	Sliding Friction f	Static Friction θ	Static Friction f
Cast iron on cast iron or bronze	Wet	17°	0.31	—	—
Earth on earth (clay)	Damp	—	—	45°	1.0
Earth on earth (clay)	Wet	—	—	17°	0.31
Hemp rope on polished wood	Dry	—	—	18°	0.33
Leather on oak	Dry	17°	0.30	26°	0.50
Leather on cast iron	Dry	29°	0.56	17°	0.30
Plastic on steel	Dry	—	0.40	—	—
Rubber (grooved) on pavement	Dry	—	0.40	—	0.55
Steel on graphite	Dry	—	0.09	—	0.21
Steel on ice	Dry	—	0.01	$1\frac{1}{2}$°	0.027
Steel on steel	Dry	—	0.50	—	0.75
Stone on concrete	Dry	—	—	37°	0.76
Stone on ground	Wet	—	—	$16\frac{1}{2}$°	0.30
Wrought iron on wrought iron	Dry	23°	0.44	—	—
Wrought iron on cast iron	Dry	10°	0.18	$10\frac{1}{2}$°	0.19

The angle of repose or angle of static friction θ is the angle with the horizontal plane at which sliding of one surface upon another will begin.

$$\tan \theta = F/N = f \tag{2-59}$$

Hydrodynamic friction, when there is a thick film of lubricant between the surfaces, is represented by the equation

$$F = \mu v A/h \tag{2-60}$$

where μ is the coefficient of viscosity, h the film thickness of the lubricant, v the relative velocity, and A the area of contact.

For a journal bearing with radius r, length L, and clearance c, this equation becomes

$$F = 4\pi^2 \mu r n L/c \tag{2-61}$$

where n is the number of revolutions per second. This equation is valid only for lightly loaded bearings where the shaft is centrally located in the clearance space.

Belt or coil friction opposes the slipping of a belt, line, or brake-band on pulley or sheave. When power is transmitted, the tension T_1 on the driving side of the belt is greater than the tension T_2 on the driven side. Neglecting centrifugal force, the tensions at incipient slipping are related by

$$T_1/T_2 = e^{f\alpha} \qquad (2\text{-}62)$$

where α is the angle in radians over the entire arc of contact between belt and pulley. Power transmitted is

$$P = (T_1 - T_2)v \qquad (2\text{-}63)$$

and maximum power is

$$P_{\max} = T_2(e^{f\alpha} - 1) = T_1\left(1 - \frac{1}{e^{f\alpha}}\right)v \qquad (2\text{-}64)$$

MECHANICS OF MATERIALS

Basic Definitions

Stress is the resultant internal force that opposes change in the shape of a body under the action of external forces. *Unit stress* (σ, τ) is the stress per unit area. *Tensile stress or tension* is the internal force that opposes increase in length and *compressive stress or compression* the internal force that opposes decrease in

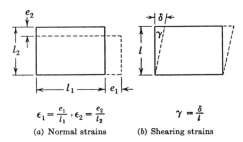

$\epsilon_1 = \dfrac{e_1}{l_1}$, $\epsilon_2 = \dfrac{e_2}{l_2}$ $\gamma = \dfrac{\delta}{l}$

(a) Normal strains   (b) Shearing strains

Fig. 2-6 Strain diagram.

length of a body acted upon by external forces. *Shearing stress or shear* is the force acting along a plane between adjacent parts of a body that opposes parallel external forces tending to slide one part of the body over another part. *Normal stress* on a section is the resultant stress which acts perpendicular to the section. *Axial stress*, a special case of normal stress, is the stress in a straight homogeneous bar when the resultant of the applied loads coincides with the axis of the bar. *True stress* is the ratio of the applied load to the smallest value of cross-sectional area existing under the applied load. *Allowable unit stress* is the maximum unit stress which it is considered safe to use in practice.

Strain or deformation is the amount of change in the shape of a body under the action of external forces. *Unit strain* (ϵ, γ) is the deformation per unit length computed as the ratio of total deformation to original length of the body. Strain may be tensile, compressive, or shear. *Elasticity* is the property of a body that permits it to undergo strain and return to its original shape upon removal

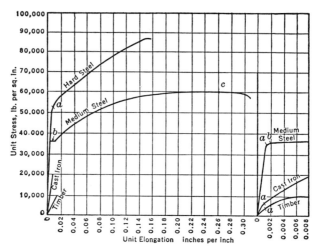

Fig. 2-7 Stress-deformation diagram.

of the external forces. *Bulk modulus* relates an increase in unit stress to the corresponding decrease in volume. *True strain* is a function of the ratio of original diameter to instantaneous diameter under stress.

$$\text{True strain} = 2 \log_e \frac{d_0}{d} \tag{2-65}$$

where d_0 is the original diameter and d the instantaneous diameter.

Poisson's ratio ν, a measure of lateral stiffness, is the ratio of lateral unit deformation to longitudinal unit deformation within the elastic limit.

Table 2-6 Equations for Basic Concepts

Unit stress	$\sigma = P/A, \quad \tau = P/A$
Unit strain	$\epsilon = \dfrac{e}{L}, \quad \gamma = \dfrac{\delta}{L}, \quad \epsilon_{th} = \alpha \Delta T$
Modulus of elasticity	$E = \dfrac{\sigma}{\epsilon} = \dfrac{P/A}{e/L} = \dfrac{PL}{Ae} = 2G(1 + \nu)$
Shear modulus	$G = \dfrac{\tau}{\gamma} = \dfrac{P/A}{\delta/L} = \dfrac{PL}{\delta A} = \dfrac{E}{2(1 + \nu)}$
Bulk modulus	$K = -\dfrac{P/A}{\Delta V/V} = \dfrac{E}{3(1 - 2\nu)}$
Hooke's law	$\epsilon = \dfrac{\sigma}{E}, \quad \gamma = \dfrac{\tau}{G}$
Poisson's ratio	$\nu = e_2/e_1 = \tfrac{1}{2} - \dfrac{E}{6K} = \dfrac{E}{2G} - 1$

Table 2-7 Properties of Structural Materials

Material	Condition	Density, lb/in³	Ultimate Strength × 10³ lb/in²		Yield Strength × 10³ lb/in²		Modulus × 10⁶ lb/in²	
			Tension	Shear	Tension	Compr.	Elasticity	Shear
Carbon steel, 1020	hot-rolled	0.284	68	58	42	42	30	13
Steel, 300-M	hardened	0.283	290	—	240	—	30	—
Stainless steel, 302	cold-rolled	0.290	190	90	165	—	28	11
Wrought iron	hot-rolled	0.278	48	40	30	30	29	12
Gray cast iron	as cast	0.260	25	20	6	20	13	6
Aluminum, 1100	annealed	0.098	13	9.5	5	5	10	4
	cold-rolled		24	13	22	22	10	4
Aluminium, 7075	heat-treated	0.101	83	49	73	—	10.4	—
Titanium	annealed	0.163	90	—	70	—	16.5	6.5
Titanium, 140A	heat treated	0.166	155	—	145	—	16.5	—
Concrete	as cast	0.084	0.3	1.0	—	1.0	2.0	—
White oak	air-dried	0.027					1.5	—
parallel to grain			7	0.5	5	4		
normal to grain			0.8	1.9	—	—		

Hooke's law, relating stress and strain within the proportional limit, states that a body acted upon by external forces will deform in proportion to the stress developed. The proportionality constant is the *modulus of elasticity* or the *shear modulus,* i.e., the slope of the stress-strain curve over the linear range. (See Fig. 2-7 and Table 2-7.)

Proportional limit is the highest unit stress at which stress is proportional to strain. Points a in Fig. 2-7 mark the proportional limit where the stress-strain relation begins to depart from a straight line. *Elastic limit* is the maximum unit stress beyond which a body will not return to its original shape upon removal of the stress. *Yield strength* is the unit stress which first produces a specified permanent set. *Yield point,* point b in Fig. 2-7, is the smallest stress at which strain increases without increase in stress. *Ultimate strength,* point c, is the highest unit stress before rupture. *Plasticity* is a state where permanent strains may occur without fracture. Lead exhibits extreme plasticity. *Ductility* refers to plasticity under tension and *malleability* to plasticity under compression.

Determination of Principal Stresses

Under some conditions of loading a body is subjected to a combination of tensile, compressive, and shear stresses. For analysis it is convenient to reduce such systems of combined stresses to a basic system of stress coordinates known as principal stresses. These stresses act on axes which differ from the axes along

122   MECHANICS

which the applied stresses are acting and represent maximum and minimum values of the normal stresses for the particular point considered.

Mohr's Circle. Let the axes x and y be chosen to represent the directions of the applied normal and shearing stresses (Fig. 2-8). Lay off to a suitable scale the distances $OA = \delta_x$, $OB = \delta_y$, and $BC = AD = \tau_{xy}$. With point E as a

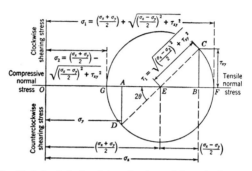

Fig. 2-8   Mohr's circle for determination of the principal stresses.

center, construct the circle DFC. Then OF and OG are the principal stresses δ_1 and δ_2, and EC is the maximum shear stress τ_1.

Beams

Flexure. In addition to compression of fibers on the upper or concave side of a simple beam and tension of fibers on the lower or convex side, shear exists along each cross section, being most intense at the two supports and zero at the middle section. The *neutral surface* is the horizontal section where there is no change in length of the fibers and no tensile or compressive forces acting on them. In homogeneous beams the intensities of the normal stresses are proportional to the distance of the fibers from the neutral surface. The *neutral axis* is the trace of the neutral surface on any cross section of the beam.

Conditions of Equilibrium. For the stresses at any cross section, (1) sum of horizontal tensile stresses equals sum of horizontal compressive stresses, (2) resisting shear equals vertical shears, and (3) resisting moment equals bending moment.

Basic Flexure Formula. If M is the bending moment, the unit stress on any fiber (usually the most remote from the neutral surface), c is the distance of that fiber from the neutral surface, and I is the moment of inertia of the section with respect to the neutral axis, within the proportional limit of the material.

$$\sigma = \frac{Mc}{I} \tag{2-66}$$

Section modulus I/c is a measure of the capacity of a section to resist any bending moment to which it may be subjected.

Bending moment M at any section, positive when the upper fibers are compressed, is the algebraic sum of the moments of the forces acting on either side of the section,

$$M = M_l + V_l x - \tfrac{1}{2}wx^2 - \sum P(x-a) \tag{2-67}$$

where M_l is the bending moment at the left support, V_l vertical shear at the left support, w the uniform load per unit length, P concentrated load upon the left of the section, x distance between left support and the section considered, and a is the distance between the left support and P.

Vertical shear V_x at any section of a horizontal beam is equal to the sum of the vertical components of the reactions to the left of the section less the sum of the vertical downward components of the loads to the left of the section. The vertical shear at the right of the left support of any span is

$$V_l = \frac{M_r - M_l}{L} + \tfrac{1}{2}wL + \sum P\left(1 - \frac{a}{L}\right) \tag{2-68}$$

where M_r and M_l are the moments at right and left supports, and L is length of span.

The maximum shear stress occurs at the neutral axis,

$$\tau_{\max} = \frac{V}{Ib}\int_0^c y\, dA \tag{2-69}$$

where b is the breadth of the section, and y the distance from the neutral axis. For a rectangular section, $\tau_{\max} = 3V/2A$ and for a circular section $\tau_{\max} = 4V/3A$.

The elastic curve is the curve formed by the neutral surface of the beam. Bending strain is a relative rotation of the sides of a beam element through an angle θ so that the length of a fiber increases from L to $L + ds$. If r is the radius of curvature of the neutral surface and c is the distance of the fiber from the neutral surface, from similar geometries $ds/c = L/r$. Since

$$\epsilon = \frac{ds}{L} = \frac{c}{r} \quad \text{and} \quad \sigma = \frac{Mc}{I} = \epsilon E$$

then
$$r = \frac{EI}{M} \tag{2-70}$$

Also a radius of curvature in general is given by

$$r = \frac{[1 + (dy/dx)^2]^{3/2}}{d^2y/dx^2} \tag{2-71}$$

If the deflection is small, higher powers of dy/dx may be ignored so that

$$r = \frac{1}{d^2y/dx^2} \tag{2-72}$$

The elastic curve is obtained by equating the two equations in r,

$$M = EI\frac{d^2y}{dx^2} \tag{2-73}$$

Table 2-8 Bending Moment, Vertical Shear, and Deflection of Beams of Uniform Cross Section under Various Conditions of Loading

P = concentrated loads, lb.
R_1, R_2 = reactions, lb.
w = uniform load per unit of length, lb per in.
W = total uniform load on beam, lb.
l = length of beam, in.
x = distance from support to any section, in.
E = modulus of elasticity, lb per sq in.

I = moment of inertia, in.^{4}
V_x = vertical shear at any section, lb.
V = maximum vertical shear, lb.
M_x = bending moment at any section, lb-in.
M = maximum bending moment, lb-in.
y = maximum deflection, in.

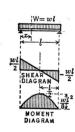

SIMPLE BEAM—UNIFORM LOAD

$R_1 = R_2 = \dfrac{wl}{2}$

$V_x = \dfrac{wl}{2} - wx$

$V = \pm \dfrac{wl}{2}$ (when $\begin{cases} x = 0 \\ x = l \end{cases}$)

$M_x = \dfrac{wlx}{2} - \dfrac{wx^2}{2}$

$M = \dfrac{wl^2}{8}$ (when $x = \dfrac{l}{2}$)

$y = \dfrac{5Wl^3}{384EI}$ (at center of span)

SIMPLE BEAM—CONCENTRATED LOAD AT ANY POINT

$R_1 = P(1 - k)$
$R_2 = Pk$
$V_x = R_1$ (when $x < kl$)
$ = R_2$ (when $x > kl$)
$V = P(1 - k)$
$ $ (when $k < 0.5$)
$ = -Pk$ (when $k > 0.5$)
$M_x = Px(1 - k)$
 (when $x < kl$)
$ = Pk(l - x)$
 (when $x > kl$)
$M = Pkl(1 - k)$ (at point of load)

$y = \dfrac{Pl^3}{3EI}(1 - k)$
$ \times (2/3k - 1/3k^2)^{3/2}$
(at $x = l\sqrt{2/3k - 1/3k^2}$)

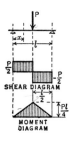

SIMPLE BEAM—CONCENTRATED LOAD AT CENTER

$R_1 = R_2 = \dfrac{P}{2}$

$V_x = V = \pm \dfrac{P}{2}$

$M_x = \dfrac{Px}{2}$

$M = \dfrac{Pl}{4}$ (when $x = \dfrac{l}{2}$)

$y = \dfrac{Pl^3}{48EI}$ (at center of span)

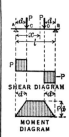

SIMPLE BEAM—TWO EQUAL CONCENTRATED LOADS AT EQUAL DISTANCES FROM SUPPORTS

$R_1 = R_2 = P$
$V_x = P$ for AC
$ = 0$ for CD
$ = -P$ for DB
$V = \pm P$
$M_x = Px$ for AC
$ = Pd$ for CD
$ = P(l - x)$ for DB
$M = Pd$

$y = \dfrac{Pd}{24EI}(3l^2 - 4d^2)$

(at center of span)

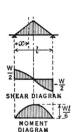

SIMPLE BEAM—LOAD INCREASING UNIFORMLY FROM SUPPORTS TO CENTER OF SPAN

$R_1 = R_2 = \dfrac{W}{2}$

$V_x = W\left(\dfrac{1}{2} - \dfrac{2x^2}{l^2}\right)$

(when $x < \dfrac{l}{2}$)

$V = \pm \dfrac{W}{2}$ (at supports)

$M_x = Wx\left(\dfrac{1}{2} - \dfrac{2x^2}{3l^2}\right)$

$M = \dfrac{Wl}{6}$ (at center of span)

$y = \dfrac{Wl^3}{60EI}$ (at center of span)

CANTILEVER BEAM—LOAD CONCENTRATED AT FREE END

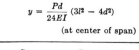

$R = P$

$V_x = V = -P$

$M_x = -P(l - x)$

$M = -Pl$ (when $x = 0$)

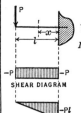

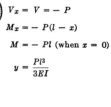

$y = \dfrac{Pl^3}{3EI}$

Table 2-8 *(Continued)*

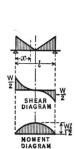 SHEAR DIAGRAM / MOMENT DIAGRAM	**Simple Beam — Load Increasing Uniformly from Center to Supports** $R_1 = R_2 = \dfrac{W}{2}$ $V_x = -W\left(\dfrac{2x}{l} - \dfrac{2x^2}{l^2} - \dfrac{1}{2}\right)$ $\left(\text{when } x < \dfrac{l}{2}\right)$ $V = \pm \dfrac{W}{2}$ $M_x = Wx\left(\dfrac{1}{2} - \dfrac{x}{l} + \dfrac{2}{3}\dfrac{x^2}{l^2}\right)$ $\left(\text{when } x < \dfrac{l}{2}\right)$ $M = \dfrac{Wl}{12}$ (at center of span) $y = \dfrac{3}{320}\dfrac{Wl^3}{EI}$ (at center of span)	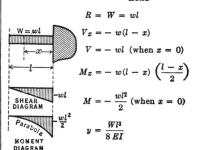 **Cantilever Beam — Uniform Load** $R = W = wl$ $V_x = -w(l - x)$ $V = -wl$ (when $x = 0$) $M_x = -w(l - x)\left(\dfrac{l - x}{2}\right)$ $M = -\dfrac{wl^2}{2}$ (when $x = 0$) $y = \dfrac{Wl^3}{8EI}$

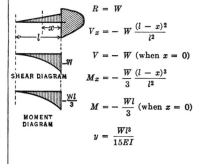 SHEAR DIAGRAM / MOMENT DIAGRAM	**Simple Beam — Load Increasing Uniformly from One Support to the Other** $R_1 = \dfrac{W}{3}; \quad R_2 = \dfrac{2}{3}W$ $V_x = W\left(\dfrac{1}{3} - \dfrac{x^2}{l^2}\right)$ $V = -\dfrac{2}{3}W$ (when $x = l$) $M_x = \dfrac{Wx}{3}\left(1 - \dfrac{x^2}{l^2}\right)$ $M = \dfrac{2}{9\sqrt{3}}Wl$ $\left(\text{when } x = \dfrac{l}{\sqrt{3}}\right)$ $y = \dfrac{0.01304}{EI}Wl^3$	**Cantilever Beam — Load Increasing Uniformly from Free End to Support** $R = W$ $V_x = -W\dfrac{(l - x)^2}{l^2}$ $V = -W$ (when $x = 0$) $M_x = -\dfrac{W}{3}\dfrac{(l - x)^3}{l^2}$ $M = -\dfrac{Wl}{3}$ (when $x = 0$) $y = \dfrac{Wl^3}{15EI}$

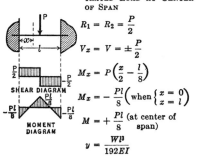 SHEAR DIAGRAM / MOMENT DIAGRAM	**Fixed Beam — Concentrated Load at Center of Span** $R_1 = R_2 = \dfrac{P}{2}$ $V_x = V = \pm \dfrac{P}{2}$ $M_x = P\left(\dfrac{x}{2} - \dfrac{l}{8}\right)$ $M_x = -\dfrac{Pl}{8}$ $\left(\text{when } \begin{cases} x = 0 \\ x = l \end{cases}\right)$ $M = +\dfrac{Pl}{8}$ (at center of span) $y = \dfrac{Wl^3}{192EI}$	**Fixed Beam — Uniform Load** $R_1 = R_2 = \dfrac{wl}{2} = \dfrac{W}{2}$ $V_x = \dfrac{wl}{2} - wx$ $V = \pm \dfrac{wl}{2}$ (at ends) $M_x = -\dfrac{wl^2}{2}\left(\dfrac{1}{6} - \dfrac{x}{l} + \dfrac{x^2}{l^2}\right)$ $M = -1/12\, wl^2$ $\left(\text{when } \begin{cases} x = 0 \\ x = l \end{cases}\right)$ $M = \dfrac{wl^2}{24}\left(\text{when } x = \dfrac{l}{2}\right)$ $y = \dfrac{Wl^3}{384EI}$

Table 2-8 (Continued)

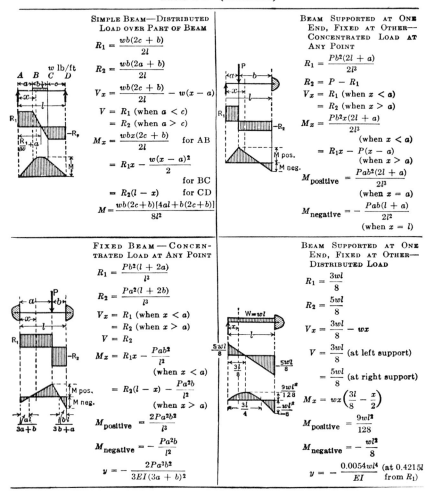

Deflection y of a statically determinate beam is obtained by double integration of the elastic curve with constants of integration established by known boundary conditions. For a simple beam supported at the ends with uniformly distributed load, the deflection is

$$y = \frac{\sigma L^2}{48 E d} \qquad (2\text{-}74)$$

Solutions for other specific conditions are given in Table 2-8.

Continuous Beams. As in simple beams, beams resting on more than two supports are governed by the expressions $M = \sigma I/c$ and $\tau = V/A$. Here the bending moments at the various sections must be determined to derive the vertical shears at the sections and the reactions at the supports. The theorem of three moments is used to obtain the bending moments at any three consecutive supports. (See Fig. 2-9.)

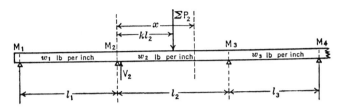

Fig. 2-9 Three-span beam.

For uniform loads:

$$M_1 l_1 + 2M_2(l_1 + l_2) + M_3 l_2 = -\tfrac{1}{4} w_1 l_1^3 - \tfrac{1}{4} w_2 l_2^3 \qquad (2\text{-}75)$$

When both spans are equally loaded and of equal length,

$$M_1 + 4M_2 + M_3 = -\tfrac{1}{2} w l^2 \qquad (2\text{-}76)$$

For concentrated loads:

$$M_1 l_1 + 2M_2(l_1 + l_2) + M_3 l_2 = P_1 l_1^2(k_1^3 - k_1) + P_2 l_2^2(3k_2^2 - k_2^3 - 2k_2) \qquad (2\text{-}77)$$

Curved Beams. The effect of curvature is to increase the stress on the inside and decrease it on the outside fibers, and to shift the neutral axis from the centroidal axis toward the convex or inner side. When the curvature is too great to be ignored, the correct value for fiber unit stress may be obtained by including a correction factor K in the flexure formula, viz. $\sigma = KMc/I$. Table 2-9 gives values of K for a number of shapes and ratios R/c. For an unstressed beam (no curvature) $R = c$.

Columns

Length of a column L is the distance between points unsupported against lateral deflection. The effective length l depends on the condition of the ends. The ratio l/L is 2 for columns with one end fixed and one free, 1 for both ends free to turn, i.e., round ends, $\tfrac{2}{3}$ for one end fixed and one round, $\tfrac{1}{2}$ for both ends fixed, and $\tfrac{3}{4}$ to $\tfrac{7}{8}$ for pinned or hinged ends.

Slenderness ratio l/k is the ratio of effective length to least radius of gyration.

Column Formulas. These relations are essentially empirical. In them $P/A =$ load per unit area, $c =$ distance from neutral axis to most compressed fiber, $l/k =$ slenderness ratio, $E =$ modulus of elasticity, and $\sigma =$ allowable unit compressive stress in design or yield strength in buckling.

Table 2-9 Values of Constant K for Curved Beams

Section	$\dfrac{R}{c}$	Values of K Inside Fiber	Values of K Outside Fiber	$\dfrac{Y_0*}{R}$	Section	$\dfrac{R}{c}$	Values of K Inside Fiber	Values of K Outside Fiber	$\dfrac{Y_0*}{R}$
(two stacked circles, h, c)	1.2	3.41	0.54	0.224	(rectangle, c, R)	1.2	2.89	0.57	0.305
	1.4	2.40	0.60	0.151		1.4	2.13	0.63	0.204
	1.6	1.96	0.65	0.108		1.6	1.79	0.67	0.149
	1.8	1.75	0.68	0.084		1.8	1.63	0.70	0.112
	2.0	1.62	0.71	0.069		2.0	1.52	0.73	0.090
	3.0	1.33	0.79	0.030		3.0	1.30	0.81	0.041
	4.0	1.23	0.84	0.016		4.0	1.20	0.85	0.021
	6.0	1.14	0.89	0.0070		6.0	1.12	0.90	0.0093
	8.0	1.10	0.91	0.0039		8.0	1.09	0.92	0.0052
	10.0	1.08	0.93	0.0025		10.0	1.07	0.94	0.0033
(trapezoid b, c, $2b$, R)	1.2	3.01	0.54	0.336	(trapezoid $3b$, c, $2b$, R)	1.2	3.09	0.56	0.336
	1.4	2.18	0.60	0.229		1.4	2.25	0.62	0.229
	1.6	1.87	0.65	0.168		1.6	1.91	0.66	0.168
	1.8	1.69	0.68	0.128		1.8	1.73	0.70	0.128
	2.0	1.58	0.71	0.102		2.0	1.61	0.73	0.102
	3.0	1.33	0.80	0.046		3.0	1.37	0.81	0.046
	4.0	1.23	0.84	0.024		4.0	1.26	0.86	0.024
	6.0	1.13	0.88	0.011		6.0	1.17	0.91	0.011
	8.0	1.10	0.91	0.0060		8.0	1.13	0.94	0.0060
	10.0	1.08	0.93	0.0039		10.0	1.11	0.95	0.0039
(triangle $5b$, c, $4b$, R)	1.2	3.14	0.52	0.352	(triangle $\tfrac{3}{5}b$, c, b, R)	1.2	3.26	0.44	0.361
	1.4	2.29	0.54	0.243		1.4	2.39	0.50	0.251
	1.6	1.93	0.62	0.179		1.6	1.99	0.54	0.186
	1.8	1.74	0.65	0.138		1.8	1.78	0.57	0.144
	2.0	1.61	0.68	0.110		2.0	1.66	0.60	0.116
	3.0	1.34	0.76	0.050		3.0	1.37	0.70	0.052
	4.0	1.24	0.82	0.028		4.0	1.27	0.75	0.029
	6.0	1.15	0.87	0.012		6.0	1.16	0.82	0.013
	8.0	1.12	0.91	0.0060		8.0	1.12	0.86	0.0060
	10.0	1.12	0.93	0.0039		10.0	1.09	0.88	0.0039
(L-section $4\tfrac{1}{2}t$, $\tfrac{3t}{2}$, $4t$, c, R)	1.2	3.63	0.58	0.418	(Z-section t, $3t$, $2t$, $4t$, t, c, R)	1.2	3.55	0.67	0.409
	1.4	2.54	0.63	0.299		1.4	2.48	0.72	0.292
	1.6	2.14	0.67	0.229		1.6	2.07	0.76	0.224
	1.8	1.89	0.70	0.183		1.8	1.83	0.78	0.178
	2.0	1.73	0.72	0.149		2.0	1.69	0.80	0.144
	3.0	1.41	0.79	0.069		3.0	1.38	0.86	0.067
	4.0	1.29	0.83	0.040		4.0	1.26	0.89	0.038
	6.0	1.18	0.88	0.018		6.0	1.15	0.92	0.018
	8.0	1.13	0.91	0.010		8.0	1.10	0.94	0.010
	10.0	1.10	0.92	0.0065		10.0	1.08	0.95	0.0065
(I-section t, $4t$, t, $3t$, c, R)	1.2	2.52	0.67	0.408	(H-section t, $4t$, t, c, R)	1.2	2.37	0.73	0.453
	1.4	1.90	0.71	0.285		1.4	1.79	0.77	0.319
	1.6	1.63	0.75	0.208		1.6	1.56	0.79	0.236
	1.8	1.50	0.77	0.160		1.8	1.44	0.81	0.183
	2.0	1.41	0.79	0.127		2.0	1.36	0.83	0.147
	3.0	1.23	0.86	0.058		3.0	1.19	0.88	0.067
	4.0	1.16	0.89	0.030		4.0	1.13	0.91	0.036
	6.0	1.10	0.92	0.013		6.0	1.08	0.94	0.016
	8.0	1.07	0.94	0.0076		8.0	1.06	0.95	0.0089
	10.0	1.05	0.95	0.0048		10.0	1.05	0.96	0.0057
(hollow circle $2d$, d, c, R)	1.2	3.28	0.58	0.269	(hollow rectangle $4t$, $4t$, $2t$, t, c, R)	1.2	2.63	0.68	0.399
	1.4	2.31	0.64	0.182		1.4	1.97	0.73	0.280
	1.6	1.89	0.68	0.134		1.6	1.66	0.76	0.205
	1.8	1.70	0.71	0.104		1.8	1.51	0.78	0.159
	2.0	1.57	0.73	0.083		2.0	1.43	0.80	0.127
	3.0	1.31	0.81	0.038		3.0	1.23	0.86	0.058
	4.0	1.21	0.85	0.020		4.0	1.15	0.89	0.031
	6.0	1.13	0.90	0.0087		6.0	1.09	0.92	0.014
	8.0	1.10	0.92	0.0049		8.0	1.07	0.94	0.0076
	10.0	1.07	0.93	0.0031		10.0	1.06	0.95	0.0048

* Y_0 is distance from centroidal axis to neutral axis, where beam is subjected to pure bending

Euler's formula for buckling load on long columns ($l/k > 200$) is

$$\frac{P}{A} = \frac{\pi^2 E}{(l/k)^2} \tag{2-78}$$

Secant formula for buckling of intermediate columns where e is an assumed initial eccentricity and ec/k^2 is usually taken as 0.25:

$$\frac{P}{A} = \frac{\sigma}{1 + (ec/k^2) \sec\left(\dfrac{l}{2k}\sqrt{P/AE}\right)} \tag{2-79}$$

The secant formula can also be used for eccentric loading if e is taken as the actual eccentricity plus the assumed initial eccentricity.

Rankine's formula, sometimes called *Gordon's formula*, for intermediate and long columns where ϕ is the reciprocal of the unit stress allowable for design or the yield strength for rupture is

$$\frac{P}{A} = \frac{\sigma}{1 + \phi(l/k)^2} \tag{2-80}$$

In many city building codes, Rankine's formula is specified with σ and $1/\phi$, each taken as 18,000 lb/sq in. for steel.

Modified Rankine formula for eccentric loads where e is the distance of the applied load from the axis is

$$\frac{P}{A} = \frac{\sigma}{1 + \phi(l/k)^2 + (ec/k)^2} \tag{2-81}$$

Transverse Loading. A column subjected to cross-bending loads may be considered as a beam with end thrust. The maximum unit stress from the bending load is Mc/I, and if deflection is neglected, the unit stress from the axial load is P/A. The interaction formula becomes

$$\frac{P/A}{\sigma_a} + \left(\frac{Mc/I}{\sigma_b}\right)_x + \left(\frac{Mc/I}{\sigma_b}\right)_y \leqq 1 \tag{2-82}$$

where σ_a is the allowable axial unit stress according to a column formula, and σ_b is the allowable bending unit stress as a beam.

Stress due to Temperature Change. For a straight member with ends restrained, where α is the coefficient of linear thermal expansion,

$$\sigma = E\alpha\,\Delta T \tag{2-83}$$

Cylinders, Plates, and Rollers

Thin-Walled Cylinders under Internal Pressure. The tensile hoop stress on a longitudinal section with diameter D and thickness t under internal pressure

P/A is

$$\sigma = \frac{P}{A}\frac{D}{2t} \tag{2-84}$$

Thick-Walled Cylinder under Internal Pressure. With maximum shear theory the criterion of failure,

$$\tau = \frac{P}{A}\frac{r_2^2 + r_1^2}{r_2^2 - r_1^2} \tag{2-85}$$

Stresses in Uniformly Loaded Plates. Circular plate, supported edge, maximum stress at center:

$$\sigma = (3 + \nu)\frac{3r^2 P/A}{8t^2} \tag{2-86}$$

Circular plate, fixed edge, maximum stress at edge:

$$\sigma = \frac{3r^2 P/A}{4t^2} \tag{2-87}$$

Rectangular plate, supported edges, maximum stress at center:

$$\sigma = \frac{L^2 b^2 P/A}{2t^2(L^2 + b^2)} \tag{2-88}$$

Rectangular plate, fixed edges, maximum stress at center of long edge:

$$\sigma = \frac{b^2 P/A}{2t^2(1 + 0.62 b^6/L^6)} \tag{2-89}$$

Pressures with Two Surfaces in Contact. Maximum stress σ at the center of contact is given below for dimensions of inches and pounds.

Two spheres: $\quad \sigma = 0.616 \sqrt[3]{PE^2\left(\dfrac{D_1 + D_2}{D_1 D_2}\right)^2} \tag{2-90}$

Sphere and plane: $\quad \sigma = 0.616 \sqrt[3]{PE^2/D^2} \tag{2-91}$

Cylinder and plane: $\quad \sigma = 0.591 \sqrt{P_1 E/D} \tag{2-92}$

Two cylinders: $\quad \sigma = 0.591 \sqrt{P_1 E\left(\dfrac{D_1 + D_2}{D_1 D_2}\right)} \tag{2-93}$

P is total load, P_1 is load per inch of length, E is modulus of elasticity, and D_1, D_2 are diameters in inches.

Shafts

Torsional Shear Stress τ in Circular Shafts. The twisting moment T is opposed by an equal resisting moment $T_r = T = \tau J/c$.

For solid shafts: $\quad \tau = 16T/\pi D^3, \quad J = \pi D^4/32 \tag{2-94}$

For hollow shafts: $\quad \tau = \dfrac{16TD}{\pi(D^4 - D_i^4)}, \quad J = \dfrac{\pi(D^4 - D_i^4)}{32}$

Torque (lb in.) = 63,030 horsepower/revolutions per minute.

MECHANICS OF MATERIALS

Angle of Twist. For a bar held fast at one end, the displacement at the other end under tortional stress is called the angle of twist θ.

For solid, circular shafts: $\quad \theta = TL/GJ \quad\quad$ (2-95)

For hollow, circular shafts: $\quad \theta = \dfrac{32TL}{\pi(D^4 - D_i^4)G} \quad$ (2-96)

J is polar moment of inertia, G is shear modulus, L is length of shaft, and D, D_i are diameters. For values of G see Table 2-7.

REFERENCES

A.I.S.C. Handbook, Am. Inst. Steel Construction, New York, 1963.
Carpenter, S. T., *Structural Mechanics*, John Wiley, New York, 1960.
Flugge, W., *Handbook of Engineering Mechanics*, McGraw-Hill, New York, 1962.
Leech, J. W., *Classical Mechanics*, John Wiley, New York, 1958.
Mabie, H. H. and D. H. Young, *Mechanics and Dynamics of Machinery*, John Wiley, New York, 1957.
Seely, F. B. and J. O. Smith, *Advanced Mechanics of Materials*, John Wiley, New York, 1952.
Timoshenko, S. P., *Strength of Materials*, third edition, D. Van Nostrand, Princeton, N.J., 1956.
Timoshenko, S. P. and D. H. Young, *Engineering Mechanics*, McGraw-Hill, New York, fourth edition, 1956.

Table 2-10 Allowable Unit Stress, Structural Steel

σ_y = yield strength, $C = (2\pi^2 E/\sigma_y)^{1/2}$

	Allowable Unit Stress
Tension	$0.6\sigma_y$
Bending	$0.6\sigma_y$
Shear	$0.4\sigma_y$
Compression Columns, $l/k < C$	$\dfrac{1 - \dfrac{(l/k)^2}{2C^2}\sigma_y}{\dfrac{5}{3} + \dfrac{3}{8}\left(\dfrac{l/k}{C}\right) - \dfrac{1}{8}\left(\dfrac{l/k}{C}\right)^3}$
Columns, $l/k > C$	$\dfrac{149{,}000{,}000}{(l/k)^2}$
Other members	$0.6\sigma_y$
Bearing plates and web of rolled shapes	$0.75\sigma_y$

Table 2-11 Allowable Unit Stress for Lumber, lb/in²

Species and Commercial Grade E = Modulus of Elasticity	Tension Parallel to Grain	Horizontal Shear	Compression	
			Parallel to Grain	Normal to Grain
Douglas fir, $E = 1.6 \times 10^6$				
Select structural	1900	120	1450	415
No. 1 structural	1450	120	1200	390
No. 2	1100	110	1075	390
Norway pine, $E = 1.2 \times 10^6$				
Prime structural	1200	75	900	360
Common structural	1100	75	775	360
Utility structural	950	75	650	360
Redwood, $E = 1.2 \times 10^6$				
Dense structural	1700	110	1450	320
Heart structural	1300	95	1100	320
Southern pine, $E = 1.6 \times 10^6$				
Select structural	2400	120	1750	455
No. 1 structural	1600	120	1150	455
No. 2	1100	85	875	390
Spruce, $E = 1.2 \times 10^6$				
Select structural	1450	110	1050	300

Table 2-12 Allowable Unit Stresses in Reinforced Concrete under Static Loads

(Requirements of ACI 318—51)

Allowable Unit Stresses

Description		For Any Strength of Concrete in Accordance with Section 302 $n = \dfrac{30{,}000}{f'_c}$	Maximum Value, psi	For Strength of Concrete Shown Below			
				$f'_c = 2000$ psi $n = 15$	$f'_c = 2500$ psi $n = 12$	$f'_c = 3000$ psi $n = 10$	$f'_c = 3750$ psi $n = 8$
Flexure: f_c							
Extreme fiber stress in compression	f_c	$0.45f'_c$		900	1125	1350	1688
Extreme fiber stress in tension in plain concrete footings	f_c	$.03f'_c$		60	75	90	113
Shear: v (as a measure of diagonal tension)							
Beams with no web reinforcement	v_c	$.03f'_c$		60	75	90	113
Beams with properly designed web reinforcement	v	$.12f'_c$		240	300	360	450
Flat slabs at distant d from edge of column capital or drop panel	v_c	$.03f'_c$		60	75	90	113
Footings	v_c	$.03f'_c$	75	60	75	75	75
Bond: u							
Deformed bars							
Top bars*	u	$.07f'_c$	245	140	175	210	245
In 2-way footings (except top bars)	u	$.08f'_c$	280	160	200	240	280
All others	u	$.10f'_c$	350	200	250	300	350
Plain bars (must be hooked)							
Top bars	u	$.03f'_c$	105	60	75	90	105
In 2-way footings (except top bars)	u	$.036f'_c$	126	72	90	108	126
All others	u	$.045f'_c$	158	90	113	135	158
Bearing: f_c							
On full area	f_c	$.25f'_c$		500	625	750	938
On one-third area or less†	f_c	$.375f'_c$		750	938	1125	1405

* Top bars are horizontal bars so placed that more than 12 in. of concrete is cast in the member below the bar.

† The allowable bearing stress on an area greater than one-third but less than the full area shall be interpolated between the values given.

Table 2-13 Properties of Wide-Flange Sections—Steel

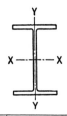

Nominal Size	Weight per Foot	Area	Depth	Flange Width	Flange Thickness	Web Thickness	Axis X-X I	Axis X-X S	Axis X-X r	Axis Y-Y I	Axis Y-Y S	Axis Y-Y r
in.	lb	in.^{2}	in.	in.	in.	in.	in.^{4}	in.^{3}	in.	in.^{4}	in.^{3}	in.
36 × 16½	300	88.17	36.72	16.655	1.680	0.945	20290.2	1105.1	15.17	1225.2	147.1	3.73
	230	67.73	35.88	16.475	1.260	.765	14988.4	835.5	14.88	870.9	105.7	3.59
36 × 12	194	57.11	36.48	12.117	1.260	.770	12103.4	663.6	14.56	355.4	58.7	2.49
	170	49.98	36.16	12.027	1.100	.680	10470.0	579.1	14.47	300.6	50.0	2.45
	150	44.16	35.84	11.972	0.940	.625	9012.1	502.9	14.29	250.4	41.8	2.38
30 × 15	210	61.78	30.38	15.105	1.315	.775	9872.4	649.9	12.64	707.9	93.7	3.38
	190	55.90	30.12	15.040	1.185	.710	8825.9	586.1	12.57	624.6	83.1	3.34
	172	50.65	29.88	14.985	1.065	.655	7891.5	528.2	12.48	550.1	73.4	3.30
27 × 14	177	52.10	27.31	14.090	1.190	.725	6728.6	492.8	11.36	518.9	73.7	3.16
	160	47.04	27.08	14.023	1.075	.658	6018.6	444.5	11.31	458.0	65.3	3.12
	145	42.68	26.88	13.965	0.975	.600	5414.3	402.9	11.26	406.9	58.3	3.09
24 × 12	120	35.29	24.31	12.088	0.930	0.556	3635.3	299.1	10.15	254.0	42.0	2.68
	110	32.36	24.16	12.042	0.855	0.510	3315.0	274.4	10.12	229.1	38.0	2.66
	100	29.43	24.00	12.000	0.775	0.468	2987.3	248.9	10.08	203.5	33.9	2.63
18 × 8¾	85	24.97	18.32	8.838	0.911	0.526	1429.9	156.1	7.57	99.4	22.5	2.00
	77	22.63	18.16	8.787	0.831	0.475	1286.8	141.7	7.54	88.6	20.2	1.98
	70	20.56	18.00	8.750	0.751	0.438	1153.9	128.2	7.49	78.5	17.9	1.95
16 × 8½	78	22.92	16.32	8.586	0.875	0.529	1042.6	127.8	6.74	87.5	20.4	1.95
	71	20.86	16.16	8.543	0.795	0.486	936.9	115.9	6.70	77.9	18.2	1.93
	64	18.80	16.00	8.500	0.715	0.443	833.8	104.2	6.66	68.4	16.1	1.91
14 × 16	426	125.25	18.69	16.695	3.033	1.875	6610.3	707.4	7.26	2359.5	282.7	4.34
	264	77.63	16.50	16.025	1.938	1.205	3526.0	427.4	6.74	1331.2	166.1	4.14
	228	67.06	16.00	15.865	1.688	1.045	2942.4	367.8	6.62	1124.8	141.8	4.10
	193	56.73	15.50	15.710	1.438	0.890	2402.4	310.0	6.51	930.1	118.4	4.05
	158	46.47	15.00	15.550	1.188	0.730	1900.6	253.4	6.40	745.0	95.8	4.00
14 × 12	84	24.71	14.18	12.023	0.778	0.451	928.4	130.9	6.13	225.5	37.5	3.02
	78	22.94	14.06	12.000	0.718	0.428	851.2	121.1	6.09	206.9	34.5	3.00
14 × 6¾	38	11.17	14.12	6.776	0.513	0.313	385.3	54.6	5.87	24.6	7.3	1.49
	34	10.00	14.00	6.750	0.453	0.287	339.2	48.5	5.83	21.3	6.3	1.46
	30	8.81	13.86	6.733	0.383	0.270	289.6	41.8	5.73	17.5	5.2	1.41
12 × 12	190	55.86	14.38	12.670	1.736	1.060	1892.5	263.2	5.82	589.7	93.1	3.25
	120	35.31	13.12	12.320	1.106	0.710	1071.7	163.4	5.51	345.1	56.0	3.13
	92	27.06	12.62	12.155	0.856	0.545	788.9	125.0	5.40	256.4	42.2	3.08
	65	19.11	12.12	12.000	0.606	0.390	533.4	88.0	5.28	174.6	29.1	3.02
12 × 6½	36	10.59	12.24	6.565	0.540	0.305	280.8	45.9	5.15	23.7	7.2	1.50
	31	9.12	12.09	6.525	0.465	0.265	238.4	39.4	5.11	19.8	6.1	1.47
	27	7.97	11.95	6.500	0.400	0.240	204.1	34.1	5.06	16.6	5.1	1.44
10 × 10	100	29.43	11.12	10.345	1.118	0.685	625.0	112.4	4.61	206.6	39.9	2.65
	72	21.18	10.50	10.170	0.808	0.510	420.7	80.1	4.46	141.8	27.9	2.59
	49	14.40	10.00	10.000	0.558	0.340	272.9	54.6	4.35	93.0	18.6	2.54
10 × 5¾	29	8.53	10.22	5.799	0.500	0.289	157.3	30.8	4.29	15.2	5.2	1.34
	25	7.35	10.08	5.762	0.430	0.252	133.2	26.4	4.26	12.7	4.4	1.31
	21	6.19	9.90	5.750	0.340	0.240	106.3	21.5	4.14	9.7	3.4	1.25
8 × 8	67	19.70	9.00	8.287	0.933	0.575	271.8	60.4	3.71	88.6	21.4	2.12
	48	14.11	8.50	8.117	0.683	0.405	183.7	43.2	3.61	60.9	15.0	2.08
	31	9.12	8.00	8.000	0.433	0.288	109.7	27.4	3.47	37.0	9.2	2.01
8 × 5¼	20	5.88	8.14	5.268	0.378	0.248	69.2	17.0	3.43	8.5	3.2	1.20
	17	5.00	8.00	5.250	0.308	0.230	56.4	14.1	3.36	6.7	2.6	1.16

MECHANICS OF MATERIALS 135

Table 2-14 Properties of American Standard Beams—Steel

Nominal Size	Weight per Foot	Area	Depth	Flange		Web Thickness	Axis X-X			Axis Y-Y		
				Width	Thickness		I	S	r	I	S	r
in.	lb	in.^{2}	in.	in.	in.	in.	in.^{4}	in.^{3}	in.	in.^{4}	in.^{3}	in.
24 × 7⅞	120.0	35.13	24.00	8.048	1.102	0.798	3010.8	250.9	9.26	84.9	21.1	1.56
	105.9	30.98	24.00	7.875	1.102	.625	2811.5	234.3	9.53	78.9	20.0	1.60
24 × 7	100.0	29.25	24.00	7.247	0.871	.747	2371.8	197.6	9.05	48.4	13.4	1.29
	90.0	26.30	24.00	7.124	0.871	.624	2230.1	185.8	9.21	45.5	12.8	1.32
	79.9	23.33	24.00	7.000	0.871	.500	2087.2	173.9	9.46	42.9	12.2	1.36
20 × 7	95.0	27.74	20.00	7.200	0.916	.800	1599.7	160.0	7.59	50.5	14.0	1.35
	85.0	24.80	20.00	7.053	0.916	.653	1501.7	150.2	7.78	47.0	13.3	1.38
20 × 6¼	75.0	21.90	20.00	6.391	0.789	.641	1263.5	126.3	7.60	30.1	9.4	1.17
	65.4	19.08	20.00	6.250	0.789	.500	1169.5	116.9	7.83	27.9	8.9	1.21
18 × 6	70.0	20.46	18.00	6.251	0.691	.711	917.5	101.9	6.70	24.5	7.8	1.09
	54.7	15.94	18.00	6.000	0.691	.460	795.5	88.4	7.07	21.2	7.1	1.15
15 × 5½	50.0	14.59	15.00	5.640	0.622	.550	481.1	64.2	5.74	16.0	5.7	1.05
	42.9	12.49	15.00	5.500	0.622	.410	441.8	58.9	5.95	14.6	5.3	1.08
12 × 5¼	50.0	14.57	12.00	5.477	0.659	.687	301.6	50.3	4.55	16.0	5.8	1.05
	40.8	11.84	12.00	5.250	0.659	.460	268.9	44.8	4.77	13.8	5.3	1.08
12 × 5	35.0	10.20	12.00	5.078	0.544	.428	227.0	37.8	4.72	10.0	3.9	0.99
	31.8	9.26	12.00	5.000	0.544	.350	215.8	36.0	4.83	9.5	3.8	1.01
10 × 4⅝	35.0	10.22	10.00	4.944	0.491	.594	145.8	29.2	3.78	8.5	3.4	0.91
	25.4	7.38	10.00	4.660	0.491	.310	122.1	24.4	4.07	6.9	3.0	0.97
8 × 4	23.0	6.71	8.00	4.171	0.425	.441	64.2	16.0	3.09	4.4	2.1	0.81
	18.4	5.34	8.00	4.000	0.425	.270	56.9	14.2	3.26	3.8	1.9	0.84
7 × 3⅝	20.0	5.83	7.00	3.860	0.392	.450	41.9	12.0	2.68	3.1	1.6	0.74
	15.3	4.43	7.00	3.660	0.392	.250	36.2	10.4	2.86	2.7	1.5	0.78
6 × 3⅜	17.25	5.02	6.00	3.565	0.359	.465	26.0	8.7	2.28	2.3	1.3	0.68
	12.5	3.61	6.00	3.330	0.359	.230	21.8	7.3	2.46	1.8	1.1	0.72
5 × 3	14.75	4.29	5.00	3.284	0.326	.494	15.0	6.0	1.87	1.7	1.0	0.63
	10.0	2.87	5.00	3.000	0.326	.210	12.1	4.8	2.05	1.2	0.82	0.65
4 × 2⅝	9.5	2.76	4.00	2.796	0.293	.326	6.7	3.3	1.56	0.91	0.65	0.58
	7.7	2.21	4.00	2.660	0.293	.190	6.0	3.0	1.64	0.77	0.58	0.59
3 × 2⅜	7.5	2.17	3.00	2.509	0.260	.349	2.9	1.9	1.15	0.59	0.47	0.52
	5.7	1.64	3.00	2.330	0.260	.170	2.5	1.7	1.23	0.46	0.40	0.53

Table 2-15 Properties of American Standard Channels—Steel

Nominal Size	Weight per Foot	Area	Depth	Flange Width	Flange Average Thickness	Web Thickness	Axis X-X I	Axis X-X S	Axis X-X r	Axis Y-Y I	Axis Y-Y S	Axis Y-Y r	x
in.	lb	in.^{2}	in.	in.	in.	in.	in.^{4}	in.^{3}	in.	in.^{4}	in.^{3}	in.	in.
18 × 4*	58.0	16.98	18.00	4.200	0.625	0.700	670.7	74.5	6.29	18.5	5.6	1.04	0.88
	51.9	15.18	18.00	4.100	.625	.600	622.1	69.1	6.40	17.1	5.3	1.06	.87
	45.8	13.38	18.00	4.000	.625	.500	573.5	63.7	6.55	15.8	5.1	1.09	.89
	42.7	12.48	18.00	3.950	.625	.450	549.2	61.0	6.64	15.0	4.9	1.10	.90
15 × 3⅜	50.0	14.64	15.00	3.716	.650	.716	401.4	53.6	5.24	11.2	3.8	0.87	.80
	40.0	11.70	15.00	3.520	.650	.520	346.3	46.2	5.44	9.3	3.4	0.89	.78
	33.9	9.90	15.00	3.400	.650	.400	312.6	41.7	5.62	8.2	3.2	0.91	.79
12 × 3	30.0	8.79	12.00	3.170	.501	.510	161.2	26.9	4.28	5.2	2.1	0.77	.68
	25.0	7.32	12.00	3.047	.501	.387	143.5	23.9	4.43	4.5	1.9	0.79	.68
	20.7	6.03	12.00	2.940	.501	.280	128.1	21.4	4.61	3.9	1.7	0.81	.70
10 × 2⅝	30.0	8.80	10.00	3.033	.436	.673	103.0	20.6	3.42	4.0	1.7	0.67	.65
	25.0	7.33	10.00	2.886	.436	.526	90.7	18.1	3.52	3.4	1.5	0.68	.62
	20.0	5.86	10.00	2.739	.436	.379	78.5	15.7	3.66	2.8	1.3	0.70	.61
	15.3	4.47	10.00	2.600	.436	.240	66.9	13.4	3.87	2.3	1.2	0.72	.64
9 × 2½	20.0	5.86	9.00	2.648	.413	.448	60.6	13.5	3.22	2.4	1.2	0.65	.59
	15.0	4.39	9.00	2.485	.413	.285	50.7	11.3	3.40	1.9	1.0	0.67	.59
	13.4	3.89	9.00	2.430	.413	.230	47.3	10.5	3.49	1.8	0.97	0.67	.61
8 × 2¼	18.75	5.49	8.00	2.527	.390	.487	43.7	10.9	2.82	2.0	1.0	0.60	.57
	13.75	4.02	8.00	2.343	.390	.303	35.8	9.0	2.99	1.5	0.86	0.62	.56
	11.5	3.36	8.00	2.260	.390	.220	32.3	8.1	3.10	1.3	0.79	0.63	.58
7 × 2⅛	14.75	4.32	7.00	2.299	.366	.419	27.1	7.7	2.51	1.4	0.79	0.57	.53
	12.25	3.58	7.00	2.194	.366	.314	24.1	6.9	2.59	1.2	0.71	0.58	.53
	9.8	2.85	7.00	2.090	.366	.210	21.1	6.0	2.72	0.98	0.63	0.59	.55
6 × 2	13.0	3.81	6.00	2.157	.343	.437	17.3	5.8	2.13	1.1	0.65	0.53	.52
	10.5	3.07	6.00	2.034	.343	.314	15.1	5.0	2.22	0.87	0.57	0.53	.50
	8.2	2.39	6.00	1.920	.343	.200	13.0	4.3	2.34	0.70	0.50	0.54	.52
5 × 1¾	9.0	2.63	5.00	1.885	.320	.325	8.8	3.5	1.83	0.64	0.45	0.49	.48
	6.7	1.95	5.00	1.750	.320	.190	7.4	3.0	1.95	0.48	0.38	0.50	.49
4 × 1⅝	7.25	2.12	4.00	1.720	.296	.320	4.5	2.3	1.47	0.44	0.35	0.46	.46
	5.4	1.56	4.00	1.580	.296	.180	3.8	1.9	1.56	0.32	0.29	0.45	.46
3 × 1½	6.0	1.75	3.00	1.596	.273	.356	2.1	1.4	1.08	0.31	0.27	0.42	.46
	5.0	1.46	3.00	1.498	.273	.258	1.8	1.2	1.12	0.25	0.24	0.41	.44
	4.1	1.19	3.00	1.410	.273	.170	1.6	1.1	1.17	0.20	0.21	0.41	.44

* Car and Shipbuilding Channel; not an American Standard.

MECHANICS OF MATERIALS

Table 2-16 Properties of Angles with Equal Legs—Steel

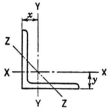

Size	Thickness	Weight per Foot	Area	Axis X-X and Axis Y-Y				Axis Z-Z
				I	S	r	x or y	r
in.	in.	lb	in.^{2}	in.^{4}	in.^{3}	in.	in.	in.
8 × 8	1⅛	56.9	16.73	98.0	17.5	2.42	2.41	1.56
	1	51.0	15.00	89.0	15.8	2.44	2.37	1.56
	⅞	45.0	13.23	79.6	14.0	2.45	2.32	1.57
	¾	38.9	11.44	69.7	12.2	2.47	2.28	1.57
	⅝	32.7	9.61	59.4	10.3	2.49	2.23	1.58
	9/16	29.6	8.68	54.1	9.3	2.50	2.21	1.58
	½	26.4	7.75	48.6	8.4	2.50	2.19	1.59
6 × 6	1	37.4	11.00	35.5	8.6	1.80	1.86	1.17
	⅞	33.1	9.73	31.9	7.6	1.81	1.82	1.17
	¾	28.7	8.44	28.2	6.7	1.83	1.78	1.17
	⅝	24.2	7.11	24.2	5.7	1.84	1.73	1.18
	9/16	21.9	6.43	22.1	5.1	1.85	1.71	1.18
	½	19.6	5.75	19.9	4.6	1.86	1.68	1.18
	7/16	17.2	5.06	17.7	4.1	1.87	1.66	1.19
	⅜	14.9	4.36	15.4	3.5	1.88	1.64	1.19
	5/16	12.5	3.66	13.0	3.0	1.89	1.61	1.19
4 × 4	¾	18.5	5.44	7.7	2.8	1.19	1.27	0.78
	⅝	15.7	4.61	6.7	2.4	1.20	1.23	0.78
	½	12.8	3.75	5.6	2.0	1.22	1.18	0.78
	7/16	11.3	3.31	5.0	1.8	1.23	1.16	0.78
	⅜	9.8	2.86	4.4	1.5	1.23	1.14	0.79
	5/16	8.2	2.40	3.7	1.3	1.24	1.12	0.79
	¼	6.6	1.94	3.0	1.1	1.25	1.09	0.80
3 × 3	½	9.4	2.75	2.2	1.1	0.90	0.93	0.58
	7/16	8.3	2.43	2.0	0.95	0.91	0.91	0.58
	⅜	7.2	2.11	1.8	0.83	0.91	0.89	0.58
	5/16	6.1	1.78	1.5	0.71	0.92	0.87	0.59
	¼	4.9	1.44	1.2	0.58	0.93	0.84	0.59
	3/16	3.71	1.09	0.96	0.44	0.94	0.82	0.59
2½ × 2½	½	7.7	2.25	1.2	0.72	0.74	0.81	0.49
	⅜	5.9	1.73	0.98	0.57	0.75	0.76	0.49
	5/16	5.0	1.47	0.85	0.48	0.76	0.74	0.49
	¼	4.1	1.19	0.70	0.39	0.77	0.72	0.49
	3/16	3.07	0.90	0.55	0.30	0.78	0.69	0.49
2 × 2	⅜	4.7	1.36	0.48	0.35	0.59	0.64	0.39
	5/16	3.92	1.15	0.42	0.30	0.60	0.61	0.39
	¼	3.19	0.94	0.35	0.25	0.61	0.59	0.39
	3/16	2.44	0.71	0.27	0.19	0.62	0.57	0.39
	⅛	1.65	0.48	0.19	0.13	0.63	0.55	0.40
1½ × 1½	¼	2.34	.69	.14	.13	.45	.47	.29
	3/16	1.80	.53	.11	.10	.46	.44	.29
	⅛	1.23	.36	.08	.07	.47	.42	.30
1 × 1	¼	1.49	.44	.04	.06	.29	.34	.20
	3/16	1.16	.34	.03	.04	.30	.32	.19
	⅛	0.80	.23	.02	.03	.30	.30	.20

Table 2-17 Properties of Angles with Unequal Legs—Steel

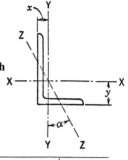

Size	Thick-ness	Weight per Foot	Area	Axis X-X				Axis Y-Y				Axis Z-Z	
				I	S	r	y	I	S	r	x	r	Tan α
in.	in.	lb	in.^{2}	in.^{4}	in.^{3}	in.	in.	in.^{4}	in.^{3}	in.	in.	in.	
9 × 4	1	40.8	12.00	97.0	17.6	2.84	3.50	12.0	4.0	1.00	1.00	0.83	0.203
	7/8	36.1	10.61	86.8	15.7	2.86	3.45	10.8	3.6	1.01	0.95	0.84	.208
	3/4	31.3	9.19	76.1	13.6	2.88	3.41	9.6	3.1	1.02	0.91	0.84	.212
	5/8	26.3	7.73	64.9	11.5	2.90	3.36	8.3	2.6	1.04	0.86	0.85	.216
	9/16	23.8	7.00	59.1	10.4	2.91	3.33	7.6	2.4	1.04	0.83	0.85	.218
	1/2	21.3	6.25	53.2	9.3	2.92	3.31	6.9	2.2	1.05	0.81	0.85	.220
8 × 6	1	44.2	13.00	80.8	15.1	2.49	2.65	38.8	8.9	1.73	1.65	1.28	.543
	7/8	39.1	11.48	72.3	13.4	2.51	2.61	34.9	7.9	1.74	1.61	1.28	.547
	3/4	33.8	9.94	63.4	11.7	2.53	2.56	30.7	6.9	1.76	1.56	1.29	.551
	5/8	28.5	8.36	54.1	9.9	2.54	2.52	26.3	5.9	1.77	1.52	1.29	.554
	9/16	25.7	7.56	49.3	9.0	2.55	2.50	24.0	5.3	1.78	1.50	1.30	.556
	1/2	23.0	6.75	44.3	8.0	2.56	2.47	21.7	4.8	1.79	1.47	1.30	.558
6 × 4	7/8	27.2	7.98	27.7	7.2	1.86	2.12	9.8	3.4	1.11	1.12	0.86	.421
	3/4	23.6	6.94	24.5	6.3	1.88	2.08	8.7	3.0	1.12	1.08	0.86	.428
	5/8	20.0	5.86	21.1	5.3	1.90	2.03	7.5	2.5	1.13	1.03	0.86	.435
	9/16	18.1	5.31	19.3	4.8	1.90	2.01	6.9	2.3	1.14	1.01	0.87	.438
	1/2	16.2	4.75	17.4	4.3	1.91	1.99	6.3	2.1	1.15	0.99	0.87	.440
	7/16	14.3	4.18	15.5	3.8	1.92	1.96	5.6	1.9	1.16	0.96	0.87	.443
	3/8	12.3	3.61	13.5	3.3	1.93	1.94	4.9	1.6	1.17	0.94	0.88	.446
5 × 3	1/2	12.8	3.75	9.5	2.9	1.59	1.75	2.6	1.1	0.83	0.75	.65	.357
	7/16	11.3	3.31	8.4	2.6	1.60	1.73	2.3	1.0	0.84	0.73	.65	.361
	3/8	9.8	2.86	7.4	2.2	1.61	1.70	2.0	0.89	0.84	0.70	.65	.364
	5/16	8.2	2.40	6.3	1.9	1.61	1.68	1.8	0.75	0.85	0.68	.66	.368
	1/4	6.6	1.94	5.1	1.5	1.62	1.66	1.4	0.61	0.86	0.66	.66	.371
4 × 3	5/8	13.6	3.98	6.0	2.3	1.23	1.37	2.9	1.4	0.85	0.87	.64	.534
	1/2	11.1	3.25	5.1	1.9	1.25	1.33	2.4	1.1	0.86	0.83	.64	.543
	7/16	9.8	2.87	4.5	1.7	1.25	1.30	2.2	1.0	0.87	0.80	.64	.547
	3/8	8.5	2.48	4.0	1.5	1.26	1.28	1.9	0.87	0.88	0.78	.64	.551
	5/16	7.2	2.09	3.4	1.2	1.27	1.26	1.7	0.73	0.89	0.76	.65	.554
	1/4	5.8	1.69	2.8	1.0	1.28	1.24	1.4	0.60	0.90	0.74	.65	.558
3½ × 2½	1/2	9.4	2.75	3.2	1.4	1.09	1.20	1.4	0.76	0.70	0.70	.53	.486
	7/16	8.3	2.43	2.9	1.3	1.09	1.18	1.2	0.68	0.71	0.68	.54	.491
	3/8	7.2	2.11	2.6	1.1	1.10	1.16	1.1	0.59	0.72	0.66	.54	.496
	5/16	6.1	1.78	2.2	0.93	1.11	1.14	0.94	0.50	0.73	0.64	.54	.501
	1/4	4.9	1.44	1.8	0.75	1.12	1.11	0.78	0.41	0.74	0.61	.54	.506
2½ × 2	3/8	5.3	1.55	0.91	0.55	0.77	0.83	0.51	0.36	0.58	0.58	.42	.614
	5/16	4.5	1.31	0.79	0.47	0.78	0.81	0.45	0.31	0.58	0.56	.42	.620
	1/4	3.62	1.06	0.65	0.38	0.78	0.79	0.37	0.25	0.59	0.54	.42	.626
	3/16	2.75	0.81	0.51	0.29	0.79	0.76	0.29	0.20	0.60	0.51	.43	.631
2 × 1½	1/4	2.77	0.81	0.32	0.24	0.62	0.66	0.15	0.14	0.43	0.41	.32	.543
	3/16	2.12	0.62	0.25	0.18	0.63	0.64	0.12	0.11	0.44	0.39	.32	.551
	1/8	1.44	0.42	0.17	0.13	0.64	0.62	0.09	0.08	0.45	0.37	.33	.558
1¾ × 1¼	1/4	2.34	0.69	0.20	0.18	0.54	0.60	0.09	0.10	0.35	0.35	.27	.486
	3/16	1.80	0.53	0.16	0.14	0.55	0.58	0.07	0.08	0.36	0.33	.27	.496
	1/8	1.23	0.36	0.11	0.09	0.56	0.56	0.05	0.05	0.37	0.31	.27	.506

MECHANICS OF MATERIALS 139

Table 2-18 Properties and Dimensions of Tees—Steel

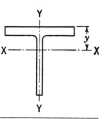

Section Number	Weight per Foot	Area	Depth of Tee	Flange		Stem Thickness	Axis X-X				Axis Y-Y		
				Width	Average Thickness		I	S	r	y	I	S	r
	lb	in.²	in.	in.	in.	in.	in.⁴	in.³	in.	in.	in.⁴	in.³	in.
ST 18 WF	150	44.09	18.36	16.655	1.680	0.945	1222.7	85.9	5.27	4.13	612.6	73.6	3.73
	115	33.86	17.94	16.475	1.260	.765	935.8	67.2	5.26	4.02	435.5	52.9	3.59
ST 15 WF	105	30.89	15.19	15.105	1.315	.775	578.0	48.7	4.33	3.31	354.0	46.9	3.38
	86	25.32	14.94	14.985	1.065	.655	471.0	40.2	4.31	3.23	275.1	36.7	3.30
ST 12 WF	60	17.64	12.16	12.088	0.930	.556	213.6	22.4	3.48	2.62	127.0	21.0	2.68
	50	14.71	12.00	12.000	0.775	.468	176.7	18.7	3.46	2.54	101.8	17.0	2.63
ST 10 WF	56	16.47	10.50	13.000	0.865	.527	136.4	16.2	2.88	2.06	144.8	22.3	2.96
	41	12.05	10.43	8.962	0.795	.499	115.4	14.5	3.09	2.48	44.8	10.0	1.93
ST 8 WF	48	14.11	8.16	11.533	0.875	.535	64.7	9.82	2.14	1.57	103.6	18.0	2.71
	32	9.40	8.00	8.500	0.715	.443	48.3	7.71	2.27	1.73	34.2	8.05	1.91
	20	5.88	8.00	7.000	0.503	.307	33.2	5.37	2.37	1.82	13.3	3.79	1.50
ST 6 WF	80.5	23.69	6.94	12.515	1.486	.905	62.6	11.5	1.63	1.47	243.1	38.9	3.20
	26.5	7.80	6.03	10.000	0.576	.345	17.7	3.54	1.51	1.02	48.0	9.60	2.48
	20	5.89	5.97	8.000	0.516	.294	14.4	2.94	1.56	1.08	22.0	5.50	1.94
	13.5	3.98	5.98	6.500	0.400	.240	11.4	2.39	1.69	1.21	8.3	2.55	1.44
ST 4 WF	33.5	9.85	4.50	8.287	.933	.575	10.94	3.07	1.05	0.94	44.3	10.7	2.12
	15.5	4.56	4.00	8.000	.433	.288	4.31	1.30	0.97	0.67	18.5	4.60	2.01
	8.5	2.50	4.00	5.250	.308	.230	3.21	1.01	1.13	0.84	3.36	1.28	1.16

Nominal Size	Weight per Foot	Area	Dimensions				Axis X-X				Axis Y-Y		
			Depth	Width of flange	Minimum thickness		I	S	r	y	I	S	r
					Flange	Stem							
in.	lb	in.²	in.	in.	in.	in.	in.⁴	in.³	in.	in.	in.⁴	in.³	in.
5 × 3	11.5	3.37	3	5	3/8	13/32	2.4	1.1	0.84	0.76	3.9	1.6	1.10
4 × 4½	11.2	3.29	4½	4	3/8	3/8	6.3	2.0	1.39	1.31	2.1	1.1	0.80
4 × 4	13.5	3.97	4	4	½	½	5.7	2.0	1.20	1.18	2.8	1.4	0.84
4 × 3	9.2	2.68	3	4	3/8	3/8	2.0	0.90	0.86	0.78	2.1	1.1	0.89
4 × 2½	8.5	2.48	2½	4	3/8	3/8	1.2	0.62	0.69	0.62	2.1	1.0	0.92
3 × 3	7.8	2.29	3	3	3/8	3/8	1.84	0.86	0.89	0.88	0.89	0.60	0.63
3 × 2½	6.1	1.77	2½	3	5/16	5/16	0.94	0.51	0.73	0.68	0.75	0.50	0.65
2½ × 2½	6.4	1.87	2½	2½	3/8	3/8	1.0	0.59	0.74	0.76	0.52	0.42	0.53
2¼ × 2¼	4.1	1.19	2¼	2¼	¼	¼	0.52	0.32	0.66	0.65	0.25	0.22	0.46
2 × 2	4.3	1.26	2	2	5/16	5/16	0.44	0.31	0.59	0.61	0.23	0.23	0.43

Table 2-19 Square and Round Bars*

Size, in.	Square Weight, lb	Square Area, sq in.	Round Weight, lb	Round Area, sq in.	Size, in.	Square Weight, lb	Square Area, sq in.	Round Weight, lb	Round Area, sq in.
0					4	54.40	16.000	42.73	12.566
1/16	0.013	0.0039	0.010	0.0031	1/16	56.11	16.504	44.07	12.962
1/8	.053	.0156	.042	.0123	1/8	57.85	17.016	45.44	13.364
3/16	.120	.0352	.094	.0276	3/16	59.62	17.535	46.83	13.772
1/4	.213	.0625	.167	.0491	1/4	61.41	18.063	48.23	14.186
5/16	.332	.0977	.261	.0767	5/16	63.23	18.598	49.66	14.607
3/8	.478	.1406	.376	.1105	3/8	65.08	19.141	51.11	15.033
7/16	.651	.1914	.511	.1503	7/16	66.95	19.691	52.58	15.466
1/2	.850	.2500	.668	.1963	1/2	68.85	20.250	54.07	15.904
9/16	1.076	.3164	.845	.2485	9/16	70.78	20.816	55.59	16.349
5/8	1.328	.3906	1.043	.3068	5/8	72.73	21.391	57.12	16 800
11/16	1.607	.4727	1.262	.3712	11/16	74.71	21.973	58.67	17.257
3/4	1.913	.5625	1.502	.4418	3/4	76.71	22.563	60.25	17.721
13/16	2.245	.6602	1.763	.5185	13/16	78.74	23.160	61.85	18.190
7/8	2.603	.7656	2.044	.6013	7/8	80.80	23.765	63.46	18.665
15/16	2.988	.8789	2.347	.6903	15/16	82.89	24.379	65.10	19.147
1	3.400	1.0000	2.670	.7854	5	85.00	25.000	66.76	19.635
1/16	3.838	1.1289	3.015	.8866	1/16	87.14	25.629	68.44	20.129
1/8	4.303	1.2656	3.380	.9940	1/8	89.30	26.266	70.14	20.629
3/16	4.795	1.4102	3.766	1.1075	3/16	91.49	26.910	71.86	21.135
1/4	5.313	1.5625	4.172	1.2272	1/4	93.71	27.563	73.60	21.648
5/16	5.857	1.7227	4.600	1.3530	5/16	95.96	28.223	75.36	22.166
3/8	6.428	1.8906	5.049	1.4849	3/8	98.23	28.891	77.15	22.691
7/16	7.026	2.0664	5.518	1.6230	7/16	100.53	29.566	78.95	23.221
1/2	7.650	2.2500	6.008	1.7671	1/2	102.85	30.250	80.78	23.758
9/16	8.301	2.4414	6.519	1.9175	9/16	105.20	30.941	82.62	24.301
5/8	8.978	2.6406	7.051	2.0739	5/8	107.58	31.641	84.49	24.850
11/16	9.682	2.8477	7.604	2.2365	11/16	109.98	32.348	86.38	25.406
3/4	10.413	3.0625	8.178	2.4053	3/4	112.41	33.063	88.29	25.967
13/16	11.170	3.2852	8.773	2.5802	13/16	114.87	33.785	90.22	26.535
7/8	11.953	3.5156	9.388	2.7612	7/8	117.35	34.516	92.17	27.109
15/16	12.763	3.7539	10.024	2.9483	15/16	119.86	35.254	94.14	27.688
2	13.600	4.0000	10.681	3.1416	6	122.40	36.000	96.13	28.274
1/16	14.463	4.2539	11.359	3.3410	1/16	124.96	36.754	98.15	28.866
1/8	15.353	4.5156	12.058	3.5466	1/8	127.55	37.516	100.18	29.465
3/16	16.270	4.7852	12.778	3.7583	3/16	130.17	38.285	102.23	30.069
1/4	17.213	5.0625	13.519	3.9761	1/4	132.81	39.063	104.31	30.680
5/16	18.182	5.3477	14.280	4.2000	5/16	135.48	39.848	106.41	31.296
3/8	19.178	5.6406	15.062	4.4301	3/8	138.18	40.641	108.53	31.919
7/16	20.201	5.9414	15.866	4.6664	7/16	140.90	41.441	110.66	32.548
1/2	21.250	6.2500	16.690	4.9087	1/2	143.65	42.250	112.82	33.183
9/16	22.326	6.5664	17.534	5.1572	9/16	146.43	43.066	115.00	33.824
5/8	23.428	6.8906	18.400	5.4119	5/8	149.23	43.891	117.20	34.472
11/16	24.557	7.2227	19.287	5.6727	11/16	152.06	44.723	119.43	35.125
3/4	25.713	7.5625	20.195	5.9396	3/4	154.91	45.563	121.67	35.785
13/16	26.895	7.9102	21.123	6.2126	13/16	157.79	46.410	123.93	36.450
7/8	28.103	8.2656	22.072	6.4918	7/8	160.70	47.266	126.22	37.122
15/16	29.338	8.6289	23.042	6.7771	15/16	163.64	48.129	128.52	37.800
3	30.60	9.000	24.03	7.069	7	166.60	49.000	130.85	38.485
1/16	31.89	9.379	25.05	7.366	1/16	169.59	49.879	133.19	39.175
1/8	33.20	9.766	26.08	7.670	1/8	172.60	50.766	135.56	39.871
3/16	34.54	10.160	27.13	7.980	3/16	175.64	51.660	137.95	40.574
1/4	35.91	10.563	28.21	8.296	1/4	178.71	52.563	140.36	41.282
5/16	37.31	10.973	29.30	8.618	5/16	181.81	53.473	142.79	41.997
3/8	38.73	11.391	30.42	8.946	3/8	184.93	54.391	145.24	42.718
7/16	40.18	11.816	31.55	9.281	7/16	188.07	55.316	147.71	43.445
1/2	41.65	12.250	32.71	9.621	1/2	191.25	56.250	150.21	44.179
9/16	43.15	12.691	33.89	9.968	9/16	194.45	57.191	152.72	44.918
5/8	44.68	13.141	35.09	10.321	5/8	197.68	58.141	155.26	45.664
11/16	46.23	13.598	36.31	10.680	11/16	200.93	59.098	157.81	46.415
3/4	47.81	14.063	37.55	11.045	3/4	204.21	60.063	160.39	47.173
13/16	49.42	14.535	38.81	11.416	13/16	207.52	61.035	162.99	47.937
7/8	51.05	15.016	40.10	11.793	7/8	210.85	62.016	165.60	48.707
15/16	52.71	15.504	41.40	12.177	15/16	214.21	63.004	168.24	49.483
4	54.40	16.000	42.73	12.566	8	217.60	64.000	170.90	50.265

* One cubic inch of rolled steel is assumed to weigh 0.2833 lb.

MECHANICS OF MATERIALS 141

Table 2-20 Pipe
(*Steel Construction*, 1949, A.I.S.C.)

	Dimensions					Couplings			Properties		
Nom. Diam., in.	Outside Diam., in.	Inside Diam., in.	Thickness, in.	Weight per Foot, lb		Threads per inch	Outside Diam., in.	Length, in.	Weight, lb	I, in.^{4}	A, in.^{2}	k, in.
				Plain Ends	Thread and Coupling							

Standard

Nom.	OD	ID	Thk	Plain	T&C	TPI	OD	L	Wt	I	A	k
1/8	0.405	0.269	0.068	0.24	0.25	27	0.562	7/8	0.03	0.001	0.072	0.12
1/4	0.540	0.364	0.088	0.42	0.43	18	0.685	1	0.04	0.003	0.125	0.16
3/8	0.675	0.493	0.091	0.57	0.57	18	0.848	1 1/8	0.07	0.007	0.167	0.21
1/2	0.840	0.622	0.109	0.85	0.85	14	1.024	1 3/8	0.12	0.017	0.250	0.26
3/4	1.050	0.824	0.113	1.13	1.13	14	1.281	1 5/8	0.21	0.037	0.333	0.33
1	1.315	1.049	0.133	1.68	1.68	11 1/2	1.576	1 7/8	0.35	0.087	0.494	0.42
1 1/4	1.660	1.380	0.140	2.27	2.28	11 1/2	1.950	2 1/8	0.55	0.195	0.669	0.54
1 1/2	1.900	1.610	0.145	2.72	2.73	11 1/2	2.218	2 3/8	0.76	0.310	0.799	0.62
2	2.375	2.067	0.154	3.65	3.68	11 1/2	2.760	2 5/8	1.23	0.666	1.075	0.79
2 1/2	2.875	2.469	0.203	5.79	5.82	8	3.276	2 7/8	1.76	1.530	1.704	0.95
3	3.500	3.068	0.216	7.58	7.62	8	3.948	3 1/8	2.55	3.017	2.228	1.16
3 1/2	4.000	3.548	0.226	9.11	9.20	8	4.591	3 5/8	4.33	4.788	2.680	1.34
4	4.500	4.026	0.237	10.79	10.89	8	5.091	3 5/8	5.41	7.233	3.174	1.51
5	5.563	5.047	0.258	14.62	14.81	8	6.296	4 1/8	9.16	15.16	4.300	1.88
6	6.625	6.065	0.280	18.97	19.19	8	7.358	4 1/8	10.82	28.14	5.581	2.25
8	8.625	8.071	0.277	24.70	25.00	8	9.420	4 5/8	15.84	63.35	7.265	2.95
8	8.625	7.981	0.322	28.55	28.81	8	9.420	4 5/8	15.84	72.49	8.399	2.94
10	10.750	10.192	0.279	31.20	32.00	8	11.721	6 1/8	33.92	125.4	9.178	3.70
10	10.750	10.136	0.307	34.24	35.00	8	11.721	6 1/8	33.92	137.4	10.07	3.69
10	10.750	10.020	0.365	40.48	41.13	8	11.721	6 1/8	33.92	160.7	11.91	3.67
12	12.750	12.090	0.330	43.77	45.00	8	13.958	6 1/8	48.27	248.5	12.88	4.39
12	12.750	12.000	0.375	49.56	50.71	8	13.958	6 1/8	48.27	279.3	14.38	4.38

Extra Strong

Nom.	OD	ID	Thk	Plain	T&C	TPI	OD	L	Wt	I	A	k
1/8	0.405	0.215	0.095	0.31	0.32	27	0.582	1 1/8	0.05	0.001	0.093	0.12
1/4	0.540	0.302	0.119	0.54	0.54	18	0.724	1 3/8	0.07	0.004	0.157	0.16
3/8	0.675	0.423	0.126	0.74	0.75	18	0.898	1 5/8	0.13	0.009	0.217	0.20
1/2	0.840	0.546	0.147	1.09	1.10	14	1.085	1 7/8	0.22	0.020	0.320	0.25
3/4	1.050	0.742	0.154	1.47	1.49	14	1.316	2 1/8	0.33	0.045	0.433	0.32
1	1.315	0.957	0.179	2.17	2.20	11 1/2	1.575	2 3/8	0.47	0.106	0.639	0.41
1 1/4	1.660	1.278	0.191	3.00	3.05	11 1/2	2.054	2 7/8	1.04	0.242	0,881	0.52
1 1/2	1.900	1.500	0.200	3.63	3.69	11 1/2	2.294	2 7/8	1.17	0.391	1.068	0.61
2	2.375	1.939	0.218	5.02	5.13	11 1/2	2.870	3 5/8	2.17	0.868	1.477	0.77
2 1/2	2.875	2.323	0.276	7.66	7.83	8	3.389	4 1/8	3.43	1.924	2.254	0.92
3	3.500	2.900	0.300	10.25	10.46	8	4.014	4 1/8	4.13	3.894	3.016	1.14
3 1/2	4.000	3.364	0.318	12.51	12.82	8	4.628	4 5/8	6.29	6.280	3.678	1.31
4	4.500	3.826	0.337	14.98	15.39	8	5.233	4 5/8	8.16	9.610	4.407	1.48
5	5.563	4.813	0.375	20.78	21.42	8	6.420	5 1/8	12.87	20.67	6.112	1.84
6	6.625	5.761	0.432	28.57	29.33	8	7.482	5 1/8	15.18	40.49	8.405	2.20
8	8.625	7.625	0.500	43.39	44.72	8	9.596	6 1/8	26.63	105.7	12.76	2.88
10	10.750	9.750	0.500	54.74	56.94	8	11.958	6 5/8	44.16	211.9	16.10	3.63
12	12.750	11.750	0.500	65.42	68.02	8	13.958	6 5/8	51.99	361.5	19.24	4.34

Double—Extra Strong

Nom.	OD	ID	Thk	Plain	T&C	TPI	OD	L	Wt	I	A	k
1/2	0.840	0.252	0.294	1.71	1.73	14	1.085	1 7/8	0.22	0.024	0.504	0.22
3/4	1.050	0.434	0.308	2.44	2.46	14	1.316	2 1/8	0.33	0.058	0.718	0.28
1	1.315	0.599	0.358	3.66	3.68	11 1/2	1.575	2 3/8	0.47	0.140	1.076	0.36
1 1/4	1.660	0.896	0.382	5.21	5.27	11 1/2	2.054	2 7/8	1.04	0.341	1.534	0.47
1 1/2	1.900	1.100	0.400	6.41	6.47	11 1/2	2.294	2 7/8	1.17	0.568	1.885	0.55
2	2.375	1.503	0.436	9.03	9.14	11 1/2	2.870	3 5/8	2.17	1.311	2.656	0.70
2 1/2	2.875	1.771	0.552	13.70	13.87	8	3.389	4 1/8	3.43	2.871	4.028	0.84
3	3.500	2.300	0.600	18.58	18.79	8	4.014	4 1/8	4.13	5.992	5.466	1.05
3 1/2	4.000	2.728	0.636	22.85	23.16	8	4.628	4 5/8	6.29	9.848	6.721	1.21
4	4.500	3.152	0.674	27.54	27.95	8	5.233	4 5/8	8.16	15.28	8 101	1.37
5	5.563	4.063	0.750	38.55	39.20	8	6.420	5 1/8	12.87	33.64	11.34	1.72
6	6.625	4.897	0.864	53.16	53.92	8	7.482	5 1/8	15.18	66.33	15.64	2.06
8	8.625	6.875	0.875	72.42	73.76	8	9.596	6 1/8	26.63	162.0	21.30	2.76

Large O. D. Pipe

Pipe 14″ and larger is sold by actual O. S. diameter and thickness.
Sizes 14″, 15″, and 16″ are available regularly in thicknesses varying by 1/16″ from 1/4 to 1″, inclusive.
All pipe is furnished random length unless otherwise ordered, viz: 12 to 22 ft with privilege of furnishing 5 per cent in 6 to 12 ft lengths. Pipe railing is most economically detailed with slip joints and random lengths between couplings.

142 MECHANICS

Table 2-21 Properties of American Standard Yard Lumber and Timber Sizes

Nominal Size, in.	American Standard Dressed Size, in.	Area of Section, A=bd, sq in.	Weight per Lineal foot, lb	Moment of Inertia, $I=\dfrac{bd^3}{12}$	Section Modulus, $S=\dfrac{bd^2}{6}$	Nominal Size, in.	American Standard Dressed Size, in.	Area of Section, A=bd, sq in.	Weight per Lineal foot, lb	Moment of Inertia, $I=\dfrac{bd^3}{12}$	Section Modulus, $S=\dfrac{bd^2}{6}$
2× 4	1⅝× 3⅝	5.89	1.6	6.45	3.56	10×20	9½×19½	185.25	51.4	5870.05	602.06
2× 6	1⅝× 5⅝	9.14	2.5	24.10	8.57	10×22	9½×21½	204.25	56.7	7867.81	731.89
2× 8	1⅝× 7½	12.19	3.4	57.13	15.32	10×24	9½×23½	223.25	62.0	10274.06	874.39
2×10	1⅝× 9½	15.44	4.3	116.09	24.44	10×26	9½×25½	242.25	67.3	13126.81	1029.56
						10×28	9½×27½	261.25	72.5	16465.24	1197.39
2×12	1⅝×11½	18.69	5.2	205.94	35.82	10×30	9½×29½	280.25	77.8	20323.79	1377.89
2×14	1⅝×13½	23.62	6.5	333.15	49.36						
2×16	1⅝×15½	25.18	7.0	504.24	65.07	12×12	11½×11½	132.25	36.7	1457.50	253.47
2×18	1⅝×17½	28.43	7.9	725.71	82.94	12×14	11½×13½	155.25	43.1	2357.85	349.31
2×20	1⅝×19½	31.69	8.8	1004.05	102.98	12×16	11½×15½	178.25	49.5	3568.70	460.48
						12×18	11½×17½	201.25	55.9	5136.49	586.98
3× 4	2⅝× 3⅝	9.51	2.6	10.42	5.75	12×20	11½×19½	224.25	62.3	7105.90	728.81
3× 6	2⅝× 5⅝	14.76	4.2	38.93	13.84						
3× 8	2⅝× 7½	19.68	5.7	92.28	24.60	12×22	11½×21½	247.25	68.7	9524.24	885.98
3×10	2⅝× 9½	24.93	7.2	187.55	39.48	12×24	11½×23½	270.25	75.0	12437.08	1058.47
						12×26	11½×25½	293.25	81.4	15890.42	1246.31
3×12	2⅝×11½	30.18	8.8	332.69	57.86	12×28	11½×27½	316.25	87.8	19932.58	1449.47
3×14	2⅝×13½	35.43	10.3	538.21	79.73	12×30	11½×29½	339.25	94.2	24602.61	1667.97
3×16	2⅝×15½	40.68	11.3	814.60	105.11						
3×18	2⅝×17½	45.94	12.8	1172.36	133.98	14×14	13½×13½	182.25	50.6	2767.92	410.06
3×20	2⅝×19½	51.19	14.2	1622.00	166.36	14×16	13½×15½	209.25	58.1	4189.36	540.56
						14×18	13½×17½	236.25	65.6	6029.29	689.06
4× 4	3⅝× 3⅝	13.14	3.6	14.38	7.94	14×20	13½×19½	263.25	73.1	8341.73	855.56
4× 6	3⅝× 5⅝	20.39	5.7	53.76	19.11	14×22	13½×21½	290.25	80.6	11180.67	1040.06
4× 8	3⅝× 7½	27.18	7.5	127.44	33.98						
4×10	3⅝× 9½	34.43	9.6	258.99	54.52	14×24	13½×23½	317.25	88.1	14600.10	1242.56
						14×26	13½×25½	344.25	95.6	18654.04	1463.06
4×12	3⅝×11½	41.68	11.6	459.42	79.90	14×28	13½×27½	371.25	103.1	23398.73	1701.56
4×14	3⅝×13½	48.93	13.6	743.23	110.11	14×30	13½×29½	398.25	110.6	28881.42	1958.06
4×16	3⅝×15½	56.18	15.6	1124.90	145.15						
4×18	3⅝×17½	63.43	17.6	1618.96	185.02	16×16	15½×15½	240.25	66.7	4809.98	620.64
4×20	3⅝×19½	70.69	19.6	2239.88	229.73	16×18	15½×17½	271.25	75.3	6922.49	791.14
						16×20	15½×19½	302.25	83.9	9577.50	982.31
6× 6	5½× 5½	30.25	8.4	76.25	27.73	16×22	15½×21½	333.25	92.5	12837.00	1194.14
6× 8	5½× 7½	41.25	11.4	193.35	51.56						
6×10	5½× 9½	52.25	14.5	392.96	82.73	16×24	15½×23½	364.25	101.2	16763.00	1426.64
6×12	5½×11½	63.25	17.5	697.06	121.23	16×26	15½×25½	395.25	109.8	21417.50	1679.81
						16×28	15½×27½	426.25	118.4	26863.78	1953.64
6×14	5½×13½	74.25	20.6	1127.66	167.06	16×30	15½×29½	457.25	127.0	33159.98	2248.14
6×16	5½×15½	85.25	23.6	1706.76	220.22						
6×18	5½×17½	96.25	26.7	2456.36	280.73	18×18	17½×17½	306.25	85.0	7815.73	893.23
6×20	5½×19½	107.25	29.8	3398.46	348.56	18×20	17½×19½	341.25	94.8	10813.33	1109.06
6×22	5½×21½	118.25	32.8	4555.05	423.73	18×22	17½×21½	376.25	104.5	14493.43	1348.23
						18×24	17½×23½	411.25	114.2	18926.02	1610.72
8× 8	7½× 7½	56.25	15.6	263.67	70.31						
8×10	7½× 9½	71.25	19.8	535.85	112.81	18×26	17½×25½	446.25	123.9	24181.11	1896.56
8×12	7½×11½	86.25	23.9	950.55	165.31	18×28	17½×27½	481.25	133.7	30331.62	2205.72
8×14	7½×13½	101.25	28.0	1537.73	227.81	18×30	17½×29½	516.25	143.4	37438.79	2538.22
						20×20	19½×19½	380.25	105.6	12049.49	1235.81
8×16	7½×15½	116.25	32.0	2327.42	300.31	20×22	19½×21½	419.25	116.4	16149.86	1502.31
8×18	7½×17½	131.25	36.4	2249.60	382.81	20×24	19½×23½	458.25	127.3	21089.04	1794.81
8×20	7½×19½	146.25	40.6	4634.30	475.31						
8×22	7½×21½	161.25	44.8	6211.48	577.81	20×26	19½×25½	497.25	138.1	26944.73	2113.31
8×24	7½×23½	176.25	48.9	8111.17	690.31	20×28	19½×27½	536.25	148.9	33798.17	2457.81
						20×30	19½×29½	575.25	159.8	41717.61	2828.31
10×10	9½× 9½	90.25	25.0	678.75	142.89						
10×12	9½×11½	109.25	30.3	1204.01	209.39	24×24	23½×23½	522.25	153.4	25414.96	2162.97
10×14	9½×13½	128.25	35.6	1947.78	288.56	24×26	23½×25½	599.25	166.4	32471.80	2546.81
10×16	9½×15½	147.25	40.9	2984.04	380.39	24×28	23½×27½	646.25	179.5	40731.06	2916.97
10×18	9½×17½	166.25	46.1	4242.80	484.89	24×30	23½×29½	693.25	192.5	50274.98	3408.47

The weights given are based on assumed average weight of 40 lb per cubic foot.

Table 2-22 Properties of Standard I-Beams—Aluminum

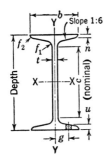

Elements of Sections

All dimensions in inches.
Weight in pounds per foot.
Area in square inches.
I = moment of inertia in in.^{4}
S = section modulus in in.^{3}
r = radius of gyration in inches.

J = torsion factor in in.^{4}
Rivet given is maximum allowable in flange.
g = usual gage.
u = nominal grip.

Size	Depth	3			4		5			6	
	t	0.170*	0.251	0.349*	0.190*	0.326*	0.210*	0.347	0.494*	0.230*	0.343*
Weight		2.02	2.31	2.67	2.72	3.38	3.53	4.36	5.25	4.43	5.25
Area		1.67	1.91	2.21	2.25	2.79	2.92	3.60	4.34	3.66	4.34
	b	2.330	2.411	2.509	2.660	2.796	3.000	3.137	3.284	3.330	3.443
	n	0.170	0.170	0.170	0.190	0.190	0.210	0.210	0.210	0.230	0.230
	f	0.27	0.27	0.27	0.29	0.29	0.31	0.31	0.31	0.33	0.33
	f_1	0.10	0.10	0.10	0.11	0.11	0.13	0.13	0.13	0.14	0.14
	c_2	1¾	1¾	1¾	2¾	2¾	3½	3½	3½	4½	4½
Axis X-X	I	2.52	2.71	2.93	6.06	6.79	12.26	13.69	15.22	22.08	24.11
	S	1.68	1.80	1.95	3.03	3.39	4.90	5.48	6.09	7.36	8.04
	r	1.23	1.19	1.15	1.64	1.56	2.05	1.95	1.87	2.46	2.36
Axis Y-Y	I	0.46	0.51	0.59	0.76	0.90	1.21	1.41	1.66	1.82	2.04
	S	0.39	0.42	0.47	0.57	0.65	0.81	0.90	1.01	1.09	1.19
	r	0.52	0.52	0.52	0.58	0.57	0.64	0.63	0.62	0.71	0.69
Rivet Data	Diam.	⅜	⅜	⅜	½	½	½	½	½	⅝	⅝
	g	¾	¾	¾	¾	¾	⅞	⅞	⅞	1	1
	u	⁵⁄₁₆	⁵⁄₁₆	⁵⁄₁₆	⁵⁄₁₆	⁵⁄₁₆	⅜	⅜	⅜	⅜	⅜
	J	0.045	0.061	0.093	0.074	0.12	0.12	0.19	0.33	0.17	0.24

Size	Depth	6 (Cont.)	7	8		9	10		12	
	t	0.465	0.345*	0.270*	0.532*	0.290	0.310*	0.594	0.350*	0.565
Weight		6.13	6.23	6.53	9.07	7.72	9.01	12.45	11.31	16.01
Area		5.07	5.15	5.40	7.49	6.38	7.45	10.29	9.35	13.23
	b	3.565	3.755	4.000	4.262	4.330	4.660	4.944	5.000	5.355
	n	0.230	0.250	0.270	0.270	0.290	0.310	0.310	0.350	0.460
	f_1	0.33	0.35	0.37	0.37	0.39	0.41	0.41	0.45	0.56
	f_2	0.14	0.15	0.16	0.16	0.17	0.19	0.19	0.21	0.28
	c	4½	5¼	6¼	6¼	7	8	8	9¾	9¼
Axis X-X	I	26.31	39.40	57.55	68.73	85.90	123.39	147.06	218.13	287.27
	S	8.77	11.26	14.39	171.8	19.09	24.68	29.41	36.35	47.88
	r	2.28	2.77	3.27	3.03	3.67	4.07	3.78	4.83	4.66
Axis Y-Y	I	2.31	2.88	3.73	4.66	5.09	6.78	8.36	9.35	14.50
	S	1.30	1.53	1.86	2.19	2.35	2.91	3.38	3.74	5.42
	r	0.68	0.75	0.83	0.79	0.89	0.95	0.90	1.00	1.05
Rivet Data	Diam.	⅝	⅝	¾	¾	¾	¾	¾	¾	¾
	g	1	1	1⅛	1⅛	1¼	1⅜	1⅜	1½	1½
	u	⅜	⅜	⁷⁄₁₆	½	½	½	½	⁹⁄₁₆	¾
	J	0.38	0.32	0.34	0.75	0.46	0.62	1.31	0.92	2.19

Table 2-23 Properties of Wide-Flange Beams—Aluminum

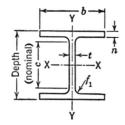

Elements of Sections

All dimensions in inches.
Weight in pounds per foot.
Area in square inches.
I = moment of inertia in in.[4]

S = section modulus in in.[3]
r = radius of gyration in inches.
J = torsion factor in in.[4]

Nominal Size		6 × 4	6 × 6	8 × 5	8 × 7	8 × 8	10 × 5¾
Actual Depth		6.00*	6.00*	8.00*	8.00*	8.00*	9.90*
t		0.230	0.240	0.230	0.245	0.288	0.240
Weight		4.28	5.56	6.07	8.56	11.04	7.51
Area		3.54	4.59	5.02	7.08	9.12	6.21
b		4.00	6.00	5.25	6.50	8.00	5.75
n		0.279	0.269	0.308	0.398	0.433	0.340
f_1		0.250	0.250	0.320	0.400	0.400	0.312
c		4⅞	4⅞	6¾	6⅜	6⅜	8½
Axis X-X	I	21.75	30.17	56.73	84.15	109.66	106.74
	S	7.25	10.06	14.18	21.04	27.41	21.56
	r	2.48	2.56	3.36	3.45	3.47	4.15
Axis Y-Y	I	2.98	9.69	7.44	18.23	36.97	10.77
	S	1.49	3.23	2.83	5.61	9.24	3.75
	r	0.92	1.45	1.22	1.61	2.01	1.32
J		0.082	0.106	0.135	0.312	0.497	0.196

Table 2-24 Properties of H-Beams—Aluminum

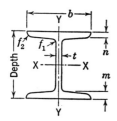

Elements of Sections

All dimensions in inches.
Weight in pounds per foot.
Area in square inches.
I = moment of inertia in in.[4]

S = section modulus in in.[3]
r = radius of gyration in inches.
J = torsion factor in in.[4]

Size	Depth	4	5	6			8		
	t	0.313*	0.313*	0.250*	0.313	0.438	0.313*	0.375	0.500*
Weight		4.85	6.63	8.04	8.49	9.40	11.51	12.11	13.32
Area		4.00	5.48	6.64	7.02	7.77	9.52	10.01	11.01
b		4.000	5.000	5.938	6.000	6.125	7.938	8.000	8.125
m		0.453	0.503	0.542	0.542	0.542	0.560	0.560	0.560
n		0.290	0.330	0.360	0.360	0.360	0.358	0.358	0.358
f_1		0.313	0.313	0.313	0.313	0.313	0.313	0.313	0.313
f_2		0.145	0.165	0.180	0.180	0.180	0.179	0.179	0.179
Axis X-X	I	10.72	23.82	44.06	45.19	47.44	112.94	115.58	120.92
	S	5.36	9.53	14.69	15.06	15.81	28.23	28.90	30.23
	r	1.64	2.08	2.58	2.54	2.47	3.45	3.40	3.31
Axis Y-Y	I	3.56	7.82	14.18	14.65	15.65	34.15	35.01	36.79
	S	1.78	3.13	4.77	4.88	5.11	8.60	8.75	9.06
	r	0.94	1.19	1.46	1.44	1.42	1.89	1.87	1.83
J		0.22	0.34	0.45	0.50	0.62	0.68	0.75	0.96

SECTION 3

FLUID MECHANICS

DEFINITIONS

Principal Symbols

D	diameter		q	volumetric flow rate
E	energy		r	radius
f	friction factor		S	surface area
F	force		t	time
ΔF	energy dissipation		V	average velocity
g	acceleration of gravity		w	mass flow rate
g_c	dimensional constant, see Tables 1-1 and 3-6		x, y	coordinates
			Z	elevation
G	mass velocity, ρV		γ	ratio of specific heats
h	fluid head		μ	absolute viscosity
k	surface roughness		ν	kinematic viscosity
L	length		ρ	density
p	pressure		τ	shear stress

A *fluid* is a substance, such as a gas, liquid, or fluidized solid powder, that undergoes continuous deformation when subjected to a shear stress. A fluid may be considered to consist of finite particles, each much larger than a molecule but infinitesimal in size compared to the total volume of fluid. The whole fluid is then regarded as continuous, and the action of forces on the particles is treated as producing relative motion which results in translation, rotation, and deformation. In a solid the deforming force depends largely on the extent of the deformation, but in a fluid the force depends mainly on the speed of deformation.

Consistency is the property of a fluid that measures resistance to deformation. Quantitatively, consistency is expressed as a coefficient which is the ratio of the shear stress to the velocity gradient (force per unit area divided by velocity change per unit distance normal to the force),

$$\text{consistency} = \tau g_c \bigg/ \frac{du}{dy} \qquad (3\text{-}1)$$

in which g_c is the dimensional constant (see page 163). The hypothetical fluid with zero consistency is called an *ideal or perfect fluid*.

A *Newtonian fluid* is one in which the consistency is independent of the shear stress and hence of the rate of shear. This constant coefficient is called the *absolute viscosity* or dynamic viscosity—in English units, lb-mass/ft sec. *Kinematic viscosity* is the ratio of absolute viscosity to density.

For *complex* or *non-Newtonian* substances the consistency is a function of the shear stress at present and perhaps past times as well. A *plastic* solid behaves like an elastic solid up to a yield point and then deforms continuously with a resistance that is a function of the shear stress. The *rigidity* of such a substance is $g_c(\tau - \tau_0)/(du/dy)$, analogous to viscosity. If deformation reduces the consistency, the substance is *thixotropic*, e.g., many gels, greases, and soaps. If deformation increases the consistency, the substance is *dilatant*, e.g., slurries. (See Fig. 3-1.)

Steady flow is motion whose velocity is independent of time at all points; it is usually implied that pressure, density, and temperature are also independent of time. *Uniform flow* is motion whose velocity is the same at every point at a given instant of time. *Incompressible flow* assumes that the density does not change throughout the path of flow. Liquids are nearly incompressible except near the critical state. Gases, although highly compressible, are usually assumed to be incompressible when the change in density produced by the motion is a small fraction of the normal density. Thus change in density during flow through an orifice may usually be neglected for velocities up to one-fifth the velocity of sound in the gas.

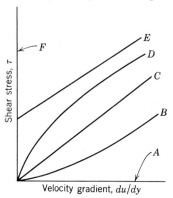

Fig. 3-1 Fluid types. (*A*) Inviscid. (*B*) Dilatant. (*C*) Newtonian. (*D*) Pseudoplastic. (*E*) Plastic. (*F*) Elastic.

FLUID STATICS

Hydrostatic Equilibrium. In a stationary mass of fluid, the pressure is constant throughout any plane normal to the gravitational force but varies with depth of fluid. For a fluid at rest the resultant of all forces acting at a point is zero; thus

$$dp + \frac{g}{g_c} \rho \, dZ = 0 \qquad (3\text{-}2)$$

where Z is measured upward from the bottom of the fluid column. See Table 1-1 for values of g_c. Where density is constant (incompressible fluid),

$$p_1 + \frac{g}{g_c} \rho Z_1 = p_2 + \frac{g}{g_c} \rho Z_2 = \text{const.} \tag{3-3}$$

In terms of equivalent fluid column or fluid head

$$\frac{g_c p}{g \rho} \tag{3-4}$$

The manometer for measuring pressure differences (Fig. 3-2) is based on hydrostatic equilibrium

$$p_a - p_b = (R - R_0)(\rho_a - \rho_b) \sin \alpha \tag{3-5}$$

where R_0 is the reading (length) when $p_a = p_b$ and R is the reading under pressure difference. The area of the tube must be negligible compared to the area of the reservoir.

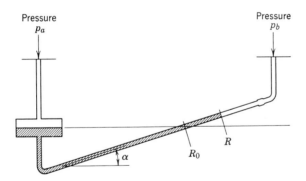

Fig. 3-2 Manometer.

Pressure on Surfaces. The pressure on the floor of a fluid container is equivalent to the maximum height of fluid column and is not equal to the weight of fluid contained.

The normal pressure at a point on a container wall is

$$p_{\text{normal}} = \frac{p + g \rho Z}{\sin \theta}$$

The horizontal component of pressure is equal to the pressure on a vertical wall at the same depth.

The *center of pressure* of a submerged plane area A is the point at which a single supporting force will balance all the pressures on the surface. The

148 FLUID MECHANICS

coordinates of the center of pressure x_p and y_p are

$$x_p = \frac{\int xy \, dA}{Ay_g} \qquad (3\text{-}6)$$

$$y_p = \frac{\text{moment of inertia about } OX}{Ay_g} \qquad (3\text{-}7)$$

where OX is the intersection of the plane of the area with the liquid surface, and y_g is the distance of the center of gravity of the area from OX.

Buoyancy. For a submerged or floating object at rest, the summation of all horizontal forces must be zero, and the gravitational force must be balanced by a buoyancy force. According to *Archimedes' principle*, the buoyancy is equal to the weight of the fluid displaced and the line of action is through the centroid of the displaced volume.

Stability of Submerged Bodies. A submerged body at rest is stable if a small displacement produces forces tending to restore the body to its original position.

For horizontal stability the center of gravity of a rigid body must be below the center of buoyancy. When the original center of buoyancy is below the center of gravity, the body is unstable unless the centers of gravity or buoyancy shift on rotation and set up a new position of stability.

FLUID DYNAMICS

Conservation Equations

Many problems of engineering interest may be solved by means of mass, momentum, and energy balances. Such balances are called conservation equations. Consider the isolated volume of fluid B shown in Fig. 3-3. The rate accumulation of mass, momentum, and energy in the volume B is the sum of their rate of generation R and their rate of transport through the surface S. Letting n denote the concentration (amount per unit volume) of any of these

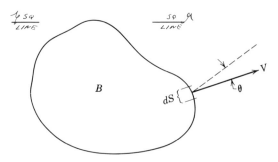

Fig. 3-3

entities,

$$\frac{\partial}{\partial t} \int n \, dB = \int R \, dB - \int nu \cos \theta \, dS - \int N \, dS \tag{3-8}$$

where the last term defines the net rate of transport across the surface by means other than flow, e.g., diffusion, radiation, conduction, work.

Continuity of Mass. For mass the concentration is density ρ, and the generation and nonflow terms vanish (except for situations of mass-energy conversions); hence Eq. 3-8 becomes

$$\int \frac{\partial \rho}{\partial t} dB = - \int \rho u \cos \theta \, dS \tag{3-9}$$

For incompressible flow or compressible flow in steady state where $\partial \rho / \partial t = 0$,

$$\int \rho u \cos \theta \, dS = 0 \tag{3-10}$$

For conduits where flow is normal to the cross section as in Fig. 3-4 and where there is negligible radial variation in density,

$$S_2 V_2 \rho_2 - S_1 V_1 \rho_1 = 0 \tag{3-11}$$

Conservation of Momentum. For the x component of momentum, concentration is $\rho V_x = G_x$ (i.e., the mass velocity) and the second and fourth terms of Eq. 3-8 are represented by a net force F_x acting in the x direction; hence

$$\int \frac{\partial G_x}{\partial t} dB = g_c F_x - \int G_x u \cos \theta \, dS \tag{3-12}$$

in steady state where $\partial G_x / \partial t = 0$,

$$F_x = \frac{1}{g_c} \int G_x u \cos \theta \, dS \tag{3-13}$$

For the special case where u and $\cos \theta$ are uniform over the areas, $S_1 u_1 \cos \theta_1 = S_2 u_2 \cos \theta_2 = q$, the volumetric flow rate, and Eq. 3-12 becomes

$$F_x = \frac{q}{g_c} (G_{x2} - G_{x1}) \tag{3-14}$$

For fully developed flow in tubes where $\cos \theta = 1$ and the velocity distribution does not change along the length of the tube, solution of Eq. 3-13 provides general relationships for the axial distribution of shear stress and velocity.

Consider the control volume isolated by the pipe sections (1) and (2) of Fig. 3-4. By applying the x-component of forces acting on the control volume, Eq. 3-12 becomes

$$-\left[\Delta p + \rho \frac{g}{g_c}(z_2 - z_1)\right] \pi r_0^2 - 2\pi r_0 \tau_0 (x_2 - x_1) = \pi r_0^2 \rho \frac{V_2^2}{\beta g_c} - \pi r_0^2 \rho \frac{V_1^2}{\beta g_c} \tag{3-12a}$$

where β is a velocity distribution parameter, constant along the pipe. Then

$$\tau = -\frac{r(\Delta p_f + \Delta p_g)}{2L} \tag{3-15}$$

where Δp_f is the change in pressure from flow, Δp_g the change in pressure from gravitational effect, and L the length of the tube; moreover, where subscript zero refers to the wall,

$$\tau = \tau_0(r/r_0) \tag{3-16}$$

i.e., shear stress is linear with the radius regardless of whether the flow is laminar or turbulent.

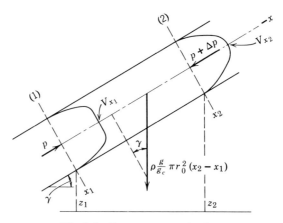

Fig. 3-4 Forces on control volume.

The velocity distribution is a function of the shear stress,

$$dV_x/dr = -g_c \psi(\tau) \tag{3-17}$$

Hence from Eqs. 3-16 and 3-17,

$$V_x = \frac{g_c r_0}{\tau_0} \int_r^{r_0} \psi(\tau) \, d\tau \tag{3-18}$$

For laminar flow of Newtonian fluids, $\psi(\tau) = \tau/\mu$; for non-Newtonian fluids $\psi(\tau)$ must be obtained experimentally; for turbulent flow Prandtl and von Karman have developed theoretical relationships.

Conservation of Energy. For energy, concentration is ρE and $N = -q - w_s + \rho u \cos \theta$, where q and w_s represent the input rates of heat and shear work per unit area. Hence Eq. 3-8 becomes

$$\frac{\partial}{\partial t}(\rho E)\, dB = Q' + W_s' - \int \left(\frac{p}{\rho} + E\right) \rho u \cos \theta \, dS \tag{3-19}$$

where E is the total energy, Q' the rate of heat input, and W_s' the input rate of shear work per unit mass.

For steady state, $[d(\rho E)/dt = 0]$, and flow normal to the control surface S, $(\cos \theta = 1)$,

$$Q + W_s + U_1 + \frac{p_1}{\rho_1} + Z_1\frac{g}{g_c} + \frac{V_1^2}{2\alpha_1 g_c} = U_2 + \frac{p_2}{\rho_2} + Z_2\frac{g}{g_c} + \frac{V_2^2}{2\alpha_2 g_c} \quad (3\text{-}20)$$

where Q is the heat input, W_s is the input of shaft work, U is the molecular internal energy per unit mass, and α is a velocity distribution parameter. For fully developed flow in a tube of constant cross section, $V_1 = V_2$ and $\alpha_1 = \alpha_2$ so that the velocity terms cancel; and if the pipe is horizontal, the gravitational terms also cancel and Eq. 3-20 reduces to the restricted thermodynamic equation,

$$Q + W_s = \Delta U + \Delta\left(\frac{p}{\rho}\right) = \Delta H \quad (3\text{-}21)$$

The Mechanical-Energy Balance. This balance is a definition of a dissipation term ΔF, which represents the loss of available energy from irreversible processes occurring during flow. Thus

$$W_s + Z_1\frac{g}{g_c} + \frac{V_1^2}{2\alpha_1 g_c} - \int_1^2 \frac{dp}{\rho} = Z_2\frac{g}{g_c} + \frac{V_2^2}{2\alpha_2 g_c} + \Delta F \quad (3\text{-}22)$$

In general, ΔF must be obtained from experiment or by comparison of the mechanical-energy balance with the momentum balance. When ΔF vanishes, however, the total- and mechanical-energy balances may be used interchangeably, the choice being a matter of convenience.

When the total dissipation loss is due only to skin friction, sudden expansions, and sudden contractions,

$$\Delta F = \Delta F_f + \Delta F_e + \Delta F_c \quad (3\text{-}23)$$

The right-hand terms of Eq. 3-23 are usually evaluated by equations derived from the momentum balance; hence for these cases the mechanical-energy balance is regarded as a restricted form of the momentum balance.

The energy loss per unit mass of fluid ΔF_f can be expressed as the dissipation part of pressure or fluid head by

$$\Delta p_f = \Delta F_f \quad (3\text{-}24)$$

and

$$\Delta h_f = \frac{g_c \Delta F_f}{g} \quad (3\text{-}25)$$

For uniform, steady flow ($\alpha = 1$) of incompressible fluids ($\rho = $ const.), Eq. 3-22 becomes the following.

In terms of energy per unit mass of fluid,

$$W_s + Z_1\frac{g}{g_c} + \frac{V_1^2}{2g_c} + \frac{p_1}{\rho} = Z_2\frac{g}{g_c} + \frac{V_2^2}{2g_c} + \frac{p_2}{\rho} + \Delta F \quad (3\text{-}26)$$

In terms of head of fluid,

$$\frac{g_c W_s}{g} + Z_1 + \frac{V_1^2}{2g} + \frac{g_c p_1}{g \rho} = Z_2 + \frac{V_2^2}{2g} + \frac{g_c p_2}{g \rho} + \frac{g_c \Delta F}{g} \quad (3\text{-}27)$$

In terms of pressure,

$$\rho W_s + \rho Z_1 \frac{g}{g_c} + \frac{\rho V_1^2}{2g_c} + p_1 = \rho Z_2 \frac{g}{g_c} + \frac{\rho V_2^2}{2g_c} + p_2 + \rho \Delta F \quad (3\text{-}28)$$

The mechanical-energy balance should not be confused with *Bernoulli's theorem* which was derived by integration along one streamline of the equations of motion simplified for an inviscid fluid to give

$$\frac{V_1^2}{2} + g_c \frac{p_1}{\rho_1} + gZ_1 = \frac{V_2^2}{2} + g_c \frac{p_2}{\rho_2} + gZ_2 \quad (3\text{-}29)$$

Flow through Constrictions

The energy losses from flow through changes in cross section can be derived theoretically from the velocity distributions and integration of the momentum or mechanical-energy equations. For incompressible flow with uniform velocity distributions (i.e., fully developed turbulence where α approaches 1), the integrated equations are given below, where the loss is expressed as energy per unit mass and subscript 1 refers to the upstream condition. Loss coefficients are shown in Table 3-1 for both turbulent flow and laminar flow in circular sections ($\alpha = 2$).

In general, from Eq. 3-26 since Z and W vanish, the differential pressure is given by

$$\frac{p_1 - p_2}{\rho} = \Delta F - \frac{V_1^2 - V_2^2}{2g_c} \quad (3\text{-}30)$$

Table 3-1 Coefficients for Constrictions

Area Ratio	0	0.2	0.4	0.6	0.8	1.0
C_c	0.61	0.63	0.67	0.71	0.78	1.0
C_d						
Nozzle		0.80	0.82	0.84	0.88	
Venturi		0.98	0.98	0.98	0.98	
Orifice		0.60	0.61	0.66	0.72	
K_e, turb.	1.0	0.64	0.36	0.16	0.04	0
K_e, laminar	1.0	0.5	0.10	−0.24	−0.5	−0.67
K_c, turb.	0.41	0.34	0.24	0.16	0.08	0
K_c, laminar	1.08	0.99	0.91	0.83	0.77	0.68
K_v		0.1	0.08	0.05	0.02	
K_0	1.0	0.86	0.67	0.40	0.14	0

Sudden Expansion

$$\frac{p_1 - p_2}{\rho} = \frac{V_1^2}{g_c}\left[\left(\frac{S_1}{S_2}\right)^2 - \frac{S_1}{S_2}\right] \quad (3\text{-}31)$$

and

$$\Delta F_e = \frac{V_1^2}{2g_c}\left(1 - \frac{S_1}{S_2}\right)^2 = K_e \frac{V_1^2}{2g_c} \quad (3\text{-}32)$$

For similar expansions through well-designed conical or trumpet diffusers, the energy loss is substantially less than shown by Eq. 3-31. Note that there is pressure recovery downstream after an expansion.

Sudden Contraction

$$\Delta F_c = \frac{V_2^2}{2g_c}\left(\frac{1}{C_c} - 1\right)^2 = K_c \frac{V_2^2}{2g_c} \quad (3\text{-}33)$$

For rounded entrances, K_c is 0.01 to 0.05; for re-entrant inlet, $K_c = 0.8$.

Nozzles, Venturi Tubes, and Orifices. The general equations for turbulent flow through these constrictions are given below, where w and q are mass and volume rates of flow, C_d is a composite discharge coefficient dependent on design flow conditions, Y is an expansion factor (for liquids $Y = 1$), β is the ratio of area of constriction to area of upstream channel, and subscript 1 refers to the upstream point and subscript c refers to the constriction. For accurate flow measurements values of $C_d Y$ should be obtained by calibration.* Table 3-1 gives approximate values of C_d.

Rate of Flow

$$w = q_1 \rho_1 = C_d Y A_c \left[\frac{2g_c(p_1 - p_c)\rho_1}{1 - \beta^4}\right]^{1/2} \quad (3\text{-}34)$$

As evident in Fig. 3-5, the relationship between C_d and $(p_1 - p_c)$ depends on the location of the pressure taps.

For ideal gases, the expansion coefficient is given by

$$Y = 1 - B\left(\frac{p_1 - p_c}{\gamma p_1}\right)(0.41 + 0.35\beta^4)$$

where B is 1 for orifice and 2 for Venturi, and γ is the ratio of specific heats at constant pressure and constant volume.

Energy Loss

$$\Delta F = \frac{p_1 - p_2}{\rho} = K\frac{V_2^2}{2g_c} \quad (3\text{-}35)$$

where subscript 2 refers to the fully developed downstream condition. Approximate values of K for turbulent flow are given in Table 3-1 where K_e applies to nozzle, K_v to Venturi, and K_o to orifice.

* See *Fluid Meters: Their Theory and Application*, 4th Ed., Am. Soc. Mech. Engrs., N.Y., 1937, for complete discussion and coefficients for standard meters.

154 FLUID MECHANICS

Maximum Discharge Rate. The velocity of flow through a construction is limited since change of pressure cannot be transferred more rapidly than the velocity of sound. Hence the upper limit of flow is the acoustical velocity U_A in the fluid flowing,

$$U_A = \left(\frac{Kg_c}{\rho}\right)^{1/2} = \left(\frac{\gamma g_c p}{\rho}\right)^{1/2} \tag{3-36}$$

where for liquids K is the bulk modulus of volume elasticity, and for gases γ is the ratio of specific heats.

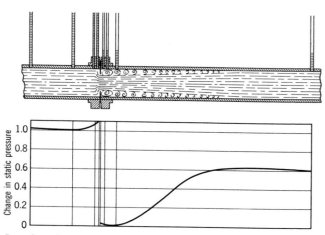

Fig. 3-5 Location of pressure taps. (From "Fluid Meters," *ASME*, New York, 1937.)

Rotameter. The general equation for rotameters is given below, where w and q are the rates of mass or volume flow, A_w is the area of minimum section of annulus, A_f is area of maximum section of float, ρ_f and ρ_w are densities of float and fluid, v_f is volume of float, and C is a coefficient of discharge which is determined by calibration:

$$w = q\rho = CA_w\left(\frac{2qv_f(\rho_f - \rho_w)\rho_w}{A_f}\right) \tag{3-37}$$

Packed Beds. For flow of an incompressible fluid through beds packed with solid particles of uniform size, the energy loss per unit mass of fluid,

$$\Delta F_f = \frac{fV_0^2 L\left(\dfrac{1}{D_e} + \dfrac{4}{D_t}\right)}{g_c \epsilon^3} \tag{3-38}$$

and

$$f = 0.3\phi + \frac{5\mu}{D_e \rho V_0} \tag{3-39}$$

FLUID DYNAMICS   155

where μ and ρ are fluid viscosity and density, V_0 is superficial velocity based on empty vessel, L is length of packed bed, D_e is ratio of total volume of packed bed to total surface area of packing [for spheres, $D_e = D_{sp}/6(1 - \epsilon)$], ϵ is fraction voids, D_t is diameter of bed, and ϕ is particle shape factor, i.e., 1.0 for spheres, 1.1 for cylinders, and 1.5 for irregular granules.

Darcy's Law of Permeability. For laminar flow through a fine porous bed or capillary passages in a solid, the velocity based on the total cross section is proportional to the ratio of pressure gradient to viscosity. The proportionality constant k, called the permeability, is a property of the bed independent of the fluid. The equivalent diameter D_e used in Eqs. 3-38 and 3-39 can be obtained experimentally by laminar flow of a gas through the bed and solution of Eqs. 3-40 and 3-41:

$$g_c \operatorname{grad} p = \frac{g_c \, dp + \rho g \, dZ}{dL} = \frac{\mu V_0}{k} \tag{3-40}$$

$$k = \frac{g_c \epsilon^3 D_e^2}{5} \tag{3-41}$$

Banks of Tubes. The pressure loss from turbulent flow across banks of tubes, such as heat exchangers, is

$$\Delta p_s = \rho \, \Delta F_f = \frac{4 f N_T \rho V_m^2}{2 g_c} \tag{3-42}$$

and

$$f = 0.75 \left(\frac{D_e V_m \rho}{\mu} \right)^{-0.2} \tag{3-43}$$

where N_T is number of rows of tubes in flow path, V_m is fluid velocity through minimum free area, and D_e is equivalent diameter, i.e., tube clearance.

Flow through Tubes and Conduits

Reynolds Analogy. The Reynolds number N_{Re} is the ratio of the inertial forces to the viscous forces and is applicable in situations where these forces are the dominant ones, as in the shear stress exerted on a pipe wall. At low Reynolds number the viscous forces are dominant and the inertial forces negligible, and at high Reynolds number the inertial forces are dominant and the viscous forces negligible:

$$N_{\mathrm{Re}} = \frac{D_e \rho V}{\mu} \tag{3-44}$$

In using the Reynolds analogy, it is necessary to select the proper representative velocity V and dimension D_e. In tubes and conduits, V is taken as the average velocity in the direction of flow and D_e as *equivalent diameter* or length. Except for laminar flow, a generalized equivalent diameter is four times the ratio of cross section to wetted perimeter, i.e., four times the *hydraulic radius* r_H. Table 3-2 gives the equivalent diameters for conduits of various shapes.

156 FLUID MECHANICS

The general equation for incompressible, steady flow in conduits, known as the Darcy-Weisbach equation, is derived from Eq. 3-22 as

$$\Delta F_f = \frac{\Delta p_f}{\rho} = \frac{g\,\Delta h_f}{g_c} = \frac{fV^2L}{2g_cD} \tag{3-45}$$

where f is the Darcy friction factor which is a function of the Reynolds number N_{Re} and the surface roughness k, as charted in Fig. 3-6 and discussed under Friction Factors below.

Table 3-2   Equivalent Diameters, $D_e = 4r_H$

Cross Section of Stream	Equivalent Diameter
Ducts running full	
Circle	D
Annulus	$D_o - D_i$
Square	L
Rectangle	$2L_1L_2/(L_1 + L_2)$
Partly filled ducts	
Rectangle, h deep and L wide	$4hL/(L + 2h)$
Semicircle	D
Wide shallow stream, h deep	$4h$
Triangular trough, h deep, L broad, s side	hL/s
Trapezoid, h deep, a top, b bottom, s side	$2h(a + b)/(b + 2s)$

Laminar Flow. The Hagen-Poiseuille equation applies to laminar, incompressible, steady flow in circular tubes where the velocity distribution is parabolic, and the mean free path of the fluid molecules is small compared to the tube diameter D.

$$q = \frac{g_c\,\Delta p_f \pi D^4}{128\mu L} \tag{3-46}$$

where q is the volumetric discharge rate, and D and L are tube diameter and length.

The energy dissipation per unit mass of fluid

$$\Delta F_f = \frac{\Delta p_f}{\rho} = \frac{g\,\Delta h_f}{g_c} = \frac{32\mu VL}{g_c\rho D^2} \tag{3-47}$$

For noncircular conduits, the distributions of shear stress must be known or derived to obtain the proper equivalent diameter. In the case of laminar flow through an annulus,

$$D_e = \tfrac{1}{2}\left[D_o^2 + D_i^2 - \frac{D_o^2 - D_i^2}{\ln(D_o/D_i)}\right]^{1/2} \tag{3-48}$$

where D_o and D_i refer to outer and inner diameters.

FLUID DYNAMICS

Friction Factors. It should be especially noted that the friction factor f used here is the Darcy factor defined by Eq. 3-45 since various quantitatively different friction factors appear in the literature. Chemical engineers use the Fanning friction factor, which is related to the Darcy factor by

$$f_{\text{Darcy}} = 4 f_{\text{Fanning}} \tag{3-49}$$

Algebraic equations for the friction factor have been derived for the various flow regimes shown in Fig. 3-6.

Laminar flow:

$$f = \frac{64}{N_{\text{Re}}} \tag{3-50}$$

Smooth pipes (von Karman):

$$\frac{1}{\sqrt{f}} = 2 \log N_{\text{Re}} \sqrt{f} - 0.8 \tag{3-51}$$

Complete turbulence (Nikuradse):

$$\frac{1}{\sqrt{f}} = 2 \log \left(3.7 \frac{D}{k} \right) \tag{3-52}$$

Transition zone (Colebrook):

$$\frac{1}{\sqrt{f}} = 1.15 + 2 \log \frac{D}{k} - 2 \log \left(1 + 4.67 \frac{D/k}{N_{\text{Re}} \sqrt{f}} \right) \tag{3-53}$$

Effective roughness of various commercial pipes is given in Fig. 3-6.

For flow that is not isothermal, the fluid properties used in the Reynolds number may be obtained at the film temperature T_f from the empirical equation

$$T_f = 0.4(T_w - T_b) + T_b \tag{3-54}$$

where T_w is the temperature of the wall and T_b is the temperature of the bulk fluid.

Losses in Valves and Fittings. Friction losses in valves and fittings must be determined by experiment. Table 3-3 gives for screwed fittings approximate data for the equivalent length of pipe and the coefficient k' in the equation for loss of fluid head,

$$h_f = \frac{k'V^2}{2g} = \frac{fL_eV^2}{D2g} \tag{3-55}$$

and

$$L_e = 50k'D \tag{3-56}$$

where k' and L_e/D are assumed to be independent of the Reynolds number.

Compressible Flow. When the pressure drop is less than 15% of the initial pressure, it is usually permissible to apply the equations for incompressible flow to gases. For accurate results with larger pressure drop in flow of gases, it is

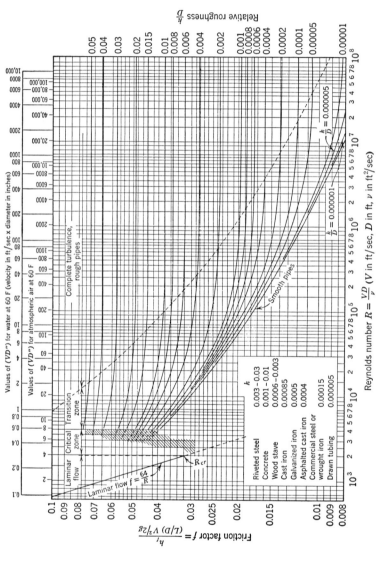

Fig. 3-6 Darcy friction factor chart. [From Moody, *Trans. ASME*, 66, 671 (1944).]

FLUID DYNAMICS

Table 3-3 Loss Coefficients for Screwed Valves and Fittings

Globe valve, open	10	45° elbow	0.5
Angle valve, open	5	90° standard elbow	1.0
Gate valve, open	0.2	90° medium-radius elbow	0.8
Gate valve, one-quarter closed	1.2	90° long-radius elbow	0.6
Gate valve, half closed	6	Standard tee	2
Gate valve, three-quarters closed	24	Close return bend	2.5

necessary to integrate over the flow path the differential form of Eq. 3-22:

$$dp + \frac{\rho g}{g_c} dZ + \frac{G^2}{\alpha g_c} d\left(\frac{1}{\rho}\right) + \frac{fG^2}{2\rho g_c D} dL = 0 \quad (3\text{-}57)$$

In situations where f is relatively insensitive to Reynolds number, average values of f and ρ can be used and Eq. 3-57 is integrated directly.

For *isothermal flow* of an ideal gas in a horizontal pipeline, the equation becomes

$$\frac{p_1^2 - p_2^2}{p_1} = \frac{fG^2 L}{\rho_1 D} \quad (3\text{-}58)$$

where subscripts 1 and 2 refer to upstream and downstream conditions.

Water hammer resulting from sudden stoppage of flow is caused by a pressure wave which moves upstream with the velocity of sound and is reflected back and forth until dissipated by friction and imperfect elasticity. The rise in head h_w and the velocity of the pressure wave u_w are given by

$$h_w = \frac{V}{\left[\frac{\rho g^2}{g_c}\left(\frac{1}{K_m} + \frac{D}{tE}\right)\right]^{1/2}} \quad (3\text{-}59)$$

$$u_w = \frac{1}{\left[\frac{\rho}{g_c}\left(\frac{1}{K_m} + \frac{D}{tE}\right)\right]^{1/2}} \quad (3\text{-}60)$$

where K_m is the bulk modulus of elasticity of the liquid, t is the thickness, and E the modulus of elasticity of the pipe wall.

Flow in Open Channels

Flow in open channels or partly filled ducts necessarily applies only to liquids and is treated as incompressible flow.

Laminar flow in open channels occurs only with very viscous liquids, very small velocities, or very shallow depth such as in a film. Turbulent flow in channels is of much wider engineering interest, but has been little studied for fluids other than water.

Weirs. Flow over weirs is treated as the conversion of potential energy to kinetic energy, $h = V^2/2g$ with suitable correction factors for velocity of approach and contraction of flow. The theoretical equation for volumetric rate of discharge for a sharp-crested weir is

$$q = c'AV = c'A(2gh)^{1/2} = cAh^{1/2} \tag{3-61}$$

where A is the upstream cross-sectional area of fluid flowing, h is the upstream head over the weir crest, and c is determined by experiment or from the following empirical formulas for various types of weirs.

Rectangular, suppressed (Francis):

$$q = 3.33Lh^{3/2} \tag{3-62}$$

Rectangular, contracted:

$$q = 3.33(L - 0.2h)h^{3/2} \qquad L > 2h \tag{3-63}$$

Circular, inflow (Gourley):

$$q = 3.0Lh^{1.4} \qquad L > 16h \tag{3-64}$$

Triangular (90° notch):

$$q = 2.53h^{5/2} \tag{3-65}$$

Cippoletti trapezoidal:

$$q = 3.37Lh^{3/2} \tag{3-66}$$

Correction for *velocity of approach* (when greater than 1 ft per sec.) is made by replacing $h^{3/2}$ with $[h + (V/2g)^{3/2} - (V/2g)^{3/2}]$.

Critical Depth. A minimum potential energy per unit mass of fluid is required for flow. This minimum head or critical depth for a rectangular channel

$$y_c = \left(\frac{q_b^2}{g}\right)^{1/3} \tag{3-67}$$

where q_b is the volumetric flow rate per unit width.

Flow Rates. For steady uniform flow in a channel, the Manning formula is commonly used to estimate average liquid velocity V and volumetric flow rate q,

$$V = \frac{1.49}{n} r_H^{2/3} \sin\theta^{1/2} \tag{3-68}$$

$$q = \frac{1.49}{n} A r_H^{2/3} \sin\theta^{1/2} \tag{3-69}$$

where A is cross-sectional area, r_H is hydraulic radius (see Table 3-2), θ is the angle between the grade line and the horizontal plane, and n is a roughness factor as given in Table 3-4.

Table 3-4 Values of Manning Roughness, n

Boundary Surface	n, (ft)$^{1/6}$
Wood	0.010–0.015
Concrete	0.011–0.016
Cast iron	0.013–0.017
Riveted steel	0.017–0.020
Brick	0.012–0.020
Earth or rubble	0.020–0.030
Gravel	0.022–0.035
Earth with weeds	0.025–0.040

Resistance of Immersed Bodies

The force on a body resulting from motion relative to a fluid is usually at an angle relative to the path of motion. The component of force normal to the path of the upstream fluid is called *lift* and is in addition to the static buoyant force which is independent of the motion. The force component parallel to the path is called *drag*. The total drag F_D is made up of two parts: the pressure drag caused by stresses normal to the surface and the friction drag caused by tangential stresses.

The *drag coefficient* is defined by

$$C_D = \frac{2g_c F_D}{\rho V_x^2 A_p} \qquad (3\text{-}70)$$

where V_x is the average relative velocity and A_p is the projected area of the body normal to V. By dimensional analysis, C_D may be a function of Reynolds number, Froude number, Mach number, and aspect ratio L/D. In most practical problems one or more of these numbers may be neglected. Thus

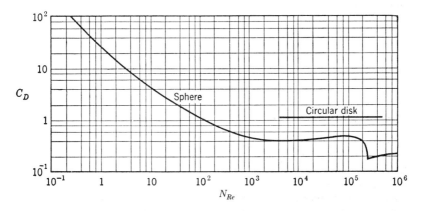

Fig. 3-7 Drag coefficients.

for incompressible flow only the Reynolds number and L/D ratio are important.

Drag coefficient as a function of Reynolds number is shown for spheres in Fig. 3-7. The sudden drop at $N_{Re} = 3 \times 10^5$ is caused by transition from laminar to turbulent flow within the boundary layer. Drag coefficients for various bodies are given in Table 3-5.

Table 3-5 Drag Coefficients, C_D for Various Bodies

Form of Body	$N_{Re} < 1.0$	$10^3 < N_{Re}$ $< 3 \times 10^5$	$10^6 < N_{Re} < 10^9$
Sphere	$24/N_{Re}$	0.44	0.2
Disk*	$24/N_{Re}$	1.12	1.12
Flat plate*	$24/N_{Re}$		
Length/breadth = 1		1.16	1.16
= 20		1.50	1.50
Circular cylinder*	$8/N_{Re}^{0.85}$		0.35
Length/diam. = 1		0.91	
= 2		0.85	
= 7		0.99	
Airfoil		0.04	0.007
Flat plate†		$1.33/(N_{Re})^{1/2}$	$0.073/(N_{Re})^{1/5}$
Circular cylinder†			
Length/diam. = 1		0.63	0.35
= 20		0.90	—
= ∞		1.20	0.33

* Flat side normal to flow.   † Flat side parallel to flow.

Stokes' Law. Since Reynolds number measures the ratio of inertial to viscous forces, at very low Reynolds number the effect of inertia may be neglected and only the viscosity effect considered. Hence the drag is proportional to $\mu D V_x$, and $C_D = \text{const.}/N_{Re}$.

For a sphere the Stokes equation is

$$g_c F_D = 3\pi \mu D V_x \qquad (3\text{-}71)$$

Hence from Eq. 3-70

$$C_D = 24/N_{Re} \qquad N_{Re} < 1 \qquad (3\text{-}72)$$

For forms other than spheres, the appropriate characteristic dimension is used to obtain C_D from Eq. 3-72, and Eq. 3-70 is then solved for drag F_D.

Newton's Drag Law. At Reynolds numbers between 10^3 and 3×10^5 where the viscous forces are negligible, the drag is independent of viscosity and the drag coefficient C_D is nearly constant over wide ranges of Reynolds number, as indicated in Table 3-5.

Between the Stokes and Newton regions, no simple statement of C_D is possible and values must be obtained from graphs like Fig. 3-7.

The *terminal settling velocity* V_t of a body relative to a surrounding fluid is obtained by equating the buoyant and drag forces acting on the body:

$$V_t = \left[\frac{2gm_p(\rho_p - \rho)}{\rho \rho_p A_p C_D}\right]^{1/2} \tag{3-73}$$

where m_p is the mass, ρ_p the density, and A_p the projected area of the particle.

For a sphere, in general,

$$V_t = \left[\frac{4gD_p(\rho_p - \rho)}{3\rho C_D}\right]^{1/2} \tag{3-74}$$

and in the Stokes region,

$$V_t = \frac{gD_p^2(\rho_p - \rho)}{18\mu} \tag{3-75}$$

High-Speed Compressible Flow. At speeds that are of the order of the velocity of sound ($N_M > 0.1$) the drag coefficient is a function of both the Reynolds number and the Mach number. At speeds above about $N_M = 0.6$, however, the Mach number dominates and the effect of Reynolds number is negligible. The drag coefficient on spheres was found to double between $N_M = 0.6$ and $N_M = 1.5$ and then decline slowly as N_M increased further. At the higher Mach numbers the friction drag is more than half the total drag, reaching as much as three-fourths of the drag at $N_M = 5$ for a well-designed airfoil. As a consequence, frictional or *aerodynamic heating* of the surfaces becomes a problem at high speeds.

The surface temperature T_s is a function of the heat flux q, the heat transfer coefficient h, the boundary layer temperature T_b, the surrounding air temperature T_a, the average relative velocity V_x, the Mach number N_M and the air properties c_p, ρ, μ, k, and γ. These variables are related through the equations

$$q = h(T_b - T_s) \tag{3-76}$$

$$T_b = T_a\left[1 + (N_{\text{Pr}})^{1/2} \frac{\gamma - 1}{2} N_M^2\right] \tag{3-77}$$

$$C_f = 2\frac{h}{c_p \rho V_x}(N_{\text{Pr}})^{2/3} \tag{3-78}$$

where C_f is the coefficient of friction drag and N_{Pr} is the Prandtl number $C_p\mu/k$.

At high Mach number the friction drag is nearly the total drag and the drag coefficient is nearly constant.

Table 3-6   Units of Absolute Viscosity

Viscosity Unit	Force Unit	Mass Unit	g_c	Conversion Factor*
Poise	dyne	gram	1	0.01
(Kilogram-force)(sec)/m²	kilogram	kilogram	9.807	1.02×10^{-4}
(Pound-force)(sec)/ft²	pound	slug	1	2.09×10^{-5}
(Poundal)(sec)/ft²	poundal	pound	1	6.72×10^{-4}
(Pound-mass)/(ft)(sec)	pound	pound	32.17	6.72×10^{-4}

* Multiply centipoises by factor to obtain other units.

REFERENCES

Elementary

Bird, R. B., W. E. Stewart, and E. H. Lightfoot, *Transport Phenomena*, John Wiley, New York, 1960.

Hunsaker, J. C. and B. G. Rightmire, *Engineering Applications of Fluid Mechanics*, McGraw-Hill, New York, 1947.

Vennard, J. K., *Elementary Fluid Mechanics*, John Wiley, New York, fourth edition, 1961.

Advanced

Batchelor, G. K., *The Theory of Homogeneous Turbulence*, Cambridge Univ. Press, London, 1955.

Lamb, H., *Hydrodynamics*, Dover, New York, sixth edition, 1945.

Rouse, Hunter, *Advanced Mechanics of Fluids*, John Wiley, New York, 1959.

Schlicting, H., *Boundary Layer Theory*, McGraw-Hill, New York, fourth edition, 1960.

SECTION 4

THERMODYNAMICS

Principal Symbols

A	Helmholtz free energy	R	gas constant, $pV = nRT$
C_p	molal heat capacity at constant pressure	S	entropy
		T	absolute temperature
C_v	molal heat capacity at constant volume	u	velocity
		V	volume
E	internal energy	W	work added to a system
f	fugacity	x, y	mole fractions, n_i/n
F_k	generalized force	X_k	generalized displacement
G	Gibbs free energy	β	volumetric coefficient of thermal expansion
H	enthalpy		
k_S	adiabatic compressibility	γ	ratio of specific heats, C_p/C_v
k_T	isothermal compressibility	μ	Gibbs chemical potential, $\left(\dfrac{\partial G}{\partial n_i}\right)$
n	number of moles		
p	pressure		
Q	heat added to a system		

DEFINITIONS

Thermodynamic processes are those where a difference is observed in any macroscopic property of the system between different times or states.

Spontaneous processes are those which occur in the direction of an intensive property or driving force contained within the system, the flow of heat from a higher to a lower temperature for example. For an adiabatically isolated spontaneous process, $dS > d\delta Q/T_{\min}$ where $T_{\min}$ is the lowest temperature in any part of the system.

Forced processes require an agency outside the system to force the process. For an infinitesimal forced process, $dS < d\delta Q/T_{\min}$ where $T_{\min}$ is the lowest temperature in any part of the system.

Reversible processes are idealized processes which represent the limiting case between a spontaneous and a forced process; i.e., for a reversible process, $dS = d\delta Q/T$ (temperature is uniform throughout). A reversible process consists of a passage of the system through a continuous series of states of equilibrium where an infinitesimal change in conditions will produce a change in either direction and, with the original conditions restored, the energy of the system is returned to the original form, location, and amount. Processes are *irreversible* when they involve a dissipation of potential work with increase in entropy, such as occurs with friction, inelasticity, or electrical resistance. All real processes are irreversible and produce some increase in entropy however small.

Generalized forces, among the *intensive properties* of a system, are those independent properties each of which is uniquely involved as the force in some mode of work. Examples are pressure, chemical potential, electric potential, field strength, mechanical stress, and surface tension. Many other intensive properties such as temperature, color, reflectivity, and conductivity are not part of a work function.

Generalized displacements, among the *extensive properties* of a system, are those independent properties each of which is uniquely involved as the displacement in some mode of work. Examples are volume, amount of mass, charge, magnetization, mechanical strain and area, but not entropy and enthalpy.

The Basic Principle of thermodynamics, sometimes called the *zeroth law*, postulates that two systems in equilibrium with a third system are in equilibrium with each other. Equilibrium is completely defined by the specification of the internal energy, the enthalpy, and the generalized displacements.

The First Law of thermodynamics is a rigorous statement of the principle of conservation of energy. In its general form it states that the change in the stored energy of a system is equal to the sum of the heat absorbed by the system from the surroundings and the work done on the system by the surroundings. Work is defined as the product of force and displacement and includes not only work of fluid expansion but chemical, elastic, magnetic, electric, surface, and all other work as well.

The Second Law establishes entropy as a fundamental property of a thermodynamic system. The second law implies that, in the absence of a constraint, the system will select the state of maximum entropy.

The Third Law introduces the concept of absolute entropy. It states that the total entropy of a pure substance approaches zero as the absolute thermodynamic temperature approaches zero. The third law implies further that the partial derivatives with respect to temperature of generalized force and generalized displacement also vanish at absolute zero.

Note that the thermodynamic equations are expressed in consistent units. If heat is expressed in Btu's, work must also be expressed in Btu's. If one of the extensive properties is related to unit mass (or moles) all other extensive properties must be related to unit mass (or moles).

THE GENERAL EQUATIONS

If the basic definitions of the thermodynamic parameters are expressed in terms that apply to all systems, we have a set of general equations of the widest possible utility. These general equations can then be simplified as need be for use in restricted systems and processes. Such simplifications of general equations call attention to the premises and assumptions involved in the specific application.

The general energy or conservation equation which applies to all processes whether reversible or irreversible states that the change in stored energy = $Q + W$

or
$$\Delta E + \Delta \left(\frac{mgZ}{g_c}\right) + \Delta \left(\frac{mV}{2g_c}\right) = Q + W \qquad (4\text{-}1)$$

The thermodynamic potentials or property functions which depend only on the initial and final properties of the system, and hence are independent of the path followed by the process, are defined as

$$H = E + pV \qquad (4\text{-}2)$$

$$A = E - TS \qquad (4\text{-}3)$$

$$G = E - TS + pV \qquad (4\text{-}4)$$

These thermodynamic potentials represent the *maximum potentials for work* in a process, i.e., ΔE for an adiabatic or isentropic process, ΔH for a process at constant pressure, ΔA for an isothermal process, and ΔG for a process at constant temperature and pressure.

The differential equations for reversible processes, which can be integrated only with knowledge of the path followed by the process, are

$$dE = T\,dS - p\,dV + \sum F_k\,dX_k + \sum \mu_j\,dn_j \qquad (4\text{-}5)$$

$$dH = T\,dS + V\,dp + \sum F_k\,dX_k + \sum \mu_j\,dn_j \qquad (4\text{-}6)$$

$$dA = -S\,dT - p\,dV + \sum F_k\,dX_k + \sum \mu_j\,dn_j \qquad (4\text{-}7)$$

$$dG = -S\,dT + V\,dp + \sum F_k\,dX_k + \sum \mu_j\,dn_j \qquad (4\text{-}8)$$

In these equations, the $\mu\,dn$ terms represent the chemical work associated with the transfer of material and the $F\,dX$ terms represent all other kinds of work except expansion or compression. For systems where the composition is constant, and pV work is the only work involved, the $\mu\,dn$ and $F\,dX$ terms disappear, of course.

Equations 4-5 through 4-8 cannot usually be applied to irreversible processes since here the path is macroscopically indeterminate.

The definitive equations for the intensive properties which express the intensive parameters in terms of the extensive parameters are

$$-p = \left(\frac{\partial E}{\partial V}\right)_{S,X,n} = \left(\frac{\partial A}{\partial V}\right)_{T,X,n} \tag{4-9}$$

$$T = \left(\frac{\partial E}{\partial S}\right)_{V,X,n} = \left(\frac{\partial H}{\partial S}\right)_{p,X,n} \tag{4-10}$$

$$F_k = \left(\frac{\partial E}{\partial X_k}\right)_{S,V,n} = \left(\frac{\partial H}{\partial X_k}\right)_{S,p,n} = \left(\frac{\partial A}{\partial X_k}\right)_{T,V,n} = \left(\frac{\partial G}{\partial X_k}\right)_{T,p,n} \tag{4-11}$$

$$\mu_j = \left(\frac{\partial E}{\partial n_j}\right)_{S,V,X,n_i} = \left(\frac{\partial H}{\partial n_j}\right)_{S,p,X,n_i} = \left(\frac{\partial A}{\partial n_j}\right)_{T,V,X,n_i} = \left(\frac{\partial G}{\partial n_j}\right)_{T,p,X,n_i} \tag{4-12}$$

In general, $F_j = F_j(X_1, \ldots, X_k)$ (4-13)

Derived functions for the extensive parameters of volume and entropy are

$$V = \left(\frac{\partial H}{\partial p}\right)_{S,X,n} = \left(\frac{\partial G}{\partial p}\right)_{T,X,n} \tag{4-14}$$

$$-S = \left(\frac{\partial A}{\partial T}\right)_{V,X,n} = \left(\frac{\partial G}{\partial T}\right)_{p,X,n} \tag{4-15}$$

The Maxwell relations, which derive from the mathematical requirement of equality between the mixed partial derivatives of a fundamental relation, may be represented generally by

$$\frac{\partial F_j}{\partial X_k} = \frac{\partial F_k}{\partial X_j}, \quad \frac{\partial X_j}{\partial X_k} = \frac{-\partial F_k}{\partial F_j}, \quad \frac{\partial X_j}{\partial F_k} = \frac{\partial X_k}{\partial F_j} \tag{4-16}$$

where X indicates an extensive parameter, and F indicates an intensive parameter.

By application of the general equalities, given by Eq. 4-16, to any of the fundamental equations, the Maxwell relations may be easily recalled. For example, from Eq. 4-5 where $E = E(S, V, X, n)$, the partial derivatives of F_j with respect to X_k are

$$\left(\frac{\partial T}{\partial V}\right)_{S,X,n} = -\left(\frac{\partial p}{\partial S}\right)_{V,X,n} \qquad \left(\frac{\partial T}{\partial X}\right)_{S,V,n} = \left(\frac{\partial F}{\partial S}\right)_{V,X,n}$$

$$\left(\frac{\partial T}{\partial n}\right)_{S,V,X} = \left(\frac{\partial \mu}{\partial S}\right)_{V,X,n} \qquad -\left(\frac{\partial p}{\partial X}\right)_{S,V,n} = \left(\frac{\partial F}{\partial V}\right)_{S,X,n}$$

$$-\left(\frac{\partial p}{\partial n}\right)_{S,V,X} = \left(\frac{\partial F}{\partial V}\right)_{S,X,n} \qquad \left(\frac{\partial F}{\partial n}\right)_{S,V,X} = \left(\frac{\partial \mu}{\partial X}\right)_{S,V,n}$$

RELATIONS FOR SIMPLE SYSTEMS

Simple systems are defined as those where the composition is constant and PV work is the only work. For such systems, the $F_k\, dX_k$ and $\mu\, dn$ terms disappear from Eqs. 4-5 through 4-8.

Physically Measurable Properties

Isothermal compressibility: $k_T = -\dfrac{1}{V}\left(\dfrac{\partial V}{\partial p}\right)_T = -\dfrac{1}{V}\left(\dfrac{\partial S}{\partial p}\right)_T$ (4-17)

Adiabatic compressibility: $k_S = -\dfrac{1}{V}\left(\dfrac{\partial V}{\partial p}\right)_S$ (4-18)

Coefficient of thermal expansion. $\beta = \dfrac{1}{V}\left(\dfrac{\partial V}{\partial T}\right)_p$ (4-19)

Heat capacity or specific heat:

at constant pressure, $C_p = \left(\dfrac{Q}{dT}\right)_p = \left(\dfrac{\partial H}{\partial T}\right)_p = T\left(\dfrac{\partial S}{\partial T}\right)_p$

$= T\left(\dfrac{\partial V}{\partial T}\right)_p\left(\dfrac{\partial p}{\partial T}\right)_S$ (4-20)

at constant volume, $C_V = \left(\dfrac{Q}{dT}\right)_V = \left(\dfrac{\partial E}{\partial T}\right)_V = T\left(\dfrac{\partial S}{\partial T}\right)_V$

$= -T\left(\dfrac{\partial p}{\partial T}\right)_V\left(\dfrac{\partial V}{\partial T}\right)_S$ (4-21)

$C_p - C_V = T\left(\dfrac{\partial p}{\partial T}\right)_V\left(\dfrac{\partial V}{\partial T}\right)_p = \dfrac{TV\beta^2}{k_T}$ (4-22)

Joule-Thomson coefficient $= \left(\dfrac{\partial T}{\partial p}\right)_H = \dfrac{1}{C_p}\left[T\left(\dfrac{\partial V}{\partial T}\right)_p - V\right]$ (4-23)

Properties at Absolute Zero

With subscripts indicating final absolute temperature (θ), zero absolute temperature (0), and all phase transitions (tr),

$$H_\theta - H_0 = \sum(\Delta H)_{\text{tr}} + \int_0^\theta C_p\, dT \qquad (4\text{-}24)$$

$$S_\theta - S_0 = \sum\left(\dfrac{\Delta H}{T}\right)_{\text{tr}} + \int_0^\theta \dfrac{C_p}{T}\, dT \qquad (4\text{-}25)$$

$$G_\theta = H_\theta - \theta S_\theta \qquad (4\text{-}26)$$

Since S_0 is zero for a pure substance according to the postulates of the Third Law, it is possible in theory to compute the absolute enthalpy, entropy, and

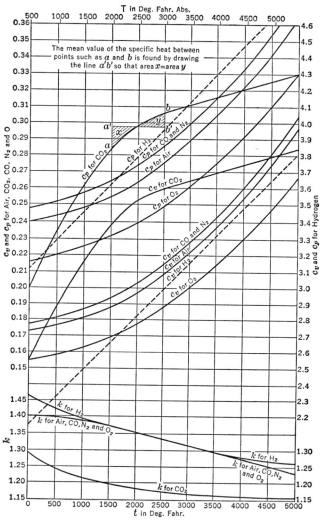

Fig. 4-1 Specific heats of diatomic gases. (Reprinted by permission from Barnard Ellenwood, and Hirshfeld, *Heat Power Engineering*, John Wiley, 1926).

free energy of any substance from knowledge of the heat capacities and heats of transitions. Since all the transitions are seldom known it is more practical to treat S_0 as an empirical constant of integration.

Also as absolute zero is approached, the coefficient of thermal expansion approaches zero as, in general, does the change of any intensive or extensive property with respect to temperature as well as the isothermal change of entropy with respect to any intensive or extensive property.

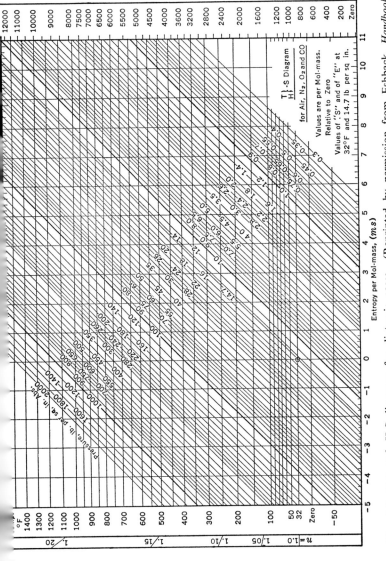

Fig. 4-2 *T-S* and *H-S* diagrams for diatomic gases. (Reprinted by permission from Eshback, *Handbook of Engineering Fundamentals*, John Wiley, 1952.)

Table 4-1 Properties of Saturated Steam, per Pound

Abs Press., lb sq in. p	Temp., F t	Specific Volume		Enthalpy			Entropy			Internal Energy		Abs Press., lb sq in. p
		Sat. Liquid v_f	Sat. Vapor v_g	Sat. Liquid h_f	Evap. h_{fg}	Sat. Vapor h_g	Sat. Liquid s_f	Evap. s_{fg}	Sat. Vapor s_g	Sat. Liquid u_f	Sat. Vapor u_g	
1.0	101.74	0.01614	333.6	69.70	1036.3	1106.0	0.1326	1.8456	1.9782	69.70	1044.3	1.0
2.0	126.08	0.01623	173.73	93.99	1022.2	1116.2	0.1749	1.7451	1.9200	93.98	1051.9	2.0
3.0	141.48	0.01630	118.71	109.37	1013.2	1122.6	0.2008	1.6855	1.8863	109.36	1056.7	3.0
4.0	152.97	0.01636	90.63	120.86	1006.4	1127.3	0.2198	1.6427	1.8625	120.85	1060.2	4.0
5.0	162.24	0.01640	73.52	130.13	1001.0	1131.1	0.2347	1.6094	1.8441	130.12	1063.1	5.0
6.0	170.06	0.01645	61.98	137.96	996.2	1134.2	0.2472	1.5820	1.8292	137.94	1065.4	6.0
7.0	176.85	0.01649	53.64	144.76	992.1	1136.9	0.2581	1.5586	1.8167	144.74	1067.4	7.0
8.0	182.86	0.01653	47.34	150.79	988.5	1139.3	0.2674	1.5383	1.8057	150.77	1069.2	8.0
9.0	188.28	0.01656	42.40	156.22	985.2	1141.4	0.2759	1.5203	1.7962	156.19	1070.8	9.0
10	193.21	0.01659	38.42	161.17	982.1	1143.3	0.2835	1.5041	1.7876	161.14	1072.2	10
14.696	212.00	0.01672	26.80	180.07	970.3	1150.4	0.3120	1.4446	1.7566	180.02	1077.5	14.696
15	213.03	0.01672	26.29	181.11	969.7	1150.8	0.3135	1.4415	1.7549	181.06	1077.8	15
20	227.96	0.01683	20.089	196.16	960.1	1156.3	0.3356	1.3962	1.7319	196.10	1081.9	20
25	240.07	0.01692	16.303	208.42	952.1	1160.6	0.3533	1.3606	1.7139	208.34	1085.1	25
30	250.33	0.01701	13.746	218.82	945.3	1164.1	0.3680	1.3313	1.6993	218.73	1087.8	30
35	259.28	0.01708	11.898	227.91	939.2	1167.1	0.3807	1.3063	1.6870	227.80	1090.1	35
40	267.25	0.01715	10.498	236.03	933.7	1169.7	0.3919	1.2844	1.6763	235.90	1092.0	40
45	274.44	0.01721	9.401	243.36	928.6	1172.0	0.4019	1.2650	1.6669	243.22	1093.7	45
50	281.01	0.01727	8.515	250.09	924.0	1174.1	0.4110	1.2474	1.6585	249.93	1095.3	50
55	287.07	0.01732	7.787	256.30	919.6	1175.9	0.4193	1.2316	1.6509	256.12	1096.7	55
60	292.71	0.01738	7.175	262.09	915.5	1177.6	0.4270	1.2168	1.6438	261.90	1097.9	60
65	297.97	0.01743	6.655	267.50	911.6	1179.1	0.4342	1.2032	1.6374	267.29	1099.1	65
70	302.92	0.01748	6.206	272.61	907.9	1180.6	0.4409	1.1906	1.6315	272.38	1100.2	70
75	307.60	0.01753	5.816	277.43	904.5	1181.9	0.4472	1.1787	1.6259	277.19	1101.2	75
80	312.03	0.01757	5.472	282.02	901.1	1183.1	0.4531	1.1676	1.6207	281.76	1102.1	80
85	316.25	0.01761	5.168	286.39	897.8	1184.2	0.4587	1.1571	1.6158	286.11	1102.9	85
90	320.27	0.01766	4.896	290.56	894.7	1185.3	0.4641	1.1471	1.6112	290.27	1103.7	90
95	324.12	0.01770	4.652	294.56	891.7	1186.2	0.4692	1.1376	1.6068	294.25	1104.5	95
100	327.81	0.01774	4.432	298.40	888.8	1187.2	0.4740	1.1286	1.6026	298.08	1105.2	100
110	334.77	0.01782	4.049	305.66	883.2	1188.9	0.4832	1.1117	1.5948	305.30	1106.5	110
120	341.25	0.01789	3.728	312.44	877.9	1190.4	0.4916	1.0962	1.5878	312.05	1107.6	120
130	347.32	0.01796	3.455	318.81	872.9	1191.7	0.4995	1.0817	1.5812	318.38	1108.6	130
140	353.02	0.01802	3.220	324.82	868.2	1193.0	0.5069	1.0682	1.5751	324.35	1109.6	140
150	358.42	0.01809	3.015	330.51	863.6	1194.1	0.5138	1.0556	1.5694	330.01	1110.5	150
160	363.53	0.01815	2.834	335.93	859.2	1195.1	0.5204	1.0436	1.5640	335.39	1111.2	160
170	368.41	0.01822	2.675	341.09	854.9	1196.0	0.5266	1.0324	1.5590	340.52	1111.9	170
180	373.06	0.01827	2.532	346.03	850.8	1196.9	0.5325	1.0217	1.5542	345.42	1112.5	180
190	377.51	0.01833	2.404	350.79	846.8	1197.6	0.5381	1.0116	1.5497	350.15	1113.1	190
200	381.79	0.01839	2.288	355.36	843.0	1198.4	0.5435	1.0018	1.5453	354.68	1113.7	200
250	400.95	0.01865	1.8438	376.00	825.1	1201.1	0.5675	0.9588	1.5263	375.14	1115.8	250
300	417.33	0.01890	1.5433	393.84	809.0	1202.8	0.5879	0.9225	1.5104	392.79	1117.1	300
350	431.72	0.01913	1.3260	409.69	794.2	1203.9	0.6056	0.8910	1.4966	408.45	1118.0	350
400	444.59	0.0193	1.1613	424.0	780.5	1204.5	0.6214	0.8630	1.4844	422.6	1118.5	400
450	456.28	0.0195	1.0320	437.2	767.4	1204.6	0.6356	0.8378	1.4734	435.5	1118.7	450
500	467.01	0.0197	0.9278	449.4	755.0	1204.4	0.6487	0.8147	1.4634	447.6	1118.6	500
550	476.94	0.0199	0.8424	460.8	743.1	1203.9	0.6608	0.7934	1.4542	458.8	1118.2	550
600	486.21	0.0201	0.7698	471.6	731.6	1203.2	0.6720	0.7734	1.4454	469.4	1117.7	600
650	494.90	0.0203	0.7083	481.8	720.5	1202.3	0.6826	0.7548	1.4374	479.4	1117.1	650
700	503.10	0.0205	0.6554	491.5	709.7	1201.2	0.6925	0.7371	1.4296	488.8	1116.3	700
750	510.86	0.0207	0.6092	500.8	699.2	1200.0	0.7019	0.7204	1.4223	498.0	1115.4	750
800	518.23	0.0209	0.5687	509.7	688.9	1198.6	0.7108	0.7045	1.4153	506.6	1114.4	800
850	525.26	0.0210	0.5327	518.3	678.8	1197.1	0.7194	0.6891	1.4085	515.0	1113.3	850
900	531.98	0.0212	0.5006	526.6	668.8	1195.4	0.7275	0.6744	1.4020	523.1	1112.1	900
950	538.43	0.0214	0.4717	534.6	659.1	1193.7	0.7355	0.6602	1.3957	530.9	1110.8	950
1000	544.61	0.0216	0.4456	542.4	649.4	1191.8	0.7430	0.6467	1.3897	538.4	1109.4	1000
1100	556.31	0.0220	0.4001	557.4	630.4	1187.8	0.7575	0.6205	1.3780	552.9	1106.4	1100
1200	567.22	0.0223	0.3619	571.7	611.7	1183.4	0.7711	0.5956	1.3667	566.7	1103.0	1200
1300	577.46	0.0227	0.3293	585.4	593.2	1178.6	0.7840	0.5719	1.3559	580.0	1099.4	1300
1400	587.10	0.0231	0.3012	598.7	574.7	1173.4	0.7963	0.5491	1.3454	592.7	1095.4	1400
1500	596.23	0.0235	0.2765	611.6	556.3	1167.9	0.8082	0.5269	1.3351	605.1	1091.2	1500
2000	635.82	0.0257	0.1878	671.7	463.4	1135.1	0.8619	0.4230	1.2849	662.2	1065.6	2000
2500	668.13	0.0287	0.1307	730.6	360.5	1091.1	0.9126	0.3197	1.2322	717.3	1030.6	2500
3000	695.36	0.0346	0.0858	802.5	217.8	1020.3	0.9731	0.1885	1.1615	783.4	972.7	3000
3206.2	705.40	0.0503	0.0503	902.7	0	902.7	1.0580	0	1.0580	872.9	872.9	3206.2

Properties of the Ideal Gas

$$pV = nRT \tag{4-27}$$

$$\left(\frac{\partial V}{\partial T}\right)_p = \frac{V}{T}, \quad \left(\frac{\partial p}{\partial T}\right)_V = \frac{p}{T}$$

$$dE = nC_V\, dT \tag{4-28}$$

$$\left(\frac{\partial E}{\partial V}\right)_T = 0, \quad \left(\frac{\partial E}{\partial p}\right)_T = 0, \quad \left(\frac{\partial C_V}{\partial V}\right)_T = 0$$

$$dH = nC_p\, dT \tag{4-29}$$

$$\left(\frac{\partial H}{\partial p}\right)_T = 0, \quad \left(\frac{\partial C_p}{\partial p}\right)_T = 0, \quad \left(\frac{\partial T}{\partial p}\right)_H = 0$$

$$C_p - C_V = R, \quad C_p = \frac{R\gamma}{\gamma - 1} \tag{4-30}$$

$$dS = \frac{nC_p}{T}\, dT - nR\, \frac{dp}{p} \tag{4-31}$$

$$dA = nC_V\, dT - nTC_V\, \frac{dT}{T} - nRT\, \frac{dV}{V} \tag{4-32}$$

$$dG = nC_p\, dT - nTC_p\, \frac{dT}{T} + nRT\, \frac{dp}{p} \tag{4-33}$$

$$k_T = 1/p \tag{4-34}$$

$$k_S = C_V/pC_p \tag{4-35}$$

$$\beta = 1/T \tag{4-36}$$

Table 4-2 Gas Constant, $R = pV/nT$

Energy	Temperature	Mole	R
lb-ft^2/sec^2	°Rankine	lb	4.969×10^4
ft lbf	°Rankine	lb	1544
cu ft atm	°Rankine	lb	0.7302
cu ft (lbf/sq in.)	°Rankine	lb	10.73
Btu	°Rankine	lb	1.987
hp-hr	°Rankine	lb	7.805×10^{-4}
kwhr	°Rankine	lb	5.819×10^{-4}
joule (abs)	°Kelvin	gm	8.314
kg-m^2/sec^2	°Kelvin	kg	8.314×10^3
kgf m	°Kelvin	kg	8.478×10^2
cu cm atm	°Kelvin	gm	82.06
calorie	°Kelvin	gm	1.987

lb = pounds-mass; lbf = pounds-force; kg = kilograms-mass; kgf = kilograms-force; °Rankine = °F + 460; °Kelvin = °C + 273.16.

Other Widely Used Relations

The thermodynamic equation of state:

$$\left(\frac{\partial E}{\partial V}\right)_T = -p + T\left(\frac{\partial p}{\partial T}\right)_V \tag{4-37}$$

The Clapeyron equation for change of state when two phases are in equilibrium:

$$\frac{dp}{dT} = \frac{\Delta H}{T\,\Delta V} \tag{4-38}$$

where ΔH is the enthalpy change and ΔV is the volume change associated with change of state.

The Maxwell relations:

From Eq. 4-5 where $E = E(S, V)$,
$$\left(\frac{\partial T}{\partial V}\right)_S = -\left(\frac{\partial p}{\partial S}\right)_V \tag{4-39}$$

From Eq. 4-6 where $H = H(S, p)$,
$$\left(\frac{\partial T}{\partial p}\right)_S = -\left(\frac{\partial V}{\partial S}\right)_p \tag{4-40}$$

From Eq. 4-7 where $A = A(T, V)$,
$$\left(\frac{\partial S}{\partial V}\right)_T = \left(\frac{\partial p}{\partial T}\right)_V \tag{4-41}$$

From Eq. 4-8 where $G = G(T, p)$,
$$-\left(\frac{\partial S}{\partial p}\right)_T = \left(\frac{\partial V}{\partial T}\right)_p \tag{4-42}$$

Table 4-3, suggested by P. W. Bridgman, gives a convenient means for obtaining any of the first partial derivatives for single-phase, simple systems in which p and T are permissible independent variables. For example, $(\partial V/\partial S)_p$ is found under the column of constant p opposite ∂V as $(\partial V/dT)_p$ and opposite ∂S as nC_p/T,

$$\left(\frac{\partial V}{\partial S}\right)_p = \left(\frac{\partial V}{\partial T}\right)_p \left(\frac{T}{nC_p}\right)$$

but
$$nC_p = T\left(\frac{\partial V}{\partial T}\right)_p \left(\frac{\partial p}{\partial T}\right)_S$$

Hence
$$\left(\frac{\partial V}{\partial S}\right)_p = \left(\frac{\partial T}{\partial p}\right)_S$$

Table 4-3 can be used for systems with other than pV changes by substituting F_j for pressure and X_j for volume.

Table 4-4 may be used in a similar manner for simple systems comprising two coexisting phases in equilibrium, e.g., saturated steam and water.

These tables can be used also to verify functions obtained elsewhere. For example, verify $nC_p/T = (\partial V/\partial T)_p(\partial p/\partial T)_S$. From Table 4-3,

$$\left(\frac{\partial V}{\partial T}\right)_p \left(\frac{\partial p}{\partial T}\right)_S = \frac{\left(\frac{\partial V}{\partial T}\right)_p \left(-\frac{nC_p}{T}\right)}{1\left[-\left(\frac{\partial V}{\partial T}\right)_p\right]} = \frac{nC_p}{T}$$

Table 4-3 Thermodynamic Formulas for Simple Systems, Single Phase

	At Constant			
	T	p	V	S
∂T	0	1	$\left(\dfrac{\partial V}{\partial p}\right)_T$	$-\left(\dfrac{\partial V}{\partial T}\right)_p$
∂p	-1	0	$-\left(\dfrac{\partial V}{\partial T}\right)_p$	$-\dfrac{nC_p}{T}$
∂V	$-\left(\dfrac{\partial V}{\partial p}\right)_T$	$\left(\dfrac{\partial V}{\partial T}\right)_p$	0	$-\dfrac{nC_p}{T}\left(\dfrac{\partial V}{\partial p}\right)_T - \left(\dfrac{\partial V}{\partial T}\right)_p^2$
∂S	$\left(\dfrac{\partial V}{\partial T}\right)_p$	$\dfrac{nC_p}{T}$	$\dfrac{nC_p}{T}\left(\dfrac{\partial V}{\partial p}\right)_T + \left(\dfrac{\partial V}{\partial T}\right)_p^2$	0
∂E	$T\left(\dfrac{\partial V}{\partial T}\right)_p + p\left(\dfrac{\partial V}{\partial T}\right)_T$	$nC_p - p\left(\dfrac{\partial V}{\partial T}\right)_p$	$nC_p\left(\dfrac{\partial V}{\partial p}\right)_T + T\left(\dfrac{\partial V}{\partial T}\right)_p^2$	$\dfrac{pnC_p}{T}\left(\dfrac{\partial V}{\partial p}\right)_T + p\left(\dfrac{\partial V}{\partial T}\right)_p^2$
∂H	$-V + T\left(\dfrac{\partial V}{\partial T}\right)_p$	nC_p	$nC_p\left(\dfrac{\partial V}{\partial p}\right)_T + T\left(\dfrac{\partial V}{\partial T}\right)_p^2 - V\left(\dfrac{\partial V}{\partial T}\right)_p$	$-\dfrac{VnC_p}{T}$
∂A	$p\left(\dfrac{\partial V}{\partial p}\right)_T$	$-S - p\left(\dfrac{\partial V}{\partial T}\right)_p$	$-S\left(\dfrac{\partial V}{\partial p}\right)_T$	$\dfrac{pnC_p}{T}\left(\dfrac{\partial V}{\partial p}\right)_T + p\left(\dfrac{\partial V}{\partial T}\right)_p^2 + S\left(\dfrac{\partial V}{\partial T}\right)_p$
∂G	$-V$	$-S$	$-V\left(\dfrac{\partial V}{\partial Tp}\right)_p - S\left(\dfrac{V\partial}{\partial p}\right)_T$	$-\dfrac{VnC_p}{T} + S\left(\dfrac{\partial V}{\partial T}\right)_p$

176 THERMODYNAMICS

Table 4-4 Thermodynamic Formulas for Simple Systems, Two Phases

At Constant

	T	p	V	S	x
∂T	0	0	-1	$-\dfrac{dp}{dT}$	$-\dfrac{1}{V_1 - V_2}$
∂p	0	0	$-\dfrac{dp}{dT}$	$-\left(\dfrac{dp}{dT}\right)^2$	$-\dfrac{dp}{dT}\left(\dfrac{1}{V_1 - V_2}\right)$
∂V	1	$\dfrac{dp}{dT}$	0	$\dfrac{nC_V}{T}$	$\dfrac{f(x)}{V_1 - V_2}$
∂S	$\dfrac{dp}{dT}$	$\left(\dfrac{dp}{dT}\right)^2$	$-\dfrac{nC_V}{T}$	0	$-\dfrac{1}{V_1 - V_2}\left[\dfrac{nC_V}{T} + \dfrac{dp}{dT}f(x)\right]$
∂E	$T\dfrac{dp}{dT} - p$	$\left(T\dfrac{dp}{dT} - p\right)\dfrac{dp}{dT}$	$-nC_V$	$-p\dfrac{nC_V}{T}$	$-\dfrac{1}{V_1 - V_2}\left[nC_V - \left(p - T\dfrac{dp}{dT}\right)f(x)\right]$
∂H	$T\dfrac{dp}{dT}$	$T\left(\dfrac{dp}{dT}\right)^2$	$-nC_V - V\dfrac{dp}{dT}$	$-V\left(\dfrac{dp}{dT}\right)^2$	$-\dfrac{1}{V_1 - V_2}\left[nC_V + V\dfrac{dp}{dT} + T\dfrac{dp}{dT}f(x)\right]$
∂A	$-p$	$-p\dfrac{dp}{dT}$	S	$-p\dfrac{nC_V}{T} + S\dfrac{dp}{dT}$	$\dfrac{1}{V_1 - V_2}[S + pf(x)]$
∂G	0	0	$S - V\dfrac{dp}{dT}$	$\dfrac{dp}{dT}\left(S - V\dfrac{dp}{dT}\right)$	$\dfrac{S - V\dfrac{dp}{dT}}{V_1 - V_2}$
∂x	$\dfrac{1}{V_1 - V_2}$	$\dfrac{dp}{dT}\dfrac{1}{V_1 - V_2}$	$\dfrac{1}{V_1 - V_2}f(x)$	$\dfrac{nC_A}{T(V_1 - V_2)} + \dfrac{dp}{dT}\left(\dfrac{f(x)}{V_1 - V_2}\right)$	0

x = fraction of system existing as pure phase 1

$$f(x) = x\left[\left(\dfrac{\partial V_1}{\partial T}\right)_p + \left(\dfrac{\partial V_1}{\partial p}\right)_T \dfrac{dp}{dT}\right] + (1-x)\left[\left(\dfrac{\partial V_2}{\partial T}\right)_p + \left(\dfrac{\partial V_2}{\partial p}\right)_T \dfrac{dp}{dT}\right]$$

RELATIONS FOR SIMPLE SYSTEMS

Table 4-5 Nonflow Processes, Simple Systems

Process		General System	Ideal Gas
Constant V:		Reversible	$p/T = $ constant
	ΔE	$\int C_V \, dT$	$\left(\dfrac{1}{\gamma - 1}\right)(V \, \Delta p)$
	ΔS	$\int C_V \dfrac{dT}{T}$	$C_V \log_e \dfrac{p_2}{p_1}$ (C_V constant)
	ΔH	$\int C_V \, dT + V \, \Delta p$	$\left(\dfrac{\gamma}{\gamma - 1}\right)(V \, \Delta p)$
	Q	ΔE	ΔE
	W	0	0
Constant p:		Reversible	$V/T = $ constant
	ΔE	$\int C_p \, dT - p \, \Delta V$	$\left(\dfrac{1}{\gamma - 1}\right)(p \, \Delta V)$
	ΔS	$\int C_p \dfrac{dT}{T}$	$C_p \log_e \dfrac{V_2}{V_1}$ (C_p constant)
	ΔH	$\int C_p \, dT$	$\left(\dfrac{\gamma}{\gamma - 1}\right)(p \, \Delta V)$
	Q	ΔH	ΔH
	W	$-p \, \Delta V$	$-p \, \Delta V$
Constant T:		Reversible, small Δp	$pV = $ constant
	ΔE	$\int (pVk_T - T\beta V) \, dp$	0
	ΔS	$-\int \beta V \, dp$	$nR \log_e (p_1/p_2)$
	ΔH	$T \Delta S + nRT \log_e (f_1/f_2)$	0
	Q	$T \Delta S$	$nRT \log_e (p_1/p_2)$
	W	$-\int p \, dV = \int pVk_T \, dp$	$-Q$

178 THERMODYNAMICS

Table 4-5 (*Continued*)

Process	General System	Ideal Gas
Constant S: (adiabatic)	Reversible	pV^γ = constant C_V constant, γ constant
ΔE	$-\int p\, dV$	$\dfrac{nRT_1}{\gamma - 1}\left[\left(\dfrac{p_2}{p_1}\right)^{(\gamma-1)/\gamma} - 1\right]$
ΔS	0	0
ΔH	$\int V\, dp$	$\gamma\, \Delta E$
Q	0	0
W	$-\int p\, dV$	ΔE
Polytropic: (nonadiabatic)	Internally reversible	pV^k = constant
ΔE	$\dfrac{nRT_1}{(\gamma - 1)}\left[\left(\dfrac{p_2}{p_1}\right)^{(k-1)/k} - 1\right]$	
ΔS	$C_V\left(\dfrac{k - \gamma}{k}\right) \log_e \dfrac{p_2}{p_1}$	$\begin{cases} k < \gamma \text{ in compression,} \\ \qquad\qquad Q \text{ negative} \\ k > \gamma \text{ in expansion,} \\ \qquad\qquad Q \text{ negative} \end{cases}$
ΔH	$\gamma\, \Delta E$	
Q	$\dfrac{n(k - \gamma)RT_1}{(\gamma - 1)(k - 1)}\left[\left(\dfrac{p_2}{p_1}\right)^{(k-1)/k} - 1\right]$	
W	$\dfrac{nRT_1}{(k - 1)}\left[\left(\dfrac{p_2}{p_1}\right)^{(k-1)/k} - 1\right]$	

General Irreversible Process:

$$\Delta E = Q + W, \qquad W \neq -\int p\, dV, \qquad dS > \delta Q/T$$

Table 4-6 Flow Processes, Simple Systems

Steady Flow: Reversible and irreversible

$$Q + W_S + H_1 + Z_1 \frac{g}{g_c} + \frac{u_1^2}{2g_c} = H_2 + Z_2 \frac{g}{g_c} + \frac{u_2^2}{2g_c}$$

where W_S is input of shaft work, Z is elevation, and u is velocity; see Table 1-1 for values of g_c.

Reversible Adiabatic Flow: $Q = 0$, $\Delta S = 0$, $\Delta H = \int V\, dp$

Expansion through enlargement, $\Delta Z = 0$, $W = 0$, $\dfrac{u_2^2 - u_1^2}{2g_c} = -\Delta H$

Expansion through engine or compression through cycle, $\Delta Z = 0$, $W_S = \Delta H$

Throttling Flow: Irreversible, $p_2 < p_1$, $S_2 > S_1$
ΔE, ΔH, Q, and W all zero

THERMODYNAMIC ENGINES

Heat Engines

A heat engine is a device for converting heat energy into work. In the most general form it operates through the input of heat Q at temperature T, the output of work W, and the rejection of heat, Q_S to a receiver or sink at temperature T_S. For a cycle, $\Delta E = 0$ and $Q - Q_S = W$, hence for any such process

$$\text{actual efficiency} = \frac{W}{Q} = \frac{Q - Q_S}{Q} = 1 - \frac{Q_S}{Q} \qquad (4\text{-}43)$$

For a *reversible cycle* operating between the constant temperatures T and T_S, $Q = T\, dS$ and $Q_S = T_S\, dS$, hence for this ideal process

$$\text{Ideal efficiency} = \frac{T - T_S}{T} \qquad (4\text{-}44)$$

Any irreversibilities in the cycle make Q_S greater than $T_S\, dS$ and hence reduce the efficiency.

The Carnot cycle (Fig. 4-3) consists of (1) a reversible isothermal expansion during which heat Q is received from a source at constant temperature T with output of work; (2) a reversible adiabatic expansion during which the temperature of the fluid decreases from T to the receiver temperature T_S with output of work; (3) a reversible isothermal compression during which heat Q_S is rejected to a receiver at temperature T_S, and (4) a reversible adiabatic compression to the original temperature T with input of work. The lower the temperature,

180   THERMODYNAMICS

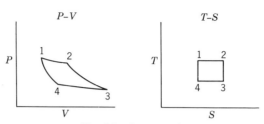

Fig. 4-3 Carnot cycle.

T_S, the greater is the efficiency of the cycle. The Carnot cycle represents the ultimate ideal in efficiency for a heat engine. Since some temperature driving force, however small, is required for heat transfer the Carnot efficiency can never be attained in practice.

The Rankine cycle (Fig. 4-4) is composed of (1) heating, vaporizing, and superheating from T_1 to T_2 at constant pressure p_1, (2) adiabatic (isentropic) expansion from p_1 to p_3 with output of work (3) isothermal condensation at

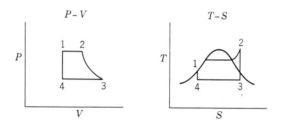

Fig. 4-4 Rankine cycle.

constant p_3 with rejection of heat, and (4) adiabatic compression of condensed fluid from p_4 to p_1 with input of work.

$$Q = H_2 - H_1, \quad Q_s = H_3 - H_4, \quad W_2 = H_2 - H_3, \quad W_4 = H_1 - H_4$$

$$\text{Ideal efficiency} = \frac{W_2 - W_4}{Q} = \frac{(H_2 - H_3) - (H_1 - H_4)}{H_2 - H_1} = 1 - \frac{Q_s}{Q} \quad (4\text{-}46)$$

Since Q_S is greater in the Rankine cycle than in the Carnot cycle, the cycle efficiency is less than unity.

The Otto cycle (Fig. 4-5) in idealized form assumes intake of fuel-air mixture without pressure loss, (1) adiabatic compression of mixture, (2) combustion at constant volume, (3) adiabatic expansion of combustion products with output of work, and (4) discharge at constant volume.

$$\text{Ideal efficiency} = 1 - \frac{1}{(r_c)^{\gamma-1}} \quad (4\text{-}47)$$

where r_c = compression ratio = $V_{\max}/V_{\min}$. For air, $\gamma = 1.4$.

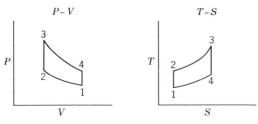

Fig. 4-5 Ideal Otto cycle.

The thermal efficiency of an actual engine operating on the Otto cycle is substantially less than the ideal efficiency because of breathing losses, friction losses, incomplete combustion, and nonadiabatic conditions.

The Diesel cycle (Fig. 4-6) ideally is composed of (1) intake of air only without pressure loss *ab* (2) adiabatic compression of air *bc* (3) liquid fuel injection at constant pressure maintained by vaporization during expansion *cd* (4) adiabatic combustion and expansion with output of work *de* and (5) discharge at end of expansion with negligible pressure loss *ef*.

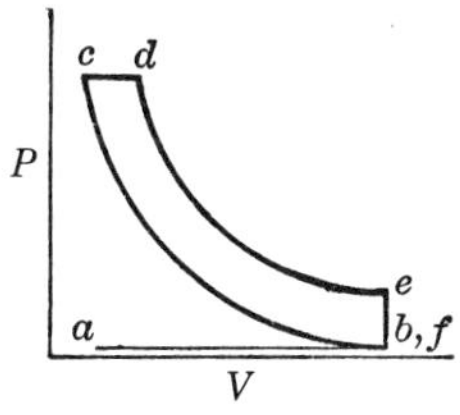

$$\text{Ideal efficiency} = 1 - \frac{(r_c/r_e)^\gamma - 1}{(r_c)^{\gamma-1}\gamma(r_c/r_e) - 1} \quad (4\text{-}48)$$

where r_c = compression ratio = V_{max}/V_{min} and r_e = expansion ratio = V_{max}/V (after injection).

Fig. 4-6 Ideal diesel cycle.

The Diesel cycle is inferior to the Otto cycle in ideal efficiency but superior in actual efficiencies especially at partial loads.

The Brayton or Joule cycle (Fig. 4-7) used in gas turbines consists of (1) adiabatic compression of fuel-air mixtures, (2) combustion at constant pressure, (3) adiabatic expansion, and (4) heat rejection at constant pressure. If the working fluid is assumed to be an ideal gas,

$$\text{Ideal efficiency} = 1 - \left(\frac{p_1}{p_2}\right)^{\gamma-1} \quad (4\text{-}49)$$

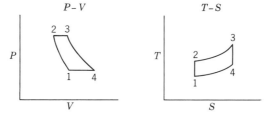

Fig. 4-7 Brayton cycle.

The thermoelectric generator or **thermopile** is based on the observation by Seebeck in 1822 that an electric current is produced in a circuit composed of two dissimilar materials when the two junctions are held at different temperatures, as in the familiar thermocouple. The thermal efficiency is given by Eq. 4-43 if Q is the heat input at the hot junction and Q_S is heat removed at the cold junction. At the optimum temperature of the hot junction,

$$\text{Ideal efficiency} = \frac{T_h - T_c}{T_h} \frac{(1 + M)^{1/2} - 1}{(1 + M)^{1/2} + T_c/T_h} \quad (4\text{-}50)$$

where T_h and T_c are the temperatures of hot and cold junctions and the *figure of merit* of the materials,

$$M = \left(\frac{T_h + T_c}{2}\right)\left[\frac{\alpha_p - \alpha_n}{(k_p \rho_p)^{1/2} + (k_n \rho_n)^{1/2}}\right] \quad (4\text{-}51)$$

where α is the Seebeck coefficient, k is the specific thermal conductivity, and ρ the specific electrical resistivity of the two materials p and n.

By *cascading* or connecting the elements in series, the thermal efficiency is increased; approximately

$$\text{Cascade thermal efficiency} = 1 - \left(\frac{T_c}{T_h}\right)^n \quad (4\text{-}52)$$

where n is the logarithmic mean value of the last term of Eq. 4–50.

Refrigeration Systems

The Carnot refrigeration cycle is obtained by operation of the Carnot cycle in reverse whereby heat Q_2 is removed from the reservoir at T_2 and with the input of work W heat Q_1 is delivered to the surroundings at higher fluid temperature T_1. For this cycle the *coefficient of performance* is the ratio of heat removed to work required, or ideally,

$$\frac{Q_2}{W} = \frac{T_2 \, \Delta S}{\Delta T \, \Delta S} = \frac{T_2}{T_1 - T_2} \quad (4\text{-}53)$$

In the practical vapor compression cycle (Fig. 4-8) the saturated liquid refrigerant at T_1, p_1 passes from a to b through a vaporizing expansion and thence from b to c through heat exchange with the cold reservoir to complete

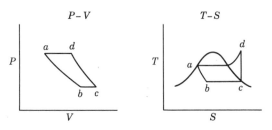

Fig. 4-8 Vapor refrigeration cycle.

the evaporation at $T_3 p_2$. The vaporized refrigerant is then compressed isentropically (in the ideal case) from c to d, and condensed by heat exchange with cooling medium from d to a. The temperature differences that must be maintained in the heat exchangers and the irreversible expansion (a to b) reduce the cycle efficiency.

$$\text{Coefficient of performance} = \frac{H_c - H_a}{H_d - H_c} \qquad (4\text{-}54)$$

The air compression system is like the vapor compression cycle except that the throttling expansion is replaced or partially replaced by an expansion engine which supplies part of the work needed for compression. For the ideal cycle with isentropic expansion and compression,

$$\text{Coefficient of performance} = \frac{T_b}{T_a - T_b} = \frac{T_c}{T_d - T_c} \qquad (4\text{-}55)$$

The cycle efficiency is relatively low and hence this cycle is not used much for refrigeration alone. As an air-conditioning installation for combined heating in winter and cooling in summer it is finding some applications in milder climates.

The vapor absorption system is similar to the vapor compression system except that the power-driven compressor is replaced by an absorption-desorption cycle for which the energy input is heat instead of mechanical energy.

Magnetic refrigeration is based on the coupling between the thermal and magnetic properties of paramagnetic materials whereby an adiabatic decrease in the applied field produces a decrease in temperature. The process, known as *adiabatic demagnetization*, is the preferred method of cooling at temperatures below 1°K. In this temperature range the cooling ratio is given approximately by

$$\frac{T_{\text{final}}}{T_{\text{initial}}} = \frac{C}{H_e} \qquad (4\text{-}56)$$

where H_e is the initial magnetic intensity and C is a constant characteristic of the paramagnetic material. The coefficient of performance is very, very small.

Thermoelectric refrigeration is based on the observation by Peltier in 1834 that heat is observed when a current flows in the proper direction across the junction of two dissimilar metals. Heat is thus transported from the cold junction to the hot junction. The heat transported is proportional to the current and the resistance losses are proportional to the square of the current, hence there is an optimum current. At the optimum current, the ideal coefficient of performance is

$$\frac{T_c}{T_h - T_c} \frac{(1+M)^{1/2} - T_h/T_c}{(1+M)^{1/2} + 1} \qquad (4\text{-}57)$$

where the symbols have the same meaning as in Eqs. 4-50 and 4-51.

Compressors

The ideal compressor cycle is composed of (1) constant pressure expansion at the intake, (2) isothermal or adiabatic compression, (3) constant pressure discharge, and (4) pressure decrease at zero volume. The cycle work then is

$$W_{\text{cycle}} = -p_1 V_1 - \int p\, dV + p_2 V_2 + 0 = \int V\, dp \qquad (4\text{-}58)$$

The total flow work, reversible or irreversible, is

$$W_{\text{flow}} = W_{\text{kinetic}} + W_{\text{cycle}} = \frac{u_2^2 - u_1^2}{2g_c} + \Delta H - Q \qquad (4\text{-}59)$$

Any heat abstracted during compression decreases both the final enthalpy and the shaft work required.

In the reversible isothermal compression of any gas, work is related to the Gibbs free energy G, and the fugacity f:

$$W_{\text{cycle}} = \int V\, dp = \Delta G = \Delta H - T\Delta S \qquad (4\text{-}60)$$

$$= nRT \log_e \frac{f_1}{f_2} \qquad (4\text{-}61)$$

For an ideal gas, $f = p$. For real gases f may be obtained from Fig. 4-10.

For single-stage, adiabatic compression of any gas, work is related to the isentropic change in enthalpy:

$$W_{\text{cycle}} = \int V\, dp = (H_2 - H_1)_S \qquad (4\text{-}62)$$

For a "dry" ideal gas,

$$W_{\text{cycle}} = \frac{\gamma}{\gamma - 1} nRT_1 \left[\left(\frac{p_2}{p_1}\right)^{(\gamma - 1)/\gamma} - 1 \right] \qquad (4\text{-}63)$$

and

$$\frac{T_2}{T_1} = \left(\frac{p_2}{p_1}\right)^{(\gamma - 1)/\gamma} \qquad (4\text{-}64)$$

For ideal multi-stage, adiabatic compression with complete recooling between stages, the minimum work for an ideal gas, is

$$W_{\text{cycle}} = \frac{N\gamma}{\gamma - 1} p_1 V_1 \left[\left(\frac{p_N}{p_1}\right)^{(\gamma - 1)/N\gamma} - 1 \right] \qquad (4\text{-}65)$$

where N is the number of stages and the intermediate pressures are selected as

$$p_2 = (p_N p_1^{N-1})^{1/N}, \quad p_3 = (p_N^2 p_1^{N-2})^{1/N}$$
$$p_4 = (p_N^3 p_1^{N-3})^{1/N}, \ldots P_{N-1} = (p_N^{N-1} p_1)^{1/N} \qquad (4\text{-}66)$$

PHASE EQUILIBRIA

Satisfactory empirical equations for the vapor pressure of liquids take the form

$$\log p = \frac{A}{T} + B \log T + C \tag{4-67}$$

where the constants A, B, and C are fitted to data over a range of temperatures.

The Clausius-Clapeyron equation relates the vapor pressure to the enthalpy of vaporization or sublimation.

$$\log_e p = -\frac{\Delta H}{RT} + \text{const.} \tag{4-68}$$

Henry's law represents the solubility of a gas in a liquid at small concentrations.

$$py = h_T x \tag{4-69}$$

where y and x are the mole fractions of the gas in the vapor and liquid phases, and h_T is the Henry's law coefficient at temperature T.

Raoult's law, applicable only to ideal vapors and liquids, equates the equilibrium partial pressures of a solution component in the coexisting phases.

$$py_1 = p_i^* x_i \tag{4-70}$$

where p_i^* is the vapor pressure of pure component i.

The vapor-liquid equilibrium ratio,

$$K_i = \frac{y}{x_i} = \left(\frac{f_2^\circ}{f_i^\circ}\right)_i \tag{4-71}$$

For ideal solutions, f_1° is the fugacity of component i at the temperature and pressure of the system, and f_2° is approximately the fugacity of the pure component at its vapor pressure for the temperature of the system.

The volatility ratio between two components,

$$\alpha = \frac{k_i}{k_j} = \frac{y_i x_j}{x_i y_j} \tag{4-72}$$

For regular binary systems where α is a constant,

$$y_i = \frac{\alpha x_i}{1 + x_i(\alpha - 1)} \tag{4-73}$$

For nonideal solutions, including most liquid-liquid systems, actual data or highly involved thermodynamic computations are required to obtain the compositions of the coexisting phases.

CHEMICAL REACTIONS

Chemical Symbols

a_i	activity of species i
A_k	symbol of chemical element k
C_i	symbol of chemical component i
$\Delta G°$	increment of free-energy at standard state
$\Delta H°$	increment of enthalpy at standard state
K	equilibrium constant
n_i	number of moles of species i
$n_i°$	original number of moles of species i
ΔN	conversion
$\mu°$	chemical potential at standard state
ν	stoichiometric coefficient

Chemical Equations

Stoichiometry. In a chemical reaction the number of moles in the system may change as a result of transfers between reactants and products. The total of each atomic species remains constant, however, except in a nuclear reaction. The relationships among the mole numbers are expressed by chemical equations of the general type

$$\nu_1 C_1 + \nu_2 C_2 \leftrightarrows \nu_s C_s + \nu_{s+1} C_{s+1} \tag{4-74}$$

where C_i is one mole of chemical compound i. By convention the *stoichiometric coefficients*, ν, are positive for the products on the right and negative for the reactants on the left. Material balance requires that

$$\sum \nu_i C_i = 0 \tag{4-75}$$

$$dn_i = \nu_i \, dN \tag{4-76}$$

and

$$n_i = n_i° + \nu_i \, \Delta N \tag{4-77}$$

where i refers to each of the chemical formulas or species taking part in the reaction.

For the reaction by which one mole of compound C_i is formed from its chemical elements

$$C_i + \sum \nu_k A_k = 0 \tag{4-78}$$

where A_k is one mole of element k, ν_k is the stoichiometric coefficient of k, and the summation is taken over all the elements in compound C_i. As reactants the ν_k's are negative.

Hence for any reaction not involving nuclear transmutation,

$$\sum \nu_i (C + \sum \nu_k A_k)_i = 0 \tag{4-79}$$

Table 4-7 1961 Atomic Weights, Basis C^{12}

	Symbol	Atomic No.	Atomic Weight		Symbol	Atomic No.	Atomic Weight
Actinium	Ac	89	(227)	Mercury	Hg	80	200.59
Aluminum	Al	13	26.9815	Molybdenum	Mo	42	95.94
Americium	Am	95	(243)	Neodymium	Nd	60	144.24
Antimony	Sb	51	121.75	Neon	Ne	10	20.183
Argon	Ar	18	39.948	Neptunium	Np	93	(237)
Arsenic	As	33	74.9216	Nickel	Ni	28	58.71
Astatine	At	85	(210)	Niobium	Nb	41	92.906
Barium	Ba	56	137.34	Nitrogen	N	7	14.0067
Berkelium	Bk	97	(249)	Nobelium	No	102	(253)
Beryllium	Be	4	9.0122	Osmium	Os	76	190.2
Bismuth	Bi	83	208.980	Oxygen	O	8	15.9994
Boron	B	5	10.811	Palladium	Pd	46	106.4
Bromine	Br	35	79.909	Phosphorus	P	15	30.9738
Cadmium	Cd	48	112.40	Platinum	Pt	78	195.09
Calcium	Ca	20	40.08	Plutonium	Pu	94	(244)
Californium	Cf	98	(249)	Polonium	Po	84	(210)
Carbon	C	6	12.01115	Potassium	K	19	39.102
Cerium	Ce	58	140.12	Praseodymium	Pr	59	140.907
Cesium	Cs	55	132.905	Promethium	Pm	61	(145)
Chlorine	Cl	17	35.453	Protactinium	Pa	91	(231)
Chromium	Cr	24	51.996	Radium	Ra	88	(226)
Cobalt	Co	27	58.9332	Radon	Rn	86	(222)
Copper	Cu	29	63.54	Rhenium	Re	75	186.2
Curium	Cm	96	(245)	Rhodium	Rh	45	102.905
Dysprosium	Dy	66	162.50	Rubidium	Rb	37	85.47
Einsteinium	Es	99	(254)	Ruthenium	Ru	44	101.07
Erbium	Er	68	167.26	Samarium	Sm	62	150.35
Europium	Eu	63	151.96	Scandium	Sc	21	44.956
Fermium	Fm	100	(252)	Selenium	Se	34	78.96
Fluorine	F	9	18.9984	Silicon	Si	14	28.086
Francium	Fr	87	(223)	Silver	Ag	47	107.870
Gadolinium	Gd	64	157.25	Sodium	Na	11	22.9898
Gallium	Ga	31	69.72	Strontium	Sr	38	87.62
Germanium	Ge	32	72.59	Sulfur	S	16	32.064
Gold	Au	79	196.967	Tantalum	Ta	73	180.948
Hafnium	Hf	72	178.49	Technetium	Tc	43	(99)
Helium	He	2	4.0026	Tellurium	Te	52	127.60
Holmium	Ho	67	164.930	Terbium	Tb	65	158.924
Hydrogen	H	1	1.00797	Thallium	Tl	81	204.37
Indium	In	49	114.82	Thorium	Th	90	232.038
Iodine	I	53	126.9044	Thulium	Tm	69	168.934
Iridium	Ir	77	192.2	Tin	Sn	50	118.69
Iron	Fe	26	55.847	Titanium	Ti	22	47.90
Krypton	Kr	36	83.80	Tungsten	W	74	183.85
Lanthanum	La	57	138.91	Uranium	U	92	238.03
Lead	Pb	82	207.19	Vanadium	V	23	50.942
Lithium	Li	3	6.939	Xenon	Xe	54	131.30
Lutetium	Lu	71	174.97	Ytterbium	Yb	70	173.04
Magnesium	Mg	12	24.312	Yttrium	Y	39	88.905
Manganese	Mn	25	54.9380	Zinc	Zn	30	65.37
Mendelevium	Md	101	(256)	Zirconium	Zr	40	91.22

Value in parenthesis denotes isotope of longest half-life.

If X is any extensive property of the system, and X_i is the corresponding molal property of pure i, for the reaction

$$\Delta X = \sum v_i X_i \tag{4-80}$$

In simultaneous or successive reactions, where the numerical subscripts refer to the different reactions,

$$dn_i = (dn_i)_1 + (dn_i)_2 + \cdots \tag{4-81}$$

and

$$n_i = n_i^\circ + (v_i \Delta N)_1 + (v_i \Delta N)_2 + \cdots \tag{4-82}$$

Extensive properties that are state functions, such as H, G, S, are additive over successive reactions carried out at constant temperature and pressure, according to the principle known as *Hess's law*:

$$\Delta X = \sum (v_i X_i)_1 + \sum (v_i X_i)_2 + \cdots \tag{4-83}$$

Thus ΔX for a reaction may be obtained from the summation of a series of simpler or better-known reactions that give the same overall chemical balance. Thus heat of a reaction may be obtained from heats of combustion of the compounds involved in the reaction.

Standard heats of reaction and formation. The principles expressed in Eqs. 4-79, Eq. 4-80, and Eq. 4-83 make it possible to obtain the increments of enthalpy of a reaction from the corresponding values for formation from the elements of the compounds taking part in the reaction.

$$\Delta H^\circ = \sum v_i (H_i^\circ)_f \tag{4-84}$$

where the superscript refers to an arbitrary standard state and subscript f refers to formation from the elements. Chemical handbooks give tabulations of heat, entropy, and free-energy of formation for a wide range of compounds usually at the standard state of 25°C and 1 atm pressure. In using these tables, it should be noted that American chemists employ the confusing convention that ΔH is *positive for endothermic* and *negative for exothermic* reactions.

Conversion from the temperature of the standard state to the actual temperature of reaction is given by

$$\left(\frac{\partial \Delta H}{\partial T} \right)_p = \sum v_i (C_p)_i = \Delta C_p \tag{4-85}$$

where C_p is the heat capacity per mole at constant pressure, which must be known as a function of temperature. An expansion of ΔC_p as a power series in T is often used, where the coefficients are obtained by summation of the coefficients of the individual $(C_p)_i$.

Equilibrium Constants

The activity of a chemical species is defined in terms of its chemical potential at a chosen temperature and pressure by,

$$\mu_i = \mu_i^\circ + RT \log_e a_i \tag{4-86}$$

where μ_i° is an arbitrary reference value chosen so as to be a function of T and p only.

The condition for chemical equilibrium in a closed system then becomes

$$\Delta G° + \sum v_i RT \log_e a_i = 0 \quad (4\text{-}87)$$

where
$$\Delta G° = \sum v_i \mu_i°$$

The increment of standard free-energy for the reaction $\Delta G°$, in terms of tabulated values of the standard free-energies of formation from the elements ΔG_f, is given by

$$\Delta G° = \sum v_i (\Delta G_f°)_i \quad (4\text{-}88)$$

The equilibrium constant for the transfer in a closed system of mole mass between chemical species or phases is defined by

$$\log_e K = -\Delta G°/RT \quad (4\text{-}89)$$

and in terms of activities from Eq. 4-87 by

$$\log_e K = \sum v_i \log_e a_i \quad (4\text{-}90)$$

For ideal gases $a_i = px_i$, for ideal liquids $a_i = x_i$, and for real gases (approximately) $a_i = f_i x_i$ where f_i is the fugacity of pure component i at the same temperature and total pressure as the mixture.

The composition of the equilibrium mixture is computed from Eq. 4-82, Eq. 4-89, and Eq. 4-90 since

$$x_i = \frac{n_i}{n} = \frac{n_i° + v_i \Delta N}{n° + \Delta N \sum v_i} \quad (4\text{-}91)$$

The temperature dependence of K is given by

$$\left(\frac{d \log_e K}{dT}\right)_p = \frac{\Delta H}{RT^2} \quad (4\text{-}92)$$

Conversion from the standard state (formation at 298°K and 1 atm) to the actual temperature of reaction is obtained by substitution of Eq. 4-85 and integration to give an equation of the form

$$\log_e K = \frac{-\Delta H_o}{RT} + \frac{a \log_e T}{R} + \frac{bT}{2R} + \frac{cT^2}{6R} + \text{const.} \quad (4\text{-}93)$$

where ΔH_o is an empirical constant determined from an observed value of K. If the heat capacity of the products equals that of the initial reactants, ΔH is constant and

$$\log_e K = \frac{-\Delta H}{RT} + \text{const.} \quad (4\text{-}94)$$

and
$$\Delta G_T = \Delta H_T - T \Delta S_T \quad (4\text{-}95)$$

The pressure dependence of K for liquids and gases is given by

$$\left(\frac{d \log_e K}{dp}\right)_T = -\frac{\Delta V°}{RT} = \frac{-\sum v_i V_i°}{RT} \quad \text{and} \quad \left(\frac{d \log_e K}{dp}\right)_T = 0 \quad (4\text{-}96)$$

respectively, where $V_i°$ is the standard partial molal volume of component i at T and p.

Additivity of Reactions. It is often convenient to consider an overall reaction as the sum of two or more intermediate reactions, thus

$$\sum (v_i C_i)_1 + \sum (v_i C_i)_2 = \sum [(B_1 v_i)_1 + (B_2 v_i)_2] C_i = 0 \tag{4-97}$$

where B_1 and B_2 are the coefficients needed to balance the equations.

The equilibrium constant, K_3, for the overall reaction is given by

$$\log_e K_3 = B_1 \log_e K_1 + B_2 \log_e K_2 \tag{4-98}$$

Tables of K for formation from the elements may thus be applied to other reactions.

GENERALIZED THERMODYNAMIC PROPERTIES

Various proposals have been made to modify the ideal gas law to obtain an equation of state that describes the behavior of real gases. In the equations below, V is the volume of one mole of gas and subscript c refers to values at the critical point.

Van der Waal's equation:

$$\left(p + \frac{a}{V^2}\right)(V - b) = RT \tag{4-99}$$

$$a = 3 p_c V_c^2, \quad b = \frac{V_c}{3}$$

Redlich-Kwong equation:

$$p = \frac{RT}{V - b} - \frac{a}{T^{1/2} V(V + b)} \tag{4-100}$$

$$a = 0.4278 \frac{R^2 T_c^{2.5}}{p_c}, \quad b = 0.0867 \frac{RT_c}{p_c}$$

The Reduced Equation of State. The pVT relations of any gas may be expressed as

$$pV = znRT \tag{4-101}$$

in which z is the *compressibility factor* or the ratio of the volume of a real gas to that of an ideal gas. If the parameters p and T are replaced by their ratios to the corresponding critical values, called *reduced properties*, all gases are in *corresponding states* and can be approximately represented by a general chart of z versus pR on lines of constant T_R such as Fig. 4-9.

Most individual gases deviate from such a general graph by 2 to 7%. Hydrogen and helium, however, deviate so much that the reduced properties for them are computed by adding 8 atm to the critical pressure and 8°K to the critical temperature.

A generalized fugacity chart (Fig. 4-10) can be constructed from the compressibility chart by integration, along each isotherm, of the equation

$$\log_e \frac{f}{p} = \int_0^{p_R} (z - 1) d \log_e p_R \tag{4-102}$$

GENERALIZED THERMODYNAMIC PROPERTIES

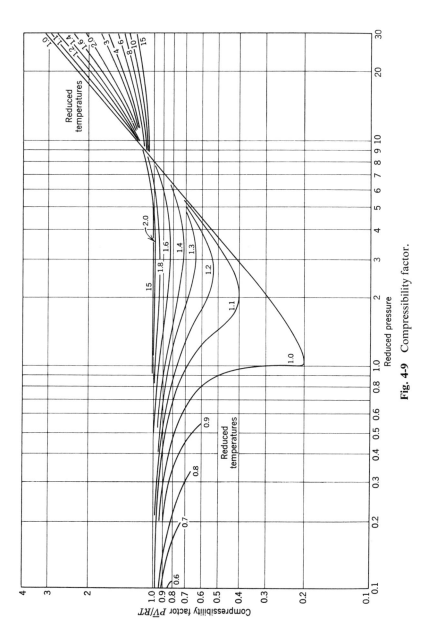

Fig. 4-9 Compressibility factor.

192 THERMODYNAMICS

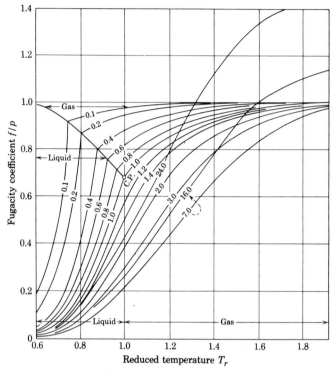

Fig. 4-10 Fugacity of gases and liquids. (Reprinted by permission from Hougen, Watson, and Ragatz, *Chemical Process Principles*, John Wiley, 1959.)

REFERENCES

General Texts

Callen, H. B., *Thermodynamics*, John Wiley, New York, 1960.
Guggenheim E. A., *Thermodynamics*, Interscience, New York, fourth edition, 1960.
Hougen, O. A., K. M. Watson, and R. A. Ragatz, *Chemical Process Principles*, Part II, John Wiley, New York, second edition, 1959.
Lewis, G. N., M. Randall, K. A. Pitzer, and L. Brewer, *Thermodynamics*, McGraw-Hill, New York, second edition, 1961.
Prigogine, I., *Introduction to Thermodynamics of Irreversible Processes*, John Wiley, New York, second edition, 1962.
Zemansky, M. W., *Heat and Thermodynamics*, McGraw-Hill, New York, fourth edition, 1957.

Thermodynamic Data

Handbook of Chemistry and Physics, Chemical Rubber Pub., Cleveland, Ohio, 44th edition, 1963.
Chemical Engineers Handbook, McGraw-Hill, New York, fourth edition, 1963.

REFERENCES

Selected Values of Properties of Hydrocarbons and Related Compounds, API Research Project 44, Carnegie Institute of Technology, Pittsburgh, 1952.

Selected Values of Chemical Thermodynamic Properties, Nat'l Bur. Standards, cir. 500, Washington, D.C.

Hougen, O. A., K. M. Watson, and R. A. Ragatz, *Chemical Process Principles Charts*, John Wiley, New York, 1960.

Keenan, J. H., and F. G. Keyes, *Thermodynamic Properties of Steam*, John Wiley, New York, 1936.

Stull, D. B., and G. C. Sinke, *Thermodynamic Properties of the Elements*, American Chemical Society, Washington, D.C., 1956.

SECTION 5

HEAT TRANSFER

Principal Symbols

G	mass velocity		***Subscripts***
h	coefficient of heat transfer	a	arithmetic mean
k	thermal conductivity	b	in bulk fluid
L	characteristic dimension	f	at film or flame
q	rate of heat flow	g	gas
t	time	i	inner
T	temperature	ln	logarithmic mean
U	overall coefficient	m	length mean
α	absorptivity	o	outer
ϵ	emissivity	s	solid or surface
σ	Stefan-Boltzmann constant	w	at wall

Dimensionless Numbers

N_{Fo} = Fourier number, $\dfrac{kt}{c_p \rho L^2}$ N_{Pe} = Peclet number, $\dfrac{LGc_p}{k}$

N_{Gz} = Graetz number, $\dfrac{wc_p}{kL}$ N_{Pr} = Prandtl number, $\dfrac{c_p \mu}{k}$

N_{Gr} = Grashof number, $\dfrac{L^3 \rho^2 g \beta \, \Delta T}{\mu^2}$ N_{Re} = Reynolds number, $\dfrac{LG}{\mu}$

N_{Nu} = Nusselt number, $\dfrac{hL}{k}$ N_{St} = Stanton number, $\dfrac{h}{c_p G}$

CONDUCTION

Thermal Conductivity

The instantaneous rate of heat conduction in the x direction q is given by the Fourier equation

$$q = -kA \frac{dT}{dx} \tag{5-1}$$

where k is the thermal conductivity, A the area of the path normal to the direction of heat flow, and $-dT/dx$ is the temperature gradient in the x direction.

Thermal conductivity is thus defined as the heat energy flowing per unit time per unit normal cross section, per unit temperature gradient. In English units,

$$k = \frac{\text{(Btu)/(hr)(sq ft)}}{\text{°F/ft}} = \frac{\text{Btu}}{\text{(hr)(ft)(°F)}} \tag{5-2}$$

Thermal conductivities vary widely with the composition of the material so that even with homogeneous solids, minor impurities often have a significant effect. With amorphous materials, the conductivity is largely influenced by the void spaces so that it becomes a function of the bulk density. At high temperatures, apparent conductivities are higher since radiation as well as conduction contributes to the internal heat transfer.

One-Dimensional Steady State Conduction

For practical applications, Eq. 5-1 must be integrated with the result expressed as:

$$q = k_{\text{avg}} A_{\text{avg}} \Delta T / x \tag{5-3}$$

Over the temperature ranges commonly encountered, k is nearly linear with temperature, hence k_{avg} may be taken as the arithmetic mean average value. When k versus T is nonlinear, k_{avg} is computed by

$$k_{\text{avg}} = \frac{1}{T_1 - T_2} \int_{T_1}^{T_2} k \, dT \tag{5-4}$$

A_{avg} is computed by a relationship corresponding to the geometry of the path. For long hollow cylinders where $A = 2\pi rL$,

$$A_{\text{avg}} = \frac{A_2 - A_1}{\log_e (A_2/A_1)} \tag{5-5}$$

For hollow spheres where $A = 4\pi r^2$,

$$A_{\text{avg}} = (A_1 A_2)^{1/2} \tag{5-6}$$

For a hollow rectangular parallelopiped with uniform thick walls, approximately,

$$A_{\text{avg}} = A + 0.54 \times \sum L + 1.2x^2$$

where A is the inside area, ΣL is the sum of the lengths of the twelve inside edges, and x, the wall thickness, is between one-half and five times the length of an inside edge.

For conduction through bodies in series,

$$q = \frac{\Delta T}{R} = \frac{\Delta T_1 + \Delta T_2 + \cdots + \Delta T_n}{R_1 + R_2 + \cdots + R_n} = \frac{\Delta T_{\text{overall}}}{\sum R_i + \sum R_j} \qquad (5\text{-}7)$$

where the thermal resistance of an individual body is $(X/kA)_i$ and a contact or interstitial resistance R_j is $(1/h_c A)_j$.

With uniform internal generation of heat, as by electrical heating of a solid body, and dissipation at only one surface (the other being adiabatic), the equation to be solved is $d^2T/dx^2 = -Q/k$ where Q is the heat input per unit volume. The result may be expressed in terms of q and the difference in temperature between the two surfaces.

For a flat plate:

$$q = 2k_a A(T_1 - T_2)/x \qquad (5\text{-}8)$$

Equations (5-9)–(5-11), where L is length and the subscript i refers to the center or axis, apply for other shapes.

Rod:

$$q_o = k_{\text{avg}} 4\pi L(T_i - T_o) \qquad (5\text{-}9)$$

Hollow cylinder, adiabatic outer surface:

$$q_2 = k_{\text{avg}} \frac{2\pi L(T_2 - T_1)}{\dfrac{r_2^2}{r_2^2 - r_1^2} \log_e\left(\dfrac{r_2}{r_1} - \dfrac{1}{2}\right)} \qquad (5\text{-}10)$$

Solid sphere:

$$q_o = k_{\text{avg}} 8\pi r_2 (T_i - T_o) \qquad (5\text{-}11)$$

For situations where the boundary surfaces are neither parallel nor concentric the solutions are complex. Schneider discusses techniques for solving such problems.

Unsteady Conduction

Solution of problems in unsteady or transient conduction requires the integration of the general differential equation

$$\frac{\partial T}{\partial t} = \frac{1}{\rho c_p}\left[\frac{\partial}{\partial x}\left(k_x \frac{\partial T}{\partial x}\right) + \frac{\partial}{\partial y}\left(k_y \frac{\partial T}{\partial y}\right) + \frac{\partial}{\partial z}\left(k_z \frac{\partial T}{\partial z}\right)\right] \qquad (5\text{-}12)$$

Solutions for a number of shapes and simplified boundary conditions have been worked out and are reviewed by Jakob and Schneider. In many cases Eq. 5-12 is nonlinear and can be solved in practice only by a complex computer program or an electrical analog or replaced by actual experiment.

CONVECTION

Film heat transfer coefficient or unit surface conductance h is defined by

$$q = hA \, \Delta T \tag{5-13}$$

and is usually expressed as Btu/hr-ft^2-°F.

Temperature difference or potential ΔT is the difference between the bulk temperature of a fluid and the temperature of a surface. The fluid temperature changes along a heat transfer surface in the direction of flow, hence the proper mean temperature difference to use in Eq. 5-13 is

$$\text{Length mean} = \Delta T_m = \int \frac{d_q}{d(hA)} \tag{5-14}$$

When the specific heat and heat transfer coefficient are constant and the operation is steady without external heat losses, this length mean temperature difference becomes

$$\text{Logarithmic mean} = \Delta T_{\ln} = \frac{\Delta T_1 - \Delta T_2}{\log_e (\Delta T_1/\Delta T_2)} \tag{5-15}$$

An approximation valid when $\Delta T_1 - \Delta T_2$ is small compared with ΔT_1 or ΔT_2 is

$$\text{Arithmetic mean} = \Delta T_a = \frac{\Delta T_1 - \Delta T_2}{2} \tag{5-16}$$

Table 5-1 Heat Transfer by Natural Convection

$$\frac{hL}{k} = C(N_{\text{Gr}} N_{\text{Pr}})^n \qquad N_{\text{Pr}} > 0.6$$

	Range of $N_{\text{Gr}} \cdot N_{\text{Pr}}$	C	n	Equation for Air
Vertical plane or cylinder where L = height:				
Laminar	10^4 to 10^9	0.59	$\tfrac{1}{4}$	$h = 0.29(\Delta T/L)^{1/4}$
Turbulent	10^9 to 10^{12}	0.13	$\tfrac{1}{3}$	$h = 0.19(\Delta T)^{1/3}$
Single horizontal cylinder where L = outside diameter	10^3 to 10^9	0.53	$\tfrac{1}{4}$	$h = 0.27(\Delta T/D_0)^{1/4}$
Horizontal plates, heated upward or cooled downward				
Laminar	10^5 to 2×10^7	0.54	$\tfrac{1}{4}$	$h = 0.27(\Delta T/L)^{1/4}$
Turbulent	2×10^7 to 3×10^{10}	0.14	$\tfrac{1}{3}$	$h = 0.22(\Delta T)^{1/3}$
Horizontal plates, heated downward or cooled upward	3×10^5 to 3×10^{10}	0.27	$\tfrac{1}{4}$	$h = 0.12(\Delta T/L)^{1/4}$

Natural Convection

Dimensionless equations for natural convection when the induced flows are laminar give the Nusselt number as a function of the product of the Grashof and Prandtl numbers, often called the Rayleigh number. Over selected ranges of the Rayleigh and Prandtl numbers this function may be represented as

$$N_{Nu} = C(N_{Gr}N_{Pr})^n \qquad (5\text{-}17)$$

in which the fluid properties ρ, β, μ, c_p, and k are evaluated at the average temperature of the film.

Table 5-1 gives values of C and n as well as simplified equations for air at ordinary atmospheric conditions.

Forced Convection

For flow parallel to a flat plate, with laminar boundary layer,

$$N_{St}(N_{Pr})^{2/3} = \tfrac{2}{3}(N_{Re})^{-1/2} \qquad (5\text{-}18)$$

For $N_{Re} > 5 \times 10^5$ the constant in Eq. 5-18 becomes 0.036 and the exponent on N_{Re} becomes $-\tfrac{1}{5}$.

For laminar flow inside horizontal tubes at moderate ΔT and small diameter,

$$\frac{hD}{k_b}\left(\frac{\mu_w}{\mu_b}\right)^{0.14} = 1.86\left[(N_{Re})_b(N_{Pr})_b\frac{D}{L}\right]^{1/3} \qquad (5\text{-}19)$$

For turbulent flow inside tubes at moderate ΔT,

$$\frac{hD}{k_b} = 0.023(N_{Re})_b^{0.8}(N_{Pr})_b^{1/3} \qquad (5\text{-}20)$$

The right-hand side of Eq. 5-20 may be multiplied by $(1 + D/L)^{0.7}$ to correct for short tubes or by $(1 + 3.5D/D_c)$ to correct for curved tubes or coils where D_c is the diameter of the coil.

For gas heaters or coolers, a simplified design equation is

$$\frac{T_2 - T_1}{\Delta T_{\ln}} = \frac{0.0576(L/D)}{(DG)^{0.2}} \qquad (5\text{-}21)$$

in which the units are inches for L and D, lb/sec-ft² for G, and °F for T.

For flow outside tubes the general equation is

$$\frac{h_m D_0}{k_f} = B_1 + B_2(N_{Re})^n(N_{Pr})_f^m \qquad (5\text{-}22)$$

Table 5-2 gives values of the constants in Eq. 5-22 for various conditions.

CONVECTION   199

Table 5.2   Heat Transfer outside Tubes

$$\frac{h_m D_0}{k_f} = B_1 + B_2(N_{Re})^n(N_{Pr})_f^m \qquad N_{Re} = \frac{D_0 G_{max}}{\mu_f}$$

	N_{Re} Range	B_1	B_2	n	m
Flow normal to single cylinder					
Air	1 to 1000	0.32	0.43	0.52	0
	1000 to 50,000	0	0.24	0.60	0
Liquids	1 to 300	0.35	0.56	.52	0
Fluids in general	>100	0	0.38	.56	0.3
Unbaffled banks of tubes					
Staggered	2000 to 32,000	0	0.33	0.6	$\frac{1}{3}$
In line	2000 to 32,000	0	0.26	0.6	$\frac{1}{3}$

For air flow normal to extended surfaces such as pins or strips,

$$\frac{h_m}{c_p G_{max}}\left(\frac{c_p \mu}{k}\right)_f^{2/3} = \left(\frac{sG_{max}}{\mu}\right)^{-1/2} \tag{5-23}$$

where s is the distance across a strip traversed by fluid and N_{Re} is between 200 and 12,000.

For packed beds, the film coefficient between fluid and solid packing is given by

$$\frac{h}{c_p G_0}\left(\frac{c_p \mu}{k}\right)^{2/3} = 1.06\left(\frac{D_p G_0}{\mu}\right)^{-0.41} \tag{5-24}$$

where D_p is the particle diameter and $D_p G_0/\mu$ is between 200 and 4000.

For fluidized solids, heat transfer to the container wall is given by

$$\frac{h_m D_p}{k_g} = 0.58\left(\frac{\rho_z}{\rho_g}\right)^{0.18}(C_r)^{0.36}\left(\frac{c_p \mu}{k}\right)_g^{0.5}\left(\frac{D_p G_0}{\mu_g}\right)^{0.45} \tag{5-25}$$

where ρ_z is the apparent density of the settled bed, C_r is the ratio of the specific heat of the solid to that of the gas, and subscripts p and g refer to solid particle and gas, respectively.

Overall Coefficients

The overall coefficient or thermal transmittance U is related to the individual film, scale, and wall coefficients by

$$\frac{1}{U} = \frac{dA}{h_h \, dA_h} + \frac{dA}{h_{dh} \, dA_{dh}} + \frac{x_w \, dA}{k_w \, dA_w} + \frac{dA}{h_{dc} \, dA_{dc}} + \frac{dA}{h_c \, dA_c} \tag{5-26}$$

       hot   hot   wall   cool   cool
       film   scale   scale   film

When the areas are approximately equal, as is the case when the thickness of a tube wall is small compared to the diameter, the area terms may be omitted and all coefficients based on the outside area of the tube, dA.

Table 5-3 Typical Coefficients for Deposits

	Temperature, °F	h_d
Treated feed water or sea water	<125	2000
Treated cooling tower water	<125	1000
Great Lakes water	<125	1000
Clean river water	<125	500
Hard water (over 15 gm/gal)	<125	330
Light hydrocarbons	<400	2000
Gasoline, organic liquids, brine, steam	<400	1000
Hydrocarbons, refrigerant vapors	<500	500
Gas oils, vegetable oil	>500	330
Crude oil, fuel oil		200
Cracked residuum, coke-oven gas		100

RADIATION

Definitions

Electromagnetic radiation covers a wide range of wavelengths from 10^{-11} centimeters to more than 1000 meters, but the part of the spectrum of importance in heat flow lies between 0.5 and 50 microns. Gases, liquids, and solids emit and receive radiation.

Absorptivity is the fraction of total radiation intercepted by a body that is absorbed by the body. Radiant energy can flow into or through a body only by transmission or by conduction of heat.

Transmissivity is the fraction of intercepted radiation that is transmitted through the body. Most solid heat-transfer surfaces are nearly opaque so that radiation may usually be treated as a surface phenomenon with zero transmissivity.

Reflectivity is the fraction of intercepted radiation that is reflected from the body. Most heat-transfer surfaces are sufficiently rough that reflection is diffuse, and the simplifying assumption can be made that reflectivity and absorptivity are independent of the angle of incidence.

For conservation of energy, the sum of absorptivity, transmissivity, and reflectivity must be unity. The radiant energy emitted by a surface is independent of that emitted by any other body in the field of view, although the energy received and reflected may not be. A surface that absorbs all incident radiation without reflection or transmission is called a *black body*.

Emissivity is the ratio of the total radiating power of a body to that of a black body at the same temperature. *A gray surface* is one for which the total absorptivity is independent of the spectral distribution of the incident radiation, so that emissivity may be substituted for absorptivity even though the temperatures of the radiation and receiver are not the same.

Kirchhoff's law implies that emissivity and absorptivity are equal for a body at temperature equilibrium with its surroundings and depend only on the temperature.

Stefan-Boltzmann law states that the total emissive power E_b of a black body is proportional to the fourth power of the absolute temperature.

$$E_b = \sigma T^4 \tag{5-27}$$

or for convenience in calculations,

$$E_b = \sigma \times 10^8 \left(\frac{T}{100}\right)^4 \tag{5-28a}$$

$\sigma = 0.1714 \times 10^{-8}$ Btu/(ft^2)(hr)($°R$)^4
$= 5.67 \times 10^{-5}$ erg/(cm^2)(sec)($°K$)^4
$= 4.88 \times 10^{-8}$ kg-cal/(m^2)(hr)($°K$)^4

Radiation between Opaque Surfaces

According to Lambert's cosine law, the radiant transfer rate between areas dA_1 and dA_2,

$$d^2E_{12} = I_1 \, dA_1 \cos \phi_1 \, dA_2 \cos \phi_2 / r^2 \tag{5-29}$$

in which I is the radiation intensity and the other terms are defined by Fig. 5-1.

The angle factor F_{12} or fraction of radiation leaving surface A_1 in all directions which is intercepted by surface A_2 is obtained by integration of Eq. 5-29 and division by the total emission of A_1. Jakob (Vol. II), McAdams, and Trinks present graphs of angle factors for various geometries. Figure 5-2 illustrates F for various parallel surfaces.

Combination of the angle factor with Eq. 5-28a gives the relationship for interchange between black surfaces,

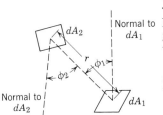

Fig. 5-1 Radiation geometry.

$$q_{12} = 0.171 AF\left[\left(\frac{T_1}{100}\right)^4 - \left(\frac{T_2}{100}\right)^4\right] \tag{5-30}$$

where q_{12} is in Btu per hour and T is °F + 460.

Allowance for refractory surfaces may be made by substituting an interchange factor $\bar{F}$ for F in Eq. 5-30. Commonly the source and sink are connected by refractory walls for which the net radiation at the wall surface is negligible. Curves 5 to 8 of Fig. 5-2 give values of $\bar{F}$ for parallel surfaces of equal areas with the assumptions that wall reflection is diffuse and the angular distribution of emission is like that of a black body.

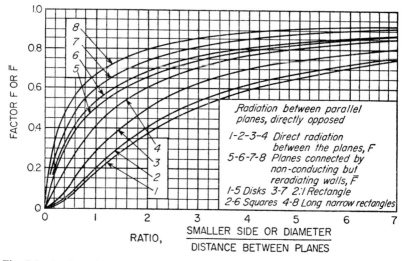

Fig. 5-2 Angle and interchange factors, parallel surfaces. (By permission of H. C. Hottel.)

With only one source and one sink, neither of which can see itself, and the assumption of uniform refractory temperature, the expression for $\bar{F}$ is

$$\bar{F}_{12} = \frac{A_2 - A_1 F_{12}^2}{A_1 + A_2 - 2A_1 F_{12}} \qquad (5\text{-}31)$$

Equation 5-31 applies to most problems in a furnace enclosure if the underlying assumptions are permissible.

Allowance for nonblack surfaces may be made by assuming all surfaces are gray and conform to the cosine law so that a factor F^* can be substituted for F in Eq. 5-30:

$$F_{12}^* = \frac{1}{\dfrac{1}{\bar{F}_{12}} + \left(\dfrac{1}{\epsilon_1} - 1\right) + \dfrac{A_1}{A_2}\left(\dfrac{1}{\epsilon_2} - 1\right)} \qquad (5\text{-}32)$$

in which ϵ_1 and ϵ_2 are emissivities of source and sink.

If no refractory walls are involved, F_{12} replaces $\bar{F}_{12}$ in Eq. 5-32. In the case of a small gray surface surrounded by a much larger surface,

$$q_{12} = \sigma A_1 \epsilon_1 (T_1^4 - T_2^4) \qquad (5\text{-}33)$$

For multiple surfaces with various emissivities, a network analysis of energy balances on all surfaces is required. This problem requires a large computing machine. [See A. K., Oppenheim, *Trans. A.S.M.E.* **78**, 725 (1956).]

Radiation from Nonluminous Gases

When black-body radiation passes through a gas, absorption occurs in regions of the infrared spectrum or, if the gas is hot, it radiates in the same region of the spectrum. Gases with significant emissivities include water vapor, the oxides of carbon and sulfur, hydrocarbons, alcohols, ammonia, and hydrogen chloride.

Consider a gas at uniform temperature T_g surrounded by a black surface at uniform temperature T_s. The radiant interchange between the two is expressed as

$$\frac{q}{A_s} = \sigma(\epsilon_g T_g^4 - \alpha_{gs} T_s^4) \tag{5-34}$$

Gas emissivity is a function of the product of the partial pressure of the radiating component p_g and the mean beam length L. The beam length L is defined as the radius of a hemisphere of gas which radiates to a black surface

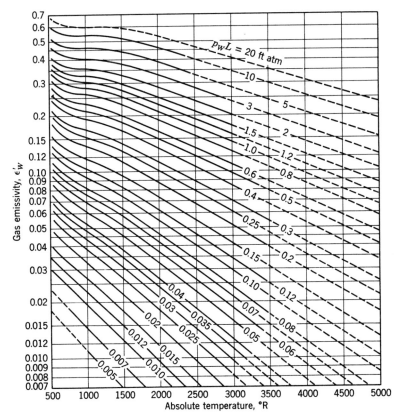

Fig. 5-3 Emissivity of water vapor. (By permission of H. C. Hottel.)

element located at the center of the base of the hemisphere. The gas emissivity at a total pressure of one atmosphere containing a radiating component at partial pressure p_g may be evaluated from graphs such as Fig. 5-3 and Fig. 5-4. At low values of $p_g L$ the mean beam length for other shaped volumes may be approximated as four times the volume divided by the surrounding surface area. For magnitudes of $p_g L$ between 0.01 to 1.0 the shielding effect of the molecules is taken into account by using L as 0.85 of this calculated value.

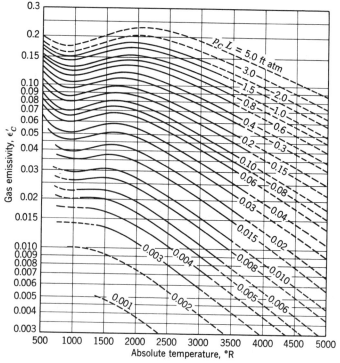

Fig. 5-4 Emissivity of carbon dioxide. (By permission of H. C. Hottel.)

When CO_2 and H_2O are both present, the total radiation from both gases is reduced due to partial opacity of each gas to radiation from the other. In this case the correction to the combined emissivity as shown in Fig. 5-5 should be included. The total gas emissivity then becomes

$$\epsilon_g = \epsilon_{CO_2} + \epsilon_{H_2O} - \Delta\epsilon \tag{5-35}$$

The accurate evaluation of α_{gs} requires a rather elaborate procedure details of which may be found in Jakob or McAdams. As an approximation α_{gs} may be evaluated as ϵ_g at T_s when the temperature of the gas is much lower than that of the surface, within an error of less than 10% in q/A.

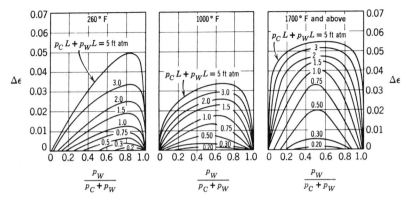

Fig. 5-5 Correction for superimposed radiation from CO_2 and H_2O. (By permission of H. C. Hottel.)

Allowance for gray surfaces may be made by multiplying the right-hand side of Eq. 5-34 by the factor $(\epsilon_s + 1)/2$ when the surface emissivity ϵ_s is above 0.8 as is the case with most high-temperature applications.

Radiation from Luminous Flames

Flames are made luminous by clouds of particles arising largely from soot from decomposition of hydrocarbons. Although it is not now possible to

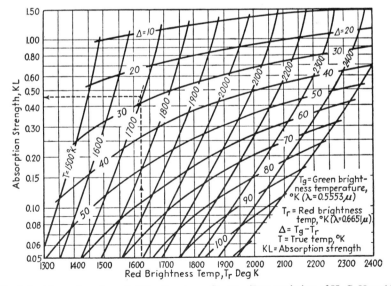

Fig. 5-6 Absorption strength KL of luminous flames. (By permission of H. C. Hottel.)

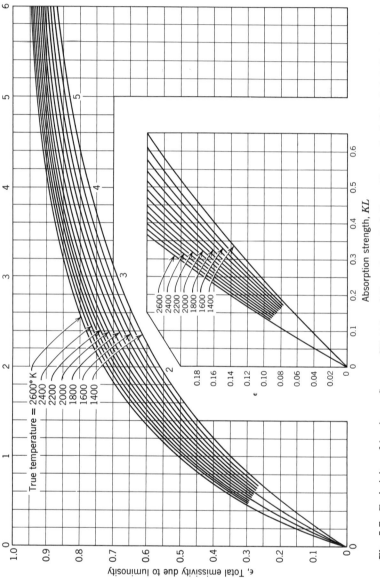

Fig. 5-7 Emissivity of luminous flames. [Hottel and Broughton, *Ind. Eng. Chem., Anal. Ed.*, **4**, 166 (1932).]

predict flame luminosity except for a limited range of fuels, it is possible to measure the brightness temperature and the corresponding emissivity by the use of an optical pyrometer.

Figure 5-6 gives the true temperature T_f and the absorption strength KL and Figure 5-7 the effective emissivity ϵ_f of the flame envelope for use in the equation for heat transfer by radiation from flame area A_f to all surrounding walls,

$$\frac{q}{A_f} = \sigma \epsilon_f \epsilon_s \left[\left(\frac{T_f}{100}\right)^4 - \left(\frac{T_s}{100}\right)^4 \right] \tag{5-36}$$

For a more complete discussion see McAdams.

Table 5-5 Representative Normal Emissivities

Surface	Temperature, °F	Emissivity
Bright polished metals	70	0.03–0.045
	1000	0.05–0.10
Aluminum roofing	100	0.21
Galvanized sheet iron	80	0.25
Polished cast or wrought iron	400	0.25
Aluminum paints	212	0.30–0.50
Molten iron or steel	2300–3300	0.30
Tungsten filament	6000	0.40
Monel metal, oxidized	1100	0.46
Polished steel sheet or casting	1400–2000	0.52–0.60
Oxidized rolled steel	70	0.65–0.80
Fire brick	1100–1800	0.65–0.75
Silica brick	2000	0.80–0.85
Paints, enamels, all colors	70–200	0.85–0.95
Plaster, rough lime	50–200	0.91
Red brick	70	0.93
Roofing paper	70	0.92
Strongly oxidized iron or steel	100–700	0.94–0.97
Hard rubber	75	0.95
Asbestos board	75	0.96
Lampblack	70	0.97

REFERENCES

Jakob, Max, *Heat Transfer*, John Wiley, New York, Vol. I, 1949; Vol. II, 1957.

Katz, D. L. and J. G. Knudsen, *Fluid Dynamics and Heat Transfer*, McGraw-Hill, New York, 1958.

Kern, D. Q., *Process Heat Transfer*, McGraw-Hill, New York, 1950.
McAdams, W. H., *Heat Transmission*, McGraw-Hill, New York, third edition, 1954.
Rohsenow, W. M. and H. Y. Choi, *Heat, Mass, and Momentum Transfer*, Prentice-Hall, Englewood Cliffs, N.J., 1961.
Schack, A., *Der Industrielle Warmeubergang*, Verlag Stahleisen, Dusseldorf, 1957.
Schneider, P. J., *Conduction Heat Transfer*, Addison-Wesley, Reading, Mass., 1955.
Trinks, W., *Industrial Furnaces*, John Wiley, New York, Vol. I, fifth edition, 1961.

SECTION 6

ELECTRICITY AND MAGNETISM

Principal Symbols

Boldface indicates vector; for units see Table 1–2.

B, B	magnetic flux density	r	radius
C	capacitance	$\mathscr{R}$	reluctance
D, D	electric flux density	S	surface
E, e	electromotive force	t	time
E, E	electric field intensity	V, v	potential, voltage
F, F	force	W	energy, work
f	frequency	X	reactance
$\mathscr{F}$	magnetomotive force	Z	impedance
G	conductance	γ	conductivity
H, H	magnetic field intensity	ϵ	permittivity, dielectric constant
I, i	current		
J, j	current density	μ	permeability
L	inductance	ρ	resistivity
l, l	length	σ	surface charge density
M	mutual induction	τ	torque
N	number of turns	ϕ	flux
Q, q	electric charge	ω	angular velocity
R	resistance		

GENERAL RELATIONS

Ohm's Law. Materials are said to obey Ohm's law when the current through a specimen is proportional to the impressed voltage. This result leads to the limited statement of Ohm's law as

$$V = IR \tag{6-1}$$

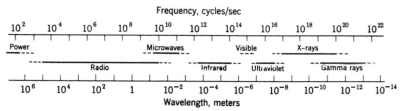

Fig. 6-1 The electromagnetic spectrum.

in which the resistance R is a constant of proportionality. An equivalent form of Ohm's law in terms of electric field strength **E**, current density **J**, and conductivity γ is

$$\mathbf{E} = \mathbf{J}/\gamma \tag{6-2}$$

A more general form which takes into account back emfs (ΣE) or the emf induced by motion in a magnetic field ($\mathbf{v} \times \mathbf{B}$) is given by

$$V = IR - \sum E \quad \text{or} \quad \mathbf{E} = \mathbf{J}/\gamma - \mathbf{v} \times \mathbf{B} \tag{6-3}$$

Kirchhoff's Laws. Charge or current is conserved so that at any junction in an electric network the currents entering equal the currents leaving.

$$\sum I = 0 \tag{6-4}$$

Kirchhoff's second law or rule, based on the conservation of energy, requires that the algebraic sum of the potential changes around any closed loop of conductors is zero.

$$\sum V = 0 \quad \text{or} \quad \sum E - \sum RI = 0 \tag{6-5}$$

Joule's Law. Flow of current through any resistance R dissipates energy as heat. For a constant current I the rate of heat dissipation

$$P_h = RI^2 \tag{6-6}$$

With a time varying current i, the *effective current* in Eq. 6-6 is the root-mean-square value

$$I_{\text{eff}} = I_{\text{rms}} = \left(\frac{1}{t}\int_0^t i^2\, dt\right)^{1/2} \tag{6-7}$$

where t is the time of one cycle. When a sinusoidal current is superimposed on a constant current I_{dc}

$$I_{\text{eff}} = (I_{dc}^2 + I_m^2/2)^{1/2} \tag{6-8}$$

in which I_m is the maximum value of the sinusoidal current.

With a varying current, losses beyond the resistance loss may occur because of (1) dielectric hysteresis accompanying the displacement current, (2) magnetic hysteresis accompanying the varying magnetic flux produced by the current, and (3) eddy currents induced in nearby conductors.

Table 6-1 Dimensional Constants in Various Systems of Units

c = velocity of light in vacuum; mksc 2.9979×10^8 meters/sec; cgs 2.9979×10^{10} cm/sec

Symbol Designation	k_e, Coulomb's Law Proportionality Constant	k_m, Ampere's Law Proportionality Constant	k_i, Faraday's Law Proportionality Constant	ϵ_0, Dielectric Constant, Permittivity of Vacuum	μ_0, Permeability of Vacuum
Mksc rationalized units	$c^2/10^7$ or 8.9875×10^9 meters/farad	2×10^{-7} henry/meter	1	$10^7/4\pi c^2$ or 8.854×10^{-12} farad/meter	$4\pi \times 10^{-7}$ or 1.257×10^{-6} henry/meter
Mksc unrationalized units	$c^2/10^7$ meters/farad	2×10^{-7} henry/meter	1	$10^7/c^2$ farad/meter	10^{-7} henry/meter
Cgs symmetric units	1 cm/statfarad	$2/c$ abhenry/cm	$1/c$	1 statfarad/cm	1 abhenry/cm
Cgs electromagnetic units	c^2 cm/statfarad	2 abhenrys/cm	1	$1/c^2$ abfarad/cm	1 abhenry/cm
Cgs electrostatic units	1 cm/statfarad	$2/c^2$ stathenry/cm	1	1 statfarad/cm	$1/c^2$ stathenry/cm

Energy and Power. For a constant current I maintained through any device by an impressed voltage V during time t, the energy input

$$W = Pt = IVt \qquad (6\text{-}9)$$

With time varying current i and voltage v, the average power input

$$P_{\text{avg}} = \frac{1}{t}\int_0^t iv\, dt \qquad (6\text{-}10)$$

Power factor is the ratio of average power input to the product of rms current and rms voltage.

$$\text{Power factor} = P_{\text{avg}}/I_{\text{eff}}V_{\text{eff}} \qquad (6\text{-}11)$$

With sinusoidal current and voltage and angle θ expressing the phase difference between them

$$\text{Power factor} = \cos\theta \qquad (6\text{-}12)$$

Table 6-2 Maxwell's Basic Equations
(with no dielectric or magnetic material present)

Designation	Equation	Describes
Gauss's law for electricity	$\epsilon_0 \phi_E = q = \epsilon_0 \oint \mathbf{E}\cdot d\mathbf{S}$	Charge and the electric field
Gauss's law for magnetism	$\phi_B = \oint B\, d\mathbf{S} = 0$	Impossibility of isolated magnetic pole
Ampere's law	$\dfrac{1}{\mu_0}\oint \mathbf{B}\cdot d\mathbf{l} = \epsilon_0 \dfrac{d\phi_E}{dt} + i_c$	The magnetic effect of a current or of a changing electric field
Faraday's law of induction	$\oint \mathbf{E}\cdot d\mathbf{l} = -\dfrac{d\phi_B}{dt}$	The electrical effect of a changing magnetic field

ELECTROSTATICS

Fields, Potentials, and Currents

Coulomb's law states that the force F exerted by electric charges on each other is directly proportional to the product of the charges Q_1 and Q_2 and inversely proportional to the square of the distance r between them.

$$F = k_e \frac{Q_1 Q_2}{\epsilon r^2} \qquad (6\text{-}13)$$

in which ϵ is the dielectric constant of the medium relative to that of vacuum. ϵ is unity for evacuated space and nearly unity for nonionized gases. See Table 6-1 for values of k_e.

Intensity of an electric field **E**, a vector, is defined as the force exerted on a unit positive charge at a point within the field.

$$\mathbf{E} = \frac{\mathbf{F}}{q} = k_e \frac{q}{\epsilon r^2} \qquad (6\text{-}14)$$

The direction of **E** is the direction of **F**, the direction in which a positive charge tends to move. For a group of point charges, **E** is obtained by vectorial addition of the field from each charge as if it existed alone.

$$\mathbf{E} = \mathbf{E}_1 + \mathbf{E}_2 + \cdots = \sum \mathbf{E}_i \qquad (6\text{-}15)$$

Electric equipotential surface is one drawn in an electric field so that it is normal at each point to the line of force through that point. The field intensity has no component along such a surface; hence no work is required to move a charge from one point to another over this surface.

Electric potential difference between two points is measured by the work required in moving a unit positive charge against the electric field intensity

$$V_1 - V_2 = \frac{W_2}{q_0} = -\int_1^2 E \cos\theta \, dl \qquad (6\text{-}16)$$

where dl is an elementary length of path and $-E \cos\theta$ is the component of force along dl. If $d\mathbf{n}$ is the differential element taken normal to electric equipotential surfaces,

$$\mathbf{E} = -dV/d\mathbf{n} = -\mathrm{grad}\, V \qquad (6\text{-}17)$$

Currents. The *direction of an electric current* is taken as the direction of movement of a positive charge and is the same as direction of the field intensity. Current I, a scalar, is related to *current density* **J**, a vector, by

$$I = \int \mathbf{J} \cdot d\mathbf{S} \qquad (6\text{-}18)$$

where $d\mathbf{S}$ is an element of surface area and the integral is taken over the surface considered.

The instantaneous current density at any point is given by

$$\mathbf{J} = \gamma \mathbf{E} + (d/dt)(\epsilon_0 \epsilon \mathbf{E}) \qquad (6\text{-}19)$$

in which γ is the conductivity of the substance. For a specific conductor at constant temperature and pressure γ is independent of the strength, distribution or time variation of the current. For a dielectric, however, γ depends on the time variation of the field intensity.

Dielectric Flux

Dielectric flux density **D** is defined as

$$\mathbf{D} = \epsilon_0 \epsilon\, \mathbf{E} \qquad (6\text{-}20)$$

The direction of **D** is chosen to be the same as that of **E**. The total dielectric flux ϕ_E through a surface S is given by

$$\phi_E = \int D \cos \alpha \, dS \tag{6-21}$$

in which dS is an element of surface area, $D \cos \alpha$ is the component of **D** normal to dS, and the integral is taken over the surface considered.

Displacement and Conduction Currents. The displacement current I_d through the surface and the momentary conduction current I_c flowing through the conductor to the surface are related to the dielectric flux ϕ by

$$I_c = I_d = d\phi/dt \quad \text{and} \quad \phi = \int I_c \, dt = Q \tag{6-22}$$

where Q is the quantity of electricity conducted through the conductor to the surface of contact between the conductor and the dielectric.

Capacitance

Capacitance and Condensers. Two conductors with dielectric between them, if their surfaces are equipotential surfaces, form a *capacitor* or an *electric condenser*. When all flux lines from one conductor end on the second conductor, as when they are connected to the two terminals of a battery, the *capacitance C* of the condenser is given by

$$C = Q/V = \epsilon_0 \epsilon \mathscr{L} \tag{6-23}$$

in which Q is the total charge on either conductor and $\mathscr{L}$ is a parameter dependent on the geometry of the condenser. $\mathscr{L}$ has the dimension of length.

$$\text{For parallel plates,} \quad \mathscr{L} = A/s \tag{6-24}$$

where A is the area of a plate and s is the distance between plates.

For concentric cylinders, $\mathscr{L} = 2\pi l/\log_e (b/a)$, where l is the length of the cylinder, a is the diameter of the inner cylinder, and b is the inside diameter of the outer shell.

General Equation for Condensers. For any part of a dielectric bounded laterally by flux lines and at the ends by equipotential conductors, the capacitance is given by

$$C = \frac{\epsilon_0 \epsilon \phi}{\int \mathbf{D} \cdot d\mathbf{l}} = \frac{\epsilon_0 \epsilon Q}{\int \mathbf{D} \cdot d\mathbf{l}} \tag{6-25}$$

in which the integral is taken along the flux line from one end surface to the other. For a specified dielectric and distribution of flux the capacitance is constant, but when any substance with different dielectric constant is introduced in the electric field of a condenser the distribution of flux and the capacitance are changed.

ELECTROSTATICS

Capacitances in Parallel and Series. When several capacitances are connected in parallel between the same pair of equipotential surfaces, the total equivalent capacitance is given by

$$C = C_1 + C_2 + C_3 + \cdots \qquad (6\text{-}26)$$

When the capacitances are connected in series end to end so that the same dielectric flux passes through each of them,

$$\frac{1}{C} = \frac{1}{C_1} + \frac{1}{C_2} + \frac{1}{C_3} + \cdots \qquad (6\text{-}27)$$

Multiple Conductors. With several charged conductors in an electric field the capacitance between any two conductors i and j is

$$C_{ij} = Q_{ij}/V_{ij} \qquad (6\text{-}28)$$

in which Q_{ij} is that portion of the charge on i which is balanced by an equal and opposite charge on j.

Transmission Line in Air. If r is radius of the wire w, h elevation above the earth g, s distance between wires, $h > 2s$, and $s > 20r$, the capacitance in microfarads per 1000 ft for wires parallel to the earth is given by

$$\text{Single wire,} \qquad C_{w-g} = \frac{7.354}{1000 \log (2h/r)} \qquad (6\text{-}29)$$

$$\text{Two parallel wires,} \qquad C_{w-w} = \frac{3.677}{1000 \log (s/r)} \qquad (6\text{-}30)$$

Charge and Discharge Current. When a switch is closed in a simple capacitor circuit, the current i as a function of time t is given by

$$\text{Charging current,} \qquad i = \frac{V - Q_0/C}{R} e^{-t/RC} \qquad (6\text{-}31)$$

$$\text{Discharging current,} \qquad i = \frac{Q_0}{RC} e^{-tRC} \qquad (6\text{-}32)$$

Relation between Capacitance and Conductance. Expressions for the capacitance C and the conductance G of the dielectric between the plates of any shape of condenser differ only by a constant coefficient.

$$G = (\gamma/\epsilon_0 \epsilon)C \qquad (6\text{-}33)$$

in which γ is the conductivity of the dielectric.

The total current through a condenser is given by

$$I = GV + C(dV/dt) \qquad (6\text{-}34)$$

in which GV is the *leakage current* and $C(dV/dt)$ is the *displacement current*.

Energy and Forces

Electrostatic Energy. When the potential difference between the plates of a condenser is increased from 0 to V, the energy input W is given by

$$W = \int_0^t GV^2 \, dt + \int_0^V CV \, dV \tag{6-35}$$

The first term on the right represents energy dissipated as heat in the dielectric. The second term on the right represents electrostatic energy W_E stored in the dielectric analogous to the energy stored in a compressed spring.

If C and ϵ are constant,

$$W_E = \frac{Q^2}{2C} = \frac{QV}{2} = \frac{\phi V}{2} = \frac{CV^2}{2} \tag{6-36}$$

The electrostatic energy per unit volume of an electric field W_E is given by

$$w_E = \frac{DE}{2} = \frac{\epsilon_0 \epsilon E^2}{2} = \frac{D^2}{2\epsilon_0 \epsilon} = \frac{D^2}{2} \tag{6-37}$$

Mechanical Forces in an Electric Field. All bodies in an electric field exert mechanical forces upon one another which tend to produce a relative motion that will decrease the energy of the field. If the motion does not change the existing electric charges in the field,

$$F = \frac{-dW}{dx} \tag{6-38}$$

in which dx is an elementary distance in the direction of the relative motion.

In the special case of the two conductors forming a condenser the force of attraction exerted by one conductor on the other is

$$F = -\frac{V^2}{2}\frac{dC}{dx} \tag{6-39}$$

in which dC represents the increase in capacitance when one conductor moves a distance dx away from the other.

For a parallel-plate condenser (see Eq. 6-23)

$$F = \frac{V^2 \epsilon_0 \epsilon A}{2s^2} \tag{6-40}$$

ELECTROMAGNETISM

Magnetic Flux

Induced Emf. If a closed coil is moved within a magnetic field or if the field is varied, a momentary emf is induced in the coil, causing a momentary current to flow in the coil. If one looks at the coil in the direction of the increase in flux, the induced emf is in the counterclockwise direction.

According to *Faraday's law* the magnitude of the induced emf e in a single turn is

$$e = -\frac{d\phi_B}{dt} \quad (6\text{-}41)$$

According to Lenz's law the induced current will appear in a direction that opposes the cause of the current, hence the minus sign before $d\phi/dt$.

When there are multiple turns the total emf is given by

$$e = -\frac{d}{dt}(n_1\phi_1 + n_2\phi_2 + \cdots) \quad (6\text{-}42)$$

in which ϕ_1, ϕ_2, etc., represent the fluxes linking the various number of turns n_1, n_2, etc. When all N turns link the same flux, as with a tightly wound coil,

$$\sum n_i\phi_i = N\phi_B \quad (6\text{-}43)$$

Magnetic Flux Density. The magnetic flux per unit area through any surface normal to the direction of the field is the *magnetic induction B*, also called the magnetic flux density. If dS is an elementary area of surface and $B \cos \alpha$ is the component of flux density normal to dS,

$$\phi_B = \int (B \cos \alpha)\, dS = \int \mathbf{B} \cdot d\mathbf{S} \quad (6\text{-}44)$$

Lines of magnetic flux form continuous loops; hence the total magnetic flux coming to any surface is equal to the total flux leaving that surface. For a closed Gaussian surface then

$$\phi_B = \oint \mathbf{B} \cdot d\mathbf{S} = 0 \quad (6\text{-}45)$$

where the integral is taken over the entire closed surface.

Directions of Current and Induced Field. Each stream line of electric current is linked with the flux lines of an induced magnetic field (see Figs. 6-2 and 6-3). The directions of the current and its magnetic flux are related by the *right-handed screw law*. If the current has the same direction as a point on the edge of a right-handed screw, the direction of the threading flux lines is the same as

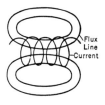

Fig. 6-2

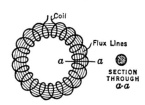

Fig. 6-3

the direction of the advancing screw. If one looks at a straight conductor in the direction of the electric current, the magnetic flux lines are in the clockwise direction around the conductor.

The Magnetic Circuit

In simple magnetic circuits there is a formal (but not physical) analogy to Ohm's law for electric circuits,

$$\mathscr{F} = \mathscr{R}\phi \qquad (6\text{-}46)$$

in which $\mathscr{F}$ is the magnetomotive force analogous to electromotive force, $\mathscr{R}$ is the reluctance of the material in the circuit, and ϕ is the flux analogous to current.

Magnetomotive force may be regarded as the agent that establishes the flux. The source or seat of $\mathscr{F}$ is usually a coil wound around a core of magnetic material with current flowing in the coil (see Fig. 6-3). If N is the number of turns of the coil and I is the current in the coil,

$$\mathscr{F} = NI \qquad (6\text{-}47)$$

If there is more than one seat of $\mathscr{F}$, the total magnetomotive force in the circuit is given by

$$\mathscr{F} = \sum NI \qquad (6\text{-}48)$$

$\mathscr{F}$ is taken as positive when the current links the flux lines in the right-handed screw direction.

Reluctance is defined as the ratio of magnetomotive force to the flux it produces through a unit of homogeneous material, $\mathscr{R} = \mathscr{F}/\phi$. The end surfaces of the unit of material must be magnetic equipotential surfaces and through every cross section the same flux must pass. For magnetic materials $\mathscr{R}$ varies with the flux density and hence is not a constant. Unlike electric resistance, reluctance does not contribute to dissipation of energy. For a straight bar of uniform cross section A and length l through which the flux lines are straight, parallel, and uniformly distributed, the reluctance is given by

$$\mathscr{R} = \frac{l}{\mu_0 \mu A} \qquad (6\text{-}49)$$

in which μ is the permeability of the material relative to that of vacuum. μ is nearly unity except for ferromagnetic materials.

Magnetic permeance $\mathscr{P}$ is the reciprocal of magnetic reluctance. The *permeability* of a material is defined as the permeance of a unit cube of the material when the flux through the cube is parallel to four edges of the cube and is uniformly distributed over a section at right angles to these four edges.

Magnetic Intensity

Magnetic field intensity or *magnetizing force*, may be regarded as the force acting on unit positive magnetic pole. Magnetic intensity **H** is related to flux density **B** by

$$\mathbf{B} = \mu_0\mu\mathbf{H} \qquad (6\text{-}50)$$

except within a permanent magnet where the direction of H due to the magnet alone is opposite to the direction of the flux lines.

By definition the line integral of H around any closed path in a magnetic field is equal to the total magnetomotive force acting around the path. From Ampere's law and Eq. 6-50

$$\sum NI = \frac{1}{\mu_0\mu}\oint \mathbf{B}\cdot d\mathbf{l} = \oint \mathbf{H}\cdot d\mathbf{l} = \int (H\cos\theta)\,dl \qquad (6\text{-}51)$$

where dl represents an elementary length along this path (see Fig. 6-4).

When the path coincides with the magnetizing force

$$\sum NI = \int H\,dl \qquad (6\text{-}52)$$

Stream Line of Electric Current (Fig. 6-5). For a closed stream line of electric current surrounded by a medium with uniform magnetic properties, each

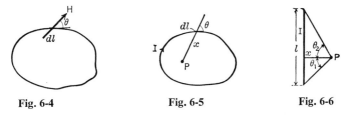

Fig. 6-4   **Fig. 6-5**   **Fig. 6-6**

elementary length dl contributes to H at any point P an amount

$$dH = \frac{(I\sin\theta)\,dl}{4\pi x^2} \qquad (6\text{-}53)$$

Straight Circular Wire (Fig. 6-6). The magnetic intensity H for length l of straight wire with circular cross section at any point P at distance x from the wire is given by

$$H = \frac{I}{4\pi x}(\sin\theta_1 + \sin\theta_2) \qquad (6\text{-}54)$$

If the wire is very long compared with x,

$$H = \frac{I}{2\pi x} \qquad (6\text{-}55)$$

For a long wire of radius r when the current density is uniform over the cross section, the usual case, the magnetic intensity inside the wire is given by

$$H_i = \frac{xI}{2\pi r^2} \qquad (6\text{-}56)$$

220   ELECTRICITY AND MAGNETISM

Flat Plate. The magnetic intensity at a point near a flat plate with breadth h is given by

$$H = \frac{I}{2h} \tag{6-57}$$

Circular Coil of N Turns. If r is the mean radius of the coil, and x is the distance of a point from the center of the circle,

$$H = \frac{NIr^2}{2(r^2 + x^2)^{3/2}} \tag{6-58}$$

Solenoid. If r is the mean radius and l is the length of the helix, at any point on the axis of the helix and at a distance x from the center,

$$H = \frac{NI}{2l}\left(\frac{0.5l + x}{\sqrt{r^2 + (0.5l + x)^2}} + \frac{0.5l - x}{\sqrt{r^2 + (0.5l - x)^2}}\right) \tag{6-59}$$

provided the thickness of the winding is small compared with r.

When l is large compared with r,

$$H = \frac{NI}{l} \tag{6-60}$$

Equation 6-60 also applies to nonaxial points within the central portion of a long solenoid where the field is uniform over the cross section.

Torus (Fig. 6-3). In the usual torus uniformly wound on a core of one homogeneous magnetic material, the lines of flux and magnetic intensity are concentric circles each of which represents a path of equal magnetic intensity. For any point P within the core, the magnetic intensity is given by Eq. 6-60 where l is the circumference of a circle through P_1.

Mechanical Forces

Magnetic Force on a Current. If θ is the angle between the current I and the flux density **B**,

$$F/l = IB \sin \theta \tag{6-61}$$

In the vector form of Ampere's law, where **l** is a displacement vector in the direction of the current,

$$\mathbf{F} = I\mathbf{B} \times \mathbf{l} \tag{6-62}$$

The relative directions of **F**, **B**, and I are conveniently determined by the *left-hand rule*. When the forefinger of the left hand is pointed in the direction of the flux and the middle finger in the direction of the current, the thumb held perpendicular to these two fingers points in the direction of the force.

Magnetic Force between Parallel Plates. From Eqs. 6-62, 6-57, and 6-50 the force on a length l of one of a pair of plate conductors of breadth h and carrying current I is given by

$$F = \frac{\mu_0 \mu I^2 l}{2h} \tag{6-63}$$

and the force per unit area of the plate by

$$\frac{F}{lh} = \frac{\mu_0 \mu}{2}\left(\frac{I}{h}\right)^2 \tag{6-64}$$

Magnetic Force on a Charged Particle. For a particle of mass m with elementary charge q_e and moving with a velocity u of constant magnitude and with no component parallel to the field, the initial force on the particle $F_e = q_e u B$ produces an acceleration $a = F_e/m$ that changes the direction of the velocity. The direction of the magnetic force remains perpendicular to the direction of the velocity u hence *the particle moves in a circle*. The radius of the path is given by

$$r = \frac{mu}{q_e B} \tag{6-65}$$

and the angular velocity ω and the frequency f by

$$\omega = \frac{q_e B}{m}, \quad f = \frac{q_e B}{2\pi m} \tag{6-66}$$

These relations for the ballistics of charged particles in magnetic fields are applied in mass spectrometers, in the focusing of particle beams, and in the cyclotron.

Force on magnetic bodies in a magnetic field with constant magnetomotive force is given by

$$F = -\tfrac{1}{2}\phi^2 (d\mathcal{R}/dx) \tag{6-67}$$

in which ϕ is the total flux threading the body and $d\mathcal{R}$ represents the increase in reluctance corresponding to a displacement dx in the direction of the flux lines.

Equation 6-67 accounts for the attraction of unlike poles of a magnet or the attraction of paramagnetic materials and the repulsion of diamagnetic materials by either pole of a magnet.

Torque on a coil is the basic operating principle of the electric meters for current or voltage. Torque on a coil with N turns and area A is

$$\tau = NIAB \sin \theta \tag{6-68}$$

where θ is the angle that the normal to the plane of the coil makes with the direction of B. When the plane of the coil is parallel with the magnetic field,

$$\tau = NIAB \tag{6-69}$$

Torque on a Rotating Coil. Consider a longitudinal coil with N turns of length l and radius r rotating about its axis in a magnetic field such as the armature of a two-pole motor. The average flux density per pole $B_{\text{avg}} = 2B_{\text{max}}/\pi$, and the average torque

$$\tau = N(IlB_{\text{avg}})r = \frac{2Nr}{\pi} IlB_{\text{max}} \tag{6-70}$$

in which I for each conductor is one-half the total armature current since the commutator divides the current.

Electromagnetic Induction

Inductance. When the current i in an electric circuit varies with time a self-induced emf e is produced according to the general relation

$$e = -L\frac{di}{dt} \tag{6-71}$$

in which the proportionality constant L is the *inductance* or *the coefficient of self-induction*. Equation 6-71 is the defining equation for L for all coils of any shape or size with or without nearby magnetic material. Inductance depends on the geometry and permeability of the coil or magnetic circuit. From Eq. 6-41 inductance may also be expressed as

$$L = \frac{\partial \phi_L}{\partial i} \quad \text{henrys} \tag{6-72}$$

in which $\phi_L = \Sigma(n\phi)_i$ weber turns is the number of magnetic linkages between the electric circuit and the flux established by the current i (see Eqs. 6-42 and 6-43).

The total induced emf is given by

$$e = -\frac{\partial \phi_L}{\partial i}\frac{di}{dt} - \frac{\partial \phi_L}{\partial t} \quad \text{volts} \tag{6-73}$$

in which the first term on the right is the self-induced emf and the second term is the emf induced by motion in the magnetic field.

For a single-layer helical coil which is not near any ferromagnetic material and which has N turns of radius r and axial length l, the self-inductance is approximately

$$L = \frac{r^2 N^2 l}{9r + 10} \times 10^{-6} \quad \text{henrys} \tag{6-74}$$

where r and l are expressed in inches and $l > r$.

The total induced emf becomes

$$e = -\frac{d(Li)}{dt} \tag{6-75}$$

When every flux line is linked by every stream line of electric current, as in toroids and at the center of long solenoids, and the permeability of the entire circuit is independent of current

$$L = N^2/\mathscr{R} \tag{6-76}$$

where N is the number of turns forming the circuit and $\mathscr{R}$ is the reluctance of the complete magnetic circuit.

For a two-wire transmission line where the length is large compared to the distance apart s, the inductance per wire

$$L = 0.01524 + 0.14037 \log \frac{2s}{D} \quad \text{millihenrys}/1000 \text{ ft} \tag{6-77}$$

Mutual Inductance. The emf e_a induced in any circuit a by a varying current i_b in any other circuit b is

$$e_a = -M_{ab}\frac{di_b}{dt} \text{ volts} \tag{6-78}$$

When permeability is independent of current and every flux line linking both a and b is linked by every turn in a and b,

$$M_{ab} = N_a N_b / \mathscr{R}_{ab} \text{ henrys} \tag{6-79}$$

where $\mathscr{R}_{ab}$ is the reluctance of that part of the magnetic circuit through which the flux from a to b passes when there is current in only one coil.

For two equal and parallel straight wires each of l meters length and s meters apart, the mutual inductance between the wires is approximately

$$M = 2l\left(\frac{s}{l} - 1 + \log_e \frac{2l}{r}\right) \times 10^{-4} \text{ millihenrys} \tag{6-80}$$

Potential and Current. At any time t after closing a circuit containing in series a source of emf V, a constant resistance R, and a constant inductance L, the current

$$i = \frac{V}{R}(1 - \exp^{-Rt/L}) \tag{6-81}$$

With two nearby coils a and b, the potential drop through one of them

$$V_a = R_a i_a + L_a \frac{di_a}{dt} + M_{ab}\frac{di_b}{dt} \tag{6-82}$$

Magnetic Energy. When the current in an electric circuit increases from zero to I, the total input of energy U into a magnetic circuit is given by

$$U = \int_0^I Li\, di \tag{6-83}$$

When permeability and L are constant

$$U = \frac{LI^2}{2} = \frac{\mathscr{F}^2}{2\mathscr{R}} \text{ joules} \tag{6-84}$$

and for a concentrated winding or a toroid

$$U = \frac{(NI)^2}{2\mathscr{R}} \text{ joules} \tag{6-85}$$

The energy per unit volume u of the magnetic field is

$$u = \frac{HB}{2} = \frac{\mu_0 \mu H^2}{2} = \frac{B^2}{2\mu_0 \mu} \text{ joules per cubic meter} \tag{6-86}$$

When the permeability is not constant, the energy transferred to unit volume of the magnetic field due to any number of currents is

$$w = \int_0^B H\, dB \text{ joules per cubic meter} \tag{6-87}$$

The *hysteresis loss* per cycle

$$\mu_h = \eta B_m^{1.6} \text{ ergs per cubic centimeter} \tag{6-88}$$

where η is a coefficient characteristic of the material (see Table 6-4), and B_m is the maximum flux density (in gauses) reached during the cycle.

The *eddy current loss* of power, for insulated metal sheets with each thickness small compared to other dimensions and with lines of force parallel to the planes of the laminations, is given by

$$P_c = \frac{(\pi a f B_m)^2}{8\rho} \text{ watts per cubic meter} \tag{6-89}$$

where a is the thickness of each sheet, f is the frequency in cycles per second, B_m is the maximum flux density in webers per square meter, and ρ is the resistivity in ohm-meters.

DIRECT-CURRENT CIRCUITS

Direct-Current Networks

Direction of an Emf. The direction of the emf in any part of a circuit is the direction that a positive charge would take if this emf were the only source of emf in the circuit. All emf's in a circuit may be added algebraically to obtain a resultant emf by assigning a positive value to each of the emf's acting around the circuit in one direction and a negative value to those acting in the opposite direction.

Potential Drop. The difference of electric potential or voltage between two points in a circuit is defined by the work function Eq. 6-16. The potential drop V from the higher potential (positive terminal) to the lower potential (negative terminal) is regarded as positive.

Resistances in Series. For several conductors connected end to end, so that the same current flows through each of them, the total resistance

$$R = R_1 + R_2 + R_3 + \cdots \tag{6-90}$$

and the equivalent emf

$$E = E_1 + E_2 + E_3 + \cdots \tag{6-91}$$

Resistances in Parallel. For several conductors connected to two common junction points the potential drop is the same through each of them. When there are no emf's in the branches the combined resistance R between the common junctions is given by

$$\frac{1}{R} = \frac{1}{R_1} + \frac{1}{R_2} + \frac{1}{R_3} + \cdots \tag{6-92}$$

In the special case of two conductors in parallel with no emf in either the combined resistance

$$R = \frac{R_1 R_2}{R_1 + R_2} \tag{6-93}$$

Kirchhoff's Laws. These two laws, Eqs. 6-4 and 6-5, provide a set of simultaneous equations for a network. In applying these equations a current leaving a point is equivalent to a negative current entering that point, and an emf in a chosen direction is equivalent to a rise of potential in that direction. At any junction point currents directed toward the point are considered as positive. For a closed mesh, currents and emf's directed around the mesh in the clockwise direction are considered as positive.

Many simpler problems can be solved by writing two independent expressions for the potential drop between each pair at points and equating them.

Solution of Networks. In solving networks of conductors there are several alternative methods, all of which are equivalent as far as results are concerned. Directions for solution by one method are given here. In Fig. 6-7 let the resistances and emf's be given. It is desired to find the currents in the five branches. There are four junction points, A, B, C, and D. Since the currents and their directions are unknown some assumption of direction must be made. Assume directions of currents in all branches and indicate by arrows and symbols I_1, I_2, I_3, I_4, I_5, and I_6. This is shown in the figure. Apply Eq. 6-4 to junctions A, B, and C as follows, remembering that a current approaching a junction is positive and a current leaving a junction is negative. Since $\Sigma I = 0$,

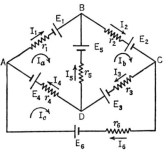

Fig. 6-7

At junction A	$I_4 + I_6 - I_1 = 0$	(6-94a)
At junction B	$I_1 + I_5 - I_2 = 0$	(6-94b)
At junction C	$I_2 - I_3 - I_6 = 0$	(6-94c)

If Eq. 6-4 is applied to the currents at the junction D, there will result an equation derivable from Eqs. 6-94. Equation 6-5 is next applied to the closed loops formed by branches. The potential drops will be summed *clockwise* about the loops. The following convention will be adopted:

1. All potential drops will be positive, and all potential rises will be negative.
2. In summing potential drops around a mesh, a given emf will be considered positive when we proceed from its positive to its negative pole, *regardless of the direction of the current through it.*
3. In summing potential drops around a mesh, a given emf will be considered negative when we proceed from its negative to its positive pole, *regardless of the direction of the current through it.*

226   ELECTRICITY AND MAGNETISM

4. In summing potential drops around a mesh, the difference of potential across a given resistance will be considered positive if the assumed direction of the current is clockwise with respect to that mesh.

5. In summing potential drops around a mesh, the difference of potential across a given resistance will be considered negative if the assumed direction of the current is counterclockwise with respect to that mesh.

Applying the above conventions to Fig. 6-7 the equations representing the second law may be written:

Mesh $ABDA$	$r_1 I_1 - E_1 + E_5 - r_5 I_5 + r_4 I_4 + E_4 = 0$	(6-95a)
Mesh $BCDB$	$r_2 I_2 + E_2 + r_3 I_3 - E_3 + r_5 I_5 - E_5 = 0$	(6-95b)
Mesh $ADCA$	$-E_4 - r_4 I_4 + E_3 - r_3 I_3 + r_6 I_6 - E_6 = 0$	(6-95c)

Any further mesh equations are derivable from Eqs. 6-95. The six equations permit the solution for currents in all branches. When these equations are solved a positive sign in front of a current indicates that the assumed direction of the current was correct. When a negative sign appears in front of a current, that current actually flows in a direction *opposite* to the direction assumed.

It is frequently convenient to use fictitious currents called *mesh currents* since the first law is automatically taken care of when writing the second-law equations by means of mesh currents. In Fig. 6-7, the circular arrows represent the mesh currents. The second-law equations may then be written

Mesh a	$r_1 I_a + r_5(I_a - I_b) + r_4(I_a - I_c) = E_1 - E_5 - E_4$	(6-96a)
Mesh b	$r_2 I_b + r_3(I_b - I_c) + r_5(I_b - I_a) = E_3 + E_5 - E_2$	(6-96b)
Mesh c	$r_6 I_c + r_4(I_c - I_a) + r_3(I_c - I_b) = E_6 + E_4 - E_3$	(6-96c)

These three equations may be solved for I_a, I_b, and I_c. The branch currents can then be found from

$$I_1 = I_a \qquad \text{(6-97a)}$$
$$I_2 = I_b \qquad \text{(6-97b)}$$
$$I_3 = I_b - I_c \qquad \text{(6-97c)}$$
$$I_4 = I_a - I_c \qquad \text{(6-97d)}$$
$$I_5 = I_b - I_a \qquad \text{(6-97e)}$$
$$I_6 = I_c \qquad \text{(6-97f)}$$

The law of the *superposition of currents and voltages* simplifies the calculation of certain types of distributing networks. If electric energy is being supplied over a network to a number of individual loads: (1) the current at any point in the network is equal to the algebraic sum of the currents which would flow if the individual load currents were considered in succession instead of simultaneously; and (2) the voltage drop from the source to any point in the network is equal to the algebraic sum of the drops to that point, each drop being calculated on the basis of individual load currents, taken successively instead of simultaneously.

SINUSOIDAL CURRENTS

Simple Sinusoidal Current and Voltage (Fig. 6-8). The instantaneous current i and voltage v are given by

$$i = I_m \sin(\omega t + \theta) \qquad (6\text{-}98)$$

$$v = V_m \sin \omega t \qquad (6\text{-}99)$$

where I_m and V_m are the maximum values of current and voltage, $\omega = 2\pi f$ and f is the frequency in cycles per second, t is the time measured from the instant when $v = 0$, and θ is the angular *phase difference* between current and voltage.

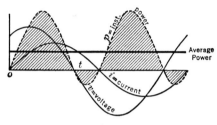

Fig. 6-8

The half-period *average* values I_{avg}, V_{avg}, and the effective meter or *root-mean-square* values I, V are given in terms of the maximum values I_m, V_m by

$$I_{\text{avg}} = 2I_m/\pi = 0.636 I_m \qquad (6\text{-}100)$$

$$V_{\text{avg}} = 2V_m/\pi = 0.636 V_m \qquad (6\text{-}101)$$

$$I = I_m/\sqrt{2} = 0.707 I_m \qquad (6\text{-}102)$$

$$V = V_m/\sqrt{2} = 0.707 V_m \qquad (6\text{-}103)$$

$$V_m = ZI_m \quad \text{or} \quad V = ZI \qquad (6\text{-}104)$$

The ratio $I/I_{\text{avg}} = 1.11$ is called the *form factor*.

Power. If current and voltage have the same frequency:

$$\text{Apparent power, } P_a = VI \text{ volt-amperes} \qquad (6\text{-}105)$$

$$\text{Active power, } P_p = VI \cos\theta \text{ watts} \qquad (6\text{-}106)$$

$$\text{Reactive power, } P_q = VI \sin\theta \text{ vars} \qquad (6\text{-}107)$$

For any one load,

$$VI = (P_p^2 + P_q^2)^{1/2} \qquad (6\text{-}108)$$

$$\text{Power factor} = P_p/P_a = \cos\theta \text{ watts per volt-ampere} \qquad (6\text{-}109)$$

$$\text{Reactive factor} = P_q/P_a = \sin\theta \text{ vars per volt-ampere} \qquad (6\text{-}110)$$

Series Circuits with Constant R, L, and C

Resistance and Inductance. When a sinusoidal voltage with f cycles per second is applied to a circuit containing a resistance R (ohms) and an inductance L (henrys), at steady state

$$i = \frac{V_m}{Z} \sin(2\pi ft - \theta) \text{ amperes} \tag{6-111}$$

$$v = ZI_m \sin 2\pi ft \text{ volts} \tag{6-112}$$

$$Z = \sqrt{R^2 + (2\pi fL)^2} \text{ ohms} \tag{6-113}$$

$$\theta = \tan^{-1} \frac{2\pi fL}{R} \text{ degrees} \tag{6-114}$$

If the resistance R is a negligible part of the impedance Z, the angle θ by which the current lags the voltage is 90°.

Resistance and Capacitance. When a sinusoidal voltage with f cycles per second is applied to a circuit containing a resistance R (ohms) and a capacitance C (farads), at steady state

$$i = \frac{V_m}{Z} \sin(2\pi ft + \theta) \text{ amperes} \tag{6-115}$$

$$v = ZI_m \sin 2\pi ft \text{ volts} \tag{6-116}$$

$$Z = \sqrt{R^2 + (1/2\pi fC)^2} \text{ ohms} \tag{6-117}$$

$$\theta = \tan^{-1} \frac{1}{2\pi fCR} \text{ degrees} \tag{6-118}$$

Here the current leads the voltage.

Resistance, Inductance, and Capacitance. Since the current lags the voltage with an inductance and leads the voltage with a capacitor, the two reactances X offset each other. Hence when both are present in the RLC series circuit

$$X_{LC} = 2\pi fL - 1/2\pi fC \tag{6-119}$$

$$Z = \sqrt{R^2 + X_{LC}^2} \tag{6-120}$$

$$\theta = \tan^{-1} \frac{X_{LC}}{R} \tag{6-121}$$

At steady state the RLC circuit behaves as a RL circuit when $2\pi fL > 1/2\pi fC$, as a RC circuit when $1/2\pi fC > 2\pi fL$, and as a simple resistance circuit when $2\pi fL = 1/2\pi fC$.

TRANSIENTS

From Kirchhoff's second law, the general equation for the impedance drop v due to flow of current i through a series circuit with constant resistance R, inductance I, and capacitance C may be expressed as

$$v = Ri + L\frac{di}{dt} + \frac{1}{C}\int i\,dt \tag{6-122}$$

and

$$Z = R + L + \frac{1}{C}\int dt \tag{6-123}$$

For many transient currents a solution is

$$i = I_0 e^{\beta t} \tag{6-124}$$

$$v = Zi = \left(R + \beta L + \frac{1}{\beta C}\right) I_0 e^{\beta t} \tag{6-125}$$

where I_0 is the initial value and β is the periodicity of the current.

The RL Circuit. When a constant emf E is applied to a series circuit with resistance R (ohms) and inductance L (henrys) in which an initial current I_0 is flowing, the current at any subsequent time t

$$i = \frac{E}{R} - \left(\frac{E}{R} - I_0\right) e^{-Rt/L} \text{ amperes} \tag{6-126}$$

The short-circuit current is obtained by putting E equal to zero.

The RC Circuit. When a constant emf E is applied to a series circuit with resistance R (ohms) and capacitance C (farads) with initial charge Q_0 (coulombs), the charge q and current i at any subsequent time are given by

$$q = CE - (CE - Q_0)e^{-t/RC} \text{ coulombs} \tag{6-127}$$

$$i = \left(\frac{E}{R} - \frac{Q_0}{RC}\right) e^{-t/RC} \text{ amperes} \tag{6-128}$$

The short-circuit current ($E = 0$) during discharge of a condenser initially charged to a potential $V_0 = Q_0/C$ is

$$i = -\frac{V_0}{R} e^{-t/RC} \tag{6-129}$$

The RLC Circuit. When a RLC series circuit is energized with a constant emf E, the current at any subsequent time t

$$i = I_0 e^{-at} \cos \omega t - \left[\frac{a}{\omega} I_0 + \frac{a^2 + \omega^2}{\omega}(Q_0 - CE)\right] e^{-at} \sin \omega t \tag{6-130}$$

in which
$$a = R/2L \tag{6-131}$$

$$\omega = \left(\frac{1}{LC} - \frac{R^2}{4L^2}\right)^{1/2} \tag{6-132}$$

If no initial current I_0 or charge Q_0 is present:

Case 1, $\dfrac{1}{LC} = \dfrac{R^2}{4L^2}$

$$i = \frac{Et}{L} e^{-R/2L} \tag{6-133}$$

Case 2, $\dfrac{1}{LC} > \dfrac{R^2}{4L^2}$

$$i = \frac{E}{\omega L} e^{-at} \sin \omega t \tag{6-134}$$

with oscillatory frequency $f = \omega/2\pi$ cycles per second

Case 3, $\dfrac{R^2}{4L^2} > \dfrac{1}{LC}$

$$i = \frac{E}{\omega L} e^{-at} \sinh \omega t \tag{6-135}$$

where $\omega = \sqrt{R^2/4L^2 - 1/LC}$

ELECTRICAL MACHINES

D-C Motors and Generators

Generated Emf between brush sets or terminals

$$E = \frac{p\phi Zn}{60 \times 10^8 m} = \frac{2f\phi Z}{10^8 m} \text{ volts} \tag{6-136}$$

in which p is the number of field poles, ϕ (Maxwells) is the flux per pole intercepted by armature conductors, Z is the total number of active conductors in the armature winding, n (revolutions per minute) is the armature speed, m is the number of parallel circuits through the armature, and f (cycles per minute) is the frequency of emf generated in each conductor.

Terminal Voltage V. If I is the armature current, R is the resistance between brushes, and E the generated emf

$$V = E - IR \tag{6-137}$$

Armature Torque τ of a Generator. With the symbols given for Eq. 6-136,

$$\tau = \frac{p\phi ZI}{8.52 \times 10^8 m} \text{ lb-ft} \tag{6-138}$$

Power Input P_i to a generator, or power output of a d-c motor

$$P_i = 0.1420\tau n \text{ watts}$$
$$= 1.903\tau n \times 10^{-4} \text{ horsepower} \quad (6\text{-}139)$$

Power Output P_0 of a generator, or power input to a d-c motor

$$P_0 = VI \text{ watts}$$
$$= 1.341 VI \times 10^{-4} \text{ horsepower} \quad (6\text{-}140)$$

Internal Generated Voltage. If f is the frequency in cycles per second, ϕ is the flux per pole in maxwells, N is the number of series-connected active conductors in each phase of the armature winding (twice the number of turns per phase), p is the number of poles, and n is the number of revolutions per minute, the average emf generated for a "full-pitch" winding is approximately

$$E = 2.1 f \phi N \times 10^{-8} \text{ volts} \quad (6\text{-}141)$$

and
$$f = \frac{pn}{120} \text{ cycles per second} \quad (6\text{-}142)$$

Induction Motors

Synchronous Speed n_s is the speed of rotation of magnetic flux.

$$n_s = 120 f/p \text{ rpm} \quad (6\text{-}143)$$

where f is the frequency of emf and current in cycles per second and p is the number of poles in the stator winding.

Slip S is the relative difference between the actual speed of the rotor n and of the rotating flux n_s

$$S = \frac{n_s - n}{n_s} \quad (6\text{-}144)$$

Shaft Torque τ. If R_r is the rotor resistance in ohms and I_r is the rotor current in amperes

$$\tau = \frac{21.12 R_r I_r^2}{n_s S_{\text{per unit}}} \text{ lb-ft} \quad (6\text{-}145)$$

Total Shaft Power and Efficiency. Excluding friction and windage losses,

$$P = 3 R_r I_r^2 \left(\frac{1-S}{S}\right) \text{ watts}$$
$$= 4.023 R_r I_r^2 \left(\frac{1-S}{S}\right) \times 10^{-4} \text{ horsepower} \quad (6\text{-}146)$$
$$\text{Efficiency} = 1 - S \quad (6\text{-}147)$$

Table 6-3 Electrical Properties of Insulating Materials

Material	Dielectric Strength		Resistivity				Specific Inductive Capacity, ϵ
	Specimen Thickness, mm	Kv per mm*	Volume, ohm-cm	Surface†			Air Unity
				Ohms, 30%	Ohms, 90%		
Asbestos paper	1.2	4.2	1.6×10^{11}				
Asphalt (Byerlyte)	3.6	14.0					2.7
Bakelite, wood molding mixture		17.7 to 21.6	1×10^{12}				4.5 to 5.5
Bakelite, asbestos molding mixture		up to 9.8	4×10^{11}				
Bakelite, Micarta-213		up to 31.4	5×10^{11}				
Cellophane	0.022	51 to 66					5
Celluloid (clear)	0.25	12 to 28	2×10^{10}	8×10^{10}	2×10^{9}		8
Cellulose acetate	0.019	48.0		8×10^{16}	8×10^{16}		5
Ceresin			over 5×10^{18}				
Empire cloth, muslin	0.38	48.0					
Fiber, vulcanized, including hard fiber, all colors	{3.2 6.4}	4.9 to 10.8 3.9 to 8.9	5 to 20 $\times 10^{9}$ 5 to 20 $\times 10^{9}$	3×10^{10} 3×10^{10}	1×10^{7} 1×10^{7}		5 5
Glass (ordinary)		8 to 9	9×10^{13}				5.5 to 9.1
Glass (plate)			2×10^{13}	3×10^{13}	2×10^{7}		5.5 to 9.1
Jute (impregnated)	6	1.2					3 to 4
Lava		3 to 10	2×10^{10}	6×10^{11}	1×10^{8}		
Marble		6.5	1 to 100 $\times 10^{9}$	8×10^{10}	2×10^{7}		8.3
Mica		21 to 28	0.04 to 200 $\times 10^{15}$	2×10^{13}	3×10^{9}		5 to 7
Micabond, plate	0.6	37.5					2.5
Micabond, flexible	1.6	23.1					2.6
Oil, insulating	1.6	10–16					2.1
Paper	2.54	8.7					4.4
Paraffin (parowax)	0.13	11.5	5×10^{4}	1.5×10^{16}	5×10^{15}		5.0
Porcelain	20	8.0	1×10^{16}	4×10^{13}	5×10^{8}		
Pressboard (oiled)	1.58	29.2	3×10^{14}				
Pressboard (varnished)	1.58	15.5					3
Rosin			5×10^{16}	8×10^{14}	2×10^{14}		2.5
Rubber (hard)	0.5	70	1×10^{18}	6×10^{15}	1×10^{9}		2.0 to 3.5
Shellac			1×10^{16}	2×10^{14}	6×10^{9}		3.0 to 3.7
Slate	10.3	1.3	1×10^{18}	2×10^{8}	1×10^{7}		6.6 to 7.4
Sulfur			1×10^{17}	1×10^{16}	1×10^{14}		2.9 to 3.2
Wood (maple), paraffined	15.2	4.6	3×10^{10}	1×10^{12}	2×10^{9}		4.1

* To obtain volts per mil multiply kilovolts per millimeter by 25.4.
† At 30 per cent and 90 per cent relative humidity.

ELECTRICAL MACHINES 233

Table 6-4 General Data on Magnetic Properties of Commercial Materials

Figures represent approximate average properties; individual samples may differ somewhat

	Supermalloy (Western Electric)	78 Permalloy (Western Electric)	Mu Metal (Allegheny Ludlum)	Hipernik (Westinghouse Electric)	Trancor 3X (oriented) (Armco)	52 Grade	Low-Carbon Steel Sheet or Shapes	Cast Steel (Annealed)
Physical								
Gage, in.	0.002	0.014	0.014	0.014	0.014	0.014		
Density, gm/cm³	8.77	8.60	8.58	8.25	7.65*	7.55*	7.85	7.80
Sheet weight, lb/sq ft		0.63	0.63	0.60	0.556	0.549		
Ultimate tensile strength, lb/sq in.			70,000	65,000	67,000	70,000	45,000	60,000
Yield point, lb/sq in.			45,000	20,000	56,000	66,000	25,000	30,000
Elongation, per cent in 2-in. gage length			23	50	12	2	18	20
Ductility, Erichsen draw, mm				8	5.5	2.5	9	
Miscellaneous								
Approximate per cent silicon					3.0	5.0		0.4
Electrical resistivity, microhm-cm	60	16	42	45	47	65	13	15
Magnetic								
Initial permeability	100,000	9,000	20,000	6,000	1,500	9,800	250	175
Maximum permeability	800,000	100,000	100,000	90,000	51,500	19,200	2,500	1,500
Saturation induction (ferric), gausses	7,900	10,700	6,500	16,000	20,200	19,200	21,200	21,000
Coercive force, oersteds (from 10,000 gauss tip)	0.002	0.05	0.04	0.06	0.1	0.22	2.0	5.0
Steinmetz hysteresis coefficient		0.0001		0.00015			0.003	0.005
Hysteresis loss, ergs/cm³/cycle	10	200	200	200		640		
Iron loss at 10,000 gausses, 60 cycles, watts/lb				0.25	0.32	0.52		
Iron loss at 15,000 gausses, 60 cycles, watts/lb					0.73	1.33		
Maximum aging, iron loss, per cent	Nil	Nil	Nil	Nil	Nil	3		
Approximate cost, per cent			1,940	1,790	220	170	52	
Typical applications	Telephone equipment, magnetic amplifiers, audio transformers, instrument transformers				Distribution and power transformers		Fields, frame of d-c and synchronous machines	Frames, solid poles

Table 6-5 Wire Table, Standard Annealed Copper American Wire Gage (B. & S.). English Units

Gage No. A.W.G.	Diameter in Mils at 20°C	Cross Section at 20°C		Ohms per 1000 ft* at 20°C (=68 F)	Pounds per 1000 ft	Feet per Pound	Feet per Ohm† at 20°C (68 F)	Ohms per Pound at 20°C (=68 F)
		Circular Mils	Square Inches					
0000	460.0	211 600.	0.1662	0.049 01	640.5	1.561	20 400.	0.000 076 52
000	409.6	167 800.	.1318	0.061 80	507.9	1.968	16 180.	0.000 1217
00	364.8	133 100.	.1045	0.077 93	402.8	2.482	12 830.	0.000 1935
0	324.9	105 500.	.082 89	0.098 27	319.5	3.130	10 180.	0.000 3076
1	289.3	83 690.	.065 73	0.1239	253.3	3.947	8070.	0.000 4891
2	257.6	66 370.	.052 13	0.1563	200.9	4.977	6400.	0.000 7778
3	229.4	52 640.	.041 34	0.1970	159.3	6.276	5075.	0.001 237
4	204.3	41 740.	.032 78	0.2485	126.4	7.914	4025.	0.001 966
5	181.9	33 100.	.026 00	0.3133	100.2	9.980	3192.	0.003 127
6	162.0	26 250.	.020 62	0.3951	79.46	12.58	2531.	0.004 972
7	144.3	20 820.	.016 35	0.4982	63.02	15.87	2007.	0.007 905
8	128.5	16 510.	.012 97	0.6282	49.98	20.01	1592.	0.012 57
9	114.4	13 090.	.010 28	0.7921	39.63	25.23	1262.	0.019 99
10	101.9	10 380.	.008 155	0.9989	31.43	31.82	1001.	0.031 78
11	90.74	8234.	.006 467	1.260	24.92	40.12	794.0	0.050 53
12	80.81	6530.	.005 129	1.588	19.77	50.59	629.6	0.080 35
13	71.96	5178.	.004 067	2.003	15.68	63.80	499.3	0.1278
14	64.08	4107.	.003 225	2.525	12.43	80.44	396.0	0.2032
15	57.07	3257.	.002 558	3.184	9.858	101.4	314.0	0.3230
16	50.82	2583.	.002 028	4.016	7.818	127.9	249.0	0.5136
17	45.26	2048.	.001 609	5.064	6.200	161.3	197.5	0.8167
18	40.30	1624.	.001 276	6.385	4.917	203.4	156.6	1.299
19	35.89	1288.	.001 012	8.051	3.899	256.5	124.2	2.065
20	31.96	1022.	.000 802 3	10.15	3.092	323.4	98.50	3.283
21	28.46	810.1	.000 636 3	12.80	2.452	407.8	78.11	5.221
22	25.35	642.4	.000 504 6	16.14	1.945	514.2	61.95	8.301
23	22.57	509.5	.000 400 2	20.36	1.542	648.4	49.13	13.20
24	20.10	404.0	.000 317 3	25.67	1.223	817.7	38.96	20.99
25	17.90	320.4	.000 251 7	32.37	0.9699	1031.	30.90	33.37
26	15.94	254.1	.000 199 6	40.81	0.7692	1300.	24.50	53.06
27	14.20	201.5	.000 158 3	51.47	0.6100	1639.	19.43	84.37
28	12.64	159.8	.000 125 5	64.90	0.4837	2067.	15.41	134.2
29	11.26	126.7	.000 099 53	81.83	0.3836	2607.	12.22	213.3
30	10.03	100.5	.000 078 94	103.2	0.3042	3287.	9.691	339.2
31	8.928	79.70	.000 062 60	130.1	0.2413	4145.	7.685	539.3
32	7.950	63.21	.000 049 64	164.1	0.1913	5227.	6.095	857.6
33	7.080	50.13	.000 039 37	206.9	0.1517	6591.	4.833	1364.
34	6.305	39.75	.000 031 22	260.9	0.1203	8310.	3.833	2168.
35	5.615	31.52	.000 024 76	329.0	0.095 42	10 480.	3.040	3448.
36	5.000	25.00	.000 019 64	414.8	0.075 68	13 210.	2.411	5482.
37	4.453	19.83	.000 015 57	523.1	0.060 01	16 660.	1.912	8717.
38	3.965	15.72	.000 012 35	659.6	0.047 59	21 010.	1.516	13 860.
39	3.531	12.47	.000 009 793	831.8	0.037 74	26 500.	1.202	22 040.
40	3.145	9.888	.000 007 766	1049.	0.029 93	33 410.	0.9534	35 040.

* Resistance at the stated temperatures of a wire whose length is 1000 ft at 20°C.
† Length at 20°C of a wire whose resistance is 1 ohm at the stated temperatures.

Table 6-6 Wire Table, Aluminum

Hard-drawn aluminum wire at 20°C (68°F)
American Wire Gage (B. & S.), English Units

Gage No.	Diameter, Mils	Cross Section		Ohms per 1000 ft	Pounds per 1000 ft	Pounds per Ohm
		Circular, Mils	Square Inches			
0000	460.	212 000.	0.166	0.0804	195.	2420.
000	410.	168 000.	.132	0.101	154.	1520.
00	365.	133 000.	.105	0.128	122.	957.
0	325.	106 000.	.0829	0.161	97.0	602.
1	289.	83 700.	.0657	0.203	76.9	379.
2	258.	66 400.	.0521	0.256	61.0	238.
3	229.	52 600.	.0413	0.323	48.4	150.
4	204.	41 700.	.0328	0.408	38.4	94.2
5	182.	33 100.	.0260	0.514	30.4	59.2
6	162.	26 300.	.0206	0.648	24.1	37.2
7	144.	20 800.	.0164	0.817	19.1	23.4
8	128.	16 500.	.0130	1.03	15.2	14.7
9	114.	13 100.	.0103	1.30	12.0	9.26
10	102.	10 400.	.008 15	1.64	9.55	5.83
11	91.	8230.	.006 47	2.07	7.57	3.66
12	81.	6530.	.005 13	2.61	6.00	2.30
13	72.	5180.	.004 07	3.29	4.76	1.45
14	64.	4110.	.003 23	4.14	3.78	0.911
15	57.	3260.	.002 56	5.22	2.99	0.573
16	51.	2580.	.002 03	6.59	2.37	0.360
17	45.	2050.	.001 61	8.31	1.88	0.227
18	40.	1620.	.001 28	10.5	1.49	0.143
19	36.	1290.	.001 01	13.2	1.18	0.0897
20	32.	1020.	.000 802	16.7	0.939	0.0564
21	28.5	810.	.000 636	21.0	0.745	0.0355
22	25.3	642.	.000 505	26.5	0.591	0.0223
23	22.6	509.	.000 400	33.4	0.468	0.0140
24	20.1	404.	.000 317	42.1	0.371	0.008 82
25	17.9	320.	.000 252	53.1	0.295	0.005 55
26	15.9	254.	.000 200	67.0	0.234	0.003 49
27	14.2	202.	.000 158	84.4	0.185	0.002 19
28	12.6	160.	.000 126	106.	0.147	0.001 38
29	11.3	127.	.000 099 5	134.	0.117	0.000 868
30	10.0	101.	.000 078 9	169.	0.0924	0.000 546
31	8.9	79.7	.000 062 6	213.	0.0733	0.000 343
32	8.0	63.2	.000 049 6	269.	0.0581	0.000 216
33	7.1	50.1	.000 039 4	339.	0.0461	0.000 136
34	6.3	39.8	.000 031 2	428.	0.0365	0.000 085 4
35	5.6	31.5	.000 024 8	540.	0.0290	0.000 053 7
36	5.0	25.0	.000 019 6	681.	0.0230	0.000 033 8
37	4.5	19.8	.000 015 6	858.	0.0182	0.000 021 2
38	4.0	15.7	.000 012 3	1080.	0.0145	0.000 013 4
39	3.5	12.5	.000 009 79	1360.	0.0115	0.000 008 40
40	3.1	9.9	.000 007 77	1720.	0.0091	0.000 005 28

ELECTRONICS

Electron Dynamics

Electron in Electric Fields. Application of the principles of particle dynamics to the electron, with charge of 1.6×10^{-19} coulomb and mass of 9.1×10^{-31} kilogram, gives the relations:

Force, $\qquad \mathbf{F} = 1.6 \times 10^{-19}\,\mathrm{grad}\,V$ newton/m $\qquad$ (6-148)

Acceleration, $\qquad \mathbf{a} = 1.76 \times 10^{11}\,\mathrm{grad}\,V$ meters/sec² $\qquad$ (6-149)

Velocity at anode in uniform electric field,

$$v = (v_i^2 \pm 35.2 \times 10^{10} V)^{1/2} \text{ meter/sec} \qquad (6\text{-}150)$$

where v_i is the initial velocity parallel to **E**.

The path in a uniform electric field is a parabola (see Eq. 2-32).

Electrostatic deflection between plates in a cathode-ray tube is given by

$$y = -V_D x^2 / f \delta V_0 \qquad (6\text{-}151)$$

in which V_D is the potential difference, δ is the distance between the deflecting plates, x is the horizontal component of the path, and V_0 is the potential difference in the electron gun. Relativistic effects become important when the electron velocity exceeds one-tenth the velocity of light or at about 3000 volts.

Electron in Magnetic Fields. In a uniform magnetic field the force on an electron is zero when it is at rest or is moving parallel to the direction of the field. When the velocity of an electron has a component normal to the magnetic field a magnetic force is developed (see Eq. 6-62). If the velocity has no component parallel to the field, the path is a circle (see Eqs. 6-65 and 6-66).

Magnetic deflection y in a cathode-ray tube is given by

$$y = Blx \left(\frac{q_e}{2mV_0} \right)^{1/2} \qquad (6\text{-}152)$$

in which l is the length of the magnetic path, x is the horizontal component of the path from the center of the field to the screen, and V_0 is the potential difference in the electron gun. Note that deflection depends on the ratio of charge to mass in a magnetic field but not in the electric field.

Parallel Electric and Magnetic Fields. If initially the electron is moving in the x-direction, the only force is that of the electric field which acts to move the electron in a straight line with constant acceleration. If the electron initially has a component of velocity normal to **B**, the projection of the path on the y-z plane is a circle of radius:

$$r = \frac{m(v_y^2 + v_z^2)^{1/2}}{q_e B} \tag{6-153}$$

The path is thus a helix of constant radius r and increasing pitch because of uniform acceleration in the x-direction.

Crossed Electric and Magnetic Fields. With electric and magnetic fields normal to each other, the path of the electron is one of a family of cycloids, i.e., the path of a point on the radius of a rolling circle. The initial velocity of the electron determines the distance of the point from the center of the circle. When the initial velocity $v_i = E/B$, the path is a straight line, i.e., point at the center of the circle. When $v_i = 0$ the path is a common cycloid, i.e., point on the circumference of the circle. For all cycloids the net drift velocity is equal to E/B.

Electron Tubes

Amplification factor:

$$\mu = \frac{dV_p}{dV_g}, \quad \text{with } I_p \text{ constant}$$

AC (dynamic) plate resistance:

$$r_p = \frac{dV_p}{dI_p}, \quad \text{with } V_g \text{ constant}$$

Mutual conductance:

$$g_m = \frac{dI_p}{dV_g}, \quad \text{with } V_p \text{ constant}$$

Gain of an amplifier stage:

$$\text{gain} = -\mu \frac{R_L}{R_L + r_p}$$

Table 6-7 gives approximate characteristics at selected operating conditions for several typical electron tubes. Values of the characteristics, obtained from manufacturers' tabulations or graphs similar to Fig. 6-10, may be used in the equations of Table 6-8 to evaluate the performance of a circuit.

Figure 6-11 shows the basic forms of amplifier coupling. Figure 6-9 gives some transistor and electron-tube equivalents.

Table 6-7 Typical Operating Conditions and Characteristic Values for Several Tube Types

Tube Type	Service	Plate Volts	Screen Volts	Negative Grid Volts	Plate Current, ma	Screen Current, ma	r_p, ohms	g_m, mhos	μ
6AK5	Class A amplifier	180	120	2.0	7.7	2.4	500,000	5100	—
		120	120	1.8	7.5	2.5	300,000	5000	—
12AT7	Class A amplifier	250	—	2.0	10	—	10,900	5500	60
		100	—	0.8	3.7	—	15,000	4000	60
12AU6	Class A amplifier	250	150	0.7	10.6	4.3	106	5200	—
		100	100	0.75	5.0	2.1	500,000	3900	—
12AU7	Class A amplifier	250	—	8.5	10.5	—	7,700	2200	17
		100	—	0	11.8	—	6,500	3100	20
12AX7	Class A Amplifier	100	—	1.0	0.5	—	80,000	1250	100
		250	—	2.0	1.2	—	62,500	1600	100
12BD6	Class A amplifier	250	100	3.0	9.0	3.5	700,000	2000	—
12BK5	Class A amplifier	250	250	5.0	35	3.5	100,000	8500	—
12BQ6	Reflection amplifier	250	150	22.5	57	2.1	14,500	5900	—
12BY7	Class A amplifier	250	180	3.2	26	5.8	93,000	11000	—
5672	Class A amplifier	67.5	67.5	6.5	3.2	1.1	—	650	—
5687	Class A amplifier	180	—	7.0	21	—	2100	8250	17.5
		250	—	12.5	12.5	—	3000	5500	16.5
5693	Class A amplifier	250	100	3.0	3.0	0.8	10^6	1650	—

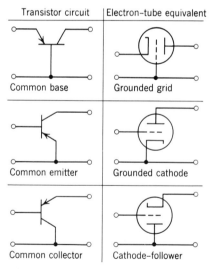

Fig. 6-9 Transistor and electron-tube equivalents. (Reproduced by permission from *Reference Data for Radio Engineers*, International Tel. and Tel. Corp., New York, fourth edition, 1956.)

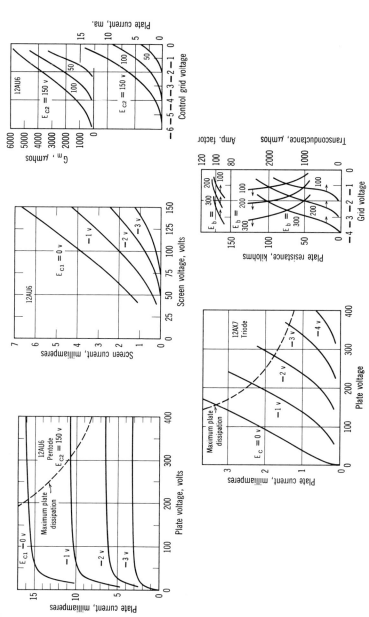

Fig. 6-10 Typical characteristics of two electron tubes.

Table 6-8 Equivalent Amplifier Circuits

	Grounded-Cathode	Grounded-Grid	Grounded-Plate or Cathode-Follower
Circuit Schematic			
Equivalent Circuit Alternating Current Component, Class-A Operation			
Voltage Gain, A for Output Load Impedance $= Z_2$ $$A = \frac{E_2}{E_1}$$	$A = \dfrac{-\mu Z_2}{r_p + Z_2}$ $= -g_m \dfrac{r_p Z_2}{r_p + Z_2}$ neglecting C_{gp} (Z_2 includes C_{pk})	$A = (1 + \mu) \dfrac{Z_2}{r_p + Z_2}$ neglecting C_{pk} (Z_2 includes C_{gp})	$A = \dfrac{\mu Z_2}{r_p + (1 + \mu) Z_2}$ neglecting C_{gk} (Z_2 includes C_{pk})
Input Admittance $$Y_1 = \frac{I_1}{E_1}$$	$Y_1 = j\omega[C_{gk} + (1 - A)C_{gp}]$	$Y_1 = j\omega[C_{gk} + (1 - A)C_{pk}] + \dfrac{1 + \mu}{r_p + Z_2}$	$Y_1 = j\omega[C_{gp} + (1 - A)C_{gk}]$
Equivalent Generator Seen by Load at Output Terminals	r_p; $-\mu E_1$; neglecting C_{gp}	r_p; $(1 + \mu)E_1$; neglecting C_{pk}	$\dfrac{r_p}{1 + \mu}$; $\dfrac{\mu}{1 + \mu} E_1$; neglecting C_{gk}

Reproduced by permission from *Reference Data for Radio Engineers*, International Tel. and Tel. Corp., New York, fourth edition, 1956.

ELECTRONICS 241

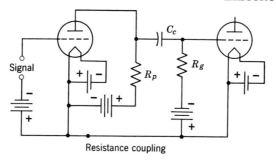

Resistance coupling

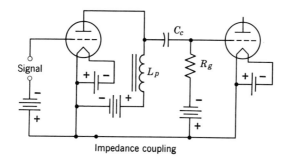

Impedance coupling

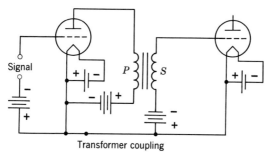

Transformer coupling

Fig. 6-11 Amplifier coupling.

Transistors*

Figures 6-12 and 6-13 show the three modes of connection for *p-n-p* and *n-p-n* transistors.

Input impedance: $$Z_i = \frac{dV_i}{dI_i}$$

Load impedance: $$Z_1 = \frac{dV_1}{dI_1}$$

* This discussion of transistors has been adapted by permission from R. F. Shea, *Transistor Applications*, John Wiley and Sons, New York, 1964.

Power gain:
$$A = \frac{P_1}{P_i}$$

Short-circuit current gain for common base:
$$\alpha = \frac{dI_c}{dI_e} = \frac{\beta}{1+\beta}$$

Short-circuit current gain for common emitter:
$$\beta = \frac{dI_c}{dI_b} = \frac{\alpha}{1-\alpha}$$

Two-Port Networks

Conventionally current flowing *into* the network is positive.

Admittance parameters:
$$I_1 = y_{11}V_1 + y_{12}V_2$$
$$I_2 = y_{21}V_1 + y_{22}V_2$$

y_{11} = input admittance with output short circuited
y_{12} = backward transfer admittance with input short circuited
y_{21} = forward transfer admittance with output short circuited
y_{22} = output admittance with input short circuited

Impedance parameters:
$$V_1 = z_{11}I_1 + z_{12}I_2$$
$$V_2 = z_{21}I_1 + z_{22}I_2$$

z_{11} = input impedance with open-circuit output
z_{12} = backward transfer impedance with open-circuit input
z_{21} = forward transfer impedance with open-circuit output
z_{22} = output impedance with open-circuit input

Hybrid parameters:
$$V_1 = h_{11}I_1 + h_{12}V_2$$
$$I_2 = h_{21}I_1 + h_{22}V_2$$

h_{11} = input impedance with output short circuited
h_{12} = reverse voltage gain with open-circuit input
h_{21} = forward current gain with output short circuited
h_{22} = output admittance with open-circuit input

Other parameters sometimes used are the *g*, *a*, and *b* parameters.
Manufacturers' specification sheets usually give values for the hybrid parameters expressed as h_i, h_r, h_f, and h_0 which correspond respectively to h_{11}, h_{12}, h_{21} and h_{22} as used in the network equations.

Table 6-9 Matrix Interrelations

In Terms of

	z	y	h	g	a	b
[z]	—	$\dfrac{y_{22}}{\Delta^y}\ \dfrac{-y_{12}}{\Delta^y}$ $\dfrac{-y_{21}}{\Delta^y}\ \dfrac{y_{11}}{\Delta^y}$	$\dfrac{\Delta^h}{h_{22}}\ \dfrac{h_{12}}{h_{22}}$ $\dfrac{-h_{21}}{h_{22}}\ \dfrac{1}{h_{22}}$	$\dfrac{1}{g_{11}}\ \dfrac{-g_{12}}{g_{11}}$ $\dfrac{g_{21}}{g_{11}}\ \dfrac{\Delta^g}{g_{11}}$	$\dfrac{a_{11}}{a_{21}}\ \dfrac{\Delta^a}{a_{21}}$ $\dfrac{1}{a_{21}}\ \dfrac{a_{22}}{a_{21}}$	$\dfrac{b_{22}}{b_{21}}\ \dfrac{1}{b_{21}}$ $\dfrac{\Delta^b}{b_{21}}\ \dfrac{b_{11}}{b_{21}}$
[y]	$\dfrac{z_{22}}{\Delta^z}\ \dfrac{-z_{12}}{\Delta^z}$ $\dfrac{-z_{21}}{\Delta^z}\ \dfrac{z_{11}}{\Delta^z}$	—	$\dfrac{1}{h_{11}}\ \dfrac{-h_{12}}{h_{11}}$ $\dfrac{h_{21}}{h_{11}}\ \dfrac{\Delta^h}{h_{11}}$	$\dfrac{\Delta^g}{g_{22}}\ \dfrac{g_{12}}{g_{22}}$ $\dfrac{-g_{21}}{g_{22}}\ \dfrac{1}{g_{22}}$	$\dfrac{a_{22}}{a_{12}}\ \dfrac{-\Delta^a}{a_{12}}$ $\dfrac{-1}{a_{12}}\ \dfrac{a_{11}}{a_{12}}$	$\dfrac{b_{11}}{b_{12}}\ \dfrac{-1}{b_{12}}$ $\dfrac{-\Delta^b}{b_{12}}\ \dfrac{b_{22}}{b_{12}}$
[h]	$\dfrac{\Delta^z}{z_{22}}\ \dfrac{z_{12}}{z_{22}}$ $\dfrac{-z_{21}}{z_{22}}\ \dfrac{1}{z_{22}}$	$\dfrac{1}{y_{11}}\ \dfrac{-y_{12}}{y_{11}}$ $\dfrac{y_{21}}{y_{11}}\ \dfrac{\Delta^y}{y_{11}}$	—	$\dfrac{g_{22}}{\Delta^g}\ \dfrac{-g_{12}}{\Delta^g}$ $\dfrac{-g_{21}}{\Delta^g}\ \dfrac{g_{11}}{\Delta^g}$	$\dfrac{a_{12}}{a_{22}}\ \dfrac{\Delta^a}{a_{22}}$ $\dfrac{-1}{a_{22}}\ \dfrac{a_{21}}{a_{22}}$	$\dfrac{b_{12}}{b_{11}}\ \dfrac{1}{b_{11}}$ $\dfrac{-\Delta^b}{b_{11}}\ \dfrac{b_{21}}{b_{11}}$
[g]	$\dfrac{1}{z_{11}}\ \dfrac{-z_{12}}{z_{11}}$ $\dfrac{z_{21}}{z_{11}}\ \dfrac{\Delta^z}{z_{11}}$	$\dfrac{\Delta^y}{y_{22}}\ \dfrac{y_{12}}{y_{22}}$ $\dfrac{-y_{21}}{y_{22}}\ \dfrac{1}{y_{22}}$	$\dfrac{h_{22}}{\Delta^h}\ \dfrac{-h_{12}}{\Delta^h}$ $\dfrac{-h_{21}}{\Delta^h}\ \dfrac{h_{11}}{\Delta^h}$	—	$\dfrac{a_{21}}{a_{11}}\ \dfrac{-\Delta^a}{a_{11}}$ $\dfrac{1}{a_{11}}\ \dfrac{a_{12}}{a_{11}}$	$\dfrac{b_{21}}{b_{22}}\ \dfrac{-1}{b_{22}}$ $\dfrac{\Delta^b}{b_{22}}\ \dfrac{b_{12}}{b_{22}}$
[a]	$\dfrac{z_{11}}{z_{21}}\ \dfrac{\Delta^z}{z_{21}}$ $\dfrac{1}{z_{21}}\ \dfrac{z_{22}}{z_{21}}$	$\dfrac{-y_{22}}{y_{21}}\ \dfrac{-1}{y_{21}}$ $\dfrac{-\Delta^y}{y_{21}}\ \dfrac{-y_{11}}{y_{21}}$	$\dfrac{-\Delta^h}{h_{21}}\ \dfrac{-h_{11}}{h_{21}}$ $\dfrac{-h_{22}}{h_{21}}\ \dfrac{-1}{h_{21}}$	$\dfrac{1}{g_{21}}\ \dfrac{g_{22}}{g_{21}}$ $\dfrac{g_{11}}{g_{21}}\ \dfrac{\Delta^g}{g_{21}}$	—	$\dfrac{b_{22}}{\Delta^b}\ \dfrac{b_{12}}{\Delta^b}$ $\dfrac{b_{21}}{\Delta^b}\ \dfrac{b_{11}}{\Delta^b}$
[b]	$\dfrac{z_{22}}{z_{12}}\ \dfrac{\Delta^z}{z_{12}}$ $\dfrac{1}{z_{12}}\ \dfrac{z_{11}}{z_{12}}$	$\dfrac{-y_{11}}{y_{12}}\ \dfrac{-1}{y_{12}}$ $\dfrac{-\Delta^y}{y_{12}}\ \dfrac{-y_{22}}{y_{12}}$	$\dfrac{1}{h_{12}}\ \dfrac{h_{11}}{h_{12}}$ $\dfrac{h_{22}}{h_{12}}\ \dfrac{\Delta^h}{h_{12}}$	$\dfrac{-\Delta^g}{g_{12}}\ \dfrac{-g_{22}}{g_{12}}$ $\dfrac{-g_{11}}{g_{12}}\ \dfrac{-1}{g_{12}}$	$\dfrac{a_{22}}{\Delta^a}\ \dfrac{a_{12}}{\Delta^a}$ $\dfrac{a_{21}}{\Delta^a}\ \dfrac{a_{11}}{\Delta^a}$	—

Table 6-10 Determinant Interrelations

	\multicolumn{6}{c}{In Terms of}
	z	y	h	g	a	b
Δ^z	—	$\dfrac{1}{\Delta^y}$	$\dfrac{h_{11}}{h_{22}}$	$\dfrac{g_{22}}{g_{11}}$	$\dfrac{a_{12}}{a_{21}}$	$\dfrac{b_{12}}{b_{21}}$
Δ^y	$\dfrac{1}{\Delta^z}$	—	$\dfrac{h_{22}}{h_{11}}$	$\dfrac{g_{11}}{g_{22}}$	$\dfrac{a_{21}}{a_{12}}$	$\dfrac{b_{21}}{b_{12}}$
Δ^h	$\dfrac{z_{11}}{z_{22}}$	$\dfrac{y_{22}}{y_{11}}$	—	$\dfrac{1}{\Delta^g}$	$\dfrac{a_{11}}{a_{22}}$	$\dfrac{b_{22}}{b_{11}}$
Δ^g	$\dfrac{z_{22}}{z_{11}}$	$\dfrac{y_{11}}{y_{22}}$	$\dfrac{1}{\Delta^h}$	—	$\dfrac{a_{22}}{a_{11}}$	$\dfrac{b_{11}}{b_{22}}$
Δ^a	$\dfrac{z_{12}}{z_{21}}$	$\dfrac{y_{12}}{y_{21}}$	$-\dfrac{h_{12}}{h_{21}}$	$-\dfrac{g_{12}}{g_{21}}$	—	$\dfrac{1}{\Delta^b}$
Δ^b	$\dfrac{z_{21}}{z_{12}}$	$\dfrac{y_{21}}{y_{12}}$	$-\dfrac{h_{21}}{h_{12}}$	$-\dfrac{g_{21}}{g_{12}}$	$\dfrac{1}{\Delta^a}$	—

Matrix determinants designated by

$$\begin{vmatrix} n_{11} & n_{12} \\ n_{21} & n_{22} \end{vmatrix}$$

are evaluated as the difference between the products of the diagonal terms

$$\Delta^n = n_{11}n_{22} - n_{12}n_{21}$$

where n is any one of the parameters $y, z, h, g, a,$ or b.

Table 6-9 gives the matrix relationships among the six different sets parameters, and Table 6-10 gives the relationships among their determinant In Table 6-9, for example, a_{11} is found to be equal to $-y_{22}/y_{21}$.

Table 6-11 gives the approximate matrix relationships among the h paramete for the three modes of transistor connection shown in Fig. 6-12. Table 6- gives the interrelations for the y parameters and Table 6-13 for the z paramete

ELECTRONICS 245

Table 6-11 Matrix Interrelations of Transistor h Parameters (Approximate) In Terms of Typical Manufacturers Parameters

	Common Base	Common Emitter	Common Collector
h_{11b}	h_{ib}	$h_{ie}/(1 + h_{fe})$	$h_{ic}/(-h_{fc})$
h_{12b}	h_{rb}	$(\Delta_e^h - h_{re})/(1 + h_{fe})$	$(\Delta_c^h + h_{fc})/(-h_{fc})$
h_{21b}	h_{fb}	$-h_{fe}/(1 + h_{fe})$	$-h_{rc}$
h_{22b}	h_{ob}	$h_{oe}/(1 + h_{fe})$	$h_{oc}/(-h_{fc})$
Δ_b^h	$\begin{vmatrix} h_{ib} & h_{rb} \\ h_{fb} & h_{ob} \end{vmatrix}$	$\Delta_e^h/(1 + h_{fe})$	$(h_{fc} + \Delta_c^h)/(-h_{fc})$
h_{11e}	$h_{ib}/(1 + h_{fb})$	h_{ie}	h_{ic}
h_{12e}	$(\Delta_b^h - h_{rb})/(1 + h_{fb})$	h_{re}	$1 - h_{rc}$
h_{21e}	$-h_{fb}/(1 + h_{fb})$	h_{fe}	$-h_{fc}$
h_{22e}	$h_{ob}/(1 + h_{fb})$	h_{oe}	h_{oc}
Δ_e^h	$\Delta_b^h/(1 + h_{fb})$	$\begin{vmatrix} h_{ie} & h_{re} \\ h_{fe} & h_{oe} \end{vmatrix}$	$h_{fc} + \Delta_c^h$
h_{11c}	$h_{ib}/(1 + h_{fb})$	h_{ie}	h_{ic}
h_{21c}	1	1	h_{rc}
h_{12c}	$-1/(1 + h_{cb})$	$-(1 + h_{fe})$	h_{fc}
h_{22c}	$h_{ob}/(1 + h_{fb})$	h_{oe}	h_{oc}
Δ_c^h	$1/(1 + h_{fb})$	$1 + h_{fe}$	$\begin{vmatrix} h_{ic} & h_{rc} \\ h_{fc} & h_{oc} \end{vmatrix}$

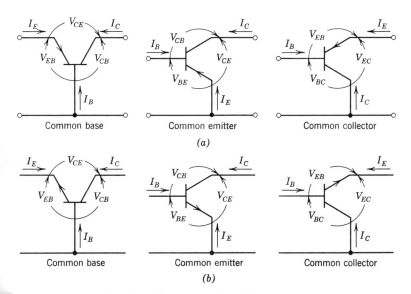

Fig. 6-12 Three modes of transistor connection (a) p-n-p type. (b) n-p-n type.

Table 6-12 Matrix Interrelations of the Transistor y Parameters

	In Terms of Common-Base Parameters	In Terms of Common-Emitter Parameters	In Terms of Common-Collector Parameters
$[y_b]$	—	$\begin{matrix} y_{ie}+y_{re}+y_{fe}+y_{oe} & -y_{oe}-y_{re} \\ -y_{oe}-y_{fe} & y_{oe} \end{matrix}$	$\begin{matrix} y_{oc} & -y_{oc}-y_{fc} \\ -y_{oc}-y_{rc} & y_{ic}+y_{rc}+y_{fc}+y_{oc} \end{matrix}$
$[y_e]$	$\begin{matrix} y_{ib}+y_{rb}+y_{fb}+y_{ob} & -y_{ob}-y_{rb} \\ -y_{ob}-y_{fb} & y_{ob} \end{matrix}$	—	$\begin{matrix} y_{ic} & -y_{ic}-y_{rc} \\ -y_{ic}-y_{fc} & y_{ic}+y_{rc}+y_{fc}+y_{oc} \end{matrix}$
$[y_c]$	$\begin{matrix} y_{ib}+y_{rb}+y_{fb}+y_{ob} & -y_{ib}-y_{fb} \\ -y_{ib}-y_{rb} & y_{ib} \end{matrix}$	$\begin{matrix} y_{ie} & -y_{ie}-y_{re} \\ -y_{ie}-y_{fe} & y_{ie}+y_{re}+y_{fe}+y_{oe} \end{matrix}$	—

Table 6-13 Matrix Interrelations of the Transistor z Parameters

	In Terms of Common-Base Parameters	In Terms of Common-Emitter Parameters	In Terms of Common-Collector Parameters
$[z_b]$	—	$\begin{matrix} z_{ie}-z_{re} & z_{re} \\ z_{ie}-z_{re}-z_{fe}+z_{oe} & z_{oe}-z_{re} \end{matrix}$	$\begin{matrix} z_{ic}-z_{rc}-z_{fc}+z_{oc} & z_{ic}-z_{fc} \\ z_{ic}-z_{rc} & z_{ic} \end{matrix}$
$[z_e]$	$\begin{matrix} z_{ib}-z_{rb} & z_{rb} \\ z_{ib}-z_{rb}-z_{fb}+z_{ob} & z_{ob}-z_{rb} \end{matrix}$	—	$\begin{matrix} z_{ic}-z_{rc}-z_{fc}+z_{oc} & z_{ic}-z_{fc} \\ z_{oc}-z_{fc} & z_{oc} \end{matrix}$
$[z_c]$	$\begin{matrix} z_{ob}-z_{rb} & z_{ob}-z_{fb} \\ z_{ib}-z_{rb}-z_{fb}+z_{ob} & z_{ob} \end{matrix}$	$\begin{matrix} z_{ie}-z_{re}-z_{fe}+z_{oe} & z_{oe}-z_{re} \\ z_{oe}-z_{fe} & z_{oe} \end{matrix}$	—

Terminated Network

When the appropriate set of parameters has been evaluated, the performance of the terminated network (see Fig. 6-13) may be found from the relations given in Table 6-14.

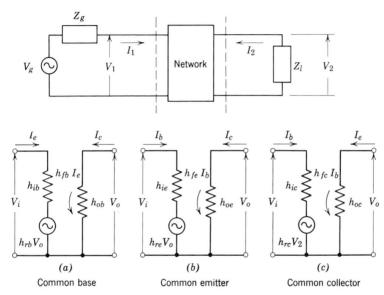

Fig. 6-13 Terminated network and equivalent circuits for three modes of transistor connection.

Figure 6-13 shows the equivalent circuits and parameters used in the analysis of low-frequency circuits. Tables 6-15, 6-16, and 6-17 give equations for the performance of terminated networks in terms of the parameters most frequently found in data tabulations. The following definitions apply to these equations.

A_i = current amplification
A_v = voltage amplification
G = power gain = $\dfrac{\text{power dissipated in load}}{\text{power dissipated in input}}$
G_t = transducer power gain = $\dfrac{\text{power dissipated in load}}{\text{power available from source}}$
G_m = maximum possible gain with matched input and output
R_g = source resistance
R_l = load resistance
R_{im} = matched source resistance for maximum gain
R_{lm} = matched load resistance for maximum gain

For analysis of high-frequency circuits an equivalent circuit which includes the parasitic elements must be used. A new set of parameters which are complex functions of the low frequency parameters and the parasitic elements must be derived. Shea's *Transistor Applications* presents a good discussion of high-frequency circuit analysis with convenient simplifications for various frequency regions.

Table 6-14 Properties of the Terminated Network

Z_i	$\dfrac{\Delta^z + z_{11}Z_l}{z_{22} + Z_l}$	$\dfrac{1 + y_{22}Z_l}{y_{11} + \Delta^y Z_l}$	$\dfrac{h_{11} + \Delta^h Z_l}{1 + h_{22}Z_l}$	$\dfrac{g_{22} + Z_l}{\Delta^g + g_{11}Z_l}$	$\dfrac{a_{12} + a_{11}Z_l}{a_{22} + a_{21}Z_l}$	$\dfrac{b_{12} + b_{22}Z_l}{b_{11} + b_{21}Z_l}$
Z_o	$\dfrac{\Delta^z + z_{22}Z_g}{z_{11} + Z_g}$	$\dfrac{1 + y_{11}Z_g}{y_{22} + \Delta^y Z_g}$	$\dfrac{h_{11} + Z_g}{\Delta^h + h_{22}Z_g}$	$\dfrac{g_{22} + \Delta^g Z_g}{1 + g_{11}Z_g}$	$\dfrac{a_{12} + a_{22}Z_g}{a_{11} + a_{21}Z_g}$	$\dfrac{b_{12} + b_{11}Z_g}{b_{22} + b_{21}Z_g}$
$\dfrac{V_2}{V_1}$	$\dfrac{z_{21}Z_l}{\Delta^z + z_{11}Z_l}$	$\dfrac{-y_{21}Z_l}{1 + y_{22}Z_l}$	$\dfrac{-h_{21}Z_l}{h_{11} + \Delta^h Z_l}$	$\dfrac{g_{21}Z_l}{g_{22} + Z_l}$	$\dfrac{Z_l}{a_{12} + a_{11}Z_l}$	$\dfrac{\Delta^b Z_l}{b_{12} + b_{22}Z_l}$
$\dfrac{I_2}{I_1}$	$\dfrac{-z_{21}}{z_{22} + Z_l}$	$\dfrac{y_{21}}{y_{11} + \Delta^y Z_l}$	$\dfrac{h_{21}}{1 + h_{22}Z_l}$	$\dfrac{-g_{21}}{\Delta^g + g_{11}Z_l}$	$\dfrac{-1}{a_{22} + a_{21}Z_l}$	$\dfrac{-\Delta^b}{b_{11} + b_{12}Z_l}$

Table 6-15 Properties of Terminated Common-Base Stage

$$Z_i = \frac{h_{ib} + [h_{ib}h_{ob} + h_{rb}h_{fe}/(1 + h_{fe})]Z_l}{1 + h_{ob}Z_l}$$

$$Z_o = \frac{h_{ib} + Z_g}{h_{ob}Z_g + h_{ib}h_{ob} + h_{rb}h_{fe}/(1 + h_{fe})}$$

$$A_i = \frac{-h_{fe}}{(1 + h_{fe})(1 + h_{ob}Z_l)}$$

$$A_v = \frac{h_{fe}Z_l}{h_{ib}(1 + h_{fe}) + (1 + h_{fe})[h_{ib}h_{ob} + h_{rb}h_{fe}/(1 + h_{fe})]Z_l}$$

$$G = \frac{(h_{fe})^2 R_l}{(1 + h_{fe})^2 \{h_{ib} + [h_{ib}h_{ob} + h_{rb}h_{fe}/(1 + h_{fe})]R_l\}(1 + h_{ob}R_l)}$$

$$G_t = \frac{4(h_{fe})^2 R_g}{R_l\{(1 + h_{fe})[R_g(h_{ob} + 1/R_l) + h_{ib}/R_l + h_{ib}h_{ob} + h_{rb}h_{fe}/(1 + h_{fe})]\}^2}$$

$$G_m = \frac{(h_{fe})^2}{\{(1 + h_{fe})[\sqrt{h_{ib}h_{ob} + h_{rb}h_{fe}/(1 + h_{fe})} + \sqrt{h_{ib}h_{ob}}]\}^2}$$

$$R_{lm} = \left\{\frac{h_{ib}}{h_{ob}[h_{ib}h_{ob} + h_{rb}h_{fe}/(1 + h_{fe})]}\right\}^{1/2}$$

$$R_{im} = \left\{\frac{h_{ib}[h_{ib}h_{ob} + h_{rb}h_{fe}/(1 + h_{fe})]}{h_{ob}}\right\}^{1/2}$$

ELECTRONICS 249

Table 6-16 Properties of the Terminated Common-Emitter Stage

$$Z_i = \frac{h_{ib} + [h_{ib}h_{ob} + h_{rb}h_{fe}/(1 + h_{fe})]Z_l}{h_{ob}Z_l + 1/(1 + h_{fe})}$$

$$Z_o = \frac{h_{ib} + Z_g/(1 + h_{fe})}{h_{ob}(h_{ib} + Z_g) + h_{rb}h_{fe}/(1 + h_{fe})}$$

$$A_i = \frac{h_{fe}}{1 + h_{ob}Z_l(1 + h_{fe})}$$

$$A_v = \frac{-h_{fe}Z_l}{h_{ib}(1 + h_{fe})(1 + h_{ob}Z_l) + h_{rb}h_{fe}Z_l}$$

$$G_t = \frac{4(h_{fe})^2 R_g}{R_l[h_{ib}(1 + h_{fe})/R_l + R_g h_{ob}(1 + h_{fe}) + R_g/R_l + h_{ib}h_{ob}(1 + h_{fe}) + h_{rb}h_{fe}]^2}$$

$$G_m = \frac{(h_{fe})^2[h_{ib}h_{ob}(1 + h_{fe}) + h_{rb}h_{fe}]}{[h_{ib}h_{ob}(1 + h_{fe}) + h_{rb}h_{fe} + (1 + h_{fe})\sqrt{(h_{ib}h_{ob})^2(1 + h_{fe}) + h_{rb}h_{fe}}]^2}$$

$$R_{lm} = \left\{\frac{h_{ib}}{h_{ob}[h_{ib}h_{ob}(1 + h_{fe}) + h_{rb}h_{fe}]}\right\}^{1/2}$$

$$R_{im} = \left\{\frac{h_{ib}[h_{ib}h_{ob}(1 + h_{fe}) + h_{rb}h_{fe}]}{h_{ob}}\right\}^{1/2}$$

Table 6-17 Properties of the Terminated Common-Collector Stage

$$Z_i = \frac{h_{ib} + Z_l}{h_{ob}Z_l + 1/(1 + h_{fe})}$$

$$Z_o = \frac{h_{ib}(1 + h_{fe}) + Z_g}{(1 + h_{ob}Z_g)(1 + h_{fe})}$$

$$A_i = \frac{-(1 + h_{fe})}{1 + h_{ob}Z_l(1 + h_{fe})}$$

$$A_v = \frac{Z_l}{h_{ib} + Z_l}$$

$$G_t = \frac{4R_g}{R_l[1 + h_{ib}/R_l + R_g(h_{ob} + 1/R_l(1 + h_{fe}))]^2}$$

$$G_m = \frac{1 + h_{fe}}{[1 + \sqrt{h_{ib}h_{ob}(1 + h_{fe})}]^2}$$

$$R_{lm} = \left[\frac{h_{ib}}{h_{ob}(1 + h_{fe})}\right]^{1/2}$$

$$R_{im} = \left[\frac{h_{ib}(1 + h_{fe})}{h_{ob}}\right]^{1/2}$$

Transistor Bias

Equations for two-battery arrangement (Fig. 6-14):

$$I_E = \frac{V_{EE} - R_2 I_{CBO}}{R_1 + R_2(1 + h_{fb})}$$

$$I_B = -\left[\frac{R_1 I_{CBO} + V_{EE}(1 + h_{fb})}{R_1 + R_2(1 + h_{fb})}\right]$$

$$I_C = \frac{(R_1 + R_2)I_{CBO} + h_{fb} V_{EE}}{R_1 + R_2(1 + h_{fb})}$$

$$V_{BG} = V_{EG} = R_2 \left[\frac{R_1 I_{CBO} + V_{EE}(1 + h_{fb})}{R_1 + R_2(1 + h_{fb})}\right]$$

$$V_{CG} = V_{CC} - R_C \left[\frac{(R_1 + R_2)I_{CBO} + h_{fb} V_{EE}}{R_1 + R_2(1 + h_{fb})}\right]$$

$$V_{CB} = V_{CC} - V_{EE}\left[\frac{R_2(1 + h_{fb}) + h_{fb} R_C}{R_1 + R_2(1 + h_{fb})}\right] - I_{CBO}\left[\frac{R_1 R_2 + R_C(R_1 + R_2)}{R_1 + R_2(1 + h_{fb})}\right]$$

$$S_{I_E} = \frac{dI_E}{dI_{CBO}} = \frac{-R_2}{R_1 + R_2(1 + h_{fb})}$$

$$S_{I_C} = \frac{dI_C}{dI_{CBO}} = \frac{R_1 + R_2}{R_1 + R_2(1 + h_{fb})}$$

$$S_V = \frac{dV_{CB}}{dI_{CBO}} = -\left[\frac{R_1 R_2 + R_C(R_1 + R_2)}{R_1 + R_2(1 + h_{fb})}\right]$$

$$S_{V_1} = \frac{dV_{CB}}{dV_{CC}} = 1$$

$$S_{V_2} = \frac{dV_{CB}}{dV_{EE}} = -\frac{R_C h_{fb} - R_2(1 + h_{fb})}{R_1 + R_2(1 + h_{fb})}$$

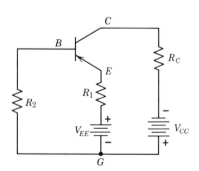

Fig. 6-14. Two-battery biasing, common-emitter. (Reproduced by permission from R. F. Shea, *Transistor Applications*, John Wiley, New York, 1964.)

Note. V_{CC} will be negative, V_{EE} positive, and I_{CBO} negative in p-n-p transistors. The reverse will be true with n-p-n transistors.

In the above equations, G refers to a virtual ground at the connection between the low side of the resistor R_2 and the common side of the two batteries.

Equations for single-battery arrangement (Fig. 6-15):

$$I_E = \frac{-R_2 V_{CC}}{R_1 R_2 + R_1 R_3 + R_2 R_3(1 + h_{fb})} - \frac{R_2 R_3 I_{CBO}}{R_1 R_2 + R_1 R_3 + R_2 R_3(1 + h_{fb})}$$

$$I_B = \frac{R_2 V_{CC}(1 + h_{fb})}{R_1 R_2 + R_1 R_3 + R_2 R_3(1 + h_{fb})} - \frac{R_1(R_2 + R_3) I_{CBO}}{R_1 R_2 + R_1 R_3 + R_2 R_3(1 + h_{fb})}$$

$$I_C = \frac{-h_{fb} R_2 V_{CC}}{R_1 R_2 + R_1 R_3 + R_2 R_3(1 + h_{fb})} + \frac{(R_1 R_2 + R_1 R_3 + R_2 R_3) I_{CBO}}{R_1 R_2 + R_1 R_3 + R_2 R_3(1 + h_{fb})}$$

$$V_{BG} = V_{EG} = \frac{R_1 R_2 V_{CC}}{R_1 R_2 + R_1 R_3 + R_2 R_3(1 + h_{fb})} + \frac{R_1 R_2 R_3 I_{CBO}}{R_1 R_2 + R_1 R_3 + R_2 R_3(1 + h_{fb})}$$

$$V_{CG} = V_{CC}\left[1 + \frac{R_2 R_C h_{fb}}{R_1 R_2 + R_1 R_3 + R_2 R_3(1 + h_{fb})}\right] - \frac{R_C(R_1 R_2 + R_1 R_3 + R_2 R_3) I_{CBO}}{R_1 R_2 + R_1 R_3 + R_2 R_3(1 + h_{fb})}$$

$$V_{CB} = V_{CC}\left[\frac{R_1 R_3 + R_2 R_3(1 + h_{fb}) + R_2 R_C h_{fb}}{R_1 R_2 + R_1 R_3 + R_2 R_3(1 + h_{fb})}\right] - I_{CBO}\left[\frac{R_1 R_2 R_3 + R_C(R_1 R_2 + R_1 R_3 + R_2 R_3)}{R_1 R_2 + R_1 R_3 + R_2 R_3(1 + h_{fb})}\right]$$

$$S_{I_E} = \frac{dI_E}{dI_{CBO}} = \frac{-R_2 R_3}{R_1 R_2 + R_1 R_3 + R_2 R_3(1 + h_{fb})}$$

$$S_{I_C} = \frac{dI_C}{dI_{CBO}} = \frac{R_1 R_2 + R_1 R_3 + R_2 R_3}{R_1 R_2 + R_1 R_3 + R_2 R_3(1 + h_{fb})}$$

$$S_V = \frac{dV_{CB}}{dI_{CBO}} = -\left[\frac{R_1 R_2 R_3 + R_C(R_1 R_2 + R_1 R_3 + R_2 R_3)}{R_1 R_2 + R_1 R_3 + R_2 R_3(1 + h_{fb})}\right]$$

$$S_{V_1} = \frac{dV_{CB}}{dV_{CC}} = \frac{R_1 R_3 + R_2 R_3(1 + h_{fb}) + R_2 R_C h_{fb}}{R_1 R_2 + R_1 R_3 + R_2 R_3(1 + h_{fb})}$$

Note. The same polarity conventions apply as for the two-battery arrangement.

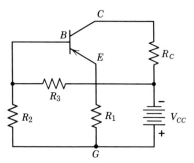

Fig. 6-15 Single-battery biasing, common-emitter. (Reproduced by permission from R. F. Shea, *Transistor Applications*, John Wiley, New York, 1964.)

REFERENCES

Corcoran, G. F. and H. R. Reed, *Introductory Electrical Engineering*, John Wiley, New York, 1957.

Ham, J. M. and G. R. Slemon, *Scientific Basis of Electrical Engineering*, John Wiley, New York, 1961.

Hemenway, C. L., R. W. Henry, and M. Coulton, *Physical Electronics*, John Wiley, New York, 1962.

Hunter, L. P., *Handbook of Semiconductor Electronics*, McGraw-Hill, New York, second edition, 1962.

Pender, H. and W. A. DelMar, *Electrical Engineers' Handbook*, Vol. 1, *Electric Power*, John Wiley, New York, fourth edition, 1949.

Pender, H. and K. McIlwain, *Electrical Engineers' Handbook*, Vol. 2, *Communications and Electronics*, John Wiley, New York, fourth edition, 1950.

Reference Data for Radio Engineers, International Tel. and Tel. Corp., New York, fourth edition, 1956.

Shea, R. F., *Transistor Applications*, John Wiley, New York, 1964.

Shea, R F., ed., *Transistor Circuit Engineering*, John Wiley, New York, 1957.

Transistor Characteristics Tabulation, Derivation and Tabulation Associates, Orange, N.J., semi-annual.

Vasseur, J. P., *Properties and Applications of Transistors*, Macmillan, New York, 1964.

SECTION 7

NUCLEAR PHYSICS

Principal Symbols

A	mass number or number of nucleons	V	velocity
		x	distance, thickness
c	velocity of light	Z	atomic number or number of protons
k	multiplication factor		
I	intensity of radiation	γ	gamma radiation
L	thermal diffusion length	λ	mean free path, decay constant
n	neutron		
N	number density of nuclei	ν	neutrons produced per fission
N_A	Avogadro's number		
p	proton	ρ	density
r	roentgen, radius	σ	microscopic cross section
t	time	Σ	macroscopic cross section
T	temperature	ϕ	neutron flux

GENERAL RELATIONS

Units and Constants

The units of time and length are the conventional second and centimeter. Since the square centimeter is very large compared to the nucleus of an atom, a more convenient unit of area is the *barn* or 10^{-24} cm². The unit of mass is the *atomic mass unit* (amu), which is one-sixteenth of the mass of the O^{16} oxygen isotope. The unit of energy is the *electron volt* (ev), million electron volts (Mev), or billion electron volts (Bev = 10^{12} ev). The electron volt is the amount of energy required to raise the potential of an electron by 1 volt. One ev equals 1.60203×10^{-12} erg or 1.60203×10^{-19} joule.

Table 7-1 Some Nuclear Constants

	amu	grams × 10^{-24}	Rest energy, Mev
Unit mass, m	1	1.65990	931.16
Electron, $_1e^0$ or β^-	0.00054862	0.00091091	0.51083
Proton, p or $_1p^1$	1.007595	1.67247	938.17
Neutron, n or $_0n^1$	1.008983	1.67472	939.43
Hydrogen atom, $_1H^1$	1.00812	1.67338	938.68
Alpha particle, α or $_2He^4$	4.00280	6.64424	3727.07

Charge on electron $(1.60206 \pm .00007) \times 10^{-19}$ coulomb
Radius of electron $(2.81784 \pm .00010) \times 10^{-13}$ cm
Radius of nucleus $(1.5 \pm .15)\sqrt{A} \times 10^{-13}$ cm

Equivalence of Energy and Mass

In macroscopic physics energy and mass are conserved independently. In nuclear physics it is the composite of mass and energy that is conserved. From *Einstein's law*, when velocity is small relative to c,

$$E = mc^2 \tag{7-1}$$

For example, complete conversion of 1 gram of matter into energy yields 9×10^{20} ergs or 2.5×10^7 kilowatt hours of energy. In the practical development of nuclear power, the yield of energy is very small, only about 0.1 % of the mass of U-235 being converted to energy in the fission process, for example.

Mass Defect. When a compound nucleus has less mass than the sum of the masses of its particles or nucleons, it is said to have a mass defect. The mass defect is equivalent to the energy radiated when the particles combined or to the energy required to separate the nucleus into its particles, which is called the *binding energy*. The binding energy per nucleon increases up to mass number of about 50 and then decreases gradually with increasing mass number. (See Fig. 7-1.)

Mass of a Moving Body. According to the theory of relativity, the mass of a body is a function of its velocity, as

$$m = \frac{m_0}{(1 - V^2/c^2)^{1/2}} \tag{7-2}$$

where m_0 is the rest mass and c is the velocity of light. At low velocities, $m = m_0$; at velocities approaching c, mass becomes very large and acceleration produces relatively large increase in mass.

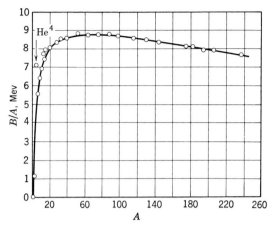

Fig. 7-1 Average binding energy per nucleon.

From combination of Eqs. 7-1 and 7-2,

$$E_{\text{tot}} = \frac{m_0 c^2}{(1 - V^2/c^2)^{1/2}} \tag{7-3}$$

and

$$E_{\text{kin}} = \frac{m_0 c^2}{(1 - V^2/c^2)^{1/2}} - m_0 c^2 \tag{7-4}$$

or

$$E_{\text{kin}} = \frac{m_0 V^2}{2}\left(1 + \frac{3V^2}{4c^2} + \frac{5V^4}{8c^4} + \cdots\right) \tag{7-5}$$

NUCLEAR REACTIONS

The conventional notation for a nuclear reaction is

$$(_zC^A)_1 + (_zC^A)_2 \rightarrow (_zC^A)_3 + (_zC^A)_4 + Q \tag{7-6}$$

in which z is the number of protons, A is the mass number, C is the chemical symbol for the atom, electron, or nucleon, and Q is the energy released.

The conservation equations are,

$$\sum Z_i = 0 \quad \text{and} \quad \sum A_i = 0 \tag{7-7}$$

$$\text{Initial mass} - \text{final mass} = Q \tag{7-8}$$

An example is the *fission reaction:*

$$_{92}U^{235} + {}_0n^1 \rightarrow {}_{92}U^{236} \rightarrow {}_{38}Sr^{94} + {}_{54}Xe^{140} + 2{}_0n^1 + Q$$

The strontium and xenon products are highly radioactive and decay further to other products. The final result is a spectrum of products. Table 7-2 gives the distribution of representative energies from fission of U-233, U-235, or Pu-239.

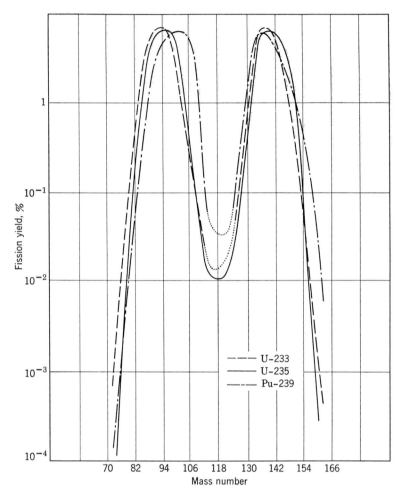

Fig. 7-2 Thermal fission yield.

Table 7-2 Average Distribution of Fission Energy

Source	Energy, Mev
Kinetic energy of fission fragments	167
Prompt gamma rays	6
Kinetic energy of fission neutrons	5
Delayed gamma rays	6
Delayed beta radiation	8
Other	12
Total energy per fission	204

Modes of Radioactive Decay

Negative beta (electron) emission:

$$_0n^1 \rightarrow {_{-1}e^0} + {_1H^1} + \text{neutrino}$$

$$_{38}Sr^{94} \xrightarrow{\beta^-} {_{39}Y^{94}} \xrightarrow{\beta^-} {_{40}Zr^{94}}$$

$$_{54}Xe^{140} \rightarrow 4(_{-1}e^0) + {_{58}Ce^{140}}$$

Positive beta (positron) emission:

$$_7N^{13} \rightarrow {_{+1}e^0} + {_6C^{13}} + \text{neutrino}$$

$$_{+1}e^0 + {_{-1}e^0} \rightarrow 2 \text{ gammas of 0.51 Mev each}$$

Alpha emission:

$$_{94}Pu^{239} \rightarrow {_2He^4} + {_{92}U^{235}}$$

Neutron emission:

$$_{53}I^{137} \xrightarrow{\beta^-} {_{54}Xe^{137}} \longrightarrow {_0n^1} + {_{54}Xe^{136}}$$

Orbital electron (K) capture:

$$_{29}Cu^{64} + {_{-1}e^0} \rightarrow {_{28}Ni^{64}}$$

Gamma emission by:

a. ejection of a gamma photon from an excited nucleus,
b. isomeric transition of a nucleus from one energy level to another,
c. annihilation of an electron following positive beta emission.

Nuclei with an excess of neutrons are usually electron emitters. Among nuclei with a deficiency of neutrons, the heavy ones usually decay by alpha emission, and the light ones by positron emission or orbital electron capture. Gamma emission often accompanies other types of decay.

Decay with Time

From statistical considerations the rate of decay (alpha emission) is proportional to the number N of radioactive nuclei present,

$$\frac{dN}{dt} = -\lambda N \tag{7-9}$$

where the proportionality constant λ is called the *disintegration constant* or the *radioactive decay constant*. Integration of Eq. 7-9 gives

$$N = N_0 e^{-\lambda t} \tag{7-10}$$

where N_0 is the initial number of radioactive nuclei. The *half-life* $t_{1/2}$ or the time required for one-half of the original atoms to decay, by substitution in Eq. 7-10, is

$$t_{1/2} = 0.693/\lambda \tag{7-11}$$

When there is more than one radioisotope in the decay chain

$$P \xrightarrow{\lambda_1} Q \xrightarrow{\lambda_2} R \xrightarrow{\lambda_3}$$

for the second member or daughter Q

$$\frac{N_Q}{N_{P_1}} = \frac{\lambda_1}{\lambda_2 - \lambda_1} e^{-\lambda_1 t} + \frac{\lambda_1}{\lambda_1 - \lambda_2} e^{-\lambda_2 t} \qquad (7\text{-}12)$$

If the half-life of the parent P is longer than the half-life of the daughter $Q(\lambda_2 > \lambda_1)$, after a lapse of time $e^{-\lambda_2 t}$ becomes negligible and

$$\frac{N_Q}{N_P} = \frac{\lambda_1}{\lambda_2 - \lambda_1} e^{-\lambda_1 t} \qquad (7\text{-}13)$$

With native radioisotopes the half-life of the parent is very long compared to that of the daughter, so that Eq. 7-13 reduces to

$$\frac{N_Q}{N_P} = \frac{\lambda_1}{\lambda_2} e^{-\lambda_1 t} \qquad (7\text{-}14)$$

Nuclear Cross Sections

Microscopic. When a neutron is directed into a material it may (1) pass through, (2) collide with a nucleus and be absorbed, or (3) be scattered. If ϕ is the neutron flux (number/sec-cm²) and N is the nuclei density (number/cm³), the rate of encounter

$$\frac{dN}{dt} = -\sigma \phi N \qquad (7\text{-}15)$$

where the proportionality constant σ is called the *microscopic cross section* per nucleus. The unit of σ is the barn or 10^{-24} cm².

Macroscopic. The total cross section for all nuclei present or the *macroscopic cross section* is

$$\Sigma = \sigma N \qquad (7\text{-}16)$$

For a chemical element of atomic mass A and density ρ

$$\Sigma = \frac{\rho \sigma N_A}{A} \qquad (7\text{-}17)$$

where N_A is Avogadro's number. If the material is a chemical compound, Σ must be summed for all its elements. The unit of Σ is cm⁻¹. The neutron flux ϕ varies with the thickness x of the target

$$\phi_x / \phi_0 = e^{-\sigma N x} \qquad (7\text{-}18)$$

where ϕ_0 is the initial neutron flux.

NUCLEAR REACTIONS 259

Table 7-3 Thermal Neutron Cross Sections

Atomic Number	Element	σ_a (barns)	σ_s (barns)	Atomic Number	Element	σ_a (barns)	σ_s (barns)
1	H	0.332	38	49	In	191	2.2
2	He		0.8	50	Sn	0.625	4
3	Li	71.0	1.4	51	Sb	5.7	4.3
4	Be	0.010	7	52	Te	4.7	5
5	B	755	4	53	I	7.0	3.6
6	C	0.0037	4.8	54	Xe		4.3
7	N	1.88	10	55	Cs	28	
8	O	0.0002	4.2	56	Ba	1.2	8
9	F	0.010	3.9	57	La	8.9	
10	Ne	2.8	2.4	58	Ce	0.73	
11	Na	0.505	4.0	59	Pr	11.3	
12	Mg	0.069	3.6	60	Nd	46	
13	Al	0.230	1.4	61	Pm-147	150	
14	Si	0.16	1.7	62	Sm	5,600	
15	P	0.20	5	63	Eu	4,300	8
16	S	0.52	1.1	64	Gd	46,000	
17	Cl	33.8	16	65	Tb	46	
18	A	0.66	1.5	66	Dy	950	100
19	K	2.07	1.5	67	Ho	9.1	
20	Ca	0.44		68	Er	173	
21	Sc	24.0	24	69	Tm	127	7
22	Ti	5.8	4	70	Yb	37	12
23	V	5.00	5	71	Lu	112	
24	Cr	3.1	3.0	71	Hf	105	8
25	Mn	13.2	2.3	73	Ta	21	5
26	Fe	2.62	11	74	W	19.2	5
27	Co	38.0	7	75	Re	86	14
28	Ni	4.6	17.5	76	Os	15.3	
29	Cu	3.85	7.5	77	Ir	440	
30	Zn	1.10	3.6	78	Pt	8.8	10
31	Ga	2.80	4	79	Au	98.8	9.3
32	Ge	2.45	3	80	Hg	380	20
33	As	4.3	6	81	Tl	3.4	14
34	Se	12.3	11	82	Pb	0.170	11
35	Br	6.7	6	83	Bi	0.034	9
36	Kr	31	7.2	89	Ac-227	795	
37	Rb	0.73		90	Th-232	7.56	12.6
38	Sr	1.21	10	92	U	7.68	9.0
39	Y	1.31		93	Np-237	170	
40	Zr	0.185	6.3	94	Pu-239	1,026	9.6
41	Nb	1.16	5	94	Pu-240	295	
42	Mo	2.70	7	94	Pu-241	1,400	
43	Tc-99	22	5	94	Pu-242	30	
44	Ru	2.56	6	95	Am-241	630	
45	Rh	149	5	95	Am-242	8,000	
46	Pd	8.0	3.6	96	Cm	250	
47	Ag	63	6	98	Cf-249	900	
48	Cd	2,450	7	99	E-254	2,700	

Source: BNL-325 and Supplements.

The *mean free path* of a particle

$$\lambda = \Sigma \int_0^\infty x e^{-\Sigma x}\, dx = \frac{1}{\Sigma} \qquad (7\text{-}19)$$

and if $x = \lambda$, from Eqs. 7-18 and 7-19

$$\phi_x/\phi_0 = e^{-\Sigma\lambda} = 1/e = 0.3679 \qquad (7\text{-}20)$$

Properties of Cross Sections. Colliding neutrons may be absorbed or scattered, but the cross section for elastic scattering σ_s is not the same as the cross section for absorption σ_a. The collision cross section is considered to be the sum

$$\sigma = \sigma_a + \sigma_s \qquad (7\text{-}21)$$

and σ_a includes all events for which the state of the target nucleus is altered such as capture (σ_c) and fission (σ_f).

The value of σ depends not only on the species of nuclei but also on the energy of the neutron. Although there is a tendency for the cross section to vary inversely with the speed of the neutrons, many materials have resonance peaks or narrow regions of large cross section, an important consideration in the design of thermal nuclear reactors. With neutron energies above 1 Mev σ_a becomes negligible compared to σ_s.

The energy of a neutron ejected after fission reaches a value in equilibrium with the material with which it is colliding. This equilibrium energy is called *thermal energy* since it is a function of temperature, about 0.0253 ev at 20°C. Table 7-3 gives cross sections of the elements for *thermal neutrons* with velocity 2200 m/sec.

Normally a population of neutrons has a range of energies. If the Maxwell-Boltzmann and the $1/V$ relations are followed, the average cross section

$$\bar{\sigma}_a = \frac{(\pi)^{1/2}}{2} \sigma_{kt} \qquad (7\text{-}22)$$

where σ_{kt} is the value for the most probable neutron velocity with energy kT. At 20°C the most probable velocity is 2200 meters per second.

NUCLEAR REACTORS

When fission produces more neutrons than are absorbed, a nuclear chain reaction is possible. A nuclear reactor consists essentially of a fuel core for the nuclear chain reaction, a system for controlling the rate of energy release, means for removing the energy such as a cooling system, and a radiation shield for the protection of personnel. Reactors may be classified as *homogeneous* when the fuel is uniformly distributed in the core, or as *heterogeneous* when fuel elements are separated by a moderator or other material. They are called *thermal reactors* when thermal neutrons are used for fission, or *fast reactors* when fast neutrons are used without moderators.

NUCLEAR REACTORS   261

Important considerations in reactor design and operation include the critical size and mass of the fuel core, the effective neutron multiplication as influenced by neutron losses, and conditions for criticality and control. The relations which follow are for the simple case of a thermal reactor with a bare homogeneous core and ignore the effects of fission products and of structural and coolant materials.

The Neutron Cycle

Figure 7-3 shows the sequence of events in a chain reaction in a thermal nuclear reactor and gives the material balances for the neutrons involved in the cycle.

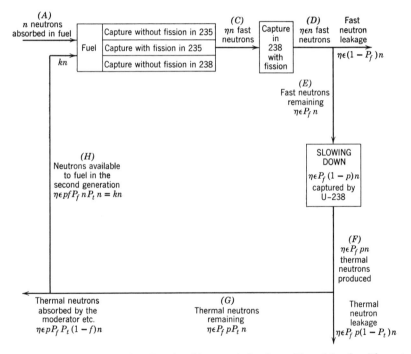

Fig. 7-3 The neutron cycle. (Reprinted by permission from Glenn Murphy, *Elements of Nuclear Engineering*, John Wiley, 1961.)

- n neutrons at start of cycle, neutrons produced per neutron absorbed, fast fission factor
- P_f fraction of fast neutrons that does not leak out
- p fraction of neutrons slowed to thermal energies without capture, 0.9 for graphite-moderated reactor
- P_t fraction of thermal neutrons not diffusing out
- f ratio of neutrons absorbed by fuel to total neutrons absorbed
- k neutron multiplication factor

Multiplication Factor. The ratio of the number of neutrons at the end of one cycle to the number at the beginning of the cycle is called the neutron multiplication factor k. When a reactor is operating at constant power, k is equal to unity. In operating any reactor k must be kept very near unity because of the sensitivity to k of the rate of change of n. See Fig. 7-3 for definitions of symbols.

For an infinitely large reactor from which there is no leakage of neutrons

$$k_\infty = \eta \epsilon p f \tag{7-23}$$

$$\eta = \frac{\sum_i (\nu N \sigma_f)_i}{\sum_i (N \sigma_a)_i} \tag{7-24}$$

where i represents each atomic species present.

$$f = \frac{\Sigma_a(\text{fuel})}{\Sigma_a(\text{fuel}) + \Sigma_a(\text{other material})} \tag{7-25}$$

See Eqs. 7-16 and 7-21 and Tables 7-4 and 7-5. The value of p varies from 0.90 for natural uranium to 1.00 for fully enriched U-235. The value of ϵ is about unity for homogeneous reactors and 1.03 for heterogeneous cores of uranium and graphite.

For a finite reactor with leakage taken into account

$$k = k_\infty P_f P_t \tag{7-26}$$

$$P_f = e^{-B^2 \tau} \tag{7-27}$$

Table 7-4 Properties of Moderators

	L_m, cm	τ, cm²	Density, gm/cm³	σ_s(epithermal), barns	σ_a(thermal), barns
H₂O	2.88	33	1.000	49	0.66
D₂O	100	120	1.10	10.5	0.0026
C	50	350	1.65	4.8	0.0045
Be	24	98	1.85	6.0	0.009

Table 7-5 Properties of Nuclear Fuels

	ν	σ_c, barns	σ_f, barns	σ_s, barns	Density, gm/cm³
U-233	2.51 ± 0.02	53 ± 6	525 ± 4		18.68
U-235	2.43 ± 0.02	101 ± 5	582 ± 4	10.0 ± 2	18.68
U-238		2.75 ± 0.03	<0.0005	8.3	18.68
Pu-239	2.89 ± 0.03	286 ± 9	742 ± 4	9.6 ± 0.5	19.6
Pu-241		400 ± 50	1010 ± 13		19.6

$\sigma_a = \sigma_f + \sigma_c$ = absorption cross section for 2200 mps neutrons.

Source: World Consistent Set of Data, from BNL-325, Supplement 1, second edition, Government Printing Office, 1960.

and
$$P_t = \frac{1}{1 + B^2 L^2} \quad (7\text{-}28)$$

in which B^2 is the buckling, τ is the square of the fast diffusion length, and L is the thermal diffusion length.

The *critical mass* of fuel in a reactor is the amount required to make k equal to unity.

Buckling. The buckling constant B^2 is called *material buckling* when calculated from the material properties and *geometric buckling* when related to the size and shape of the reactor core. For an exactly critical reactor ($k = 1$) material buckling must equal geometric buckling. Values for geometrical buckling for various reactor shapes are:

spherical (radius r), $\quad B^2 = (\pi/r)^2 \quad (7\text{-}29)$

cylindrical (radius r, height h),

$$B^2 = \left(\frac{2.405}{r}\right)^2 + \left(\frac{\pi}{h}\right)^2 \quad (7\text{-}30)$$

cubical (side a), $\quad B^2 = 3(\pi/a)^2 \quad (7\text{-}31)$

The value for material buckling is given by Eq. 7-26 and by

$$B^2 = \frac{-\Sigma_a(1 - k_\infty)}{\lambda_t/3} \quad (7\text{-}32)$$

where
$$\lambda_t = \frac{1}{\Sigma_s(1 - 2/3A)} \quad (7\text{-}33)$$

in which λ_t is the transport mean free path of a neutron, and Σ_a, Σ_s, and A refer to the reactor material.

Diffusion Length. The neutron flux in a homogeneous medium decreases exponentially with distance from the source, and the diffusion length is a measure of the average distance a neutron travels from formation to capture. Moderators, materials with large scattering cross section and small absorption cross section, are used to slow fast neutrons down to thermal neutrons by elastic collisions. For fast neutrons τ is one-sixth the mean square displacement while being slowed to thermal neutrons. For thermal neutrons L^2 is one-sixth the mean square displacement from thermalization to capture. (See Table 7-4.)

For homogeneous reactors in which fuel is mixed in the moderator,

$$L^2 = L_m^2(1 - f) \quad (7\text{-}34)$$

where f is the ratio of neutrons absorbed by fuel to the total neutrons absorbed. (See Eq. 7-25.)

The value of τ is affected only slightly by fuel in the moderator, so that the value in Table 7-4 may be used for the mixture.

Units of Radiation Exposure

The *roentgen* is the standard unit for measuring exposure to ionizing radiation. It represents the quantity of gamma or X-radiation resulting in the absorption of 83.9 ergs per gram of dry air at standard conditions. The *roentgen equivalent physical* or rep is an obsolescent unit for the absorbed radiation that has the same effect as one roentgen of gamma radiation. The *rad* is the conventional unit of absorbed radiation dose corresponding to the absorption of 100 ergs per gram of any material. It can be applied to any type of radiation that produces ionization.

Relative biological effectiveness (RBE) is the ratio of gamma radiation to another type of radiation that produces the same biological effect, both expressed in rads. (See Table 7-6.) The *roentgen equivalent man* (rem) is the unit of absorbed radiation dose for biological effects. It is defined as

$$\text{rem} = \text{RBE} \times (\text{number of rad})$$

The maximum permissible external radiation dose for adult workers in the radiation industries is 0.1 rem per week (8 hr day). The corresponding fluxes are given in Table 7-6. A single chest X-ray is about twice this dose (0.2 rem) and continuous exposure of man to natural background radiation at sea level and 50 degrees latitude is about one-thirtieth of this rate (0.0033 rem per week).

The ingestion of radioactive materials in air, water, or food presents hazards which depend not only on the concentration and absorption but also on the rate of elimination. For example, Sr-90 with a long half-life is absorbed in bone structure and persists for years.

Table 7-6 Maximum Permissible Flux

Dose rate 0.1 rem per week

Type of Radiation	RBE	Exposure, rad/week	Energy Level	Flux Units/sec-cm^2
Gamma and X-rays	1	0.100	Various	1400/Mev
Electrons	1	0.100	1 Mev	15
Thermal neutrons	2.5	0.040	0.025 ev	667
Fast neutrons	10	0.010	2 Mev	13
Alpha particles	10	0.010	5 Mev	0.005
Protons	10	0.010	5 Mev	0.06
Heavy ions	20	0.005	5 Mev	0.0002

Shielding

Since alpha and beta particles have relatively short range, shielding is provided mainly for neutron and gamma radiation. Shielding for neutrons usually involves thermalization of the neutrons followed by their absorption in a material with high absorption cross section. Neutron shielding involves also provision for photons from neutron-gamma reactions in the shield. Gamma shielding normally is provided by a material with a high absorption coefficient or by large thickness. Design considerations include geometrical arrangement and balance of thickness and effectiveness of materials against cost. Effectiveness depends both on the material and on the energy of radiation as shown in Table 7-7.

The attenuation with distance x of the intensity I of gamma radiation from a plane source is given by

$$I_x/I_0 = be^{-\mu x} \tag{7-37}$$

and from a point source surrounded by a spherical shield by

$$J = \frac{bI_0 e^{-\mu x}}{4\pi x^2} \tag{7-38}$$

where I_0 is the intensity at the source (Mev/cm²-sec), μ is the attenuation coefficient (see Table 7-7), b is the build-up factor for shields of appreciable thickness, and J is the current density (Mev/cm²-sec). The factor b, determined experimentally, is found to depend on μx and I_0 as well as on the shield material.

Equations 7-37 and 7-38 can be applied to the attenuation of thermal neutrons where $\Sigma = \mu$. Attenuation of fast neutrons is a much more complex problem and requires specialized knowledge.

The energy of radiation attenuated by a shield is nearly all converted to heat energy. With the assumption that all the energy is absorbed within one free-path, the maximum temperature rise in a plane shield is given by

$$\Delta T = I_0/K\mu \tag{7-39}$$

and in a spherical shield by

$$\Delta T = \frac{I_0}{4\pi K}\left(\frac{1}{r_1} - \frac{1}{r_2}\right) \tag{7-40}$$

where K is the thermal conductivity of the shield (in Mev/sec-cm-°C).

Table 7-7 Shielding Properties of Materials

	Water	Iron	Lead	Portland Concrete	Barite Concrete
Density, g/cm³	1.00	7.78	11.3	2.37	3.49
Thermal neutrons					
Σ, cm⁻¹	0.100	0.156	0.113	0.094	0.094
γ at 0.5 Mev					
μ, cm⁻¹	0.096	0.653	1.72	0.20	0.30
Build-up b					
At $\mu x = 1$	2.46	2.80	1.51		
At $\mu x = 10$	71.5	34.2	3.01		
γ at 6 Mev					
μ, cm⁻¹	0.027	0.237	0.503	0.065	0.109
Build-up b					
At $\mu x = 1$	1.46	1.30	1.14		
At $\mu x = 10$	5.18	7.10	4.20		

REFERENCES

Etherington, Harold, editor, *Nuclear Engineering Handbook*, McGraw-Hill, New York, 1958.
Halliday, David, *Introduction to Nuclear Physics*, John Wiley, New York, second edition, 1955.
Murphy, Glenn, *Elements of Nuclear Engineering*, John Wiley, New York, 1961.
Stephenson, Richard, *Introduction to Nuclear Engineering*, McGraw-Hill, New York, second edition, 1958.
Permissible Dose from External Sources of Ionizing Radiation, NBS Handbooks, U.S. Government Printing Office, Washington.
TID-4575, *Guide to Atomic Energy for the Civilian Application Program*, Office of Technical Services, U.S. Department of Commerce, Washington.

Table 7-8 Selected Radioisotopes

Element	Mass Number	β (Mev)	γ (Mev)	Half-Life
Aluminum	29	2.5	—	6.7 m
Antimony	122	1.36, 1.94	0.57	2.8 d
	124	0.74, 2.45	1.72	60 d
	125	0.3, 0.7	0.55	2.7 y
Arsenic	77	0.8	None	40 h
Beryllium	10	0.56	None	3×10^6 y
Bismuth	210	1.17	None	5 d
Bromine	82	0.465	0.547, 0.787, 1.35	34 h
Cadmium	115	0.6, 1.11	0.65	2.8 d
Calcium	45	0.2, 0.9	None	180 d
	49	2.3	0.8	2.5 h
Carbon	11	$0.97e^+$	—	20.5 m
	14	0.145	None	5100 y
Cerium	141	0.6	0.21	28 d
Chlorine	36	0.66	None	10^6 y
	38	1.1, 2.8, 5.0	1.65, 2.15	37 m
Cobalt	60	0.31	1.10, 1.30	5.3 y
Columbium	95	0.15	0.75	35 d
Copper	61	0.9, 1.23	None	3.4 h
Europium	154	0.9	1.4	5.4 y
	155	0.23	0.084	2 y
Fluorine	18	0.7	—	1.86 h
Gallium	72	0.8, 3.1	0.84, 2.25	14.1 h
Germanium	71	1.2	—	40 h
	77	1.9	—	12 h
Gold	198	0.98	0.12, 0.41	2.7 d
	199	1.01	0.45	3.3 d
Hafnium	181	0.8	0.5	46 d
Hydrogen	3	0.011, 0.015	None	12 y
Iodine	131	0.315, 0.600	0.367, 0.080, 0.284, 0.638	8.0 d
Iridium	192	0.67	$0.137 \to 0.651(12\gamma)$	74.7 d
	194	0.48, 2.18	0.38, 1.43	19 h
Iron	59	0.26, 0.46	1.1, 1.3	46.3 d
Krypton	85	1.0	0.17, 0.37	4.5 h

Table 7-8 (*continued*)

Element	Mass Number	β (Mev)	γ (Mev)	Half-Life
Lanthanum	140	1.32, 1.67, 2.26	0.093 → 2.5	40 h
Magnesium	27	0.9, 1.80	0.64, 0.84, 1.02	9.58 m
Mercury	203	0.205	0.286	43.5 d
Molybdenum	99	0.445, 1.23	0.04, 0.741, 0.780	68.3 h
Neodymium	147	0.17, 0.78	0.035, 0.58	11 d
Nitrogen	13	$1.23e^+$	None	10.1 m
Osmium	191	0.142	0.039, 0.127	15.0 d
	193	1.15	1.58	32 h
Phosphorus	32	1.718	None	14.3 d
Platinum	197	0.65	None	18 h
Potassium	40	1.40	1.45, K	9.9×10^8 y
	42	3.5	None	12.4 h
Praeseodymium	142	0.636, 2.154	1.57	19.1 h
	143	0.932	None	13.8 d
Promethium	147	0.229	None	2.26 y
Rhenium	186	0.64, 0.95, 1.09	0.132, 0.275, 1.70	92.8 h
Rhodium	105	0.57	0.33	36.2 h
Rubidium	86	0.72, 1.80	1.08	19.5 d
Ruthenium	103	0.205, 0.670	0.494	42 d
Samarium	153	0.68, 0.80	0.070, 0.103, 0.61	47 h
Scandium	46	0.36, 1.49	0.89, 1.12	85 d
Silicon	31	1.5	None	2.59 h
Silver	110	0.087, 0.53	0.885, 0.935, 0.1389, 1.516	270 d
	111	1.06	None	7.5 d
Sodium	22	$0.58e^+$	1.3	2.6 y
	24	1.390	1.380, 2.758	15.0 h
Strontium	89	1.50	None	53 d
	90	0.54	None	19.9 y
Sulfur	35	0.167	None	87.1 d
Tantalum	182	0.52 m, 1.1	0.05 → 1.24(33γ)	115 d
Technetium	97	IT	0.097	90 d > 10^3 y
	99	0.30	0.140	2.1×10^5 y
Tellurium	127	IT ~ 0.8	0.089	90 d, 9.3 h
	129	IT 1.46	0.102, 0.3, 0.8	32 d, 72 m
	131	IT > 1.8	0.177	30 h, 25 m
Thallium	206	0.58	None	2.7 y
Titanium	51	0.36	1.0	72 d
Tungsten	185	0.428	None	73.2 d
	187	0.627, 1.318	0.086, 0.70	24.1 h
Yttrium	90	2.24	None	61 h

SECTION 8

ENGINEERING ECONOMY

Principal Symbols

D	net disbursement	P	present amount or worth
i	periodic interest rate	q	project interest rate
k	incremental interest rate	r	continuous interest rate
L	life of project	R	net receipt
n	number of interest periods	S	future amount or worth
p	minimum acceptable return		

TIME VALUE OF MONEY

Periodic Interest

By definition interest is the difference between the sum S received at the end and the deposit P made at the beginning of a period of time. The periodic rate of interest i is then $(S_1 - P)/P$ and

$$S_1/P = 1 + i \tag{8-1}$$

Amount at Compound Interest. The future amount S_n of an initial principal deposit P_0 accumulated over n periods at rate of interest i credited at the end of each period is given by

$$S_n/P_0 = (1 + i)^n \tag{8-2}$$

in which S_n/P_0 is known as the *compound amount factor* for a single payment. (See Table 8-2.)

Present Worth. The reciprocal of Eq. 8-2 gives the present worth P of an amount S_n payable at the end of n periods,

$$\frac{P}{S_n} = \frac{1}{(1 + i)^n} \tag{8-3}$$

in which P/S_n is known as the *present worth factor* for a single payment. (See Table 8-3.)

Continuous Interest

Interest which is applied at every instant or continuous interest r is often a mathematical convenience. Two different conventions are used with continuous interest. One of these assumes that receipts are concentrated at an instant of time such as at the end of a period. When continuous interest is used with this convention, the convenient tables of the exponential functions e^x and e^{-x} serve as the interest tables with $x = rn$.

The second convention assumes that cash flow is uniformly distributed throughout the period, an assumption that simulates business experience. This convention eliminates much of the mathematical simplicity of continuous interest and may not be worth the extra labor, especially when future cash flows are uncertain.

Compound Amount Factor. If S is the compound amount at any time n and r is the continuous rate of interest per period, usually one year,

$$dS = Sr\,dn \quad \text{and} \quad \int_{P_0}^{S_n} \frac{dS}{S} = \int_0^n r\,dn$$

or

$$\log_e \frac{S_n}{P_0} = rn \quad \text{and} \quad \frac{S_n}{P_0} = e^{rn} \qquad (8\text{-}4)$$

in which S_n is the accumulated amount, with continuous interest r from a principal deposit P_0 at zero time.

If the receipt R_1 is distributed uniformly throughout any one period the accumulated amount S_1 at the end of the period is given by

$$\frac{S_1}{R_1} = \frac{e^r - 1}{r} \qquad (8\text{-}5)$$

Present Worth Factor. The reciprocal of Eq. 8-4 gives the present worth P of an amount S_n payable at time n with continuous interest at the rate r,

$$P/S_n = e^{-rn} \qquad (8\text{-}6)$$

For a receipt or deposit R_c distributed uniformly throughout the period the present worth P is given by

$$\frac{P}{R_c} = \frac{(e^r - 1)e^{-rn}}{r} \qquad (8\text{-}7)$$

Table 8-7 gives solutions of Eq. 8-7 for a series of periods with various interest rates r.

Equivalent Values of i and r. Equations 8-2 and 8-4 do not yield the same numerical results since $(S_n)_r > (S_n)_i$ if $i = r$ and $i > r$ if $(S_n)_i = (S_n)_r$ except in the limit. The equivalent values of i and r per annum for $(S_n)_i = (S_n)_r$ are found by equating Eqs. 8-2 and 8-4 and substituting $n = 1$, from which

$$i = e^r - 1 \qquad (8\text{-}8)$$

Uniform Series

For a uniform series of receipts R per n there are four simple general formulas for solving interest problems. Table 8-1 shows these formulas for (1) periodic interest with R at the end of each period, (2) continuous interest with R at the end of each period, and (3) continuous interest with R uniformly distributed throughout each period. These three different conventions should be clearly distinguished when applying the formulas.

The Series Compound Amount Factor S/R or the *amount of unit annuity* is the accumulated amount of periodic deposits of unity with interest. This amount is the summation of the series of $n - 1$ terms from Eq. 8-2 for $n = 1$ through $n = n - 1$ or the similar series of terms from Eqs. 8-4 and 8-5.

The Series Present Worth Factor P/R or the *present worth of unit annuity* is the discounted value of a series of periodic deposits of unity. This value is the summation of the series of n terms from Eqs. 8-3, 8-6, and 8-7 for $n = 1$ through $n = n$.

The Capital Recovery Factor R/P or the *annuity which has a present worth of unity* is the reciprocal of the series present worth factor. The capital recovery factor is also equal to the sinking fund factor plus the interest rate.

The Sinking Fund Deposit Factor R/S or the periodic deposit that with interest will accumulate to unity is the reciprocal of the series compound amount factor.

Table 8-1 Interest Formulas for Uniform Annual Series

	Periodic Interest i	Continuous Interest r	
	R end of year	R end of year	R continuous
S/R Compound amount factor Amount of unit annuity	$\dfrac{(1+i)^n - 1}{i}$	$\dfrac{e^{rn} - 1}{e^r - 1}$	$\dfrac{e^{rn} - 1}{r}$
P/R Present worth factor Present worth of unit annuity	$\dfrac{1 - (1+i)^{-n}}{i}$	$\dfrac{1 - e^{-rn}}{e^r - 1}$	$\dfrac{1 - e^{-rn}}{r}$
R/P Capital recovery factor Annuity with present worth of unity	$\dfrac{i}{1 - (1+i)^{-n}}$	$\dfrac{e^r - 1}{1 - e^{-rn}}$	$\dfrac{r}{1 - e^{-rn}}$
R/S Sinking fund deposit factor Annuity that will amount to unity	$\dfrac{i}{(1+i)^n - 1}$	$\dfrac{e^r - 1}{e^{rn} - 1}$	$\dfrac{r}{e^{rn} - 1}$

ECONOMIC CHOICE

Minimum Cost Point

Time Independent. When only immediate expenditures are relevant to the choice among alternatives, the time value of money is not involved. The problem then is to obtain the design with the minimum first cost. One such problem is the design of a power line where the variables are the pole height and the span length. Another one is the design of a multispan bridge where the variables are the costs of trusses and piers as functions of span length. These problems are solved graphically or analytically by expressing the costs as functions of the design variable (length of span), adding the separate costs, and locating the point of minimum total cost.

Economic Balance. When increase in annual capital charges results in decreased annual operating charges, both expressed in terms of a common design variable, the solution for minimum total annual cost is called economic balance. Choice of wire size for electrical transmission, insulation for a steam line, diameter of a pipe line, or lot size for a production run are all problems in economic balance. Although graphical or tabular constructions are generally applicable, many of these problems may be solved analytically. For example, if x is the common design variable, a, b, c, and d are constants related to the specific problem, and

$$\text{annual capital cost} = ax + c$$
$$\text{annual operating cost} = b/x + d$$
$$\text{total annual cost} = ax + b/x + (c + d)$$

differentiating total cost and equating to zero gives the optimum solution as

$$x = (b/a)^{1/2} \tag{8-9}$$

When there are two common design variables x and y, and

$$\text{total annual cost} = ax + b/xy + cy + d$$

taking the partial derivatives with respect to x and y and equating them gives

$$x/y = c/a \tag{8-10}$$

from which a plot of total annual cost versus x at various values of y gives a cost curve with minimum point as the optimum. For limitations and other forms see Allen, *Mathematical Analysis for Economists*.

Measures of Profitability

In the choice among alternative investments the objective is to make the best possible use of a limited resource, capital. Decision criteria are thus required, among which the principal ones are described here using the convention of continuous interest r with periodic receipts and disbursements. To convert these equations to periodic interest substitute $(1 + i)^n$ for e^{rn}.

Periodic and continuous interest should not be intermixed in economic comparisons. All results should be converted to one form of interest by Eq. 8-8 before comparing profitabilities.

Minimum Acceptable Return. In the array of prospective investments, some more profitable than others, a minimum standard of comparison is needed. Minimum acceptable return p may be looked upon as the interest rate of the marginal investment that is generally available to the investor. For the state it is the cost of borrowed money. For a public utility it is the rate set by a regulatory agency, in theory the rate required to attract new capital. For a competitive enterprise it is commonly taken as the average rate of earnings on the total assets of the company, often called the *pool rate*. Rates of return are usually computed after income tax.

Internal Rate Method. This well-known method, recently referred to as *discounted cash flow* and *interest rate of return*, serves adequately for the great majority of investment and budgetary decisions. It may be applied to the incremental cash flow from an incremental investment in a project as well as to an array of unrelated projects. The internal rate method postulates that the algebraic sum of the compound amounts of all cash flows for a project is zero at some internal rate of return r found by trial-and-error solution of Eq. 8-11.

or

$$\sum_{n=1}^{n=L} R_n e^{r(L-n)} - \sum_{n=0}^{n=L} D_n e^{r(L-n)} = 0$$

$$\sum_{n=1}^{n=L} R_n e^{-rn} - \sum_{n=0}^{n=L} D_n e^{-rn} = 0 \quad (8\text{-}11)$$

The decision criterion is the internal rate r, larger r being preferred.

This method is not entirely satisfactory because of two inherent defects: (1) it is based on the questionable assumption that the receipts from a project will be reinvested in an equally profitable investment, and (2) the solution for r may be indeterminate (imaginary or multiple roots) when there is more than one reversal in the direction of annual net cash flow.

Proportional Gain Method. This method, attributed to Bernoulli, avoids trial-and-error solutions and is suitable for choosing between mutually exclusive alternates or for ranking an array of investment opportunities. This formulation postulates that the net receipts are accumulated in one account and the net investments in another account both at interest rate p. When the project is terminated at time L, the relative gain G is the ratio of the two accounts or the ratio of their present worths:

$$\frac{\sum_{n=1}^{n=L} R_n e^{p(L-n)}}{\sum_{n=0}^{n=L} D_n e^{p(L-n)}} = \frac{\sum_{n=1}^{n=L} R_n e^{-pn}}{\sum_{n=0}^{n=L} D_n e^{-pn}} = G = e^{kL} \quad (8\text{-}12)$$

The proper decision criterion is G, not k as erroneously used by some authors. Since this method is biased in favor of long-term investments, it is generally reliable only for comparing investments with nearly equal lives as originally proposed by Bernoulli.

Present Worth Method. This method, also referred to as venture worth and incremental present worth, is restricted to comparison of projects that have identical lives L or that cover the same total time span. If it can be assumed that the costs and returns of replacements will repeat those of the original asset, multiple cycles of a short-term project may be compared with a single long-term project covering the same time span, e.g., three 5-year lives considered equivalent to one 15-year life.

With this method present worth is the decision criterion, larger present worth being preferred. The present worth P of each project is computed, with r equal to the minimum acceptable return, as the algebraic sum of the present worths of the annual net cash flows with salvage value taken as a receipt at the end of the project life.

$$P = \sum_{n=1}^{n=L} R_n e^{-rn} - \sum_{n=0}^{n=L} D_n e^{-rn} \tag{8-13}$$

This method is suitable for projects that have no positive receipts, a situation in which the other two methods are indeterminate.

REFERENCES

Allen, R. G. C., *Mathematical Analysis for Economists*, Macmillan, London, 1947.
Barish, N. N., *Economic Analysis for Engineering and Managerial Decision Making*, McGraw-Hill, New York, 1963.
Grant, E. L. and W. G. Ireson, *Principles of Engineering Economy*, Ronald, New York, fourth edition, 1960.
Happel, J., *Chemical Process Economics*, John Wiley, New York, 1958.
Rautenstrauch, W., *The Economics of Business Enterprise*, John Wiley, New York, 1939.

Table 8-2 Amount at Compound Interest, $(1 + i)^n$

n	$1\frac{1}{2}\%$	2%	$2\frac{1}{2}\%$	3%	$3\frac{1}{2}\%$	4%	$4\frac{1}{2}\%$	5%
1	1.01500	1.02000	1.02500	1.03000	1.03500	1.04000	1.04500	1.05000
2	1.03023	1.04040	1.05062	1.06090	1.07122	1.08160	1.09203	1.10250
3	1.04568	1.06121	1.07689	1.09273	1.10872	1.12486	1.14117	1.15763
4	1.06136	1.08243	1.10381	1.12551	1.14752	1.16986	1.19252	1.21551
5	1.07728	1.10408	1.13141	1.15927	1.18769	1.21665	1.24618	1.27628
6	1.09344	1.12616	1.15969	1.19405	1.22926	1.26532	1.30226	1.34010
7	1.10984	1.14869	1.18869	1.22987	1.27228	1.31593	1.36086	1.40710
8	1.12649	1.17166	1.21840	1.26677	1.31681	1.36857	1.42210	1.47746
9	1.14339	1.19509	1.24886	1.30477	1.36290	1.42331	1.48610	1.55133
10	1.16054	1.21899	1.28008	1.34392	1.41060	1.48024	1.55297	1.62889
11	1.17795	1.24337	1.31209	1.38423	1.45997	1.53945	1.62285	1.71034
12	1.19362	1.26824	1.34489	1.42576	1.51107	1.60103	1.69588	1.79586
13	1.21355	1.29361	1.37851	1.46853	1.56396	1.66507	1.77220	1.88565
14	1.23176	1.31948	1.41297	1.51259	1.61869	1.73168	1.85194	1.97993
15	1.25023	1.34587	1.44830	1.55797	1.67535	1.80094	1.93528	2.07893
16	1.26899	1.37279	1.48451	1.60471	1.73399	1.87298	2.02237	2.18287
17	1.28802	1.40024	1.52162	1.65285	1.79468	1.94790	2.11338	2.29202
18	1.30734	1.42825	1.55966	1.70243	1.85749	2.02582	2.20848	2.40662
19	1.32695	1.45681	1.59865	1.75351	1.92250	2.10685	2.30786	2.52695
20	1.34685	1.48595	1.63862	1.80611	1.98979	2.19112	2.41171	2.65330
21	1.36706	1.51567	1.67958	1.86029	2.05943	2.27877	2.52024	2.78596
22	1.38756	1.54598	1.72157	1.91610	2.13151	2.36992	2.63365	2.92526
23	1.40838	1.57690	1.76461	1.97359	2.20611	2.46472	2.75217	3.07152
24	1.42950	1.60844	1.80873	2.03279	2.28333	2.56330	2.87601	3.22510
25	1.45095	1.64061	1.85394	2.09378	2.36324	2.66584	3.00543	3.38635
26	1.47271	1.67342	1.90029	2.15659	2.44596	2.77247	3.14068	3.55567
27	1.49480	1.70689	1.94780	2.22129	2.53157	2.88337	3.28201	3.73346
28	1.51722	1.74102	1.99650	2.28793	2.62017	2.99870	3.42970	3.92013
29	1.53998	1.77584	2.04640	2.35657	2.71188	3.11865	3.58404	4.11614
30	1.56308	1.81136	2.09757	2.42726	2.80679	3.24340	3.74532	4.32194
31	1.58653	1.84759	2.15001	2.50008	2.90503	3.37313	3.91386	4.53804
32	1.61032	1.88454	2.20376	2.57508	3.00671	3.50806	4.08998	4.76494
33	1.63448	1.92223	2.25885	2.65234	3.11194	3.64838	4.27403	5.00319
34	1.65900	1.96068	2.31532	2.73191	3.22086	3.79432	4.46636	5.25335
35	1.68388	1.99989	2.37321	2.81386	3.33359	3.94609	4.66735	5.51602
36	1.70914	2.03989	2.43254	2.89828	3.45027	4.10393	4.87738	5.79182
37	1.73478	2.08069	2.49335	2.98523	3.57103	4.26809	5.09686	6.08141
38	1.76080	2.12230	2.55568	3.07478	3.69601	4.43881	5.32622	6.38548
39	1.78721	2.16474	2.61957	3.16703	3.82537	4.61637	5.56590	6.70475
40	1.81402	2.20804	2.68506	3.26204	3.95926	4.80102	5.81636	7.03999
41	1.84123	2.25220	2.75219	3.35990	4.09783	4.99306	6.07810	7.39199
42	1.86885	2.29724	2.82100	3.46070	4.24126	5.19278	6.35162	7.76159
43	1.89688	2.34319	2.89152	3.56452	4.38970	5.40050	6.63744	8.14967
44	1.92533	2.39005	2.96381	3.67145	4.54334	5.61652	6.93612	8.55715
45	1.95241	2.43785	3.03790	3.78160	4.70236	5.84118	7.24825	8.98501
46	1.98353	2.48661	3.11385	3.89504	4.86694	6.07482	7.57442	9.43426
47	2.01328	2.53634	3.19170	4.01190	5.03728	6.31782	7.91527	9.90597
48	2.04348	2.58707	3.27149	4.13225	5.21359	6.57053	8.27146	10.4013
49	2.07413	2.63881	3.35328	4.25622	5.39606	6.83335	8.64367	10.9213
50	2.10524	2.69159	3.43711	4.38391	5.58493	7.10668	9.03264	11.4674

Table 8-2 (*continued*)

n	5½%	6%	7%	8%	10%	12%	15%	20%
1	1.055	1.060	1.070	1.080	1.100	1.120	1.150	1.200
2	1.113	1.124	1.145	1.166	1.210	1.254	1.322	1.440
3	1.174	1.191	1.225	1.260	1.331	1.405	1.521	1.728
4	1.239	1.262	1.311	1.360	1.464	1.574	1.749	2.074
5	1.307	1.338	1.403	1.469	1.611	1.762	2.011	2.488
6	1.379	1.419	1.501	1.587	1.772	1.974	2.313	2.986
7	1.455	1.504	1.606	1.714	1.949	2.211	2.660	3.583
8	1.535	1.594	1.718	1.851	2.144	2.476	3.059	4.300
9	1.619	1.689	1.838	1.999	2.358	2.773	3.518	5.160
10	1.708	1.791	1.967	2.159	2.594	3.106	4.046	6.192
11	1.802	1.898	2.105	2.332	2.853	3.479	4.652	7.430
12	1.901	2.012	2.252	2.518	3.138	3.896	5.350	8.916
13	2.006	2.133	2.410	2.720	3.452	4.363	6.153	10.699
14	2.116	2.261	2.579	2.937	3.797	4.887	7.076	12.839
15	2.232	2.397	2.759	3.172	4.177	5.474	8.137	15.407
16	2.355	2.540	2.952	3.426	4.595	6.130	9.358	18.488
17	2.485	2.693	3.159	3.700	5.054	6.866	10.761	22.186
18	2.621	2.854	3.380	3.996	5.560	7.690	12.375	26.623
19	2.766	3.026	3.617	4.316	6.116	8.613	14.232	31.948
20	2.918	3.207	3.870	4.661	6.727	9.646	16.367	38.338
21	3.078	3.400	4.141	5.034	7.400	10.804	18.821	46.005
22	3.248	3.604	4.430	5.437	8.140	12.100	21.645	55.206
23	3.426	3.820	4.741	5.871	8.954	13.552	24.891	66.247
24	3.615	4.049	5.072	6.341	9.850	15.179	28.625	79.497
25	3.813	4.292	5.427	6.848	10.835	17.000	32.919	95.396
26	4.023	4.549	5.807	7.396	11.918	19.040	37.857	114.475
27	4.244	4.822	6.214	7.988	13.110	21.325	43.535	137.370
28	4.478	5.112	6.649	8.627	14.421	23.884	50.065	164.845
29	4.724	5.418	7.114	9.317	15.863	26.750	57.575	197.813
30	4.984	5.743	7.612	10.063	17.449	29.960	66.212	237.376
31	5.258	6.088	8.145	10.868	19.194	33.555	76.143	284.851
32	5.547	6.453	8.715	11.737	21.114	37.582	87.565	341.822
33	5.852	6.841	9.325	12.676	23.225	42.091	100.700	410.186
34	6.174	7.251	9.978	13.690	25.548	47.142	115.805	492.223
35	6.514	7.686	10.677	14.785	28.102	52.799	133.175	590.668

Table 8-3 Present Worth, $(1+i)^{-n}$

n	2%	3%	4%	5%	6%	7%	8%	10%
1	0.98039	0.97087	0.96154	0.95238	0.94340	0.9346	0.9259	0.9091
2	0.96117	0.94260	0.92456	0.90703	0.89000	0.8734	0.8573	0.8264
3	0.94232	0.91514	0.88900	0.86384	0.83962	0.8163	0.7938	0.7513
4	0.92385	0.88849	0.85480	0.82270	0.79209	0.7629	0.7350	0.6830
5	0.90573	0.86261	0.82193	0.78353	0.74726	0.7130	0.6806	0.6209
6	0.88797	0.83748	0.79031	0.74622	0.70496	0.6663	0.6302	0.5645
7	0.87056	0.81309	0.75992	0.71068	0.66506	0.6227	0.5835	0.5132
8	0.85349	0.78941	0.73069	0.67684	0.62741	0.5820	0.5403	0.4665
9	0.83676	0.76642	0.70259	0.64461	0.59190	0.5439	0.5002	0.4241
10	0.82035	0.74409	0.67556	0.61391	0.55839	0.5083	0.4632	0.3855
11	0.80426	0.72242	0.64958	0.58468	0.52679	0.4751	0.4289	0.3505
12	0.78849	0.70138	0.62460	0.55684	0.49697	0.4440	0.3971	0.3186
13	0.77303	0.68095	0.60057	0.53032	0.46884	0.4150	0.3677	0.2897
14	0.75788	0.66112	0.57748	0.50507	0.44230	0.3878	0.3405	0.2633
15	0.74301	0.64186	0.55526	0.48102	0.41727	0.3624	0.3152	0.2394
16	0.72845	0.62317	0.53391	0.45811	0.39365	0.3387	0.2919	0.2176
17	0.71416	0.60502	0.51337	0.43630	0.37136	0.3166	0.2703	0.1978
18	0.70016	0.58739	0.49363	0.41552	0.35034	0.2959	0.2502	0.1799
19	0.68643	0.57029	0.47464	0.39573	0.33051	0.2765	0.2317	0.1635
20	0.67297	0.55368	0.45639	0.37689	0.31180	0.2584	0.2145	0.1486
21	0.65978	0.53755	0.43883	0.35894	0.29416	0.2415	0.1987	0.1351
22	0.64684	0.52189	0.42196	0.34185	0.27751	0.2257	0.1839	0.1228
23	0.63416	0.50669	0.40573	0.32557	0.26180	0.2109	0.1703	0.1117
24	0.62172	0.49193	0.39012	0.31007	0.24698	0.1971	0.1577	0.1015
25	0.60953	0.47761	0.37512	0.29530	0.23300	0.1842	0.1460	0.0923
26	0.59758	0.46369	0.36069	0.28124	0.21981	0.1722	0.1352	0.0839
27	0.58586	0.45019	0.34682	0.26785	0.20737	0.1609	0.1252	0.0763
28	0.57437	0.43708	0.33348	0.25509	0.19563	0.1504	0.1159	0.0693
29	0.56311	0.42435	0.32065	0.24295	0.18456	0.1406	0.1073	0.0630
30	0.55207	0.41199	0.30832	0.23138	0.17411	0.1314	0.0994	0.0573
31	0.54125	0.39999	0.29646	0.22036	0.16425	0.1228	0.0920	0.0521
32	0.53063	0.38834	0.28506	0.20987	0.15496	0.1147	0.0852	0.0474
33	0.52023	0.37703	0.27409	0.19987	0.14619	0.1072	0.0789	0.0431
34	0.51003	0.36604	0.26355	0.19035	0.13791	0.1002	0.0730	0.0391
35	0.50003	0.35538	0.25342	0.18129	0.13011	0.0937	0.0676	0.0356
36	0.49022	0.34503	0.24367	0.17266	0.12274	0.0875	0.0626	
37	0.48061	0.33498	0.23430	0.16444	0.11579	0.0818	0.0580	
38	0.47119	0.32523	0.22529	0.15661	0.10924	0.0765	0.0537	
39	0.46195	0.31575	0.21662	0.14915	0.10306	0.0715	0.0497	
40	0.45289	0.30656	0.20829	0.14205	0.09722	0.0668	0.0460	0.0221
41	0.44401	0.29763	0.20028	0.13528	0.09172	0.0624	0.0426	
42	0.43530	0.28896	0.19257	0.12884	0.08653	0.0583	0.0395	
43	0.42677	0.28054	0.18517	0.12270	0.08163	0.0545	0.0365	
44	0.41840	0.27237	0.17805	0.11686	0.07701	0.0509	0.0338	
45	0.41020	0.26444	0.17120	0.11130	0.07265	0.0476	0.0313	0.0137
46	0.40215	0.25674	0.16461	0.10600	0.06854	0.0445	0.0290	
47	0.39427	0.24926	0.15828	0.10095	0.06466	0.0416	0.0269	
48	0.38654	0.24200	0.15219	0.09614	0.06100	0.0389	0.0249	
49	0.37896	0.23495	0.14634	0.09156	0.05755	0.0363	0.0230	
50	0.37153	0.22811	0.14071	0.08720	0.05429	0.0339	0.0213	0.0085

Table 8-3 (*Continued*)

n	12%	15%	20%	25%	30%	35%	40%	45%
1	0.8929	0.8696	0.8333	0.8000	0.7692	0.7407	0.7143	0.6897
2	0.7972	0.7561	0.6944	0.6400	0.5917	0.5487	0.5102	0.4756
3	0.7118	0.6575	0.5787	0.5120	0.4552	0.4064	0.3644	0.3280
4	0.6355	0.5718	0.4823	0.4096	0.3501	0.3011	0.2603	0.2262
5	0.5674	0.4972	0.4019	0.3277	0.2693	0.2230	0.1859	0.1560
6	0.5066	0.4323	0.3349	0.2621	0.2072	0.1652	0.1328	0.1076
7	0.4523	0.3759	0.2791	0.2097	0.1594	0.1224	0.0949	0.0742
8	0.4039	0.3269	0.2326	0.1678	0.1226	0.0906	0.0678	0.0512
9	0.3606	0.2843	0.1938	0.1342	0.0943	0.0671	0.0484	0.0353
10	0.3220	0.2472	0.1615	0.1074	0.0725	0.0497	0.0346	0.0243
11	0.2875	0.2149	0.1346	0.0859	0.0558	0.0368	0.0247	0.0168
12	0.2567	0.1869	0.1122	0.0687	0.0429	0.0273	0.0176	0.0116
13	0.2292	0.1625	0.0935	0.0550	0.0330	0.0202	0.0126	0.0080
14	0.2046	0.1413	0.0779	0.0440	0.0254	0.0150	0.0090	0.0055
15	0.1827	0.1229	0.0649	0.0352	0.0195	0.0111	0.0064	0.0038
16	0.1631	0.1069	0.0541	0.0281	0.0150	0.0082	0.0046	0.0026
17	0.1456	0.0929	0.0451	0.0225	0.0116	0.0061	0.0033	0.0018
18	0.1300	0.0808	0.0376	0.0180	0.0089	0.0045	0.0023	0.0012
19	0.1161	0.0703	0.0313	0.0144	0.0068	0.0033	0.0017	0.0009
20	0.1037	0.0611	0.0261	0.0115	0.0053	0.0025	0.0012	0.0006
21	0.0926	0.0531	0.0217	0.0092	0.0040	0.0018	0.0009	0.0004
22	0.0826	0.0462	0.0181	0.0074	0.0031	0.0014	0.0006	0.0003
23	0.0738	0.0402	0.0151	0.0059	0.0024	0.0010	0.0004	0.0002
24	0.0659	0.0349	0.0126	0.0047	0.0018	0.0007	0.0003	0.0001
25	0.0588	0.0304	0.0105	0.0038	0.0014	0.0006	0.0002	0.0001
26	0.0525	0.0264	0.0087	0.0030	0.0011	0.0004	0.0002	0.0001
27	0.0469	0.0230	0.0073	0.0024	0.0008	0.0003	0.0001	
28	0.0419	0.0200	0.0061	0.0019	0.0006	0.0002	0.0001	
29	0.0374	0.0174	0.0051	0.0015	0.0005	0.0002	0.0001	
30	0.0334	0.0151	0.0042	0.0012	0.0004	0.0001		
31	0.0298	0.0131	0.0035	0.0010	0.0003	0.0001		
32	0.0266	0.0114	0.0029	0.0008	0.0002	0.0001		
33	0.0238	0.0099	0.0024	0.0006	0.0002	0.0001		
34	0.0212	0.0086	0.0020	0.0005	0.0001			
35	0.0189	0.0075	0.0017	0.0004	0.0001			

278 ENGINEERING ECONOMY

Table 8-4 Amount of Annuity, $\dfrac{(1+i)^n - 1}{i}$

n	$1\frac{1}{2}\%$	2%	$2\frac{1}{2}\%$	3%	$3\frac{1}{2}\%$	4%	$4\frac{1}{2}\%$	5%
1	1.00000	1.00000	1.00000	1.00000	1.00000	1.00000	1.00000	1.00000
2	2.01500	2.02000	2.02500	2.03000	2.03500	2.04000	2.04500	2.05000
3	3.04522	3.06040	3.07562	3.09090	3.10623	3.12160	3.13702	3.15250
4	4.09090	4.12161	4.15252	4.18363	4.21494	4.24646	4.27819	4.31013
5	5.15227	5.20404	5.25633	5.30914	5.36247	5.41632	5.47071	5.52563
6	6.22955	6.30812	6.38774	6.46841	6.55015	6.63298	6.71689	6.80191
7	7.32299	7.43428	7.54743	7.66246	7.77941	7.89829	8.01915	8.14201
8	8.43284	8.58297	8.73612	8.89234	9.05169	9.21423	9.38001	9.54911
9	9.55933	9.75463	9.95452	10.1591	10.3685	10.5828	10.8021	11.0266
10	10.70272	10.94972	11.2034	11.4639	11.7314	12.0061	12.2882	12.5779
11	11.86326	12.16872	12.4835	12.8078	13.1420	13.4864	13.8412	14.2068
12	13.04121	13.41209	13.7956	14.1920	14.6020	15.0258	15.4640	15.9171
13	14.23683	14.68033	15.1404	15.6178	16.1130	16.6268	17.1599	17.7130
14	15.45038	15.97394	16.5190	17.0863	17.6770	18.2919	18.9321	19.5986
15	16.68214	17.29342	17.9319	18.5989	19.2957	20.0236	20.7841	21.5786
16	17.93237	18.63929	19.3802	20.1569	20.9710	21.8245	22.7193	23.6575
17	19.20136	20.01207	20.8647	21.7616	22.7050	23.6975	24.7417	25.8404
18	20.48938	21.41231	22.3863	23.4144	24.4997	25.6454	26.8551	28.1324
19	21.79672	22.84056	23.9460	25.1169	26.3572	27.6712	29.0636	30.5390
20	23.12367	24.29737	25.5447	26.8704	28.2797	29.7781	31.3714	33.0660
21	24.47052	25.78332	27.1833	28.6765	30.2695	31.9692	33.7831	35.7193
22	25.83758	27.29899	28.8629	30.5368	32.3289	34.2480	36.3034	38.5052
23	27.22514	28.84496	30.5844	32.4529	34.4604	36.6179	38.9370	41.4305
24	28.63352	30.42186	32.3490	34.4265	36.6665	39.0826	41.6892	44.5020
25	30.06302	32.03030	34.1578	36.4593	38.9499	41.6459	44.5652	47.7271
26	31.51397	33.67091	36.0117	38.5530	41.3131	44.3117	47.5706	51.1135
27	32.98668	35.34432	37.9120	40.7096	43.7591	47.0842	50.7113	54.6691
28	34.48148	37.05121	39.8598	42.9309	46.2906	49.9676	53.9933	58.4026
29	35.99870	38.79223	41.8563	45.2189	48.9180	52.9663	57.4230	62.3227
30	37.53868	40.56808	43.9027	47.5754	51.6227	56.0849	61.0071	66.4388
31	39.10176	42.37944	46.0003	50.0027	54.4295	59.3283	64.7524	70.7608
32	40.68829	44.22703	48.1503	52.5028	57.3345	62.7015	68.6662	75.2988
33	42.29861	46.11157	50.3540	55.0778	60.3412	66.2095	72.7562	80.0638
34	43.93309	48.03380	52.6129	57.7302	63.4532	69.8579	77.0303	85.0670
35	45.59208	49.99448	54.9282	60.4621	66.6740	73.6522	81.4966	90.3203
36	47.27597	51.99437	57.3014	63.2759	70.0076	77.5983	86.1640	95.8363
37	48.98511	54.03425	59.7339	66.1742	73.4579	81.7022	91.0413	101.628
38	50.71989	56.11494	62.2273	69.1594	77.0289	85.9703	96.1382	107.710
39	52.48068	58.23724	64.7830	72.2342	80.7249	90.4091	101.464	114.095
40	54.26789	60.40198	67.4026	75.4013	84.5503	95.0255	107.030	120.800
41	56.08191	62.61002	70.0876	78.6633	88.5095	99.8265	112.847	127.840
42	57.92314	64.86222	72.8398	82.0232	92.6074	104.820	118.925	135.232
43	59.79199	67.15947	75.6608	85.4839	96.8486	110.012	125.276	142.993
44	61.68887	69.50266	78.5523	89.0484	101.238	115.413	131.914	151.143
45	63.61420	71.89271	81.5161	92.7199	105.782	121.029	138.850	159.700
46	65.56841	74.33056	84.5540	96.5015	110.484	126.871	146.098	168.685
47	67.55194	76.81718	87.6679	100.397	115.351	132.945	153.673	178.119
48	69.56522	79.35352	90.8596	104.408	120.388	139.263	161.588	188.025
49	71.60870	81.94059	94.1311	108.541	125.602	145.834	169.859	198.427
50	73.68283	84.57940	97.4843	112.797	130.998	152.667	178.503	209.348

Table 8-4 (*Continued*)

n	5½%	6%	7%	8%	10%	12%	15%	20%
1	1.00000	1.00000	1.000	1.000	1.000	1.000	1.000	1.000
2	2.05500	2.06000	2.070	2.080	2.100	2.120	2.150	2.200
3	3.16803	3.18360	3.215	3.246	3.310	3.374	3.472	3.640
4	4.34227	4.37462	4.440	4.506	4.641	4.779	4.993	5.368
5	5.58109	5.63709	5.751	5.867	6.105	6.353	6.742	7.442
6	6.88805	6.97532	7.153	7.336	7.716	8.115	8.754	9.930
7	8.26689	8.39384	8.654	8.923	9.487	10.089	11.067	12.916
8	9.72157	9.89747	10.260	10.637	11.436	12.300	13.727	16.499
9	11.2563	11.4913	11.978	12.488	13.579	14.776	16.786	20.799
10	12.8754	13.1808	13.816	14.487	15.937	17.549	20.304	25.959
11	14.5835	14.9716	15.784	16.645	18.531	20.655	24.349	32.150
12	16.3856	16.8699	17.888	18.977	21.384	24.133	29.002	39.580
13	18.2868	18.8821	20.141	21.495	24.523	28.029	34.352	48.497
14	20.2926	21.0151	22.550	24.215	27.975	32.393	40.505	59.196
15	22.4087	23.2760	25.129	27.152	31.772	37.280	47.580	72.035
16	24.6411	25.6725	27.888	30.324	35.950	42.753	55.717	87.442
17	26.9964	28.2129	30.840	33.750	40.545	48.884	65.075	105.931
18	29.4812	30.9057	33.999	37.450	45.599	55.750	75.836	128.117
19	32.1027	33.7600	37.379	41.446	51.159	63.440	88.212	154.740
20	34.8683	36.7856	40.995	45.762	57.275	72.052	102.443	186.688
21	37.7861	39.9927	44.865	50.423	64.002	81.699	118.810	225.025
22	40.8643	43.3923	49.006	55.457	71.403	92.502	137.631	271.031
23	44.1118	46.9958	53.436	60.893	79.543	104.603	159.276	326.237
24	47.5380	50.8156	58.177	66.765	88.497	118.155	184.167	392.484
25	51.1526	54.8645	63.249	73.106	98.347	133.334	212.793	471.981
26	54.9660	59.1564	68.676	79.954	109.182	150.334	245.711	567.377
27	58.9891	63.7058	74.484	87.351	121.100	169.374	283.568	681.852
28	63.2335	68.5281	80.698	95.339	134.210	190.699	327.103	819.223
29	67.7114	73.6398	87.347	103.966	148.631	214.582	377.169	984.067
30	72.4355	79.0582	94.461	113.283	164.494	241.332	434.744	1181.881
31	77.4194	84.8017	102.073	123.346	181.943	271.292	500.956	1419.257
32	82.6775	90.8898	110.218	134.214	201.138	304.847	577.099	1704.108
33	88.2248	97.3432	118.933	145.951	222.252	342.429	664.664	2045.930
34	94.0071	104.184	128.259	158.627	245.477	384.520	765.364	2456.116
35	100.251	111.435	138.237	172.317	271.024	431.663	881.168	2948.339

Table 8-5 Present Worth of Annuity, $\dfrac{1-(1+i)^{-n}}{i}$

n	2%	3%	4%	5%	6%	7%	8%	10%
1	0.98039	0.97087	0.96154	0.95238	0.94340	0.935	0.926	0.909
2	1.94156	1.91347	1.88609	1.85941	1.83339	1.808	1.783	1.736
3	2.88388	2.82861	2.77509	2.72325	2.67301	2.624	2.577	2.487
4	3.80773	3.71710	3.62990	3.54595	3.46511	3.387	3.312	3.170
5	4.71346	4.57971	4.45182	4.32948	4.21236	4.100	3.993	3.791
6	5.60143	5.41719	5.24214	5.07569	4.91732	4.767	4.623	4.355
7	6.47199	6.23028	6.00205	5.78637	5.58238	5.389	5.206	4.868
8	7.32548	7.01969	6.73274	6.46321	6.20979	5.971	5.747	5.335
9	8.16224	7.78611	7.43533	7.10782	6.80169	6.515	6.247	5.759
10	8.98258	8.53020	8.11090	7.72173	7.36009	7.024	6.710	6.144
11	9.78685	9.25262	8.76048	8.30641	7.88687	7.499	7.139	6.495
12	10.57534	9.95400	9.38507	8.86325	8.38384	7.943	7.536	6.814
13	11.34837	10.6350	9.98565	9.39357	8.85268	8.358	7.904	7.103
14	12.10625	11.2961	10.5631	9.89864	9.29498	8.745	8.244	7.367
15	12.84926	11.9379	11.1184	10.3797	9.71225	9.108	8.559	7.606
16	13.57771	12.5611	11.6523	10.8378	10.1059	9.447	8.851	7.824
17	14.29187	13.1661	12.1657	11.2741	10.4773	9.763	9.122	8.022
18	14.99203	13.7535	12.6593	11.6896	10.8276	10.059	9.372	8.201
19	15.67846	14.3238	13.1339	12.0853	11.1581	10.336	9.604	8.365
20	16.35143	14.8775	13.5903	12.4622	11.4699	10.594	9.818	8.514
21	17.01121	15.4150	14.0292	12.8212	11.7641	10.836	10.017	8.649
22	17.65805	15.9369	14.4511	13.1630	12.0416	11.061	10.201	8.772
23	18.29220	16.4436	14.8568	13.4886	12.3034	11.272	10.371	8.883
24	18.91393	16.9355	15.2470	13.7986	12.5504	11.469	10.529	8.985
25	19.52346	17.4131	15.6221	14.0939	12.7834	11.654	10.675	9.077
26	20.12104	17.8768	15.9828	14.3752	13.0032	11.826	10.810	9.161
27	20.70690	18.3270	16.3296	14.6430	13.2105	11.987	10.935	9.237
28	21.28127	18.7641	16.6631	14.8981	13.4062	12.137	11.051	9.307
29	21.84438	19.1885	16.9837	15.1411	13.5907	12.278	11.158	9.370
30	22.39646	19.6004	17.2920	15.3725	13.7648	12.409	11.258	9.427
31	22.93770	20.0004	17.5885	15.5928	13.9291	12.532	11.350	9.479
32	23.46833	20.3888	17.8736	15.8027	14.0840	12.647	11.435	9.526
33	23.98856	20.7658	18.1476	16.0025	14.2302	12.754	11.514	9.569
34	24.49859	21.1318	18.4112	16.1929	14.3681	12.854	11.587	9.609
35	24.99862	21.4872	18.6646	16.3742	14.4982	12.948	11.655	9.644
36	25.48884	21.8323	18.9083	16.5469	14.6210			
37	25.96945	22.1672	19.1426	16.7113	14.7368			
38	26.44064	22.4925	19.3679	16.8679	14.8460			
39	26.90259	22.8082	19.5845	17.0170	14.9491			
40	27.35548	23.1148	19.7928	17.1591	15.0463			
41	27.79949	23.4124	19.9931	17.2944	15.1380			
42	28.23479	23.7014	20.1856	17.4232	15.2245			
43	28.66156	23.9819	20.3708	17.5459	15.3062			
44	29.07996	24.2543	20.5488	17.6628	15.3832			
45	29.49016	24.5187	20.7200	17.7741	15.4558			
46	29.89231	24.7754	20.8847	17.8801	15.5244			
47	30.28658	25.0247	21.0429	17.9810	15.5890			
48	30.67312	25.2667	21.1951	18.0772	15.6500			
49	31.05208	25.5017	21.3415	18.1687	15.7076			
50	31.42361	25.7298	21.4822	18.2559	15.7619			

Table 8-5 (*Continued*)

n	12%	15%	20%	25%	30%	35%	40%	45%
1	0.893	0.870	0.833	0.800	0.769	0.741	0.714	0.690
2	1.690	1.626	1.528	1.440	1.361	1.289	1.224	1.165
3	2.402	2.283	2.106	1.952	1.816	1.696	1.589	1.493
4	3.037	2.855	2.589	2.362	2.166	1.997	1.849	1.720
5	3.605	3.352	2.991	2.689	2.436	2.220	2.035	1.876
6	4.111	3.784	3.326	2.951	2.643	2.385	2.168	1.983
7	4.564	4.160	3.605	3.161	2.802	2.507	2.263	2.057
8	4.968	4.487	3.837	3.329	2.925	2.598	2.331	2.109
9	5.328	4.772	4.031	3.463	3.019	2.665	2.379	2.144
10	5.650	5.019	4.192	3.571	3.092	2.715	2.414	2.168
11	5.938	5.234	4.327	3.656	3.147	2.752	2.438	2.185
12	6.194	5.421	4.439	3.725	3.190	2.779	2.456	2.196
13	6.424	5.583	4.533	3.780	3.223	2.799	2.469	2.204
14	6.628	5.724	4.611	3.824	3.249	2.814	2.478	2.210
15	6.811	5.847	4.675	3.859	3.268	2.825	2.484	2.214
16	6.974	5.945	4.730	3.887	3.283	2.834	2.489	2.216
17	7.120	6.047	4.775	3.910	3.295	2.840	2.492	2.218
18	7.250	6.128	4.812	3.928	3.304	2.844	2.494	2.219
19	7.366	6.198	4.844	3.942	3.311	2.848	2.496	2.220
20	7.469	6.259	4.870	3.954	3.316	2.850	2.497	2.221
21	7.562	6.312	4.891	3.963	3.320	2.852	2.498	2.221
22	7.645	6.359	4.909	3.970	3.323	2.853	2.498	2.222
23	7.718	6.399	4.925	3.976	3.325	2.854	2.499	2.222
24	7.784	6.434	4.937	3.981	3.327	2.855	2.499	2.222
25	7.843	6.464	4.948	3.985	3.329	2.856	2.499	2.222
26	7.896	6.491	4.956	3.988	3.330	2.856	2.500	2.222
27	7.943	6.514	4.964	3.990	3.331	2.856	2.500	2.222
28	7.984	6.534	4.970	3.992	3.331	2.857	2.500	2.222
29	8.022	6.551	4.975	3.994	3.332	2.857	2.500	2.222
30	8.055	6.566	4.979	3.995	3.332	2.875	2.500	2.222
31	8.085	6.579	4.982	3.996	3.332	2.857	2.500	2.222
32	8.112	6.591	4.985	3.997	3.333	2.857	2.500	2.222
33	8.135	6.600	4.988	3.997	3.333	2.857	2.500	2.222
34	8.157	6.609	4.990	3.998	3.333	2.857	2.500	2.222
35	8.176	6.617	4.992	3.998	3.333	2.857	2.500	2.222
∞	8.333	6.667	5.000	4.000	3.333	2.857	2.500	2.222

Table 8-6 Annuity with Present Worth of Unity, $\dfrac{i}{1 - (1 + i)^{-n}}$

n	2%	3%	4%	5%	6%	7%	8%	10%
1	1.020000	1.030000	1.040000	1.050000	1.060000	1.07000	1.08000	1.10000
2	0.515050	0.522611	0.530196	0.537805	0.545437	0.55309	0.56077	0.57619
3	0.346755	0.353530	0.360349	0.367209	0.374110	0.38105	0.38803	0.40211
4	0.262624	0.269027	0.275490	0.282012	0.288591	0.29523	0.30192	0.31547
5	0.212158	0.218355	0.224627	0.230975	0.237396	0.34389	0.25046	0.26380
6	0.178526	0.184598	0.190762	0.197017	0.203363	0.20980	0.21632	0.22961
7	0.154512	0.160506	0.166610	0.172820	0.179135	0.18555	0.19207	0.20541
8	0.136510	0.142456	0.148528	0.154722	0.161036	0.16747	0.17401	0.18744
9	0.122515	0.128434	0.134493	0.140690	0.147022	0.15349	0.16008	0.17364
10	0.111327	0.117231	0.123291	0.129505	0.135868	0.14238	0.14903	0.16275
11	0.102178	0.108077	0.114149	0.120389	0.126793	0.13336	0.14008	0.15396
12	0.094560	0.100462	0.106552	0.112825	0.119277	0.12590	0.13270	0.14676
13	0.088118	0.094030	0.100144	0.106456	0.112960	0.11965	0.12652	0.14078
14	0.082602	0.088526	0.094669	0.101024	0.107585	0.11434	0.12130	0.13575
15	0.077826	0.083767	0.089941	0.096342	0.102963	0.10979	0.11683	0.13147
16	0.073650	0.079611	0.085820	0.092270	0.098952	0.10586	0.11298	0.12782
17	0.069970	0.075953	0.082199	0.088699	0.095445	0.10243	0.10963	0.12466
18	0.066702	0.072709	0.078993	0.085546	0.092357	0.09941	0.10670	0.12193
19	0.063782	0.069814	0.076139	0.082745	0.089621	0.09675	0.10413	0.11955
20	0.061157	0.067216	0.073582	0.080243	0.087185	0.09439	0.10185	0.11746
21	0.058785	0.064872	0.071280	0.077996	0.085005	0.09229	0.09983	0.11562
22	0.056631	0.062747	0.069199	0.075971	0.083046	0.09041	0.09803	0.11401
23	0.054668	0.060814	0.067309	0.074137	0.081278	0.08871	0.09642	0.11257
24	0.052871	0.059047	0.065587	0.072471	0.079679	0.08719	0.09498	0.11130
25	0.051220	0.057428	0.064012	0.070952	0.078227	0.08581	0.09368	0.11017
26	0.049699	0.055938	0.062567	0.069564	0.076904	0.08456	0.09251	0.10916
27	0.048293	0.054564	0.061239	0.068292	0.075697	0.08343	0.09145	0.10826
28	0.046990	0.053293	0.060013	0.067123	0.074593	0.08239	0.09049	0.10745
29	0.045778	0.052115	0.058880	0.066046	0.073580	0.08145	0.08962	0.10673
30	0.044650	0.051019	0.057830	0.065051	0.072649	0.08059	0.08883	0.10608

Table 8-6 (*Continued*)

n	12%	15%	20%	25%	30%	35%	40%	45%
1	1.12000	1.15000	1.20000	1.25000	1.30000	1.35000	1.40000	1.45000
2	0.59170	0.61512	0.65455	0.69444	0.73478	0.77553	0.81667	0.85816
3	0.41635	0.43798	0.47473	0.51230	0.55063	0.58966	0.62936	0.66966
4	0.32923	0.35027	0.38629	0.42344	0.46163	0.50076	0.54077	0.58156
5	0.27741	0.29832	0.33438	0.37185	0.41058	0.45046	0.49136	0.53318
6	0.24323	0.26424	0.30071	0.33882	0.37839	0.41926	0.46126	0.50426
7	0.21912	0.24036	0.27742	0.31634	0.35687	0.39880	0.44192	0.48607
8	0.20130	0.22285	0.26061	0.30040	0.34192	0.38489	0.42907	0.47427
9	0.18768	0.20957	0.24808	0.28876	0.33124	0.37519	0.42034	0.46646
10	0.17698	0.19925	0.23852	0.28007	0.32346	0.36832	0.41432	0.46123
11	0.16842	0.19107	0.23110	0.27349	0.31773	0.36339	0.41013	0.45768
12	0.16144	0.18448	0.22526	0.26845	0.31345	0.35982	0.40718	0.45527
13	0.15568	0.17911	0.22062	0.26454	0.31024	0.35722	0.40510	0.45362
14	0.15087	0.17469	0.21689	0.26150	0.30782	0.35532	0.40363	0.45249
15	0.14682	0.17102	0.21388	0.25912	0.30598	0.35393	0.40259	0.45172
16	0.14339	0.16795	0.21144	0.25724	0.30458	0.35290	0.40185	0.45118
17	0.14046	0.16537	0.20944	0.25576	0.30351	0.35214	0.40132	0.45081
18	0.13794	0.16319	0.20781	0.25459	0.30269	0.35158	0.40094	0.45056
19	0.13576	0.16134	0.20646	0.25366	0.30207	0.35117	0.40067	0.45039
20	0.13388	0.15976	0.20536	0.25292	0.30159	0.35087	0.40048	0.45027

Table 8-7 Present Worth of Unit R Uniformly Distributed Over Period With Continuous Interest, $\dfrac{(e^r - 1)}{r} e^{-rn}$

Period	1%	5%	10%	15%	20%	25%	30%	35%	40%	50%
0–1	0.995	0.975	0.952	0.929	0.906	0.885	0.864	0.844	0.824	0.787
1–2	0.985	0.928	0.861	0.799	0.742	0.689	0.640	0.595	0.552	0.477
2–3	0.975	0.883	0.779	0.688	0.608	0.537	0.474	0.419	0.370	0.290
3–4	0.966	0.840	0.705	0.592	0.497	0.418	0.351	0.295	0.248	0.176
4–5	0.956	0.799	0.638	0.510	0.407	0.326	0.260	0.208	0.166	0.106
5–6	0.946	0.760	0.577	0.439	0.333	0.254	0.193	0.147	0.112	0.065
6–7	0.937	0.723	0.522	0.378	0.273	0.197	0.143	0.103	0.075	0.039
7–8	0.928	0.687	0.473	0.325	0.224	0.154	0.106	0.073	0.050	0.024
8–9	0.918	0.654	0.428	0.280	0.183	0.120	0.078	0.051	0.034	0.014
9–10	0.909	0.622	0.387	0.241	0.150	0.093	0.058	0.036	0.022	0.009

Table 8-8 Equivalent Values of Periodic and Continuous Interest

Periodic $i\%$	Nominal i or $r\%$	Continuous $r\%$	Periodic $i\%$	Nominal i or $r\%$	Continuous $r\%$
1.010	1	0.995	12.75	12	11.333
2.020	2	1.980	16.18	15	13.976
3.045	3	2.956	22.14	20	18.232
4.081	4	3.922	28.40	25	22.314
5.127	5	4.879	34.99	30	26.236
6.184	6	5.827	41.91	35	30.104
7.251	7	6.766	49.18	40	33.647
8.328	8	7.696	56.83	45	37.156
10.52	10	9.531	64.87	50	40.546

Table 8-9 1941 CSO Mortality Table

l_x number living, d_x number of deaths, p_x yearly probability of living, $1 - p_x$ net term premium, $\overset{\circ}{e}_x$ years life expectancy

x	l_x	d_x	p_x	$\overset{\circ}{e}_x$	x	l_x	d_x	p_x	$\overset{\circ}{e}_x$
0	1 023 102	23 102	0.977 42	62.33	50	810 900	9 990	0.987 68	21.37
1	1 000 000	5 770	.994 23	62.76	51	800 910	10 628	.986 73	20.64
2	994 230	4 116	.995 86	62.12	52	790 282	11 301	.985 70	19.91
3	990 114	3 347	.996 62	61.37	53	778 981	12 020	.984 57	19.19
4	986 767	2 950	.997 01	60.58	54	766 961	12 770	.983 35	18.48
5	983 817	2 715	.997 24	59.76	55	754 191	13 560	.982 02	17.78
6	981 102	2 561	.997 39	58.92	56	740 631	14 390	.980 57	17.10
7	978 541	2 417	.997 53	58.08	57	726 241	15 251	.979 00	16.43
8	976 124	2 255	.997 69	57.22	58	710 990	16 147	.977 29	15.77
9	973 869	2 065	.997 88	56.35	59	694 843	17 072	.975 43	15.13
10	971 804	1 914	.998 03	55.47	60	677 771	18 022	.973 41	14.50
11	969 890	1 852	.998 09	54.58	61	659 749	18 988	.971 22	13.88
12	968 038	1 859	.998 08	53.68	62	640 761	19 979	.968 82	13.27
13	966 179	1 913	.998 02	52.78	63	620 782	20 958	.966 24	12.69
14	964 266	1 996	.997 93	51.89	64	599 824	21 942	.963 42	12.11
15	962 270	2 069	.997 85	50.99	65	577 882	22 907	.960 36	11.55
16	960 201	2 103	.997 81	50.10	66	554 975	23 842	.957 04	11.01
17	958 098	2 156	.997 75	49.21	67	531 133	24 730	.953 44	10.48
18	955 942	2 199	.997 70	48.32	68	506 403	25 553	.949 54	9.97
19	953 743	2 260	.997 63	47.43	69	480 850	26 302	.945 30	9.47
20	951 483	2 312	.997 57	46.54	70	454 548	26 955	.940 70	8.99
21	949 171	2 382	.997 49	45.66	71	427 593	27 481	.935 73	8.52
22	946 789	2 452	.997 41	44.77	72	400 112	27 872	.930 34	8.08
23	944 337	2 531	.997 32	43 88	73	372 240	28 104	.924 50	7.64
24	941 806	2 609	.997 23	43.00	74	344 136	28 154	.918 19	7.23
25	939 197	2 705	.997 12	42.12	75	315 982	28 009	.911 36	6.82
26	936 492	2 800	.997 01	41.24	76	287 973	27 651	.903 98	6.44
27	933 692	2 904	.996 89	40.36	77	260 322	27 071	.896 01	6.07
28	930 788	3 025	.996 75	39.49	78	233 251	26 262	.887 41	5.72
29	927 763	3 154	.996 60	38.61	79	206 989	25 224	.878 14	5.38
30	924 609	3 292	.996 44	37.74	80	181 765	23 966	.868 15	5.06
31	921 317	3 437	.996 27	36.88	81	157 799	22 502	.857 40	4.75
32	917 880	3 598	.996 08	36.01	82	135 297	20 857	.845 84	4.46
33	914 282	3 767	.995 88	35.15	83	114 440	19 062	.833 43	4.18
34	910 515	3 961	.995 65	34.29	84	95 378	17 157	.820 12	3.91
35	906 554	4 161	.995 41	33.44	85	78 221	15 185	.805 87	3.66
36	902 393	4 386	.995 14	32.59	86	63 036	13 198	.790 63	3.42
37	898 007	4 625	.994 85	31.75	87	49 838	11 245	.774 37	3.19
38	893 382	4 878	.994 54	30.91	88	38 593	9 378	.757 00	2.98
39	888 504	5 162	.994 19	30.08	89	29 215	7 638	.738 56	2.77
40	883 342	5 459	.993 82	29.25	90	21 577	6 063	.719 01	2.58
41	877 883	5 785	.993 41	28.43	91	15 514	4 681	.698 27	2.39
42	872 098	6 131	.992 97	27.62	92	10 833	3 506	.676 36	2.21
43	865 967	6 503	.992 49	26.81	93	7 327	2 540	.653 34	2.03
44	859 464	6 910	.991 96	26.01	94	4 787	1 776	.629 00	1.84
45	852 554	7 340	.991 39	25.21	95	3 011	1 193	.603 79	1.63
46	845 214	7 801	.990 77	24.43	96	1 818	813	.552 81	1.37
47	837 413	8 299	.900 09	23.65	97	1 005	551	.451 74	1.08
48	829 114	8 822	.989 36	22.88	98	454	329	.275 33	.78
49	820 292	9 392	.988 55	22.12	99	125	125	.000 00	.50

Reproduced by permission of the Society of Actuaries

Table 8-10 Cost Index Numbers

Year	Consumer Prices[a]	Wholesale Commodities[a]	Construction[b]	Building[b]	Labor[a] Avg. Hourly
1930	71	56	203	185	
1935	59	52	196	166	
1940	60	51	242	203	
1945	77	69	308	239	
1950	103	103	510	376	110
1955	114	111	660	469	163
1960	126	120	824	559	205
1965	134	121	974	636	237

[a] Bureau of Labor Statistics, 1947–49 = 100
[b] 1913 = 100; quoted by permission of *Engineering News-Record*, copyright 1965 by McGraw-Hill Book Co.

SECTION 9
MATHEMATICAL AND PHYSICAL TABLES

Defined and Fundamental Constants

Yard (U.S. statute)	3600/3937 meters
Foot (U.S.)	0.30480070 meters
Gallon (U.S. statute)	231 cubic inches
Gallon (U.S. statute)	3785.4344 cubic centimeters
Pound (U.S. avdp)	453.59237 grams
Second	1/86,400 mean solar days
Temperature 0°C, ice point	273.15°K or 491.67°R
Avogadro number	6.02252×10^{23} molecules/mol
Boltzmann constant	1.38054×10^{-23} joules/°K
Elementary charge	1.60210×10^{-19} coulombs
Electron rest mass	9.1091×10^{-28} grams
Faraday constant (electromag)	9.64870×10^4 coulombs/mol
Gravitational constant	6.670×10^{-8} dyne-cm²/gm²
Ideal gas constant	8.3143 joules/°K-mol
Light speed in vacuum	2.997925×10^8 meters/second
Planck constant	6.6256×10^{-34} joule-seconds
Standard acceleration of gravity	9.80665 meters/sec²
Standard acceleration of gravity	32.1740 feet/sec²
Standard atmospheric pressure	1013250 dynes/cm²
Stefan-Boltzmann constant	5.6697×10^{-5} ergs/cm²-sec-°K⁴
Volume ideal gas at 0°C, 1 atmos	22413.6 cm³/gram mol

Solar System

Body	Mean dist. from sun 10^6 km	Orbital period sidereal days	Axial period sidereal hr	Equatorial diameter km	Mass 10^{24} kg	Density g/cm^3	g m/sec^2	Escape velocity km/sec
Sun	—	—	609 h, 6'	1.392×10^6	1980000	1.39	271	—
Mercury	57.85	87.97	—	4800	0.32	5.3	3.33	4.2
Venus	108.2	224.70	30 h	12400	4.9	4.95	8.52	10.3
Earth	149.6	365.26	23 h, 56', 4.1"	12756.6	6.0	5.52	9.81	11.2
Moon	0.38*	27.32	655 h, 43', 11"	3478	0.074	3.39	1.62	2.4
Mars	227.9	686.98	24 h, 37', 23"	6783	0.64	3.95	3.77	5.1
Jupiter	778.3	4332.6	9 h, 50', 30"	142600	1900	1.33	25.1	61
Saturn	1428	10759	10 h, 14'	119000	570	0.69	10.72	37
Uranus	2872	30687	10 h, 49'	51500	87	1.56	8.83	22
Neptune	4498	60184	15 h, 40'	49900	103	2.27	11.00	25
Pluto	5910	90700	16 h	12800	5.6	5	9.1	10

* Distance from earth

Common Logarithms of Numbers

N	0	1	2	3	4	5	6	7	8	9	Diff.
100	000000	000434	000868	001301	001734	002166	002598	003029	003461	003891	432
1	004321	004751	005181	005609	006038	006466	006894	007321	007748	008174	428
2	008600	009026	009451	009876	010300	010724	011147	011570	011993	012415	424
3	012837	013259	013680	014100	014521	014940	015360	015779	016197	016616	420
4	017033	017451	017868	018284	018700	019116	019532	019947	020361	020775	416
5	021189	021603	022016	022428	022841	023252	023664	024075	024486	024896	412
6	025306	025715	026125	026533	026942	027350	027757	028164	028571	028978	408
7	029384	029789	030195	030600	031004	031408	031812	032216	032619	033021	404
8	033424	033826	034227	034628	035029	035430	035830	036230	036629	037028	400
9	037426	037825	038223	038620	039017	039414	039811	040207	040602	040998	397
110	041393	041787	042182	042576	042969	043362	043755	044148	044540	044932	393
1	045323	045714	046105	046495	046885	047275	047664	048053	048442	048830	390
2	049218	049606	049993	050380	050766	051153	051538	051924	052309	052694	386
3	053078	053463	053846	054230	054613	054996	055378	055760	056142	056524	383
4	056905	057286	057666	058046	058426	058805	059185	059563	059942	060320	379
5	060698	061075	061452	061829	062206	062582	062958	063333	063709	064083	376
6	064458	064832	065206	065580	065953	066326	066699	067071	067443	067815	373
7	068186	068557	068928	069298	069668	070038	070407	070776	071145	071514	370
8	071882	072250	072617	072985	073352	073718	074085	074451	074816	075182	366
9	075547	075912	076276	076640	077004	077368	077731	078094	078457	078819	363
120	079181	079543	079904	080266	080626	080987	081347	081707	082067	082426	360

PROPORTIONAL PARTS

Diff.	1	2	3	4	5	6	7	8	9
434	43.4	86.8	130.2	173.6	217.0	260.4	303.8	347.2	390.6
432	43.2	86.4	129.6	172.8	216.0	259.2	302.4	345.6	388.8
430	43.0	86.0	129.0	172.0	215.0	258.0	301.0	344.0	387.0
428	42.8	85.6	128.4	171.2	214.0	256.8	299.6	342.4	385.2
426	42.6	85.2	127.8	170.4	213.0	255.6	298.2	340.8	383.4
424	42.4	84.8	127.2	169.6	212.0	254.4	296.8	339.2	381.6
422	42.2	84.4	126.6	168.8	211.0	253.2	295.4	337.6	379.8
420	42.0	84.0	126.0	168.0	210.0	252.0	294.0	336.0	378.0
418	41.8	83.6	125.4	167.2	209.0	250.8	292.6	334.4	376.2
416	41.6	83.2	124.8	166.4	208.0	249.6	291.2	332.8	374.4
414	41.4	82.8	124.2	165.6	207.0	248.4	289.8	331.2	372.6
412	41.2	82.4	123.6	164.8	206.0	247.2	288.4	329.6	370.8
410	41.0	82.0	123.0	164.0	205.0	246.0	287.0	328.0	369.0
408	40.8	81.6	122.4	163.2	204.0	244.8	285.6	326.4	367.2
406	40.6	81.2	121.8	162.4	203.0	243.6	284.2	324.8	365.4
404	40.4	80.8	121.2	161.6	202.0	242.4	282.8	323.2	363.6
402	40.2	80.4	120.6	160.8	201.0	241.2	281.4	321.6	361.8
400	40.0	80.0	120.0	160.0	200.0	240.0	280.0	320.0	360.0
398	39.8	79.6	119.4	159.2	199.0	238.8	278.6	318.4	358.2
396	39.6	79.2	118.8	158.4	198.0	237.6	277.2	316.8	356.4
394	39.4	78.8	118.2	157.6	197.0	236.4	275.8	315.2	354.6
392	39.2	78.4	117.6	156.8	196.0	235.2	274.4	313.6	352.8
390	39.0	78.0	117.0	156.0	195.0	234.0	273.0	312.0	351.0
388	38.8	77.6	116.4	155.2	194.0	232.8	271.6	310.4	349.2
386	38.6	77.2	115.8	154.4	193.0	231.6	270.2	308.8	347.4
384	38.4	76.8	115.2	153.6	192.0	230.4	268.8	307.2	345.6
382	38.2	76.4	114.6	152.8	191.0	229.2	267.4	305.6	343.8
380	38.0	76.0	114.0	152.0	190.0	228.0	266.0	304.0	342.0
378	37.8	75.6	113.4	151.2	189.0	226.8	264.6	302.4	340.2
376	37.6	75.2	112.8	150.4	188.0	225.6	263.2	300.8	338.4
374	37.4	74.8	112.2	149.6	187.0	224.4	261.8	299.2	336.6
372	37.2	74.4	111.6	148.8	186.0	223.2	260.4	297.6	334.8
370	37.0	74.0	111.0	148.0	185.0	222.0	259.0	296.0	333.0

COMMON LOGARITHMS

N	0	1	2	3	4	5	6	7	8	9	Diff.
120	079181	079543	079904	080266	080626	080987	081347	081707	082067	082426	360
1	082785	083144	083503	083861	084219	084576	084934	085291	085647	086004	357
2	086360	086716	087071	087426	087781	088136	088490	088845	089198	089552	355
3	089905	090258	090611	090963	091315	091667	092018	092370	092721	093071	352
4	093422	093772	094122	094471	094820	095169	095518	095866	096215	096562	349
5	096910	097257	097604	097951	098298	098644	098990	099335	099681	100026	346
6	100371	100715	101059	101403	101747	102091	102434	102777	103119	103462	343
7	103804	104146	104487	104828	105169	105510	105851	106191	106531	106871	341
8	107210	107549	107888	108227	108565	108903	109241	109579	109916	110253	338
9	110590	110926	111263	111599	111934	112270	112605	112940	113275	113609	335
130	113943	114277	114611	114944	115278	115611	115943	116276	116608	116940	333
1	117271	117603	117934	118265	118595	118926	119256	119586	119915	120245	330
2	120574	120903	121231	121560	121888	122216	122544	122871	123198	123525	328
3	123852	124178	124504	124830	125156	125481	125806	126131	126456	126781	325
4	127105	127429	127753	128076	128399	128722	129045	129368	129690	130012	323
5	130334	130655	130977	131298	131619	131939	132260	132580	132900	133219	321
6	133539	133858	134177	134496	134814	135133	135451	135769	136086	136403	318
7	136721	137037	137354	137671	137987	138303	138618	138934	139249	139564	316
8	139879	140194	140508	140822	141136	141450	141763	142076	142389	142702	314
9	143015	143327	143639	143951	144263	144574	144885	145196	145507	145818	311
140	146128	146438	146748	147058	147367	147676	147985	148294	148603	148911	309

PROPORTIONAL PARTS

Diff.	1	2	3	4	5	6	7	8	9
370	37.0	74.0	111.0	148.0	185.0	222.0	259.0	296.0	333.0
368	36.8	73.6	110.4	147.2	184.0	220.8	257.6	294.4	331.2
366	36.6	73.2	109.8	146.4	183.0	219.6	256.2	292.8	329.4
36	36.4	72.8	109.2	145.6	182.0	218.4	254.8	291.2	327.6
362	36.2	72.4	108.6	144.8	181.0	217.2	253.4	289.6	325.8
360	36.0	72.0	108.0	144.0	180.0	216.0	252.0	288.0	324.0
358	35.8	71.6	107.4	143.2	179.0	214.8	250.6	286.4	322.2
356	35.6	71.2	106.8	142.4	178.0	213.6	249.2	284.8	320.4
354	35.4	70.8	106.2	141.6	177.0	212.4	247.8	283.2	318.6
352	35.2	70.4	105.6	140.8	176.0	211.2	246.4	281.6	316.8
350	35.0	70.0	105.0	140.0	175.0	210.0	245.0	280.0	315.0
348	34.8	69.6	104.4	139.2	174.0	208.8	243.6	278.4	313.2
346	34.6	69.2	103.8	138.4	173.0	207.6	242.2	276.8	311.4
344	34.4	68.8	103.2	137.6	172.0	206.4	240.8	275.2	309.6
342	34.2	68.4	102.6	136.8	171.0	205.2	239.4	273.6	307.8
340	34.0	68.0	102.0	136.0	170.0	204.0	238.0	272.0	306.0
338	33.8	67.6	101.4	135.2	169.0	202.8	236.6	270.4	304.2
336	33.6	67.2	100.8	134.4	168.0	201.6	235.2	268.8	302.4
334	33.4	66.8	100.2	133.6	167.0	200.4	233.8	267.2	300.6
332	33.2	66.4	99.6	132.8	166.0	199.2	232.4	265.6	298.8
330	33.0	66.0	99.0	132.0	165.0	198.0	231.0	264.0	297.0
328	32.8	65.6	98.4	131.2	164.0	196.8	229.6	262.4	295.2
326	32.6	65.2	97.8	130.4	163.0	195.6	228.2	260.8	293.4
324	32.4	64.8	97.2	129.6	162.0	194.4	226.8	259.2	291.6
322	32.2	64.4	96.6	128.8	161.0	193.2	225.4	257.6	289.8
320	32.0	64.0	96.0	128.0	160.0	192.0	224.0	256.0	288.0
318	31.8	63.6	95.4	127.2	159.0	190.8	222.6	254.4	286.2
316	31.6	63.2	94.8	126.4	158.0	189.6	221.2	252.8	284.4
314	31.4	62.8	94.2	125.6	157.0	188.4	219.8	251.2	282.6
312	31.2	62.4	93.6	124.8	156.0	187.2	218.4	249.6	280.8
310	31.0	62.0	93.0	124.0	155.0	186.0	217.0	248.0	279.0
308	30.8	61.6	92.4	123.2	154.0	184.8	215.6	246.4	277.2

290 MATHEMATICAL AND PHYSICAL TABLES

N	0	1	2	3	4	5	6	7	8	9	Diff.
140	146128	146438	146748	147058	147367	147676	147985	148294	148603	148911	309
1	149219	149527	149835	150142	150449	150756	151063	151370	151676	151982	307
2	152288	152594	152900	153205	153510	153815	154120	154424	154728	155032	305
3	155336	155640	155943	156246	156549	156852	157154	157457	157759	158061	303
4	158362	158664	158965	159266	159567	159868	160168	160469	160769	161068	301
5	161368	161667	161967	162266	162564	162863	163161	163460	163758	164055	299
6	164353	164650	164947	165244	165541	165838	166134	166430	166726	167022	297
7	167317	167613	167908	168203	168497	168792	169086	169380	169674	169968	295
8	170262	170555	170848	171141	171434	171726	172019	172311	172603	172895	293
9	173186	173478	173769	174060	174351	174641	174932	175222	175512	175802	291
150	176091	176381	176670	176959	177248	177536	177825	178113	178401	178689	289
1	178977	179264	179552	179839	180126	180413	180699	180986	181272	181558	287
2	181844	182129	182415	182700	182985	183270	183555	183839	184123	184407	285
3	184691	184975	185259	185542	185825	186108	186391	186674	186956	187239	283
4	187521	187803	188084	188366	188647	188928	189209	189490	189771	190051	281
5	190332	190612	190892	191171	191451	191730	192010	192289	192567	192846	279
6	193125	193403	193681	193959	194237	194514	194792	195069	195346	195623	278
7	195900	196176	196453	196729	197005	197281	197556	197832	198107	198382	276
8	198657	198932	199206	199481	199755	200029	200303	200577	200850	201124	274
9	201397	201670	201943	202216	202488	202761	203033	203305	203577	203848	272
160	204120	204391	204663	204934	205204	205475	205746	206016	206286	206556	271
1	206826	207096	207365	207634	207904	208173	208441	208710	208979	209247	269
2	209515	209783	210051	210319	210586	210853	211121	211388	211654	211921	267
3	212188	212454	212720	212986	213252	213518	213783	214049	214314	214579	266
4	214844	215109	215373	215638	215902	216166	216430	216694	216957	217221	264
5	217484	217747	218010	218273	218536	218798	219060	219323	219585	219846	262
6	220108	220370	220631	220892	221153	221414	221675	221936	222196	222456	261
7	222716	222976	223236	223496	223755	224015	224274	224533	224792	225051	259
8	225309	225568	225826	226084	226342	226600	226858	227115	227372	227630	258
9	227887	228144	228400	228657	228913	229170	229426	229682	229938	230193	256
170	230449	230704	230960	231215	231470	231724	231979	232234	232488	232742	255

PROPORTIONAL PARTS

Diff.	1	2	3	4	5	6	7	8	9
310	31.0	62.0	93.0	124.0	155.0	186.0	217.0	248.0	279.0
308	30.8	61.6	92.4	123.2	154.0	184.8	215.6	246.4	277.2
306	30.6	61.2	91.8	122.4	153.0	183.6	214.2	244.8	275.4
304	30.4	60.8	91.2	121.6	152.0	182.4	212.8	243.2	273.6
302	30.2	60.4	90.6	120.8	151.0	181.2	211.4	241.6	271.8
300	30.0	60.0	90.0	120.0	150.0	180.0	210.0	240.0	270.0
298	29.8	59.6	89.4	119.2	149.0	178.8	208.6	238.4	268.2
296	29.6	59.2	88.8	118.4	148.0	177.6	207.2	236.8	266.4
294	29.4	58.8	88.2	117.6	147.0	176.4	205.8	235.2	264.6
292	29.2	58.4	87.6	116.8	146.0	175.2	204.4	233.6	262.8
290	29.0	58.0	87.0	116.0	145.0	174.0	203.0	232.0	261.0
288	28.8	57.6	86.4	115.2	144.0	172.8	201.6	230.4	259.2
286	28.6	57.2	85.8	114.4	143.0	171.6	200.2	228.8	257.4
284	28.4	56.8	85.2	113.6	142.0	170.4	198.8	227.2	255.6
282	28.2	56.4	84.6	112.8	141.0	169.2	197.4	225.6	253.8
280	28.0	56.0	84.0	112.0	140.0	168.0	196.0	224.0	252.0
278	27.8	55.6	83.4	111.2	139.0	166.8	194.6	222.4	250.2
276	27.6	55.2	82.8	110.4	138.0	165.6	193.2	220.8	248.4
274	27.4	54.8	82.2	109.6	137.0	164.4	191.8	219.2	246.6
272	27.2	54.4	81.6	108.8	136.0	163.2	190.4	217.6	244.8
270	27.0	54.0	81.0	108.0	135.0	162.0	189.0	216.0	243.0
268	26.8	53.6	80.4	107.2	134.0	160.8	187.6	214.4	241.2
266	26.6	53.2	79.8	106.4	133.0	159.6	186.2	212.8	239.4
264	26.4	52.8	79.2	105.6	132.0	158.4	184.8	211.2	237.6

COMMON LOGARITHMS

N	0	1	2	3	4	5	6	7	8	9	Diff.
170	230449	230704	230960	231215	231470	231724	231979	232234	232488	232742	255
1	232996	233250	233504	233757	234011	234264	234517	234770	235023	235276	253
2	235528	235781	236033	236285	236537	236789	237041	237292	237544	237795	252
3	238046	238297	238548	238799	239049	239299	239550	239800	240050	240300	250
4	240549	240799	241048	241297	241546	241795	242044	242293	242541	242790	249
5	243038	243286	243534	243782	244030	244277	244525	244772	245019	245266	248
6	245513	245759	246006	246252	246499	246745	246991	247237	247482	247728	246
7	247973	248219	248464	248709	248954	249198	249443	249687	249932	250176	245
8	250420	250664	250908	251151	251395	251638	251881	252125	252368	252610	243
9	252853	253096	253338	253580	253822	254064	254306	254548	254790	255031	242
180	255273	255514	255755	255996	256237	256477	256718	256958	257198	257439	241
1	257679	257918	258158	258398	258637	258877	259116	259355	259594	259833	239
2	260071	260310	260548	260787	261025	261263	261501	261739	261976	262214	238
3	262451	262688	262925	263162	263399	263636	263873	264109	264346	264582	237
4	264818	265054	265290	265525	265761	265996	266232	266467	266702	266937	235
5	267172	267406	267641	267875	268110	268344	268578	268812	269046	269279	234
6	269513	269746	269980	270213	270446	270679	270912	271144	271377	271609	233
7	271842	272074	272306	272538	272770	273001	273233	273464	273696	273927	232
8	274158	274389	274620	274850	275081	275311	275542	275772	276002	276232	230
9	276462	276692	276921	277151	277380	277609	277838	278067	278296	278525	229
190	278754	278982	279211	279439	279667	279895	280123	280351	280578	280806	228
1	281033	281261	281488	281715	281942	282169	282396	282622	282849	283075	227
2	283301	283527	283753	283979	284205	284431	284656	284882	285107	285332	226
3	285557	285782	286007	286232	286456	286681	286905	287130	287354	287578	225
4	287802	288026	288249	288473	288696	288920	289143	289366	289589	289812	223
5	290035	290257	290480	290702	290925	291147	291369	291591	291813	292034	222
6	292256	292478	292699	292920	293141	293363	293584	293804	294025	294246	221
7	294466	294687	294907	295127	295347	295567	295787	296007	296226	296446	220
8	296665	296884	297104	297323	297542	297761	297979	298198	298416	298635	219
9	298853	299071	299289	299507	299725	299943	300161	300378	300595	300813	218
200	301030	301247	301464	301681	301898	302114	302331	302547	302764	302980	217

PROPORTIONAL PARTS

Diff.	1	2	3	4	5	6	7	8	9
262	26.2	52.4	78.6	104.8	131.0	157.2	183.4	209.6	235.8
260	26.0	52.0	78.0	104.0	130.0	156.0	182.0	208.0	234.0
258	25.8	51.6	77.4	103.2	129.0	154.8	180.6	206.4	232.2
256	25.6	51.2	76.8	102.4	128.0	153.6	179.2	204.8	230.4
254	25.4	50.8	76.2	101.6	127.0	152.4	177.8	203.2	228.6
252	25.2	50.4	75.6	100.8	126.0	151.2	176.4	201.6	226.8
250	25.0	50.0	75.0	100.0	125.0	150.0	175.0	200.0	225.0
248	24.8	49.6	74.4	99.2	124.0	148.8	173.6	198.4	223.2
246	24.6	49.2	73.8	98.4	123.0	147.6	172.2	196.8	221.4
244	24.4	48.8	73.2	97.6	122.0	146.4	170.8	195.2	219.6
242	24.2	48.4	72.6	96.8	121.0	145.2	169.4	193.6	217.8
240	24.0	48.0	72.0	96.0	120.0	144.0	168.0	192.0	216.0
238	23.8	47.6	71.4	95.2	119.0	142.8	166.6	190.4	214.2
236	23.6	47.2	70.8	94.4	118.0	141.6	165.2	188.8	212.4
234	23.4	46.8	70.2	93.6	117.0	140.4	163.8	187.2	210.6
232	23.2	46.4	69.6	92.8	116.0	139.2	162.4	185.6	208.8
230	23.0	46.0	69.0	92.0	115.0	138.0	161.0	184.0	207.0
228	22.8	45.6	68.4	91.2	114.0	136.8	159.6	182.4	205.2
226	22.6	45.2	67.8	90.4	113.0	135.6	158.2	180.8	203.4
224	22.4	44.8	67.2	89.6	112.0	134.4	156.8	179.2	201.6
222	22.2	44.4	66.6	88.8	111.0	133.2	155.4	177.6	199.8
220	22.0	44.0	66.0	88.0	110.0	132.0	154.0	176.0	198.0
218	21.8	43.6	65.4	87.2	109.0	130.8	152.6	174.4	196.2
216	21.6	43.2	64.8	86.4	108.0	129.6	151.2	172.8	194.4

MATHEMATICAL AND PHYSICAL TABLES

N	0	1	2	3	4	5	6	7	8	9	Diff.
200	**301030**	**301247**	**301464**	**301681**	**301898**	**302114**	**302331**	**302547**	**302764**	**302980**	217
1	303196	303412	303628	303844	304059	304275	304491	304706	304921	305136	216
2	305351	305566	305781	305996	306211	306425	306639	306854	307068	307282	215
3	307496	307710	307924	308137	308351	308564	308778	308991	309204	309417	213
4	309630	309843	310056	310268	310481	310693	310906	311118	311330	311542	212
5	311754	311966	312177	312389	312600	312812	313023	313234	313445	313656	211
6	313867	314078	314289	314499	314710	314920	315130	315340	315551	315760	210
7	315970	316180	316390	316599	316809	317018	317227	317436	317646	317854	209
8	318063	318272	318481	318689	318898	319106	319314	319522	319730	319938	208
9	320146	320354	320562	320769	320977	321184	321391	321598	321805	322012	207
210	**322219**	**322426**	**322633**	**322839**	**323046**	**323252**	**323458**	**323665**	**323871**	**324077**	206
1	324282	324488	324694	324899	325105	325310	325516	325721	325926	326131	205
2	326336	326541	326745	326950	327155	327359	327563	327767	327972	328176	204
3	328380	328583	328787	328991	329194	329398	329601	329805	330008	330211	203
4	330414	330617	330819	331022	331225	331427	331630	331832	332034	332236	202
5	332438	332640	332842	333044	333246	333447	333649	333850	334051	334253	
6	334454	334655	334856	335057	335257	335458	335658	335859	336059	336260	201
7	336460	336660	336860	337060	337260	337459	337659	337858	338058	338257	200
8	338456	338656	338855	339054	339253	339451	339650	339849	340047	340246	199
9	340444	340642	340841	341039	341237	341435	341632	341830	342028	342225	198
220	**342423**	**342620**	**342817**	**343014**	**343212**	**343409**	**343606**	**343802**	**343999**	**344196**	197
1	344392	344589	344785	344981	345178	345374	345570	345766	345962	346157	196
2	346353	346549	346744	346939	347135	347330	347525	347720	347915	348110	195
3	348305	348500	348694	348889	349083	349278	349472	349666	349860	350054	194
4	350248	350442	350636	350829	351023	351216	351410	351603	351796	351989	193
5	352183	352375	352568	352761	352954	353147	353339	353532	353724	353916	
6	354108	354301	354493	354685	354876	355068	355260	355452	355643	355834	192
7	356026	356217	356408	356599	356790	356981	357172	357363	357554	357744	191
8	357935	358125	358316	358506	358696	358886	359076	359266	359456	359646	190
9	359835	360025	360215	360404	360593	360783	360972	361161	361350	361539	189
230	**361728**	**361917**	**362105**	**362294**	**362482**	**362671**	**362859**	**363048**	**363236**	**363424**	188
1	363612	363800	363988	364176	364363	364551	364739	364926	365113	365301	
2	365488	365675	365862	366049	366236	366423	366610	366796	366983	367169	187
3	367356	367542	367729	367915	368101	368287	368473	368659	368845	369030	186
4	369216	369401	369587	369772	369958	370143	370328	370513	370698	370883	185
5	371068	371253	371437	371622	371806	371991	372175	372360	372544	372728	184
6	372912	373096	373280	373464	373647	373831	374015	374198	374382	374565	
7	374748	374932	375115	375298	375481	375664	375846	376029	376212	376394	183
8	376577	376759	376942	377124	377306	377488	377670	377852	378034	378216	182
9	378398	378580	378761	378943	379124	379306	379487	379668	379849	380030	181
240	**380211**	**380392**	**380573**	**380754**	**380934**	**381115**	**381296**	**381476**	**381656**	**381837**	

PROPORTIONAL PARTS

Diff.	1	2	3	4	5	6	7	8	9
216	21.6	43.2	64.8	86.4	108.0	129.6	151.2	172.8	194.4
214	21.4	42.8	64.2	85.6	107.0	128.4	149.8	171.2	192.6
212	21.2	42.4	63.6	84.8	106.0	127.2	148.4	169.6	190.8
210	21.0	42.0	63.0	84.0	105.0	126.0	147.0	168.0	189.0
208	20.8	41.6	62.4	83.2	104.0	124.8	145.6	166.4	187.2
206	20.6	41.2	61.8	82.4	103.0	123.6	144.2	164.8	185.4
204	20.4	40.8	61.2	81.6	102.0	122.4	142.8	163.2	183.6
202	20.2	40.4	60.6	80.8	101.0	121.2	141.4	161.6	181.8
200	20.0	40.0	60.0	80.0	100.0	120.0	140.0	160.0	180.0
198	19.8	39.6	59.4	79.2	99.0	118.8	138.6	158.4	178.2
196	19.6	39.2	58.8	78.4	98.0	117.6	137.2	156.8	176.4
194	19.4	38.8	58.2	77.6	97.0	116.4	135.8	155.2	174.6
192	19.2	38.4	57.6	76.8	96.0	115.2	134.4	153.6	172.8
190	19.0	38.0	57.0	76.0	95.0	114.0	133.0	152.0	171.0
188	18.8	37.6	56.4	75.2	94.0	112.8	131.6	150.4	169.2
186	18.6	37.2	55.8	74.4	93.0	111.6	130.2	148.8	167.4

COMMON LOGARITHMS

N	0	1	2	3	4	5	6	7	8	9	Diff.
240	380211	380392	380573	380754	380934	381115	381296	381476	381656	381837	181
1	382017	382197	382377	382557	382737	382917	383097	383277	383456	383636	180
2	383815	383995	384174	384353	384533	384712	384891	385070	385249	385428	179
3	385606	385785	385964	386142	386321	386499	386677	386856	387034	387212	178
4	387390	387568	387746	387924	388101	388279	388456	388634	388811	388989	
5	389166	389343	389520	389698	389875	390051	390228	390405	390582	390759	177
6	390935	391112	391288	391464	391641	391817	391993	392169	392345	392521	176
7	392697	392873	393048	393224	393400	393575	393751	393926	394101	394277	
8	394452	394627	394802	394977	395152	395326	395501	395676	395850	396025	175
9	396199	396374	396548	396722	396896	397071	397245	397419	397592	397766	174
250	397940	398114	398287	398461	398634	398808	398981	399154	399328	399501	173
1	399674	399847	400020	400192	400365	400538	400711	400883	401056	401228	
2	401401	401573	401745	401917	402089	402261	402433	402605	402777	402949	172
3	403121	403292	403464	403635	403807	403978	404149	404320	404492	404663	171
4	404834	405005	405176	405346	405517	405688	405858	406029	406199	406370	
5	406540	406710	406881	407051	407221	407391	407561	407731	407901	408070	170
6	408240	408410	408579	408749	408918	409087	409257	409426	409595	409764	169
7	409933	410102	410271	410440	410609	410777	410946	411114	411283	411451	
8	411620	411788	411956	412124	412293	412461	412629	412796	412964	413132	168
9	413300	413467	413635	413803	413970	414137	414305	414472	414639	414806	167
260	414973	415140	415307	415474	415641	415808	415974	416141	416308	416474	
1	416641	416807	416973	417139	417306	417472	417638	417804	417970	418135	166
2	418301	418467	418633	418798	418964	419129	419295	419460	419625	419791	165
3	419956	420121	420286	420451	420616	420781	420945	421110	421275	421439	
4	421604	421768	421933	422097	422261	422426	422590	422754	422918	423082	164
5	423246	423410	423574	423737	423901	424065	424228	424392	424555	424718	
6	424882	425045	425208	425371	425534	425697	425860	426023	426186	426349	163
7	426511	426674	426836	426999	427161	427324	427486	427648	427811	427973	162
8	428135	428297	428459	428621	428783	428944	429106	429268	429429	429591	
9	429752	429914	430075	430236	430398	430559	430720	430881	431042	431203	161
270	431364	431525	431685	431846	432007	432167	432328	432488	432649	432809	
1	432969	433130	433290	433450	433610	433770	433930	434090	434249	434409	160
2	434569	434729	434888	435048	435207	435367	435526	435685	435844	436004	159
3	436163	436322	436481	436640	436799	436957	437116	437275	437433	437592	
4	437751	437909	438067	438226	438384	438542	438701	438859	439017	439175	158
5	439333	439491	439648	439806	439964	440122	440279	440437	440594	440752	
6	440909	441066	441224	441381	441538	441695	441852	442009	442166	442323	157
7	442480	442637	442793	442950	443106	443263	443419	443576	443732	443889	
8	444045	444201	444357	444513	444669	444825	444981	445137	445293	445449	156
9	445604	445760	445915	446071	446226	446382	446537	446692	446848	447003	155
280	447158	447313	447468	447623	447778	447933	448088	448242	448397	448552	

PROPORTIONAL PARTS

Diff.	1	2	3	4	5	6	7	8	9
184	18.4	36.8	55.2	73.6	92.0	110.4	128.8	147.2	165.6
182	18.2	36.4	54.6	72.8	91.0	109.2	127.4	145.6	163.8
180	18.0	36.0	54.0	72.0	90.0	108.0	126.0	144.0	162.0
178	17.8	35.6	53.4	71.2	89.0	106.8	124.6	142.4	160.2
176	17.6	35.2	52.8	70.4	88.0	105.6	123.2	140.8	158.4
174	17.4	34.8	52.2	69.6	87.0	104.4	121.8	139.2	156.6
172	17.2	34.4	51.6	68.8	86.0	103.2	120.4	137.6	154.8
170	17.0	34.0	51.0	68.0	85.0	102.0	119.0	136.0	153.0
168	16.8	33.6	50.4	67.2	84.0	100.8	117.6	134.4	151.2
166	16.6	33.2	49.8	66.4	83.0	99.6	116.2	132.8	149.4
164	16.4	32.8	49.2	65.6	82.0	98.4	114.8	131.2	147.6
162	16.2	32.4	48.6	64.8	81.0	97.2	113.4	129.6	145.8
160	16.0	32.0	48.0	64.0	80.0	96.0	112.0	128.0	144.0
158	15.8	31.6	47.4	63.2	79.0	94.8	110.6	126.4	142.2
156	15.6	31.2	46.8	62.4	78.0	93.6	109.2	124.8	140.4
154	15.4	30.8	46.2	61.6	77.0	92.4	107.8	123.2	138.6

MATHEMATICAL AND PHYSICAL TABLES

N	0	1	2	3	4	5	6	7	8	9	Diff.
280	447158	447313	447468	447623	447778	447933	448088	448242	448397	448552	
1	448706	448861	449015	449170	449324	449478	449633	449787	449941	450095	154
2	450249	450403	450557	450711	450865	451018	451172	451326	451479	451633	
3	451786	451940	452093	452247	452400	452553	452706	452859	453012	453165	153
4	453318	453471	453624	453777	453930	454082	454235	454387	454540	454692	
5	454845	454997	455150	455302	455454	455606	455758	455910	456062	456214	152
6	456366	456518	456670	456821	456973	457125	457276	457428	457579	457731	
7	457882	458033	458184	458336	458487	458638	458789	458940	459091	459242	151
8	459392	459543	459694	459845	459995	460146	460296	460447	460597	460748	
9	460898	461048	461198	461348	461499	461649	461799	461948	462098	462248	150
290	462398	462548	462697	462847	462997	463146	463296	463445	463594	463744	
1	463893	464042	464191	464340	464490	464639	464788	464936	465085	465234	149
2	465383	465532	465680	465829	465977	466126	466274	466423	466571	466719	
3	466868	467016	467164	467312	467460	467608	467756	467904	468052	468200	148
4	468347	468495	468643	468790	468938	469085	469233	469380	469527	469675	
5	469822	469969	470116	470263	470410	470557	470704	470851	470998	471145	147
6	471292	471438	471585	471732	471878	472025	472171	472318	472464	472610	146
7	472756	472903	473049	473195	473341	473487	473633	473779	473925	474071	
8	474216	474362	474508	474653	474799	474944	475090	475235	475381	475526	
9	475671	475816	475962	476107	476252	476397	476542	476687	476832	476976	145
300	477121	477266	477411	477555	477700	477844	477989	478133	478278	478422	
1	478566	478711	478855	478999	479143	479287	479431	479575	479719	479863	144
2	480007	480151	480294	480438	480582	480725	480869	481012	481156	481299	
3	481443	481586	481729	481872	482016	482159	482302	482445	482588	482731	143
4	482874	483016	483159	483302	483445	483587	483730	483872	484015	484157	
5	484300	484442	484585	484727	484869	485011	485153	485295	485437	485579	142
6	485721	485863	486005	486147	486289	486430	486572	486714	486855	486997	
7	487138	487280	487421	487563	487704	487845	487986	488127	488269	488410	141
8	488551	488692	488833	488974	489114	489255	489396	489537	489677	489818	
9	489958	490099	490239	490380	490520	490661	490801	490941	491081	491222	140
310	491362	491502	491642	491782	491922	492062	492201	492341	492481	492621	
1	492760	492900	493040	493179	493319	493458	493597	493737	493876	494015	139
2	494155	494294	494433	494572	494711	494850	494989	495128	495267	495406	
3	495544	495683	495822	495960	496099	496238	496376	496515	496653	496791	
4	496930	497068	497206	497344	497483	497621	497759	497897	498035	498173	138
5	498311	498448	498586	498724	498862	498999	499137	499275	499412	499550	
6	499687	499824	499962	500099	500236	500374	500511	500648	500785	500922	137
7	501059	501196	501333	501470	501607	501744	501880	502017	502154	502291	
8	502427	502564	502700	502837	502973	503109	503246	503382	503518	503655	136
9	503791	503927	504063	504199	504335	504471	504607	504743	504879	505014	
320	505150	505286	505421	505557	505693	505828	505964	506099	506234	506370	
1	506505	506640	506776	506911	507046	507181	507316	507451	507586	507721	135
2	507856	507991	508126	508260	508395	508530	508664	508799	508934	509068	
3	509203	509337	509471	509606	509740	509874	510009	510143	510277	510411	134
4	510545	510679	510813	510947	511081	511215	511349	511482	511616	511750	
5	511883	512017	512151	512284	512418	512551	512684	512818	512951	513084	133
6	513218	513351	513484	513617	513750	513883	514016	514149	514282	514415	
7	514548	514681	514813	514946	515079	515211	515344	515476	515609	515741	
8	515874	516006	516139	516271	516403	516535	516668	516800	516932	517064	132
9	517196	517328	517460	517592	517724	517855	517987	518119	518251	518382	
330	518514	518646	518777	518909	519040	519171	519303	519434	519566	519697	131

PROPORTIONAL PARTS

Diff.	1	2	3	4	5	6	7	8	9
154	15.4	30.8	46.2	61.6	77.0	92.4	107.8	123.2	138.6
152	15.2	30.4	45.6	60.8	76.0	91.2	106.4	121.6	136.8
150	15.0	30.0	45.0	60.0	75.0	90.0	105.0	120.0	135.0
148	14.8	29.6	44.4	59.2	74.0	88.8	103.6	118.4	133.2
146	14.6	29.2	43.8	58.4	73.0	87.6	102.2	116.8	131.4
144	14.4	28.8	43.2	57.6	72.0	86.4	100.8	115.2	129.6
142	14.2	28.4	42.6	56.8	71.0	85.2	99.4	113.6	127.8
140	14.0	28.0	42.0	56.0	70.0	84.0	98.0	112.0	126.0
138	13.8	27.6	41.4	55.2	69.0	82.8	96.6	110.4	124.2
136	13.6	27.2	40.8	54.4	68.0	81.6	95.2	108.8	122.4

COMMON LOGARITHMS

N	0	1	2	3	4	5	6	7	8	9	Diff.
330	**518514**	**518646**	**518777**	**518909**	**519040**	**519171**	**519303**	**519434**	**519566**	**519697**	
1	519828	519959	520090	520221	520353	520484	520615	520745	520876	521007	
2	521138	521269	521400	521530	521661	521792	521922	522053	522183	522314	
3	522444	522575	522705	522835	522966	523096	523226	523356	523486	523616	130
4	523746	523876	524006	524136	524266	524396	524526	524656	524785	524915	
5	525045	525174	525304	525434	525563	525693	525822	525951	526081	526210	129
6	526339	526469	526598	526727	526856	526985	527114	527243	527372	527501	
7	527630	527759	527888	528016	528145	528274	528402	528531	528660	528788	
8	528917	529045	529174	529302	529430	529559	529687	529815	529943	530072	128
9	530200	530328	530456	530584	530712	530840	530968	531096	531223	531351	
340	**531479**	**531607**	**531734**	**531862**	**531990**	**532117**	**532245**	**532372**	**532500**	**532627**	
1	532754	532882	533009	533136	533264	533391	533518	533645	533772	533899	127
2	534026	534153	534280	534407	534534	534661	534787	534914	535041	535167	
3	535294	535421	535547	535674	535800	535927	536053	536180	536306	536432	126
4	536558	536685	536811	536937	537063	537189	537315	537441	537567	537693	
5	537819	537945	538071	538197	538322	538448	538574	538699	538825	538951	
6	539076	539202	539327	539452	539578	539703	539829	539954	540079	540204	125
7	540329	540455	540580	540705	540830	540955	541080	541205	541330	541454	
8	541579	541704	541829	541953	542078	542203	542327	542452	542576	542701	
9	542825	542950	543074	543199	543323	543447	543571	543696	543820	543944	124
350	**544068**	**544192**	**544316**	**544440**	**544564**	**544688**	**544812**	**544936**	**545060**	**545183**	
1	545307	545431	545555	545678	545802	545925	546049	546172	546296	546419	
2	546543	546666	546789	546913	547036	547159	547282	547405	547529	547652	123
3	547775	547898	548021	548144	548267	548389	548512	548635	548758	548881	
4	549003	549126	549249	549371	549494	549616	549739	549861	549984	550106	
5	550228	550351	550473	550595	550717	550840	550962	551084	551206	551328	122
6	551450	551572	551694	551816	551938	552060	552181	552303	552425	552547	
7	552668	552790	552911	553033	553155	553276	553398	553519	553640	553762	121
8	553883	554004	554126	554247	554368	554489	554610	554731	554852	554973	
9	555094	555215	555336	555457	555578	555699	555820	555940	556061	556182	
360	**556303**	**556423**	**556544**	**556664**	**556785**	**556905**	**557026**	**557146**	**557267**	**557387**	120
1	557507	557627	557748	557868	557988	558108	558228	558349	558469	558589	
2	558709	558829	558948	559068	559188	559308	559428	559548	559667	559787	
3	559907	560026	560146	560265	560385	560504	560624	560743	560863	560982	119
4	561101	561221	561340	561459	561578	561698	561817	561936	562055	562174	
5	562293	562412	562531	562650	562769	562887	563006	563125	563244	563362	
6	563481	563600	563718	563837	563955	564074	564192	564311	564429	564548	
7	564666	564784	564903	565021	565139	565257	565376	565494	565612	565730	118
8	565848	565966	566084	566202	566320	566437	566555	566673	566791	566909	
9	567026	567144	567262	567379	567497	567614	567732	567849	567967	568084	
370	**568202**	**568319**	**568436**	**568554**	**568671**	**568788**	**568905**	**569023**	**569140**	**569257**	117
1	569374	569491	569608	569725	569842	569959	570076	570193	570309	570426	
2	570543	570660	570776	570893	571010	571126	571243	571359	571476	571592	
3	571709	571825	571942	572058	572174	572291	572407	572523	572639	572755	116
4	572872	572988	573104	573220	573336	573452	573568	573684	573800	573915	
5	574031	574147	574263	574379	574494	574610	574726	574841	574957	575072	
6	575188	575303	575419	575534	575650	575765	575880	575996	576111	576226	115
7	576341	576457	576572	576687	576802	576917	577032	577147	577262	577377	
8	577492	577607	577722	577836	577951	578066	578181	578295	578410	578525	
9	578639	578754	578868	578983	579097	579212	579326	579441	579555	579669	114
380	**579784**	**579898**	**580012**	**580126**	**580241**	**580355**	**580469**	**580583**	**580697**	**580811**	

PROPORTIONAL PARTS

Diff.	1	2	3	4	5	6	7	8	9
134	13.4	26.8	40.2	53.6	67.0	80.4	93.8	107.2	120.6
132	13.2	26.4	39.6	52.8	66.0	79.2	92.4	105.6	118.8
130	13.0	26.0	39.0	52.0	65.0	78.0	91.0	104.0	117.0
128	12.8	25.6	38.4	51.2	64.0	76.8	89.6	102.4	115.2
126	12.6	25.2	37.8	50.4	63.0	75.6	88.2	100.8	113.4
124	12.4	24.8	37.2	49.6	62.0	74.4	86.8	99.2	111.6
122	12.2	24.4	36.6	48.8	61.0	73.2	85.4	97.6	109.8
120	12.0	24.0	36.0	48.0	60.0	72.0	84.0	96.0	108.0
118	11.8	23.6	35.4	47.2	59.0	70.8	82.6	94.4	106.2
116	11.6	23.2	34.8	46.4	58.0	69.6	81.2	92.8	104.4
114	11.4	22.8	34.2	45.6	57.0	68.4	79.8	91.2	102.6

MATHEMATICAL AND PHYSICAL TABLES

N	0	1	2	3	4	5	6	7	8	9	Diff.
380	**579784**	**579898**	**580012**	**580126**	**580241**	**580355**	**580469**	**580583**	**580697**	**580811**	114
1	580925	581039	581153	581267	581381	581495	581608	581722	581836	581950	
2	582063	582177	582291	582404	582518	582631	582745	582858	582972	583085	
3	583199	583312	583426	583539	583652	583765	583879	583992	584105	584218	
4	584331	584444	584557	584670	584783	584896	585009	585122	585235	585348	113
5	585461	585574	585686	585799	585912	586024	586137	586250	586362	586475	
6	586587	586700	586812	586925	587037	587149	587262	587374	587486	587599	
7	587711	587823	587935	588047	588160	588272	588384	588496	588608	588720	112
8	588832	588944	589056	589167	589279	589391	589503	589615	589726	589838	
9	589950	590061	590173	590284	590396	590507	590619	590730	590842	590953	
390	**591065**	**591176**	**591287**	**591399**	**591510**	**591621**	**591732**	**591843**	**591955**	**592066**	
1	592177	592288	592399	592510	592621	592732	592843	592954	593064	593175	111
2	593286	593397	593508	593618	593729	593840	593950	594061	594171	594282	
3	594393	594503	594614	594724	594834	594945	595055	595165	595275	595386	
4	595496	595606	595717	595827	595937	596047	596157	596267	596377	596487	
5	596597	596707	596817	596927	597037	597146	597256	597366	597476	597586	110
6	597695	597805	597914	598024	598134	598243	598353	598462	598572	598681	
7	598791	598900	599009	599119	599228	599337	599446	599556	599665	599774	
8	599883	599992	600101	600210	600319	600428	600537	600646	600755	600864	109
9	600973	601082	601191	601299	601408	601517	601625	601734	601843	601951	
400	**602060**	**602169**	**602277**	**602386**	**602494**	**602603**	**602711**	**602819**	**602928**	**603036**	
1	603144	603253	603361	603469	603577	603686	603794	603902	604010	604118	108
2	604226	604334	604442	604550	604658	604766	604874	604982	605089	605197	
3	605305	605413	605521	605628	605736	605844	605951	606059	606166	606274	
4	606381	606489	606596	606704	606811	606919	607026	607133	607241	607348	
5	607455	607562	607669	607777	607884	607991	608098	608205	608312	608419	107
6	608526	608633	608740	608847	608954	609061	609167	609274	609381	609488	
7	609594	609701	609808	609914	610021	610128	610234	610341	610447	610554	
8	610660	610767	610873	610979	611036	611192	611298	611405	611511	611617	
9	611723	611829	611936	612042	612148	612254	612360	612466	612572	612678	106
410	**612784**	**612890**	**612996**	**613102**	**613207**	**613313**	**613419**	**613525**	**613630**	**613736**	
1	613842	613947	614053	614159	614264	614370	614475	614581	614686	614792	
2	614897	615003	615108	615213	615319	615424	615529	615634	615740	615845	
3	615950	616055	616160	616265	616370	616476	616581	616686	616790	616895	105
4	617000	617105	617210	617315	617420	617525	617629	617734	617839	617943	
5	618048	618153	618257	618362	618466	618571	618676	618780	618884	618989	
6	619093	619198	619302	619406	619511	619615	619719	619824	619928	620032	
7	620136	620240	620344	620448	620552	620656	620760	620864	620968	621072	104
8	621176	621280	621384	621488	621592	621695	621799	621903	622007	622110	
9	622214	622318	622421	622525	622628	622732	622835	622939	623042	623146	
420	**623249**	**623353**	**623456**	**623559**	**623663**	**623766**	**623869**	**623973**	**624076**	**624179**	
1	624282	624385	624488	624591	624695	624798	624901	625004	625107	625210	103
2	625312	625415	625518	625621	625724	625827	625929	626032	626135	626238	
3	626340	626443	626546	626648	626751	626853	626956	627058	627161	627263	
4	627366	627468	627571	627673	627775	627878	627980	628082	628185	628287	
5	628389	628491	628593	628695	628797	628900	629002	629104	629206	629308	102
6	629410	629512	629613	629715	629817	629919	630021	630123	630224	630326	
7	630428	630530	630631	630733	630835	630936	631038	631139	631241	631342	
8	631444	631545	631647	631748	631849	631951	632052	632153	632255	632356	
9	632457	632559	632660	632761	632862	632963	633064	633165	633266	633367	
430	**633468**	**633569**	**633670**	**633771**	**633872**	**633973**	**634074**	**634175**	**634276**	**634376**	101

PROPORTIONAL PARTS

Diff.	1	2	3	4	5	6	7	8	9
114	11.4	22.8	34.2	45.6	57.0	68.4	79.8	91.2	102.6
112	11.2	22.4	33.6	44.8	56.0	67.2	78.4	89.6	100.8
110	11.0	22.0	33.0	44.0	55.0	66.0	77.0	88.0	99.0
108	10.8	21.6	32.4	43.2	54.0	64.8	75.6	86.4	97.2
106	10.6	21.2	31.8	42.4	53.0	63.6	74.2	84.8	95.4
104	10.4	20.8	31.2	41.6	52.0	62.4	72.8	83.2	93.6
102	10.2	20.4	30.6	40.8	51.0	61.2	71.4	81.6	91.8

COMMON LOGARITHMS

N	0	1	2	3	4	5	6	7	8	9	Diff.
430	**633468**	**633569**	**633670**	**633771**	**633872**	**633973**	**634074**	**634175**	**634276**	**634376**	
1	634477	634578	634679	634779	634880	634981	635081	635182	635283	635383	
2	635484	635584	635685	635785	635886	635986	636087	636187	636287	636388	
3	636488	636588	636688	636789	636889	636989	637089	637189	637290	637390	
4	637490	637590	637690	637790	637890	637990	638090	638190	638290	638389	100
5	638489	638589	638689	638789	638888	638988	639088	639188	639287	639387	
6	639486	639586	639686	639785	639885	639984	640084	640183	640283	640382	
7	640481	640581	640680	640779	640879	640978	641077	641177	641276	641375	
8	641474	641573	641672	641771	641871	641970	642069	642168	642267	642366	
9	642465	642563	642662	642761	642860	642959	643058	643156	643255	643354	99
440	**643453**	**643551**	**643650**	**643749**	**643847**	**643946**	**644044**	**644143**	**644242**	**644340**	
1	644439	644537	644636	644734	644832	644931	645029	645127	645226	645324	
2	645422	645521	645619	645717	645815	645913	646011	646110	646208	646306	
3	646404	646502	646600	646698	646796	646894	646992	647089	647187	647285	98
4	647383	647481	647579	647676	647774	647872	647969	648067	648165	648262	
5	648360	648458	648555	648653	648750	648848	648945	649043	649140	649237	
6	649335	649432	649530	649627	649724	649821	649919	650016	650113	650210	
7	650308	650405	650502	650599	650696	650793	650890	650987	651084	651181	
8	651278	651375	651472	651569	651666	651762	651859	651956	652053	652150	97
9	652246	652343	652440	652536	652633	652730	652826	652923	653019	653116	
450	**653213**	**653309**	**653405**	**653502**	**653598**	**653695**	**653791**	**653888**	**653984**	**654080**	
1	654177	654273	654369	654465	654562	654658	654754	654850	654946	655042	
2	655138	655235	655331	655427	655523	655619	655715	655810	655906	656002	96
3	656098	656194	656290	656386	656482	656577	656673	656769	656864	656960	
4	657056	657152	657247	657343	657438	657534	657629	657725	657820	657916	
5	658011	658107	658202	658298	658393	658488	658584	658679	658774	658870	
6	658965	659060	659155	659250	659346	659441	659536	659631	659726	659821	
7	659916	660011	660106	660201	660296	660391	660486	660581	660676	660771	95
8	660865	660960	661055	661150	661245	661339	661434	661529	661623	661718	
9	661813	661907	662002	662096	662191	662286	662380	662475	662569	662663	
460	**662758**	**662852**	**662947**	**663041**	**663135**	**663230**	**663324**	**663418**	**663512**	**663607**	
1	663701	663795	663889	663983	664078	664172	664266	664360	664454	664548	
2	664642	664736	664830	664924	665018	665112	665206	665299	665393	665487	94
3	665581	665675	665769	665862	665956	666050	666143	666237	666331	666424	
4	666518	666612	666705	666799	666892	666986	667079	667173	667266	667360	
5	667453	667546	667640	667733	667826	667920	668013	668106	668199	668293	
6	668386	668479	668572	668665	668759	668852	668945	669038	669131	669224	
7	669317	669410	669503	669596	669689	669782	669875	669967	670060	670153	93
8	670246	670339	670431	670524	670617	670710	670802	670895	670988	671080	
9	671173	671265	671358	671451	671543	671636	671728	671821	671913	672005	
470	**672098**	**672190**	**672283**	**672375**	**672467**	**672560**	**672652**	**672744**	**672836**	**672929**	
1	673021	673113	673205	673297	673390	673482	673574	673666	673758	673850	
2	673942	674034	674126	674218	674310	674402	674494	674586	674677	674769	92
3	674861	674953	675045	675137	675228	675320	675412	675503	675595	675687	
4	675778	675870	675962	676053	676145	676236	676328	676419	676511	676602	
5	676694	676785	676876	676968	677059	677151	677242	677333	677424	677516	
6	677607	677698	677789	677881	677972	678063	678154	678245	678336	678427	
7	678518	678609	678700	678791	678882	678973	679064	679155	679246	679337	91
8	679428	679519	679610	679700	679791	679882	679973	680063	680154	680245	
9	680336	680426	680517	680607	680698	680789	680879	680970	681060	681151	
480	**681241**	**681332**	**681422**	**681513**	**681603**	**681693**	**681784**	**681874**	**681964**	**682055**	

PROPORTIONAL PARTS

Diff.	1	2	3	4	5	6	7	8	9
102	10.2	20.4	30.6	40.8	51.0	61.2	71.4	81.6	91.8
100	10.0	20.0	30.0	40.0	50.0	60.0	70.0	80.0	90.0
98	9.8	19.6	29.4	39.2	49.0	58.8	68.6	78.4	88.2
96	9.6	19.2	28.8	38.4	48.0	57.6	67.2	76.8	86.4
94	9.4	18.8	28.2	37.6	47.0	56.4	65.8	75.2	84.6
92	9.2	18.4	27.6	36.8	46.0	55.2	64.4	73.6	82.8
90	9.0	18.0	27.0	36.0	45.0	54.0	63.0	72.0	81.0

MATHEMATICAL AND PHYSICAL TABLES

N	0	1	2	3	4	5	6	7	8	9	Diff.
480	**681241**	**681332**	**681422**	**681513**	**681603**	**681693**	**681784**	**681874**	**681964**	**682055**	
1	682145	682235	682326	682416	682506	682596	682686	682777	682867	682957	
2	683047	683137	683227	683317	683407	683497	683587	683677	683767	683857	90
3	683947	684037	684127	684217	684307	684396	684486	684576	684666	684756	
4	684845	684935	685025	685114	685204	685294	685383	685473	685563	685652	
5	685742	685831	685921	686010	686100	686189	686279	686368	686458	686547	
6	686636	686726	686815	686904	686994	687083	687172	687261	687351	687440	
7	687529	687618	687707	687796	687886	687975	688064	688153	688242	688331	
8	688420	688509	688598	688687	688776	688865	688953	689042	689131	689220	89
9	689309	689398	689486	689575	689664	689753	689841	689930	690019	690107	
490	**690196**	**690285**	**690373**	**690462**	**690550**	**690639**	**690728**	**690816**	**690905**	**690993**	
1	691081	691170	691258	691347	691435	691524	691612	691700	691789	691877	
2	691965	692053	692142	692230	692318	692406	692494	692583	692671	692759	
3	692847	692935	693023	693111	693199	693287	693375	693463	693551	693639	88
4	693727	693815	693903	693991	694078	694166	694254	694342	694430	694517	
5	694605	694693	694781	694868	694956	695044	695131	695219	695307	695394	
6	695482	695569	695657	695744	695832	695919	696007	696094	696182	696269	
7	696356	696444	696531	696618	696706	696793	696880	696968	697055	697142	
8	697229	697317	697404	697491	697578	697665	697752	697839	697926	698014	87
9	698100	698188	698275	698362	698449	698535	698622	698709	698796	698883	
500	**698970**	**699057**	**699144**	**699231**	**699317**	**699404**	**699491**	**699578**	**699664**	**699751**	
1	699838	699924	700011	700098	700184	700271	700358	700444	700531	700617	
2	700704	700790	700877	700963	701050	701136	701222	701309	701395	701482	
3	701568	701654	701741	701827	701913	701999	702086	702172	702258	702344	
4	702431	702517	702603	702689	702775	702861	702947	703033	703119	703205	
5	703291	703377	703463	703549	703635	703721	703807	703893	703979	704065	86
6	704151	704236	704322	704408	704494	704579	704665	704751	704837	704922	
7	705008	705094	705179	705265	705350	705436	705522	705607	705693	705778	
8	705864	705949	706035	706120	706206	706291	706376	706462	706547	706632	
9	706718	706803	706888	706974	707059	707144	707229	707315	707400	707485	
510	**707570**	**707655**	**707740**	**707826**	**707911**	**707996**	**708081**	**708166**	**708251**	**708336**	
1	708421	708506	708591	708676	708761	708846	708931	709015	709100	709185	85
2	709270	709355	709440	709524	709609	709694	709779	709863	709948	710033	
3	710117	710202	710287	710371	710456	710540	710625	710710	710794	710879	
4	710963	711048	711132	711217	711301	711385	711470	711554	711639	711723	
5	711807	711892	711976	712060	712144	712229	712313	712397	712481	712566	
6	712650	712734	712818	712902	712986	713070	713154	713238	713323	713407	
7	713491	713575	713659	713742	713826	713910	713994	714078	714162	714246	84
8	714330	714414	714497	714581	714665	714749	714833	714916	715000	715084	
9	715167	715251	715335	715418	715502	715586	715669	715753	715836	715920	
520	**716003**	**716087**	**716170**	**716254**	**716337**	**716421**	**716504**	**716588**	**716671**	**716754**	
1	716838	716921	717004	717088	717171	717254	717338	717421	717504	717587	
2	717671	717754	717837	717920	718003	718086	718169	718253	718336	718419	83
3	718502	718585	718668	718751	718834	718917	719000	719083	719165	719248	
4	719331	719414	719497	719580	719663	719745	719828	719911	719994	720077	
5	720159	720242	720325	720407	720490	720573	720655	720738	720821	720903	
6	720986	721068	721151	721233	721316	721398	721481	721563	721646	721728	
7	721811	721893	721975	722058	722140	722222	722305	722387	722469	722552	
8	722634	722716	722798	722881	722963	723045	723127	723209	723291	723374	
9	723456	723538	723620	723702	723784	723866	723948	724030	724112	724194	82
530	**724276**	**724358**	**724440**	**724522**	**724604**	**724685**	**724767**	**724849**	**724931**	**725013**	

PROPORTIONAL PARTS

Diff.	1	2	3	4	5	6	7	8	9
90	9.0	18.0	27.0	36.0	45.0	54.0	63.0	72.0	81.0
88	8.8	17.6	26.4	35.2	44.0	52.8	61.6	70.4	79.2
86	8.6	17.2	25.8	34.4	43.0	51.6	60.2	68.8	77.4
84	8.4	16.8	25.2	33.6	42.0	50.4	58.8	67.2	75.6
82	8.2	16.4	24.6	32.8	41.0	49.2	57.4	65.6	73.8

COMMON LOGARITHMS

N	0	1	2	3	4	5	6	7	8	9	Diff.
530	**724276**	**724358**	**724440**	**724522**	**724604**	**724685**	**724767**	**724849**	**724931**	**725013**	
1	725095	725176	725258	725340	725422	725503	725585	725667	725748	725830	
2	725912	725993	726075	726156	726238	726320	726401	726483	726564	726646	
3	726727	726809	726890	726972	727053	727134	727216	727297	727379	727460	
4	727541	727623	727704	727785	727866	727948	728029	728110	728191	728273	
5	728354	728435	728516	728597	728678	728759	728841	728922	729003	729084	
6	729165	729246	729327	729408	729489	729570	729651	729732	729813	729893	81
7	729974	730055	730136	730217	730298	730378	730459	730540	730621	730702	
8	730782	730863	730944	731024	731105	731186	731266	731347	731428	731508	
9	731589	731669	731750	731830	731911	731991	732072	732152	732233	732313	
540	**732394**	**732474**	**732555**	**732635**	**732715**	**732796**	**732876**	**732956**	**733037**	**733117**	
1	733197	733278	733358	733438	733518	733598	733679	733759	733839	733919	
2	733999	734079	734160	734240	734320	734400	734480	734560	734640	734720	80
3	734800	734880	734960	735040	735120	735200	735279	735359	735439	735519	
4	735599	735679	735759	735838	735918	735998	736078	736157	736237	736317	
5	736397	736476	736556	736635	736715	736795	736874	736954	737034	737113	
6	737193	737272	737352	737431	737511	737590	737670	737749	737829	737908	
7	737987	738067	738146	738225	738305	738384	738463	738543	738622	738701	
8	738781	738860	738939	739018	739097	739177	739256	739335	739414	739493	
9	739572	739651	739731	739810	739889	739968	740047	740126	740205	740284	79
550	**740363**	**740442**	**740521**	**740600**	**740678**	**740757**	**740836**	**740915**	**740994**	**741073**	
1	741152	741230	741309	741388	741467	741546	741624	741703	741782	741860	
2	741939	742018	742096	742175	742254	742332	742411	742489	742568	742647	
3	742725	742804	742882	742961	743039	743118	743196	743275	743353	743431	
4	743510	743588	743667	743745	743823	743902	743980	744058	744136	744215	
5	744293	744371	744449	744528	744606	744684	744762	744840	744919	744997	
6	745075	745153	745231	745309	745387	745465	745543	745621	745699	745777	78
7	745855	745933	746011	746089	746167	746245	746323	746401	746479	746556	
8	746634	746712	746790	746868	746945	747023	747101	747179	747256	747334	
9	747412	747489	747567	747645	747722	747800	747878	747955	748033	748110	
560	**748188**	**748266**	**748343**	**748421**	**748498**	**748576**	**748653**	**748731**	**748808**	**748885**	
1	748963	749040	749118	749195	749272	749350	749427	749504	749582	749659	
2	749736	749814	749891	749968	750045	750123	750200	750277	750354	750431	
3	750508	750586	750663	750740	750817	750894	750971	751048	751125	751202	
4	751279	751356	751433	751510	751587	751664	751741	751818	751895	751972	77
5	752048	752125	752202	752279	752356	752433	752509	752586	752663	752740	
6	752816	752893	752970	753047	753123	753200	753277	753353	753430	753506	
7	753583	753660	753736	753813	753889	753966	754042	754119	754195	754272	
8	754348	754425	754501	754578	754654	754730	754807	754883	754960	755036	
9	755112	755189	755265	755341	755417	755494	755570	755646	755722	755799	
570	**755875**	**755951**	**756027**	**756103**	**756180**	**756256**	**756332**	**756408**	**756484**	**756560**	
1	756636	756712	756788	756864	756940	757016	757092	757168	757244	757320	76
2	757396	757472	757548	757624	757700	757775	757851	757927	758003	758079	
3	758155	758230	758306	758382	758458	758533	758609	758685	758761	758836	
4	758912	758988	759063	759139	759214	759290	759366	759441	759517	759592	
5	759668	759743	759819	759894	759970	760045	760121	760196	760272	760347	
6	760422	760498	760573	760649	760724	760799	760875	760950	761025	761101	
7	761176	761251	761326	761402	761477	761552	761627	761702	761778	761853	
8	761928	762003	762078	762153	762228	762303	762378	762453	762529	762604	75
9	762679	762754	762829	762904	762978	763053	763128	763203	763278	763353	
580	**763428**	**763503**	**763578**	**763653**	**763727**	**763802**	**763877**	**763952**	**764027**	**764101**	

Proportional Parts

Diff.	1	2	3	4	5	6	7	8	9
82	8.2	16.4	24.6	32.8	41.0	49.2	57.4	65.6	73.8
80	8.0	16.0	24.0	32.0	40.0	48.0	56.0	64.0	72.0
78	7.8	15.6	23.4	31.2	39.0	46.8	54.6	62.4	70.2
76	7.6	15.2	22.8	30.4	38.0	45.6	53.2	60.8	68.4
74	7.4	14.8	22.2	29.6	37.0	44.4	51.8	59.2	66.6

N	0	1	2	3	4	5	6	7	8	9	Diff.
580	**763428**	**763503**	**763578**	**763653**	**763727**	**763802**	**763877**	**763952**	**764027**	**764101**	
1	764176	764251	764326	764400	764475	764550	764624	764699	764774	764848	
2	764923	764998	765072	765147	765221	765296	765370	765445	765520	765594	
3	765669	765743	765818	765892	765966	766041	766115	766190	766264	766338	
4	766413	766487	766562	766636	766710	766785	766859	766933	767007	767082	
5	767156	767230	767304	767379	767453	767527	767601	767675	767749	767823	
6	767898	767972	768046	768120	768194	768268	768342	768416	768490	768564	74
7	768638	768712	768786	768860	768934	769008	769082	769156	769230	769303	
8	769377	769451	769525	769599	769673	769746	769820	769894	769968	770042	
9	770115	770189	770263	770336	770410	770484	770557	770631	770705	770778	
590	**770852**	**770926**	**770999**	**771073**	**771146**	**771220**	**771293**	**771367**	**771440**	**771514**	
1	771587	771661	771734	771808	771881	771955	772028	772102	772175	772248	
2	772322	772395	772468	772542	772615	772688	772762	772835	772908	772981	
3	773055	773128	773201	773274	773348	773421	773494	773567	773640	773713	
4	773786	773860	773933	774006	774079	774152	774225	774298	774371	774444	73
5	774517	774590	774663	774736	774809	774882	774955	775028	775100	775173	
6	775246	775319	775392	775465	775538	775610	775683	775756	775829	775902	
7	775974	776047	776120	776193	776265	776338	776411	776483	776556	776629	
8	776701	776774	776846	776919	776992	777064	777137	777209	777282	777354	
9	777427	777499	777572	777644	777717	777789	777862	777934	778006	778079	
600	**778151**	**778224**	**778296**	**778368**	**778441**	**778513**	**778585**	**778658**	**778730**	**778802**	
1	778874	778947	779019	779091	779163	779236	779308	779380	779452	779524	
2	779596	779669	779741	779813	779885	779957	780029	780101	780173	780245	
3	780317	780389	780461	780533	780605	780677	780749	780821	780893	780965	72
4	781037	781109	781181	781253	781324	781396	781468	781540	781612	781684	
5	781755	781827	781899	781971	782042	782114	782186	782258	782329	782401	
6	782473	782544	782616	782688	782759	782831	782902	782974	783046	783117	
7	783189	783260	783332	783403	783475	783546	783618	783689	783761	783832	
8	783904	783975	784046	784118	784189	784261	784332	784403	784475	784546	
9	784617	784689	784760	784831	784902	784974	785045	785116	785187	785259	
610	**785330**	**785401**	**785472**	**785543**	**785615**	**785686**	**785757**	**785828**	**785899**	**785970**	
1	786041	786112	786183	786254	786325	786396	786467	786538	786609	786680	71
2	786751	786822	786893	786964	787035	787106	787177	787248	787319	787390	
3	787460	787531	787602	787673	787744	787815	787885	787956	788027	788098	
4	788168	788239	788310	788381	788451	788522	788593	788663	788734	788804	
5	788875	788946	789016	789087	789157	789228	789299	789369	789440	789510	
6	789581	789651	789722	789792	789863	789933	790004	790074	790144	790215	
7	790285	790356	790426	790496	790567	790637	790707	790778	790848	790918	
8	790988	791059	791129	791199	791269	791340	791410	791480	791550	791620	
9	791691	791761	791831	791901	791971	792041	792111	792181	792252	792322	
620	**792392**	**792462**	**792532**	**792602**	**792672**	**792742**	**792812**	**792882**	**792952**	**793022**	70
1	793092	793162	793231	793301	793371	793441	793511	793581	793651	793721	
2	793790	793860	793930	794000	794070	794139	794209	794279	794349	794418	
3	794488	794558	794627	794697	794767	794836	794906	794976	795045	795115	
4	795185	795254	795324	795393	795463	795532	795602	795672	795741	795811	
5	795880	795949	796019	796088	796158	796227	796297	796366	796436	796505	
6	796574	796644	796713	796782	796852	796921	796990	797060	797129	797198	
7	797268	797337	797406	797475	797545	797614	797683	797752	797821	797890	
8	797960	798029	798098	798167	798236	798305	798374	798443	798513	798582	
9	798651	798720	798789	798858	798927	798996	799065	799134	799203	799272	69
630	**799341**	**799409**	**799478**	**799547**	**799616**	**799685**	**799754**	**799823**	**799892**	**799961**	

PROPORTIONAL PARTS

Diff.	1	2	3	4	5	6	7	8	9
76	7.6	15.2	22.8	30.4	38.0	45.6	53.2	60.8	68.4
74	7.4	14.8	22.2	29.6	37.0	44.4	51.8	59.2	66.6
72	7.2	14.4	21.6	28.8	36.0	43.2	50.4	57.6	64.8
70	7.0	14.0	21.0	28.0	35.0	42.0	49.0	56.0	63.0
68	6.8	13.6	20.4	27.2	34.0	40.8	47.6	54.4	61.2

COMMON LOGARITHMS

N	0	1	2	3	4	5	6	7	8	9	Diff.
630	**799341**	**799409**	**799478**	**799547**	**799616**	**799685**	**799754**	**799823**	**799892**	**799961**	
1	800029	800098	800167	800236	800305	800373	800442	800511	800580	800648	
2	800717	800786	800854	800923	800992	801061	801129	801198	801266	801335	
3	801404	801472	801541	801609	801678	801747	801815	801884	801952	802021	
4	802089	802158	802226	802295	802363	802432	802500	802568	802637	802705	
5	802774	802842	802910	802979	803047	803116	803184	803252	803321	803389	
6	803457	803525	803594	803662	803730	803798	803867	803935	804003	804071	
7	804139	804208	804276	804344	804412	804480	804548	804616	804685	804753	
8	804821	804889	804957	805025	805093	805161	805229	805297	805365	805433	68
9	805501	805569	805637	805705	805773	805841	805908	805976	806044	806112	
640	**806180**	**806248**	**806316**	**806384**	**806451**	**806519**	**806587**	**806655**	**806723**	**806790**	
1	806858	806926	806994	807061	807129	807197	807264	807332	807400	807467	
2	807535	807603	807670	807738	807806	807873	807941	808008	808076	808143	
3	808211	808279	808346	808414	808481	808549	808616	808684	808751	808818	
4	808886	808953	809021	809088	809156	809223	809290	809358	809425	809492	
5	809560	809627	809694	809762	809829	809896	809964	810031	810098	810165	
6	810233	810300	810367	810434	810501	810569	810636	810703	810770	810837	
7	810904	810971	811039	811106	811173	811240	811307	811374	811441	811508	67
8	811575	811642	811709	811776	811843	811910	811977	812044	812111	812178	
9	812245	812312	812379	812445	812512	812579	812646	812713	812780	812847	
650	**812913**	**812980**	**813047**	**813114**	**813181**	**813247**	**813314**	**813381**	**813448**	**813514**	
1	813581	813648	813714	813781	813848	813914	813981	814048	814114	814181	
2	814248	814314	814381	814447	814514	814581	814647	814714	814780	814847	
3	814913	814980	815046	815113	815179	815246	815312	815378	815445	815511	
4	815578	815644	815711	815777	815843	815910	815976	816042	816109	816175	
5	816241	816308	816374	816440	816506	816573	816639	816705	816771	816838	
6	816904	816970	817036	817102	817169	817235	817301	817367	817433	817499	
7	817565	817631	817698	817764	817830	817896	817962	818028	818094	818160	
8	818226	818292	818358	818424	818490	818556	818622	818688	818754	818820	66
9	818885	818951	819017	819083	819149	819215	819281	819346	819412	819478	
660	**819544**	**819610**	**819676**	**819741**	**819807**	**819873**	**819939**	**820004**	**820070**	**820136**	
1	820201	820267	820333	820399	820464	820530	820595	820661	820727	820792	
2	820858	820924	820989	821055	821120	821186	821251	821317	821382	821448	
3	821514	821579	821645	821710	821775	821841	821906	821972	822037	822103	
4	822168	822233	822299	822364	822430	822495	822560	822626	822691	822756	
5	822822	822887	822952	823018	823083	823148	823213	823279	823344	823409	
6	823474	823539	823605	823670	823735	823800	823865	823930	823996	824061	
7	824126	824191	824256	824321	824386	824451	824516	824581	824646	824711	
8	824776	824841	824906	824971	825036	825101	825166	825231	825296	825361	65
9	825426	825491	825556	825621	825686	825751	825815	825880	825945	826010	
670	**826075**	**826140**	**826204**	**826269**	**826334**	**826399**	**826464**	**826528**	**826593**	**826658**	
1	826723	826787	826852	826917	826981	827046	827111	827175	827240	827305	
2	827369	827434	827499	827563	827628	827692	827757	827821	827886	827951	
3	828015	828080	828144	828209	828273	828338	828402	828467	828531	828595	
4	828660	828724	828789	828853	828918	828982	829046	829111	829175	829239	
5	829304	829368	829432	829497	829561	829625	829690	829754	829818	829882	
6	829947	830011	830075	830139	830204	830268	830332	830396	830460	830525	
7	830589	830653	830717	830781	830845	830909	830973	831037	831102	831166	
8	831230	831294	831358	831422	831486	831550	831614	831678	831742	831806	64
9	831870	831934	831998	832062	832126	832189	832253	832317	832381	832445	
680	**832509**	**832573**	**832637**	**832700**	**832764**	**832828**	**832892**	**832956**	**833020**	**833083**	

PROPORTIONAL PARTS

Diff.	1	2	3	4	5	6	7	8	9
70	7.0	14.0	21.0	28.0	35.0	42.0	49.0	56.0	63.0
68	6.8	13.6	20.4	27.2	34.0	40.8	47.6	54.4	61.2
66	6.6	13.2	19.8	26.4	33.0	39.6	46.2	52.8	59.4
64	6.4	12.8	19.2	25.6	32.0	38.4	44.8	51.2	57.6
62	6.2	12.4	18.6	24.8	31.0	37.2	43.4	49.6	55.8

302 MATHEMATICAL AND PHYSICAL TABLES

N	0	1	2	3	4	5	6	7	8	9	Diff.
680	**832509**	**832573**	**832637**	**832700**	**832764**	**832828**	**832892**	**832956**	**833020**	**833083**	
1	833147	833211	833275	833338	833402	833466	833530	833593	833657	833721	
2	833784	833848	833912	833975	834039	834103	834166	834230	834294	834357	
3	834421	834484	834548	834611	834675	834739	834802	834866	834929	834993	
4	835056	835120	835183	835247	835310	835373	835437	835500	835564	835627	
5	835691	835754	835817	835881	835944	836007	836071	836134	836197	836261	
6	836324	836387	836451	836514	836577	836641	836704	836767	836830	836894	
7	836957	837020	837083	837146	837210	837273	837336	837399	837462	837525	
8	837588	837652	837715	837778	837841	837904	837967	838030	838093	838156	63
9	838219	838282	838345	838408	838471	838534	838597	838660	838723	838786	
690	**838849**	**838912**	**838975**	**839038**	**839101**	**839164**	**839227**	**839289**	**839352**	**839415**	
1	839478	839541	839604	839667	839729	839792	839855	839918	839981	840043	
2	840106	840169	840232	840294	840357	840420	840482	840545	840608	840671	
3	840733	840796	840859	840921	840984	841046	841109	841172	841234	841297	
4	841359	841422	841485	841547	841610	841672	841735	841797	841860	841922	
5	841985	842047	842110	842172	842235	842297	842360	842422	842484	842547	
6	842609	842672	842734	842796	842859	842921	842983	843046	843108	843170	
7	843233	843295	843357	843420	843482	843544	843606	843669	843731	843793	
8	843855	843918	843980	844042	844104	844166	844229	844291	844353	844415	
9	844477	844539	844601	844664	844726	844788	844850	844912	844974	845036	
700	**845098**	**845160**	**845222**	**845284**	**845346**	**845408**	**845470**	**845532**	**845594**	**845656**	62
1	845718	845780	845842	845904	845966	846028	846090	846151	846213	846275	
2	846337	846399	846461	846523	846585	846646	846708	846770	846832	846894	
3	846955	847017	847079	847141	847202	847264	847326	847388	847449	847511	
4	847573	847634	847696	847758	847819	847881	847943	848004	848066	848128	
5	848189	848251	848312	848374	848435	848497	848559	848620	848682	848743	
6	848805	848866	848928	848989	849051	849112	849174	849235	849297	849358	
7	849419	849481	849542	849604	849665	849726	849788	849849	849911	849972	
8	850033	850095	850156	850217	850279	850340	850401	850462	850524	850585	
9	850646	850707	850769	850830	850891	850952	851014	851075	851136	851197	
710	**851258**	**851320**	**851381**	**851442**	**851503**	**851564**	**851625**	**851686**	**851747**	**851809**	
1	851870	851931	851992	852053	852114	852175	852236	852297	852358	852419	61
2	852480	852541	852602	852663	852724	852785	852846	852907	852968	853029	
3	853090	853150	853211	853272	853333	853394	853455	853516	853577	853637	
4	853698	853759	853820	853881	853941	854002	854063	854124	854185	854245	
5	854306	854367	854428	854488	854549	854610	854670	854731	854792	854852	
6	854913	854974	855034	855095	855156	855216	855277	855337	855398	855459	
7	855519	855580	855640	855701	855761	855822	855882	855943	856003	856064	
8	856124	856185	856245	856306	856366	856427	856487	856548	856608	856668	
9	856729	856789	856850	856910	856970	857031	857091	857152	857212	857272	
720	**857332**	**857393**	**857453**	**857513**	**857574**	**857634**	**857694**	**857755**	**857815**	**857875**	
1	857935	857995	858056	858116	858176	858236	858297	858357	858417	858477	
2	858537	858597	858657	858718	858778	858838	858898	858958	859018	859078	
3	859138	859198	859258	859318	859379	859439	859499	859559	859619	859679	60
4	859739	859799	859859	859918	859978	860038	860098	860158	860218	860278	
5	860338	860398	860458	860518	860578	860637	860697	860757	860817	860877	
6	860937	860996	861056	861116	861176	861236	861295	861355	861415	861475	
7	861534	861594	861654	861714	861773	861833	861893	861952	862012	862072	
8	862131	862191	862251	862310	862370	862430	862489	862549	862608	862668	
9	862728	862787	862847	862906	862966	863025	863085	863144	863204	863263	
730	**863323**	**863382**	**863442**	**863501**	**863561**	**863620**	**863680**	**863739**	**863799**	**863858**	

Proportional Parts

Diff.	1	2	3	4	5	6	7	8	9
64	6.4	12.8	19.2	25.6	32.0	38.4	44.8	51.2	57.6
62	6.2	12.4	18.6	24.8	31.0	37.2	43.4	49.6	55.8
60	6.0	12.0	18.0	24.0	30.0	36.0	42.0	48.0	54.0
58	5.8	11.6	17.4	23.2	29.0	34.8	40.6	46.4	52.2

COMMON LOGARITHMS 303

N	0	1	2	3	4	5	6	7	8	9	Diff.
730	**863323**	**863382**	**863442**	**863501**	**863561**	**863620**	**863680**	**863739**	**863799**	**863858**	
1	863917	863977	864036	864096	864155	864214	864274	864333	864392	864452	
2	864511	864570	864630	864689	864748	864808	864867	864926	864985	865045	
3	865104	865163	865222	865282	865341	865400	865459	865519	865578	865637	
4	865696	865755	865814	865874	865933	865992	866051	866110	866169	866228	
5	866287	866346	866405	866465	866524	866583	866642	866701	866760	866819	
6	866878	866937	866996	867055	867114	867173	867232	867291	867350	867409	59
7	867467	867526	867585	867644	867703	867762	867821	867880	867939	867998	
8	868056	868115	868174	868233	868292	868350	868409	868468	868527	868586	
9	868644	868703	868762	868821	868879	868938	868997	869056	869114	869173	
740	**869232**	**869290**	**869349**	**869408**	**869466**	**869525**	**869584**	**869642**	**869701**	**869760**	
1	869818	869877	869935	869994	870053	870111	870170	870228	870287	870345	
2	870404	870462	870521	870579	870638	870696	870755	870813	870872	870930	
3	870989	871047	871106	871164	871223	871281	871339	871398	871456	871515	
4	871573	871631	871690	871748	871806	871865	871923	871981	872040	872098	
5	872156	872215	872273	872331	872389	872448	872506	872564	872622	872681	
6	872739	872797	872855	872913	872972	873030	873088	873146	873204	873262	
7	873321	873379	873437	873495	873553	873611	873669	873727	873785	873844	
8	873902	873960	874018	874076	874134	874192	874250	874308	874366	874424	58
9	874482	874540	874598	874656	874714	874772	874830	874888	874945	875003	
750	**875061**	**875119**	**875177**	**875235**	**875293**	**875351**	**875409**	**875466**	**875524**	**875582**	
1	875640	875698	875756	875813	875871	875929	875987	876045	876102	876160	
2	876218	876276	876333	876391	876449	876507	876564	876622	876680	876737	
3	876795	876853	876910	876968	877026	877083	877141	877199	877256	877314	
4	877371	877429	877487	877544	877602	877659	877717	877774	877832	877889	
5	877947	878004	878062	878119	878177	878234	878292	878349	878407	878464	
6	878522	878579	878637	878694	878752	878809	878866	878924	878981	879039	
7	879096	879153	879211	879268	879325	879383	879440	879497	879555	879612	
8	879669	879726	879784	879841	879898	879956	880013	880070	880127	880185	
9	880242	880299	880356	880413	880471	880528	880585	880642	880699	880756	
760	**880814**	**880871**	**880928**	**880985**	**881042**	**881099**	**881156**	**881213**	**881271**	**881328**	
1	881385	881442	881499	881556	881613	881670	881727	881784	881841	881898	
2	881955	882012	882069	882126	882183	882240	882297	882354	882411	882468	57
3	882525	882581	882638	882695	882752	882809	882866	882923	882980	883037	
4	883093	883150	883207	883264	883321	883377	883434	883491	883548	883605	
5	883661	883718	883775	883832	883888	883945	884002	884059	884115	884172	
6	884229	884285	884342	884399	884455	884512	884569	884625	884682	884739	
7	884795	884852	884909	884965	885022	885078	885135	885192	885248	885305	
8	885361	885418	885474	885531	885587	885644	885700	885757	885813	885870	
9	885926	885983	886039	886096	886152	886209	886265	886321	886378	886434	
770	**886491**	**886547**	**886604**	**886660**	**886716**	**886773**	**886829**	**886885**	**886942**	**886998**	
1	887054	887111	887167	887223	887280	887336	887392	887449	887505	887561	
2	887617	887674	887730	887786	887842	887898	887955	888011	888067	888123	
3	888179	888236	888292	888348	888404	888460	888516	888573	888629	888685	
4	888741	888797	888853	888909	888965	889021	889077	889134	889190	889246	
5	889302	889358	889414	889470	889526	889582	889638	889694	889750	889806	56
6	889862	889918	889974	890030	890086	890141	890197	890253	890309	890365	
7	890421	890477	890533	890589	890645	890700	890756	890812	890368	890924	
8	890980	891035	891091	891147	891203	891259	891314	891370	891426	891482	
9	891537	891593	891649	891705	891760	891816	891872	891928	891983	892039	
780	**892095**	**892150**	**892206**	**892262**	**892317**	**892373**	**892429**	**892484**	**892540**	**892595**	

PROPORTIONAL PARTS

Diff.	1	2	3	4	5	6	7	8	9
60	6.0	12.0	18.0	24.0	30.0	36.0	42.0	48.0	54.0
58	5.8	11.6	17.4	23.2	29.0	34.8	40.6	46.4	52.2
56	5.6	11.2	16.8	22.4	28.0	33.6	39.2	44.8	50.4
54	5.4	10.8	16.2	21.6	27.0	32.4	37.8	43.2	48.6

MATHEMATICAL AND PHYSICAL TABLES

N	0	1	2	3	4	5	6	7	8	9	Diff.
780	892095	892150	892206	892262	892317	892373	892429	892484	892540	892595	
1	892651	892707	892762	892818	892873	892929	892985	893040	893096	893151	
2	893207	893262	893318	893373	893429	893484	893540	893595	893651	893706	
3	893762	893817	893873	893928	893984	894039	894094	894150	894205	894261	
4	894316	894371	894427	894482	894538	894593	894648	894704	894759	894814	
5	894870	894925	894980	895036	895091	895146	895201	895257	895312	895367	
6	895423	895478	895533	895588	895644	895699	895754	895809	895864	895920	
7	895975	896030	896085	896140	896195	896251	896306	896361	896416	896471	
8	896526	896581	896636	896692	896747	896802	896857	896912	896967	897022	
9	897077	897132	897187	897242	897297	897352	897407	897462	897517	897572	
790	897627	897682	897737	897792	897847	897902	897957	898012	898067	898122	55
1	898176	898231	898286	898341	898396	898451	898506	898561	898615	898670	
2	898725	898780	898835	898890	898944	898999	899054	899109	899164	899218	
3	899273	899328	899383	899437	899492	899547	899602	899656	899711	899766	
4	899821	899875	899930	899985	900039	900094	900149	900203	900258	900312	
5	900367	900422	900476	900531	900586	900640	900695	900749	900804	900859	
6	900913	900968	901022	901077	901131	901186	901240	901295	901349	901404	
7	901458	901513	901567	901622	901676	901731	901785	901840	901894	901948	
8	902003	902057	902112	902166	902221	902275	902329	902384	902438	902492	
9	902547	902601	902655	902710	902764	902818	902873	902927	902981	903036	
800	903090	903144	903199	903253	903307	903361	903416	903470	903524	903578	
1	903633	903687	903741	903795	903849	903904	903958	904012	904066	904120	
2	904174	904229	904283	904337	904391	904445	904499	904553	904607	904661	
3	904716	904770	904824	904878	904932	904986	905040	905094	905148	905202	
4	905256	905310	905364	905418	905472	905526	905580	905634	905688	905742	54
5	905796	905850	905904	905958	906012	906066	906119	906173	906227	906281	
6	906335	906389	906443	906497	906551	906604	906658	906712	906766	906820	
7	906874	906927	906981	907035	907089	907143	907196	907250	907304	907358	
8	907411	907465	907519	907573	907626	907680	907734	907787	907841	907895	
9	907949	908002	908056	908110	908163	908217	908270	908324	908378	908431	
810	908485	908539	908592	908646	908699	908753	908807	908860	908914	908967	
1	909021	909074	909128	909181	909235	909289	909342	909396	909449	909503	
2	909556	909610	909663	909716	909770	909823	909877	909930	909984	910037	
3	910091	910144	910197	910251	910304	910358	910411	910464	910518	910571	
4	910624	910678	910731	910784	910838	910891	910944	910998	911051	911104	
5	911158	911211	911264	911317	911371	911424	911477	911530	911584	911637	
6	911690	911743	911797	911850	911903	911956	912009	912063	912116	912169	
7	912222	912275	912328	912381	912435	912488	912541	912594	912647	912700	
8	912753	912806	912859	912913	912966	913019	913072	913125	913178	913231	
9	913284	913337	913390	913443	913496	913549	913602	913655	913708	913761	53
820	913814	913867	913920	913973	914026	914079	914132	914184	914237	914290	
1	914343	914396	914449	914502	914555	914608	914660	914713	914766	914819	
2	914872	914925	914977	915030	915083	915136	915189	915241	915294	915347	
3	915400	915453	915505	915558	915611	915664	915716	915769	915822	915875	
4	915927	915980	916033	916085	916138	916191	916243	916296	916349	916401	
5	916454	916507	916559	916612	916664	916717	916770	916822	916875	916927	
6	916980	917033	917085	917138	917190	917243	917295	917348	917400	917453	
7	917506	917558	917611	917663	917716	917768	917820	917873	917925	917978	
8	918030	918083	918135	918188	918240	918293	918345	918397	918450	918502	
9	918555	918607	918659	918712	918764	918816	918869	918921	918973	919026	
830	919078	919130	919183	919235	919287	919340	919392	919444	919496	919549	
1	919601	919653	919706	919758	919810	919862	919914	919967	920019	920071	
2	920123	920176	920228	920280	920332	920384	920436	920489	920541	920593	
3	920645	920697	920749	920801	920853	920906	920958	921010	921062	921114	52
4	921166	921218	921270	921322	921374	921426	921478	921530	921582	921634	
5	921686	921738	921790	921842	921894	921946	921998	922050	922102	922154	

Proportional Parts

Diff.	1	2	3	4	5	6	7	8	9
56	5.6	11.2	16.8	22.4	28.0	33.6	39.2	44.8	50.4
54	5.4	10.8	16.2	21.6	27.0	32.4	37.8	43.2	48.6
52	5.2	10.4	15.6	20.8	26.0	31.2	36.4	41.6	46.8

COMMON LOGARITHMS 305

N	0	1	2	3	4	5	6	7	8	9	Diff.
835	921686	921738	921790	921842	921894	921946	921998	922050	922102	922154	
6	922206	922258	922310	922362	922414	922466	922518	922570	922622	922674	
7	922725	922777	922829	922881	922933	922985	923037	923089	923140	923192	
8	923244	923296	923348	923399	923451	923503	923555	923607	923658	923710	
9	923762	923814	923865	923917	923969	924021	924072	924124	924176	924228	
840	924279	924331	924383	924434	924486	924538	924589	924641	924693	924744	
1	924796	924848	924899	924951	925003	925054	925106	925157	925209	925261	
2	925312	925364	925415	925467	925518	925570	925621	925673	925725	925776	
3	925828	925879	925931	925982	926034	926085	926137	926188	926240	926291	
4	926342	926394	926445	926497	926548	926600	926651	926702	926754	926805	
5	926857	926908	926959	927011	927062	927114	927165	927216	927268	927319	
6	927370	927422	927473	927524	927576	927627	927678	927730	927781	927832	
7	927883	927935	927986	928037	928088	928140	928191	928242	928293	928345	
8	928396	928447	928498	928549	928601	928652	928703	928754	928805	928857	
9	928908	928959	929010	929061	929112	929163	929215	929266	929317	929368	
850	929419	929470	929521	929572	929623	929674	929725	929776	929827	929879	
1	929930	929981	930032	930083	930134	930185	930236	930287	930338	930389	51
2	930440	930491	930542	930592	930643	930694	930745	930796	930847	930898	
3	930949	931000	931051	931102	931153	931204	931254	931305	931356	931407	
4	931458	931509	931560	931610	931661	931712	931763	931814	931865	931915	
5	931966	932017	932068	932118	932169	932220	932271	932322	932372	932423	
6	932474	932524	932575	932626	932677	932727	932778	932829	932879	932930	
7	932981	933031	933082	933133	933183	933234	933285	933335	933386	933437	
8	933487	933538	933589	933639	933690	933740	933791	933841	933892	933943	
9	933993	934044	934094	934145	934195	934246	934296	934347	934397	934448	
860	934498	934549	934599	934650	934700	934751	934801	934852	934902	934953	
1	935003	935054	935104	935154	935205	935255	935306	935356	935406	935457	
2	935507	935558	935608	935658	935709	935759	935809	935860	935910	935960	
3	936011	936061	936111	936162	936212	936262	936313	936363	936413	936463	
4	936514	936564	936614	936665	936715	936765	936815	936865	936916	936966	
5	937016	937066	937116	937167	937217	937267	937317	937367	937418	937468	
6	937518	937568	937618	937668	937718	937769	937819	937869	937919	937969	
7	938019	938069	938119	938169	938219	938269	938320	938370	938420	938470	50
8	938520	938570	938620	938670	938720	938770	938820	938870	938920	938970	
9	939020	939070	939120	939170	939220	939270	939320	939369	939419	939469	
870	939519	939569	939619	939669	939719	939769	939819	939869	939919	939968	
1	940018	940068	940118	940168	940218	940267	940317	940367	940417	940467	
2	940516	940566	940616	940666	940716	940765	940815	940865	940915	940964	
3	941014	941064	941114	941163	941213	941263	941313	941362	941412	941462	
4	941511	941561	941611	941660	941710	941760	941809	941859	941909	941958	
5	942008	942058	942107	942157	942207	942256	942306	942355	942405	942455	
6	942504	942554	942603	942653	942702	942752	942801	942851	942901	942950	
7	943000	943049	943099	943148	943198	943247	943297	943346	943396	943445	
8	943495	943544	943593	943643	943692	943742	943791	943841	943890	943939	
9	943989	944038	944088	944137	944186	944236	944285	944335	944384	944433	
880	944483	944532	944581	944631	944680	944729	944779	944828	944877	944927	
1	944976	945025	945074	945124	945173	945222	945272	945321	945370	945419	
2	945469	945518	945567	945616	945665	945715	945764	945813	945862	945912	
3	945961	946010	946059	946108	946157	946207	946256	946305	946354	946403	
4	946452	946501	946551	946600	946649	946698	946747	946796	946845	946894	
5	946943	946992	947041	947090	947140	947189	947238	947287	947336	947385	
6	947434	947483	947532	947581	947630	947679	947728	947777	947826	947875	49
7	947924	947973	948022	948070	948119	948168	948217	948266	948315	948364	
8	948413	948462	948511	948560	948608	948657	948706	948755	948804	948853	
9	948902	948951	948999	949048	949097	949146	949195	949244	949292	949341	
890	949390	949439	949488	949536	949585	949634	949683	949731	949780	949829	

PROPORTIONAL PARTS

Diff.	1	2	3	4	5	6	7	8	9
52	5.2	10.4	15.6	20.8	26.0	31.2	36.4	41.6	46.8
50	5.0	10.0	15.0	20.0	25.0	30.0	35.0	40.0	45.0
48	4.8	9.6	14.4	19.2	24.0	28.8	33.6	38.4	43.2

MATHEMATICAL AND PHYSICAL TABLES

N	0	1	2	3	4	5	6	7	8	9	Diff.
890	**949390**	**949439**	**949488**	**949536**	**949585**	**949634**	**949683**	**949731**	**949780**	**949829**	
1	949878	949926	949975	950024	950073	950121	950170	950219	950267	950316	
2	950365	950414	950462	950511	950560	950608	950657	950706	950754	950803	
3	950851	950900	950949	950997	951046	951095	951143	951192	951240	951289	
4	951338	951386	951435	951483	951532	951580	951629	951677	951726	951775	
5	951823	951872	951920	951969	952017	952066	952114	952163	952211	952260	
6	952308	952356	952405	952453	952502	952550	952599	952647	952696	952744	
7	952792	952841	952889	952938	952986	953034	953083	953131	953180	953228	
8	953276	953325	953373	953421	953470	953518	953566	953615	953663	953711	
9	953760	953808	953856	953905	953953	954001	954049	954098	954146	954194	
900	**954243**	**954291**	**954339**	**954387**	**954435**	**954484**	**954532**	**954580**	**954628**	**954677**	
1	954725	954773	954821	954869	954918	954966	955014	955062	955110	955158	
2	955207	955255	955303	955351	955399	955447	955495	955543	955592	955640	
3	955688	955736	955784	955832	955880	955928	955976	956024	956072	956120	
4	956168	956216	956265	956313	956361	956409	956457	956505	956553	956601	
5	956649	956697	956745	956793	956840	956888	956936	956984	957032	957080	48
6	957128	957176	957224	957272	957320	957368	957416	957464	957512	957559	
7	957607	957655	957703	957751	957799	957847	957894	957942	957990	958038	
8	958086	958134	958181	958229	958277	958325	958373	958421	958468	958516	
9	958564	958612	958659	958707	958755	958803	958850	958898	958946	958994	
910	**959041**	**959089**	**959137**	**959185**	**959232**	**959280**	**959328**	**959375**	**959423**	**959471**	
1	959518	959566	959614	959661	959709	959757	959804	959852	959900	959947	
2	959995	960042	960090	960138	960185	960233	960280	960328	960376	960423	
3	960471	960518	960566	960613	960661	960709	960756	960804	960851	960899	
4	960946	960994	961041	961089	961136	961184	961231	961279	961326	961374	
5	961421	961469	961516	961563	961611	961658	961706	961753	961801	961848	
6	961895	961943	961990	962038	962085	962132	962180	962227	962275	962322	
7	962369	962417	962464	962511	962559	962606	962653	962701	962748	962795	
8	962843	962890	962937	962985	963032	963079	963126	963174	963221	963268	
9	963316	963363	963410	963457	963504	963552	963599	963646	963693	963741	
920	**963788**	**963835**	**963882**	**963929**	**963977**	**964024**	**964071**	**964118**	**964165**	**964212**	
1	964260	964307	964354	964401	964448	964495	964542	964590	964637	964684	
2	964731	964778	964825	964872	964919	964966	965013	965061	965108	965155	
3	965202	965249	965296	965343	965390	965437	965484	965531	965578	965625	
4	965672	965719	965766	965813	965860	965907	965954	966001	966048	966095	47
5	966142	966189	966236	966283	966329	966376	966423	966470	966517	966564	
6	966611	966658	966705	966752	966799	966845	966892	966939	966986	967033	
7	967080	967127	967173	967220	967267	967314	967361	967408	967454	967501	
8	967548	967595	967642	967688	967735	967782	967829	967875	967922	967969	
9	968016	968062	968109	968156	968203	968249	968296	968343	968390	968436	
930	**968483**	**968530**	**968576**	**968623**	**968670**	**968716**	**968763**	**968810**	**968856**	**968903**	
1	968950	968996	969043	969090	969136	969183	969229	969276	969323	969369	
2	969416	969463	969509	969556	969602	969649	969695	969742	969789	969835	
3	969882	969928	969975	970021	970068	970114	970161	970207	970254	970300	
4	970347	970393	970440	970486	970533	970579	970626	970672	970719	970765	
5	970812	970858	970904	970951	970997	971044	971090	971137	971183	971229	
6	971276	971322	971369	971415	971461	971508	971554	971601	971647	971693	
7	971740	971786	971832	971879	971925	971971	972018	972064	972110	972157	
8	972203	972249	972295	972342	972388	972434	972481	972527	972573	972619	
9	972666	972712	972758	972804	972851	972897	972943	972989	973035	973082	
940	**973128**	**973174**	**973220**	**973266**	**973313**	**973359**	**973405**	**973451**	**973497**	**973543**	
1	973590	973636	973682	973728	973774	973820	973866	973913	973959	974005	
2	974051	974097	974143	974189	974235	974281	974327	974374	974420	974466	
3	974512	974558	974604	974650	974696	974742	974788	974834	974880	974926	
4	974972	975018	975064	975110	975156	975202	975248	975294	975340	975386	46
5	975432	975478	975524	975570	975616	975662	975707	975753	975799	975845	

Proportional Parts

Diff.	1	2	3	4	5	6	7	8	9
50	5.0	10.0	15.0	20.0	25.0	30.0	35.0	40.0	45.0
48	4.8	9.6	14.4	19.2	24.0	28.8	33.6	38.4	43.2
46	4.6	9.2	13.8	18.4	23.0	27.6	32.2	36.8	41.4

COMMON LOGARITHMS

N	0	1	2	3	4	5	6	7	8	9	Diff.
945	975432	975478	975524	975570	975616	975662	975707	975753	975799	975845	
6	975891	975937	975983	976029	976075	976121	976167	976212	976258	976304	
7	976350	976396	976442	976488	976533	976579	976625	976671	976717	976763	
8	976808	976854	976900	976946	976992	977037	977083	977129	977175	977220	
9	977266	977312	977358	977403	977449	977495	977541	977586	977632	977678	
950	**977724**	**977769**	**977815**	**977861**	**977906**	**977952**	**977998**	**978043**	**978089**	**978135**	
1	978181	978226	978272	978317	978363	978409	978454	978500	978546	978591	
2	978637	978683	978728	978774	978819	978865	978911	978956	979002	979047	
3	979093	979138	979184	979230	979275	979321	979366	979412	979457	979503	
4	979548	979594	979639	979685	979730	979776	979821	979867	979912	979958	
5	980003	980049	980094	980140	980185	980231	980276	980322	980367	980412	
6	980458	980503	980549	980594	980640	980685	980730	980776	980821	980867	
7	980912	980957	981003	981048	981093	981139	981184	981229	981275	981320	
8	981366	981411	981456	981501	981547	981592	981637	981683	981728	981773	
9	981819	981864	981909	981954	982000	982045	982090	982135	982181	982226	
960	**982271**	**982316**	**982362**	**982407**	**982452**	**982497**	**982543**	**982588**	**982633**	**982678**	
1	982723	982769	982814	982859	982904	982949	982994	983040	983085	983130	
2	983175	983220	983265	983310	983356	983401	983446	983491	983536	983581	
3	983626	983671	983716	983762	983807	983852	983897	983942	983987	984032	
4	984077	984122	984167	984212	984257	984302	984347	984392	984437	984482	
5	984527	984572	984617	984662	984707	984752	984797	984842	984887	984932	45
6	984977	985022	985067	985112	985157	985202	985247	985292	985337	985382	
7	985426	985471	985516	985561	985606	985651	985696	985741	985786	985830	
8	985875	985920	985965	986010	986055	986100	986144	986189	986234	986279	
9	986324	986369	986413	986458	986503	986548	986593	986637	986682	986727	
970	**986772**	**986817**	**986861**	**986906**	**986951**	**986996**	**987040**	**987085**	**987130**	**987175**	
1	987219	987264	987309	987353	987398	987443	987488	987532	987577	987622	
2	987666	987711	987756	987800	987845	987890	987934	987979	988024	988068	
3	988113	988157	988202	988247	988291	988336	988381	988425	988470	988514	
4	988559	988604	988648	988693	988737	988782	988826	988871	988916	988960	
5	989005	989049	989094	989138	989183	989227	989272	989316	989361	989405	
6	989450	989494	989539	989583	989628	989672	989717	989761	989806	989850	
7	989895	989939	989983	990028	990072	990117	990161	990206	990250	990294	
8	990339	990383	990428	990472	990516	990561	990605	990650	990694	990738	
9	990783	990827	990871	990916	990960	991004	991049	991093	991137	991182	
980	**991226**	**991270**	**991315**	**991359**	**991403**	**991448**	**991492**	**991536**	**991580**	**991625**	
1	991669	991713	991758	991802	991846	991890	991935	991979	992023	992067	
2	992111	992156	992200	992244	992288	992333	992377	992421	992465	992509	
3	992554	992598	992642	992686	992730	992774	992819	992863	992907	992951	
4	992995	993039	993083	993127	993172	993216	993260	993304	993348	993392	
5	993436	993480	993524	993568	993613	993657	993701	993745	993789	993833	
6	993877	993921	993965	994009	994053	994097	994141	994185	994229	994273	
7	994317	994361	994405	994449	994493	994537	994581	994625	994669	994713	
8	994757	994801	994845	994889	994933	994977	995021	995065	995108	995152	
9	995196	995240	995284	995328	995372	995416	995460	995504	995547	995591	
990	**995635**	**995679**	**995723**	**995767**	**995811**	**995854**	**995898**	**995942**	**995986**	**996030**	
1	996074	996117	996161	996205	996249	996293	996337	996380	996424	996468	
2	996512	996555	996599	996643	996687	996731	996774	996818	996862	996906	
3	996949	996993	997037	997080	997124	997168	997212	997255	997299	997343	
4	997386	997430	997474	997517	997561	997605	997648	997692	997736	997779	
5	997823	997867	997910	997954	997998	998041	998085	998129	998172	998216	
6	998259	998303	998347	998390	998434	998477	998521	998564	998608	998652	
7	998695	998739	998782	998826	998869	998913	998956	999000	999043	999087	
8	999131	999174	999218	999261	999305	999348	999392	999435	999479	999522	
9	999565	999609	999652	999696	999739	999783	999826	999870	999913	999957	
1000	**000000**	**000043**	**000087**	**000130**	**000174**	**000217**	**000260**	**000304**	**000347**	**000391**	43

PROPORTIONAL PARTS

Diff.	1	2	3	4	5	6	7	8	9
46	4.6	9.2	13.8	18.4	23.0	27.6	32.2	36.8	41.4
44	4.4	8.8	13.2	17.6	22.0	26.4	30.8	35.2	39.6
42	4.2	8.4	12.6	16.8	21.0	25.2	29.4	33.6	37.8

Natural (Naperian) Logarithms of Numbers

The table gives the natural logarithms of numbers from 1.00 to 9.99 directly, and permits the finding of the logarithms of numbers outside of that range by the addition or subtraction of the natural logarithms of powers of 10.

EXAMPLES: $\log_e 679. = \log_e 6.79 + \log_e 10^2 = 1.9155 + 4.6052 = 6.5207$.
$\log_e .0679 = \log_e 6.79 - \log_e 10^2 = 1.9155 - 4.6052 = -2.6897$.

Natural Logarithms of Powers of 10

$\log_e 10 = 2.302\ 585$ $\log_e 10^4 = 9.210\ 340$ $\log_e 10^7 = 16.118\ 096$
$\log_e 10^2 = 4.605\ 170$ $\log_e 10^5 = 11.512\ 925$ $\log_e 10^8 = 18.420\ 681$
$\log_e 10^3 = 6.907\ 755$ $\log_e 10^6 = 13.815\ 511$ $\log_e 10^9 = 20.723\ 266$

To obtain the common logarithm, the natural logarithm is multiplied by $\log_{10} e$, which is 0.434 294, or $\log_{10} N = 0.434\ 294 \log_e N$.

A negative number or number less than zero has no real logarithm.

N	0	1	2	3	4	5	6	7	8	9
1.0	0.0000	0.0100	0.0198	0.0296	0.0392	0.0488	0.0583	0.0677	0.0770	0.0862
1.1	0.0953	0.1044	0.1133	0.1222	0.1310	0.1398	0.1484	0.1570	0.1655	0.1740
1.2	0.1823	0.1906	0.1989	0.2070	0.2151	0.2231	0.2311	0.2390	0.2469	0.2546
1.3	0.2624	0.2700	0.2776	0.2852	0.2927	0.3001	0.3075	0.3148	0.3221	0.3293
1.4	0.3365	0.3436	0.3507	0.3577	0.3646	0.3716	0.3784	0.3853	0.3920	0.3988
1.5	0.4055	0.4121	0.4187	0.4253	0.4318	0.4383	0.4447	0.4511	0.4574	0.4637
1.6	0.4700	0.4762	0.4824	0.4886	0.4947	0.5008	0.5068	0.5128	0.5188	0.5247
1.7	0.5306	0.5365	0.5423	0.5481	0.5539	0.5596	0.5653	0.5710	0.5766	0.5822
1.8	0.5878	0.5933	0.5988	0.6043	0.6098	0.6152	0.6206	0.6259	0.6313	0.6366
1.9	0.6419	0.6471	0.6523	0.6575	0.6627	0.6678	0.6729	0.6780	0.6831	0.6881
2.0	0.6931	0.6981	0.7031	0.7080	0.7129	0.7178	0.7227	0.7275	0.7324	0.7372
2.1	0.7419	0.7467	0.7514	0.7561	0.7608	0.7655	0.7701	0.7747	0.7793	0.7839
2.2	0.7885	0.7930	0.7975	0.8020	0.8065	0.8109	0.8154	0.8198	0.8242	0.8286
2.3	0.8329	0.8372	0.8416	0.8459	0.8502	0.8544	0.8587	0.8629	0.8671	0.8713
2.4	0.8755	0.8796	0.8838	0.8879	0.8920	0.8961	0.9002	0.9042	0.9083	0.9123
2.5	0.9163	0.9203	0.9243	0.9282	0.9322	0.9361	0.9400	0.9439	0.9478	0.9517
2.6	0.9555	0.9594	0.9632	0.9670	0.9708	0.9746	0.9783	0.9821	0.9858	0.9895
2.7	0.9933	0.9969	1.0006	1.0043	1.0080	1.0116	1.0152	1.0188	1.0225	1.0260
2.8	1.0296	1.0332	1.0367	1.0403	1.0438	1.0473	1.0508	1.0543	1.0578	1.0613
2.9	1.0647	1.0682	1.0716	1.0750	1.0784	1.0818	1.0852	1.0886	1.0919	1.0953
3.0	1.0986	1.1019	1.1053	1.1086	1.1119	1.1151	1.1184	1.1217	1.1249	1.1282
3.1	1.1314	1.1346	1.1378	1.1410	1.1442	1.1474	1.1506	1.1537	1.1569	1.1600
3.2	1.1632	1.1663	1.1694	1.1725	1.1756	1.1787	1.1817	1.1848	1.1878	1.1909
3.3	1.1939	1.1969	1.2000	1.2030	1.2060	1.2090	1.2119	1.2149	1.2179	1.2208
3.4	1.2238	1.2267	1.2296	1.2326	1.2355	1.2384	1.2413	1.2442	1.2470	1.2499
3.5	1.2528	1.2556	1.2585	1.2613	1.2641	1.2669	1.2698	1.2726	1.2754	1.2782
3.6	1.2809	1.2837	1.2865	1.2892	1.2920	1.2947	1.2975	1.3002	1.3029	1.3056
3.7	1.3083	1.3110	1.3137	1.3164	1.3191	1.3218	1.3244	1.3271	1.3297	1.3324
3.8	1.3350	1.3376	1.3403	1.3429	1.3455	1.3481	1.3507	1.3533	1.3558	1.3584
3.9	1.3610	1.3635	1.3661	1.3686	1.3712	1.3737	1.3762	1.3788	1.3813	1.3838
4.0	1.3863	1.3888	1.3913	1.3938	1.3962	1.3987	1.4012	1.4036	1.4061	1.4085
4.1	1.4110	1.4134	1.4159	1.4183	1.4207	1.4231	1.4255	1.4279	1.4303	1.4327
4.2	1.4351	1.4375	1.4398	1.4422	1.4446	1.4469	1.4493	1.4516	1.4540	1.4563
4.3	1.4586	1.4609	1.4633	1.4656	1.4679	1.4702	1.4725	1.4748	1.4770	1.4793
4.4	1.4816	1.4839	1.4861	1.4884	1.4907	1.4929	1.4951	1.4974	1.4996	1.5019
4.5	1.5041	1.5063	1.5085	1.5107	1.5129	1.5151	1.5173	1.5195	1.5217	1.5239
4.6	1.5261	1.5282	1.5304	1.5326	1.5347	1.5369	1.5390	1.5412	1.5433	1.5454
4.7	1.5476	1.5497	1.5518	1.5539	1.5560	1.5581	1.5602	1.5623	1.5644	1.5665
4.8	1.5686	1.5707	1.5728	1.5748	1.5769	1.5790	1.5810	1.5831	1.5851	1.5872
4.9	1.5892	1.5913	1.5933	1.5953	1.5974	1.5994	1.6014	1.6034	1.6054	1.6074

NATURAL LOGARITHMS

N	0	1	2	3	4	5	6	7	8	9
5.0	**1.6094**	**1.6114**	**1.6134**	**1.6154**	**1.6174**	**1.6194**	**1.6214**	**1.6233**	**1.6253**	**1.6273**
5.1	1.6292	1.6312	1.6332	1.6351	1.6371	1.6390	1.6409	1.6429	1.6448	1.6467
5.2	1.6487	1.6506	1.6525	1.6544	1.6563	1.6582	1.6601	1.6620	1.6639	1.6658
5.3	1.6677	1.6696	1.6715	1.6734	1.6752	1.6771	1.6790	1.6808	1.6827	1.6845
5.4	1.6864	1.6882	1.6901	1.6919	1.6938	1.6956	1.6974	1.6993	1.7011	1.7029
5.5	1.7047	1.7066	1.7084	1.7102	1.7120	1.7138	1.7156	1.7174	1.7192	1.7210
5.6	1.7228	1.7246	1.7263	1.7281	1.7299	1.7317	1.7334	1.7352	1.7370	1.7387
5.7	1.7405	1.7422	1.7440	1.7457	1.7475	1.7492	1.7509	1.7527	1.7544	1.7561
5.8	1.7579	1.7596	1.7613	1.7630	1.7647	1.7664	1.7681	1.7699	1.7716	1.7733
5.9	1.7750	1.7766	1.7783	1.7800	1.7817	1.7834	1.7851	1.7867	1.7884	1.7901
6.0	**1.7918**	**1.7934**	**1.7951**	**1.7967**	**1.7984**	**1.8001**	**1.8017**	**1.8034**	**1.8050**	**1.8066**
6.1	1.8083	1.8099	1.8116	1.8132	1.8148	1.8165	1.8181	1.8197	1.8213	1.8229
6.2	1.8245	1.8262	1.8278	1.8294	1.8310	1.8326	1.8342	1.8358	1.8374	1.8390
6.3	1.8405	1.8421	1.8437	1.8453	1.8469	1.8485	1.8500	1.8516	1.8532	1.8547
6.4	1.8563	1.8579	1.8594	1.8610	1.8625	1.8641	1.8656	1.8672	1.8687	1.8703
6.5	1.8718	1.8733	1.8749	1.8764	1.8779	1.8795	1.8810	1.8825	1.8840	1.8856
6.6	1.8871	1.8886	1.8901	1.8916	1.8931	1.8946	1.8961	1.8976	1.8991	1.9006
6.7	1.9021	1.9036	1.9051	1.9066	1.9081	1.9095	1.9110	1.9125	1.9140	1.9155
6.8	1.9169	1.9184	1.9199	1.9213	1.9228	1.9242	1.9257	1.9272	1.9286	1.9301
6.9	1.9315	1.9330	1.9344	1.9359	1.9373	1.9387	1.9402	1.9416	1.9430	1.9445
7.0	**1.9459**	**1.9473**	**1.9488**	**1.9502**	**1.9516**	**1.9530**	**1.9544**	**1.9559**	**1.9573**	**1.9587**
7.1	1.9601	1.9615	1.9629	1.9643	1.9657	1.9671	1.9685	1.9699	1.9713	1.9727
7.2	1.9741	1.9755	1.9769	1.9782	1.9796	1.9810	1.9824	1.9838	1.9851	1.9865
7.3	1.9879	1.9892	1.9906	1.9920	1.9933	1.9947	1.9961	1.9974	1.9988	2.0001
7.4	2.0015	2.0028	2.0042	2.0055	2.0069	2.0082	2.0096	2.0109	2.0122	2.0136
7.5	2.0149	2.0162	2.0176	2.0189	2.0202	2.0215	2.0229	2.0242	2.0255	2.0268
7.6	2.0281	2.0295	2.0308	2.0321	2.0334	2.0347	2.0360	2.0373	2.0386	2.0399
7.7	2.0412	2.0425	2.0438	2.0451	2.0464	2.0477	2.0490	2.0503	2.0516	2.0528
7.8	2.0541	2.0554	2.0567	2.0580	2.0592	2.0605	2.0618	2.0631	2.0643	2.0656
7.9	2.0669	2.0681	2.0694	2.0707	2.0719	2.0732	2.0744	2.0757	2.0769	2.0782
8.0	**2.0794**	**2.0807**	**2.0819**	**2.0832**	**2.0844**	**2.0857**	**2.0869**	**2.0882**	**2.0894**	**2.0906**
8.1	2.0919	2.0931	2.0943	2.0956	2.0968	2.0980	2.0992	2.1005	2.1017	2.1029
8.2	2.1041	2.1054	2.1066	2.1078	2.1090	2.1102	2.1114	2.1126	2.1138	2.1150
8.3	2.1163	2.1175	2.1187	2.1199	2.1211	2.1223	2.1235	2.1247	2.1258	2.1270
8.4	2.1282	2.1294	2.1306	2.1318	2.1330	2.1342	2.1353	2.1365	2.1377	2.1389
8.5	2.1401	2.1412	2.1424	2.1436	2.1448	2.1459	2.1471	2.1483	2.1494	2.1506
8.6	2.1518	2.1529	2.1541	2.1552	2.1564	2.1576	2.1587	2.1599	2.1610	2.1622
8.7	2.1633	2.1645	2.1656	2.1668	2.1679	2.1691	2.1702	2.1713	2.1725	2.1736
8.8	2.1748	2.1759	2.1770	2.1782	2.1793	2.1804	2.1815	2.1827	2.1838	2.1849
8.9	2.1861	2.1872	2.1883	2.1894	2.1905	2.1917	2.1928	2.1939	2.1950	2.1961
9.0	**2.1972**	**2.1983**	**2.1994**	**2.2006**	**2.2017**	**2.2028**	**2.2039**	**2.2050**	**2.2061**	**2.2072**
9.1	2.2083	2.2094	2.2105	2.2116	2.2127	2.2138	2.2148	2.2159	2.2170	2.2181
9.2	2.2192	2.2203	2.2214	2.2225	2.2235	2.2246	2.2257	2.2268	2.2279	2.2289
9.3	2.2300	2.2311	2.2322	2.2332	2.2343	2.2354	2.2364	2.2375	2.2386	2.2395
9.4	2.2407	2.2418	2.2428	2.2439	2.2450	2.2460	2.2471	2.2481	2.2492	2.2502
9.5	2.2513	2.2523	2.2534	2.2544	2.2555	2.2565	2.2576	2.2586	2.2597	2.2607
9.6	2.2618	2.2628	2.2638	2.2649	2.2659	2.2670	2.2680	2.2690	2.2701	2.2711
9.7	2.2721	2.2732	2.2742	2.2752	2.2762	2.2773	2.2783	2.2793	2.2803	2.2814
9.8	2.2824	2.2834	2.2844	2.2854	2.2865	2.2875	2.2885	2.2895	2.2905	2.2915
9.9	2.2925	2.2935	2.2946	2.2956	2.2966	2.2976	2.2986	2.2996	2.3006	2.3016

Exponential and Hyperbolic Functions

	Natural Values					Common Logarithms			
x	e^x	e^{-x}	Sinh x	Cosh x	Tanh x	e^x	Sinh x	Cosh x	Tanh x
0.00	1.0000	1.0000	0.0000	1.0000	.00000	0.00000	$-\infty$	0.00000	$-\infty$
0.01	1.0101	.99005	0.0100	1.0001	.01000	.00434	$\bar{2}$.00001	.00002	$\bar{3}$.99999
0.02	1.0202	.98020	0.0200	1.0002	.02000	.00869	.30106	.00009	$\bar{2}$.30097
0.03	1.0305	.97045	0.0300	1.0005	.02999	.01303	.47719	.00020	.47699
0.04	1.0408	.96079	0.0400	1.0008	.03998	.01737	.60218	.00035	.60183
0.05	1.0513	.95123	0.0500	1.0013	.04996	.02171	.69915	.00054	.69861
0.06	1.0618	.94176	0.0600	1.0018	.05993	.02606	.77841	.00078	.77763
0.07	1.0725	.93239	0.0701	1.0025	.06989	.03040	.84545	.00106	.84439
0.08	1.0833	.92312	0.0801	1.0032	.07983	.03474	.90355	.00139	.90216
0.09	1.0942	.91393	0.0901	1.0041	.08976	.03909	.95483	.00176	.95307
0.10	1.1052	.90484	0.1002	1.0050	.09967	0.04343	$\bar{1}$.00072	0.00217	$\bar{2}$.99856
0.11	1.1163	.89583	0.1102	1.0061	.10956	.04777	.04227	.00262	$\bar{1}$.03965
0.12	1.1275	.88692	0.1203	1.0072	.11943	.05212	.08022	.00312	.07710
0.13	1.1388	.87810	0.1304	1.0085	.12927	.05646	.11517	.00366	.11151
0.14	1.1503	.86936	0.1405	1.0098	.13909	.06080	.14755	.00424	.14330
0.15	1.1618	.86071	0.1506	1.0113	.14889	.06514	.17772	.00487	.17285
0.16	1.1735	.85214	0.1607	1.0128	.15865	.06949	.20597	.00554	.20044
0.17	1.1853	.84366	0.1708	1.0145	.16838	.07383	.23254	.00625	.22629
0.18	1.1972	.83527	0.1810	1.0162	.17808	.07817	.25762	.00700	.25062
0.19	1.2092	.82696	0.1911	1.0181	.18775	.08252	.28136	.00779	.27357
0.20	1.2214	.81873	0.2013	1.0201	.19738	0.08686	$\bar{1}$.30392	0.00863	$\bar{1}$.29529
0.21	1.2337	.81058	0.2115	1.0221	.20697	.09120	.32541	.00951	.31590
0.22	1.2461	.80252	0.2218	1.0243	.21652	.09554	.34592	.01043	.33549
0.23	1.2586	.79453	0.2320	1.0266	.22603	.09989	.36555	.01139	.35416
0.24	1.2712	.78663	0.2423	1.0289	.23550	.10423	.38437	.01239	.37198
0.25	1.2840	.77880	0.2526	1.0314	.24492	.10857	.40245	.01343	.38902
0.26	1.2969	.77105	0.2629	1.0340	.25430	.11292	.41986	.01452	.40534
0.27	1.3100	.76338	0.2733	1.0367	.26362	.11726	.43663	.01564	.42099
0.28	1.3231	.75578	0.2837	1.0395	.27291	.12160	.45282	.01681	.43601
0.29	1.3364	.74826	0.2941	1.0423	.28213	.12595	.46847	.01801	.45046
0.30	1.3499	.74082	0.3045	1.0453	.29131	0.13029	1.48362	0.01926	$\bar{1}$.46436
0.31	1.3634	.73345	0.3150	1.0484	.30044	.13463	.49830	.02054	.47775
0.32	1.3771	.72615	0.3255	1.0516	.30951	.13897	.51254	.02107	.49067
0.33	1.3910	.71892	0.3360	1.0549	.31852	.14332	.52637	.02323	.50314
0.34	1.4049	.71177	0.3466	1.0584	.32748	.14766	.53981	.02463	.51518
0.35	1.4191	.70469	0.3572	1.0619	.33638	.15200	.55290	.02607	.52682
0.36	1.4333	.69768	0.3678	1.0655	.34521	.15635	.56564	.02755	.53809
0.37	1.4477	.69073	0.3785	1.0692	.35399	.16069	.57807	.02907	.54899
0.38	1.4623	.68386	0.3892	1.0731	.36271	.16503	.59019	.03063	.55956
0.39	1.4770	.67706	0.4000	1.0770	.37136	.16937	.60202	.03222	.56980
0.40	1.4918	.67032	0.4108	1.0811	.37995	0.17372	$\bar{1}$.61358	0.03385	$\bar{1}$.57973
0.41	1.5068	.66365	0.4216	1.0852	.38847	.17806	.62488	.03552	.58936
0.42	1.5220	.65705	0.4325	1.0895	.39693	.18240	.63594	.03723	.59871
0.43	1.5373	.65051	0.4434	1.0939	.40532	.18675	.64677	.03897	.60780
0.44	1.5527	.64404	0.4543	1.0984	.41364	.19109	.65738	.04075	.61663
0.45	1.5683	.63763	0.4653	1.1030	.42190	.19543	.66777	.04256	.62521
0.46	1.5841	.63128	0.4764	1.1077	.43008	.19978	.67797	.04441	.63355
0.47	1.6000	.62500	0.4875	1.1125	.43820	.20412	.68797	.04630	.64167
0.48	1.6161	.61878	0.4986	1.1174	.44624	.20846	.69779	.04822	.64957
0.49	1.6323	.61263	0.5098	1.1225	.45422	.21280	.70744	.05018	.65726
0.50	1.6487	.60653	0.5211	1.1276	.46212	0.21715	1.71692	0.05217	1.66475
0.51	1.6653	.60050	0.5324	1.1329	.46995	.22149	.72624	.05419	.67205
0.52	1.6820	.59452	0.5438	1.1383	.47770	.22583	.73540	.05625	.67916
0.53	1.6989	.58860	0.5552	1.1438	.48538	.23018	.74442	.05834	.68608
0.54	1.7160	.58275	0.5666	1.1494	.49299	.23452	.75330	.06046	.69284
0.55	1.7333	.57695	0.5782	1.1551	.50052	.23886	.76204	.06262	.69942
0.56	1.7507	.57121	0.5897	1.1609	.50798	.24320	.77065	.06481	.70584
0.57	1.7683	.56553	0.6014	1.1669	.51536	.24755	.77914	.06703	.71211
0.58	1.7860	.55990	0.6131	1.1730	.52267	.25189	.78751	.06929	.71822
0.59	1.8040	.55433	0.6248	1.1792	.52990	.25623	.79576	.07157	.72419
0.60	1.8221	.54881	0.6367	1.1855	.53705	0.26058	$\bar{1}$.80390	0.07389	$\bar{1}$.73001

EXPONENTIAL AND HYPERBOLIC FUNCTIONS 311

x	Natural Values					Common Logarithms			
	e^x	e^{-x}	Sinh x	Cosh x	Tanh x	e^x	Sinh x	Cosh x	Tanh x
0.60	1.8221	.54881	0.6367	1.1855	.53705	0.26058	$\bar{1}$.80390	0.07389	$\bar{1}$.73001
0.61	1.8404	.54335	0.6485	1.1919	.54413	.26492	.81194	.07624	.73570
0.62	1.8589	.53794	0.6605	1.1984	.55113	.26926	.81987	.07861	.74125
0.63	1.8776	.53259	0.6725	1.2051	.55805	.27361	.82770	.08102	.74667
0.64	1.8965	.52729	0.6846	1.2119	.56490	.27795	.83543	.08346	.75197
0.65	1.9155	.52205	0.6967	1.2188	.57167	.28229	.84308	.08593	.75715
0.66	1.9348	.51685	0.7090	1.2258	.57836	.28663	.85063	.08843	.76220
0.67	1.9542	.51171	0.7213	1.2330	.58498	.29098	.85809	.09095	.76714
0.68	1.9739	.50662	0.7336	1.2402	.59152	.29532	.86548	.09351	.77197
0.69	1.9937	.50158	0.7461	1.2476	.59798	.29966	.87278	.09609	.77669
0.70	2.0138	.49659	0.7586	1.2552	.60437	0.30401	$\bar{1}$.88000	0.09870	$\bar{1}$.78130
0.71	2.0340	.49164	0.7712	1.2628	.61068	.30835	.88715	.10134	.78581
0.72	2.0544	.48675	0.7838	1.2706	.61691	.31269	.89423	.10401	.79022
0.73	2.0751	.48191	0.7966	1.2785	.62307	.31704	.90123	.10670	.79453
0.74	2.0959	.47711	0.8094	1.2865	.62915	.32138	.90817	.10942	.79875
0.75	2.1170	.47237	0.8223	1.2947	.63515	.32572	.91504	.11216	.80288
0.76	2.1383	.46767	0.8353	1.3030	.64108	.33006	.92185	.11493	.80691
0.77	2.1598	.46301	0.8484	1.3114	.64693	.33441	.92859	.11773	.81086
0.78	2.1815	.45841	0.8615	1.3199	.65271	.33875	.93527	.12055	.81472
0.79	2.2034	.45384	0.8748	1.3286	.65841	.34309	.94190	.12340	.81850
0.80	2.2255	.44933	0.8881	1.3374	.66404	0.34744	$\bar{1}$.94846	0.12627	1.82219
0.81	2.2479	.44486	0.9015	1.3464	.66959	.35178	.95498	.12917	.82581
0.82	2.2705	.44043	0.9150	1.3555	.67507	.35612	.96144	.13209	.82935
0.83	2.2933	.43605	0.9286	1.3647	.68048	.36046	.96784	.13503	.83281
0.84	2.3164	.43171	0.9423	1.3740	.68581	.36481	.97420	.13800	.83620
0.85	2.3396	.42741	0.9561	1.3835	.69107	.36915	.98051	.14099	.83952
0.86	2.3632	.42316	0.9700	1.3932	.69626	.37349	.98677	.14400	.84277
0.87	2.3869	.41895	0.9840	1.4029	.70137	.37784	.99299	.14704	.84595
0.88	2.4109	.41478	0.9981	1.4128	.70642	.38218	.99916	.15009	.84906
0.89	2.4351	.41066	1.0122	1.4229	.71139	.38652	0.00528	.15317	.85211
0.90	2.4596	.40657	1.0265	1.4331	.71630	0.39087	0.01137	0.15627	$\bar{1}$.85509
0.91	2.4843	.40252	1.0409	1.4434	.72113	.39521	.01741	.15939	.85801
0.92	2.5093	.39852	1.0554	1.4539	.72590	.39955	.02341	.16254	.86088
0.93	2.5345	.39455	1.0700	1.4645	.73059	.40389	.02937	.16570	.86368
0.94	2.5600	.39063	1.0847	1.4753	.73522	.40824	.03530	.16888	.86642
0.95	2.5857	.38674	1.0995	1.4862	.73978	.41258	.04119	.17208	.86910
0.96	2.6117	.38289	1.1144	1.4973	.74428	.41692	.04704	.17531	.87173
0.97	2.6379	.37908	1.1294	1.5085	.74870	.42127	.05286	.17855	.87431
0.98	2.6645	.37531	1.1446	1.5199	.75307	.42561	.05864	.18181	.87683
0.99	2.6912	.37158	1.1598	1.5314	.75736	.42995	.06439	.18509	.87930
1.00	2.7183	.36788	1.1752	1.5431	.76159	0.43429	0.07011	0.18839	$\bar{1}$.88172
1.01	2.7456	.36422	1.1907	1.5549	.76576	.43864	.07580	.19171	.88409
1.02	2.7732	.36059	1.2063	1.5669	.76987	.44298	.08146	.19504	.88642
1.03	2.8011	.35701	1.2220	1.5790	.77391	.44732	.08708	.19839	.88869
1.04	2.8292	.35345	1.2379	1.5913	.77789	.45167	.09268	.20176	.89092
1.05	2.8577	.34994	1.2539	1.6038	.78181	.45601	.09825	.20515	.89310
1.06	2.8864	.34646	1.2700	1.6164	.78566	.46035	.10379	.20855	.89524
1.07	2.9154	.34301	1.2862	1.6292	.78946	.46470	.10930	.21197	.89733
1.08	2.9447	.33960	1.3025	1.6421	.79320	.46904	.11479	.21541	.89938
1.09	2.9743	.33622	1.3190	1.6552	.79688	.47338	.12025	.21886	.90139
1.10	3.0042	.33287	1.3356	1.6685	.80050	0.47772	0.12569	0.22233	$\bar{1}$.90336
1.11	3.0344	.32956	1.3524	1.6820	.80406	.48207	.13111	.22582	.90529
1.12	3.0649	.32628	1.3693	1.6956	.80757	.48641	.13649	.22931	.90718
1.13	3.0957	.32303	1.3863	1.7093	.81102	.49075	.14186	.23283	.90903
1.14	3.1268	.31982	1.4035	1.7233	.81441	.49510	.14720	.23636	.91085
1.15	3.1582	.31664	1.4208	1.7374	.81775	.49944	.15253	.23990	.91262
1.16	3.1899	.31349	1.4382	1.7517	.82104	.50378	.15783	.24346	.91436
1.17	3.2220	.31037	1.4558	1.7662	.82427	.50812	.16311	.24703	.91607
1.18	3.2544	.30728	1.4735	1.7808	.82745	.51247	.16836	.25062	.91774
1.19	3.2871	.30422	1.4914	1.7957	.83058	.51681	.17360	.25422	.91938
1.20	3.3201	.30119	1.5095	1.8107	.83365	0.52115	0.17882	0.25784	$\bar{1}$.92099

x	Natural Values					Common Logarithms			
	e^x	e^{-x}	Sinh x	Cosh x	Tanh x	e^x	Sinh x	Cosh x	Tanh x
1.20	3.3201	.30119	1.5095	1.8107	.83365	0.52115	0.17882	0.25784	1.92099
1.21	3.3535	.29820	1.5276	1.8258	.83668	.52550	.18402	.26146	.92256
1.22	3.3872	.29523	1.5460	1.8412	.83965	.52984	.18920	.26510	.92410
1.23	3.4212	.29229	1.5645	1.8568	.84258	.53418	.19437	.26876	.92561
1.24	3.4556	.28938	1.5831	1.8725	.84546	.53853	.19951	.27242	.92709
1.25	3.4903	.28650	1.6019	1.8884	.84828	.54287	.20464	.27610	.92854
1.26	3.5254	.28365	1.6209	1.9045	.85106	.54721	.20975	.27979	.92996
1.27	3.5609	.28083	1.6400	1.9208	.85380	.55155	.21485	.28349	.93135
1.28	3.5966	.27804	1.6593	1.9373	.85648	.55590	.21993	.28721	.93272
1.29	3.6328	.27527	1.6788	1.9540	.85913	.56024	.22499	.29093	.93406
1.30	3.6693	.27253	1.6984	1.9709	.86172	0.56458	0.23004	0.29467	$\bar{1}$.93537
1.31	3.7062	.26982	1.7182	1.9880	.86428	.56893	.23507	.29842	.93665
1.32	3.7434	.26714	1.7381	2.0053	.86678	.57327	.24009	.30217	.93791
1.33	3.7810	.26448	1.7583	2.0228	.86925	.57761	.24509	.30594	.93914
1.34	3.8190	.26185	1.7786	2.0404	.87167	.58195	.25008	.30972	.94035
1.35	3.8574	.25924	1.7991	2.0583	.87405	.58630	.25505	.31352	.94154
1.36	3.8962	.25666	1.8198	2.0764	.87639	.59064	.26002	.31732	.94270
1.37	3.9354	.25411	1.8406	2.0947	.87869	.59498	.26496	.32113	.94384
1.38	3.9749	.25158	1.8617	2.1132	.88095	.59933	.26990	.32495	.94495
1.39	4.0149	.24908	1.8829	2.1320	.88317	.60367	.27482	.32878	.94604
1.40	4.0552	.24660	1.9043	2.1509	.88535	0.60801	0.27974	0.33262	$\bar{1}$.94712
1.41	4.0960	.24414	1.9259	2.1700	.88749	.61236	.28464	.33647	.94817
1.42	4.1371	.24171	1.9477	2.1894	.88960	.61670	.28952	.34033	.94919
1.43	4.1787	.23931	1.9697	2.2090	.89167	.62104	.29440	.34420	.95020
1.44	4.2207	.23693	1.9919	2.2288	.89370	.62538	.29926	.34807	.95119
1.45	4.2631	.23457	2.0143	2.2488	.89569	.62973	.30412	.35196	.95216
1.46	4.3060	.23224	2.0369	2.2691	.89765	.63407	.30896	.35585	.95311
1.47	4.3492	.22993	2.0597	2.2896	.89958	.63841	.31379	.35976	.95404
1.48	4.3929	.22764	2.0827	2.3103	.90147	.64276	.31862	.36367	.95495
1.49	4.4371	.22537	2.1059	2.3312	.90332	.64710	.32343	.36759	.95584
1.50	4.4817	.22313	2.1293	2.3524	.90515	0.65144	0.32823	0.37151	$\bar{1}$.95672
1.51	4.5267	.22091	2.1529	2.3738	.90694	.65578	.33303	.37545	.95758
1.52	4.5722	.21871	2.1768	2.3955	.90870	.66013	.33781	.37939	.95842
1.53	4.6182	.21654	2.2008	2.4174	.91042	.66447	.34258	.38334	.95924
1.54	4.6646	.21438	2.2251	2.4395	.91212	.66881	.34735	.38730	.96005
1.55	4.7115	.21225	2.2496	2.4619	.91379	.67316	.35211	.39126	.96084
1.56	4.7588	.21014	2.2743	2.4845	.91542	.67750	.35686	.39524	.96162
1.57	4.8066	.20805	2.2993	2.5073	.91703	.68184	.36160	.39921	.96238
1.58	4.8550	.20598	2.3245	2.5305	.91860	.68619	.36633	.40320	.96313
1.59	4.9037	.20393	2.3499	2.5538	.92015	.69053	.37105	.40719	.96386
1.60	4.9530	.20190	2.3756	2.5775	.92167	0.69487	0.37577	0.41119	$\bar{1}$.96457
1.61	5.0028	.19989	2.4015	2.6013	.92316	.69921	.38048	.41520	.96528
1.62	5.0531	.19790	2.4276	2.6255	.92462	.70356	.38518	.41921	.96597
1.63	5.1039	.19593	2.4540	2.6499	.92606	.70790	.38987	.42323	.96664
1.64	5.1552	.19398	2.4806	2.6746	.92747	.71224	.39456	.42725	.96730
1.65	5.2070	.19205	2.5075	2.6995	.92886	.71659	.39923	.43129	.96795
1.66	5.2593	.19014	2.5346	2.7247	.93022	.72093	.40391	.43532	.96858
1.67	5.3122	.18825	2.5620	2.7502	.93155	.72527	.40857	.43937	.96921
1.68	5.3656	.18637	2.5896	2.7760	.93286	.72961	.41323	.44341	.96982
1.69	5.4195	.18452	2.6175	2.8020	.93415	.73396	.41788	.44747	.97042
1.70	5.4739	.18268	2.6456	2.8283	.93541	0.73830	0.42253	0.45153	$\bar{1}$.97100
1.71	5.5290	.18087	2.6740	2.8549	.93665	.74264	.42717	.45559	.97158
1.72	5.5845	.17907	2.7027	2.8818	.93786	.74699	.43180	.45966	.97214
1.73	5.6407	.17728	2.7317	2.9090	.93906	.75133	.43643	.46374	.97269
1.74	5.6973	.17552	2.7609	2.9364	.94023	.75567	.44105	.46782	.97323
1.75	5.7546	.17377	2.7904	2.9642	.94138	.76002	.44567	.47191	.97376
1.76	5.8124	.17204	2.8202	2.9922	.94250	.76436	.45028	.47600	.97428
1.77	5.8709	.17033	2.8503	3.0206	.94361	.76870	.45488	.48009	.97479
1.78	5.9299	.16864	2.8806	3.0492	.94470	.77304	.45948	.48419	.97529
1.79	5.9895	.16696	2.9112	3.0782	.94576	.77739	.46408	.48830	.97578
1.80	6.0496	.16530	2.9422	3.1075	.94681	0.78173	0.46867	0.49241	$\bar{1}$.97626

EXPONENTIAL AND HYPERBOLIC FUNCTIONS 313

x	Natural Values					Common Logarithms			
	e^x	e^{-x}	Sinh x	Cosh x	Tanh x	e^x	Sinh x	Cosh x	Tanh x
1.80	6.0496	.16530	2.9422	3.1075	.94681	0.78173	0.46867	0.49241	$\bar{1}$.97626
1.81	6.1104	.16365	2.9734	3.1371	.94783	.78607	.47325	.49652	.97673
1.82	6.1719	.16203	3.0049	3.1669	.94884	.79042	.47783	.50064	.97719
1.83	6.2339	.16041	3.0367	3.1972	.94983	.79476	.48241	.50476	.97764
1.84	6.2965	.15882	3.0689	3.2277	.95080	.79910	.48698	.50889	.97809
1.85	6.3598	.15724	3.1013	3.2585	.95175	.80344	.49154	.51302	.97852
1.86	6.4237	.15567	3.1340	3.2897	.95268	.80779	.49610	.51716	.97895
1.87	6.4883	.15412	3.1671	3.3212	.95359	.81213	.50066	.52130	.97936
1.88	6.5535	.15259	3.2005	3.3530	.95449	.81647	.50521	.52544	.97977
1.89	6.6194	.15107	3.2341	3.3852	.95537	.82082	.50976	.52959	.98017
1.90	6.6859	.14957	3.2682	3.4177	.95624	0.82516	0.51430	0.53374	$\bar{1}$.98057
1.91	6.7531	.14808	3.3025	3.4506	.95709	.82950	.51884	.53789	.98095
1.92	6.8210	.14661	3.3372	3.4838	.95792	.83385	.52338	.54205	.98133
1.93	6.8895	.14515	3.3722	3.5173	.95873	.83819	.52791	.54621	.98170
1.94	6.9588	.14370	3.4075	3.5512	.95953	.84253	.53244	.55038	.98206
1.95	7.0287	.14227	3.4432	3.5855	.96032	.84687	.53696	.55455	.98242
1.96	7.0993	.14086	3.4792	3.6201	.96109	.85122	.54148	.55872	.98272
1.97	7.1707	.13946	3.5156	3.6551	.96185	.85556	.54600	.56290	.98311
1.98	7.2427	.13807	3.5523	3.6904	.96259	.85990	.55051	.56707	.98344
1.99	7.3155	.13670	3.5894	3.7261	.96331	.86425	.55502	.57126	.98377
2.00	7.3891	.13534	3.6269	3.7622	.96403	0.86859	0.55953	0.57544	$\bar{1}$.98409
2.01	7.4633	.13399	3.6647	3.7987	.96473	.87293	.56403	.57963	.98440
2.02	7.5383	.13266	3.7028	3.8355	.96541	.87727	.56853	.58382	.98471
2.03	7.6141	.13134	3.7414	3.8727	.96609	.88162	.57303	.58802	.98502
2.04	7.6906	.13003	3.7803	3.9103	.96675	.88596	.57753	.59221	.98531
2.05	7.7679	.12873	3.8196	3.9483	.96740	.89030	.58202	.59641	.98560
2.06	7.8460	.12745	3.8593	3.9867	.96803	.89465	.58650	.60061	.98589
2.07	7.9248	.12619	3.8993	4.0255	.96865	.89899	.59099	.60482	.98617
2.08	8.0045	.12493	3.9398	4.0647	.96926	.90333	.59547	.60903	.98644
2.09	8.0849	.12369	3.9806	4.1043	.96986	.90768	.59995	.61324	.98671
2.10	8.1662	.12246	4.0219	4.1443	.97045	0.91202	0.60443	0.61745	$\bar{1}$.98697
2.11	8.2482	.12124	4.0635	4.1847	.97103	.91636	.60890	.62167	.98723
2.12	8.3311	.12003	4.1056	4.2256	.97159	.92070	.61337	.62589	.98748
2.13	8.4149	.11884	4.1480	4.2669	.97215	.92505	.61784	.63011	.98773
2.14	8.4994	.11765	4.1909	4.3085	.97269	.92939	.62231	.63433	.98798
2.15	8.5849	.11648	4.2342	4.3507	.97323	.93373	.62677	.63856	.98821
2.16	8.6711	.11533	4.2779	4.3932	.97375	.93808	.63123	.64278	.98845
2.17	8.7583	.11418	4.3221	4.4362	.97426	.94242	.63569	.64701	.98868
2.18	8.8463	.11304	4.3666	4.4797	.97477	.94676	.64015	.65125	.98890
2.19	8.9352	.11192	4.4116	4.5236	.97526	.95110	.64460	.65548	.98912
2.20	9.0250	.11080	4.4571	4.5679	.97574	0.95545	0.64905	0.65972	$\bar{1}$.98934
2.21	9.1157	.10970	4.5030	4.6127	.97622	.95979	.65350	.66396	.98955
2.22	9.2073	.10861	4.5494	4.6580	.97668	.96413	.65795	.66820	.98975
2.23	9.2999	.10753	4.5962	4.7037	.97714	.96848	.66240	.67244	.98996
2.24	9.3933	.10646	4.6434	4.7499	.97759	.97282	.66684	.67668	.99016
2.25	9.4877	.10540	4.6912	4.7966	.97803	.97716	.67128	.68093	.99035
2.26	9.5831	.10435	4.7394	4.8437	.97846	.98151	.67572	.68518	.99054
2.27	9.6794	.10331	4.7880	4.8914	.97888	.98585	.68016	.68943	.99073
2.28	9.7767	.10228	4.8372	4.9395	.97929	.99019	.68459	.69368	.99091
2.29	9.8749	.10127	4.8868	4.9881	.97970	.99453	.68903	.69794	.99109
2.30	9.9742	.10026	4.9370	5.0372	.98010	0.99888	0.69346	0.70219	$\bar{1}$.99127
2.31	10.074	.09926	4.9876	5.0868	.98049	1.00322	.69789	.70645	.99144
2.32	10.176	.09827	5.0387	5.1370	.98087	.00756	.70232	.71071	.99161
2.33	10.278	.09730	5.0903	5.1876	.98124	.01191	.70675	.71497	.99178
2.34	10.381	.09633	5.1425	5.2388	.98161	.01625	.71117	.71923	.99194
2.35	10.486	.09537	5.1951	5.2905	.98197	.02059	.71559	.72349	.99210
2.36	10.591	.09442	5.2483	5.3427	.98233	.02493	.72002	.72776	.99226
2.37	10.697	.09348	5.3020	5.3954	.98267	.02928	.72444	.73203	.99241
2.38	10.805	.09255	5.3562	5.4487	.98301	.03362	.72885	.73630	.99256
2.39	10.913	.09163	5.4109	5.5026	.98335	.03796	.73327	.74056	.99271
2.40	11.023	.09072	5.4662	5.5569	.98367	1.04231	0.73769	0.74484	$\bar{1}$.99285

MATHEMATICAL AND PHYSICAL TABLES

x	Natural Values					Common Logarithms			
	e^x	e^{-x}	Sinh x	Cosh x	Tanh x	e^x	Sinh x	Cosh x	Tanh x
2.40	11.023	.09072	5.4662	5.5569	.98367	1.04231	0.73769	0.74484	1̄.99285
2.41	11.134	.08982	5.5221	5.6119	.98400	.04665	.74210	.74911	.99299
2.42	11.246	.08892	5.5785	5.6674	.98431	.05099	.74652	.75338	.99313
2.43	11.359	.08804	5.6354	5.7235	.98462	.05534	.75093	.75766	.99327
2.44	11.473	.08716	5.6929	5.7801	.98492	.05968	.75534	.76194	.99340
2.45	11.588	.08629	5.7510	5.8373	.98522	.06402	.75975	.76621	.99353
2.46	11.705	.08543	5.8097	5.8951	.98551	.06836	.76415	.77049	.99366
2.47	11.822	.08458	5.8689	5.9535	.98579	.07271	.76856	.77477	.99379
2.48	11.941	.08374	5.9288	6.0125	.98607	.07705	.77296	.77906	.99391
2.49	12.061	.08291	5.9892	6.0721	.98635	.08139	.77737	.78334	.99403
2.50	12.182	.08208	6.0502	6.1323	.98661	1.08574	0.78177	0.78762	1̄.99415
2.51	12.305	.08127	6.1118	6.1931	.98688	.09008	.78617	.79191	.99426
2.52	12.429	.08046	6.1741	6.2545	.98714	.09442	.79057	.79619	.99438
2.53	12.554	.07966	6.2369	6.3166	.98739	.09877	.79497	.80048	.99449
2.54	12.680	.07887	6.3004	6.3793	.98764	.10311	.79937	.80477	.99460
2.55	12.807	.07808	6.3645	6.4426	.98788	.10745	.80377	.80906	.99470
2.56	12.936	.07730	6.4293	6.5066	.98812	.11179	.80816	.81335	.99481
2.57	13.066	.07654	6.4946	6.5712	.98835	.11614	.81256	.81764	.99491
2.58	13.197	.07577	6.5607	6.6365	.98858	.12048	.81695	.82194	.99501
2.59	13.330	.07502	6.6274	6.7024	.98881	.12482	.82134	.82623	.99511
2.60	13.464	.07427	6.6947	6.7690	.98903	1.12917	0.82573	0.83052	1̄.99521
2.61	13.599	.07353	6.7628	6.8363	.98924	.13351	.83012	.83482	.99530
2.62	13.736	.07280	6.8315	6.9043	.98946	.13785	.83451	.83912	.99540
2.63	13.874	.07208	6.9008	6.9729	.98966	.14219	.83890	.84341	.99549
2.64	14.013	.07136	6.9709	7.0423	.98987	.14654	.84329	.84771	.99558
2.65	14.154	.07065	7.0417	7.1123	.99007	.15088	.84768	.85201	.99566
2.66	14.296	.06995	7.1132	7.1831	.99026	.15522	.85206	.85631	.99575
2.67	14.440	.06925	7.1854	7.2546	.99045	.15957	.85645	.86061	.99583
2.68	14.585	.06856	7.2583	7.3268	.99064	.16391	.86083	.86492	.99592
2.69	14.732	.06788	7.3319	7.3998	.99083	.16825	.86522	.86922	.99600
2.70	14.880	.06721	7.4063	7.4735	.99101	1.17260	0.86960	0.87352	1̄.99608
2.71	15.029	.06654	7.4814	7.5479	.99118	.17694	.87398	.87783	.99615
2.72	15.180	.06587	7.5572	7.6231	.99136	.18128	.87836	.88213	.99623
2.73	15.333	.06522	7.6338	7.6991	.99153	.18562	.88274	.88644	.99631
2.74	15.487	.06457	7.7112	7.7758	.99170	.18997	.88712	.89074	.99638
2.75	15.643	.06393	7.7894	7.8533	.99186	.19431	.89150	.89505	.99645
2.76	15.800	.06329	7.8683	7.9316	.99202	.19865	.89588	.89936	.99652
2.77	15.959	.06266	7.9480	8.0106	.99218	.20300	.90026	.90367	.99659
2.78	16.119	.06204	8.0285	8.0905	.99233	.20734	.90463	.90798	.99666
2.79	16.281	.06142	8.1098	8.1712	.99248	.21168	.90901	.91229	.99672
2.80	16.445	.06081	8.1919	8.2527	.99263	1.21602	0.91339	0.91660	1̄.99679
2.81	16.610	.06020	8.2749	8.3351	.99278	.22037	.91776	.92091	.99685
2.82	16.777	.05961	8.3586	8.4182	.99292	.22471	.92213	.92522	.99691
2.83	16.945	.05901	8.4432	8.5022	.99306	.22905	.92651	.92953	.99698
2.84	17.116	.05843	8.5287	8.5871	.99320	.23340	.93088	.93385	.99704
2.85	17.288	.05784	8.6150	8.6728	.99333	.23774	.93525	.93816	.99709
2.86	17.462	.05727	8.7021	8.7594	.99346	.24208	.93963	.94247	.99715
2.87	17.637	.05670	8.7902	8.8469	.99359	.24643	.94400	.94679	.99721
2.88	17.814	.05613	8.8791	8.9352	.99372	.25077	.94837	.95110	.99726
2.89	17.993	.05558	8.9689	9.0244	.99384	.25511	.95274	.95542	.99732
2.90	18.174	.05502	9.0596	9.1146	.99396	1.25945	0.95711	0.95974	1̄.99737
2.91	18.357	.05448	9.1512	9.2056	.99408	.26380	.96148	.96405	.99742
2.92	18.541	.05393	9.2437	9.2976	.99420	.26814	.96584	.96837	.99747
2.93	18.728	.05340	9.3371	9.3905	.99431	.27248	.97021	.97269	.99752
2.94	18.916	.05287	9.4315	9.4844	.99443	.27683	.97458	.97701	.99757
2.95	19.106	.05234	9.5268	9.5791	.99454	.28117	.97895	.98133	.99762
2.96	19.298	.05182	9.6231	9.6749	.99464	.28551	.98331	.98565	.99767
2.97	19.492	.05130	9.7203	9.7716	.99475	.28985	.98768	.98997	.99771
2.98	19.688	.05079	9.8185	9.8693	.99485	.29420	.99205	.99429	.99776
2.99	19.886	.05029	9.9177	9.9680	.99496	.29854	.99641	.99861	.99780
3.00	20.086	.04979	10.018	10.068	.99505	1.30288	1.00078	1.00293	1̄.99785

EXPONENTIAL AND HYPERBOLIC FUNCTIONS

x	Natural Values					Common Logarithms			
	e^x	e^{-x}	Sinh x	Cosh x	Tanh x	e^x	Sinh x	Cosh x	Tanh x
3.00	20.086	.04979	10.018	10.068	.99505	1.30288	1.00078	1.00293	$\bar{1}$.99785
3.01	20.287	.04929	10.119	10.168	.99515	.30723	.00514	.00725	.99789
3.02	20.491	.04880	10.221	10.270	.99525	.31157	.00950	.01157	.99793
3.03	20.697	.04832	10.325	10.373	.99534	.31591	.01387	.01589	.99797
3.04	20.905	.04783	10.429	10.477	.99543	.32026	.01823	.02022	.99801
3.05	21.115	.04736	10.534	10.581	.99552	.32460	.02259	.02454	.99805
3.06	21.328	.04689	10.640	10.687	.99561	.32894	.02696	.02886	.99809
3.07	21.542	.04642	10.748	10.794	.99570	.33328	.03132	.03319	.99813
3.08	21.758	.04596	10.856	10.902	.99578	.33763	.03568	.03751	.99817
3.09	21.977	.04550	10.966	11.011	.99587	.34197	.04004	.04184	.99820
3.10	22.198	.04505	11.077	11.122	.99595	1.34631	1.04440	1.04616	$\bar{1}$.99824
3.11	22.421	.04460	11.188	11.233	.99603	.35066	.04876	.05049	.99827
3.12	22.646	.04416	11.301	11.345	.99611	.35500	.05312	.05481	.99831
3.13	22.874	.04372	11.415	11.459	.99618	.35934	.05748	.05914	.99834
3.14	23.104	.04328	11.530	11.574	.99626	.36368	.06184	.06347	.99837
3.15	23.336	.04285	11.647	11.689	.99633	.36803	.06620	.06779	.99841
3.16	23.571	.04243	11.764	11.807	.99641	.37237	.07056	.07212	.99844
3.17	23.807	.04200	11.883	11.925	.99648	.37671	.07492	.07645	.99847
3.18	24.047	.04159	12.003	12.044	.99655	.38106	.07927	.08078	.99850
3.19	24.288	.04117	12.124	12.165	.99662	.38540	.08363	.08510	.99853
3.20	24.533	.04076	12.246	12.287	.99668	1.38974	1.08799	1.08943	$\bar{1}$.99856
3.21	24.779	.04036	12.369	12.410	.99675	.39409	.09235	.09376	.99859
3.22	25.028	.03996	12.494	12.534	.99681	.39843	.09670	.09809	.99861
3.23	25.280	.03956	12.620	12.660	.99688	.40277	.10106	.10242	.99864
3.24	25.534	.03916	12.747	12.786	.99694	.40711	.10542	.10675	.99867
3.25	25.790	.03877	12.876	12.915	.99700	.41146	.10977	.11108	.99869
3.26	26.050	.03839	13.006	13.044	.99706	.41580	.11413	.11541	.99872
3.27	26.311	.03801	13.137	13.175	.99712	.42014	.11849	.11974	.99875
3.28	26.576	.03763	13.269	13.307	.99717	.42449	.12284	.12407	.99877
3.29	26.843	.03725	13.403	13.440	.99723	.42883	.12720	.12840	.99879
3.30	27.113	.03688	13.538	13.575	.99728	1.43317	1.13155	1.13273	$\bar{1}$.99882
3.31	27.385	.03652	13.674	13.711	.99734	.43751	.13591	.13706	.99884
3.32	27.660	.03615	13.812	13.848	.99739	.44186	.14026	.14139	.99886
3.33	27.938	.03579	13.951	13.987	.99744	.44620	.14461	.14573	.99889
3.34	28.219	.03544	14.092	14.127	.99749	.45054	.14897	.15006	.99891
3.35	28.503	.03508	14.234	14.269	.99754	.45489	.15332	.15439	.99893
3.36	28.789	.03474	14.377	14.412	.99759	.45923	.15768	.15872	.99895
3.37	29.079	.03439	14.522	14.556	.99764	.46357	.16203	.16306	.99897
3.38	29.371	.03405	14.668	14.702	.99768	.46792	.16638	.16739	.99899
3.39	29.666	.03371	14.816	14.850	.99773	.47226	.17073	.17172	.99901
3.40	29.964	.03337	14.965	14.999	.99777	1.47660	1.17509	1.17605	$\bar{1}$.99903
3.41	30.265	.03304	15.116	15.149	.99782	.48094	.17944	.18039	.99905
3.42	30.569	.03271	15.268	15.301	.99786	.48529	.18379	.18472	.99907
3.43	30.877	.03239	15.422	15.455	.99790	.48963	.18814	.18906	.99909
3.44	31.187	.03206	15.577	15.610	.99795	.49397	.19250	.19339	.99911
3.45	31.500	.03175	15.734	15.766	.99799	.49832	.19685	.19772	.99912
3.46	31.817	.03143	15.893	15.924	.99803	.50266	.20120	.20206	.99914
3.47	32.137	.03112	16.053	16.084	.99807	.50700	.20555	.20639	.99916
3.48	32.460	.03081	16.215	16.245	.99810	.51134	.20990	.21073	.99918
3.49	32.786	.03050	16.378	16.408	.99814	.51569	.21425	.21506	.99919
3.50	33.115	.03020	16.543	16.573	.99818	1.52003	1.21860	1.21940	$\bar{1}$.99921
3.51	33.448	.02990	16.709	16.739	.99821	.52437	.22296	.22373	.99922
3.52	33.784	.02960	16.877	16.907	.99825	.52872	.22731	.22807	.99924
3.53	34.124	.02930	17.047	17.077	.99828	.53306	.23166	.23240	.99925
3.54	34.467	.02901	17.219	17.248	.99832	.53740	.23601	.23674	.99927
3.55	34.813	.02872	17.392	17.421	.99835	.54175	.24036	.24107	.99928
3.56	35.163	.02844	17.567	17.596	.99838	.54609	.24471	.24541	.99930
3.57	35.517	.02816	17.744	17.772	.99842	.55043	.24906	.24975	.99931
3.58	35.874	.02788	17.923	17.951	.99845	.55477	.25341	.25408	.99933
3.59	36.234	.02760	18.103	18.131	.99848	.55912	.25776	.25842	.99934
3.60	36.598	.02732	18.285	18.313	.99851	1.56346	1.26211	1.26275	$\bar{1}$.99935

MATHEMATICAL AND PHYSICAL TABLES

x	Natural Values					Common Logarithms			
	e^x	e^{-x}	Sinh x	Cosh x	Tanh x	e^x	Sinh x	Cosh x	Tanh x
3.60	36.598	.02732	18.285	18.313	.99851	1.56346	1.26211	1.26275	$\bar{1}$.99935
3.61	36.966	.02705	18.470	18.497	.99854	.56780	.26646	.26709	.99936
3.62	37.338	.02678	18.655	18.682	.99857	.57215	.27080	.27143	.99938
3.63	37.713	.02652	18.843	18.870	.99859	.57649	.27515	.27576	.99939
3.64	38.092	.02625	19.033	19.059	.99862	.58083	.27950	.28010	.99940
3.65	38.475	.02599	19.224	19.250	.99865	.58517	.28385	.28444	.99941
3.66	38.861	.02573	19.418	19.444	.99868	.58952	.28820	.28878	.99942
3.67	39.252	.02548	19.613	19.639	.99870	.59386	.29255	.29311	.99944
3.68	39.646	.02522	19.811	19.836	.99873	.59820	.29690	.29745	.99945
3.69	40.045	.02497	20.010	20.035	.99875	.60255	.30125	.30179	.99946
3.70	40.447	.02472	20.211	20.236	.99878	1.60689	1.30559	1.30612	$\bar{1}$.99947
3.71	40.854	.02448	20.415	20.439	.99880	.61123	.30994	.31046	.99948
3.72	41.264	.02423	20.620	20.644	.99883	.61558	.31429	.31480	.99949
3.73	41.679	.02399	20.828	20.852	.99885	.61992	.31864	.31914	.99950
3.74	42.098	.02375	21.037	21.061	.99887	.62426	.32299	.32348	.99951
3.75	42.521	.02352	21.249	21.272	.99889	.62860	.32733	.32781	.99952
3.76	42.948	.02328	21.463	21.486	.99892	.63295	.33168	.33215	.99953
3.77	43.380	.02305	21.679	21.702	.99894	.63729	.33603	.33649	.99954
3.78	43.816	.02282	21.897	21.919	.99896	.64163	.34038	.34083	.99955
3.79	44.256	.02260	22.117	22.140	.99898	.64598	.34472	.34517	.99956
3.80	44.701	.02237	22.339	22.362	.99900	1.65032	1.34907	1.34951	$\bar{1}$.99957
3.81	45.150	.02215	22.564	22.586	.99902	.65466	.35342	.35384	.99957
3.82	45.604	.02193	22.791	22.813	.99904	.65900	.35777	.35818	.99958
3.83	46.063	.02171	23.020	23.042	.99906	.66335	.36211	.36252	.99959
3.84	46.525	.02149	23.252	23.274	.99908	.66769	.36646	.36686	.99960
3.85	46.993	.02128	23.486	23.507	.99909	.67203	.37081	.37120	.99961
3.86	47.465	.02107	23.722	23.743	.99911	.67638	.37515	.37554	.99961
3.87	47.942	.02086	23.961	23.982	.99913	.68072	.37950	.37988	.99962
3.88	48.424	.02065	24.202	24.222	.99915	.68506	.38385	.38422	.99963
3.89	48.911	.02045	24.445	24.466	.99916	.68941	.38819	.38856	.99964
3.90	49.402	.02024	24.691	24.711	.99918	1.69375	1.39254	1.39290	$\bar{1}$.99964
3.91	49.899	.02004	24.939	24.960	.99920	.69809	.39689	.39724	.99965
3.92	50.400	.01984	25.190	25.210	.99921	.70243	.40123	.40158	.99966
3.93	50.907	.01964	25.444	25.463	.99923	.70678	.40558	.40591	.99966
3.94	51.419	.01945	25.700	25.719	.99924	.71112	.40993	.41025	.99967
3.95	51.935	.01925	25.958	25.977	.99926	.71546	.41427	.41459	.99968
3.96	52.457	.01906	26.219	26.238	.99927	.71981	.41862	.41893	.99968
3.97	52.985	.01887	26.483	26.502	.99929	.72415	.42296	.42327	.99969
3.98	53.517	.01869	26.749	26.768	.99930	.72849	.42731	.42761	.99970
3.99	54.055	.01850	27.018	27.037	.99932	.73284	.43166	.43195	.99970
4.00	54.598	.01832	27.290	27.308	.99933	1.73718	1.43600	1.43629	$\bar{1}$.99971
4.01	55.147	.01813	27.564	27.583	.99934	.74152	.44035	.44063	.99971
4.02	55.701	.01795	27.842	27.860	.99936	.74586	.44469	.44497	.99972
4.03	56.261	.01777	28.122	28.139	.99937	.75021	.44904	.44931	.99973
4.04	56.826	.01760	28.404	28.422	.99938	.75455	.45339	.45365	.99973
4.05	57.397	.01742	28.690	28.707	.99939	.75889	.45773	.45799	.99974
4.06	57.974	.01725	28.979	28.996	.99941	.76324	.46208	.46233	.99974
4.07	58.557	.01708	29.270	29.287	.99942	.76758	.46642	.46668	.99975
4.08	59.145	.01691	29.564	29.581	.99943	.77192	.47077	.47102	.99975
4.09	59.740	.01674	29.862	29.878	.99944	.77626	.47511	.47536	.99976
4.10	60.340	.01657	30.162	30.178	.99945	1.78061	1.47946	1.47970	$\bar{1}$.99976
4.11	60.947	.01641	30.465	30.482	.99946	.78495	.48380	.48404	.99977
4.12	61.559	.01624	30.772	30.788	.99947	.78929	.48815	.48838	.99977
4.13	62.178	.01608	31.081	31.097	.99948	.79364	.49249	.49272	.99978
4.14	62.803	.01592	31.393	31.409	.99949	.79798	.49684	.49706	.99978
4.15	63.434	.01576	31.709	31.725	.99950	.80232	.50118	.50140	.99978
4.16	64.072	.01561	32.028	32.044	.99951	.80667	.50553	.50574	.99979
4.17	64.715	.01545	32.350	32.365	.99952	.81101	.50987	.51008	.99979
4.18	65.366	.01530	32.675	32.691	.99953	.81535	.51422	.51442	.99980
4.19	66.023	.01515	33.004	33.019	.99954	.81969	.51856	.51876	.99980
4.20	66.686	.01500	33.336	33.351	.99955	1.82404	1.52291	1.52310	$\bar{1}$.99980

EXPONENTIAL AND HYPERBOLIC FUNCTIONS

x	Natural Values					Common Logarithms			
	e^x	e^{-x}	Sinh x	Cosh x	Tanh x	e^x	Sinh x	Cosh x	Tanh x
4.20	66.686	.01500	33.336	33.351	.99955	1.82404	1.52291	1.52310	$\bar{1}$.99980
4.21	67.357	.01485	33.671	33.686	.99956	.82838	.52725	.52745	.99981
4.22	68.033	.01470	34.009	34.024	.99957	.83272	.53160	.53179	.99981
4.23	68.717	.01455	34.351	34.366	.99958	.83707	.53594	.53613	.99982
4.24	69.408	.01441	34.697	34.711	.99958	.84141	.54029	.54047	.99982
4.25	70.105	.01426	35.046	35.060	.99959	.84575	.54463	.54481	.99982
4.26	70.810	.01412	35.398	35.412	.99960	.85009	.54898	.54915	.99983
4.27	71.522	.01398	35.754	35.768	.99961	.85444	.55332	.55349	.99983
4.28	72.240	.01384	36.113	36.127	.99962	.85878	.55767	.55783	.99983
4.29	72.966	.01370	36.476	36.490	.99962	.86312	.56201	.56217	.99984
4.30	73.700	.01357	36.843	36.857	.99963	1.86747	1.56636	1.56652	$\bar{1}$.99984
4.31	74.440	.01343	37.214	37.227	.99964	.87181	.57070	.57086	.99984
4.32	75.189	.01330	37.588	37.601	.99965	.87615	.57505	.57520	.99985
4.33	75.944	.01317	37.966	37.979	.99965	.88050	.57939	.57954	.99985
4.34	76.708	.01304	38.347	38.360	.99966	.88484	.58373	.58388	.99985
4.35	77.478	.01291	38.733	38.746	.99967	.88918	.58808	.58822	.99986
4.36	78.257	.01278	39.122	39.135	.99967	.89352	.59242	.59256	.99986
4.37	79.044	.01265	39.515	39.528	.99968	.89787	.59677	.59691	.99986
4.38	79.838	.01253	39.913	39.925	.99969	.90221	.60111	.60125	.99986
4.39	80.640	.01240	40.314	40.326	.99969	.90655	.60546	.60559	.99987
4.40	81.451	.01228	40.719	40.732	.99970	1.91090	1.60980	1.60993	$\bar{1}$.99987
4.41	82.269	.01216	41.129	41.141	.99970	.91524	.61414	.61427	.99987
4.42	83.096	.01203	41.542	41.554	.99971	.91958	.61849	.61861	.99987
4.43	83.931	.01191	41.960	41.972	.99972	.92392	.62283	.62296	.99988
4.44	84.775	.01180	42.382	42.393	.99972	.92827	.62718	.62730	.99988
4.45	85.627	.01168	42.808	42.819	.99973	.93261	.63152	.63164	.99988
4.46	86.488	.01156	43.238	43.250	.99973	.93695	.63587	.63598	.99988
4.47	87.357	.01145	43.673	43.684	.99974	.94130	.64021	.64032	.99989
4.48	88.235	.01133	44.112	44.123	.99974	.94564	.64455	.64467	.99989
4.49	89.121	.01122	44.555	44.566	.99975	.94998	.64890	.64901	.99989
4.50	90.017	.01111	45.003	45.014	.99975	1.95433	1.65324	1.65335	$\bar{1}$.99989
4.51	90.922	.01100	45.455	45.466	.99976	.95867	.65759	.65769	.99989
4.52	91.836	.01089	45.912	45.923	.99976	.96301	.66193	.66203	.99990
4.53	92.759	.01078	46.374	46.385	.99977	.96735	.66627	.66637	.99990
4.54	93.691	.01067	46.840	46.851	.99977	.97170	.67062	.67072	.99990
4.55	94.632	.01057	47.311	47.321	.99978	.97604	.67496	.67506	.99990
4.56	95.583	.01046	47.787	47.797	.99978	.98038	.67931	.67940	.99990
4.57	96.544	.01036	48.267	48.277	.99979	.98473	.68365	.68374	.99991
4.58	97.514	.01025	48.752	48.762	.99979	.98907	.68799	.68808	.99991
4.59	98.494	.01015	49.242	49.252	.99979	.99341	.69234	.69243	.99991
4.60	99.484	.01005	49.737	49.747	.99980	1.99775	1.69668	1.69677	$\bar{1}$.99991
4.61	100.48	.00995	50.237	50.247	.99980	2.00210	.70102	.70111	.99991
4.62	101.49	.00985	50.742	50.752	.99981	.00644	.70537	.70545	.99992
4.63	102.51	.00975	51.252	51.262	.99981	.01078	.70971	.70979	.99992
4.64	103.54	.00966	51.767	51.777	.99981	.01513	.71406	.71414	.99992
4.65	104.58	.00956	52.288	52.297	.99982	.01947	.71840	.71848	.99992
4.66	105.64	.00947	52.813	52.823	.99982	.02381	.72274	.72282	.99992
4.67	106.70	.00937	53.344	53.354	.99982	.02816	.72709	.72716	.99992
4.68	107.77	.00928	53.880	53.890	.99983	.03250	.73143	.73151	.99993
4.69	108.85	.00919	54.422	54.431	.99983	.03684	.73577	.73585	.99993
4.70	109.95	.00910	54.969	54.978	.99983	2.04118	1.74012	1.74019	$\bar{1}$.99993
4.71	111.05	.00900	55.522	55.531	.99984	.04553	.74446	.74453	.99993
4.72	112.17	.00892	56.080	56.089	.99984	.04987	.74881	.74887	.99993
4.73	113.30	.00883	56.643	56.652	.99984	.05421	.75315	.75322	.99993
4.74	114.43	.00874	57.213	57.222	.99985	.05856	.75749	.75756	.99993
4.75	115.58	.00865	57.788	57.796	.99985	.06290	.76184	.76190	.99993
4.76	116.75	.00857	58.369	58.377	.99985	.06724	.76618	.76624	.99994
4.77	117.92	.00848	58.955	58.964	.99986	.07158	.77052	.77059	.99994
4.78	119.10	.00840	59.548	59.556	.99986	.07593	.77487	.77493	.99994
4.79	120.30	.00831	60.147	60.155	.99986	.08027	.77921	.77927	.99994
4.80	121.51	.00823	60.751	60.759	.99985	2.08461	1.78355	1.78361	$\bar{1}$.99994

MATHEMATICAL AND PHYSICAL TABLES

x	Natural Values					Common Logarithms			
	e^x	e^{-x}	Sinh x	Cosh x	Tanh x	e^x	Sinh x	Cosh x	Tanh x
4.80	121.51	.00823	60.751	60.760	.99986	2.08461	1.78355	1.78361	$\bar{1}$.99994
4.81	122.73	.00815	61.362	61.370	.99987	.08896	.78790	.78796	.99994
4.82	123.97	.00807	61.979	61.987	.99987	.09330	.79224	.79230	.99994
4.83	125.21	.00799	62.601	62.609	.99987	.09764	.79658	.79664	.99994
4.84	126.47	.00791	63.231	63.239	.99987	.10199	.80093	.80098	.99995
4.85	127.74	.00783	63.866	63.874	.99988	.10633	.80527	.80532	.99995
4.86	129.02	.00775	64.508	64.516	.99988	.11067	.80962	.80967	.99995
4.87	130.32	.00767	65.157	65.164	.99988	.11501	.81396	.81401	.99995
4.88	131.63	.00760	65.812	65.819	.99988	.11936	.81830	.81835	.99995
4.89	132.95	.00752	66.473	66.481	.99989	.12370	.82265	.82269	.99995
4.90	134.29	.00745	67.141	67.149	.99989	2.12804	1.82699	1.82704	$\bar{1}$.99995
4.91	135.64	.00737	67.816	67.823	.99989	.13239	.83133	.83138	.99995
4.92	137.00	.00730	68.498	68.505	.99989	.13673	.83568	.83572	.99995
4.93	138.38	.00723	69.186	69.193	.99990	.14107	.84002	.84006	.99995
4.94	139.77	.00715	69.882	69.889	.99990	.14541	.84436	.84441	.99996
4.95	141.17	.00708	70.584	70.591	.99990	.14976	.84871	.84875	.99996
4.96	142.59	.00701	71.293	71.300	.99990	.15410	.85305	.85309	.99996
4.97	144.03	.00694	72.010	72.017	.99990	.15844	.85739	.85743	.99996
4.98	145.47	.00687	72.734	72.741	.99991	.16279	.86174	.86178	.99996
4.99	146.94	.00681	73.465	73.472	.99991	.16713	.86608	.86612	.99996
5.00	148.41	.00674	74.203	74.210	.99991	2.17147	1.87042	1.87046	$\bar{1}$.99996
5.01	149.90	.00667	74.949	74.956	.99991	.17582	.87477	.87480	.99996
5.02	151.41	.00660	75.702	75.710	.99991	.18016	.87911	.87915	.99996
5.03	152.93	.00654	76.463	76.470	.99991	.18450	.88345	.88349	.99996
5.04	154.47	.00647	77.232	77.238	.99992	.18884	.88780	.88783	.99996
5.05	156.02	.00641	78.008	78.014	.99992	.19319	.89214	.89217	.99996
5.06	157.59	.00635	78.792	78.798	.99992	.19753	.89648	.89652	.99997
5.07	159.17	.00628	79.584	79.590	.99992	.20187	.90083	.90086	.99997
5.08	160.77	.00622	80.384	80.390	.99992	.20622	.90517	.90520	.99997
5.09	162.39	.00616	81.192	81.198	.99992	.21056	.90951	.90955	.99997
5.10	164.02	.00610	82.008	82.014	.99993	2.21490	1.91386	1.91389	$\bar{1}$.99997
5.11	165.67	.00604	82.832	82.838	.99993	.21924	.91820	.91823	.99997
5.12	167.34	.00598	83.665	83.671	.99993	.22359	.92254	.92257	.99997
5.13	169.02	.00592	84.506	84.512	.99993	.22793	.92689	.92692	.99997
5.14	170.72	.00586	85.355	85.361	.99993	.23227	.93123	.93126	.99997
5.15	172.43	.00580	86.213	86.219	.99993	.23662	.93557	.93560	.99997
5.16	174.16	.00574	87.079	87.085	.99993	.24096	.93992	.93994	.99997
5.17	175.91	.00568	87.955	87.960	.99994	.24530	.94426	.94429	.99997
5.18	177.68	.00563	88.839	88.844	.99994	.24965	.94860	.94863	.99997
5.19	179.47	.00557	89.732	89.737	.99994	.25399	.95294	.95297	.99997
5.20	181.27	.00552	90.633	90.639	.99994	2.25833	1.95729	1.95731	$\bar{1}$.99997
5.21	183.09	.00546	91.544	91.550	.99994	.26267	.96163	.96166	.99997
5.22	184.93	.00541	92.464	92.470	.99994	.26702	.96597	.96600	.99997
5.23	186.79	.00535	93.394	93.399	.99994	.27136	.97032	.97034	.99998
5.24	188.67	.00530	94.332	94.338	.99994	.27570	.97466	.97469	.99998
5.25	190.57	.00525	95.281	95.286	.99994	.28005	.97900	.97903	.99998
5.26	192.48	.00520	96.238	96.243	.99995	.28439	.98335	.98337	.99998
5.27	194.42	.00514	97.205	97.211	.99995	.28873	.98769	.98771	.99998
5.28	196.37	.00509	98.182	98.188	.99995	.29307	.99203	.99206	.99998
5.29	198.34	.00504	99.169	99.174	.99995	.29742	.99638	.99640	.99998
5.30	200.34	.00499	100.17	100.17	.99995	2.30176	2.00072	2.00074	$\bar{1}$.99998
5.31	202.35	.00494	101.17	101.18	.99995	.30610	.00506	.00508	.99998
5.32	204.38	.00489	102.19	102.19	.99995	.31045	.00941	.00943	.99998
5.33	206.44	.00484	103.22	103.22	.99995	.31479	.01375	.01377	.99998
5.34	208.51	.00480	104.25	104.26	.99995	.31913	.01809	.01811	.99998
5.35	210.61	.00475	105.30	105.31	.99995	.32348	.02244	.02246	.99998
5.36	212.72	.00470	106.36	106.36	.99996	.32782	.02678	.02680	.99998
5.37	214.86	.00465	107.43	107.43	.99996	.33216	.03112	.03114	.99998
5.38	217.02	.00461	108.51	108.51	.99996	.33650	.03547	.03548	.99998
5.39	219.20	.00456	109.60	109.60	.99996	.34085	.03981	.03983	.99998
5.40	221.41	.00452	110.70	110.71	.99996	2.34519	2.04415	2.04417	$\bar{1}$.99998

EXPONENTIAL AND HYPERBOLIC FUNCTIONS

x	Natural Values					Common Logarithms			
	e^x	e^{-x}	Sinh x	Cosh x	Tanh x	e^x	Sinh x	Cosh x	Tanh x
5.40	221.41	.00452	110.70	110.71	.99996	2.34519	2.04415	2.04417	$\bar{1}$.99998
5.41	223.63	.00447	111.81	111.82	.99996	.34953	.04849	.04851	.99998
5.42	225.88	.00443	112.94	112.94	.99996	.35388	.05284	.05285	.99998
5.43	228.15	.00438	114.07	114.08	.99996	.35822	.05718	.05720	.99998
5.44	230.44	.00434	115.22	115.22	.99996	.36256	.06152	.06154	.99998
5.45	232.76	.00430	116.38	116.38	.99996	.36690	.06587	.06588	.99998
5.46	235.10	.00425	117.55	117.55	.99996	.37125	.07021	.07023	.99998
5.47	237.46	.00421	118.73	118.73	.99996	.37559	.07455	.07457	.99998
5.48	239.85	.00417	119.92	119.93	.99997	.37993	.07890	.07891	.99998
5.49	242.26	.00413	121.13	121.13	.99997	.38428	.08324	.08325	.99999
5.50	244.69	.00409	122.34	122.35	.99997	2.38862	2.08758	2.08760	$\bar{1}$.99999
5.51	247.15	.00405	123.57	123.58	.99997	.39296	.09193	.09194	.99999
5.52	249.64	.00401	124.82	124.82	.99997	.39731	.09627	.09628	.99999
5.53	252.14	.00397	126.07	126.07	.99997	.40165	.10061	.10063	.99999
5.54	254.68	.00393	127.34	127.34	.99997	.40599	.10495	.10497	.99999
5.55	257.24	.00389	128.62	128.62	.99997	.41033	.10930	.10931	.99999
5.56	259.82	.00385	129.91	129.91	.99997	.41468	.11364	.11365	.99999
5.57	262.43	.00381	131.22	131.22	.99997	.41902	.11798	.11800	.99999
5.58	265.07	.00377	132.53	132.54	.99997	.42336	.12233	.12234	.99999
5.59	267.74	.00374	133.87	133.87	.99997	.42771	.12667	.12668	.99999
5.60	270.43	.00370	135.21	135.22	.99997	2.43205	2.13101	2.13103	$\bar{1}$.99999
5.61	273.14	.00366	136.57	136.57	.99997	.43639	.13536	.13537	.99999
5.62	275.89	.00362	137.94	137.95	.99997	.44074	.13970	.13971	.99999
5.63	278.66	.00359	139.33	139.33	.99997	.44508	.14404	.14405	.99999
5.64	281.46	.00355	140.73	140.73	.99997	.44942	.14839	.14840	.99999
5.65	284.29	.00352	142.14	142.15	.99998	.45376	.15273	.15274	.99999
5.66	287.15	.00348	143.57	143.58	.99998	.45811	.15707	.15708	.99999
5.67	290.03	.00345	145.02	145.02	.99998	.46245	.16141	.16142	.99999
5.68	292.95	.00341	146.47	146.48	.99998	.46679	.16576	.16577	.99999
5.69	295.89	.00338	147.95	147.95	.99998	.47114	.17010	.17011	.99999
5.70	298.87	.00335	149.43	149.44	.99998	2.47548	2.17444	2.17445	$\bar{1}$.99999
5.71	301.87	.00331	150.93	150.94	.99998	.47982	.17879	.17880	.99999
5.72	304.90	.00328	152.45	152.45	.99998	.48416	.18313	.18314	.99999
5.73	307.97	.00325	153.98	153.99	.99998	.48851	.18747	.18748	.99999
5.74	311.06	.00321	155.53	155.53	.99998	.49285	.19182	.19182	.99999
5.75	314.19	.00318	157.09	157.10	.99998	.49719	.19616	.19617	.99999
5.76	317.35	.00315	158.67	158.68	.99998	.50154	.20050	.20051	.99999
5.77	320.54	.00312	160.27	160.27	.99998	.50588	.20484	.20485	.99999
5.78	323.76	.00309	161.88	161.88	.99998	.51022	.20919	.20920	.99999
5.79	327.01	.00306	163.51	163.51	.99998	.51457	.21353	.21354	.99999
5.80	330.30	.00303	165.15	165.15	.99998	2.51891	2.21787	2.21788	$\bar{1}$.99999
5.81	333.62	.00300	166.81	166.81	.99998	.52325	.22222	.22222	.99999
5.82	336.97	.00297	168.48	168.49	.99998	.52759	.22656	.22657	.99999
5.83	340.36	.00294	170.18	170.18	.99998	.53194	.23090	.23091	.99999
5.84	343.78	.00291	171.89	171.89	.99998	.53628	.23525	.23525	.99999
5.85	347.23	.00288	173.62	173.62	.99998	.54062	.23959	.23960	.99999
5.86	350.72	.00285	175.36	175.36	.99998	.54497	.24393	.24394	.99999
5.87	354.25	.00282	177.12	177.13	.99998	.54931	.24828	.24828	.99999
5.88	357.81	.00279	178.90	178.91	.99998	.55365	.25262	.25262	.99999
5.89	361.41	.00277	180.70	180.70	.99998	.55799	.25696	.25697	.99999
5.90	365.04	.00274	182.52	182.52	.99998	2.56234	2.26130	2.26131	$\bar{1}$.99999
5.91	368.71	.00271	184.35	184.35	.99999	.56668	.26565	.26565	.99999
5.92	372.41	.00269	186.20	186.21	.99999	.57102	.26999	.27000	.99999
5.93	376.15	.00266	188.08	188.08	.99999	.57537	.27433	.27434	.99999
5.94	379.93	.00263	189.97	189.97	.99999	.57971	.27868	.27868	.99999
5.95	383.75	.00261	191.88	191.88	.99999	.58405	.28302	.28303	.99999
5.96	387.61	.00258	193.80	193.81	.99999	.58840	.28736	.28737	.99999
5.97	391.51	.00255	195.75	195.75	.99999	.59274	.29171	.29171	.99999
5.98	395.44	.00253	197.72	197.72	.99999	.59708	.29605	.29605	.99999
5.99	399.41	.00250	199.71	199.71	.99999	.60142	.30039	.30040	.99999
6.00	403.43	.00248	201.71	201.72	.99999	2.60577	2.30473	2.30474	$\bar{1}$.99999

Powers, Roots, Reciprocals and Circles

Number, N		N^2	N^3	$\sqrt{N}$	$\sqrt[3]{N}$	$N^{3/2}$	$\sqrt[5]{N}$	$\frac{1}{N}$	Circle ($N = D$)	
Fraction	Decimal								Circum.	Area
1/64	.015625	0.000244	$.381 \times 10^{-5}$	.1250	.2500	.00195	.4353	64.0	.04909	.00019
1/32	.03125	.000977	$.305 \times 10^{-4}$	.1768	.3150	.00552	.5000	32.0	.09818	.00077
3/64	.046875	.002197	$.103 \times 10^{-3}$	.2165	.3606	.01015	.5422	18.8235	.14726	.00173
1/16	.0625	.003906	$.244 \times 10^{-3}$	.2500	.3969	.01563	.5744	16.0	.19635	.00307
5/64	.078125	.006104	$.477 \times 10^{-3}$	.2795	.4275	.02184	.6006	12.80	.24544	.00479
3/32	.09375	.008789	$.824 \times 10^{-3}$	.3062	.4543	.02871	.6229	10.6667	.29452	.00690
	.10	.010	.00100	.3162	.4642	.03162	.6310	10.0	.31416	.00785
7/64	.109375	.01196	.001308	.3307	.4782	.03617	.6424	9.1429	.34361	.00939
1/8	.125	.01563	.001953	.3536	.5000	.04419	.6598	8.0	.39270	.01227
9/64	.140625	.01978	.002782	.3750	.5200	.05273	.6755	7.1111	.44179	.01554
5/32	.15625	.02441	.003814	.3953	.5386	.06176	.6899	6.40	.49087	.01917
11/64	.171875	.02954	.005077	.4146	.5560	.07126	.7031	5.8182	.53996	.02320
3/16	.1875	.03516	.006592	.4330	.5724	.08119	.7155	5.3333	.58905	.02761
	.20	.040	.0080	.4472	.5848	.08944	.7248	5.0	.62832	.03142
13/64	.203125	.04126	.008381	.4507	.5878	.09155	.7270	4.9231	.63814	.03241
7/32	.21875	.04785	.01047	.4677	.6025	.10231	.7379	4.5714	.68722	.03758
15/64	.234375	.05493	.01287	.4841	.6166	.11347	.7481	4.2667	.73631	.04314
1/4	.250	.0625	.01563	.5000	.6300	.12500	.7579	4.0	.78540	.04909
17/64	.265625	.07056	.01874	.5154	.6428	.13690	.7671	3.7647	.83448	.05542
9/32	.28125	.07910	.02225	.5303	.6552	.14916	.7759	3.5556	.88357	.06213
19/64	.296875	.08813	.02616	.5449	.6671	.16176	.7844	3.3684	.93266	.06922
	.30	.090	.0270	.5477	.6694	.16432	.7860	3.3333	.94248	.07069
5/16	.3125	.09766	.03052	.5590	.6786	.17469	.7925	3.2000	.98175	.07670
21/64	.328125	.10767	.03533	.5728	.6897	.18796	.8002	3.0476	1.0308	.08456
11/32	.34375	.11816	.04062	.5863	.7005	.20154	.8077	2.9091	1.0799	.09281
23/64	.359375	.12915	.04641	.5995	.7110	.21544	.8149	2.7826	1.1290	.10143
3/8	.375	.14063	.05273	.6124	.7211	.22964	.8219	2.6667	1.1781	.11045
25/64	.390625	.15259	.05961	.6250	.7310	.24414	.8286	2.5600	1.2272	.11984
	.40	.16	.0640	.6325	.7368	.25298	.8326	2.50	1.2566	.12566
13/32	.40625	.16504	.06705	.6374	.7406	.25894	.8351	2.4615	1.2763	.12962
27/64	.421875	.17798	.07508	.6495	.7500	.27402	.8415	2.3704	1.3254	.13979
7/16	.4375	.19141	.08374	.6614	.7592	.28938	.8476	2.2857	1.3744	.15033
29/64	.453125	.20532	.09304	.6732	.7681	.30502	.8536	2.2069	1.4235	.16126
15/32	.46875	.21973	.10300	.6847	.7768	.32093	.8594	2.1333	1.4726	.17257
31/64	.484375	.23462	.11364	.6960	.7854	.33711	.8650	2.0645	1.5217	.18427
1/2	.50	.2500	.12500	.7071	.7937	.35355	.8706	2.0	1.5708	.19635
33/64	.515625	.26587	.13709	.7181	.8019	.37025	.8759	1.9394	1.6199	.20881
17/32	.53125	.28223	.14993	.7289	.8099	.38721	.8812	1.8824	1.6690	.22166
35/64	.546875	.29907	.16355	.7395	.8178	.40442	.8863	1.8286	1.7181	.23489
9/16	.5625	.31641	.17798	.7500	.8255	.42188	.8913	1.7778	1.7671	.24850
37/64	.578125	.33423	.19323	.7604	.8331	.43957	.8962	1.7297	1.8162	.26250
19/32	.59375	.35254	.20932	.7706	.8405	.45751	.9010	1.6842	1.8653	.27688
	.60	.3600	.21600	.7746	.8434	.46476	.9029	1.6667	1.8850	.28274
39/64	.609375	.37134	.22628	.7806	.8478	.47569	.9057	1.6410	1.9144	.29165
5/8	.625	.39063	.24414	.7906	.8550	.49410	.9103	1.6000	1.9635	.30680
41/64	.640625	.41040	.26291	.8004	.8621	.51275	.9148	1.5610	2.0126	.32233
21/32	.65625	.43066	.28262	.8101	.8690	.53162	.9192	1.5238	2.0617	.33824
43/64	.671875	.45142	.30330	.8197	.8759	.55072	.9235	1.4884	2.1108	.35454
11/16	.6875	.47266	.32495	.8297	.8826	.57005	.9278	1.4545	2.1598	.37122
	.70	.4900	.34300	.8367	.8879	.58566	.9312	1.4286	2.1991	.38485
45/64	.703125	.49438	.34761	.8385	.8892	.58959	.9320	1.4222	2.2089	.38829
23/32	.71875	.51660	.37131	.8478	.8958	.60935	.9361	1.3913	2.2580	.40574
47/64	.734375	.53931	.39605	.8570	.9022	.62933	.9401	1.3617	2.3071	.42357
3/4	.750	.56250	.42188	.8660	.9086	.64952	.9441	1.3333	2.3562	.44179
49/64	.765625	.58618	.44879	.8750	.9148	.66992	.9480	1.3061	2.4053	.46038
25/32	.78125	.61035	.47684	.8839	.9210	.69053	.9518	1.2800	2.4544	.47937
51/64	.796875	.63501	.50602	.8927	.9271	.71135	.9556	1.2549	2.5035	.49874
	.80	.6400	.51200	.8944	.9283	.71554	.9564	1.2500	2.5133	.50265
13/16	.8125	.66016	.53638	.9014	.9331	.73238	.9593	1.2308	2.5525	.51849
53/64	.828125	.68579	.56792	.9100	.9391	.75361	.9630	1.2075	2.6016	.53862
27/32	.84375	.71191	.60067	.9186	.9449	.77503	.9666	1.1852	2.6507	.55917
55/64	.859375	.73853	.63467	.9270	.9507	.79666	.9702	1.1636	2.6998	.58002
7/8	.875	.76563	.66992	.9354	.9565	.81849	.9737	1.1429	2.7489	.60132
57/64	.890625	.79321	.70645	.9437	.9621	.84051	.9771	1.1228	2.7980	.62296
	.90	.81000	.72900	.9487	.9655	.85435	.9792	1.1111	2.8274	.63617
29/32	.90625	.82129	.74429	.9520	.9677	.86272	.9805	1.1034	2.8471	.64504
59/64	.921875	.84985	.78346	.9601	.9733	.88513	.9839	1.0847	2.8962	.66746
15/16	.9375	.87891	.82398	.9683	.9787	.90773	.9872	1.0667	2.9452	.69029
61/64	.953125	.90845	.86587	.9763	.9841	.93053	.9905	1.0492	2.9943	.71347
31/32	.96875	.93848	.90915	.9843	.9895	.95349	.9937	1.0323	3.0434	.73708
63/64	.984375	.96899	.95385	.9922	.9948	.97666	.9969	1.0159	3.0925	.76108

POWERS, ROOTS, RECIPROCALS, CIRCLES 321

N	N^2	N^3	$\sqrt{N}$	$\sqrt[3]{N}$	$N^{3/2}$	$\sqrt[5]{N}$	$\dfrac{1}{N}$	Circle ($N = D$)	
								Circum.	Area
1.	1.0000	1.0000	1.0000	1.0000	1.0000	1.0000	1.0000000	3.1416	.7854
1.125	1.2656	1.4238	1.0606	1.0400	1.1932	1.0238	.8888888	3.5343	.9940
1.25	1.5625	1.9531	1.1180	1.0772	1.3975	1.0456	.80000000	3.9270	1.2272
1.375	1.8906	2.5996	1.1726	1.1120	1.6123	1.0658	.72727272	4.3197	1.4849
1.5	2.25	3.3750	1.2247	1.1447	1.8371	1.0845	.66666666	4.7124	1.7671
1.625	2.6406	4.2910	1.2748	1.1757	2.0715	1.1020	.61538462	5.1051	2.0739
1.75	3.0625	5.3594	1.3229	1.2051	2.3150	1.1186	.57142857	5.4978	2.4053
1.875	3.5156	6.5918	1.3693	1.2331	2.5675	1.1340	.53333333	5.8905	2.7612
2.	4.0000	8.0000	1.4142	1.2599	2.8284	1.1487	.50000000	6.2832	3.1416
2.125	4.5156	9.5957	1.4577	1.2856	3.0977	1.1627	.47058823	6.6759	3.5466
2.25	5.0625	11.3906	1.5000	1.3104	3.3750	1.1761	.44444444	7.0686	3.9761
2.375	5.6406	13.3965	1.5411	1.3342	3.6601	1.1889	.42105263	7.4613	4.4301
2.5	6.2500	15.6250	1.5811	1.3572	3.9529	1.2011	.40000000	7.8540	4.9087
2.625	6.8906	18.0879	1.6202	1.3795	4.2530	1.2129	.38095231	8.2467	5.4119
2.75	7.5625	20.7969	1.6583	1.4011	4.5604	1.2242	.36363636	8.6394	5.9396
2.875	8.2656	23.7637	1.6956	1.4219	4.8748	1.2352	.34782609	9.0321	6.4918
3.	9.000	27.0000	1.7321	1.4422	5.1962	1.2457	.33333333	9.4248	7.0686
3.125	9.7656	30.5176	1.7678	1.4620	5.5243	1.2559	.32000000	9.8175	7.6699
3.25	10.5625	34.3281	1.8028	1.4813	5.8590	1.2658	.30769231	10.2102	8.2958
3.375	11.3906	38.4434	1.8371	1.5000	6.2003	1.2754	.29629629	10.6029	8.9462
3.5	12.2500	42.8750	1.8708	1.5183	6.5479	1.2847	.28571429	10.9956	9.6211
3.625	13.1406	47.6348	1.9039	1.5362	6.9018	1.2938	.27586207	11.3883	10.3206
3.75	14.0625	52.7344	1.9365	1.5536	7.2619	1.3026	.26666666	11.7810	11.0447
3.875	15.0156	58.1856	1.9685	1.5707	7.6279	1.3112	.25806452	12.1737	11.7932
4.	16.0000	64.0000	2.0000	1.5874	8.0000	1.3195	.25000000	12.5664	12.5664
4.125	17.0156	70.1895	2.0310	1.6038	8.3779	1.3277	.24242424	12.9591	13.3640
4.25	18.0625	76.7656	2.0616	1.6198	8.7616	1.3356	.23529412	13.3518	14.1863
4.375	19.1406	83.7402	2.0916	1.6355	9.1510	1.3434	.22857143	13.7445	15.0330
4.5	20.2500	91.1250	2.1213	1.6510	9.5460	1.3510	.22222222	14.1372	15.9043
4.625	21.3906	98.9317	2.1506	1.6661	9.9465	1.3584	.21621622	14.5299	16.8001
4.75	22.5625	107.1719	2.1795	1.6810	10.3524	1.3656	.21052632	14.9226	17.7205
4.875	23.7656	115.8574	2.2079	1.6956	10.7637	1.3728	.20512821	15.3153	18.6655
5.	25.0000	125.0000	2.2361	1.7100	11.1803	1.3799	.20000000	15.7080	19.6350
5.125	26.2656	134.6113	2.2638	1.7241	11.6022	1.3866	.19512195	16.1006	20.6289
5.25	27.5625	144.7031	2.2913	1.7380	12.0293	1.3933	.19047619	16.4933	21.6475
5.375	28.8906	155.2871	2.3184	1.7517	12.4614	1.3998	.18604651	16.8860	22.6906
5.5	30.2500	166.3750	2.3452	1.7652	12.8987	1.4063	.18181818	17.2787	23.7583
5.625	31.6406	177.9785	2.3727	1.7784	13.3409	1.4126	.17777777	17.6714	24.8505
5.75	33.0625	190.1094	2.3979	1.7915	13.7880	1.4188	.17391304	18.0641	25.9672
5.875	34.5156	202.7793	2.4238	1.8044	14.2400	1.4250	.17021277	18.4568	27.1085
6.	36.0000	216.0000	2.4495	1.8171	14.6969	1.4310	.16666666	18.8495	28.2743
6.125	37.5156	229.7832	2.4749	1.8297	15.1586	1.4369	.16326531	19.2422	29.4647
6.25	39.0625	244.1406	2.5000	1.8420	15.6250	1.4427	.16000000	19.6349	30.6796
6.375	40.6406	259.0840	2.5249	1.8542	16.0961	1.4484	.15686275	20.0276	31.9190
6.5	42.2500	274.6250	2.5495	1.8663	16.5718	1.4542	.15384615	20.4203	33.1831
6.625	43.8906	290.7754	2.5739	1.8781	17.0522	1.4596	.15094339	20.8130	34.4716
6.75	45.5625	307.5469	2.5981	1.8899	17.5370	1.4651	.14814815	21.2057	35.7847
6.875	47.2656	324.9512	2.6220	1.9015	18.0264	1.4705	.14545454	21.5984	37.1223
7.	49.0000	343.0000	2.6458	1.9129	18.5203	1.4758	.14285714	21.9911	38.4845
7.125	50.7656	361.7051	2.6693	1.9243	19.0186	1.4810	.14035088	22.3838	39.8712
7.25	52.5625	381.0781	2.6926	1.9354	19.5212	1.4862	.13793103	22.7765	41.2825
7.375	54.3906	401.1309	2.7157	1.9465	20.0283	1.4913	.13559322	23.1692	42.7183
7.5	56.2500	421.8750	2.7386	1.9574	20.5396	1.4963	.13333333	23.5619	44.1786
7.625	58.1406	443.3223	2.7613	1.9683	21.0552	1.5012	.13114754	23.9546	45.6635
7.75	60.0625	465.4844	2.7839	1.9789	21.5751	1.5061	.12903226	24.3473	47.1730
7.875	62.0156	488.3731	2.8063	1.9895	22.0992	1.5110	.12698413	24.7400	48.7069
8.	64.0000	512.0000	2.8284	2.0000	22.6274	1.5157	.12500000	25.1327	50.2655
8.125	66.0156	536.3770	2.8504	2.0104	23.1598	1.5204	.12307692	25.5254	51.8485
8.25	68.0625	561.5156	2.8723	2.0206	23.6963	1.5251	.12121212	25.9181	53.4562
8.375	70.1406	587.4278	2.8940	2.0308	24.2369	1.5297	.11940298	26.3108	55.0883
8.5	72.2500	614.1250	2.9155	2.0408	24.7816	1.5342	.11764706	26.7035	56.7450
8.625	74.3906	641.6192	2.9368	2.0508	25.3301	1.5387	.11594203	27.0962	58.4262
8.75	76.5625	669.9219	2.9580	2.0606	25.8828	1.5431	.11428571	27.4889	60.1320
8.875	78.7656	699.0450	2.9791	2.0704	26.4394	1.5475	.11267605	27.8816	61.8623
9.	81.0000	729.0000	3.0000	2.0801	27.0000	1.5518	.11111111	28.2743	63.6172
9.125	83.2656	759.7989	3.0207	2.0897	27.5645	1.5561	.10958904	28.6670	65.3966
9.25	85.5625	791.4531	3.0414	2.0992	28.1328	1.5604	.10810811	29.0597	67.2006
9.375	87.8906	823.9746	3.0619	2.1086	28.7050	1.5646	.10666666	29.4524	69.0291
9.5	90.2500	857.3750	3.0822	2.1179	29.2810	1.5687	.10526316	29.8451	70.8822
9.625	92.6406	891.6660	3.1024	2.1272	29.8608	1.5728	.10389610	30.2378	72.7597
9.75	95.0625	926.8594	3.1225	2.1363	30.4444	1.5769	.10256410	30.6305	74.6619
9.875	97.5156	962.9668	3.1425	2.1454	31.0317	1.5809	.10126582	31.0232	76.5886

MATHEMATICAL AND PHYSICAL TABLES

N	N^2	N^3	$\sqrt{N}$	$\sqrt[3]{N}$	$N^{3/2}$	$\sqrt[5]{N}$	$\dfrac{1}{N}$	Circle ($N = D$)	
								Circum.	Area
10	100	1000	3.1623	2.1544	31.623	1.5849	.10000000	31.4159	78.5398
11	121	1331	3.3166	2.2240	36.483	1.6154	.09090909	34.5575	95.0332
12	144	1728	3.4641	2.2894	41.569	1.6438	.08333333	37.6991	113.0973
13	169	2197	3.6056	2.3513	46.873	1.6703	.07692308	40.8407	132.7323
14	196	2744	3.7417	2.4101	52.384	1.6953	.07142857	43.9823	153.9380
15	225	3375	3.8730	2.4662	58.095	1.7188	.06666667	47.1239	176.7146
16	256	4096	4.0000	2.5198	64.000	1.7411	.06250000	50.2654	201.0619
17	289	4913	4.1231	2.5713	70.093	1.7623	.05882353	53.4070	226.9801
18	324	5832	4.2426	2.6207	76.367	1.7826	.05555556	56.5486	254.4690
19	361	6859	4.3589	2.6684	82.819	1.8020	.05263158	59.6902	283.5287
20	400	8000	4.4721	2.7144	89.442	1.8206	.05000000	62.8318	314.1593
21	441	9261	4.5826	2.7589	96.235	1.8384	.04761905	65.9734	346.3606
22	484	10648	4.6904	2.8020	103.19	1.8556	.04545455	69.1150	380.1327
23	529	12167	4.7958	2.8439	110.30	1.8722	.04347826	72.2566	415.4756
24	576	13824	4.8990	2.8845	117.58	1.8882	.04166667	75.3982	452.3893
25	625	15625	5.0000	2.9240	125.00	1.9037	.04000000	78.5398	490.8739
26	676	17576	5.0990	2.9625	132.57	1.9186	.03846154	81.6813	530.9292
27	729	19683	5.1962	3.0000	140.30	1.9332	.03703704	84.8229	572.5553
28	784	21952	5.2915	3.0366	148.16	1.9473	.03571429	87.9645	615.7522
29	841	24389	5.3852	3.0723	156.17	1.9610	.03448276	91.1061	660.5198
30	900	27000	5.4772	3.1072	164.32	1.9744	.03333333	94.2477	706.8583
31	961	29791	5.5678	3.1414	172.60	1.9873	.03225806	97.3893	754.7676
32	1024	32768	5.6569	3.1748	181.02	2.0000	.03125000	100.5309	804.2477
33	1089	35937	5.7446	3.2075	189.57	2.0123	.03030303	103.6725	855.2986
34	1156	39304	5.8310	3.2396	198.25	2.0244	.02941176	106.8141	907.9203
35	1225	42875	5.9161	3.2711	207.06	2.0362	.02857143	109.9557	962.1127
36	1296	46656	6.0000	3.3019	216.00	2.0477	.02777778	113.0972	1017.8760
37	1369	50653	6.0828	3.3322	225.06	2.0589	.02702703	116.2388	1075.2101
38	1444	54872	6.1644	3.3620	234.25	2.0699	.02631579	119.3804	1134.1149
39	1521	59319	6.2450	3.3912	243.56	2.0807	.02564103	122.5220	1194.5906
40	1600	64000	6.3246	3.4200	252.98	2.0913	.02500000	125.6636	1256.6371
41	1681	68921	6.4031	3.4482	262.53	2.1016	.02439024	128.8052	1320.2543
42	1764	74088	6.4807	3.4760	272.19	2.1118	.02380952	131.9468	1385.4424
43	1849	79507	6.5574	3.5034	281.97	2.1218	.02325581	135.0884	1452.2012
44	1936	85184	6.6332	3.5303	291.86	2.1315	.02272727	138.2300	1520.5308
45	2025	91125	6.7082	3.5569	301.87	2.1411	.02222222	141.3716	1590.4313
46	2116	97336	6.7823	3.5830	311.99	2.1506	.02173913	144.5131	1661.9025
47	2209	103823	6.8557	3.6088	322.22	2.1598	.02127660	147.6547	1734.9445
48	2304	110592	6.9282	3.6342	332.55	2.1689	.02083333	150.7963	1809.5574
49	2401	117649	7.0000	3.6593	343.00	2.1779	.02040816	153.9379	1885.7410
50	2500	125000	7.0711	3.6840	353.55	2.1867	.02000000	157.0795	1963.500
51	2601	132651	7.1414	3.7084	364.21	2.1954	.01960784	160.2211	2042.820
52	2704	140608	7.2111	3.7325	374.98	2.2039	.01923077	163.3627	2123.716
53	2809	148877	7.2801	3.7563	385.85	2.2124	.01886792	166.5043	2206.183
54	2916	157464	7.3485	3.7798	396.82	2.2206	.01851852	169.6459	2290.221
55	3025	166375	7.4162	3.8030	407.89	2.2288	.01818182	172.7875	2375.829
56	3136	175616	7.4833	3.8259	419.07	2.2369	.01785714	175.9290	2463.008
57	3249	185193	7.5498	3.8485	430.35	2.2448	.01754386	179.0706	2551.758
58	3364	195112	7.6158	3.8709	441.72	2.2526	.01724138	182.2122	2642.079
59	3481	205379	7.6811	3.8930	453.19	2.2603	.01694915	185.3538	2733.970
60	3600	216000	7.7460	3.9149	464.76	2.2679	.01666667	188.4954	2827.433
61	3721	226981	7.8102	3.9365	476.43	2.2755	.01639344	191.6370	2922.466
62	3844	238328	7.8740	3.9579	488.19	2.2829	.01612903	194.7786	3019.070
63	3969	250047	7.9373	3.9791	500.05	2.2902	.01587302	197.9202	3117.245
64	4096	262144	8.0000	4.0000	512.00	2.2974	.01562500	201.0618	3216.990
65	4225	274625	8.0623	4.0207	524.05	2.3045	.01538462	204.2034	3318.307
66	4356	287496	8.1240	4.0412	536.19	2.3116	.01515152	207.3449	3421.194
67	4489	300763	8.1854	4.0615	548.42	2.3186	.01492537	210.4865	3525.652
68	4624	314432	8.2462	4.0817	560.74	2.3254	.01470588	213.6281	3631.680
69	4761	328509	8.3066	4.1016	573.16	2.3322	.01449275	216.7697	3739.280
70	4900	343000	8.3666	4.1213	585.66	2.3389	.01428571	219.9113	3848.450
71	5041	357911	8.4261	4.1408	598.26	2.3456	.01408451	223.0529	3959.191
72	5184	373248	8.4853	4.1602	610.94	2.3522	.01388889	226.1945	4071.503
73	5329	389017	8.5440	4.1793	623.71	2.3587	.01369863	229.3361	4185.386
74	5476	405224	8.6023	4.1983	636.57	2.3651	.01351351	232.4777	4300.839
75	5625	421875	8.6603	4.2172	649.52	2.3714	.01333333	235.6193	4417.864
76	5776	438976	8.7178	4.2358	662.55	2.3777	.01315789	238.7608	4536.459
77	5929	456533	8.7750	4.2543	675.68	2.3840	.01298701	241.9024	4656.625
78	6084	474552	8.8318	4.2727	688.88	2.3901	.01282051	245.0440	4778.361
79	6241	493039	8.8882	4.2908	702.17	2.3962	.01265823	248.1856	4901.669

POWERS, ROOTS, RECIPROCALS, CIRCLES

N	N^2	N^3	$\sqrt{N}$	$\sqrt[3]{N}$	$N^{3/2}$	$\sqrt[5]{N}$	$\frac{1}{N}$	Circle ($N = D$) Circum.	Area
80	6400	512000	8.9443	4.3089	715.54	2.4022	.01250000	251.327	5026.547
81	6561	531441	9.0000	4.3267	729.00	2.4082	.01234568	254.469	5152.998
82	6724	551368	9.0554	4.3445	742.54	2.4141	.01219512	257.610	5281.016
83	6889	571787	9.1104	4.3621	756.17	2.4200	.01204819	260.752	5410.607
84	7056	592704	9.1652	4.3795	769.88	2.4258	.01190476	263.894	5541.770
85	7225	614125	9.2195	4.3968	783.66	2.4315	.01176471	267.035	5674.501
86	7396	636056	9.2736	4.4140	797.53	2.4372	.01162791	270.177	5808.805
87	7569	658503	9.3274	4.4310	811.49	2.4429	.01149425	273.318	5944.679
88	7744	681472	9.3808	4.4480	825.52	2.4485	.01136364	276.460	6082.124
89	7921	704969	9.4340	4.4647	839.63	2.4540	.01123596	279.602	6221.138
90	8100	729000	9.4868	4.4814	853.82	2.4595	.01111111	282.743	6361.725
91	8281	753571	9.5394	4.4979	868.09	2.4650	.01098901	285.885	6503.882
92	8464	778688	9.5917	4.5144	882.44	2.4705	.01086957	289.026	6647.610
93	8649	804357	9.6437	4.5307	896.86	2.4758	.01075269	292.168	6792.909
94	8836	830584	9.6954	4.5468	911.36	2.4810	.01063830	295.309	6939.778
95	9025	857375	9.7468	4.5629	925.95	2.4863	.01052632	298.451	7088.219
96	9216	884736	9.7980	4.5789	940.61	2.4915	.01041667	301.593	7238.230
97	9409	912673	9.8489	4.5947	955.34	2.4966	.01030928	304.734	7389.812
98	9604	941192	9.8995	4.6104	970.15	2.5018	.01020408	307.876	7542.962
99	9801	970299	9.9499	4.6261	985.04	2.5069	.01010101	311.017	7697.688
100	10000	1000000	10.0000	4.6416	1000.0	2.5119	.01000000	314.159	7853.982
101	10201	1030301	10.0499	4.6570	1015.0	2.5169	.00990099	317.301	8011.85
102	10404	1061208	10.0995	4.6723	1030.1	2.5219	.00980392	320.442	8171.28
103	10609	1092727	10.1489	4.6875	1045.3	2.5268	.00970874	323.584	8332.29
104	10816	1124864	10.1980	4.7027	1060.6	2.5317	.00961538	326.725	8494.87
105	11025	1157625	10.2470	4.7177	1075.9	2.5365	.00952381	329.867	8659.01
106	11236	1191016	10.2956	4.7326	1091.3	2.5413	.00943396	333.009	8824.73
107	11449	1225043	10.3441	4.7475	1106.8	2.5461	.00934579	336.150	8992.02
108	11664	1259712	10.3923	4.7622	1122.4	2.5509	.00925926	339.292	9160.88
109	11881	1295029	10.4403	4.7769	1138.0	2.5556	.00917431	342.433	9331.32
110	12100	1331000	10.4881	4.7914	1153.7	2.5602	.00909091	345.575	9503.32
111	12321	1367631	10.5357	4.8059	1169.5	2.5649	.00900901	348.716	9676.89
112	12544	1404928	10.5830	4.8203	1185.3	2.5695	.00892857	351.858	9852.03
113	12769	1442897	10.6301	4.8346	1201.2	2.5740	.00884956	355.000	10028.75
114	12996	1481544	10.6771	4.8488	1217.2	2.5786	.00877193	358.141	10207.03
115	13225	1520875	10.7238	4.8629	1233.2	2.5831	.00869565	361.283	10386.89
116	13456	1560896	10.7703	4.8770	1249.4	2.5876	.00862069	364.424	10568.32
117	13689	1601613	10.8167	4.8910	1265.5	2.5920	.00854701	367.566	10751.31
118	13924	1643032	10.8628	4.9049	1281.8	2.5964	.00847458	370.708	10935.88
119	14161	1685159	10.9087	4.9187	1298.1	2.6008	.00840336	373.849	11122.02
120	14400	1728000	10.9545	4.9324	1314.5	2.6052	.00833333	376.991	11309.73
121	14641	1771561	11.0000	4.9461	1331.0	2.6095	.00826446	380.132	11499.01
122	14884	1815848	11.0454	4.9597	1347.5	2.6138	.00819672	383.274	11689.86
123	15129	1860867	11.0905	4.9732	1364.1	2.6181	.00813008	386.416	11882.29
124	15376	1906624	11.1355	4.9866	1380.8	2.6223	.00806452	389.557	12076.28
125	15625	1953125	11.1803	5.0000	1397.5	2.6265	.00800000	392.699	12271.84
126	15876	2000376	11.2250	5.0133	1414.4	2.6307	.00793651	395.840	12468.98
127	16129	2048383	11.2694	5.0265	1431.2	2.6349	.00787402	398.982	12667.68
128	16384	2097152	11.3137	5.0397	1448.2	2.6390	.00781250	402.124	12867.96
129	16641	2146689	11.3578	5.0528	1465.2	2.6431	.00775194	405.265	13069.81
130	16900	2197000	11.4018	5.0658	1482.3	2.6472	.00769231	408.407	13273.23
131	17161	2248091	11.4455	5.0788	1499.4	2.6513	.00763359	411.548	13478.22
132	17424	2299968	11.4891	5.0916	1516.6	2.6553	.00757576	414.690	13684.77
133	17689	2352637	11.5326	5.1045	1533.8	2.6593	.00751880	417.831	13892.91
134	17956	2406104	11.5758	5.1172	1551.2	2.6633	.00746269	420.973	14102.61
135	18225	2460375	11.6190	5.1299	1568.6	2.6673	.00740741	424.115	14313.88
136	18496	2515456	11.6619	5.1426	1586.0	2.6712	.00735294	427.256	14526.72
137	18769	2571353	11.7047	5.1551	1603.6	2.6751	.00729927	430.398	14741.14
138	19044	2628072	11.7473	5.1676	1621.1	2.6790	.00724638	433.539	14957.12
139	19321	2685619	11.7898	5.1801	1638.8	2.6829	.00719424	436.681	15174.67
140	19600	2744000	11.8322	5.1925	1656.5	2.6867	.00714286	439.823	15393.80
141	19881	2803221	11.8743	5.2048	1674.3	2.6906	.00709220	442.964	15614.50
142	20164	2863288	11.9164	5.2171	1692.1	2.6944	.00704225	446.106	15836.77
143	20449	2924207	11.9583	5.2293	1710.0	2.6981	.00699301	449.247	16060.60
144	20736	2985984	12.0000	5.2415	1728.0	2.7019	.00694444	452.389	16286.01
145	21025	3048625	12.0416	5.2536	1746.0	2.7057	.00689655	455.531	16512.99
146	21316	3112136	12.0830	5.2656	1764.1	2.7094	.00684932	458.672	16741.54
147	21609	3176523	12.1244	5.2776	1782.2	2.7131	.00680272	461.814	16971.67
148	21904	3241792	12.1655	5.2896	1800.5	2.7168	.00675676	464.955	17203.36
149	22201	3307949	12.2066	5.3015	1818.8	2.7204	.00671141	468.097	17436.62

324   MATHEMATICAL AND PHYSICAL TABLES

N	N^2	N^3	$\sqrt{N}$	$\sqrt[3]{N}$	$N^{3/2}$	$\sqrt[5]{N}$	$\dfrac{1}{N}$	Circle ($N = D$)	
								Circum.	Area
150	22500	3375000	12.2474	5.3133	1837.1	2.7241	.00666667	471.239	17671.46
151	22801	3442951	12.2882	5.3251	1855.5	2.7277	.00662252	474.380	17907.86
152	23104	3511808	12.3288	5.3368	1874.0	2.7314	.00657895	477.522	18145.84
153	23409	3581577	12.3693	5.3485	1892.5	2.7349	.00653595	480.663	18385.38
154	23716	3652264	12.4097	5.3601	1911.1	2.7385	.00649351	483.805	18626.50
155	24025	3723875	12.4499	5.3717	1929.7	2.7420	.00645161	486.946	18869.19
156	24336	3796416	12.4900	5.3832	1948.4	2.7455	.00641026	490.088	19113.45
157	24649	3869893	12.5300	5.3947	1967.2	2.7490	.00636943	493.230	19359.28
158	24964	3944312	12.5698	5.4061	1986.0	2.7525	.00632911	496.371	19606.68
159	25281	4019679	12.6095	5.4175	2004.9	2.7560	.00628931	499.513	19855.65
160	25600	4096000	12.6491	5.4288	2023.9	2.7595	.00625000	502.654	20106.19
161	25921	4173281	12.6886	5.4401	2042.9	2.7629	.00621118	505.796	20358.30
162	26244	4251528	12.7279	5.4514	2061.9	2.7663	.00617284	508.938	20611.99
163	26569	4330747	12.7671	5.4626	2081.0	2.7697	.00613497	512.079	20867.24
164	26896	4410944	12.8062	5.4737	2100.2	2.7731	.00609756	515.221	21124.06
165	27225	4492125	12.8452	5.4848	2119.5	2.7765	.00606061	518.362	21382.46
166	27556	4574296	12.8841	5.4959	2138.8	2.7799	.00602410	521.504	21642.43
167	27889	4657463	12.9228	5.5069	2158.1	2.7832	.00598802	524.646	21903.96
168	28224	4741632	12.9615	5.5178	2177.5	2.7865	.00595238	527.787	22167.07
169	28561	4826809	13.0000	5.5288	2197.0	2.7898	.00591716	530.929	22431.75
170	28900	4913000	13.0384	5.5397	2216.5	2.7931	.00588235	534.070	22698.00
171	29241	5000211	13.0767	5.5505	2236.1	2.7964	.00584795	537.212	22965.82
172	29584	5088448	13.1149	5.5613	2255.8	2.7997	.00581395	540.353	23235.21
173	29929	5177717	13.1529	5.5721	2275.5	2.8029	.00578035	543.495	23506.18
174	30276	5268024	13.1909	5.5828	2295.2	2.8061	.00574713	546.637	23778.71
175	30625	5359375	13.2288	5.5934	2315.0	2.8094	.00571429	549.778	24052.81
176	30976	5451776	13.2665	5.6041	2334.9	2.8126	.00568182	552.920	24328.49
177	31329	5545233	13.3041	5.6147	2354.8	2.8158	.00564972	556.061	24605.73
178	31684	5639752	13.3417	5.6252	2374.8	2.8189	.00561798	559.203	24884.55
179	32041	5735339	13.3791	5.6357	2394.9	2.8221	.00558659	562.345	25164.94
180	32400	5832000	13.4164	5.6462	2415.0	2.8252	.00555556	565.486	25446.90
181	32761	5929741	13.4536	5.6567	2435.1	2.8284	.00552486	568.628	25730.42
182	33124	6028568	13.4907	5.6671	2455.3	2.8315	.00549451	571.769	26015.52
183	33489	6128487	13.5277	5.6774	2475.6	2.8346	.00546448	574.911	26302.19
184	33856	6229504	13.5647	5.6877	2495.9	2.8377	.00543478	578.053	26590.43
185	34225	6331625	13.6015	5.6980	2516.3	2.8408	.00540541	581.194	26880.25
186	34596	6434856	13.6382	5.7083	2536.7	2.8438	.00537634	584.336	27171.63
187	34969	6539203	13.6748	5.7185	2557.2	2.8469	.00534759	587.477	27464.58
188	35344	6644672	13.7113	5.7287	2577.7	2.8499	.00531915	590.619	27759.11
189	35721	6751269	13.7477	5.7388	2598.3	2.8529	.00529101	593.761	28055.20
190	36100	6859000	13.7840	5.7489	2619.0	2.8560	.00526316	596.902	28352.87
191	36481	6967871	13.8203	5.7590	2639.7	2.8590	.00523560	600.044	28652.10
192	36864	7077888	13.8564	5.7690	2660.4	2.8619	.00520833	603.185	28952.91
193	37249	7189057	13.8924	5.7790	2681.2	2.8649	.00518135	606.327	29255.29
194	37636	7301384	13.9284	5.7890	2702.1	2.8679	.00515464	609.468	29559.24
195	38025	7414875	13.9642	5.7989	2723.0	2.8708	.00512821	612.610	29864.76
196	38416	7529536	14.0000	5.8088	2744.0	2.8738	.00510204	615.752	30171.85
197	38809	7645373	14.0357	5.8186	2765.0	2.8767	.00507614	618.893	30480.51
198	39204	7762392	14.0712	5.8285	2786.1	2.8796	.00505051	622.035	30790.74
199	39601	7880599	14.1067	5.8383	2807.2	2.8825	.00502513	625.176	31102.55
200	40000	8000000	14.1421	5.8480	2828.4	2.8854	.00500000	628.318	31415.93
201	40401	8120601	14.1774	5.8578	2849.7	2.8883	.00497512	631.460	31730.87
202	40804	8242408	14.2127	5.8675	2871.0	2.8911	.00495050	634.601	32047.39
203	41209	8365427	14.2478	5.8771	2892.3	2.8940	.00492611	637.743	32365.47
204	41616	8489664	14.2829	5.8868	2913.7	2.8968	.00490196	640.884	32685.13
205	42025	8615125	14.3178	5.8964	2935.2	2.8997	.00487805	644.026	33006.36
206	42436	8741816	14.3527	5.9059	2956.7	2.9025	.00485437	647.168	33329.16
207	42849	8869743	14.3875	5.9155	2978.2	2.9053	.00483092	650.309	33653.53
208	43264	8998912	14.4222	5.9250	2999.8	2.9081	.00480769	653.451	33979.47
209	43681	9129329	14.4568	5.9345	3021.5	2.9109	.00478469	656.592	34306.98
210	44100	9261000	14.4914	5.9439	3043.2	2.9137	.00476190	659.734	34636.06
211	44521	9393931	14.5258	5.9533	3065.0	2.9165	.00473934	662.875	34966.71
212	44944	9528128	14.5602	5.9627	3086.8	2.9192	.00471698	666.017	35298.94
213	45369	9663597	14.5945	5.9721	3108.7	2.9220	.00469484	669.159	35632.73
214	45796	9800344	14.6287	5.9814	3130.6	2.9247	.00467290	672.300	35968.09
215	46225	9938375	14.6629	5.9907	3152.5	2.9274	.00465116	675.442	36305.03
216	46656	10077696	14.6969	6.0000	3174.5	2.9302	.00462963	678.583	36643.54
217	47089	10218313	14.7309	6.0092	3196.6	2.9329	.00460829	681.725	36983.61
218	47524	10360232	14.7648	6.0185	3218.7	2.9356	.00458716	684.867	37325.26
219	47961	10503459	14.7986	6.0277	3240.9	2.9383	.00456621	688.008	37668.48

POWERS, ROOTS, RECIPROCALS, CIRCLES   325

N	N^2	N^3	$\sqrt{N}$	$\sqrt[3]{N}$	$N^{3/2}$	$\sqrt[5]{N}$	$\frac{1}{N}$	Circle ($N = D$)	
								Circum.	Area
220	48400	10648000	14.8324	6.0368	3263.1	2.9409	.00454545	691.150	38013.27
221	48841	10793861	14.8661	6.0459	3285.4	2.9436	.00452489	694.291	38359.63
222	49284	10941048	14.8997	6.0550	3307.7	2.9463	.00450450	697.433	38707.56
223	49729	11089567	14.9332	6.0641	3330.1	2.9489	.00448430	700.575	39057.07
224	50176	11239424	14.9666	6.0732	3352.5	2.9516	.00446429	703.716	39408.14
225	50625	11390625	15.0000	6.0822	3375.0	2.9542	.00444444	706.858	39760.78
226	51076	11543176	15.0333	6.0912	3397.5	2.9568	.00442478	709.999	40115.00
227	51529	11697083	15.0665	6.1002	3420.1	2.9594	.00440529	713.141	40470.78
228	51984	11852352	15.0997	6.1091	3442.7	2.9620	.00438596	716.283	40828.14
229	52441	12008989	15.1327	6.1180	3465.4	2.9646	.00436681	719.424	41187.07
230	52900	12167000	15.1658	6.1269	3488.1	2.9672	.00434783	722.566	41547.56
231	53361	12326391	15.1987	6.1358	3510.9	2.9698	.00432900	725.707	41909.63
232	53824	12487168	15.2315	6.1446	3533.7	2.9723	.00431034	728.849	42273.27
233	54289	12649337	15.2643	6.1534	3556.6	2.9749	.00429185	731.990	42638.48
234	54756	12812904	15.2971	6.1622	3579.5	2.9774	.00427350	735.132	43005.26
235	55225	12977875	15.3297	6.1710	3602.5	2.9800	.00425532	738.274	43373.61
236	55696	13144256	15.3623	6.1797	3625.5	2.9825	.00423729	741.415	43743.54
237	56169	13312053	15.3948	6.1885	3648.6	2.9850	.00421941	744.557	44115.03
238	56644	13481272	15.4272	6.1972	3671.7	2.9875	.00420168	747.698	44488.09
239	57121	13651919	15.4596	6.2058	3694.8	2.9900	.00418410	750.840	44862.73
240	57600	13824000	15.4919	6.2145	3718.0	2.9925	.00416667	753.982	45238.93
241	58081	13997521	15.5242	6.2231	3741.3	2.9950	.00414938	757.123	45616.71
242	58564	14172488	15.5563	6.2317	3764.6	2.9975	.00413223	760.265	45996.06
243	59049	14348907	15.5885	6.2403	3788.0	3.0000	.00411523	763.406	46376.98
244	59536	14526784	15.6205	6.2488	3811.4	3.0025	.00409836	766.548	46759.47
245	60025	14706125	15.6525	6.2573	3834.9	3.0049	.00408163	769.690	47143.52
246	60516	14886936	15.6844	6.2658	3858.4	3.0074	.00406504	772.831	47529.16
247	61009	15069223	15.7162	6.2743	3881.9	3.0098	.00404858	775.973	47916.36
248	61504	15252992	15.7480	6.2828	3905.5	3.0122	.00403226	779.114	48305.13
249	62001	15438249	15.7797	6.2912	3929.2	3.0147	.00401606	782.256	48695.47
250	62500	15625000	15.8114	6.2996	3952.9	3.0171	.00400000	785.398	49087.39
251	63001	15813251	15.8430	6.3080	3976.6	3.0195	.00398406	788.539	49480.87
252	63504	16003008	15.8745	6.3164	4000.4	3.0219	.00396825	791.681	49875.92
253	64009	16194277	15.9060	6.3247	4024.2	3.0243	.00395257	794.822	50272.55
254	64516	16387064	15.9374	6.3330	4048.1	3.0267	.00393701	797.964	50670.75
255	65025	16581375	15.9687	6.3413	4072.0	3.0291	.00392157	801.105	51070.52
256	65536	16777216	16.0000	6.3496	4096.0	3.0314	.00390625	804.247	51471.85
257	66049	16974593	16.0312	6.3579	4120.0	3.0338	.00389105	807.389	51874.76
258	66564	17173512	16.0624	6.3661	4144.1	3.0362	.00387597	810.530	52279.24
259	67081	17373979	16.0935	6.3743	4168.2	3.0385	.00386100	813.672	52685.29
260	67600	17576000	16.1245	6.3825	4192.4	3.0418	.00384615	816.813	53092.92
261	68121	17779581	16.1555	6.3907	4216.6	3.0432	.00383142	819.955	53502.11
262	68644	17984728	16.1864	6.3988	4240.8	3.0455	.00381679	823.097	53912.87
263	69169	18191447	16.2173	6.4070	4265.1	3.0478	.00380228	826.238	54325.21
264	69696	18399744	16.2481	6.4151	4289.5	3.0501	.00378788	829.380	54739.11
265	70225	18609625	16.2788	6.4232	4313.9	3.0524	.00377358	832.521	55154.59
266	70756	18821096	16.3095	6.4312	4338.3	3.0547	.00375940	835.663	55571.63
267	71289	19034163	16.3401	6.4393	4362.8	3.0570	.00374532	838.805	55990.25
268	71824	19248832	16.3707	6.4473	4387.3	3.0593	.00373134	841.946	56410.44
269	72361	19465109	16.4012	6.4553	4411.9	3.0616	.00371747	845.088	56832.20
270	72900	19683000	16.4317	6.4633	4436.5	3.0639	.00370370	848.229	57255.53
271	73441	19902511	16.4621	6.4713	4461.2	3.0662	.00369004	851.371	57680.43
272	73984	20123648	16.4924	6.4792	4485.9	3.0684	.00367647	854.512	58106.90
273	74529	20346417	16.5227	6.4872	4510.7	3.0707	.00366300	857.654	58534.94
274	75076	20570824	16.5529	6.4951	4535.5	3.0729	.00364964	860.796	58964.55
275	75625	20796875	16.5831	6.5030	4560.4	3.0752	.00363636	863.937	59395.74
276	76176	21024576	16.6132	6.5108	4585.3	3.0774	.00362319	867.079	59828.49
277	76729	21253933	16.6433	6.5187	4610.2	3.0796	.00361011	870.220	60262.82
278	77284	21484952	16.6733	6.5265	4635.2	3.0818	.00359712	873.362	60698.71
279	77841	21717639	16.7033	6.5343	4660.2	3.0840	.00358423	876.504	61136.18
280	78400	21952000	16.7332	6.5421	4685.3	3.0863	.00357143	879.645	61575.22
281	78961	22188041	16.7631	6.5499	4710.4	3.0885	.00355872	882.787	62015.82
282	79524	22425768	16.7929	6.5577	4735.6	3.0907	.00354610	885.928	62458.00
283	80089	22665187	16.8226	6.5654	4760.8	3.0928	.00353357	889.070	62901.75
284	80656	22906304	16.8523	6.5731	4786.0	3.0950	.00352113	892.212	63347.07
285	81225	23149125	16.8819	6.5808	4811.3	3.0972	.00350877	895.353	63793.97
286	81796	23393656	16.9115	6.5885	4836.7	3.0994	.00349650	898.495	64242.43
287	82369	23639903	16.9411	6.5962	4862.1	3.1015	.00348432	901.636	64692.46
288	82944	23887872	16.9706	6.6039	4887.5	3.1037	.00347222	904.778	65144.07
289	83521	24137569	17.0000	6.6115	4913.0	3.1058	.00346021	907.920	65597.24

326 MATHEMATICAL AND PHYSICAL TABLES

N	N^2	N^3	$\sqrt{N}$	$\sqrt[3]{N}$	$N^{3/2}$	$\sqrt[5]{N}$	$\frac{1}{N}$	Circle ($N = D$)	
								Circum.	Area
290	84100	24389000	17.0294	6.6191	4938.5	3.1080	.00344828	911.061	66051.99
291	84681	24642171	17.0587	6.6267	4964.1	3.1101	.00343643	914.203	66508.30
292	85264	24897088	17.0880	6.6343	4989.7	3.1123	.00342466	917.344	66966.19
293	85849	25153757	17.1172	6.6419	5015.4	3.1144	.00341297	920.486	67425.65
294	86436	25412184	17.1464	6.6494	5041.1	3.1165	.00340136	923.627	67886.68
295	87025	25672375	17.1756	6.6569	5066.8	3.1186	.00338983	926.769	68349.28
296	87616	25934336	17.2047	6.6644	5092.6	3.1207	.00337838	929.911	68813.45
297	88209	26198073	17.2337	6.6719	5118.4	3.1228	.00336700	933.052	69279.19
298	88804	26463592	17.2627	6.6794	5144.3	3.1249	.00335570	936.194	69746.50
299	89401	26730899	17.2916	6.6869	5170.2	3.1270	.00334448	939.335	70215.38
300	90000	27000000	17.3205	6.6943	5196.2	3.1291	.00333333	942.477	70685.83
301	90601	27270901	17.3494	6.7018	5222.2	3.1312	.00332226	945.619	71157.86
302	91204	27543608	17.3781	6.7092	5248.2	3.1333	.00331126	948.760	71631.45
303	91809	27818127	17.4069	6.7166	5274.3	3.1354	.00330033	951.902	72106.62
304	92416	28094464	17.4356	6.7240	5300.4	3.1374	.00328947	955.043	72583.36
305	93025	28372625	17.4642	6.7313	5326.6	3.1395	.00327869	958.185	73061.66
306	93636	28652616	17.4929	6.7387	5352.8	3.1416	.00326797	961.327	73541.54
307	94249	28934443	17.5214	6.7460	5379.1	3.1436	.00325733	964.468	74022.99
308	94864	29218112	17.5499	6.7533	5405.4	3.1456	.00324675	967.610	74506.01
309	95481	29503629	17.5784	6.7606	5431.7	3.1477	.00323625	970.751	74990.60
310	96100	29791000	17.6068	6.7679	5458.1	3.1497	.00322581	973.893	75476.76
311	96721	30080231	17.6352	6.7752	5484.5	3.1518	.00321543	977.034	75964.50
312	97344	30371328	17.6635	6.7824	5511.0	3.1538	.00320513	980.176	76453.80
313	97969	30664297	17.6918	6.7897	5537.5	3.1558	.00319489	983.318	76944.67
314	98596	30959144	17.7200	6.7969	5564.1	3.1578	.00318471	986.459	77437.12
315	99225	31255875	17.7482	6.8041	5590.7	3.1598	.00317460	989.601	77931.13
316	99856	31554496	17.7764	6.8113	5617.3	3.1618	.00316456	992.742	78426.72
317	100489	31855013	17.8045	6.8185	5644.0	3.1638	.00315457	995.884	78923.88
318	101124	32157432	17.8326	6.8256	5670.7	3.1658	.00314465	999.026	79422.60
319	101761	32461759	17.8606	6.8328	5697.5	3.1678	.00313480	1002.167	79922.90
320	102400	32768000	17.8885	6.8399	5724.3	3.1698	.00312500	1005.309	80424.77
321	103041	33076161	17.9165	6.8470	5751.2	3.1718	.00311526	1008.450	80928.21
322	103684	33386248	17.9444	6.8541	5778.1	3.1737	.00310559	1011.592	81433.22
323	104329	33698267	17.9722	6.8612	5805.0	3.1757	.00309598	1014.734	81939.80
324	104976	34012224	18.0000	6.8683	5832.0	3.1777	.00308642	1017.875	82447.96
325	105625	34328125	18.0278	6.8753	5859.0	3.1796	.00307692	1021.017	82957.68
326	106276	34645976	18.0555	6.8824	5886.1	3.1816	.00306748	1024.158	83468.97
327	106929	34965783	18.0831	6.8894	5913.2	3.1835	.00305810	1027.300	83981.84
328	107584	35287552	18.1108	6.8964	5940.3	3.1855	.00304878	1030.442	84496.28
329	108241	35611289	18.1384	6.9034	5967.5	3.1874	.00303951	1033.583	85012.28
330	108900	35937000	18.1659	6.9104	5994.7	3.1894	.00303030	1036.725	85529.86
331	109561	36264691	18.1934	6.9174	6022.0	3.1913	.00302115	1039.866	86049.01
332	110224	36594368	18.2209	6.9244	6049.3	3.1932	.00301205	1043.008	86569.73
333	110889	36926037	18.2483	6.9313	6076.7	3.1951	.00300300	1046.149	87092.02
334	111556	37259704	18.2757	6.9382	6104.1	3.1970	.00299401	1049.291	87615.88
335	112225	37595375	18.3030	6.9451	6131.5	3.1989	.00298507	1052.433	88141.31
336	112896	37933056	18.3303	6.9521	6159.0	3.2009	.00297619	1055.574	88668.31
337	113569	38272753	18.3576	6.9589	6186.5	3.2028	.00296736	1058.716	89196.88
338	114244	38614472	18.3848	6.9658	6214.1	3.2047	.00295858	1061.857	89727.03
339	114921	38958219	18.4120	6.9727	6241.7	3.2066	.00294985	1064.999	90258.74
340	115600	39304000	18.4391	6.9795	6269.3	3.2085	.00294118	1068.141	90792.03
341	116281	39651821	18.4662	6.9864	6297.0	3.2103	.00293255	1071.282	91326.88
342	116964	40001688	18.4932	6.9932	6324.7	3.2122	.00292398	1074.424	91863.31
343	117649	40353607	18.5203	7.0000	6352.4	3.2141	.00291545	1077.565	92401.31
344	118336	40707584	18.5472	7.0068	6380.2	3.2160	.00290698	1080.707	92940.88
345	119025	41063625	18.5742	7.0136	6408.1	3.2178	.00289855	1083.849	93482.02
346	119716	41421736	18.6011	7.0203	6436.0	3.2197	.00289017	1086.990	94024.73
347	120409	41781923	18.6279	7.0271	6463.9	3.2216	.00288184	1090.132	94569.01
348	121104	42144192	18.6548	7.0338	6491.9	3.2234	.00287356	1093.273	95114.86
349	121801	42508549	18.6815	7.0406	6519.9	3.2253	.00286533	1096.415	95662.28
350	122500	42875000	18.7083	7.0473	6547.9	3.2271	.00285714	1099.557	96211.28
351	123201	43243551	18.7350	7.0540	6576.0	3.2289	.00284900	1102.698	96761.84
352	123904	43614208	18.7617	7.0607	6604.1	3.2308	.00284091	1105.840	97313.97
353	124609	43986977	18.7883	7.0674	6632.3	3.2326	.00283286	1108.981	97867.68
354	125316	44361864	18.8149	7.0740	6660.5	3.2345	.00282486	1112.123	98422.96
355	126025	44738875	18.8414	7.0807	6688.7	3.2363	.00281690	1115.264	98979.80
356	126736	45118016	18.8680	7.0873	6717.0	3.2381	.00280899	1118.406	99538.22
357	127449	45499293	18.8944	7.0940	6745.3	3.2399	.00280112	1121.548	100098.21
358	128164	45882712	18.9209	7.1006	6773.7	3.2417	.00279330	1124.689	100659.77
359	128881	46268279	18.9473	7.1072	6802.1	3.2435	.00278552	1127.831	101222.90

POWERS, ROOTS, RECIPROCALS, CIRCLES

N	N^2	N^3	$\sqrt{N}$	$\sqrt[3]{N}$	$N^{3/2}$	$\sqrt[5]{N}$	$\dfrac{1}{N}$	Circle ($N = D$)	
								Circum.	Area
360	129600	46656000	18.9737	7.1138	6830.5	3.2453	.00277778	1130.972	101787.60
361	130321	47045881	19.0000	7.1204	6859.0	3.2471	.00277008	1134.114	102353.87
362	131044	47437928	19.0263	7.1269	6887.5	3.2489	.00276243	1137.256	102921.72
363	131769	47832147	19.0526	7.1335	6916.1	3.2507	.00275482	1140.397	103491.13
364	132496	48228544	19.0788	7.1400	6944.7	3.2525	.00274725	1143.539	104062.12
365	133225	48627125	19.1050	7.1466	6973.3	3.2543	.00273973	1146.680	104634.67
366	133956	49027896	19.1311	7.1531	7002.0	3.2561	.00273224	1149.822	105208.80
367	134689	49430863	19.1572	7.1596	7030.7	3.2579	.00272480	1152.964	105784.49
368	135424	49836032	19.1833	7.1661	7059.5	3.2597	.00271739	1156.105	106361.76
369	136161	50243409	19.2094	7.1726	7088.3	3.2614	.00271003	1159.247	106940.60
370	136900	50653000	19.2354	7.1791	7117.1	3.2632	.00270270	1162.388	107521.01
371	137641	51064811	19.2614	7.1855	7146.0	3.2650	.00269542	1165.530	108102.99
372	138384	51478848	19.2873	7.1920	7174.9	3.2668	.00268817	1168.671	108686.54
373	139129	51895117	19.3132	7.1984	7203.9	3.2685	.00268097	1171.813	109271.66
374	139876	52313624	19.3391	7.2048	7232.8	3.2702	.00267380	1174.955	109858.35
375	140625	52734375	19.3649	7.2112	7261.8	3.2719	.00266667	1178.096	110446.62
376	141376	53157376	19.3907	7.2177	7290.9	3.2737	.00265957	1181.238	111036.45
377	142129	53582633	19.4165	7.2240	7320.0	3.2754	.00265252	1184.379	111627.86
378	142884	54010152	19.4422	7.2304	7349.2	3.2772	.00264550	1187.521	112220.83
379	143641	54439939	19.4679	7.2368	7378.4	3.2789	.00263852	1190.663	112815.38
380	144400	54872000	19.4936	7.2432	7407.6	3.2807	.00263158	1193.804	113411.49
381	145161	55306341	19.5192	7.2495	7436.8	3.2824	.00262467	1196.946	114009.18
382	145924	55742968	19.5448	7.2558	7466.1	3.2841	.00261780	1200.087	114608.44
383	146689	56181887	19.5704	7.2622	7495.4	3.2858	.00261097	1203.229	115209.27
384	147456	56623104	19.5959	7.2685	7524.8	3.2875	.00260417	1206.371	115811.67
385	148225	57066625	19.6214	7.2748	7554.2	3.2892	.00259740	1209.512	116415.64
386	148996	57512456	19.6469	7.2811	7583.7	3.2909	.00259067	1212.654	117021.18
387	149769	57960603	19.6723	7.2874	7613.2	3.2926	.00258398	1215.795	117628.30
388	150544	58411072	19.6977	7.2936	7642.7	3.2943	.00257732	1218.937	118236.98
389	151321	58863869	19.7231	7.2999	7672.3	3.2960	.00257069	1222.079	118847.24
390	152100	59319000	19.7484	7.3061	7701.9	3.2977	.00256410	1225.220	119459.06
391	152881	59776471	19.7737	7.3124	7731.5	3.2994	.00255754	1228.362	120072.46
392	153664	60236288	19.7990	7.3186	7761.2	3.3011	.00255102	1231.503	120687.42
393	154449	60698457	19.8242	7.3248	7790.9	3.3028	.00254453	1234.645	121303.96
394	155236	61162984	19.8494	7.3310	7820.7	3.3045	.00253807	1237.786	121922.07
395	156025	61629875	19.8746	7.3372	7850.5	3.3061	.00253165	1240.928	122541.75
396	156816	62099136	19.8997	7.3434	7880.3	3.3078	.00252525	1244.070	123163.00
397	157609	62570773	19.9249	7.3496	7910.2	3.3095	.00251889	1247.211	123785.82
398	158404	63044792	19.9499	7.3558	7940.1	3.3111	.00251256	1250.353	124410.21
399	159201	63521199	19.9750	7.3619	7970.0	3.3128	.00250627	1253.494	125036.17
400	160000	64000000	20.0000	7.3681	8000.0	3.3145	.00250000	1256.636	125663.71
401	160801	64481201	20.0250	7.3742	8030.0	3.3161	.00249377	1259.778	126292.81
402	161604	64964808	20.0499	7.3803	8061.1	3.3178	.00248756	1262.919	126923.48
403	162409	65450827	20.0749	7.3864	8090.2	3.3194	.00248139	1266.061	127555.73
404	163216	65939264	20.0998	7.3925	8120.3	3.3211	.00247525	1269.202	128189.55
405	164025	66430125	20.1246	7.3986	8150.5	3.3227	.00246914	1272.344	128824.93
406	164836	66923416	20.1494	7.4047	8180.7	3.3243	.00246305	1275.486	129461.89
407	165649	67419143	20.1742	7.4108	8210.9	3.3260	.00245700	1278.627	130100.42
408	166464	67917312	20.1990	7.4169	8241.2	3.3276	.00245098	1281.769	130740.52
409	167281	68417929	20.2237	7.4229	8271.5	3.3292	.00244499	1284.910	131382.19
410	168100	68921000	20.2485	7.4290	8301.9	3.3308	.00243902	1288.052	132025.43
411	168921	69426531	20.2731	7.4350	8332.3	3.3325	.00243309	1291.193	132670.24
412	169744	69934528	20.2978	7.4410	8362.7	3.3341	.00242718	1294.335	133316.63
413	170569	70444997	20.3224	7.4470	8393.2	3.3357	.00242131	1297.477	133964.58
414	171396	70957944	20.3470	7.4530	8423.7	3.3373	.00241546	1300.618	134614.10
415	172225	71473375	20.3715	7.4590	8454.2	3.3390	.00240964	1303.760	135265.20
416	173056	71991296	20.3961	7.4650	8484.8	3.3406	.00240385	1306.901	135917.86
417	173889	72511713	20.4206	7.4710	8515.4	3.3422	.00239808	1310.043	136572.10
418	174724	73034632	20.4450	7.4770	8546.0	3.3438	.00239234	1313.185	137227.91
419	175561	73560059	20.4695	7.4829	8576.7	3.3454	.00238663	1316.326	137885.29
420	176400	74088000	20.4939	7.4889	8607.4	3.3470	.00238095	1319.468	138544.24
421	177241	74618461	20.5183	7.4948	8638.2	3.3485	.00237530	1322.609	139204.76
422	178084	75151448	20.5426	7.5007	8669.0	3.3501	.00236967	1325.751	139866.85
423	178929	75686967	20.5670	7.5067	8699.8	3.3517	.00236407	1328.893	140530.51
424	179776	76225024	20.5913	7.5126	8730.7	3.3533	.00235849	1332.034	141195.74
425	180625	76765625	20.6155	7.5185	8761.6	3.3549	.00235294	1335.176	141862.54
426	181476	77308776	20.6398	7.5244	8792.5	3.3564	.00234742	1338.317	142530.92
427	182329	77854483	20.6640	7.5302	8823.5	3.3580	.00234192	1341.459	143200.86
428	183184	78402752	20.6882	7.5361	8854.5	3.3596	.00233645	1344.601	143872.38
429	184041	78953589	20.7123	7.5420	8885.6	3.3612	.00233100	1347.742	144545.46

328 MATHEMATICAL AND PHYSICAL TABLES

N	N^2	N^3	$\sqrt{N}$	$\sqrt[3]{N}$	$N^{3/2}$	$\sqrt[5]{N}$	$\frac{1}{N}$	Circle ($N = D$)	
								Circum.	Area
430	184900	79507000	20.7364	7.5478	8916.7	3.3627	.00232558	1350.884	145220.12
431	185761	80062991	20.7605	7.5537	8947.8	3.3643	.00232019	1354.025	145896.35
432	186624	80621568	20.7846	7.5595	8979.0	3.3659	.00231481	1357.167	146574.15
433	187489	81182737	20.8087	7.5654	9010.1	3.3674	.00230947	1360.308	147253.52
434	188356	81746504	20.8327	7.5712	9041.4	3.3690	.00230415	1363.450	147934.46
435	189225	82312875	20.8567	7.5770	9072.7	3.3705	.00229885	1366.592	148616.97
436	190096	82881856	20.8806	7.5828	9104.0	3.3720	.00229358	1369.733	149301.05
437	190969	83453453	20.9045	7.5886	9135.3	3.3736	.00228833	1372.875	149986.70
438	191844	84027672	20.9284	7.5944	9166.7	3.3752	.00228311	1376.016	150673.92
439	192721	84604519	20.9523	7.6001	9198.1	3.3767	.00227790	1379.158	151362.72
440	193600	85184000	20.9762	7.6059	9229.5	3.3783	.00227273	1382.300	152053.08
441	194481	85766121	21.0000	7.6117	9261.0	3.3798	.00226757	1385.441	152745.02
442	195364	86350888	21.0238	7.6174	9292.5	3.3813	.00226244	1388.583	153438.53
443	196249	86938307	21.0476	7.6232	9324.1	3.3828	.00225734	1391.724	154133.60
444	197136	87528384	21.0713	7.6289	9355.7	3.3844	.00225225	1394.866	154830.25
445	198025	88121125	21.0950	7.6346	9387.3	3.3859	.00224719	1398.008	155528.47
446	198916	88716536	21.1187	7.6403	9419.0	3.3874	.00224215	1401.149	156228.26
447	199809	89314623	21.1424	7.6460	9450.7	3.3889	.00223714	1404.291	156929.62
448	200704	89915392	21.1660	7.6517	9482.4	3.3904	.00223214	1407.432	157632.55
449	201601	90518849	21.1896	7.6574	9514.2	3.3919	.00222717	1410.574	158337.05
450	202500	91125000	21.2132	7.6631	9546.0	3.3935	.00222222	1413.716	159043.13
451	203401	91733851	21.2368	7.6688	9577.8	3.3950	.00221729	1416.857	159750.77
452	204304	92345408	21.2603	7.6744	9609.6	3.3965	.00221239	1419.999	160459.99
453	205209	92959677	21.2838	7.6801	9641.5	3.3980	.00220751	1423.140	161170.77
454	206116	93576664	21.3073	7.6857	9673.5	3.3995	.00220264	1426.282	161883.13
455	207025	94196375	21.3307	7.6914	9705.5	3.4010	.00219780	1429.423	162597.05
456	207936	94818816	21.3542	7.6970	9737.5	3.4025	.00219298	1432.565	163312.55
457	208849	95443993	21.3776	7.7026	9769.5	3.4039	.00218818	1435.707	164029.62
458	209764	96071912	21.4009	7.7082	9801.6	3.4054	.00218341	1438.848	164748.26
459	210681	96702579	21.4243	7.7138	9833.8	3.4069	.00217865	1441.990	165468.47
460	211600	97336000	21.4476	7.7194	9865.9	3.4084	.00217391	1445.131	166190.25
461	212521	97972181	21.4709	7.7250	9898.1	3.4099	.00216920	1448.273	166913.60
462	213444	98611128	21.4942	7.7306	9930.3	3.4113	.00216450	1451.415	167638.52
463	214369	99252847	21.5174	7.7362	9962.6	3.4128	.00215983	1454.556	168365.02
464	215296	99897344	21.5407	7.7418	9994.8	3.4143	.00215517	1457.698	169093.08
465	216225	100544625	21.5639	7.7473	10027.	3.4158	.00215054	1460.839	169822.72
466	217156	101194696	21.5870	7.7529	10060.	3.4173	.00214592	1463.981	170553.92
467	218089	101847563	21.6102	7.7584	10092.	3.4187	.00214133	1467.123	171286.70
468	219024	102503232	21.6333	7.7639	10124.	3.4202	.00213675	1470.264	172021.05
469	219961	103161709	21.6564	7.7695	10157.	3.4217	.00213220	1473.406	172756.96
470	220900	103823000	21.6795	7.7750	10189.	3.4231	.00212766	1476.547	173494.45
471	221841	104487111	21.7025	7.7805	10222.	3.4246	.00212314	1479.689	174233.51
472	222784	105154048	21.7256	7.7860	10255.	3.4260	.00211864	1482.830	174974.14
473	223729	105823817	21.7486	7.7915	10287.	3.4275	.00211416	1485.972	175716.34
474	224676	106496424	21.7715	7.7970	10320.	3.4289	.00210970	1489.114	176460.12
475	225625	107171875	21.7945	7.8025	10352.	3.4304	.00210526	1492.255	177205.46
476	226576	107850176	21.8174	7.8079	10385.	3.4318	.00210084	1495.397	177952.37
477	227529	108531333	21.8403	7.8134	10418.	3.4332	.00209644	1498.538	178700.86
478	228484	109215352	21.8632	7.8188	10450.	3.4347	.00209205	1501.680	179450.91
479	229441	109902239	21.8861	7.8243	10483.	3.4361	.00208768	1504.822	180202.54
480	230400	110592000	21.9089	7.8297	10516.	3.4375	.00208333	1507.963	180955.74
481	231361	111284641	21.9317	7.8352	10549.	3.4390	.00207900	1511.105	181710.50
482	232324	111980168	21.9545	7.8406	10582.	3.4404	.00207469	1514.246	182466.84
483	233289	112678587	21.9773	7.8460	10615.	3.4418	.00207039	1517.388	183224.75
484	234256	113379904	22.0000	7.8514	10648.	3.4433	.00206612	1520.530	183984.23
485	235225	114084125	22.0227	7.8568	10681.	3.4447	.00206186	1523.671	184745.28
486	236196	114791256	22.0454	7.8622	10714.	3.4461	.00205761	1526.813	185507.90
487	237169	115501303	22.0681	7.8676	10747.	3.4475	.00205339	1529.954	186272.10
488	238144	116214272	22.0907	7.8730	10780.	3.4489	.00204918	1533.096	187037.86
489	239121	116930169	22.1133	7.8784	10813.	3.4504	.00204499	1536.238	187805.19
490	240100	117649000	22.1359	7.8837	10847.	3.4518	.00204082	1539.379	188574.10
491	241081	118370771	22.1585	7.8891	10880.	3.4532	.00203666	1542.521	189344.57
492	242064	119095488	22.1811	7.8944	10913.	3.4546	.00203252	1545.662	190116.62
493	243049	119823157	22.2036	7.8998	10946.	3.4560	.00202840	1548.804	190890.24
494	244036	120553784	22.2261	7.9051	10980.	3.4574	.00202429	1551.945	191665.43
495	245025	121287375	22.2486	7.9105	11013.	3.4588	.00202020	1555.087	192442.18
496	246016	122023936	22.2711	7.9158	11046.	3.4602	.00201613	1558.229	193220.51
497	247009	122763473	22.2935	7.9211	11080.	3.4616	.00201207	1561.370	194000.41
498	248004	123505992	22.3159	7.9264	11113.	3.4630	.00200803	1564.512	194781.89
499	249001	124251499	22.3383	7.9317	11147.	3.4643	.00200401	1567.653	195564.93

POWERS, ROOTS, RECIPROCALS, CIRCLES

N	N^2	N^3	$\sqrt{N}$	$\sqrt[3]{N}$	$N^{3/2}$	$\sqrt[5]{N}$	$\dfrac{1}{N}$	Circle ($N = D$) Circum.	Area
500	250000	125000000	22.3607	7.9370	11180	3.4657	.00200000	1570.795	196349.54
501	251001	125751501	22.3830	7.9423	11214	3.4671	.00199601	1573.937	197135.72
502	252004	126506008	22.4054	7.9476	11247	3.4685	.00199203	1577.078	197923.48
503	253009	127263527	22.4277	7.9528	11281	3.4699	.00198807	1580.220	198712.80
504	254016	128024064	22.4499	7.9581	11315	3.4713	.00198413	1583.361	199503.70
505	255025	128787625	22.4722	7.9634	11348	3.4726	.00198020	1586.503	200296.17
506	256036	129554216	22.4944	7.9686	11382	3.4740	.00197628	1589.645	201090.20
507	257049	130323843	22.5167	7.9739	11416	3.4754	.00197239	1592.786	201885.81
508	258064	131096512	22.5389	7.9791	11450	3.4768	.00196850	1595.928	202682.99
509	259081	131872229	22.5610	7.9843	11484	3.4781	.00196464	1599.069	203481.74
510	260100	132651000	22.5832	7.9896	11517	3.4795	.00196078	1602.211	204282.06
511	261121	133432831	22.6053	7.9948	11551	3.4808	.00195695	1605.352	205083.95
512	262144	134217728	22.6274	8.0000	11585	3.4822	.00195313	1608.494	205887.42
513	263169	135005697	22.6495	8.0052	11619	3.4836	.00194932	1611.636	206692.45
514	264196	135796744	22.6716	8.0104	11653	3.4849	.00194553	1614.777	207499.05
515	265225	136590875	22.6936	8.0156	11687	3.4863	.00194175	1617.919	208307.23
516	266256	137388096	22.7156	8.0208	11721	3.4876	.00193798	1621.060	209116.97
517	267289	138188413	22.7376	8.0260	11755	3.4890	.00193424	1624.202	209928.29
518	268324	138991832	22.7596	8.0311	11789	3.4904	.00193050	1627.344	210741.18
519	269361	139798359	22.7816	8.0363	11824	3.4917	.00192678	1630.485	211555.63
520	270400	140608000	22.8035	8.0415	11858	3.4930	.00192308	1633.627	212371.66
521	271441	141420761	22.8254	8.0466	11892	3.4944	.00191939	1636.768	213189.26
522	272484	142236648	22.8473	8.0517	11926	3.4957	.00191571	1639.910	214008.43
523	273529	143055667	22.8692	8.0569	11960	3.4970	.00191205	1643.052	214829.17
524	274576	143877824	22.8910	8.0620	11995	3.4984	.00190840	1646.193	215651.49
525	275625	144703125	22.9129	8.0671	12029	3.4997	.00190476	1649.335	216475.37
526	276676	145531576	22.9347	8.0723	12064	3.5010	.00190114	1652.476	217300.82
527	277729	146363183	22.9565	8.0774	12098	3.5024	.00189753	1655.618	218127.85
528	278784	147197952	22.9783	8.0825	12133	3.5037	.00189394	1658.760	218956.44
529	279841	148035889	23.0000	8.0876	12167	3.5050	.00189036	1661.901	219786.61
530	280900	148877000	23.0217	8.0927	12202	3.5064	.00188679	1665.043	220618.34
531	281961	149721291	23.0434	8.0978	12236	3.5077	.00188324	1668.184	221451.65
532	283024	150568768	23.0651	8.1028	12271	3.5090	.00187970	1671.326	222286.53
533	284089	151419437	23.0868	8.1079	12305	3.5103	.00187617	1674.467	223122.98
534	285156	152273304	23.1084	8.1130	12340	3.5116	.00187266	1677.609	223961.00
535	286225	153130375	23.1301	8.1180	12375	3.5130	.00186916	1680.751	224800.59
536	287296	153990656	23.1517	8.1231	12410	3.5143	.00186567	1683.892	225641.75
537	288369	154854153	23.1733	8.1281	12444	3.5156	.00186220	1687.034	226484.48
538	289444	155720872	23.1948	8.1332	12479	3.5169	.00185874	1690.175	227328.79
539	290521	156590819	23.2164	8.1382	12514	3.5182	.00185529	1693.317	228174.66
540	291600	157464000	23.2379	8.1433	12549	3.5195	.00185185	1696.459	229022.10
541	292681	158340421	23.2594	8.1483	12583	3.5208	.00184843	1699.600	229871.12
542	293764	159220088	23.2809	8.1533	12618	3.5221	.00184502	1702.742	230721.71
543	294849	160103007	23.3024	8.1583	12653	3.5234	.00184162	1705.883	231573.86
544	295936	160989184	23.3238	8.1633	12688	3.5247	.00183824	1709.025	232427.59
545	297025	161878625	23.3452	8.1683	12723	3.5260	.00183486	1712.167	233282.89
546	298116	162771336	23.3666	8.1733	12758	3.5273	.00183150	1715.308	234139.76
547	299209	163667323	23.3880	8.1783	12793	3.5286	.00182815	1718.450	234998.20
548	300304	164566592	23.4094	8.1833	12828	3.5299	.00182482	1721.591	235858.21
549	301401	165469149	23.4307	8.1882	12863	3.5311	.00182149	1724.733	236719.79
550	302500	166375000	23.4521	8.1932	12899	3.5324	.00181818	1727.875	237582.94
551	303601	167284151	23.4734	8.1982	12934	3.5337	.00181488	1731.016	238447.67
552	304704	168196608	23.4947	8.2031	12969	3.5350	.00181159	1734.158	239313.96
553	305809	169112377	23.5160	8.2081	13004	3.5363	.00180832	1737.299	240181.83
554	306916	170031464	23.5372	8.2130	13040	3.5376	.00180505	1740.441	241051.26
555	308025	170953875	23.5584	8.2180	13075	3.5388	.00180180	1743.582	241922.27
556	309136	171879616	23.5797	8.2229	13110	3.5401	.00179856	1746.724	242794.85
557	310249	172808693	23.6008	8.2278	13146	3.5414	.00179533	1749.866	243668.99
558	311364	173741112	23.6220	8.2327	13181	3.5426	.00179211	1753.007	244544.71
559	312481	174676879	23.6432	8.2377	13217	3.5439	.00178891	1756.149	245422.00
560	313600	175616000	23.6643	8.2426	13252	3.5451	.00178571	1759.290	246300.86
561	314721	176558481	23.6854	8.2475	13288	3.5464	.00178253	1762.432	247181.30
562	315844	177504328	23.7065	8.2524	13323	3.5477	.00177936	1765.574	248063.30
563	316969	178453547	23.7276	8.2573	13359	3.5490	.00177620	1768.715	248946.87
564	318096	179406144	23.7487	8.2621	13394	3.5502	.00177305	1771.857	249832.01
565	319225	180362125	23.7697	8.2670	13430	3.5515	.00176991	1774.998	250718.73
566	320356	181321496	23.7908	8.2719	13466	3.5527	.00176678	1778.140	251607.01
567	321489	182284263	23.8118	8.2768	13501	3.5540	.00176367	1781.282	252496.87
568	322624	183250432	23.8328	8.2816	13537	3.5553	.00176056	1784.423	253388.30
569	323761	184220009	23.8537	8.2865	13573	3.5565	.00175747	1787.565	254281.29

330   MATHEMATICAL AND PHYSICAL TABLES

N	N^2	N^3	$\sqrt{N}$	$\sqrt[3]{N}$	$N^{3/2}$	$\sqrt[5]{N}$	$\frac{1}{N}$	Circle ($N = D$)	
								Circum.	Area
570	324900	185193000	23.8747	8.2913	13609	3.5577	.00175439	1790.706	255175.86
571	326041	186169411	23.8956	8.2962	13644	3.5590	.00175131	1793.848	256072.00
572	327184	187149248	23.9165	8.3010	13680	3.5602	.00174825	1796.989	256969.71
573	328329	188132517	23.9374	8.3059	13716	3.5615	.00174520	1800.131	257868.99
574	329476	189119224	23.9583	8.3107	13752	3.5627	.00174216	1803.273	258769.85
575	330625	190109375	23.9792	8.3155	13788	3.5640	.00173913	1806.414	259672.27
576	331776	191102976	24.0000	8.3203	13824	3.5652	.00173611	1809.556	260576.26
577	332929	192100033	24.0208	8.3251	13860	3.5664	.00173310	1812.697	261481.83
578	334084	193100552	24.0416	8.3300	13896	3.5677	.00173010	1815.839	262388.96
579	335241	194104539	24.0624	8.3348	13932	3.5689	.00172712	1818.981	263297.67
580	336400	195112000	24.0832	8.3396	13968	3.5702	.00172414	1822.122	264207.94
581	337561	196122941	24.1039	8.3443	14004	3.5714	.00172117	1825.264	265119.79
582	338724	197137368	24.1247	8.3491	14040	3.5726	.00171821	1828.405	266033.21
583	339889	198155287	24.1454	8.3539	14077	3.5738	.00171527	1831.547	266948.20
584	341056	199176704	24.1661	8.3587	14113	3.5751	.00171233	1834.689	267864.76
585	342225	200201625	24.1868	8.3634	14149	3.5763	.00170940	1837.830	268782.89
586	343396	201230056	24.2074	8.3682	14186	3.5775	.00170648	1840.972	269702.59
587	344569	202262003	24.2281	8.3730	14222	3.5787	.00170358	1844.113	270623.86
588	345744	203297472	24.2487	8.3777	14258	3.5799	.00170068	1847.255	271546.70
589	346921	204336469	24.2693	8.3825	14295	3.5812	.00169779	1850.397	272471.12
590	348100	205379000	24.2899	8.3872	14331	3.5824	.00169492	1853.538	273397.10
591	349281	206425071	24.3105	8.3919	14368	3.5836	.00169205	1856.680	274324.66
592	350464	207474688	24.3311	8.3967	14404	3.5848	.00168919	1859.821	275253.78
593	351649	208527857	24.3516	8.4014	14440	3.5860	.00168634	1862.963	276184.48
594	352836	209584584	24.3721	8.4061	14477	3.5872	.00168350	1866.104	277116.75
595	354025	210644875	24.3926	8.4108	14514	3.5884	.00168067	1869.246	278050.58
596	355216	211708736	24.4131	8.4155	14550	3.5896	.00167785	1872.388	278985.99
597	356409	212776173	24.4336	8.4202	14587	3.5908	.00167504	1875.529	279922.97
598	357604	213847192	24.4540	8.4249	14624	3.5920	.00167224	1878.671	280861.52
599	358801	214921799	24.4745	8.4296	14660	3.5932	.00166945	1881.812	281801.65
600	360000	216000000	24.4949	8.4343	14697	3.5944	.00166667	1884.954	282743.34
601	361201	217081801	24.5153	8.4390	14734	3.5956	.00166389	1888.096	283686.60
602	362404	218167208	24.5357	8.4437	14770	3.5958	.00166113	1891.237	284631.44
603	363609	219256227	24.5561	8.4484	14807	3.5980	.00165837	1894.379	285577.84
604	364816	220348864	24.5764	8.4530	14844	3.5992	.00165563	1897.520	286525.82
605	366025	221445125	24.5967	8.4577	14881	3.6004	.00165289	1900.662	287475.36
606	367236	222545016	24.6171	8.4623	14918	3.6016	.00165017	1903.804	288426.48
607	368449	223648543	24.6374	8.4670	14955	3.6028	.00164745	1906.945	289379.17
608	369664	224755712	24.6577	8.4716	14992	3.6040	.00164474	1910.087	290333.43
609	370881	225866529	24.6779	8.4763	15029	3.6052	.00164204	1913.228	291289.26
610	372100	226981000	24.6982	8.4809	15066	3.6063	.00163934	1916.370	292246.66
611	373321	228099131	24.7184	8.4856	15103	3.6075	.00163666	1919.511	293205.63
612	374544	229220928	24.7386	8.4902	15140	3.6087	.00163399	1922.653	294166.17
613	375769	230346397	24.7588	8.4948	15177	3.6099	.00163132	1925.795	295128.28
614	376996	231475544	24.7790	8.4994	15214	3.6111	.00162866	1928.936	296091.97
615	378225	232608375	24.7992	8.5040	15252	3.6122	.00162602	1932.078	297057.22
616	379456	233744896	24.8193	8.5086	15289	3.6134	.00162338	1935.219	298024.05
617	380689	234885113	24.8395	8.5132	15326	3.6146	.00162075	1938.361	298992.44
618	381924	236029032	24.8596	8.5178	15363	3.6158	.00161812	1941.503	299962.41
619	383161	237176659	24.8797	8.5224	15400	3.6169	.00161551	1944.644	300933.95
620	384400	238328000	24.8998	8.5270	15437	3.6181	.00161290	1947.786	301907.05
621	385641	239483061	24.9199	8.5316	15475	3.6192	.00161031	1950.927	302881.73
622	386884	240641848	24.9399	8.5362	15513	3.6204	.00160772	1954.069	303857.98
623	388129	241804367	24.9600	8.5408	15550	3.6216	.00160514	1957.211	304835.80
624	389376	242970624	24.9800	8.5453	15588	3.6227	.00160256	1960.352	305815.20
625	390625	244140625	25.0000	8.5499	15625	3.6239	.00160000	1963.494	306796.16
626	391876	245314376	25.0200	8.5544	15663	3.6250	.00159744	1966.635	307778.69
627	393129	246491883	25.0400	8.5590	15700	3.6262	.00159490	1969.777	308762.79
628	394384	247673152	25.0599	8.5635	15738	3.6274	.00159236	1972.919	309748.47
629	395641	248858189	25.0799	8.5681	15775	3.6285	.00158983	1976.060	310735.71
630	396900	250047000	25.0998	8.5726	15813	3.6297	.00158730	1979.202	311724.53
631	398161	251239591	25.1197	8.5772	15850	3.6309	.00158479	1982.343	312714.92
632	399424	252435968	25.1396	8.5817	15888	3.6320	.00158228	1985.485	313706.88
633	400689	253636137	25.1595	8.5862	15926	3.6331	.00157978	1988.626	314700.40
634	401956	254840104	25.1794	8.5907	15964	3.6343	.00157729	1991.768	315695.50
635	403225	256047875	25.1992	8.5952	16002	3.6354	.00157480	1994.910	316692.17
636	404496	257259456	25.2190	8.5997	16040	3.6366	.00157233	1998.051	317690.42
637	405769	258474853	25.2389	8.6043	16077	3.6377	.00156986	2001.193	318690.23
638	407044	259694072	25.2587	8.6088	16115	3.6389	.00156740	2004.334	319691.61
639	408321	260917119	25.2784	8.6132	16153	3.6400	.00156495	2007.476	320694.56

POWERS, ROOTS, RECIPROCALS, CIRCLES 331

N	N^2	N^3	$\sqrt{N}$	$\sqrt[3]{N}$	$N^{3/2}$	$\sqrt[5]{N}$	$\dfrac{1}{N}$	Circle ($N = D$)	
								Circum.	Area
640	409600	262144000	25.2982	8.6177	16191	3.6411	.00156250	2010.618	321699.09
641	410881	263374721	25.3180	8.6222	16229	3.6423	.00156006	2013.759	322705.18
642	412164	264609288	25.3377	8.6267	16267	3.6435	.00155763	2016.901	323712.85
643	413449	265847707	25.3574	8.6312	16305	3.6446	.00155521	2020.042	324722.09
644	414736	267089984	25.3772	8.6357	16343	3.6457	.00155280	2023.184	325732.89
645	416025	268336125	25.3969	8.6401	16381	3.6468	.00155039	2026.326	326745.27
646	417316	269586136	25.4165	8.6446	16419	3.6479	.00154799	2029.467	327759.22
647	418609	270840023	25.4362	8.6490	16457	3.6499	.00154560	2032.609	328774.74
648	419904	272097792	25.4558	8.6535	16495	3.6502	.00154321	2035.750	329791.83
649	421201	273359449	25.4755	8.6579	16534	3.6513	.00154083	2038.892	330810.49
650	422500	274625000	25.4951	8.6624	16572	3.6524	.00153846	2042.034	331830.72
651	423801	275894451	25.5147	8.6668	16610	3.6536	.00153610	2045.175	332852.53
652	425104	277167808	25.5343	8.6713	16648	3.6547	.00153374	2048.317	333875.90
653	426409	278445077	25.5539	8.6757	16687	3.6558	.00153139	2051.458	334900.85
654	427716	279726264	25.5734	8.6801	16725	3.6569	.00152905	2054.600	335927.36
655	429025	281011375	25.5930	8.6845	16764	3.6580	.00152672	2057.741	336955.45
656	430336	282300416	25.6125	8.6890	16802	3.6592	.00152439	2060.883	337985.10
657	431649	283593393	25.6320	8.6934	16840	3.6603	.00152207	2064.025	339016.33
658	432964	284890312	25.6515	8.6978	16879	3.6614	.00151976	2067.166	340049.13
659	434281	286191179	25.6710	8.7022	16917	3.6625	.00151745	2070.308	341083.50
660	435600	287496000	25.6905	8.7066	16956	3.6636	.00151515	2073.449	342119.44
661	436921	288804781	25.7099	8.7110	16994	3.6647	.00151286	2076.591	343156.95
662	438244	290117528	25.7294	8.7154	17033	3.6658	.00151057	2079.733	344196.03
663	439569	291434247	25.7488	8.7198	17071	3.6669	.00150830	2082.874	345236.69
664	440896	292754944	25.7682	8.7241	17110	3.6680	.00150602	2086.016	346278.91
665	442225	294079625	25.7876	8.7285	17149	3.6691	.00150376	2089.157	347322.70
666	443556	295408296	25.8070	8.7329	17187	3.6702	.00150150	2092.299	348368.07
667	444889	296740963	25.8263	8.7373	17226	3.6713	.00149925	2095.441	349415.00
668	446224	298077632	25.8457	8.7416	17265	3.6724	.00149701	2098.582	350463.51
669	447561	299418309	25.8650	8.7460	17304	3.6735	.00149477	2101.724	351513.59
670	448900	300763000	25.8844	8.7503	17343	3.6746	.00149254	2104.865	352565.24
671	450241	302111711	25.9037	8.7547	17381	3.6757	.00149031	2108.007	353618.45
672	451584	303464448	25.9230	8.7590	17420	3.6768	.00148810	2111.148	354673.24
673	452929	304821217	25.9422	8.7634	17459	3.6779	.00148588	2114.290	355729.60
674	454276	306182024	25.9615	8.7677	17498	3.6790	.00148368	2117.432	356787.54
675	455625	307546875	25.9808	8.7721	17537	3.6801	.00148148	2120.573	357847.04
676	456976	308915776	26.0000	8.7764	17576	3.6812	.00147929	2123.715	358908.11
677	458329	310288733	26.0192	8.7807	17615	3.6823	.00147710	2126.856	359970.75
678	459684	311665752	26.0384	8.7850	17654	3.6834	.00147493	2129.998	361034.97
679	461041	313046839	26.0576	8.7893	17693	3.6845	.00147275	2133.140	362100.75
680	462400	314432000	26.0768	8.7937	17732	3.6856	.00147059	2136.281	363168.11
681	463761	315821241	26.0960	8.7980	17771	3.6866	.00146843	2139.423	364237.04
682	465124	317214568	26.1151	8.8023	17810	3.6877	.00146628	2142.564	365307.54
683	466489	318611987	26.1343	8.8066	17850	3.6888	.00146413	2145.706	366379.60
684	467856	320013504	26.1534	8.8109	17889	3.6899	.00146199	2148.848	367453.24
685	469225	321419125	26.1725	8.8152	17928	3.6909	.00145985	2151.989	368528.45
686	470596	322828856	26.1916	8.8194	17967	3.6920	.00145773	2155.131	369605.23
687	471969	324242703	26.2107	8.8237	18007	3.6931	.00145560	2158.272	370683.59
688	473344	325660672	26.2298	8.8280	18046	3.6942	.00145349	2161.414	371763.51
689	474721	327082769	26.2488	8.8323	18085	3.6953	.00145138	2164.556	372845.00
690	476100	328509000	26.2679	8.8366	18125	3.6963	.00144928	2167.697	373928.07
691	477481	329939371	26.2869	8.8408	18164	3.6974	.00144718	2170.839	375012.70
692	478864	331373888	26.3059	8.8451	18204	3.6985	.00144509	2173.980	376098.91
693	480249	332812557	26.3249	8.8493	18243	3.6995	.00144300	2177.122	377186.68
694	481636	334255384	26.3439	8.8536	18283	3.7006	.00144092	2180.263	378276.03
695	483025	335702375	26.3629	8.8578	18322	3.7016	.00143885	2183.405	379366.95
696	484416	337153536	26.3818	8.8621	18362	3.7027	.00143678	2186.547	380459.44
697	485809	338608873	26.4008	8.8663	18401	3.7038	.00143472	2189.688	381553.50
698	487204	340068392	26.4197	8.8706	18441	3.7049	.00143266	2192.830	382649.13
699	488601	341532099	26.4386	8.8748	18480	3.7059	.00143062	2195.971	383746.33
700	490000	343000000	26.4575	8.8790	18520	3.7070	.00142857	2199.113	384845.10
701	491401	344472101	26.4764	8.8833	18560	3.7080	.00142653	2202.255	385945.44
702	492804	345948408	26.4953	8.8875	18600	3.7091	.00142450	2205.396	387047.36
703	494209	347428927	26.5141	8.8917	18640	3.7101	.00142248	2208.538	388150.84
704	495616	348913664	26.5330	8.8959	18679	3.7112	.00142045	2211.679	389255.90
705	497025	350402625	26.5518	8.9001	18719	3.7123	.00141844	2214.821	390362.52
706	498436	351895816	26.5707	8.9043	18759	3.7133	.00141643	2217.963	391470.72
707	499849	353393243	26.5895	8.9085	18799	3.7144	.00141443	2221.104	392580.49
708	501264	354894912	26.6083	8.9127	18839	3.7154	.00141243	2224.246	393691.82
709	502681	356400829	26.6271	8.9169	18879	3.7165	.00141044	2227.387	394804.73

332 MATHEMATICAL AND PHYSICAL TABLES

N	N^2	N^3	$\sqrt{N}$	$\sqrt[3]{N}$	$N^{3/2}$	$\sqrt[5]{N}$	$\frac{1}{N}$	Circle ($N = D$)	
								Circum.	Area
710	504100	357911000	26.6458	8.9211	18919	3.7175	.00140845	2230.529	395919.21
711	505521	359425431	26.6646	8.9253	18959	3.7185	.00140647	2233.670	397035.26
712	506944	360944128	26.6833	8.9295	18999	3.7196	.00140449	2236.812	398152.89
713	508369	362467097	26.7021	8.9337	19039	3.7206	.00140252	2239.954	399272.08
714	509796	363994344	26.7208	8.9378	19079	3.7217	.00140056	2243.095	400392.84
715	511225	365525875	26.7395	8.9420	19119	3.7227	.00139860	2246.237	401515.18
716	512656	367061696	26.7582	8.9462	19159	3.7238	.00139665	2249.378	402639.08
717	514089	368601813	26.7769	8.9503	19199	3.7248	.00139470	2252.520	403764.56
718	515524	370146232	26.7955	8.9545	19239	3.7258	.00139276	2255.662	404891.60
719	516961	371694959	26.8142	8.9587	19280	3.7269	.00139082	2258.803	406020.22
720	518400	373248000	26.8328	8.9628	19320	3.7279	.00138889	2261.945	407150.41
721	519841	374805361	26.8514	8.9670	19360	3.7290	.00138696	2265.086	408282.17
722	521284	376367048	26.8701	8.9711	19400	3.7300	.00138504	2268.228	409415.50
723	522729	377933067	26.8887	8.9752	19440	3.7310	.00138313	2271.370	410550.40
724	524176	379503424	26.9072	8.9794	19481	3.7321	.00138122	2274.511	411686.87
725	525625	381078125	26.9258	8.9835	19521	3.7331	.00137931	2277.653	412824.91
726	527076	382657176	26.9444	8.9876	19562	3.7341	.00137741	2280.794	413964.52
727	528529	384240583	26.9629	8.9918	19602	3.7351	.00137552	2283.936	415105.71
728	529984	385828352	26.9815	8.9959	19643	3.7362	.00137363	2287.078	416248.46
729	531441	387420489	27.0000	9.0000	19683	3.7372	.00137174	2290.219	417392.79
730	532900	389017000	27.0185	9.0041	19724	3.7382	.00136986	2293.361	418538.68
731	534361	390617891	27.0370	9.0082	19764	3.7392	.00136799	2296.502	419686.15
732	535824	392223168	27.0555	9.0123	19805	3.7403	.00136612	2299.644	420835.19
733	537289	393832837	27.0740	9.0164	19845	3.7413	.00136426	2302.785	421985.79
734	538756	395446904	27.0924	9.0205	19886	3.7423	.00136240	2305.927	423137.97
735	540225	397065375	27.1109	9.0246	19927	3.7433	.00136054	2309.069	424291.72
736	541696	398688256	27.1293	9.0287	19967	3.7443	.00135870	2312.210	425447.04
737	543169	400315553	27.1477	9.0328	20008	3.7454	.00135685	2315.352	426603.94
738	544644	401947272	27.1662	9.0369	20049	3.7464	.00135501	2318.493	427762.40
739	546121	403583419	27.1846	9.0410	20090	3.7474	.00135318	2321.635	428922.43
740	547600	405224000	27.2029	9.0450	20130	3.7484	.00135135	2324.777	430084.03
741	549081	406869021	27.2213	9.0491	20171	3.7494	.00134953	2327.918	431247.21
742	550564	408518488	27.2397	9.0532	20212	3.7504	.00134771	2331.060	432411.95
743	552049	410172407	27.2580	9.0572	20253	3.7514	.00134590	2334.201	433578.27
744	553536	411830784	27.2764	9.0613	20294	3.7524	.00134409	2337.343	434746.16
745	555025	413493625	27.2947	9.0654	20335	3.7534	.00134228	2340.485	435915.62
746	556516	415160936	27.3130	9.0694	20376	3.7545	.00134048	2343.626	437086.64
747	558009	416832723	27.3313	9.0735	20417	3.7555	.00133869	2346.768	438259.24
748	559504	418508992	27.3496	9.0775	20458	3.7565	.00133690	2349.909	439433.41
749	561001	420189749	27.3679	9.0816	20499	3.7575	.00133511	2353.051	440609.16
750	562500	421875000	27.3861	9.0856	20540	3.7585	.00133333	2356.193	441786.47
751	564001	423564751	27.4044	9.0896	20581	3.7595	.00133156	2359.334	442965.35
752	565504	425259008	27.4226	9.0937	20622	3.7605	.00132979	2362.476	444145.80
753	567009	426957777	27.4408	9.0977	20663	3.7615	.00132802	2365.617	445327.83
754	568516	428661064	27.4591	9.1017	20704	3.7625	.00132626	2368.759	446511.42
755	570025	430368875	27.4773	9.1057	20745	3.7635	.00132450	2371.900	447696.59
756	571536	432081216	27.4955	9.1098	20787	3.7645	.00132275	2375.042	448883.32
757	573049	433798093	27.5136	9.1138	20828	3.7655	.00132100	2378.184	450071.63
758	574564	435519512	27.5318	9.1178	20869	3.7665	.00131926	2381.325	451261.51
759	576081	437245479	27.5500	9.1218	20910	3.7675	.00131752	2384.467	452452.96
760	577600	438976000	27.5681	9.1258	20952	3.7685	.00131579	2387.608	453645.98
761	579121	440711081	27.5862	9.1298	20993	3.7694	.00131406	2390.750	454840.57
762	580644	442450728	27.6043	9.1338	21035	3.7704	.00131234	2393.892	456036.73
763	582169	444194947	27.6225	9.1378	21076	3.7714	.00131062	2397.033	457234.46
764	583696	445943744	27.6405	9.1418	21117	3.7724	.00130890	2400.175	458433.77
765	585225	447697125	27.6586	9.1458	21159	3.7734	.00130719	2403.316	459634.64
766	586756	449455096	27.6767	9.1498	21200	3.7744	.00130548	2406.458	460837.08
767	588289	451217663	27.6948	9.1537	21242	3.7754	.00130378	2409.600	462041.10
768	589824	452984832	27.7128	9.1577	21283	3.7764	.00130208	2412.741	463246.69
769	591361	454756609	27.7308	9.1617	21325	3.7774	.00130039	2415.883	464453.84
770	592900	456533000	27.7489	9.1657	21367	3.7784	.00129870	2419.024	465662.57
771	594441	458314011	27.7669	9.1696	21408	3.7793	.00129702	2422.166	466872.87
772	595984	460099648	27.7849	9.1736	21450	3.7803	.00129534	2425.307	468084.74
773	597529	461889917	27.8029	9.1775	21492	3.7813	.00129366	2428.449	469298.18
774	599076	463684824	27.8209	9.1815	21533	3.7822	.00129199	2431.591	470513.19
775	600625	465484375	27.8388	9.1855	21575	3.7832	.00129032	2434.732	471729.77
776	602176	467288576	27.8568	9.1894	21617	3.7842	.00128866	2437.874	472947.92
777	603729	469097433	27.8747	9.1933	21658	3.7852	.00128700	2441.015	474167.65
778	605284	470910952	27.8927	9.1973	21700	3.7861	.00128535	2444.157	475388.94
779	606841	472729139	27.9106	9.2012	21742	3.7871	.00128370	2447.299	476611.81

POWERS, ROOTS, RECIPROCALS, CIRCLES 333

N	N^2	N^3	$\sqrt{N}$	$\sqrt[3]{N}$	$N^{3/2}$	$\sqrt[5]{N}$	$\dfrac{1}{N}$	Circle ($N = D$)	
								Circum.	Area
780	608400	474552000	27.9285	9.2052	21784	3.7881	.00128205	2450.440	477836.24
781	609961	476379541	27.9464	9.2091	21826	3.7890	.00128041	2453.582	479062.25
782	611524	478211768	27.9643	9.2130	21868	3.7900	.00127877	2456.723	480289.83
783	613209	480048687	27.9821	9.2170	21910	3.7910	.00127714	2459.865	481518.97
784	614656	481890304	28.0000	9.2209	21952	3.7920	.00127551	2463.007	482749.69
785	616225	483736625	28.0179	9.2248	21994	3.7929	.00127389	2466.148	483981.98
786	617796	485587656	28.0357	9.2287	22036	3.7939	.00127226	2469.290	485215.84
787	619369	487443403	28.0535	9.2326	22078	3.7949	.00127065	2472.431	486451.28
788	620944	489303872	28.0713	9.2365	22120	3.7959	.00126904	2475.573	487688.28
789	622521	491169069	28.0891	9.2404	22162	3.7969	.00126743	2478.715	488926.85
790	624100	493039000	28.1069	9.2443	22205	3.7978	.00126582	2481.856	490166.99
791	625681	494913671	28.1247	9.2482	22247	3.7987	.00126422	2484.998	491408.71
792	627264	496793088	28.1425	9.2521	22289	3.7997	.00126263	2488.139	492651.99
793	628849	498677257	28.1603	9.2560	22331	3.8006	.00126103	2491.281	493896.85
794	630436	500566184	28.1780	9.2599	22373	3.8016	.00125945	2494.422	495143.28
795	632025	502459875	28.1957	9.2638	22416	3.8025	.00125786	2497.564	496391.27
796	633616	504358336	28.2135	9.2677	22458	3.8035	.00125628	2500.706	497640.84
797	635209	506261573	28.2312	9.2716	22500	3.8044	.00125471	2503.847	498891.98
798	636804	508169592	28.2489	9.2754	22543	3.8054	.00125313	2506.989	500144.69
799	638401	510082399	28.2666	9.2793	22585	3.8064	.00125156	2510.130	501398.97
800	640000	512000000	28.2843	9.2832	22627	3.8073	.00125000	2513.272	502654.82
801	641601	513922401	28.3019	9.2870	22670	3.8083	.00124844	2516.414	503912.25
802	643204	515849608	28.3196	9.2909	22712	3.8092	.00124688	2519.555	505171.24
803	644809	517781627	28.3373	9.2948	22755	3.8102	.00124533	2522.697	506431.80
804	646416	519718464	28.3549	9.2986	22797	3.8111	.00124378	2525.838	507693.94
805	648025	521660125	28.3725	9.3025	22840	3.8121	.00124224	2528.980	508957.64
806	649636	523606616	28.3901	9.3063	22883	3.8130	.00124069	2532.122	510222.92
807	651249	525557943	28.4077	9.3102	22925	3.8139	.00123916	2535.263	511489.77
808	652864	527514112	28.4253	9.3140	22968	3.8149	.00123762	2538.405	512758.19
809	654481	529475129	28.4429	9.3179	23010	3.8158	.00123609	2541.546	514028.18
810	656100	531441000	28.4605	9.3217	23053	3.8168	.00123457	2544.688	515299.74
811	657721	533411731	28.4781	9.3255	23096	3.8177	.00123305	2547.829	516572.87
812	659344	535387328	28.4956	9.3294	23138	3.8186	.00123153	2550.971	517847.57
813	660969	537367797	28.5132	9.3332	23181	3.8196	.00123001	2554.113	519123.84
814	662596	539353144	28.5307	9.3370	23224	3.8205	.00122850	2557.254	520401.68
815	664225	541343375	28.5482	9.3408	23267	3.8215	.00122699	2560.396	521681.10
816	665856	543338496	28.5657	9.3447	23310	3.8224	.00122549	2563.537	522962.08
817	667489	545338513	28.5832	9.3485	23352	3.8234	.00122399	2566.679	524244.63
818	669124	547343432	28.6007	9.3523	23395	3.8243	.00122249	2569.821	525528.76
819	670761	549353259	28.6182	9.3561	23438	3.8252	.00122100	2572.962	526814.46
820	672400	551368000	28.6356	9.3599	23481	3.8262	.00121951	2576.104	528101.73
821	674041	553387661	28.6531	9.3637	23524	3.8271	.00121803	2579.245	529390.56
822	675684	555412248	28.6705	9.3675	23567	3.8280	.00121655	2582.387	530680.97
823	677329	557441767	28.6880	9.3713	23610	3.8290	.00121507	2585.529	531972.95
824	678976	559476224	28.7054	9.3751	23653	3.8299	.00121359	2588.670	533266.50
825	680625	561515625	28.7228	9.3789	23696	3.8308	.00121212	2591.812	534561.62
826	682276	563559976	28.7402	9.3827	23740	3.8317	.00121065	2594.953	535858.32
827	683929	565609283	28.7576	9.3865	23783	3.8327	.00120919	2598.095	537156.58
828	685584	567663552	28.7750	9.3902	23826	3.8336	.00120773	2601.237	538456.41
829	687241	569722789	28.7924	9.3940	23869	3.8345	.00120627	2604.378	539757.82
830	688900	571787000	28.8097	9.3978	23912	3.8355	.00120482	2607.520	541060.79
831	690561	573856191	28.8271	9.4016	23955	3.8364	.00120337	2610.661	542365.34
832	692224	575930368	28.8444	9.4053	23999	3.8373	.00120192	2613.803	543671.46
833	693889	578009537	28.8617	9.4091	24042	3.8382	.00120048	2616.944	544979.15
834	695556	580093704	28.8791	9.4129	24085	3.8391	.00119904	2620.086	546288.40
835	697225	582182875	28.8964	9.4166	24128	3.8401	.00119760	2623.228	547599.23
836	698896	584277056	28.9137	9.4204	24172	3.8410	.00119617	2626.369	548911.63
837	700569	586376253	28.9310	9.4241	24215	3.8419	.00119474	2629.511	550225.61
838	702244	588480472	28.9482	9.4279	24259	3.8428	.00119332	2632.652	551541.15
839	703921	590589719	28.9655	9.4316	24302	3.8437	.00119190	2635.794	552858.26
840	705600	592704000	28.9828	9.4354	24346	3.8446	.00119048	2638.936	554176.94
841	707281	594823321	29.0000	9.4391	24389	3.8456	.00118906	2642.077	555497.20
842	708964	596947688	29.0172	9.4429	24432	3.8465	.00118765	2645.219	556819.02
843	710649	599077107	29.0345	9.4466	24476	3.8474	.00118624	2648.360	558142.42
844	712336	601211584	29.0517	9.4503	24520	3.8483	.00118483	2651.502	559467.39
845	714025	603351125	29.0689	9.4541	24563	3.8492	.00118343	2654.644	560793.92
846	715716	605495736	29.0861	9.4578	24607	3.8501	.00118203	2657.785	562122.03
847	717409	607645423	29.1033	9.4615	24650	3.8510	.00118064	2660.927	563451.71
848	719104	609800192	29.1204	9.4652	24694	3.8519	.00117925	2664.068	564782.96
849	720801	611960049	29.1376	9.4690	24738	3.8528	.00117786	2667.210	566115.78

334 MATHEMATICAL AND PHYSICAL TABLES

N	N^2	N^3	$\sqrt{N}$	$\sqrt[3]{N}$	$N^{3/2}$	$\sqrt[5]{N}$	$\frac{1}{N}$	Circle ($N = D$) Circum.	Area
850	722500	614125000	29.1548	9.4727	24782	3.8558	.00117647	2670.352	567450.17
851	724201	616295051	29.1719	9.4764	24825	3.8547	.00117509	2673.493	568786.14
852	725904	618470208	29.1890	9.4801	24869	3.8556	.00117371	2676.635	570123.67
853	727609	620650477	29.2062	9.4838	24913	3.8565	.00117233	2679.776	571462.77
854	729316	622835864	29.2233	9.4875	24957	3.8574	.00117096	2682.918	572803.45
855	731025	625026375	29.2404	9.4912	25000	3.8582	.00116959	2686.059	574145.69
856	732736	627222016	29.2575	9.4949	25044	3.8592	.00116822	2689.201	575489.51
857	734449	629422793	29.2746	9.4986	25088	3.8601	.00116686	2692.343	576834.90
858	736164	631628712	29.2916	9.5023	25132	3.8610	.00116550	2695.484	578181.85
859	737881	633839779	29.3087	9.5060	25176	3.8619	.00116414	2698.626	579530.38
860	739600	636056000	29.3258	9.5097	25220	3.8628	.00116279	2701.767	580880.48
861	741321	638277381	29.3428	9.5134	25264	3.8637	.00116144	2704.909	582232.15
862	743044	640503928	29.3598	9.5171	25308	3.8646	.00116009	2708.051	583585.39
863	744769	642735647	29.3769	9.5207	25352	3.8655	.00115875	2711.192	584940.20
864	746496	644972544	29.3939	9.5244	25396	3.8664	.00115741	2714.334	586296.59
865	748225	647214625	29.4109	9.5281	25440	3.8673	.00115607	2717.475	587654.54
866	749956	649461896	29.4279	9.5317	25485	3.8682	.00115473	2720.617	589014.07
867	751689	651714363	29.4449	9.5354	25529	3.8691	.00115340	2723.759	590375.16
868	753424	653972032	29.4618	9.5391	25573	3.8700	.00115207	2726.900	591737.83
869	755161	656234909	29.4788	9.5427	25617	3.8708	.00115075	2730.042	593102.06
870	756900	658503000	29.4958	9.5464	25661	3.8717	.00114943	2733.183	594467.87
871	758641	660776311	29.5127	9.5501	25706	3.8726	.00114811	2736.325	595835.25
872	760384	663054848	29.5296	9.5537	25750	3.8735	.00114679	2739.466	597204.20
873	762129	665338617	29.5466	9.5574	25794	3.8744	.00114548	2742.608	598574.72
874	763876	667627624	29.5635	9.5610	25839	3.8753	.00114416	2745.750	599946.81
875	765625	669921875	29.5804	9.5647	25883	3.8762	.00114286	2748.891	601320.47
876	767376	672221376	29.5973	9.5683	25927	3.8771	.00114155	2752.033	602695.70
877	769129	674526133	29.6142	9.5719	25972	3.8780	.00114025	2755.174	604072.50
878	770884	676836152	29.6311	9.5756	26016	3.8789	.00113895	2758.316	605450.88
879	772641	679151439	29.6479	9.5792	26061	3.8797	.00113766	2761.458	606830.82
880	774400	681472000	29.6648	9.5828	26105	3.8806	.00113636	2764.599	608212.34
881	776161	683797841	29.6816	9.5865	26150	3.8815	.00113507	2767.741	609595.42
882	777924	686128968	29.6985	9.5901	26194	3.8823	.00113379	2770.882	610980.08
883	779689	688465387	29.7153	9.5937	26239	3.8832	.00113250	2774.024	612366.31
884	781456	690807104	29.7321	9.5973	26283	3.8841	.00113122	2777.166	613754.11
885	783225	693154125	29.7489	9.6010	26328	3.8850	.00112994	2780.307	615143.48
886	784996	695506456	29.7658	9.6046	26373	3.8859	.00112867	2783.449	616534.42
887	786769	697864103	29.7825	9.6082	26417	3.8868	.00112740	2786.590	617926.93
888	788544	700227072	29.7993	9.6118	26462	3.8877	.00112613	2789.732	619321.01
889	790321	702595369	29.8161	9.6154	26507	3.8885	.00112486	2792.874	620716.66
890	792100	704969000	29.8329	9.6190	26551	3.8894	.00112360	2796.015	622113.89
891	793881	707347971	29.8496	9.6225	26596	3.8902	.00112233	2799.157	623512.68
892	795664	709732288	29.8664	9.6262	26641	3.8911	.00112108	2802.298	624913.04
893	797449	712121957	29.8831	9.6298	26686	3.8920	.00111982	2805.440	626314.98
894	799236	714516984	29.8998	9.6334	26730	3.8929	.00111857	2808.581	627718.49
895	801025	716917375	29.9166	9.6370	26775	3.8937	.00111732	2811.723	629123.56
896	802816	719323136	29.9333	9.6406	26820	3.8946	.00111607	2814.865	630530.21
897	804609	721734273	29.9500	9.6442	26865	3.8955	.00111483	2818.006	631938.43
898	806404	724150792	29.9666	9.6477	26910	3.8963	.00111359	2821.148	633348.22
899	808201	726572699	29.9833	9.6513	26955	3.8972	.00111235	2824.289	634759.58
900	810000	729000000	30.0000	9.6549	27000	3.8981	.00111111	2827.431	636172.51
901	811801	731432701	30.0167	9.6585	27045	3.8989	.00110988	2830.573	637587.01
902	813604	733870808	30.0333	9.6620	27090	3.8998	.00110865	2833.714	639003.09
903	815409	736314327	30.0500	9.6656	27135	3.9007	.00110742	2836.856	640420.73
904	817216	738763264	30.0666	9.6692	27180	3.9015	.00110619	2839.997	641839.95
905	819025	741217625	30.0832	9.6727	27225	3.9024	.00110497	2843.139	643260.73
906	820836	743677416	30.0998	9.6763	27270	3.9032	.00110375	2846.281	644683.09
907	822649	746142643	30.1164	9.6799	27316	3.9041	.00110254	2849.422	646107.01
908	824464	748613312	30.1330	9.6834	27361	3.9050	.00110132	2852.564	647532.51
909	826281	751089429	30.1496	9.6870	27406	3.9059	.00110011	2855.705	648959.58
910	828100	753571000	30.1662	9.6905	27451	3.9067	.00109890	2858.847	650388.22
911	829921	756058031	30.1828	9.6941	27497	3.9076	.00109769	2861.988	651818.43
912	831744	758550528	30.1993	9.6976	27542	3.9084	.00109649	2865.130	653250.21
913	833569	761048497	30.2159	9.7012	27587	3.9093	.00109529	2868.272	654683.56
914	835396	763551944	30.2324	9.7047	27632	3.9101	.00109409	2871.413	656118.48
915	837225	766060875	30.2490	9.7082	27678	3.9110	.00109290	2874.555	657554.98
916	839056	768575296	30.2655	9.7118	27723	3.9118	.00109170	2877.696	658993.04
917	840889	771095213	30.2820	9.7153	27769	3.9127	.00109051	2880.838	660432.68
918	842724	773620632	30.2985	9.7188	27814	3.9135	.00108932	2883.980	661873.88
919	844561	776151559	30.3150	9.7224	27859	3.9144	.00108814	2887.121	663316.66

POWERS, ROOTS, RECIPROCALS, CIRCLES

N	N^2	N^3	$\sqrt{N}$	$\sqrt[3]{N}$	$N^{3/2}$	$\sqrt[5]{N}$	$\frac{1}{N}$	Circle ($N = D$)	
								Circum.	Area
920	846400	778688000	30.3315	9.7259	27905	3.9153	.00108696	2890.263	664761.01
921	848241	781229961	30.3480	9.7294	27950	3.9161	.00108578	2893.404	666206.92
922	850084	783777448	30.3645	9.7329	27996	3.9169	.00108460	2896.546	667654.41
923	851929	786330467	30.3809	9.7364	28042	3.9178	.00108342	2899.688	669103.47
924	853776	788889024	30.3974	9.7400	28087	3.9186	.00108225	2902.829	670554.10
925	855625	791453125	30.4138	9.7435	28133	3.9194	.00108108	2905.971	672006.30
926	857476	794022776	30.4302	9.7470	28179	3.9203	.00107991	2909.112	673460.08
927	859329	796597983	30.4467	9.7505	28224	3.9212	.00107875	2912.254	674915.42
928	861184	799178752	30.4631	9.7540	28270	3.9220	.00107759	2915.396	676372.33
929	863041	801765089	30.4795	9.7575	28315	3.9229	.00107643	2918.537	677830.82
930	864900	804357000	30.4959	9.7610	28361	3.9237	.00107527	2921.679	679290.87
931	866761	806954491	30.5123	9.7645	28407	3.9246	.00107411	2924.820	680752.50
932	868624	809557568	30.5287	9.7680	28453	3.9254	.00107296	2927.962	682215.69
933	870489	812166237	30.5450	9.7715	28499	3.9262	.00107181	2931.103	683680.46
934	872356	814780504	30.5614	9.7750	28544	3.9271	.00107066	2934.245	685146.80
935	874225	817400375	30.5778	9.7785	28590	3.9279	.00106952	2937.387	686614.71
936	876096	820025856	30.5941	9.7819	28636	3.9288	.00106838	2940.528	688084.19
937	877969	822656953	30.6105	9.7854	28682	3.9296	.00106724	2943.670	689555.24
938	879844	825293672	30.6268	9.7889	28728	3.9304	.00106610	2946.811	691027.86
939	881721	827936019	30.6431	9.7924	28774	3.9313	.00106496	2949.953	692502.05
940	883600	830584000	30.6594	9.7959	28820	3.9321	.00106383	2953.095	693977.82
941	885481	833237621	30.6757	9.7993	28866	3.9329	.00106270	2956.236	695455.15
942	887364	835896888	30.6920	9.8028	28912	3.9338	.00106157	2959.378	696934.06
943	889249	838561807	30.7083	9.8063	28958	3.9346	.00106045	2962.519	698414.53
944	891136	841232384	30.7246	9.8097	29004	3.9354	.00105932	2965.661	699896.58
945	893025	843908625	30.7409	9.8132	29050	3.9363	.00105820	2968.803	701380.19
946	894916	846590536	30.7571	9.8167	29096	3.9371	.00105708	2971.944	702865.38
947	896809	849278123	30.7734	9.8201	29142	3.9379	.00105597	2975.086	704352.14
948	898704	851971392	30.7896	9.8236	29189	3.9388	.00105485	2978.227	705840.47
949	900601	854670349	30.8058	9.8270	29235	3.9396	.00105374	2981.369	707330.37
950	902500	857375000	30.8221	9.8305	29281	3.9404	.00105263	2984.511	708821.84
951	904401	860085351	30.8383	9.8339	29327	3.9413	.00105152	2987.652	710314.88
952	906304	862801408	30.8545	9.8374	29374	3.9421	.00105042	2990.794	711809.50
953	908209	865523177	30.8707	9.8408	29420	3.9429	.00104932	2993.935	713305.68
954	910116	868250664	30.8869	9.8443	29466	3.9438	.00104822	2997.077	714803.43
955	912025	870983875	30.9031	9.8477	29513	3.9446	.00104712	3000.218	716302.76
956	913936	873722816	30.9192	9.8511	29559	3.9454	.00104603	3003.360	717803.66
957	915849	876467493	30.9354	9.8546	29605	3.9462	.00104493	3006.502	719306.12
958	917764	879217912	30.9516	9.8580	29652	3.9471	.00104384	3009.643	720810.16
959	919681	881974079	30.9677	9.8614	29698	3.9479	.00104275	3012.785	722315.77
960	921600	884736000	30.9839	9.8648	29745	3.9487	.00104167	3015.926	723822.95
961	923521	887503681	31.0000	9.8683	29791	3.9495	.00104058	3019.068	725331.70
962	925444	890277128	31.0161	9.8717	29838	3.9503	.00103950	3022.210	726842.02
963	927369	893056347	31.0322	9.8751	29884	3.9512	.00103842	3025.351	728353.91
964	929296	895841344	31.0483	9.8785	29931	3.9520	.00103734	3028.493	729867.37
965	931225	898632125	31.0644	9.8819	29977	3.9528	.00103627	3031.634	731382.40
966	933156	901428696	31.0805	9.8854	30024	3.9536	.00103520	3034.776	732899.01
967	935089	904231063	31.0966	9.8888	30070	3.9544	.00103413	3037.918	734417.18
968	937024	907039232	31.1127	9.8922	30117	3.9553	.00103306	3041.059	735936.93
969	938961	909853209	31.1288	9.8956	30164	3.9561	.00103199	3044.201	737458.24
970	940900	912673000	31.1448	9.8990	30210	3.9569	.00103093	3047.342	738981.13
971	942841	915498611	31.1609	9.9024	30257	3.9577	.00102987	3050.484	740505.59
972	944784	918330048	31.1769	9.9058	30304	3.9585	.00102881	3053.625	742031.62
973	946729	921167317	31.1929	9.9092	30351	3.9593	.00102775	3056.767	743559.22
974	948676	924010424	31.2090	9.9126	30398	3.9602	.00102669	3059.909	745088.39
975	950625	926859375	31.2250	9.9160	30444	3.9610	.00102564	3063.050	746619.13
976	952576	929714176	31.2410	9.9194	30491	3.9618	.00102459	3066.192	748151.44
977	954529	932574833	31.2570	9.9227	30538	3.9626	.00102354	3069.333	749685.32
978	956484	935441352	31.2730	9.9261	30585	3.9634	.00102249	3072.475	751220.78
979	958441	938313739	31.2890	9.9295	30632	3.9642	.00102145	3075.617	752757.80
980	960400	941192000	31.3050	9.9329	30679	3.9650	.00102041	3078.758	754296.40
981	962361	944076141	31.3209	9.9363	30726	3.9658	.00101937	3081.900	755836.59
982	964324	946966168	31.3369	9.9396	30773	3.9666	.00101833	3085.041	757378.30
983	966289	949862087	31.3528	9.9430	30820	3.9674	.00101729	3088.183	758921.61
984	968256	952763904	31.3688	9.9464	30867	3.9682	.00101626	3091.325	760466.48
985	970225	955671625	31.3847	9.9497	30914	3.9691	.00101523	3094.466	762012.93
986	972196	958585256	31.4006	9.9531	30961	3.9699	.00101420	3097.608	763560.95
987	974169	961504803	31.4166	9.9565	31008	3.9707	.00101317	3100.749	765110.54
988	976144	964430272	31.4325	9.9598	31055	3.9715	.00101215	3103.891	766661.70
989	978121	967361669	31.4484	9.9632	31102	3.9723	.00101112	3107.033	768214.44

MATHEMATICAL AND PHYSICAL TABLES

N	N^2	N^3	$\sqrt{N}$	$\sqrt[3]{N}$	$N^{3/2}$	$\sqrt[5]{N}$	$\dfrac{1}{N}$	Circle ($N = D$)	
								Circum.	Area
990	980100	970299000	31.4643	9.9666	31150	3.9731	.00101010	3110.174	769768.74
991	982081	973242271	31.4802	9.9699	31197	3.9739	.00100908	3113.316	771324.61
992	984064	976191488	31.4960	9.9733	31244	3.9747	.00100806	3116.457	772882.06
993	986049	979146657	31.5119	9.9766	31291	3.9755	.00100705	3119.599	774441.07
994	988036	982107784	31.5278	9.9800	31339	3.9763	.00100604	3122.740	776001.66
995	990025	985074875	31.5436	9.9833	31386	3.9771	.00100503	3125.882	777563.82
996	992016	988047936	31.5595	9.9866	31433	3.9779	.00100402	3129.024	779127.54
997	994009	991026973	31.5753	9.9900	31480	3.9787	.00100301	3132.165	780692.84
998	996004	994011992	31.5911	9.9933	31528	3.9795	.00100200	3135.307	782259.71
999	998001	997002999	31.6070	9.9967	31575	3.9803	.00100100	3138.448	783828.15
1000	1000000	1000000000	31.6228	10.0000	31623	3.9811	.00100000	3141.593	785398.16

Inches to Decimals of a Foot

In.	Ft.	In.	Ft.	In.	Ft.	In.	Ft.	In.	Ft.	In.	Ft.	In.	Ft.
1/16	.0052	5/16	.0260	9/16	.0469	13/16	.0677	1	.0833	5	.4167	9	.7500
1/8	.0104	3/8	.0313	5/8	.0521	7/8	.0729	2	.1667	6	.5000	10	.8333
3/16	.0156	7/16	.0364	11/16	.0573	15/16	.0781	3	.2500	7	.5833	11	.9167
1/4	.0208	1/2	.0417	3/4	.0625	1	.0833	4	.3333	8	.6667	12	1.0000

Factorials

n	$n! = 1 \cdot 2 \cdot 3 \ldots n$	$1/n!$	n	$n! = 1 \cdot 2 \cdot 3 \ldots n$	$1/n!$
1	1	1.	11	$399,168 \times 10^2$	0.250521×10^{-7}
2	2	0.5	12	$479,002 \times 10^3$	$.208768 \times 10^{-8}$
3	6	.166667	13	$622,702 \times 10^4$	$.160590 \times 10^{-9}$
4	24	$.416667 \times 10^{-1}$	14	$871,783 \times 10^5$	$.114707 \times 10^{-10}$
5	120	$.833333 \times 10^{-2}$	15	$130,767 \times 10^7$	$.764716 \times 10^{-12}$
6	720	$.138889 \times 10^{-2}$	16	$209,228 \times 10^8$	$.477948 \times 10^{-13}$
7	5,040	$.198413 \times 10^{-3}$	17	$355,687 \times 10^9$	$.281146 \times 10^{-14}$
8	40,320	$.248016 \times 10^{-4}$	18	$640,237 \times 10^{10}$	$.156192 \times 10^{-15}$
9	362,880	$.275573 \times 10^{-5}$	19	$121,645 \times 10^{12}$	$.822064 \times 10^{-17}$
10	3,628,800	$.275573 \times 10^{-6}$	20	$243,290 \times 10^{13}$	$.411032 \times 10^{-19}$

Higher Powers of Numbers

N	N^4	N^5	N^6	N^7	N^8
1	1	1	1	1	1
2	16	32	64	128	256
3	81	243	729	2187	6561
4	256	1024	4096	16384	65536
5	625	3125	15625	78125	390625
6	1296	7776	46656	279936	1679616
7	2401	16807	117649	823543	5764801
8	4096	32768	262144	2097152	16777216
9	6561	59049	531441	4782969	43046721
					$\times 10^8$
10	10000	100000	1000000	10000000	1.000000
11	14641	161051	1771561	19487171	2.143589
12	20736	248832	2985984	35831808	4.299817
13	28561	371293	4826809	62748517	8.157307
14	38416	537824	7529536	105413504	14.757891
15	50625	759375	11390625	170859375	25.628906
16	65536	1048576	16777216	268435456	42.949673
17	83521	1419857	24137569	410338673	69.757574
18	104976	1889568	34012224	612220032	110.199606
19	130321	2476099	47045881	893871739	169.835630
				$\times 10^9$	$\times 10^{10}$
20	160000	3200000	64000000	1.280000	2.560000
21	194481	4084101	85766121	1.801089	3.782286
22	234256	5153632	113379904	2.494358	5.487587
23	279841	6436343	148035889	3.404825	7.831099
24	331776	7962624	191102976	4.586471	11.007531
25	390625	9765625	244140625	6.103516	15.258789
26	456976	11881376	308915776	8.031810	20.882706
27	531441	14348907	387420489	10.460353	28.242954
28	614656	17210368	481890304	13.492929	37.780200
29	707281	20511149	594823321	17.249876	50.024641
			$\times 10^8$	$\times 10^{10}$	$\times 10^{11}$
30	810000	24300000	7.290000	2.187000	6.561000
31	923521	28629151	8.875037	2.751261	8.528910
32	1048576	33554432	10.737418	3.435974	10.995116
33	1185921	39135393	12.914680	4.261844	14.064086
34	1336336	45435424	15.448044	5.252335	17.857939
35	1500625	52521875	18.382656	6.433930	22.518754
36	1679616	60466176	21.767823	7.836416	28.211099

Higher Powers of Numbers (Continued)

N	N^4	N^5	N^6	N^7	N^8
37	1874161	69343957	25.657264	9.493188	35.124795
38	2085136	79235168	30.109364	11.441558	43.477921
39	2313441	90224199	35.187438	13.723101	53.520093
			$\times 10^9$	$\times 10^{10}$	$\times 10^{12}$
40	2560000	102400000	4.096000	16.384000	6.553600
41	2825761	115856201	4.750104	19.475427	7.984925
42	3111696	130691232	5.489032	23.053933	9.682652
43	3418801	147008443	6.321363	27.181861	11.688200
44	3748096	164916224	7.256314	31.927781	14.048224
45	4100625	184528125	8.303766	37.366945	16.815125
46	4477456	205962976	9.474297	43.581766	20.047612
47	4879681	229345007	10.779215	50.662312	23.811287
48	5308416	254803968	12.230590	58.706834	28.179280
49	5764801	282475249	13.841287	67.822307	33.232931
50	6250000	312500000	15.625000	78.125000	39.062500
			$\times 10^9$	$\times 10^{11}$	$\times 10^{13}$
50	6250000	312500000	15.625000	7.812500	3.906250
51	6765201	345025251	17.596288	8.974107	4.576794
52	7311616	380204032	19.770610	10.280717	5.345973
53	7890481	418195493	22.164361	11.747111	6.225969
54	8503056	459165024	24.794911	13.389252	7.230196
55	9150625	503284375	27.680641	15.224352	8.373394
56	9834496	550731776	30.840979	17.270948	9.671731
57	10556001	601692057	34.296447	19.548975	11.142916
58	11316496	656356768	38.068693	22.079842	12.806308
59	12117361	714924299	42.180534	24.886515	14.683044
		$\times 10^8$	$\times 10^{10}$	$\times 10^{11}$	$\times 10^{13}$
60	12960000	7.776000	4.665600	27.993600	16.796160
61	13845841	8.445963	5.152037	31.427428	19.170731
62	14776336	9.161328	5.680024	35.216146	21.834011
63	15752961	9.924365	6.252350	39.389806	24.815578
64	16777216	10.737418	6.871948	43.980465	28.147498
65	17850625	11.602906	7.541889	49.022279	31.864481
66	18974736	12.523326	8.265395	54.551607	36.004061
67	20151121	13.501251	9.045838	60.607116	40.606768
68	21381376	14.539336	9.886748	67.229888	45.716324
69	22667121	15.640313	10.791816	74.463533	51.379837

Higher Powers of Numbers (*Continued*)

N	N^4	N^5	N^6	N^7	N^8
		$\times\ 10^8$	$\times\ 10^{10}$	$\times\ 10^{12}$	$\times\ 10^{14}$
70	24010000	16.807000	11.764900	8.235430	5.764801
71	25411681	18.042294	12.810028	9.095120	6.457535
72	26873856	19.349176	13.931407	10.030613	7.222041
73	28398241	20.730716	15.133423	11.047399	8.064601
74	29986576	22.190066	16.420649	12.151280	8.991947
75	31640625	23.730469	17.797852	13.348389	10.011292
76	33362176	25.355254	19.269993	14.645195	11.130348
77	35153041	27.067842	20.842238	16.048523	12.357363
78	37015056	28.871744	22.519960	17.565569	13.701144
79	38950081	30.770564	24.308746	19.203909	15.171088
		$\times\ 10^8$	$\times\ 10^{10}$	$\times\ 10^{12}$	$\times\ 10^{14}$
80	40960000	32.768000	26.214400	20.971520	16.777216
81	43046721	34.867844	28.242954	22.876792	18.530202
82	45212176	37.073984	30.400667	24.928547	20.441409
83	47458321	39.390406	32.694037	27.136051	22.522922
84	49787136	41.821194	35.129803	29.509035	24.787589
85	52200625	44.370531	37.714952	32.057709	27.249053
86	54700816	47.042702	40.456724	34.792782	29.921793
87	57289761	49.842092	43.362620	37.725479	32.821167
88	59969536	52.772192	46.440409	40.867560	35.963452
89	62742241	55.840594	49.698129	44.231335	39.365888
		$\times\ 10^9$	$\times\ 10^{11}$	$\times\ 10^{13}$	$\times\ 10^{15}$
90	65610000	5.904900	5.314410	4.782969	4.304672
91	68574961	6.240321	5.678693	5.167610	4.702525
92	71639296	6.590815	6.063550	5.578466	5.132189
93	74805201	6.956884	6.469902	6.017009	5.595818
94	78074896	7.339040	6.898698	6.484776	6.095689
95	81450625	7.737809	7.350919	6.983373	6.634204
96	84934656	8.153727	7.827578	7.514475	7.213896
97	88529281	8.587240	8.329720	8.079828	7.837434
98	92236816	9.039208	8.858424	8.681255	8.507630
99	96059601	9.509900	9.414801	9.320653	9.227447
100	100000000	10.000000	10.000000	10.000000	10.000000

Fractional Powers of Numbers

N	Power of N								
	0.1	0.2	0.3	0.4	0.5	0.6	0.7	0.8	0.9
0.01	0.6310	0.3981	0.2512	0.1585	0.1000	0.0631	0.0398	0.0251	0.0159
.02	.6762	.4573	.3093	.2091	.1414	.0956	.0647	.0437	.0296
.03	.7042	.4959	.3492	.2460	.1732	.1220	.0859	.0605	.0426
.04	.7248	.5253	.3807	.2760	.2000	.1450	.1051	.0762	.0552
.05	.7411	.5493	.4071	.3017	.2236	.1657	.1228	.0910	.0675
.06	.7548	.5697	.4300	.3245	.2449	.1849	.1395	.1053	.0795
.07	.7665	.5875	.4503	.3452	.2646	.2028	.1554	.1192	.0913
.08	.7768	.6034	.4687	.3641	.2828	.2197	.1707	.1326	.1030
.09	.7860	.6178	.4856	.3817	.3000	.2358	.1853	.1457	.1145
.10	.7943	.6310	.5012	.3981	.3162	.2512	.1995	.1585	.1259
.11	.8019	.6431	.5157	.4136	.3317	.2660	.2133	.1711	.1372
.12	.8089	.6544	.5294	.4282	.3464	.2802	.2267	.1834	.1483
.13	.8154	.6650	.5422	.4422	.3606	.2940	.2398	.1955	.1594
.14	.8215	.6749	.5544	.4555	.3742	.3074	.2525	.2074	.1704
.15	.8272	.6843	.5660	.4682	.3873	.3204	.2650	.2192	.1813
.16	.8326	.6931	.5771	.4805	.4000	.3330	.2773	.2308	.1922
.17	.8376	.7016	.5877	.4922	.4123	.3454	.2893	.2423	.2030
.18	.8424	.7097	.5978	.5036	.4243	.3574	.3011	.2536	.2137
.19	.8470	.7174	.6076	.5146	.4359	.3692	.3127	.2649	.2243
.20	.8513	.7248	.6170	.5253	.4472	.3807	.3241	.2760	.2349
.21	.8555	.7319	.6261	.5357	.4583	.3920	.3354	.2869	.2455
.22	.8595	.7387	.6349	.5457	.4690	.4031	.3465	.2978	.2560
.23	.8633	.7453	.6435	.5555	.4796	.4140	.3575	.3086	.2664
.24	.8670	.7517	.6517	.5650	.4899	.4248	.3683	.3193	.2768
.25	.8706	.7579	.6598	.5744	.5000	.4353	.3789	.3299	.2872
.26	.8740	.7638	.6676	.5834	.5099	.4456	.3895	.3404	.2975
.27	.8773	.7696	.6752	.5923	.5196	.4559	.3999	.3508	.3078
.28	.8805	.7752	.6826	.6010	.5292	.4659	.4102	.3612	.3180
.29	.8836	.7807	.6898	.6095	.5385	.4758	.4204	.3715	.3282
.30	.8866	.7860	.6969	.6178	.5477	.4856	.4305	.3817	.3384
.31	.8895	.7912	.7037	.6260	.5568	.4952	.4405	.3918	.3485
.32	.8923	.7962	.7105	.6340	.5657	.5048	.4504	.4019	.3586
.33	.8951	.8011	.7171	.6418	.5745	.5142	.4602	.4119	.3687
.34	.8977	.8059	.7235	.6495	.5831	.5235	.4699	.4219	.3787
.35	.9003	.8106	.7298	.6571	.5916	.5327	.4796	.4318	.3887
.36	.9029	.8152	.7360	.6645	.6000	.5417	.4891	.4416	.3987
.37	.9054	.8197	.7421	.6719	.6083	.5507	.4986	.4514	.4087
.38	.9078	.8241	.7481	.6791	.6164	.5596	.5080	.4611	.4186
.39	.9101	.8284	.7539	.6862	.6245	.5684	.5173	.4708	.4285

Fractional Powers of Numbers (*Continued*)

N	_	_	_	_	Power of N	_	_	_	_
	0.1	0.2	0.3	0.4	0.5	0.6	0.7	0.8	0.9
.40	.9124	.8326	.7597	.6932	.6325	.5771	.5266	.4805	.4384
.41	.9147	.8367	.7653	.7000	.6403	.5857	.5357	.4900	.4482
.42	.9169	.8407	.7709	.7068	.6481	.5942	.5449	.4996	.4581
.43	.9191	.8447	.7763	.7135	.6558	.6027	.5539	.5091	.4679
.44	.9212	.8486	.7817	.7201	.6633	.6110	.5629	.5185	.4777
.45	.9233	.8524	.7870	.7266	.6708	.6193	.5718	.5279	.4874
.46	.9253	.8562	.7922	.7330	.6782	.6276	.5807	.5373	.4971
.47	.9273	.8598	.7973	.7393	.6856	.6357	.5895	.5466	.5069
.48	.9292	.8635	.8023	.7456	.6928	.6438	.5982	.5559	.5166
.49	.9312	.8670	.8073	.7518	.7000	.6518	.6069	.5651	.5262
0.50	.9330	.8706	.8123	.7579	.7071	.6598	.6156	.5743	.5359
.51	.9349	.8740	.8171	.7639	.7141	.6676	.6242	.5835	.5455
.52	.9367	.8774	.8219	.7698	.7211	.6755	.6327	.5927	.5551
.53	.9385	.8808	.8266	.7757	.7280	.6832	.6412	.6018	.5647
.54	.9402	.8841	.8312	.7816	.7349	.6909	.6497	.6108	.5743
.55	.9420	.8873	.8358	.7873	.7416	.6986	.6580	.6199	.5839
.56	.9437	.8905	.8403	.7930	.7483	.7062	.6664	.6289	.5934
.57	.9453	.8937	.8448	.7986	.7550	.7137	.6747	.6378	.6030
.58	.9470	.8968	.8492	.8042	.7616	.7212	.6830	.6467	.6125
.59	.9486	.8999	.8536	.8098	.7681	.7286	.6912	.6557	.6220
.60	.9502	.9029	.8579	.8152	.7746	.7360	.6994	.6645	.6315
.61	.9518	.9059	.8622	.8206	.7810	.7434	.7075	.6734	.6409
.62	.9533	.9088	.8664	.8260	.7874	.7506	.7156	.6822	.6504
.63	.9549	.9117	.8706	.8313	.7937	.7579	.7237	.6910	.6598
.64	.9564	.9146	.8747	.8365	.8000	.7651	.7317	.6998	.6692
.65	.9578	.9175	.8788	.8417	.8062	.7722	.7397	.7085	.6786
.66	.9593	.9203	.8828	.8469	.8124	.7793	.7476	.7172	.6880
.67	.9608	.9230	.8868	.8520	.8185	.7864	.7555	.7259	.6974
.68	.9622	.9258	.8907	.8571	.8246	.7934	.7634	.7345	.7067
.69	.9636	.9285	.8947	.8621	.8307	.8004	.7713	.7432	.7161
.70	.9650	.9312	.8985	.8670	.8367	.8074	.7791	.7518	.7254
.71	.9663	.9338	.9024	.8720	.8426	.8143	.7868	.7603	.7347
.72	.9677	.9364	.9062	.8769	.8485	.8211	.7946	.7689	.7441
.73	.9690	.9390	.9099	.8817	.8544	.8279	.8023	.7774	.7533
.74	.9703	.9416	.9136	.8865	.8602	.8347	.8100	.7859	.7626
.75	.9716	.9441	.9173	.8913	.8660	.8415	.8176	.7944	.7719
.76	.9729	.9466	.9210	.8960	.8718	.8482	.8252	.8029	.7811
.77	.9742	.9491	.9246	.9007	.8775	.8549	.8328	.8113	.7904
.78	.9755	.9515	.9282	.9054	.8832	.8615	.8404	.8197	.7996
.79	.9767	.9540	.9317	.9100	.8888	.8681	.8479	.8281	.8088

Fractional Powers of Numbers (*Continued*)

N	Power of N								
	0.1	0.2	0.3	0.4	0.5	0.6	0.7	0.8	0.9
.80	.9779	.9564	.9353	.9146	.8944	.8747	.8554	.8365	.8181
.81	.9792	.9587	.9388	.9192	.9000	.8812	.8629	.8449	.8273
.82	.9804	.9611	.9422	.9237	.9055	.8877	.8703	.8532	.8364
.83	.9816	.9634	.9456	.9282	.9110	.8942	.8777	.8615	.8456
.84	.9827	.9657	.9490	.9326	.9165	.9007	.8851	.8698	.8548
.85	.9839	.9680	.9524	.9371	.9220	.9071	.8925	.8781	.8639
.86	.9850	.9703	.9558	.9415	.9274	.9135	.8998	.8863	.8731
.87	.9862	.9725	.9591	.9458	.9327	.9198	.9071	.8946	.8822
.88	.9873	.9748	.9624	.9502	.9381	.9262	.9144	.9028	.8913
.89	.9884	.9770	.9656	.9545	.9434	.9325	.9217	.9110	.9004
.90	.9895	.9792	.9689	.9587	.9487	.9387	.9289	.9192	.9095
.91	.9906	.9813	.9721	.9630	.9539	.9450	.9361	.9273	.9186
.92	.9917	.9835	.9753	.9672	.9592	.9512	.9433	.9355	.9277
.93	.9928	.9856	.9785	.9714	.9644	.9574	.9505	.9436	.9368
.94	.9938	.9877	.9816	.9756	.9695	.9636	.9576	.9517	.9458
.95	.9949	.9898	.9847	.9797	.9747	.9697	.9647	.9598	.9549
.96	.9959	.9919	.9878	.9838	.9798	.9758	.9718	.9679	.9639
.97	.9970	.9939	.9909	.9879	.9849	.9819	.9789	.9759	.9730
.98	.9980	.9960	.9940	.9920	.9900	.9880	.9860	.9840	.9820
.99	.9990	.9980	.9970	.9960	.9950	.9940	.9930	.9920	.9910

FACTORS FOR COMPUTING PROBABLE ERRORS

n	$\dfrac{0.6745}{\sqrt{n-1}}$	$\dfrac{0.6745}{\sqrt{n(n-1)}}$	$\dfrac{0.8453}{\sqrt{n(n-1)}}$	$\dfrac{0.9453}{n\sqrt{n-1}}$	n	$\dfrac{0.6745}{\sqrt{n-1}}$	$\dfrac{0.6745}{\sqrt{n(n-1)}}$	$\dfrac{0.8453}{\sqrt{n(n-1)}}$	$\dfrac{0.8453}{n\sqrt{n-1}}$
1					51	0.0954	0.0134	0.0167	0.0023
2	0.6745	0.4769	0.5978	0.4227	52	.0944	.0131	.0164	.0023
3	.4769	.2754	.3451	.1993	53	.0935	.0128	.0161	.0022
4	.3894	.1947	.2440	.1220	54	.0926	.0126	.0158	.0022
5	.3372	.1508	.1890	.0845	55	.0918	.0124	.0155	.0021
6	.3016	.1231	.1543	.0630	56	.0909	.0122	.0152	.0020
7	.2754	.1041	.1304	.0493	57	.0901	.0119	.0150	.0020
8	.2549	.0901	.1130	.0399	58	.0893	.0117	.0147	.0019
9	.2385	.0795	.0996	.0332	59	.0886	.0115	.0145	.0019
10	0.2248	0.0711	0.0891	0.0282	60	0.0878	0.0113	0.0142	0.0018
11	.2133	.0643	.0806	.0243	61	.0871	.0111	.0140	.0018
12	.2034	.0587	.0736	.0212	62	.0864	.0110	.0137	.0017
13	.1947	.0540	.0677	.0188	63	.0857	.0108	.0135	.0017
14	.1871	.0500	.0627	.0167	64	.0850	.0106	.0133	.0017
15	.1803	.0465	.0583	.0151	65	.0843	.0105	.0131	.0016
16	.1742	.0435	.0546	.0136	66	.0837	.0103	.0129	.0016
17	.1686	.0409	.0513	.0124	67	.0830	.0101	.0127	.0016
18	.1636	.0386	.0483	.0114	68	.0824	.0100	.0125	.0015
19	.1590	.0365	.0457	.0105	69	.0818	.0098	.0123	.0015
20	0.1547	0.0346	0.0434	0.0097	70	0.0812	0.0097	0.0122	0.0015
21	.1508	.0329	.0412	.0090	71	.0806	.0096	.0120	.0014
22	.1472	.0314	.0393	.0084	72	.0800	.0094	.0118	.0014
23	.1438	.0300	.0376	.0078	73	.0795	.0093	.0117	.0014
24	.1406	.0287	.0360	.0073	74	.0789	.0092	.0115	.0013
25	.1377	.0275	.0345	.0069	75	.0784	.0091	.0113	.0013
26	.1349	.0265	.0332	.0065	76	.0779	.0089	.0112	.0013
27	.1323	.0255	.0319	.0061	77	.0774	.0088	.0111	.0013
28	.1298	.0245	.0307	.0058	78	.0769	.0087	.0109	.0012
29	.1275	.0237	.0297	.0055	79	.0764	.0086	.0108	.0012
30	0.1252	0.0229	0.0287	0.0052	80	0.0759	0.0085	0.0106	0.0012
31	.1231	.0221	.0277	.0050	81	.0754	.0084	.0105	.0012
32	.1211	.0214	.0268	.0047	82	.0749	.0083	.0104	.0011
33	.1192	.0208	.0260	.0045	83	.0745	.0082	.0102	.0011
34	.1174	.0201	.0252	.0043	84	.0740	.0081	.0101	.0011
35	.1157	.0196	.0245	.0041	85	.0736	.0080	.0100	.0011
36	.1140	.0190	.0238	.0040	86	.0732	.0079	.0099	.0011
37	.1124	.0185	.0232	.0038	87	.0727	.0078	.0098	.0010
38	.1109	.0180	.0225	.0037	88	.0723	.0077	.0097	.0010
39	.1094	.0175	.0220	.0035	89	.0719	.0076	.0096	.0010
40	0.1080	0.0171	0.0214	0.0034	90	0.0715	0.0075	0.0094	0.0010
41	.1066	.0167	.0209	.0033	91	.0711	.0075	.0093	.0010
42	.1053	.0163	.0204	.0031	92	.0707	.0074	.0092	.0010
43	.1041	.0159	.0199	.0030	93	.0703	.0073	.0091	.0009
44	.1029	.0155	.0194	.0029	94	.0699	.0072	.0090	.0009
45	.1017	.0152	.0190	.0028	95	.0696	.0071	.0089	.0009
46	.1005	.0148	.0186	.0027	96	.0692	.0071	.0089	.0009
47	.0994	.0145	.0182	.0027	97	.0688	.0070	.0088	.0009
48	.0984	.0142	.0178	.0026	98	.0685	.0069	.0087	.0009
49	.0974	.0139	.0174	.0025	99	.0681	.0068	.0086	.0009
50	0.0964	0.0136	0.0171	0.0024	100	0.0678	0.0068	0.0085	0.0009

Constants Containing e and π

$e = 2.7182818285$ $\qquad M = \log_{10} e = 0.4342944819$
$\pi = 3.1415926536$ $\qquad M^{-1} = \log_e 10 = 2.3025850930$

Powers of e			Multiples of π			Fractions of π		
e^n	Value	Logarithm	$n\pi$	Value	Logarithm	π/n	Value	Logarithm
e	2.718282	0.434294	π	3.141593	0.497150	$\pi/2$	1.570780	0.196120
e^{-1}	0.367879	$\bar{1}$.565706	2π	6.283185	0.798180	$\pi/3$	1.047198	0.020029
e^2	7.389057	0.868589	3π	9.424778	0.974271	$\pi/4$	0.785398	$\bar{1}$.895090
e^{-2}	0.135335	$\bar{1}$.131411	4π	12.566371	1.099210	$\pi/180$	0.017453*	$\bar{2}$.241877
$e^{1/2}$	1.648721	0.217147	5π	15.707963	1.196120			

Reciprocals of π			Powers of π			Roots of π		
n/π	Value	Logarithm	$\pi^{\pm n}$	Value	Logarithm	$\pi^{\pm 1/n}$	Value	Logarithm
$1/\pi$	0.318310	$\bar{1}$.502850	π^2	9.869604	0.994300	$\sqrt{\pi}$	1.772454	0.248575
$2/\pi$	0.636620	$\bar{1}$.803880	$1/\pi^2$	0.101321	$\bar{1}$.005700	$1/\sqrt{\pi}$	0.564190	$\bar{1}$.751425
$3/\pi$	0.954930	$\bar{1}$.979971	π^3	31.006277	1.491450	$\sqrt[3]{\pi}$	1.464592	0.165717
$180/\pi$	57.295780†	1.758123	$1/\pi^3$	0.032252	$\bar{2}$.508550	$1/\sqrt[3]{\pi}$	0.682784	$\bar{1}$.834283

* Number of radians per degree. † Number of degrees per radian.

Circular Arcs, Chords, and Segments

Central Angle in Degrees	Arc R	Height R	Chord R	Height Chord	Area R^2	Central Angle in Degrees	Arc R	Height R	Chord R	Height Chord	Area R^2
1	0.0175	0.0000	0.0175	0.0022	0.00000	31	0.5411	0.0364	0.5345	0.0680	0.01301
2	.0349	.0002	.0349	.0044	.00000	32	.5585	.0387	.5513	.0703	.01429
3	.0524	.0003	.0524	.0066	.00001	33	.5760	.0412	.5680	.0725	.01566
4	.0698	.0006	.0698	.0087	.00003	34	.5934	.0437	.5847	.0747	.01711
5	.0873	.0010	.0872	.0109	.00006	35	.6109	.0463	.6014	.0770	.01864
6	.1047	.0014	.1047	.0131	.00010	36	.6283	.0489	.6180	.0792	.02027
7	.1222	.0019	.1221	.0153	.00015	37	.6458	.0517	.6346	.0814	.02198
8	.1396	.0024	.1395	.0175	.00023	38	.6632	.0545	.6511	.0837	.02378
9	.1571	.0031	.1569	.0196	.00032	39	.6807	.0574	.6676	.0859	.02568
10	.1745	.0038	.1743	.0218	.00044	40	.6981	.0603	.6840	.0882	.02767
11	.1920	.0046	.1917	.0240	.00059	41	.7156	.0633	.7004	.0904	.02976
12	.2094	.0055	.2091	.0262	.00076	42	.7330	.0664	.7167	.0927	.03195
13	.2269	.0064	.2264	.0284	.00097	43	.7505	.0696	.7330	.0949	.03425
14	.2443	.0075	.2437	.0306	.00121	44	.7679	.0728	.7492	.0972	.03664
15	.2618	.0086	.2611	.0328	.00149	45	.7854	.0761	.7654	.0995	.03915
16	.2793	.0097	.2783	.0350	.00181	46	.8029	.0795	.7815	.1017	.04176
17	.2967	.0110	.2956	.0372	.00217	47	.8203	.0829	.7975	.1040	.04448
18	.3142	.0123	.3129	.0394	.00257	48	.8378	.0865	.8135	.1063	.04731
19	.3316	.0137	.3301	.0415	.00302	49	.8552	.0900	.8294	.1086	.05025
20	.3491	.0152	.3473	.0437	.00352	50	.8727	.0937	.8452	.1108	.05331
21	.3665	.0167	.3645	.0459	.00408	51	.8901	.0974	.8610	.1131	.05649
22	.3840	.0184	.3816	.0481	.00468	52	.9076	.1012	.8767	.1154	.05978
23	.4014	.0201	.3987	.0503	.00535	53	.9250	.1051	.8924	.1177	.06319
24	.4189	.0219	.4158	.0526	.00607	54	.9425	.1090	.9080	.1200	.06673
25	.4363	.0237	.4329	.0548	.00686	55	.9599	.1130	.9235	.1223	.07039
26	.4538	.0256	.4499	.0570	.00771	56	.9774	.1171	.9389	.1247	.07417
27	.4712	.0276	.4669	.0592	.00862	57	.9948	.1212	.9543	.1270	.07808
28	.4887	.0297	.4838	.0614	.00961	58	1.0123	.1254	.9696	.1293	.08212
29	.5061	.0319	.5008	.0636	.01067	59	1.0297	.1296	.9848	.1316	.08629
30	.5236	.0341	.5176	.0658	.01180	60	1.0472	.1340	1.0000	.1340	.09059

CIRCULAR ARCS, CHORDS, SEGMENTS 345

Central Angle in Degrees	Arc R	Height R	Chord R	Height Chord	Area R^2	Central Angle in Degrees	Arc R	Height R	Chord R	Height Chord	Area R^2
61	1.0647	.1384	1.015	.1363	.09502	121	2.1118	.5076	1.741	.2916	.62734
62	1.0821	.1428	1.030	.1387	.09958	122	2.1293	.5152	1.749	.2945	.64063
63	1.0996	.1474	1.045	.1410	.10428	123	2.1468	.5228	1.758	.2975	.65404
64	1.1170	.1520	1.060	.1434	.10911	124	2.1642	.5305	1.766	.3004	.66759
65	1.1345	.1566	1.075	.1457	.11408	125	2.1817	.5383	1.774	.3034	.68125
66	1.1519	.1613	1.089	.1481	.11919	126	2.1991	.5460	1.782	.3064	.69505
67	1.1694	.1661	1.104	.1505	.12443	127	2.2166	.5538	1.790	.3094	.70897
68	1.1868	.1710	1.118	.1529	.12982	128	2.2340	.5616	1.798	.3124	.72301
69	1.2043	.1759	1.133	.1553	.13535	129	2.2515	.5695	1.805	.3155	.73716
70	**1.2217**	**.1808**	**1.147**	**.1576**	**.14102**	**130**	**2.2689**	**.5774**	**1.813**	**.3185**	**.75143**
71	1.2392	.1859	1.161	.1601	.14683	131	2.2864	.5853	1.820	.3216	.76584
72	1.2566	.1910	1.176	.1625	.15279	132	2.3038	.5933	1.827	.3247	.78034
73	1.2741	.1961	1.190	.1649	.15889	133	2.3213	.6013	1.834	.3278	.79497
74	1.2915	.2014	1.204	.1673	.16514	134	2.3387	.6093	1.841	.3309	.80970
75	1.3090	.2066	1.218	.1697	.17154	135	2.3562	.6173	1.848	.3341	.82454
76	1.3265	.2120	1.231	.1722	.17808	136	2.3736	.6254	1.854	.3373	.83949
77	1.3439	.2174	1.245	.1746	.18477	137	2.3911	.6335	1.861	.3404	.85455
78	1.3614	.2229	1.259	.1771	.19160	138	2.4086	.6416	1.867	.3436	.86971
79	1.3788	2284	1.272	.1795	.19859	139	2.4260	.6498	1.873	.3469	.88497
80	**1.3963**	**.2340**	**1.286**	**.1820**	**.20573**	**140**	**2.4435**	**.6580**	**1.879**	**.3501**	**.90034**
81	1.4137	.2396	1.299	.1845	.21301	141	2.4609	.6662	1.885	.3534	.91580
82	1.4312	.2453	1.312	.1869	.22045	142	2.4784	.6744	1.891	.3566	.93135
83	1.4486	.2510	1.325	.1894	.22804	143	2.4958	.6827	1.897	.3599	.94700
84	1.4661	.2569	1.338	.1919	.23578	144	2.5133	.6910	1.902	.3633	.96274
85	1.4835	.2627	1.351	.1944	.24367	145	2.5307	.6993	1.907	.3666	.97858
86	1.5010	.2686	1.364	.1970	.25171	146	2.5482	.7076	1.913	.3700	.99449
87	1.5184	.2746	1.377	.1995	.25990	147	2.5656	.7160	1.918	.3734	1.0105
88	1.5359	.2807	1.389	.2020	.26825	148	2.5831	.7244	1.923	.3768	1.0266
89	1.5533	.2867	1.402	.2046	.27675	149	2.6005	.7328	1.927	.3802	1.0428
90	**1.5708**	**.2929**	**1.414**	**.2071**	**.28540**	**150**	**2.6180**	**.7412**	**1.932**	**.3837**	**1.0590**
91	1.5882	.2991	1.427	.2097	.29420	151	2.6354	.7496	1.936	.3871	1.0753
92	1.6057	.3053	1.439	.2122	.30316	152	2.6529	.7581	1.941	.3906	1.0917
93	1.6232	.3116	1.451	.2148	.31226	153	2.6704	.7666	1.945	.3942	1.1082
94	1.6406	.3180	1.463	.2174	.32152	154	2.6878	.7750	1.949	.3977	1.1247
95	1.6581	.3244	1.475	.2200	.33093	155	2.7053	.7836	1.953	.4013	1.1413
96	1.6755	.3309	1.486	.2226	.34050	156	2.7227	.7921	1.956	.4049	1.1580
97	1.6930	.3374	1.498	.2252	.35021	157	2.7402	.8006	1.960	.4085	1.1747
98	1.7104	.3439	1.509	.2279	.36008	158	2.7576	.8092	1.963	.4122	1.1915
99	1.7279	.3506	1.521	.2305	.37009	159	2.7751	.8178	1.967	.4158	1.2084
100	**1.7453**	**.3572**	**1.532**	**.2332**	**.38026**	**160**	**2.7925**	**.8264**	**1.970**	**.4195**	**1.2253**
101	1.7628	.3639	1.543	.2358	.39058	161	2.8100	.8350	1.973	.4233	1.2422
102	1.7802	.3707	1.554	.2385	.40104	162	2.8274	.8436	1.975	.4270	1.2592
103	1.7977	.3775	1.565	.2412	.41166	163	2.8449	.8522	1.978	.4308	1.2763
104	1.8151	.3843	1.576	.2439	.42242	164	2.8623	.8608	1.981	.4346	1.2934
105	1.8326	.3912	1.587	.2466	.43333	165	2.8798	.8695	1.983	.4385	1.3105
106	1.8500	.3982	1.597	.2493	.44439	166	2.8972	.8781	1.985	.4424	1.3277
107	1.8675	.4052	1.608	.2520	.45560	167	2.9147	.8868	1.987	.4463	1.3449
108	1.8850	.4122	1.618	.2548	.46695	168	2.9322	.8955	1.989	.4502	1.3621
109	1.9024	.4193	1.628	.2575	.47844	169	2.9496	.9042	1.991	.4542	1.3794
110	**1.9199**	**.4264**	**1.638**	**.2603**	**.49008**	**170**	**2.9671**	**.9128**	**1.992**	**.4582**	**1.3967**
111	1.9373	.4336	1.648	.2631	.50187	171	2.9845	.9215	1.994	.4622	1.4140
112	1.9548	.4408	1.658	.2659	.51379	172	3.0020	.9302	1.995	.4663	1.4314
113	1.9722	.4481	1.668	.2687	.52586	173	3.0194	.9390	1.996	.4704	1.4488
114	1.9897	.4554	1.677	.2715	.53807	174	3.0369	.9477	1.997	.4745	1.4662
115	2.0071	.4627	1.687	.2743	.55041	175	3.0543	.9564	1.998	.4786	1.4836
116	2.0246	.4701	1.696	.2772	.56289	176	3.0718	.9651	1.999	.4828	1.5010
117	2.0420	.4775	1.705	.2800	.57551	177	3.0892	.9738	1.999	.4871	1.5184
118	2.0595	.4850	1.714	.2829	.58827	178	3.1067	.9825	2.000	.4914	1.5359
119	2.0769	.4925	1.723	.2858	.60116	179	3.1241	.9913	2.000	.4957	1.5533
120	**2.0944**	**.5000**	**1.732**	**.2887**	**.61418**	**180**	**3.1416**	**1.0000**	**2.000**	**.5000**	**1.5708**

Values of Degrees, Minutes, and Seconds in Radians

Lengths of Circular Arcs, Radius Unity

Lengths of Circular Arcs, Radius Unity
Example. θ = 30° 20' 10''
  30° = 0.52359878
  20' = 0.00581776
  10'' = 0.00004848
  Arc length = 0.52946502

Degrees	Radians Arc Length R = 1	Degrees	Radians Arc Length R = 1	Degrees	Radians Arc Length R = 1		Radians Arc Length R = 1	
							Minutes	Seconds
0		60	1.04719755	120	2.09439510	0		
1	0.01745329	61	1.06465084	121	2.11184840	1	0.00029089	0.00000485
2	0.03490659	62	1.08210414	122	2.12930169	2	.00058178	.00000970
3	0.05235988	63	1.09955743	123	2.14675498	3	.00087266	.00001454
4	0.06981317	64	1.11701072	124	2.16420828	4	.00116355	.00001939
5	0.08726646	65	1.13446401	125	2.18166157	5	.00145444	.00002424
6	0.10471976	66	1.15191731	126	2.19911486	6	.00174533	.00002909
7	0.12217305	67	1.16937060	127	2.21656815	7	.00203622	.00003394
8	0.13962634	68	1.18682389	128	2.23402145	8	.00232711	.00003879
9	0.15707963	69	1.20427718	129	2.25147474	9	.00261799	.00004363
10	0.17453293	70	1.22173048	130	2.26892803	10	.00290888	.00004848
11	0.19198622	71	1.23918377	131	2.28638133	11	.00319977	.00005333
12	0.20943951	72	1.25663706	132	2.30383462	12	.00349066	.00005818
13	0.22689280	73	1.27409035	133	2.32128791	13	.00378155	.00006303
14	0.24434610	74	1.29154365	134	2.33874121	14	.00407243	.00006787
15	0.26179939	75	1.30899694	135	2.35619450	15	.00436332	.00007272
16	0.27925268	76	1.32645023	136	2.37364780	16	.00465421	.00007757
17	0.29670597	77	1.34390352	137	2.39110107	17	.00494510	.00008242
18	0.31415927	78	1.36135682	138	2.40855436	18	.00523599	.00008727
19	0.33161256	79	1.37881011	139	2.42600766	19	.00552688	.00009211
20	0.34906585	80	1.39626340	140	2.44346095	20	.00581776	.00009696
21	0.36651914	81	1.41371669	141	2.46091424	21	.00610865	.00010181
22	0.38397244	82	1.43116999	142	2.47836754	22	.00639954	.00010666
23	0.40142573	83	1.44862328	143	2.49582083	23	.00669043	.00011151
24	0.41887902	84	1.46607657	144	2.51327413	24	.00698132	.00011636
25	0.43633231	85	1.48352986	145	2.53072742	25	.00727221	.00012120
26	0.45378561	86	1.50098316	146	2.54818071	26	.00756309	.00012605
27	0.47123890	87	1.51843645	147	2.56563401	27	.00785398	.00013090
28	0.48869219	88	1.53588974	148	2.58308729	28	.00814487	.00013575
29	0.50614548	89	1.55334303	149	2.60054058	29	.00843576	.00014060
30	0.52359878	90	1.57079633	150	2.61799388	30	.00872665	.00014544
31	0.54105207	91	1.58824962	151	2.63544717	31	.00901753	.00015029
32	0.55850536	92	1.60570291	152	2.65290046	32	.00930842	.00015514
33	0.57595865	93	1.62315620	153	2.67035375	33	.00959931	.00015999
34	0.59341195	94	1.64060950	154	2.68780705	34	.00989020	.00016484
35	0.61086524	95	1.65806279	155	2.70526034	35	.01018109	.00016968
36	0.62831853	96	1.67551608	156	2.72271363	36	.01047198	.00017453
37	0.64577182	97	1.69296937	157	2.74016693	37	.01076286	.00017938
38	0.66322512	98	1.71042267	158	2.75762022	38	.01105375	.00018423
39	0.68067841	99	1.72787596	159	2.77507351	39	.01134464	.00018908
40	0.69813170	100	1.74532925	160	2.79252680	40	.01163553	.00019393
41	0.71558499	101	1.76278254	161	2.80998009	41	.01192642	.00019877
42	0.73303829	102	1.78023584	162	2.82743338	42	.01221730	.00020362
43	0.75049158	103	1.79768913	163	2.84488668	43	.01250819	.00020847
44	0.76794487	104	1.81514242	164	2.86233997	44	.01279908	.00021332
45	0.78539816	105	1.83259571	165	2.87979327	45	.01308997	.00021817
46	0.80285146	106	1.85004901	166	2.89724655	46	.01338086	.00022301
47	0.82030475	107	1.86750230	167	2.91469985	47	.01367175	.00022786
48	0.83775804	108	1.88495559	168	2.93215314	48	.01396263	.00023271
49	0.85521133	109	1.90240888	169	2.94960643	49	.01425352	.00023756
50	0.87266463	110	1.91986218	170	2.96705972	50	.01454441	.00024241
51	0.89011792	111	1.93731547	171	2.98451302	51	.01483530	.00024725
52	0.90757121	112	1.95476876	172	3.00196631	52	.01512619	.00025210
53	0.92502450	113	1.97222205	173	3.01941961	53	.01541707	.00025695
54	0.94247780	114	1.98967535	174	3.03687289	54	.01570796	.00026180
55	0.95993109	115	2.00712864	175	3.05432619	55	.01599885	.00026665
56	0.97738438	116	2.02458193	176	3.07177948	56	.01628974	.00027150
57	0.99483767	117	2.04203522	177	3.08923277	57	.01658063	.00027634
58	1.01229097	118	2.05948852	178	3.10668607	58	.01687152	.00028119
59	1.02974426	119	2.07694181	179	3.12413962	59	.01716240	.00028604
				180	3.14159265			

DEGREES TO RADIANS   347

Values of Radians in Degrees

Rad.	.00	.01	.02	.03	.04	.05	.06	.07	.08	.09
	Deg	Deg	Deg	Deg	Deg	Deg	Deg	Deg	Deg	Deg
0.0	0.0000	0.5730	1.1459	1.7189	2.2918	2.8648	3.4377	4.0107	4.5837	5.1566
.1	5.7296	6.3025	6.8755	7.4485	8.0214	8.5944	9.1673	9.7403	10.3132	10.8862
.2	11.4591	12.0321	12.6051	13.1780	13.7510	14.3239	14.8969	15.4699	16.0428	16.6158
.3	17.1887	17.7617	18.3346	18.9076	19.4806	20.0535	20.6265	21.1994	21.7724	22.3454
.4	22.9183	23.4913	24.0642	24.6372	25.2101	25.7831	26.3561	26.9290	27.5020	28.0749
.5	28.6479	29.2208	29.7938	30.3668	30.9397	31.1527	32.0856	32.6586	33.2316	33.8045
.6	34.3775	34.9504	35.5234	36.0963	36.6693	37.2423	37.8152	38.3882	38.9611	39.5341
.7	40.1070	40.6800	41.2530	41.8259	42.3989	42.9718	43.5448	44.1178	44.6907	45.2637
.8	45.8366	46.4096	46.9825	47.5555	48.1285	48.7014	49.2744	49.8473	50.4203	50.9932
.9	51.5662	52.1392	52.7121	53.2851	53.8580	54.4310	55.0039	55.5769	56.1499	56.7228

1 Radian = 57.29578 deg | 2 Radians = 114.59156 deg | 3 Radians = 171.88734 deg

Decimals of a Degree in Minute and Seconds

Decimal	.00		.01		.02		.03		.04		.05		.06		.07		.08		.09	
	Min	Sec	Min	Sec	Min	Sec	Min	Sec	Min	Sec	Min	Sec	Min	Sec	Min	Sec	Min	Sec	Min	Sec
0.0	0	0	0	36	1	12	1	48	2	24	3	0	3	36	4	12	4	48	5	24
.1	6	0	6	36	7	12	7	48	8	24	9	0	9	36	10	12	10	48	11	24
.2	12	0	12	36	13	12	13	48	14	24	15	0	15	36	16	12	16	48	17	24
.3	18	0	18	36	19	12	19	48	20	24	21	0	21	36	22	12	22	48	23	24
.4	24	0	24	36	25	12	25	48	26	24	27	0	27	36	28	12	28	48	29	24
.5	30	0	30	36	31	12	31	48	32	24	33	0	33	36	34	12	24	48	35	24
.6	36	0	36	36	37	12	37	48	38	24	39	0	39	36	40	12	40	48	41	24
.7	42	0	42	36	43	12	43	48	44	24	45	0	45	36	46	12	46	48	47	24
.8	48	0	48	36	49	12	49	48	50	24	51	0	51	36	52	12	52	48	53	24
.9	54	0	54	36	55	12	55	48	56	24	57	0	57	36	58	12	58	48	59	24

Minutes in Decimals of a Degree

Minutes	0	1	2	3	4	5	6	7	8	9
	Degrees	Degrees	Degrees	Degrees	Degrees	Degrees	Degrees	Degrees	Degrees	Degrees
0	0.00000	0.01667	0.03333	0.05000	0.06667	0.08333	0.10000	0.11667	0.13333	0.15000
10	.16667	.18333	.20000	.21667	.23333	.25000	.26667	.28333	.30000	.31667
20	.33333	.35000	.36667	.38333	.40000	.41667	.43333	.45000	.46667	.48333
30	.50000	.51667	.53333	.55000	.56667	.58333	.60000	.61667	.63333	.65000
40	.66667	.68333	.70000	.71667	.73333	.75000	.76667	.78333	.80000	.81667
50	.83333	.85000	.86667	.88333	.90000	.91667	.93333	.95000	.96667	.98333

Seconds in Decimals of a Degree

Seconds	0	1	2	3	4
	Degrees	Degrees	Degrees	Degrees	Degrees
0	0	0.0002778	0.0005555	0.0008333	0.0011111
10	0.0027778	.0030555	.0033333	.0036111	.0038888
20	.0055555	.0058333	.0061111	.0063888	.0066667
30	.0083333	.0086111	.0088888	.0091667	.0094444
40	.0111111	.0113888	.0116667	.0119444	.0122222
50	.0138888	.0141667	.0144444	.0147222	.0150000

Seconds	5	6	7	8	9
	Degrees	Degrees	Degrees	Degrees	Degrees
0	0.0013888	0.0016667	0.0019444	0.0022222	0.0024999
10	.0041667	.0044444	.0047222	.0050000	.0052778
20	.0069444	.0072222	.0075000	.0077778	.0080555
30	.0097222	.0100000	.0102778	.0105555	.0108333
40	.0125000	.0127778	.0130555	.0133333	.0136111
50	.0152778	.0155555	.0158333	.0161111	.0163888

Values and Logarithms of Trigonometric Functions

0°

Decimals	Minutes	Natural Values				Common Logarithms				Minutes	Decimals
		Sin	Cos	Tan	Cot	Sin	Cos	Tan	Cot		
.00	0	.00000	1.00000	.00000	+∞	−∞	10.000000	−∞	+∞	60	1.00
	1	.00029	1.00000	.00029	3437.75	6.463726	.000000	6.463726	13.536274	59	
	2	.00058	1.00000	.00058	1718.87	.764756	.000000	.764756	.235244	58	
.05	3	.00087	1.00000	.00087	1145.92	.940847	.000000	.940847	.059153	57	.95
	4	.00116	1.00000	.00116	859.436	7.065786	.000000	7.065786	12.934214	56	
	5	.00145	1.00000	.00145	687.549	.162696	.000000	.162696	.837304	55	
.10	6	.00175	1.00000	.00175	572.957	.241877	9.999999	.241878	.758122	54	.90
	7	.00204	1.00000	.00204	491.106	.308824	.999999	.308825	.691175	53	
	8	.00233	1.00000	.00233	429.718	.366816	.999999	.366817	.633183	52	
.15	9	.00262	1.00000	.00262	381.971	.417968	.999999	.417970	.582030	51	.85
	10	.00291	1.00000	.00291	343.774	.463726	.999998	.463727	.536273	50	
	11	.00320	.99999	.00320	312.521	.505118	.999998	.505120	.494880	49	
.20	12	.00349	.99999	.00349	286.478	.542906	.999997	.542909	.457091	48	.80
	13	.00378	.99999	.00378	264.441	.577668	.999997	.577672	.422328	47	
	14	.00407	.99999	.00407	245.552	.609853	.999996	.609857	.390143	46	
.25	15	.00436	.99999	.00436	229.182	.639816	.999996	.639820	.360180	45	.75
	16	.00465	.99999	.00465	214.858	.667845	.999995	.667849	.332151	44	
	17	.00495	.99999	.00495	202.219	.694173	.999995	.694179	.305821	43	
.30	18	.00524	.99999	.00524	190.984	.718997	.999994	.719003	.280997	42	.70
	19	.00553	.99998	.00553	180.932	.742478	.999993	.742484	.257516	41	
	20	.00582	.99998	.00582	171.885	.764754	.999993	.764761	.235239	40	
.35	21	.00611	.99998	.00611	163.700	.785943	.999992	.785951	.214049	39	.65
	22	.00640	.99998	.00640	156.259	.806146	.999991	.806155	.193845	38	
	23	.00669	.99998	.00669	149.465	.825451	.999990	.825460	.174540	37	
.40	24	.00698	.99998	.00698	143.237	.843934	.999989	.843944	.156056	36	.60
	25	.00727	.99997	.00727	137.507	.861662	.999989	.861674	.138326	35	
	26	.00756	.99997	.00756	132.219	.878695	.999988	.878708	.121292	34	
.45	27	.00785	.99997	.00785	127.321	.895085	.999987	.895099	.104901	33	.55
	28	.00814	.99997	.00815	122.774	.910879	.999986	.910894	.089106	32	
	29	.00844	.99996	.00844	118.540	.926119	.999985	.926134	.073866	31	
.50	30	.00873	.99996	.00873	114.589	.940842	.999983	.940858	.059142	30	.50
	31	.00902	.99996	.00902	110.892	.955082	.999982	.955100	.044900	29	
	32	.00931	.99996	.00931	107.426	.968870	.999981	.968889	.031111	28	
.55	33	.00960	.99995	.00960	104.171	.982233	.999980	.982253	.017747	27	.45
	34	.00989	.99995	.00989	101.107	.995198	.999979	.995219	.004781	26	
	35	.01018	.99995	.01018	98.2179	8.007787	.999977	8.007809	11.992191	25	
.60	36	.01047	.99995	.01047	95.4895	.020021	.999976	.020044	.979956	24	.40
	37	.01076	.99994	.01076	92.9085	.031919	.999975	.031945	.968055	23	
	38	.01105	.99994	.01105	90.4633	.043501	.999973	.043527	.956473	22	
.65	39	.01134	.99994	.01135	88.1436	.054781	.999972	.054809	.945191	21	.35
	40	.01164	.99993	.01164	85.9398	.065776	.999971	.065806	.934194	20	
	41	.01193	.99993	.01193	83.8435	.076500	.999969	.076531	.923469	19	
.70	42	.01222	.99993	.01222	81.8470	.086965	.999968	.086997	.913003	18	.30
	43	.01251	.99992	.01251	79.9434	.097183	.999966	.097217	.902783	17	
	44	.01280	.99992	.01280	78.1263	.107167	.999964	.107203	.892797	16	
.75	45	.01309	.99991	.01309	76.3900	.116926	.999963	.116963	.883037	15	.25
	46	.01338	.99991	.01338	74.7292	.126471	.999961	.126510	.873490	14	
	47	.01367	.99991	.01367	73.1390	.135810	.999959	.135851	.864149	13	
.80	48	.01396	.99990	.01396	71.6151	.144953	.999958	.144996	.855004	12	.20
	49	.01425	.99990	.01425	70.1533	.153907	.999956	.153952	.846048	11	
	50	.01454	.99989	.01455	68.7501	.162681	.999954	.162727	.837273	10	
.85	51	.01483	.99989	.01484	67.4019	.171280	.999952	.171328	.828672	9	.15
	52	.01513	.99989	.01513	66.1055	.179713	.999950	.179763	.820237	8	
	53	.01542	.99988	.01542	64.8580	.187985	.999948	.188036	.811964	7	
.90	54	.01571	.99988	.01571	63.6567	.196102	.999946	.196156	.803844	6	.10
	55	.01600	.99987	.01600	62.4992	.204070	.999944	.204126	.795874	5	
	56	.01629	.99987	.01629	61.3829	.211895	.999942	.211953	.788047	4	
.95	57	.01658	.99986	.01658	60.3058	.219581	.999940	.219641	.780359	3	.05
	58	.01687	.99986	.01687	59.2659	.227134	.999938	.227195	.772805	2	
	59	.01716	.99985	.01716	58.2612	.234557	.999936	.234621	.765379	1	
1.00	60	.01745	.99985	.01746	57.2900	8.241855	9.999934	8.241921	11.758079	0	.00
Decimals	Minutes	Cos	Sin	Cot	Tan	Cos	Sin	Cot	Tan	Minutes	Decimals
		Natural Values				Common Logarithms					

89°

VALUES AND LOGARITHMS OF TRIGONOMETRIC FUNCTIONS

1°

Decimals	Minutes	Natural Values				Common Logarithms				Minutes	Decimals
		Sin	Cos	Tan	Cot	Sin	Cos	Tan	Cot		
.00	0	.01745	.99985	.01746	57.2900	8.241855	9.999934	8.241921	11.758079	60	1.00
	1	.01774	.99984	.01775	56.3506	.249033	.999932	.249102	.750898	59	
	2	.01803	.99984	.01804	55.4415	.256094	.999929	.256165	.743835	58	
.05	3	.01832	.99983	.01833	54.5613	.263042	.999927	.263115	.736885	57	.95
	4	.01862	.99983	.01862	53.7086	.269881	.999925	.269956	.730044	56	
	5	.01891	.99982	.01891	52.8821	.276614	.999922	.276691	.723309	55	
.10	6	.01920	.99982	.01920	52.0807	.283243	.999920	.283323	.716677	54	.90
	7	.01949	.99981	.01949	51.3032	.289773	.999918	.289856	.710144	53	
	8	.01978	.99980	.01978	50.5485	.296207	.999915	.296292	.703708	52	
.15	9	.02007	.99980	.02007	49.8157	.302546	.999913	.302634	.697366	51	.85
	10	.02036	.99979	.02036	49.1039	.308794	.999910	.308884	.691116	50	
	11	.02065	.99979	.02066	48.4121	.314954	.999907	.315046	.684954	49	
.20	12	.02094	.99978	.02095	47.7395	.321027	.999905	.321122	.678878	48	.80
	13	.02123	.99977	.02124	47.0853	.327016	.999902	.327114	.672886	47	
	14	.02152	.99977	.02153	46.4489	.332924	.999899	.333025	.666975	46	
.25	15	.02181	.99976	.02182	45.8294	.338753	.999897	.338856	.661144	45	.75
	16	.02211	.99976	.02211	45.2261	.344504	.999894	.344610	.655390	44	
	17	.02240	.99975	.02240	44.6386	.350181	.999891	.350289	.649711	43	
.30	18	.02269	.99974	.02269	44.0661	.355783	.999888	.355895	.644105	42	.70
	19	.02298	.99974	.02298	43.5081	.361315	.999885	.361430	.638570	41	
	20	.02327	.99973	.02328	42.9641	.366777	.999882	.366895	.633105	40	
.35	21	.02356	.99972	.02357	42.4335	.372171	.999879	.372292	.627708	39	.65
	22	.02385	.99972	.02386	41.9158	.377499	.999876	.377622	.622378	38	
	23	.02414	.99971	.02415	41.4106	.382762	.999873	.382889	.617111	37	
.40	24	.02443	.99970	.02444	40.9174	.387962	.999870	.388092	.611908	36	.60
	25	.02472	.99969	.02473	40.4358	.393101	.999867	.393234	.606766	35	
	26	.02501	.99969	.02502	39.9655	.398179	.999864	.398315	.601685	34	
.45	27	.02530	.99968	.02531	39.5059	.403199	.999861	.403338	.596662	33	.55
	28	.02560	.99967	.02560	39.0568	.408161	.999858	.408304	.591696	32	
	29	.02589	.99966	.02589	38.6177	.413068	.999854	.413213	.586787	31	
.50	30	.02618	.99966	.02619	38.1885	.417919	.999851	.418068	.581932	30	.50
	31	.02647	.99965	.02648	37.7686	.422717	.999848	.422869	.577131	29	
	32	.02676	.99964	.02677	37.3579	.427462	.999844	.427618	.572382	28	
.55	33	.02705	.99963	.02706	36.9560	.432156	.999841	.432315	.567685	27	.45
	34	.02734	.99963	.02735	36.5627	.436800	.999838	.436962	.563038	26	
	35	.02763	.99962	.02764	36.1776	.441394	.999834	.441560	.558440	25	
.60	36	.02792	.99961	.02793	35.8006	.445941	.999831	.446110	.553890	24	.40
	37	.02821	.99960	.02822	35.4313	.450440	.999827	.450613	.549387	23	
	38	.02850	.99959	.02851	35.0695	.454893	.999824	.455070	.544930	22	
.65	39	.02879	.99959	.02881	34.7151	.459301	.999820	.459481	.540519	21	.35
	40	.02908	.99958	.02910	34.3678	.463665	.999816	.463849	.536151	20	
	41	.02938	.99957	.02939	34.0273	.467985	.999813	.468172	.531828	19	
.70	42	.02967	.99956	.02968	33.6935	.472263	.999809	.472454	.527546	18	.30
	43	.02996	.99955	.02997	33.3662	.476498	.999805	.476693	.523307	17	
	44	.03025	.99954	.03026	33.0452	.480693	.999801	.480892	.519108	16	
.75	45	.03054	.99953	.03055	32.7303	.484848	.999797	.485050	.514950	15	.25
	46	.03083	.99952	.03084	32.4213	.488963	.999794	.489170	.510830	14	
	47	.03112	.99952	.03114	32.1181	.493040	.999790	.493250	.506750	13	
.80	48	.03141	.99951	.03143	31.8205	.497078	.999786	.497293	.502707	12	.20
	46	.03170	.99950	.03172	31.5284	.501080	.999782	.501298	.498702	11	
	50	.03199	.99949	.03201	31.2416	.505045	.999778	.505267	.494733	10	
.85	51	.03228	.99948	.03230	30.9599	.508974	.999774	.509200	.490800	9	.15
	52	.03257	.99947	.03259	30.6833	.512867	.999769	.513098	.486902	8	
	53	.03286	.99946	.03288	30.4116	.516726	.999765	.516961	.483039	7	
.90	54	.03316	.99945	.03317	30.1446	.520551	.999761	.520790	.479210	6	.10
	55	.03345	.99944	.03346	29.8823	.524343	.999757	.524586	.475414	5	
	56	.03374	.99943	.03376	29.6245	.528102	.999753	.528349	.471651	4	
.95	57	.03403	.99942	.03405	29.3711	.531828	.999748	.532080	.467920	3	.05
	58	.03432	.99941	.03434	29.1220	.535523	.999744	.535779	.464221	2	
	59	.03461	.99940	.03463	28.8771	.539186	.999740	.539447	.460553	1	
1.00	60	.03490	.99939	.03492	28.6363	8.542819	9.999735	8.543084	11.456916	0	.00
Decimals	Minutes	Cos	Sin	Cot	Tan	Cos	Sin	Cot	Tan	Minutes	Decimals
		Natural Values				Common Logarithms					

88°

2°

Decimals	Minutes	Natural Values				Common Logarithms				Minutes	Decimals
		Sin	Cos	Tan	Cot	Sin	Cos	Tan	Cot		
.00	0	.03490	.99939	.03492	28.6363	8.542819	9.999735	8.543084	11.456916	60	1.00
	1	.03519	.99938	.03521	28.3994	.546422	.999731	.546691	.453309	59	
	2	.03548	.99937	.03550	28.1664	.549995	.999726	.550268	.449732	58	
.05	3	.03577	.99936	.03579	27.9372	.553539	.999722	.553817	.446183	57	.95
	4	.03606	.99935	.03609	27.7117	.557054	.999717	.557336	.442664	56	
	5	.03635	.99934	.03638	27.4899	.560540	.999713	.560828	.439172	55	
.10	6	.03664	.99933	.03667	27.2715	.563999	.999708	.564291	.435709	54	.90
	7	.03693	.99932	.03696	27.0566	.567431	.999704	.567727	.432273	53	
	8	.03723	.99931	.03725	26.8450	.570836	.999699	.571137	.428863	52	
.15	9	.03752	.99930	.03754	26.6367	.574214	.999694	.574520	.425480	51	.85
	10	.03781	.99929	.03783	26.4316	.577566	.999689	.577877	.422123	50	
	11	.03810	.99927	.03812	26.2296	.580892	.999685	.581208	.418792	49	
.20	12	.03839	.99926	.03842	26.0307	.584193	.999680	.584514	.415486	48	.80
	13	.03868	.99925	.03871	25.8348	.587469	.999675	.587795	.412205	47	
	14	.03897	.99924	.03900	25.6418	.590721	.999670	.591051	.408949	46	
.25	15	.03926	.99923	.03929	25.4517	.593948	.999665	.594283	.405717	45	.75
	16	.03955	.99922	.03958	25.2644	.597152	.999660	.597492	.402508	44	
	17	.03984	.99921	.03987	25.0798	.600332	.999655	.600677	.399323	43	
.30	18	.04013	.99919	.04016	24.8978	.603489	.999650	.603839	.396161	42	.70
	19	.04042	.99918	.04046	24.7185	.606623	.999645	.606978	.393022	41	
	20	.04071	.99917	.04075	24.5418	.609734	.999640	.610094	.389906	40	
.35	21	.04100	.99916	.04104	24.3675	.612823	.999635	.613189	.386811	39	.65
	22	.04129	.99915	.04133	24.1957	.615891	.999629	.616262	.383738	38	
	23	.04159	.99913	.04162	24.0263	.618937	.999624	.619313	.380687	37	
.40	24	.04188	.99912	.04191	23.8593	.621962	.999619	.622343	.377657	36	.60
	25	.04217	.99911	.04220	23.6945	.624965	.999614	.625352	.374648	35	
	26	.04246	.99910	.04250	23.5321	.627948	.999608	.628340	.371660	34	
.45	27	.04275	.99909	.04279	23.3718	.630911	.999603	.631308	.368692	33	.55
	28	.04304	.99907	.04308	23.2137	.633854	.999597	.634256	.365744	32	
	29	.04333	.99906	.04337	23.0577	.636776	.999592	.637184	.362816	31	
.50	30	.04362	.99905	.04366	22.9038	.639680	.999586	.640093	.359907	30	.50
	31	.04391	.99904	.04395	22.7519	.642563	.999581	.642982	.357018	29	
	32	.04420	.99902	.04424	22.6020	.645428	.999575	.645853	.354147	28	
.55	33	.04449	.99901	.04454	22.4541	.648274	.999570	.648704	.351296	27	.45
	34	.04478	.99900	.04483	22.3081	.651102	.999564	.651537	.348463	26	
	35	.04507	.99898	.04512	22.1640	.653911	.999558	.654352	.345648	25	
.60	36	.04536	.99897	.04541	22.0217	.656702	.999553	.657149	.342851	24	.40
	37	.04565	.99896	.04570	21.8813	.659475	.999547	.659928	.340072	23	
	38	.04594	.99894	.04599	21.7426	.662230	.999541	.662689	.337311	22	
.65	39	.04623	.99893	.04628	21.6056	.664968	.999535	.665433	.334567	21	.35
	40	.04653	.99892	.04658	21.4704	.667689	.999529	.668160	.331840	20	
	41	.04682	.99890	.04687	21.3369	.670393	.999524	.670870	.329130	19	
.70	42	.04711	.99889	.04716	21.2049	.673080	.999518	.673563	.326437	18	.30
	43	.04740	.99888	.04745	21.0747	.675751	.999512	.676239	.323761	17	
	44	.04769	.99886	.04774	20.9460	.678405	.999506	.678900	.321100	16	
.75	45	.04798	.99885	.04803	20.8188	.681043	.999500	.681544	.318456	15	.25
	46	.04827	.99883	.04833	20.6932	.683665	.999493	.684172	.315828	14	
	47	.04856	.99882	.04862	20.5691	.686272	.999487	.686784	.313216	13	
.80	48	.04885	.99881	.04891	20.4465	.688863	.999481	.689381	.310619	12	.20
	49	.04914	.99879	.04920	20.3253	.691438	.999475	.691963	.308037	11	
	50	.04943	.99878	.04949	20.2056	.693998	.999469	.694529	.305471	10	
.85	51	.04972	.99876	.04978	20.0872	.696543	.999463	.697081	.302919	9	.15
	52	.05001	.99875	.05007	19.9702	.699073	.999456	.699617	.300383	8	
	53	.05030	.99873	.05037	19.8546	.701589	.999450	.702139	.297861	7	
.90	54	.05059	.99872	.05066	19.7403	.704090	.999443	.704646	.295354	6	.10
	55	.05088	.99870	.05095	19.6273	.706577	.999437	.707140	.292860	5	
	56	.05117	.99869	.05124	19.5156	.709049	.999431	.709618	.290382	4	
.95	57	.05146	.99867	.05153	19.4051	.711507	.999424	.712083	.287917	3	.05
	58	.05175	.99866	.05182	19.2959	.713952	.999418	.714534	.285466	2	
	59	.05205	.99864	.05212	19.1879	.716383	.999411	.716972	.283028	1	
1.00	60	.05234	.99863	.05241	19.0811	8.718800	9.999404	8.719396	11.280604	0	.00
Decimals	Minutes	Cos	Sin	Cot	Tan	Cos	Sin	Cot	Tan	Minutes	Decimals
		Natural Values				Common Logarithms					

87°

VALUES AND LOGARITHMS OF TRIGONOMETRIC FUNCTIONS

3°

Decimals	Minutes	Natural Values				Common Logarithms				Minutes	Decimals
		Sin	Cos	Tan	Cot	Sin	Cos	Tan	Cot		
.00	0	.05234	.99863	.05241	19.0811	8.718800	9.999404	8.719396	11.280604	60	1.00
	1	.05263	.99861	.05270	18.9755	.721204	.999398	.721806	.278194	59	
	2	.05292	.99860	.05299	18.8711	.723595	.999391	.724204	.275796	58	
.05	3	.05321	.99858	.05328	18.7678	.725972	.999384	.726588	.273412	57	.95
	4	.05350	.99857	.05357	18.6656	.728337	.999378	.728959	.271041	56	
	5	.05379	.99855	.05387	18.5645	.730688	.999371	.731317	.268683	55	
.10	6	.05408	.99854	.05416	18.4645	.733027	.999364	.733663	.266337	54	.90
	7	.05437	.99852	.05445	18.3655	.735354	.999357	.735996	.264004	53	
	8	.05466	.99851	.05474	18.2677	.737667	.999350	.738317	.261683	52	
.15	9	.05495	.99849	.05503	18.1708	.739969	.999343	.740626	.259374	51	.85
	10	.05524	.99847	.05533	18.0750	.742259	.999336	.742922	.257078	50	
	11	.05553	.99846	.05562	17.9802	.744536	.999329	.745207	.254793	49	
.20	12	.05582	.99844	.05591	17.8863	.746802	.999322	.747479	.252521	48	.80
	13	.05611	.99842	.05620	17.7934	.749055	.999315	.749740	.250260	47	
	14	.05640	.99841	.05649	17.7015	.751297	.999308	.751989	.248011	46	
.25	15	.05669	.99839	.05678	17.6106	.753528	.999301	.754227	.245773	45	.75
	16	.05698	.99838	.05708	17.5205	.755747	.999294	.756453	.243547	44	
	17	.05727	.99836	.05737	17.4314	.757955	.999287	.758668	.241332	43	
.30	18	.05756	.99834	.05766	17.3432	.760151	.999279	.760872	.239128	42	.70
	19	.05785	.99833	.05795	17.2558	.762337	.999272	.763065	.236935	41	
	20	.05814	.99831	.05824	17.1693	.764511	.999265	.765246	.234754	40	
.35	21	.05844	.99829	.05854	17.0837	.766675	.999257	.767417	.232583	39	.65
	22	.05873	.99827	.05883	16.9990	.768828	.999250	.769578	.230422	38	
	23	.05902	.99826	.05912	16.9150	.770970	.999242	.771727	.228273	37	
.40	24	.05931	.99824	.05941	16.8319	.773101	.999235	.773866	.226134	36	.60
	25	.05960	.99822	.05970	16.7496	.775223	.999227	.775995	.224005	35	
	26	.05989	.99821	.05999	16.6681	.777333	.999220	.778114	.221886	34	
.45	27	.06018	.99819	.06029	16.5874	.779434	.999212	.780222	.219778	33	.55
	28	.06047	.99817	.06058	16.5075	.781524	.999205	.782320	.217680	32	
	29	.06076	.99815	.06087	16.4283	.783605	.999197	.784408	.215592	31	
.50	30	.06105	.99813	.06116	16.3499	.785675	.999189	.786486	.213514	30	.50
	31	.06134	.99812	.06145	16.2722	.787736	.999181	.788554	.211446	29	
	32	.06163	.99810	.06175	16.1952	.789787	.999174	.790613	.209387	28	
.55	33	.06192	.99808	.06204	16.1190	.791828	.999166	.792662	.207338	27	.45
	34	.06221	.99806	.06233	16.0435	.793859	.999158	.794701	.205299	26	
	35	.06250	.99804	.06262	15.9687	.795881	.999150	.796731	.203269	25	
.60	36	.06279	.99803	.06291	15.8945	.797894	.999142	.798752	.201248	24	.40
	37	.06308	.99801	.06321	15.8211	.799897	.999134	.800763	.199237	23	
	38	.06337	.99799	.06350	15.7483	.801892	.999126	.802765	.197235	22	
.65	39	.06366	.99797	.06379	15.6762	.803876	.999118	.804758	.195242	21	.35
	40	.06395	.99795	.06408	15.6048	.805852	.999110	.806742	.193258	20	
	41	.06424	.99793	.06437	15.5340	.807819	.999102	.808717	.191283	19	
.70	42	.06453	.99792	.06467	15.4638	.809777	.999094	.810683	.189317	18	.30
	43	.06482	.99790	.06496	15.3943	.811726	.999086	.812641	.187359	17	
	44	.06511	.99788	.06525	15.3254	.813667	.999077	.814589	.185411	16	
.75	45	.06540	.99786	.06554	15.2571	.815599	.999069	.816529	.183471	15	.25
	46	.06569	.99784	.06584	15.1893	.817522	.999061	.818461	.181539	14	
	47	.06598	.99782	.06613	15.1222	.819436	.999053	.820384	.179616	13	
.80	48	.06627	.99780	.06642	15.0557	.821343	.999044	.822298	.177702	12	.20
	49	.06656	.99778	.06671	14.9898	.823240	.999036	.824205	.175795	11	
	50	.06685	.99776	.06700	14.9244	.825130	.999027	.826103	.173897	10	
.85	51	.06714	.99774	.06730	14.8596	.827011	.999019	.827992	.172008	9	.15
	52	.06743	.99772	.06759	14.7954	.828884	.999010	.829874	.170126	8	
	53	.06773	.99770	.06788	14.7317	.830749	.999002	.831748	.168252	7	
.90	54	.06802	.99768	.06817	14.6685	.832607	.998993	.833613	.166387	6	.10
	55	.06831	.99766	.06847	14.6059	.834456	.998984	.835471	.164529	5	
	56	.06860	.99764	.06876	14.5438	.836297	.998976	.837321	.162679	4	
.95	57	.06889	.99762	.06905	14.4823	.838130	.998967	.839163	.160837	3	.05
	58	.06918	.99760	.06934	14.4212	.839956	.998958	.840998	.159002	2	
	59	.06947	.99758	.06963	14.3607	.841774	.998950	.842825	.157175	1	
1.00	60	.06976	.99756	.06993	14.3007	8.843585	9.998941	8.844644	11.155356	0	.00
Decimals	Minutes	Cos	Sin	Cot	Tan	Cos	Sin	Cot	Tan	Minutes	Decimals
		Natural Values				Common Logarithms					

86°

4°

Decimals	Minutes	Natural Values				Common Logarithms				Minutes	Decimals
		Sin	Cos	Tan	Cot	Sin	Cos	Tan	Cot		
.00	0	.06976	.99756	.06993	14.3007	8.843585	9.998941	8.844644	11.155356	60	1.00
	1	.07005	.99754	.07022	14.2411	.845387	.998932	.846455	.153545	59	
	2	.07034	.99752	.07051	14.1821	.847183	.998923	.848260	.151740	58	
.05	3	.07063	.99750	.07080	14.1235	.848971	.998914	.850057	.149943	57	.95
	4	.07092	.99748	.07110	14.0655	.850751	.998905	.851846	.148154	56	
	5	.07121	.99746	.07139	14.0079	.852525	.998896	.853628	.146372	55	
.10	6	.07150	.99744	.07168	13.9507	.854291	.998887	.855403	.144597	54	.90
	7	.07179	.99742	.07197	13.8940	.856049	.998878	.857171	.142829	53	
	8	.07208	.99740	.07227	13.8378	.857801	.998869	.858932	.141068	52	
.15	9	.07237	.99738	.07256	13.7821	.859546	.998860	.860686	.139314	51	.85
	10	.07266	.99736	.07285	13.7267	.861283	.998851	.862433	.137567	50	
	11	.07295	.99734	.07314	13.6719	.863014	.998841	.864173	.135827	49	
.20	12	.07324	.99731	.07344	13.6174	.864738	.998832	.865906	.134094	48	.80
	13	.07353	.99729	.07373	13.5634	.866455	.998823	.867632	.132368	47	
	14	.07382	.99727	.07402	13.5098	.868165	.998813	.869351	.130649	46	
.25	15	.07411	.99725	.07431	13.4566	.869868	.998804	.871064	.128936	45	.75
	16	.07440	.99723	.07461	13.4039	.871565	.998795	.872770	.127230	44	
	17	.07469	.99721	.07490	13.3515	.873255	.998785	.874469	.125531	43	
.30	18	.07498	.99719	.07519	13.2996	.874938	.998776	.876162	.123838	42	.70
	19	.07527	.99716	.07548	13.2480	.876615	.998766	.877849	.122151	41	
	20	.07556	.99714	.07578	13.1969	.878285	.998757	.879529	.120471	40	
.35	21	.07585	.99712	.07607	13.1461	.879949	.998747	.881202	.118798	39	.65
	22	.07614	.99710	.07636	13.0958	.881607	.998738	.882869	.117131	38	
	23	.07643	.99708	.07665	13.0458	.883258	.998728	.884530	.115470	37	
.40	24	.07672	.99705	.07695	12.9962	.884903	.998718	.886185	.113815	36	.60
	25	.07701	.99703	.07724	12.9469	.886542	.998708	.887833	.112167	35	
	26	.07730	.99701	.07753	12.8981	.888174	.998699	.889476	.110524	34	
.45	27	.07759	.99699	.07782	12.8496	.889801	.998689	.891112	.108888	33	.55
	28	.07788	.99696	.07812	12.8014	.891421	.998679	.892742	.107258	32	
	29	.07817	.99694	.07841	12.7536	.893035	.998669	.894366	.105634	31	
.50	30	.07846	.99692	.07870	12.7062	.894643	.998659	.895984	.104016	30	.50
	31	.07875	.99689	.07899	12.6591	.896246	.998649	.897596	.102404	29	
	32	.07904	.99687	.07929	12.6124	.897842	.998639	.899203	.100797	28	
.55	33	.07933	.99685	.07958	12.5660	.899432	.998629	.900803	.099197	27	.45
	34	.07962	.99683	.07987	12.5199	.901017	.998619	.902398	.097602	26	
	35	.07991	.99680	.08017	12.4742	.902596	.998609	.903987	.096013	25	
.60	36	.08020	.99678	.08046	12.4288	.904169	.998599	.905570	.094430	24	.40
	37	.08049	.99676	.08075	12.3838	.905736	.998589	.907147	.092853	23	
	38	.08078	.99673	.08104	12.3390	.907297	.998578	.908719	.091281	22	
.65	39	.08107	.99671	.08134	12.2946	.908853	.998568	.910285	.089715	21	.35
	40	.08136	.99668	.08163	12.2505	.910404	.998558	.911846	.088154	20	
	41	.08165	.99666	.08192	12.2067	.911949	.998548	.913401	.086599	19	
.70	42	.08194	.99664	.08221	12.1632	.913488	.998537	.914951	.085049	18	.30
	43	.08223	.99661	.08251	12.1201	.915022	.998527	.916495	.083505	17	
	44	.08252	.99659	.08280	12.0772	.916550	.998516	.918034	.081966	16	
.75	45	.08281	.99657	.08309	12.0346	.918073	.998506	.919568	.080432	15	.25
	46	.08310	.99654	.08339	11.9923	.919591	.998495	.921096	.078904	14	
	47	.08339	.99652	.08368	11.9504	.921103	.998485	.922619	.077381	13	
.80	48	.08368	.99649	.08397	11.9087	.922610	.998474	.924136	.075864	12	.20
	49	.08397	.99647	.08427	11.8673	.924112	.998464	.925649	.074351	11	
	50	.08426	.99644	.08456	11.8262	.925609	.998453	.927156	.072844	10	
.85	51	.08455	.99642	.08485	11.7853	.927100	.998442	.928658	.071342	9	.15
	52	.08484	.99639	.08514	11.7448	.928587	.998431	.930155	.069845	8	
	53	.08513	.99637	.08544	11.7045	.930068	.998421	.931647	.068353	7	
.90	54	.08542	.99635	.08573	11.6645	.931544	.998410	.933134	.066866	6	.10
	55	.08571	.99632	.08602	11.6248	.933015	.998399	.934616	.065384	5	
	56	.08600	.99630	.08632	11.5853	.934481	.998388	.936093	.063907	4	
.95	57	.08629	.99627	.08661	11.5461	.935942	.998377	.937565	.062435	3	.05
	58	.08658	.99625	.08690	11.5072	.937398	.998366	.939032	.060968	2	
	59	.08687	.99622	.08720	11.4685	.938850	.998355	.940494	.059506	1	
1.00	60	.08716	.99619	.08749	11.4301	8.940296	9.998344	8.941952	11.058048	0	.00
Decimals	Minutes	Cos	Sin	Cot	Tan	Cos	Sin	Cot	Tan	Minutes	Decimals
		Natural Values				Common Logarithms					

85°

VALUES AND LOGARITHMS OF TRIGONOMETRIC FUNCTIONS 353

5°

Decimals	Minutes	Natural Values				Common Logarithms				Minutes	Decimals
		Sin	Cos	Tan	Cot	Sin	Cos	Tan	Cot		
.00	0	.08716	.99619	.08749	11.4301	8.940296	9.998344	8.941952	11.058048	60	1.00
	1	.08745	.99617	.08778	11.3919	.941738	.998333	.943404	.056596	59	
	2	.08774	.99614	.08807	11.3540	.943174	.998322	.944852	.055148	58	
.05	3	.08803	.99612	.08837	11.3163	.944606	.998311	.946295	.053705	57	.95
	4	.08831	.99609	.08866	11.2789	.946034	.998300	.947734	.052266	56	
	5	.08860	.99607	.08895	11.2417	.947456	.998289	.949168	.050832	55	
.10	6	.08889	.99604	.08925	11.2048	.948874	.998277	.950597	.049403	54	.90
	7	.08918	.99602	.08954	11.1681	.950287	.998266	.952021	.047979	53	
	8	.08947	.99599	.08983	11.1316	.951696	.998255	.953441	.046559	52	
.15	9	.08976	.99596	.09013	11.0954	.953100	.998243	.954856	.045144	51	.85
	10	.09005	.99594	.09042	11.0594	.954499	.998232	.956267	.043733	50	
	11	.09034	.99591	.09071	11.0237	.955894	.998220	.957674	.042326	49	
.20	12	.09063	.99588	.09101	10.9882	.957284	.998209	.959075	.040925	48	.80
	13	.09092	.99586	.09130	10.9529	.958670	.998197	.960473	.039527	47	
	14	.09121	.99583	.09159	10.9178	.960052	.998186	.961866	.038134	46	
.25	15	.09150	.99580	.09189	10.8829	.961429	.998174	.963255	.036745	45	.75
	16	.09179	.99578	.09218	10.8483	.962801	.998163	.964639	.035361	44	
	17	.09208	.99575	.09247	10.8139	.964170	.998151	.966019	.033981	43	
.30	18	.09237	.99572	.09277	10.7797	.965534	.998139	.967394	.032606	42	.70
	19	.09266	.99570	.09306	10.7457	.966893	.998128	.968766	.031234	41	
	20	.09295	.99567	.09335	10.7119	.968249	.998116	.970133	.029867	40	
.35	21	.09324	.99564	.09365	10.6783	.969600	.998104	.971496	.028504	39	.65
	22	.09353	.99562	.09394	10.6450	.970947	.998092	.972855	.027145	38	
	23	.09382	.99559	.09423	10.6118	.972289	.998080	.974209	.025791	37	
.40	24	.09411	.99556	.09453	10.5789	.973628	.998068	.975560	.024440	36	.60
	25	.09440	.99553	.09482	10.5462	.974962	.998056	.976906	.023094	35	
	26	.09469	.99551	.09511	10.5136	.976293	.998044	.978248	.021752	34	
.45	27	.09498	.99548	.09541	10.4813	.977619	.998032	.979586	.020414	33	.55
	28	.09527	.99545	.09570	10.4491	.978941	.998020	.980921	.019079	32	
	29	.09556	.99542	.09600	10.4172	.980259	.998008	.982251	.017749	31	
.50	30	.09585	.99540	.09629	10.3854	.981573	.997996	.983577	.016423	30	.50
	31	.09614	.99537	.09658	10.3538	.982883	.997984	.984899	.015101	29	
	32	.09642	.99534	.09688	10.3224	.984189	.997972	.986217	.013783	28	
.55	33	.09671	.99531	.09717	10.2913	.985491	.997959	.987532	.012468	27	.45
	34	.09700	.99528	.09746	10.2602	.986789	.997947	.988842	.011158	26	
	35	.09729	.99526	.09776	10.2294	.988083	.997935	.990149	.009851	25	
.60	36	.09758	.99523	.09805	10.1988	.989374	.997922	.991451	.008549	24	.40
	37	.09787	.99520	.09834	10.1683	.990660	.997910	.992750	.007250	23	
	38	.09816	.99517	.09864	10.1381	.991943	.997897	.994045	.005955	22	
.65	39	.09845	.99514	.09893	10.1080	.993222	.997885	.995337	.004663	21	.35
	40	.09874	.99511	.09923	10.0780	.994497	.997872	.996624	.003376	20	
	41	.09903	.99508	.09952	10.0483	.995768	.997860	.997908	.002092	19	
.70	42	.09932	.99506	.09981	10.0187	.997036	.997847	.999188	.000812	18	.30
	43	.09961	.99503	.10011	9.98931	.998299	.997835	9.000465	10.999535	17	
	44	.09990	.99500	.10040	9.96007	.999560	.997822	.001738	.998262	16	
.75	45	.10019	.99497	.10069	9.93101	9.000816	.997809	.003007	.996993	15	.25
	46	.10048	.99494	.10099	9.90211	.002069	.997797	.004272	.995728	14	
	47	.10077	.99491	.10128	9.87338	.003318	.997784	.005534	.994466	13	
.80	48	.10106	.99488	.10158	9.84482	.004563	.997771	.006792	.993208	12	.20
	49	.10135	.99485	.10187	9.81641	.005805	.997758	.008047	.991953	11	
	50	.10164	.99482	.10216	9.78817	.007044	.997745	.009298	.990702	10	
.85	51	.10192	.99479	.10246	9.76009	.008273	.997732	.010546	.989454	9	.15
	52	.10221	.99476	.10275	9.73217	.009510	.997719	.011790	.988210	8	
	53	.10250	.99473	.10305	9.70441	.010737	.997706	.013031	.986969	7	
.90	54	.10279	.99470	.10334	9.67680	.011962	.997693	.014268	.985732	6	.10
	55	.10308	.99467	.10363	9.64935	.013182	.997680	.015502	.984498	5	
	56	.10337	.99464	.10393	9.62205	.014400	.997667	.016732	.983268	4	
.95	57	.10366	.99461	.10422	9.59490	.015613	.997654	.017959	.982041	3	.05
	58	.10395	.99458	.10452	9.56791	.016824	.997641	.019183	.980817	2	
	59	.10424	.99455	.10481	9.54106	.018031	.997628	.020403	.979597	1	
1.00	60	.10453	.99452	.10510	9.51436	9.019235	9.997614	9.021620	10.978380	0	.00
Decimals	Minutes	Cos	Sin	Cot	Tan	Cos	Sin	Cot	Tan	Minutes	Decimals
		Natural Values				Common Logarithms					

84°

6°

Decimals	Minutes	Natural Values				Common Logarithms				Minutes	Decimals
		Sin	Cos	Tan	Cot	Sin	Cos	Tan	Cot		
.00	0	.10453	.99452	.10510	9.51436	9.019235	9.997614	9.021620	10.978380	60	1.00
	1	.10482	.99449	.10540	9.48781	.020435	.997601	.022834	.977166	59	
	2	.10511	.99446	.10569	9.46141	.021632	.997588	.024044	.975956	58	
.05	3	.10540	.99443	.10599	9.43515	.022825	.997574	.025251	.974749	57	.95
	4	.10569	.99440	.10628	9.40904	.024016	.997561	.026455	.973545	56	
	5	.10597	.99437	.10657	9.38307	.025203	.997547	.027655	.972345	55	
.10	6	.10626	.99434	.10687	9.35724	.026386	.997534	.028852	.971148	54	.90
	7	.10655	.99431	.10716	9.33155	.027567	.997520	.030046	.969954	53	
	8	.10684	.99428	.10746	9.30599	.028744	.997507	.031237	.968763	52	
.15	9	.10713	.99424	.10775	9.28058	.029918	.997493	.032425	.967575	51	.85
	10	.10742	.99421	.10805	9.25530	.031089	.997480	.033609	.966391	50	
	11	.10771	.99418	.10834	9.23016	.032257	.997466	.034791	.965209	49	
.20	12	.10800	.99415	.10863	9.20516	.033421	.997452	.035969	.964031	48	.80
	13	.10829	.99412	.10893	9.18028	.034582	.997439	.037144	.962856	47	
	14	.10858	.99409	.10922	9.15554	.035741	.997425	.038316	.961684	46	
.25	15	.10887	.99406	.10952	9.13093	.036896	.997411	.039485	.960515	45	.75
	16	.10916	.99402	.10981	9.10646	.038048	.997397	.040651	.959349	44	
	17	.10945	.99399	.11011	9.08211	.039197	.997383	.041813	.958187	43	
.30	18	.10973	.99396	.11040	9.05789	.040342	.997369	.042973	.957027	42	.70
	19	.11002	.99393	.11070	9.03379	.041485	.997355	.044130	.955870	41	
	20	.11031	.99390	.11099	9.00983	.042625	.997341	.045284	.954716	40	
.35	21	.11060	.99386	.11128	8.98598	.043762	.997327	.046434	.953566	39	.65
	22	.11089	.99383	.11158	8.96227	.044895	.997313	.047582	.952418	38	
	23	.11118	.99380	.11187	8.93867	.046026	.997299	.048727	.951273	37	
.40	24	.11147	.99377	.11217	8.91520	.047154	.997285	.049869	.950131	36	.60
	25	.11176	.99374	.11246	8.89185	.048279	.997271	.051008	.948992	35	
	26	.11205	.99370	.11276	8.86862	.049400	.997257	.052144	.947856	34	
.45	27	.11234	.99367	.11305	8.84551	.050519	.997242	.053277	.946723	33	.55
	28	.11263	.99364	.11335	8.82252	.051635	.997228	.054407	.945593	32	
	29	.11291	.99360	.11364	8.79964	.052749	.997214	.055535	.944465	31	
.50	30	.11320	.99357	.11394	8.77689	.053859	.997199	.056659	.943341	30	.50
	31	.11349	.99354	.11423	8.75425	.054966	.997185	.057781	.942219	29	
	32	.11378	.99351	.11452	8.73172	.056071	.997170	.058900	.941100	28	
.55	33	.11407	.99347	.11482	8.70931	.057172	.997156	.060016	.939984	27	.45
	34	.11436	.99344	.11511	8.68701	.058271	.997141	.061130	.938870	26	
	35	.11465	.99341	.11541	8.66482	.059367	.997127	.062240	.937760	25	
.60	36	.11494	.99337	.11570	8.64275	.060460	.997112	.063348	.936652	24	.40
	37	.11523	.99334	.11600	8.62078	.061551	.997098	.064453	.935547	23	
	38	.11552	.99331	.11629	8.59893	.062639	.997083	.065556	.934444	22	
.65	39	.11580	.99327	.11659	8.57718	.063724	.997068	.066655	.933345	21	.35
	40	.11609	.99324	.11688	8.55555	.064806	.997053	.067752	.932248	20	
	41	.11638	.99320	.11718	8.53402	.065885	.997039	.068846	.931154	19	
.70	42	.11667	.99317	.11747	8.51259	.066962	.997024	.069938	.930062	18	.30
	43	.11696	.99314	.11777	8.49128	.068036	.997009	.071027	.928973	17	
	44	.11725	.99310	.11806	8.47007	.069107	.996994	.072113	.927887	16	
.75	45	.11754	.99307	.11836	8.44896	.070176	.996979	.073197	.926803	15	.25
	46	.11783	.99303	.11865	8.42795	.071242	.996964	.074278	.925722	14	
	47	.11812	.99300	.11895	8.40705	.072306	.996949	.075356	.924644	13	
.80	48	.11840	.99297	.11924	8.38625	.073366	.996934	.076432	.923568	12	.20
	49	.11869	.99293	.11954	8.36555	.074424	.996919	.077505	.922495	11	
	50	.11898	.99290	.11983	8.34496	.075480	.996904	.078576	.921424	10	
.85	51	.11927	.99286	.12013	8.32446	.076533	.996889	.079644	.920356	9	.15
	52	.11956	.99283	.12042	8.30406	.077583	.996874	.080710	.919290	8	
	53	.11985	.99279	.12072	8.28376	.078631	.996858	.081773	.918227	7	
.90	54	.12014	.99276	.12101	8.26355	.079676	.996843	.082833	.917167	6	.10
	55	.12043	.99272	.12131	8.24345	.080719	.996828	.083891	.916109	5	
	56	.12071	.99269	.12160	8.22344	.081759	.996812	.084947	.915053	4	
.95	57	.12100	.99265	.12190	8.20352	.082797	.996797	.086000	.914000	3	.05
	58	.12129	.99262	.12219	8.18370	.083832	.996782	.087050	.912950	2	
	59	.12158	.99258	.12249	8.16398	.084864	.996766	.088098	.911902	1	
1.00	60	.12187	.99255	.12278	8.14435	9.085894	9.996751	9.089144	10.910856	0	.00
Decimals	Minutes	Cos	Sin	Cot	Tan	Cos	Sin	Cot	Tan	Minutes	Decimals
		Natural Values				Common Logarithms					

83°

VALUES AND LOGARITHMS OF TRIGONOMETRIC FUNCTIONS 355

7°

Decimals	Minutes	Natural Values				Common Logarithms				Minutes	Decimals
		Sin	Cos	Tan	Cot	Sin	Cos	Tan	Cot		
.00	0	.12187	.99255	.12278	8.14435	9.085894	9.996751	9.089144	10.910856	60	1.00
	1	.12216	.99251	.12308	8.12481	.086922	.996735	.090187	.909813	59	
	2	.12245	.99248	.12338	8.10536	.087947	.996720	.091228	.908772	58	
.05	3	.12274	.99244	.12367	8.08600	.088970	.996704	.092266	.907734	57	.95
	4	.12302	.99240	.12397	8.06674	.089990	.996688	.093302	.906698	56	
	5	.12331	.99237	.12426	8.04756	.091008	.996673	.094336	.905664	55	
.10	6	.12360	.99233	.12456	8.02848	.092024	.996657	.095367	.904633	54	.90
	7	.12389	.99230	.12485	8.00948	.093037	.996641	.096395	.903605	53	
	8	.12418	.99226	.12515	7.99058	.094047	.996625	.097422	.902578	52	
.15	9	.12447	.99222	.12544	7.97176	.095056	.996610	.098446	.901554	51	.85
	10	.12476	.99219	.12574	7.95302	.096062	.996594	.099468	.900532	50	
	11	.12504	.99215	.12603	7.93438	.097065	.996578	.100487	.899513	49	
.20	12	.12533	.99211	.12633	7.91582	.098066	.996562	.101504	.898496	48	.80
	13	.12562	.99208	.12662	7.89734	.099065	.996546	.102519	.897481	47	
	14	.12591	.99204	.12692	7.87895	.100062	.996530	.103532	.896468	46	
.25	15	.12620	.99200	.12722	7.86064	.101056	.996514	.104542	.895458	45	.75
	16	.12649	.99197	.12751	7.84242	.102048	.996498	.105550	.894450	44	
	17	.12678	.99193	.12781	7.82428	.103037	.996482	.106556	.893444	43	
.30	18	.12706	.99189	.12810	7.80622	.104025	.996465	.107559	.892441	42	.70
	19	.12735	.99186	.12840	7.78825	.105010	.996449	.108560	.891440	41	
	20	.12764	.99182	.12869	7.77035	.105992	.996433	.109559	.890441	40	
.35	21	.12793	.99178	.12899	7.75254	.106973	.996417	.110556	.889444	39	.65
	22	.12822	.99175	.12929	7.73480	.107951	.996400	.111551	.888449	38	
	23	.12851	.99171	.12958	7.71715	.108927	.996384	.112543	.887457	37	
.40	24	.12880	.99167	.12988	7.69957	.109901	.996368	.113533	.886467	36	.60
	25	.12908	.99163	.13017	7.68208	.110873	.996351	.114521	.885479	35	
	26	.12937	.99160	.13047	7.66466	.111842	.996335	.115507	.884493	34	
.45	27	.12966	.99156	.13076	7.64732	.112809	.996318	.116491	.883509	33	.55
	28	.12995	.99152	.13106	7.63005	.113774	.996302	.117472	.882528	32	
	29	.13024	.99148	.13136	7.61287	.114737	.996285	.118452	.881548	31	
.50	30	.13053	.99144	.13165	7.59575	.115698	.996269	.119429	.880571	30	.50
	31	.13081	.99141	.13195	7.57872	.116656	.996252	.120404	.879596	29	
	32	.13110	.99137	.13224	7.56176	.117613	.996235	.121377	.878623	28	
.55	33	.13139	.99133	.13254	7.54487	.118567	.996219	.122348	.877652	27	.45
	34	.13168	.99129	.13284	7.52806	.119519	.996202	.123317	.876683	26	
	35	.13197	.99125	.13313	7.51132	.120469	.996185	.124284	.875716	25	
.60	36	.13226	.99122	.13343	7.49465	.121417	.996168	.125249	.874751	24	.40
	37	.13254	.99118	.13372	7.47806	.122362	.996151	.126211	.873789	23	
	38	.13283	.99114	.13402	7.46154	.123306	.996134	.127172	.872828	22	
.65	39	.13312	.99110	.13432	7.44509	.124248	.996117	.128130	.871870	21	.35
	40	.13341	.99106	.13461	7.42871	.125187	.996100	.129087	.870913	20	
	41	.13370	.99102	.13491	7.41240	.126125	.996083	.130041	.869959	19	
.70	42	.13399	.99098	.13521	7.39616	.127060	.996066	.130994	.869006	18	.30
	43	.13427	.99094	.13550	7.37999	.127993	.996049	.131944	.868056	17	
	44	.13456	.99091	.13580	7.36389	.128925	.996032	.132893	.867107	16	
.75	45	.13485	.99087	.13609	7.34786	.129854	.996015	.133839	.866161	15	.25
	46	.13514	.99083	.13639	7.33190	.130781	.995998	.134784	.865216	14	
	47	.13543	.99079	.13669	7.31600	.131706	.995980	.135726	.864274	13	
.80	48	.13572	.99075	.13698	7.30018	.132630	.995963	.136667	.863333	12	.20
	49	.13600	.99071	.13728	7.28442	.133551	.995946	.137605	.862395	11	
	50	.13629	.99067	.13758	7.26873	.134470	.995928	.138542	.861458	10	
.85	51	.13658	.99063	.13787	7.25310	.135387	.995911	.139476	.860524	9	.15
	52	.13687	.99059	.13817	7.23754	.136303	.995894	.140409	.859591	8	
	53	.13716	.99055	.13846	7.22204	.137216	.995876	.141340	.858660	7	
.90	54	.13744	.99051	.13876	7.20661	.138128	.995859	.142269	.857731	6	.10
	55	.13773	.99047	.13906	7.19125	.139037	.995841	.143196	.856804	5	
	56	.13802	.99043	.13935	7.17594	.139944	.995823	.144121	.855879	4	
.95	57	.13831	.99039	.13965	7.16071	.140850	.995806	.145044	.854956	3	.05
	58	.13860	.99035	.13995	7.14553	.141754	.995788	.145966	.854034	2	
	59	.13889	.99031	.14024	7.13042	.142655	.995771	.146885	.853115	1	
1.00	60	.13917	.99027	.14054	7.11537	9.143555	9.995753	9.147803	10.852197	0	.00
Decimals	Minutes	Cos	Sin	Cot	Tan	Cos	Sin	Cot	Tan	Minutes	Decimals
		Natural Values				Common Logarithms					

82°

8°

Decimals	Minutes	Natural Values				Common Logarithms				Minutes	Decimals
		Sin	Cos	Tan	Cot	Sin	Cos	Tan	Cot		
.00	0	.13917	.99027	.14054	7.11537	9.143555	9.995753	9.147803	10.852197	60	1.00
	1	.13946	.99023	.14084	7.10038	.144453	.995735	.148718	.851282	59	
	2	.13975	.99019	.14113	7.08546	.145349	.995717	.149632	.850368	58	
.05	3	.14004	.99015	.14143	7.07059	.146243	.995699	.150544	.849456	57	.95
	4	.14033	.99011	.14173	7.05579	.147136	.995681	.151454	.848546	56	
	5	.14061	.99006	.14202	7.04105	.148026	.995664	.152363	.847637	55	
.10	6	.14090	.99002	.14232	7.02637	.148915	.995646	.153269	.846731	54	.90
	7	.14119	.98998	.14262	7.01174	.149802	.995628	.154174	.845826	53	
	8	.14148	.98994	.14291	6.99718	.150686	.995610	.155077	.844923	52	
.15	9	.14177	.98990	.14321	6.98268	.151569	.995591	.155978	.844022	51	.85
	10	.14205	.98986	.14351	6.96823	.152451	.995573	.156877	.843123	50	
	11	.14234	.98982	.14381	6.95385	.153330	.995555	.157775	.842225	49	
.20	12	.14263	.98978	.14410	6.93952	.154208	.995537	.158671	.841329	48	.80
	13	.14292	.98973	.14440	6.92525	.155083	.995519	.159565	.840435	47	
	14	.14320	.98969	.14470	6.91104	.155957	.995501	.160457	.839543	46	
.25	15	.14349	.98965	.14499	6.89388	.156830	.995482	.161347	.838653	45	.75
	16	.14378	.98961	.14529	6.88278	.157700	.995464	.162236	.837764	44	
	17	.14407	.98957	.14559	6.86874	.158569	.995446	.163123	.836877	43	
.30	18	.14436	.98953	.14588	6.85475	.159435	.995427	.164008	.835992	42	.70
	19	.14464	.98948	.14618	6.84082	.160301	.995409	.164892	.835108	41	
	20	.14493	.98944	.14648	6.82694	.161164	.995390	.165774	.834226	40	
.35	21	.14522	.98940	.14678	6.81312	.162025	.995372	.166654	.833346	39	.65
	22	.14551	.98936	.14707	6.79936	.162885	.995353	.167532	.832468	38	
	23	.14580	.98931	.14737	6.78564	.163743	.995334	.168409	.831591	37	
.40	24	.14608	.98927	.14767	6.77199	.164600	.995316	.169284	.830716	36	.60
	25	.14637	.98923	.14796	6.75838	.165454	.995297	.170157	.829843	35	
	26	.14666	.98919	.14826	6.74483	.166307	.995278	.171029	.828971	34	
.45	27	.14695	.98914	.14856	6.73133	.167159	.995260	.171899	.828101	33	.55
	28	.14723	.98910	.14886	6.71789	.168008	.995241	.172767	.827233	32	
	29	.14752	.98906	.14915	6.70450	.168856	.995222	.173634	.826366	31	
.50	30	.14781	.98902	.14945	6.69116	.169702	.995203	.174499	.825501	30	.50
	31	.14810	.98897	.14975	6.67787	.170547	.995184	.175362	.824638	29	
	32	.14838	.98893	.15005	6.66463	.171389	.995165	.176224	.823776	28	
.55	33	.14867	.98889	.15034	6.65144	.172230	.995146	.177084	.822916	27	.45
	34	.14896	.98884	.15064	6.63831	.173070	.995127	.177942	.822058	26	
	35	.14925	.98880	.15094	6.62523	.173908	.995108	.178799	.821201	25	
.60	36	.14954	.98876	.15124	6.61219	.174744	.995089	.179655	.820345	24	.40
	37	.14982	.98871	.15153	6.59921	.175578	.995070	.180508	.819492	23	
	38	.15011	.98867	.15183	6.58627	.176411	.995051	.181360	.818640	22	
.65	39	.15040	.98863	.15213	6.57339	.177242	.995032	.182211	.817789	21	.35
	40	.15069	.98858	.15243	6.56055	.178072	.995013	.183059	.816941	20	
	41	.15097	.98854	.15272	6.54777	.178900	.994993	.183907	.816093	19	
.70	42	.15126	.98849	.15302	6.53503	.179726	.994974	.184752	.815248	18	.30
	43	.15155	.98845	.15332	6.52234	.180551	.994955	.185597	.814403	17	
	44	.15184	.98841	.15362	6.50970	.181374	.994935	.186439	.813561	16	
.75	45	.15212	.98836	.15391	6.49710	.182196	.994916	.187280	.812720	15	.25
	46	.15241	.98832	.15421	6.48456	.183016	.994896	.188120	.811880	14	
	47	.15270	.98827	.15451	6.47206	.183834	.994877	.188958	.811042	13	
.80	48	.15299	.98823	.15481	6.45961	.184651	.994857	.189794	.810206	12	.20
	49	.15327	.98818	.15511	6.44720	.185466	.994838	.190629	.809371	11	
	50	.15356	.98814	.15540	6.43484	.186280	.994818	.191462	.808538	10	
.85	51	.15385	.98809	.15570	6.42253	.187092	.994798	.192294	.807706	9	.15
	52	.15414	.98805	.15600	6.41026	.187903	.994779	.193124	.806876	8	
	53	.15442	.98800	.15630	6.39804	.188712	.994759	.193953	.806047	7	
.90	54	.15471	.98796	.15660	6.38587	.189519	.994739	.194780	.805220	6	.10
	55	.15500	.98791	.15689	6.37374	.190325	.994720	.195606	.804394	5	
	56	.15529	.98787	.15719	6.36165	.191130	.994700	.196430	.803570	4	
.95	57	.15557	.98782	.15749	6.34961	.191933	.994680	.197253	.802747	3	.05
	58	.15586	.98778	.15779	6.33761	.192734	.994660	.198074	.801926	2	
	59	.15615	.98773	.15809	6.32566	.193534	.994640	.198894	.801106	1	
1.00	60	.15643	.98769	.15838	6.31375	9.194332	9.994620	9.199713	10.800287	0	.00
Decimals	Minutes	Cos	Sin	Cot	Tan	Cos	Sin	Cot	Tan	Minutes	Decimals
		Natural Values				Common Logarithms					

81°

VALUES AND LOGARITHMS OF TRIGONOMETRIC FUNCTIONS

9°

Decimals	Minutes	Natural Values				Common Logarithms				Minutes	Decimals
		Sin	Cos	Tan	Cot	Sin	Cos	Tan	Cot		
.00	0	.15643	.98769	.15838	6.31375	9.194332	9.994620	9.199713	10.800287	60	1.00
	1	.15672	.98764	.15868	6.30189	.195129	.994600	.200529	.799471	59	
	2	.15701	.98760	.15898	6.29007	.195925	.994580	.201345	.798655	58	
.05	3	.15730	.98755	.15928	6.27829	.196719	.994560	.202159	.797841	57	.95
	4	.15758	.98751	.15958	6.26655	.197511	.994540	.202971	.797029	56	
	5	.15787	.98746	.15988	6.25486	.198302	.994519	.203782	.796218	55	
.10	6	.15816	.98741	.16017	6.24321	.199091	.994499	.204592	.795408	54	.90
	7	.15845	.98737	.16047	6.23160	.199879	.994479	.205400	.794600	53	
	8	.15873	.98732	.16077	6.22003	.200666	.994459	.206207	.793793	52	
.15	9	.15902	.98728	.16107	6.20851	.201451	.994438	.207013	.792987	51	.85
	10	.15931	.98723	.16137	6.19703	.202234	.994418	.207817	.792183	50	
	11	.15959	.98718	.16167	6.18559	.203017	.994398	.208619	.791381	49	
.20	12	.15988	.98714	.16196	6.17419	.203797	.994377	.209420	.790580	48	.80
	13	.16017	.98709	.16226	6.16283	.204577	.994357	.210220	.789780	47	
	14	.16046	.98704	.16256	6.15151	.205354	.994336	.211018	.788982	46	
.25	15	.16074	.98700	.16286	6.14023	.206131	.994316	.211815	.788185	45	.75
	16	.16103	.98695	.16316	6.12899	.206906	.994295	.212611	.787389	44	
	17	.16132	.98690	.16346	6.11779	.207679	.994274	.213405	.786595	43	
.30	18	.16160	.98686	.16376	6.10664	.208452	.994254	.214198	.785802	42	.70
	19	.16189	.98681	.16405	6.09552	.209222	.994233	.214989	.785011	41	
	20	.16218	.98676	.16435	6.08444	.209992	.994212	.215780	.784220	40	
.35	21	.16246	.98671	.16465	6.07340	.210760	.994191	.216568	.783432	39	.65
	22	.16275	.98667	.16495	6.06240	.211526	.994171	.217356	.782644	38	
	23	.16304	.98662	.16525	6.05143	.212291	.994150	.218142	.781858	37	
.40	24	.16333	.98657	.16555	6.04051	.213055	.994129	.218926	.781074	36	.60
	25	.16361	.98652	.16585	6.02962	.213818	.994108	.219710	.780290	35	
	26	.16390	.98648	.16615	6.01878	.214579	.994087	.220492	.779508	34	
.45	27	.16419	.98643	.16645	6.00797	.215338	.994066	.221272	.778728	33	.55
	28	.16447	.98638	.16674	5.99720	.216097	.994045	.222052	.777948	32	
	29	.16476	.98633	.16704	5.98646	.216854	.994024	.222830	.777170	31	
.50	30	.16505	.98629	.16734	5.97576	.217609	.994003	.223607	.776393	30	.50
	31	.16533	.98624	.16764	5.96510	.218363	.993982	.224382	.775618	29	
	32	.16562	.98619	.16794	5.95448	.219116	.993960	.225156	.774844	28	
.55	33	.16591	.98614	.16824	5.94390	.219868	.993939	.225929	.774071	27	.45
	34	.16620	.98609	.16854	5.93335	.220618	.993918	.226700	.773300	26	
	35	.16648	.98604	.16884	5.92283	.221367	.993897	.227471	.772529	25	
.60	36	.16677	.98600	.16914	5.91236	.222115	.993875	.228239	.771761	24	.40
	37	.16706	.98595	.16944	5.90191	.222861	.993854	.229007	.770993	23	
	38	.16734	.98590	.16974	5.89151	.223606	.993832	.229773	.770227	22	
.65	39	.16763	.98585	.17004	5.88114	.224349	.993811	.230539	.769461	21	.35
	40	.16792	.98580	.17033	5.87080	.225092	.993789	.231302	.768698	20	
	41	.16820	.98575	.17063	5.86051	.225833	.993768	.232065	.767935	19	
.70	42	.16849	.98570	.17093	5.85024	.226573	.993746	.232826	.767174	18	.30
	43	.16878	.98565	.17123	5.84001	.227311	.993725	.233586	.766414	17	
	44	.16906	.98561	.17153	5.82982	.228048	.993703	.234345	.765655	16	
.75	45	.16935	.98556	.17183	5.81966	.228784	.993681	.235103	.764897	15	.25
	46	.16964	.98551	.17213	5.80953	.229518	.993660	.235859	.764141	14	
	47	.16992	.98546	.17243	5.79944	.230252	.993638	.236614	.763386	13	
.80	48	.17021	.98541	.17273	5.78938	.230984	.993616	.237368	.762632	12	.20
	49	.17050	.98536	.17303	5.77936	.231715	.993594	.238120	.761880	11	
	50	.17078	.98531	.17333	5.76937	.232444	.993572	.238872	.761128	10	
.85	51	.17107	.98526	.17363	5.75941	.233172	.993550	.239622	.760378	9	.15
	52	.17136	.98521	.17393	5.74949	.233899	.993528	.240371	.759629	8	
	53	.17164	.98516	.17423	5.73960	.234625	.993506	.241118	.758882	7	
.90	54	.17193	.98511	.17453	5.72974	.235349	.993484	.241865	.758135	6	.10
	55	.17222	.98506	.17483	5.71992	.236073	.993462	.242610	.757390	5	
	56	.17250	.98501	.17513	5.71013	.236795	.993440	.243354	.756646	4	
.95	57	.17279	.98496	.17543	5.70037	.237515	.993418	.244097	.755903	3	.05
	58	.17308	.98491	.17573	5.69064	.238235	.993396	.244839	.755161	2	
	59	.17336	.98486	.17603	5.68094	.238953	.993374	.245579	.754421	1	
1.00	60	.17365	.98481	.17633	5.67128	9.239670	9.993351	9.246319	10.753681	0	.00
Decimals	Minutes	Cos	Sin	Cot	Tan	Cos	Sin	Cot	Tan	Minutes	Decimals
		Natural Values				Common Logarithms					

80°

10°

Decimals	Minutes	Natural Values				Common Logarithms				Minutes	Decimals
		Sin	Cos	Tan	Cot	Sin	Cos	Tan	Cot		
.00	0	.17365	.98481	.17633	5.67128	9.239670	9.993351	9.246319	10.753681	60	1.00
	1	.17393	.98476	.17663	5.66165	.240386	.993329	.247057	.752943	59	
	2	.17422	.98471	.17693	5.65205	.241101	.993307	.247794	.752206	58	
.05	3	.17451	.98466	.17723	5.64248	.241814	.993284	.248530	.751470	57	.95
	4	.17479	.98461	.17753	5.63295	.242526	.993262	.249264	.750736	56	
	5	.17508	.98455	.17783	5.62344	.243237	.993240	.249998	.750002	55	
.10	6	.17537	.98450	.17813	5.61397	.243947	.993217	.250730	.749270	54	.90
	7	.17565	.98445	.17843	5.60452	.244656	.993195	.251461	.748539	53	
	8	.17594	.98440	.17873	5.59511	.245363	.993172	.252191	.747809	52	
.15	9	.17623	.98435	.17903	5.58573	.246069	.993149	.252920	.747080	51	.85
	10	.17651	.98430	.17933	5.57638	.246775	.993127	.253648	.746352	50	
	11	.17680	.98425	.17963	5.56706	.247478	.993104	.254374	.745626	49	
.20	12	.17708	.98420	.17993	5.55777	.248181	.993081	.255100	.744900	48	.80
	13	.17737	.98414	.18023	5.54851	.248883	.993059	.255824	.744176	47	
	14	.17766	.98409	.18053	5.53927	.249583	.993036	.256547	.743453	46	
.25	15	.17794	.98404	.18083	5.53007	.250282	.993013	.257269	.742731	45	.75
	16	.17823	.98399	.18113	5.52090	.250980	.992990	.257990	.742010	44	
	17	.17852	.98394	.18143	5.51176	.251677	.992967	.258710	.741290	43	
.30	18	.17880	.98389	.18173	5.50264	.252373	.992944	.259429	.740571	42	.70
	19	.17909	.98383	.18203	5.49356	.253067	.992921	.260146	.739854	41	
	20	.17937	.98378	.18233	5.48451	.253761	.992898	.260863	.739137	40	
.35	21	.17966	.98373	.18263	5.47548	.254453	.992875	.261578	.738422	39	.65
	22	.17995	.98368	.18293	5.46648	.255144	.992852	.262292	.737708	38	
	23	.18023	.98362	.18323	5.45751	.255834	.992829	.263005	.736995	37	
.40	24	.18052	.98357	.18353	5.44857	.256523	.992806	.263717	.736283	36	.60
	25	.18081	.98352	.18384	5.43966	.257211	.992783	.264428	.735572	35	
	26	.18109	.98347	.18414	5.43077	.257898	.992759	.265138	.734862	34	
.45	27	.18138	.98341	.18444	5.42192	.258583	.992736	.265847	.734153	33	.55
	28	.18166	.98336	.18474	5.41309	.259268	.992713	.266555	.733445	32	
	29	.18195	.98331	.18504	5.40429	.259951	.992690	.267261	.732739	31	
.50	30	.18224	.98325	.18534	5.39552	.260633	.992666	.267967	.732033	30	.50
	31	.18252	.98320	.18564	5.38677	.261314	.992643	.268671	.731329	29	
	32	.18281	.98315	.18594	5.37805	.261994	.992619	.269375	.730625	28	
.55	33	.18309	.98310	.18624	5.36936	.262673	.992596	.270077	.729923	27	.45
	34	.18338	.98304	.18654	5.36070	.263351	.992572	.270779	.729221	26	
	35	.18367	.98299	.18684	5.35206	.264027	.992549	.271479	.728521	25	
.60	36	.18395	.98294	.18714	5.34345	.264703	.992525	.272178	.727822	24	.40
	37	.18424	.98288	.18745	5.33487	.265377	.992501	.272876	.727124	23	
	38	.18452	.98283	.18775	5.32631	.266051	.992478	.273573	.726427	22	
.65	39	.18481	.98277	.18805	5.31778	.266723	.992454	.274269	.725731	21	.35
	40	.18509	.98272	.18835	5.30928	.267395	.992430	.274964	.725036	20	
	41	.18538	.98267	.18865	5.30080	.268065	.992406	.275658	.724342	19	
.70	42	.18567	.98261	.18895	5.29235	.268734	.992382	.276351	.723649	18	.30
	43	.18595	.98256	.18925	5.28393	.269402	.992359	.277043	.722957	17	
	44	.18624	.98250	.18955	5.27553	.270069	.992335	.277734	.722266	16	
.75	45	.18652	.98245	.18986	5.26715	.270735	.992311	.278424	.721576	15	.25
	46	.18681	.98240	.19016	5.25880	.271400	.992287	.279113	.720887	14	
	47	.18710	.98234	.19046	5.25048	.272064	.992263	.279801	.720199	13	
.80	48	.18738	.98229	.19076	5.24218	.272726	.992239	.280488	.719512	12	.20
	49	.18767	.98223	.19106	5.23391	.273388	.992214	.281174	.718826	11	
	50	.18795	.98218	.19136	5.22566	.274049	.992190	.281858	.718142	10	
.85	51	.18824	.98212	.19166	5.21744	.274708	.992166	.282542	.717458	9	.15
	52	.18852	.98207	.19197	5.20925	.275367	.992142	.283225	.716775	8	
	53	.18881	.98201	.19227	5.20107	.276025	.992118	.283907	.716093	7	
.90	54	.18910	.98196	.19257	5.19293	.276681	.992093	.284588	.715412	6	.10
	55	.18938	.98190	.19287	5.18480	.277337	.992069	.285268	.714732	5	
	56	.18967	.98185	.19317	5.17671	.277991	.992044	.285947	.714053	4	
.95	57	.18995	.98179	.19347	5.16863	.278645	.992020	.286624	.713376	3	.05
	58	.19024	.98174	.19378	5.16058	.279297	.991996	.287301	.712699	2	
	59	.19052	.98168	.19408	5.15256	.279948	.991971	.287977	.712023	1	
1.00	60	.19081	.98163	.19438	5.14455	9.280599	9.991947	9.288652	10.711348	0	.00
Decimals	Minutes	Cos	Sin	Cot	Tan	Cos	Sin	Cot	Tan	Minutes	Decimals
		Natural Values				Common Logarithms					

79°

VALUES AND LOGARITHMS OF TRIGONOMETRIC FUNCTIONS   359

11°

Decimals	Minutes	Natural Values				Common Logarithms				Minutes	Decimals
		Sin	Cos	Tan	Cot	Sin	Cos	Tan	Cot		
.00	0	.19081	.98163	.19438	5.14455	9.280599	9.991947	9.288652	10.711348	60	1.00
	1	.19109	.98157	.19468	5.13658	.281248	.991922	.289326	.710674	59	
	2	.19138	.98152	.19498	5.12862	.281897	.991897	.289999	.710001	58	
.05	3	.19167	.98146	.19529	5.12069	.282544	.991873	.290671	.709329	57	.95
	4	.19195	.98140	.19559	5.11279	.283190	.991848	.291342	.708658	56	
	5	.19224	.98135	.19589	5.10490	.283836	.991823	.292013	.707987	55	
.10	6	.19252	.98129	.19619	5.09704	.284480	.991799	.292682	.707318	54	.90
	7	.19281	.98124	.19649	5.08921	.285124	.991774	.293350	.706650	53	
	8	.19309	.98118	.19680	5.08139	.285766	.991749	.294017	.705983	52	
.15	9	.19338	.98112	.19710	5.07360	.286408	.991724	.294684	.705316	51	.85
	10	.19366	.98107	.19740	5.06584	.287048	.991699	.295349	.704651	50	
	11	.19395	.98101	.19770	5.05809	.287688	.991674	.296013	.703987	49	
.20	12	.19423	.98096	.19801	5.05037	.288326	.991649	.296677	.703323	48	.80
	13	.19452	.98090	.19831	5.04267	.288964	.991624	.297339	.702661	47	
	14	.19481	.98084	.19861	5.03499	.289600	.991599	.298001	.701999	46	
.25	15	.19509	.98079	.19891	5.02734	.290236	.991574	.298662	.701338	45	.75
	16	.19538	.98073	.19921	5.01971	.290870	.991549	.299322	.700678	44	
	17	.19566	.98067	.19952	5.01210	.291504	.991524	.299980	.700020	43	
.30	18	.19595	.98061	.19982	5.00451	.292137	.991498	.300638	.699362	42	.70
	19	.19623	.98056	.20012	4.99695	.292768	.991473	.301295	.698705	41	
	20	.19652	.98050	.20042	4.98940	.293399	.991448	.301951	.698049	40	
.35	21	.19680	.98044	.20073	4.98188	.294029	.991422	.302607	.697393	39	.65
	22	.19709	.98039	.20103	4.97438	.294658	.991397	.303261	.696739	38	
	23	.19737	.98033	.20133	4.96690	.295286	.991372	.303914	.696086	37	
.40	24	.19766	.98027	.20164	4.95945	.295913	.991346	.304567	.695433	36	.60
	25	.19794	.98021	.20194	4.95201	.296539	.991321	.305218	.694782	35	
	26	.19823	.98016	.20224	4.94460	.297164	.991295	.305869	.694131	34	
.45	27	.19851	.98010	.20254	4.93721	.297788	.991270	.306519	.693481	33	.55
	28	.19880	.98004	.20285	4.92984	.298412	.991244	.307168	.692832	32	
	29	.19908	.97998	.20315	4.92249	.299034	.991218	.307816	.692184	31	
.50	30	.19937	.97992	.20345	4.91516	.299655	.991193	.308463	.691537	30	.50
	31	.19965	.97987	.20376	4.90785	.300276	.991167	.309109	.690891	29	
	32	.19994	.97981	.20406	4.90056	.300895	.991141	.309754	.690246	28	
.55	33	.20022	.97975	.20436	4.89330	.301514	.991115	.310399	.689601	27	.45
	34	.20051	.97969	.20466	4.88605	.302132	.991090	.311042	.688958	26	
	35	.20079	.97963	.20497	4.87882	.302748	.991064	.311685	.688315	25	
.60	36	.20108	.97958	.20527	4.87162	.303364	.991038	.312327	.687673	24	.40
	37	.20136	.97952	.20557	4.86444	.303979	.991012	.312968	.687032	23	
	38	.20165	.97946	.20588	4.85727	.304593	.990986	.313608	.686392	22	
.65	39	.20193	.97940	.20618	4.85013	.305207	.990960	.314247	.685753	21	.35
	40	.20222	.97934	.20648	4.84300	.305819	.990934	.314885	.685115	20	
	41	.20250	.97928	.20679	4.83590	.306430	.990908	.315523	.684477	19	
.70	42	.20279	.97922	.20709	4.82882	.307041	.990882	.316159	.683841	18	.30
	43	.20307	.97916	.20739	4.82175	.307650	.990855	.316795	.683205	17	
	44	.20336	.97910	.20770	4.81471	.308259	.990829	.317430	.682570	16	
.75	45	.20364	.97905	.20800	4.80769	.308867	.990803	.318064	.681936	15	.25
	46	.20393	.97899	.20830	4.80068	.309474	.990777	.318697	.681303	14	
	47	.20421	.97893	.20861	4.79370	.310080	.990750	.319330	.680670	13	
.80	48	.20450	.97887	.20891	4.78673	.310685	.990724	.319961	.680039	12	.20
	49	.20478	.97881	.20921	4.77978	.311289	.990697	.320592	.679408	11	
	50	.20507	.97875	.20952	4.77286	.311893	.990671	.321222	.678778	10	
.85	51	.20535	.97869	.20982	4.76595	.312495	.990645	.321851	.678149	9	.15
	52	.20563	.97863	.21013	4.75906	.313097	.990618	.322479	.677521	8	
	53	.20592	.97857	.21043	4.75219	.313698	.990591	.323106	.676894	7	
.90	54	.20620	.97851	.21073	4.74534	.314297	.990565	.323733	.676267	6	.10
	55	.20649	.97845	.21104	4.73851	.314897	.990538	.324358	.675642	5	
	56	.20677	.97839	.21134	4.73170	.315495	.990511	.324983	.675017	4	
.95	57	.20706	.97833	.21164	4.72490	.316092	.990485	.325607	.674393	3	.05
	58	.20734	.97827	.21195	4.71813	.316689	.990458	.326231	.673769	2	
	59	.20763	.97821	.21225	4.71137	.317284	.990431	.326853	.673147	1	
1.00	60	.20791	.97815	.21256	4.70463	9.317879	9.990404	9.327475	10.672525	0	.00
Decimals	Minutes	Cos	Sin	Cot	Tan	Cos	Sin	Cot	Tan	Minutes	Decimals
		Natural Values				Common Logarithms					

78°

12°

Decimals	Minutes	Natural Values				Common Logarithms				Minutes	Decimals
		Sin	Cos	Tan	Cot	Sin	Cos	Tan	Cot		
.00	0	.20791	.97815	.21256	4.70463	9.317879	9.990404	9.327475	10.672525	60	1.00
	1	.20820	.97809	.21286	4.69791	.318473	.990378	.328095	.671905	59	
	2	.20848	.97803	.21316	4.69121	.319066	.990351	.328715	.671285	58	
.05	3	.20877	.97797	.21347	4.68452	.319658	.990324	.329334	.670666	57	.95
	4	.20905	.97791	.21377	4.67786	.320249	.990297	.329953	.670047	56	
	5	.20933	.97784	.21408	4.67121	.320840	.990270	.330570	.669430	55	
.10	6	.20962	.97778	.21438	4.66458	.321430	.990243	.331187	.668813	54	.90
	7	.20990	.97772	.21469	4.65797	.322019	.990215	.331803	.668197	53	
	8	.21019	.97766	.21499	4.65138	.322607	.990188	.332418	.667582	52	
.15	9	.21047	.97760	.21529	4.64480	.323194	.990161	.333033	.666967	51	.85
	10	.21076	.97754	.21560	4.63825	.323780	.990134	.333646	.666354	50	
	11	.21104	.97748	.21590	4.63171	.324366	.990107	.334259	.665741	49	
.20	12	.21132	.97742	.21621	4.62518	.324950	.990079	.334871	.665129	48	.80
	13	.21161	.97735	.21651	4.61868	.325534	.990052	.335482	.664518	47	
	14	.21189	.97729	.21682	4.61219	.326117	.990025	.336093	.663907	46	
.25	15	.21218	.97723	.21712	4.60572	.326700	.989997	.336702	.663298	45	.75
	16	.21246	.97717	.21743	4.59927	.327281	.989970	.337311	.662689	44	
	17	.21275	.97711	.21773	4.59283	.327862	.989942	.337919	.662081	43	
.30	18	.21303	.97705	.21804	4.58641	.328442	.989915	.338527	.661473	42	.70
	19	.21331	.97698	.21834	4.58001	.329021	.989887	.339133	.660867	41	
	20	.21360	.97692	.21864	4.57363	.329599	.989860	.339739	.660261	40	
.35	21	.21388	.97686	.21895	4.56726	.330176	.989832	.340344	.659656	39	.65
	22	.21417	.97680	.21925	4.56091	.330753	.989804	.340948	.659052	38	
	23	.21445	.97673	.21956	4.55458	.331329	.989777	.341552	.658448	37	
.40	24	.21474	.97667	.21986	4.54826	.331903	.989749	.342155	.657845	36	.60
	25	.21502	.97661	.22017	4.54196	.332478	.989721	.342757	.657243	35	
	26	.21530	.97655	.22047	4.53568	.333051	.989693	.343358	.656642	34	
.45	27	.21559	.97648	.22078	4.52941	.333624	.989665	.343958	.656042	33	.55
	28	.21587	.97642	.22108	4.52316	.334195	.989637	.344558	.655442	32	
	29	.21616	.97636	.22139	4.51693	.334767	.989610	.345157	.654843	31	
.50	30	.21644	.97630	.22169	4.51071	.335337	.989582	.345755	.654245	30	.50
	31	.21672	.97623	.22200	4.50451	.335906	.989553	.346353	.653647	29	
	32	.21701	.97617	.22231	4.49832	.336475	.989525	.346949	.653051	28	
.55	33	.21729	.97611	.22261	4.49215	.337043	.989497	.347545	.652455	27	.45
	34	.21758	.97604	.22292	4.48600	.337610	.989469	.348141	.651859	26	
	35	.21786	.97598	.22322	4.47986	.338176	.989441	.348735	.651265	25	
.60	36	.21814	.97592	.22353	4.47374	.338742	.989413	.349329	.650671	24	.40
	37	.21843	.97585	.22383	4.46764	.339307	.989385	.349922	.650078	23	
	38	.21871	.97579	.22414	4.46155	.339871	.989356	.350514	.649486	22	
.65	39	.21899	.97573	.22444	4.45548	.340434	.989328	.351106	.648894	21	.35
	40	.21928	.97566	.22475	4.44942	.340996	.989300	.351697	.648303	20	
	41	.21956	.97560	.22505	4.44338	.341558	.989271	.352287	.647713	19	
.70	42	.21985	.97553	.22536	4.43735	.342119	.989243	.352876	.647124	18	.30
	43	.22013	.97547	.22567	4.43134	.342679	.989214	.353465	.646535	17	
	44	.22041	.97541	.22597	4.42534	.343239	.989186	.354053	.645947	16	
.75	45	.22070	.97534	.22628	4.41936	.343797	.989157	.354640	.645360	15	.25
	46	.22098	.97528	.22658	4.41340	.344355	.989128	.355227	.644773	14	
	47	.22126	.97521	.22689	4.40745	.344912	.989100	.355813	.644187	13	
.80	48	.22155	.97515	.22719	4.40152	.345469	.989071	.356398	.643602	12	.20
	49	.22183	.97508	.22750	4.39560	.346024	.989042	.356982	.643018	11	
	50	.22212	.97502	.22781	4.38969	.346579	.989014	.357566	.642434	10	
.85	51	.22240	.97496	.22811	4.38381	.347134	.988985	.358149	.641851	9	.15
	52	.22268	.97489	.22842	4.37793	.347687	.988956	.358731	.641269	8	
	53	.22297	.97483	.22872	4.37207	.348240	.988927	.359313	.640687	7	
.90	54	.22325	.97476	.22903	4.36623	.348792	.988898	.359893	.640107	6	.10
	55	.22353	.97470	.22934	4.36040	.349343	.988869	.360474	.639526	5	
	56	.22382	.97463	.22964	4.35459	.349893	.988840	.361053	.638947	4	
.95	57	.22410	.97457	.22995	4.34879	.350443	.988811	.361632	.638368	3	.05
	58	.22438	.97450	.23026	4.34300	.350992	.988782	.362210	.637790	2	
	59	.22467	.97444	.23056	4.33723	.351540	.988753	.362787	.637213	1	
1.00	60	.22495	.97437	.23087	4.33148	9.352088	9.988724	9.363364	10.635636	0	.00
		Cos	Sin	Cot	Tan	Cos	Sin	Cot	Tan		
Decimals	Minutes	Natural Values				Common Logarithms				Minutes	Decimals

77°

VALUES AND LOGARITHMS OF TRIGONOMETRIC FUNCTIONS 361

13°

Decimals	Minutes	Natural Values				Common Logarithms				Minutes	Decimals
		Sin	Cos	Tan	Cot	Sin	Cos	Tan	Cot		
.00	0	.22495	.97437	.23087	4.33148	9.352088	9.988724	9.363364	10.636636	60	1.00
	1	.22523	.97430	.23117	4.32573	.352635	.988695	.363940	.636060	59	
	2	.22552	.97424	.23148	4.32001	.353181	.988666	.364515	.635485	58	
.05	3	.22580	.97417	.23179	4.31430	.353726	.988636	.365090	.634910	57	.95
	4	.22608	.97411	.23209	4.30860	.354271	.988607	.365664	.634336	56	
	5	.22637	.97404	.23240	4.30291	.354815	.988578	.366237	.633763	55	
.10	6	.22665	.97398	.23271	4.29724	.355358	.988548	.366810	.633190	54	.90
	7	.22693	.97391	.23301	4.29159	.355901	.988519	.367382	.632618	53	
	8	.22722	.97384	.23332	4.28595	.356443	.988489	.367953	.632047	52	
.15	9	.22750	.97378	.23363	4.28032	.356984	.988460	.368524	.631476	51	.85
	10	.22778	.97371	.23393	4.27471	.357524	.988430	.369094	.630906	50	
	11	.22807	.97365	.23424	4.26911	.358064	.988401	.369663	.630337	49	
.20	12	.22835	.97358	.23455	4.26352	.358603	.988371	.370232	.629768	48	.80
	13	.22863	.97351	.23485	4.25795	.359141	.988342	.370799	.629201	47	
	14	.22892	.97345	.23516	4.25239	.359678	.988312	.371367	.628633	46	
.25	15	.22920	.97338	.23547	4.24685	.360215	.988282	.371933	.628067	45	.75
	16	.22948	.97331	.23578	4.24132	.360752	.988252	.372499	.627501	44	
	17	.22977	.97325	.23608	4.23580	.361287	.988223	.373064	.626936	43	
.30	18	.23005	.97318	.23639	4.23030	.361822	.988193	.373629	.626371	42	.70
	19	.23033	.97311	.23670	4.22481	.362356	.988163	.374193	.625807	41	
	20	.23062	.97304	.23700	4.21933	.362889	.988133	.374756	.625244	40	
.35	21	.23090	.97298	.23731	4.21387	.363422	.988103	.375319	.624681	39	.65
	22	.23118	.97291	.23762	4.20842	.363954	.988073	.375881	.624119	38	
	23	.23146	.97284	.23793	4.20298	.364485	.988043	.376442	.623558	37	
.40	24	.23175	.97278	.23823	4.19756	.365016	.988013	.377003	.622997	36	.60
	25	.23203	.97271	.23854	4.19215	.365546	.987983	.377563	.622437	35	
	26	.23231	.97264	.23885	4.18675	.366075	.987953	.378122	.621878	34	
.45	27	.23260	.97257	.23916	4.18137	.366604	.987922	.378681	.621319	33	.55
	28	.23288	.97251	.23946	4.17600	.367131	.987892	.379239	.620761	32	
	29	.23316	.97244	.23977	4.17064	.367659	.987862	.379797	.620203	31	
.50	30	.23345	.97237	.24008	4.16530	.368185	.987832	.380354	.619646	30	.50
	31	.23373	.97230	.24039	4.15997	.368711	.987801	.380910	.619090	29	
	32	.23401	.97223	.24069	4.15465	.369236	.987771	.381466	.618534	28	
.55	33	.23429	.97217	.24100	4.14934	.369761	.987740	.382020	.617980	27	.45
	34	.23458	.97210	.24131	4.14405	.370285	.987710	.382575	.617425	26	
	35	.23486	.97203	.24162	4.13877	.370808	.987679	.383129	.616871	25	
.60	36	.23514	.97196	.24193	4.13350	.371330	.987649	.383682	.616318	24	.40
	37	.23542	.97189	.24223	4.12825	.371852	.987618	.384234	.615766	23	
	38	.23571	.97182	.24254	4.12301	.372373	.987588	.384786	.615214	22	
.65	39	.23599	.97176	.24285	4.11778	.372894	.987557	.385337	.614663	21	.35
	40	.23627	.97169	.24316	4.11256	.373414	.987526	.385888	.614112	20	
	41	.23656	.97162	.24347	4.10736	.373933	.987496	.386438	.613562	19	
.70	42	.23684	.97155	.24377	4.10216	.374452	.987465	.386987	.613013	18	.30
	43	.23712	.97148	.24408	4.09699	.374970	.987434	.387536	.612464	17	
	44	.23740	.97141	.24439	4.09182	.375487	.987403	.388084	.611916	16	
.75	45	.23769	.97134	.24470	4.08666	.376003	.987372	.388631	.611369	15	.25
	46	.23797	.97127	.24501	4.08152	.376519	.987341	.389178	.610822	14	
	47	.23825	.97120	.24532	4.07639	.377035	.987310	.389724	.610276	13	
.80	48	.23853	.97113	.24562	4.07127	.377549	.987279	.390270	.609730	12	.20
	49	.23882	.97106	.24593	4.06616	.378063	.987248	.390815	.609185	11	
	50	.23910	.97100	.24624	4.06107	.378577	.987217	.391360	.608640	10	
.85	51	.23938	.97093	.24655	4.05599	.379089	.987186	.391903	.608097	9	.15
	52	.23966	.97086	.24686	4.05092	.379601	.987155	.392447	.607553	8	
	53	.23995	.97079	.24717	4.04586	.380113	.987124	.392989	.607011	7	
.90	54	.24023	.97072	.24747	4.04081	.380624	.987092	.393531	.606469	6	.10
	55	.24051	.97065	.24778	4.03578	.381134	.987061	.394073	.605927	5	
	56	.24079	.97058	.24809	4.03076	.381643	.987030	.394614	.605386	4	
.95	57	.24108	.97051	.24840	4.02574	.382152	.986998	.395154	.604846	3	.05
	58	.24136	.97044	.24871	4.02074	.382661	.986967	.395694	.604306	2	
	59	.24164	.97037	.24902	4.01576	.383168	.986936	.396233	.603767	1	
1.00	60	.24192	.97030	.24933	4.01078	9.383675	9.986904	9.396771	10.603229	0	.00
Decimals	Minutes	Cos	Sin	Cot	Tan	Cos	Sin	Cot	Tan	Minutes	Decimals
		Natural Values				Common Logarithms					

76°

14°

Decimals	Minutes	Natural Values				Common Logarithms				Minutes	Decimals
		Sin	Cos	Tan	Cot	Sin	Cos	Tan	Cot		
.00	0	.24192	.97030	.24933	4.01078	9.383675	9.986904	9.396771	10.603229	60	1.00
	1	.24220	.97023	.24964	4.00582	.384182	.986873	.397309	.602691	59	
	2	.24249	.97015	.24995	4.00086	.384687	.986841	.397846	.602154	58	
.05	3	.24277	.97008	.25026	3.99592	.385192	.986809	.398383	.601617	57	.95
	4	.24305	.97001	.25056	3.99099	.385697	.986778	.398919	.601081	56	
	5	.24333	.96994	.25087	3.98607	.386201	.986746	.399455	.600545	55	
.10	6	.24362	.96987	.25118	3.98117	.386704	.986714	.399990	.600010	54	.90
	7	.24390	.96980	.25149	3.97627	.387207	.986683	.400524	.599476	53	
	8	.24418	.96973	.25180	3.97139	.387709	.986651	.401058	.598942	52	
.15	9	.24446	.96966	.25211	3.96651	.388210	.986619	.401591	.598409	51	.85
	10	.24474	.96959	.25242	3.96165	.388711	.986587	.402124	.597876	50	
	11	.24503	.96952	.25273	3.95680	.389211	.986555	.402656	.597344	49	
.20	12	.24531	.96945	.25304	3.95196	.389711	.986523	.403187	.596813	48	.80
	13	.24559	.96937	.25335	3.94713	.390210	.986491	.403718	.596282	47	
	14	.24587	.96930	.25366	3.94232	.390708	.986459	.404249	.595751	46	
.25	15	.24615	.96923	.25397	3.93751	.391206	.986427	.404778	.595222	45	.75
	16	.24644	.96916	.25428	3.93271	.391703	.986395	.405308	.594692	44	
	17	.24672	.96909	.25459	3.92793	.392199	.986363	.405836	.594164	43	
.30	18	.24700	.96902	.25490	3.92316	.392695	.986331	.406364	.593636	42	.70
	19	.24728	.96894	.25521	3.91839	.393191	.986299	.406892	.593108	41	
	20	.24756	.96887	.25552	3.91364	.393685	.986266	.407419	.592581	40	
.35	21	.24784	.96880	.25583	3.90890	.394179	.986234	.407945	.592055	39	.65
	22	.24813	.96873	.25614	3.90417	.394673	.986202	.408471	.591529	38	
	23	.24841	.96866	.25645	3.89945	.395166	.986169	.408996	.591004	37	
.40	24	.24869	.96858	.25676	3.89474	.395658	.986137	.409521	.590479	36	.60
	25	.24897	.96851	.25707	3.89004	.396150	.986104	.410045	.589955	35	
	26	.24925	.96844	.25738	3.88536	.396641	.986072	.410569	.589431	34	
.45	27	.24954	.96837	.25769	3.88068	.397132	.986039	.411092	.588908	33	.55
	28	.24982	.96829	.25800	3.87601	.397621	.986007	.411615	.588385	32	
	29	.25010	.96822	.25831	3.87136	.398111	.985974	.412137	.587863	31	
.50	30	.25038	.96815	.25862	3.86671	.398600	.985942	.412658	.587342	30	.50
	31	.25066	.96807	.25893	3.86208	.399088	.985909	.413179	.586821	29	
	32	.25094	.96800	.25924	3.85745	.399575	.985876	.413699	.586301	28	
.55	33	.25122	.96793	.25955	3.85284	.400062	.985843	.414219	.585781	27	.45
	34	.25151	.96786	.25986	3.84824	.400549	.985811	.414738	.585262	26	
	35	.25179	.96778	.26017	3.84364	.401035	.985778	.415257	.584743	25	
.60	36	.25207	.96771	.26048	3.83906	.401520	.985745	.415775	.584225	24	.40
	37	.25235	.96764	.26079	3.83449	.402005	.985712	.416293	.583707	23	
	38	.25263	.96756	.26110	3.82992	.402489	.985679	.416810	.583190	22	
.65	39	.25291	.96749	.26141	3.82537	.402972	.985646	.417326	.582674	21	.35
	40	.25320	.96742	.26172	3.82083	.403455	.985613	.417842	.582158	20	
	41	.25348	.96734	.26203	3.81630	.403938	.985580	.418358	.581642	19	
.70	42	.25376	.96727	.26235	3.81177	.404420	.985547	.418873	.581127	18	.30
	43	.25404	.96719	.26266	3.80726	.404901	.985514	.419387	.580613	17	
	44	.25432	.96712	.26297	3.80276	.405382	.985480	.419901	.580099	16	
.75	45	.25460	.96705	.26328	3.79827	.405862	.985447	.420415	.579585	15	.25
	46	.25488	.96697	.26359	3.79378	.406341	.985414	.420927	.579073	14	
	47	.25516	.96690	.26390	3.78931	.406820	.985381	.421440	.578560	13	
.80	48	.25545	.96682	.26421	3.78485	.407299	.985347	.421952	.578048	12	.20
	49	.25573	.96675	.26452	3.78040	.407777	.985314	.422463	.577537	11	
	50	.25601	.96667	.26483	3.77595	.408254	.985280	.422974	.577026	10	
.85	51	.25629	.96660	.26515	3.77152	.408731	.985247	.423484	.576516	9	.15
	52	.25657	.96653	.26546	3.76709	.409207	.985213	.423993	.576007	8	
	53	.25685	.96645	.26577	3.76268	.409682	.985180	.424503	.575497	7	
.90	54	.25713	.96638	.26608	3.75828	.410157	.985146	.425011	.574989	6	.10
	55	.25741	.96630	.26639	3.75388	.410632	.985113	.425519	.574481	5	
	56	.25769	.96623	.26670	3.74950	.411106	.985079	.426027	.573973	4	
.95	57	.25798	.96615	.26701	3.74512	.411579	.985045	.426534	.573466	3	.05
	58	.25826	.96608	.26733	3.74075	.412052	.985011	.427041	.572959	2	
	59	.25854	.96600	.26764	3.73640	.412524	.984978	.427547	.572453	1	
1.00	60	.25882	.96593	.26795	3.73205	9.412996	9.984944	9.428052	10.571948	0	.00
Decimals	Minutes	Cos	Sin	Cot	Tan	Cos	Sin	Cot	Tan	Minutes	Decimals
		Natural Values				Common Logarithms					

75°

15°

Decimals	Minutes	Natural Values				Common Logarithms				Minutes	Decimals
		Sin	Cos	Tan	Cot	Sin	Cos	Tan	Cot		
.00	0	.25882	.96593	.26795	3.73205	9.412996	9.984944	9.428052	10.571948	60	1.00
	1	.25910	.96585	.26826	3.72771	.413467	.984910	.428558	.571442	59	
	2	.25938	.96578	.26857	3.72338	.413938	.984876	.429062	.570938	58	
.05	3	.25966	.96570	.26888	3.71907	.414408	.984842	.429566	.570434	57	.95
	4	.25994	.96562	.26920	3.71476	.414878	.984808	.430070	.569930	56	
	5	.26022	.96555	.26951	3.71046	.415347	.984774	.430573	.569427	55	
.10	6	.26050	.96547	.26982	3.70616	.415815	.984740	.431075	.568925	54	.90
	7	.26079	.96540	.27013	3.70188	.416283	.984706	.431577	.568423	53	
	8	.26107	.96532	.27044	3.69761	.416751	.984672	.432079	.567921	52	
.15	9	.26135	.96524	.27076	3.69335	.417217	.984638	.432580	.567420	51	.85
	10	.26163	.96517	.27107	3.68909	.417684	.984603	.433080	.566920	50	
	11	.26191	.96509	.27138	3.68485	.418150	.984569	.433580	.566420	49	
.20	12	.26219	.96502	.27169	3.68061	.418615	.984535	.434080	.565920	48	.80
	13	.26247	.96494	.27201	3.67638	.419079	.984500	.434579	.565421	47	
	14	.26275	.96486	.27232	3.67217	.419544	.984466	.435078	.564922	46	
.25	15	.26303	.96479	.27263	3.66796	.420007	.984432	.435576	.564424	45	.75
	16	.26331	.96471	.27294	3.66376	.420470	.984397	.436073	.563927	44	
	17	.26359	.96463	.27326	3.65957	.420933	.984363	.436570	.563430	43	
.30	18	.26387	.96456	.27357	3.65538	.421395	.984328	.437067	.562933	42	.70
	19	.26415	.96448	.27388	3.65121	.421857	.984294	.437563	.562437	41	
	20	.26443	.96440	.27419	3.64705	.422318	.984259	.438059	.561941	40	
.35	21	.26471	.96433	.27451	3.64289	.422778	.984224	.438554	.561446	39	.65
	22	.26500	.96425	.27482	3.63874	.423238	.984190	.439048	.560952	38	
	23	.26528	.96417	.27513	3.63461	.423697	.984155	.439543	.560457	37	
.40	24	.26556	.96410	.27545	3.63048	.424156	.984120	.440036	.559964	36	.60
	25	.26584	.96402	.27576	3.62636	.424615	.984085	.440529	.559471	35	
	26	.26612	.96394	.27607	3.62224	.425073	.984050	.441022	.558978	34	
.45	27	.26640	.96386	.27638	3.61814	.425530	.984015	.441514	.558486	33	.55
	28	.26668	.96379	.27670	3.61405	.425987	.983981	.442006	.557994	32	
	29	.26696	.96371	.27701	3.60996	.426443	.983946	.442497	.557503	31	
.50	30	.26724	.96363	.27732	3.60588	.426899	.983911	.442988	.557012	30	.50
	31	.26752	.96355	.27764	3.60181	.427354	.983875	.443479	.556521	29	
	32	.26780	.96347	.27795	3.59775	.427809	.983840	.443968	.556032	28	
.55	33	.26808	.96340	.27826	3.59370	.428263	.983805	.444458	.555542	27	.45
	34	.26836	.96332	.27858	3.58966	.428717	.983770	.444947	.555053	26	
	35	.26864	.96324	.27889	3.58562	.429170	.983735	.445435	.554565	25	
.60	36	.26892	.96316	.27921	3.58160	.429623	.983700	.445923	.554077	24	.40
	37	.26920	.96308	.27952	3.57758	.430075	.983664	.446411	.553589	23	
	38	.26948	.96301	.27983	3.57357	.430527	.983629	.446898	.553102	22	
.65	39	.26976	.96293	.28015	3.56957	.430978	.983594	.447384	.552616	21	.35
	40	.27004	.96285	.28046	3.56557	.431429	.983558	.447870	.552130	20	
	41	.27032	.96277	.28077	3.56159	.431879	.983523	.448356	.551644	19	
.70	42	.27060	.96269	.28109	3.55761	.432329	.983487	.448841	.551159	18	.30
	43	.27088	.96261	.28140	3.55364	.432778	.983452	.449326	.550674	17	
	44	.27116	.96253	.28172	3.54968	.433226	.983416	.449810	.550190	16	
.75	45	.27144	.96246	.28203	3.54573	.433675	.983381	.450294	.549706	15	.25
	46	.27172	.96238	.28234	3.54179	.434122	.983345	.450777	.549223	14	
	47	.27200	.96230	.28266	3.53785	.434569	.983309	.451260	.548740	13	
.80	48	.27228	.96222	.28297	3.53393	.435016	.983273	.451743	.548257	12	.20
	49	.27256	.96214	.28329	3.53001	.435462	.983238	.452225	.547775	11	
	50	.27284	.96206	.28360	3.52609	.435908	.983202	.452706	.547294	10	
.85	51	.27312	.96198	.28391	3.52219	.436353	.983166	.453187	.546813	9	.15
	52	.27340	.96190	.28423	3.51829	.436798	.983130	.453668	.546332	8	
	53	.27368	.96182	.28454	3.51441	.437242	.983094	.454148	.545852	7	
.90	54	.27396	.96174	.28486	3.51053	.437686	.983058	.454628	.545372	6	.10
	55	.27424	.96166	.28517	3.50666	.438129	.983022	.455107	.544893	5	
	56	.27452	.96158	.28549	3.50279	.438572	.982986	.455586	.544414	4	
.95	57	.27480	.96150	.28580	3.49894	.439014	.982950	.456064	.543936	3	.05
	58	.27508	.96142	.28612	3.49509	.439456	.982914	.456542	.543458	2	
	59	.27536	.96134	.28643	3.49125	.439897	.982878	.457019	.542981	1	
1.00	60	.27564	.96126	.28675	3.48741	9.440338	9.982842	9.457496	10.542504	0	.00
Decimals	Minutes	Cos	Sin	Cot	Tan	Cos	Sin	Cot	Tan	Minutes	Decimals
		Natural Values				Common Logarithms					

74°

16°

Decimals	Minutes	Natural Values				Common Logarithms				Minutes	Decimals
		Sin	Cos	Tan	Cot	Sin	Cos	Tan	Cot		
.00	0	.27564	.96126	.28675	3.48741	9.440338	9.982842	9.457496	10.542504	60	1.00
	1	.27592	.96118	.28706	3.48359	.440778	.982805	.457973	.542027	59	
	2	.27620	.96110	.28738	3.47977	.441218	.982769	.458449	.541551	58	
.05	3	.27648	.96102	.28769	3.47596	.441658	.982733	.458925	.541075	57	.95
	4	.27676	.96094	.28800	3.47216	.442096	.982696	.459400	.540600	56	
	5	.27704	.96086	.28832	3.46837	.442535	.982660	.459875	.540125	55	
.10	6	.27731	.96078	.28864	3.46458	.442973	.982624	.460349	.539651	54	.90
	7	.27759	.96070	.28895	3.46080	.443410	.982587	.460823	.539177	53	
	8	.27787	.96062	.28927	3.45703	.443847	.982551	.461297	.538703	52	
.15	9	.27815	.96054	.28958	3.45327	.444284	.982514	.461770	.538230	51	.85
	10	.27843	.96046	.28990	3.44951	.444720	.982477	.462242	.537758	50	
	11	.27871	.96037	.29021	3.44576	.445155	.982441	.462715	.537285	49	
.20	12	.27899	.96029	.29053	3.44202	.445590	.982404	.463186	.536814	48	.80
	13	.27927	.96021	.29084	3.43829	.446025	.982367	.463658	.536342	47	
	14	.27955	.96013	.29116	3.43456	.446459	.982331	.464128	.535872	46	
.25	15	.27983	.96005	.29147	3.43084	.446893	.982294	.464599	.535401	45	.75
	16	.28011	.95997	.29179	3.42713	.447326	.982257	.465069	.534931	44	
	17	.28039	.95989	.29210	3.42343	.447759	.982220	.465539	.534461	43	
.30	18	.28067	.95981	.29242	3.41973	.448191	.982183	.466008	.533992	42	.70
	19	.28095	.95972	.29274	3.41604	.448623	.982146	.466477	.533523	41	
	20	.28123	.95964	.29305	3.41236	.449054	.982109	.466945	.533055	40	
.35	21	.28150	.95956	.29337	3.40869	.449485	.982072	.467413	.532587	39	.65
	22	.28178	.95948	.29368	3.40502	.449915	.982035	.467880	.532120	38	
	23	.28206	.95940	.29400	3.40136	.450345	.981998	.468347	.531653	37	
.40	24	.28234	.95931	.29432	3.39771	.450775	.981961	.468814	.531186	36	.60
	25	.28262	.95923	.29463	3.39406	.451204	.981924	.469280	.530720	35	
	26	.28290	.95915	.29495	3.39042	.451632	.981886	.469746	.530254	34	
.45	27	.28318	.95907	.29526	3.38679	.452060	.981849	.470211	.529789	33	.55
	28	.28346	.95898	.29558	3.38317	.452488	.981812	.470676	.529324	32	
	29	.28374	.95890	.29590	3.37955	.452915	.981774	.471141	.528859	31	
.50	30	.28402	.95882	.29621	3.37594	.453342	.981737	.471605	.528395	30	.50
	31	.28429	.95874	.29653	3.37234	.453768	.981700	.472069	.527931	29	
	32	.28457	.95865	.29685	3.36875	.454194	.981662	.472532	.527468	28	
.55	33	.28485	.95857	.29716	3.36516	.454619	.981625	.472995	.527005	27	.45
	34	.28513	.95849	.29748	3.36158	.455044	.981587	.473457	.526543	26	
	35	.28541	.95841	.29780	3.35800	.455469	.981549	.473919	.526081	25	
.60	36	.28569	.95832	.29811	3.35443	.455893	.981512	.474381	.525619	24	.40
	37	.28597	.95824	.29843	3.35087	.456316	.981474	.474842	.525158	23	
	38	.28625	.95816	.29875	3.34732	.456739	.981436	.475303	.524697	22	
.65	39	.28652	.95807	.29906	3.34377	.457162	.981399	.475763	.524237	21	.35
	40	.28680	.95799	.29938	3.34023	.457584	.981361	.476223	.523777	20	
	41	.28708	.95791	.29970	3.33670	.458006	.981323	.476683	.523317	19	
.70	42	.28736	.95782	.30001	3.33317	.458427	.981285	.477142	.522858	18	.30
	43	.28764	.95774	.30033	3.32965	.458848	.981247	.477601	.522399	17	
	44	.28792	.95766	.30065	3.32614	.459268	.981209	.478059	.521941	16	
.75	45	.28820	.95757	.30097	3.32264	.459688	.981171	.478517	.521483	15	.25
	46	.28847	.95749	.30128	3.31914	.460108	.981133	.478975	.521025	14	
	47	.28875	.95740	.30160	3.31565	.460527	.981095	.479432	.520568	13	
.80	48	.28903	.95732	.30192	3.31216	.460946	.981057	.479889	.520111	12	.20
	49	.28931	.95724	.30224	3.30868	.461364	.981019	.480345	.519655	11	
	50	.28959	.95715	.30255	3.30521	.461782	.980981	.480801	.519199	10	
.85	51	.28987	.95707	.30287	3.30174	.462199	.980942	.481257	.518743	9	.15
	52	.29015	.95698	.30319	3.29829	.462616	.980904	.481712	.518288	8	
	53	.29042	.95690	.30351	3.29483	.463032	.980866	.482167	.517833	7	
.90	54	.29070	.95681	.30382	3.29139	.463448	.980827	.482621	.517379	6	.10
	55	.29098	.95673	.30414	3.28795	.463864	.980789	.483075	.516925	5	
	56	.29126	.95664	.30446	3.28452	.464279	.980750	.483529	.516471	4	
.95	57	.29154	.95656	.30478	3.28109	.464694	.980712	.483982	.516018	3	.05
	58	.29182	.95647	.30509	3.27767	.465108	.980673	.484435	.515565	2	
	59	.29209	.95639	.30541	3.27426	.465521	.980635	.484887	.515113	1	
1.00	60	.29237	.95630	.30573	3.27085	9.465935	9.980596	9.485339	10.514661	0	.00
Decimals	Minutes	Cos	Sin	Cot	Tan	Cos	Sin	Cot	Tan	Minutes	Decimals
		Natural Values				Common Logarithms					

73°

VALUES AND LOGARITHMS OF TRIGONOMETRIC FUNCTIONS

17°

Decimals	Minutes	Natural Values				Common Logarithms				Minutes	Decimals
		Sin	Cos	Tan	Cot	Sin	Cos	Tan	Cot		
.00	0	.29237	.95630	.30573	3.27085	9.465935	9.980596	9.485339	10.514661	60	1.00
	1	.29265	.95622	.30605	3.26745	.466348	.980558	.485791	.514209	59	
	2	.29293	.95613	.30637	3.26406	.466761	.980519	.486242	.513758	58	
.05	3	.29321	.95605	.30669	3.26067	.467173	.980480	.486693	.513307	57	.95
	4	.29348	.95596	.30700	3.25729	.467585	.980442	.487143	.512857	56	
	5	.29376	.95588	.30732	3.25392	.467996	.980403	.487593	.512407	55	
.10	6	.29404	.95579	.30764	3.25055	.468407	.980364	.488043	.511957	54	.90
	7	.29432	.95571	.30796	3.24719	.468817	.980325	.488492	.511508	53	
	8	.29460	.95562	.30828	3.24383	.469227	.980286	.488941	.511059	52	
.15	9	.29487	.95554	.30860	3.24049	.469637	.980247	.489390	.510610	51	.85
	10	.29515	.95545	.30891	3.23714	.470046	.980208	.489838	.510162	50	
	11	.29543	.95536	.30923	3.23381	.470455	.980169	.490286	.509714	49	
.20	12	.29571	.95528	.30955	3.23048	.470863	.980130	.490733	.509267	48	.80
	13	.29599	.95519	.30987	3.22715	.471271	.980091	.491180	.508820	47	
	14	.29626	.95511	.31019	3.22384	.471679	.980052	.491627	.508373	46	
.25	15	.29654	.95502	.31051	3.22053	.472086	.980012	.492073	.507927	45	.75
	16	.29682	.95493	.31083	3.21722	.472492	.979973	.492519	.507481	44	
	17	.29710	.95485	.31115	3.21392	.472898	.979934	.492965	.507035	43	
.30	18	.29737	.95476	.31147	3.21053	.473304	.979895	.493410	.506590	42	.70
	19	.29765	.95467	.31178	3.20734	.473710	.979855	.493854	.506146	41	
	20	.29793	.95459	.31210	3.20406	.474115	.979816	.494299	.505701	40	
.35	21	.29821	.95450	.31242	3.20079	.474519	.979776	.494743	.505257	39	.65
	22	.29849	.95441	.31274	3.19752	.474923	.979737	.495186	.504814	38	
	23	.29876	.95433	.31306	3.19426	.475327	.979697	.495630	.504370	37	
.40	24	.29904	.95424	.31338	3.19100	.475730	.979658	.496073	.503927	36	.60
	25	.29932	.95415	.31370	3.18775	.476133	.979618	.496515	.503485	35	
	26	.29960	.95407	.31402	3.18451	.476536	.979579	.496957	.503043	34	
.45	27	.29987	.95398	.31434	3.18127	.476938	.979539	.497399	.502601	33	.55
	28	.30015	.95389	.31466	3.17804	.477340	.979499	.497841	.502159	32	
	29	.30043	.95380	.31498	3.17481	.477741	.979459	.498282	.501718	31	
.50	30	.30071	.95372	.31530	3.17159	.478142	.979420	.498722	.501278	30	.50
	31	.30098	.95363	.31562	3.16838	.478542	.979380	.499163	.500837	29	
	32	.30126	.95354	.31594	3.16517	.478942	.979340	.499603	.500397	28	
.55	33	.30154	.95345	.31626	3.16197	.479342	.979300	.500042	.499958	27	.45
	34	.30182	.95337	.31658	3.15877	.479741	.979260	.500481	.499519	26	
	35	.30209	.95328	.31690	3.15558	.480140	.979220	.500920	.499080	25	
.60	36	.30237	.95319	.31722	3.15240	.480539	.979180	.501359	.498641	24	.40
	37	.30265	.95310	.31754	3.14922	.480937	.979140	.501797	.498203	23	
	38	.30292	.95301	.31786	3.14605	.481334	.979100	.502235	.497765	22	
.65	39	.30320	.95293	.31818	3.14288	.481731	.979059	.502672	.497328	21	.35
	40	.30348	.95284	.31850	3.13972	.482128	.979019	.503109	.496891	20	
	41	.30376	.95275	.31882	3.13656	.482525	.978979	.503546	.496454	19	
.70	42	.30403	.95266	.31914	3.13341	.482921	.978939	.503982	.496018	18	.30
	43	.30431	.95257	.31946	3.13027	.483316	.978898	.504418	.495582	17	
	44	.30459	.95248	.31978	3.12713	.483712	.978858	.504854	.495146	16	
.75	45	.30486	.95240	.32010	3.12400	.484107	.978817	.505289	.494711	15	.25
	46	.30514	.95231	.32042	3.12087	.484501	.978777	.505724	.494276	14	
	47	.30542	.95222	.32074	3.11775	.484895	.978737	.506159	.493841	13	
.80	48	.30570	.95213	.32106	3.11464	.485289	.978696	.506593	.493407	12	.20
	49	.30597	.95204	.32139	3.11153	.485682	.978655	.507027	.492973	11	
	50	.30625	.95195	.32171	3.10842	.486075	.978615	.507460	.492540	10	
.85	51	.30653	.95186	.32203	3.10532	.486467	.978574	.507893	.492107	9	.15
	52	.30680	.95177	.32235	3.10223	.486860	.978533	.508326	.491674	8	
	53	.30708	.95168	.32267	3.09914	.487251	.978493	.508759	.491241	7	
.90	54	.30736	.95159	.32299	3.09606	.487643	.978452	.509191	.490809	6	.10
	55	.30763	.95150	.32331	3.09298	.488034	.978411	.509622	.490378	5	
	56	.30791	.95142	.32363	3.08991	.488424	.978370	.510054	.489946	4	
.95	57	.30819	.95133	.32396	3.08685	.488814	.978329	.510485	.489515	3	.05
	58	.30846	.95124	.32428	3.08379	.489204	.978288	.510916	.489084	2	
	59	.30874	.95115	.32460	3.08073	.489593	.978247	.511346	.488654	1	
1.00	60	.30902	.95106	.32492	3.07768	9.489982	9.978206	9.511776	10.488224	0	.00
Decimals	Minutes	Cos	Sin	Cot	Tan	Cos	Sin	Cot	Tan	Minutes	Decimals
		Natural Values				Common Logarithms					

72°

18°

Decimals	Minutes	Natural Values				Common Logarithms				Minutes	Decimals
		Sin	Cos	Tan	Cot	Sin	Cos	Tan	Cot		
.00	0	.30902	.95106	.32492	3.07768	9.489982	9.978206	9.511776	10.488224	60	1.00
	1	.30929	.95097	.32524	3.07464	.490371	.978165	.512206	.487794	59	
	2	.30957	.95088	.32556	3.07160	.490759	.978124	.512635	.487365	58	
.05	3	.30985	.95079	.32588	3.06857	.491147	.978083	.513064	.486936	57	.95
	4	.31012	.95070	.32621	3.06554	.491535	.978042	.513493	.486507	56	
	5	.31040	.95061	.32653	3.06252	.491922	.978001	.513921	.486079	55	
.10	6	.31068	.95052	.32685	3.05950	.492308	.977959	.514349	.485651	54	.90
	7	.31095	.95043	.32717	3.05649	.492695	.977918	.514777	.485223	53	
	8	.31123	.95033	.32749	3.05349	.493081	.977877	.515204	.484796	52	
.15	9	.31151	.95024	.32782	3.05049	.493466	.977835	.515631	.484369	51	.85
	10	.31178	.95015	.32814	3.04749	.493851	.977794	.516057	.483943	50	
	11	.31206	.95006	.32846	3.04450	.494236	.977752	.516484	.483516	49	
.20	12	.31233	.94997	.32878	3.04152	.494621	.977711	.516910	.483090	48	.80
	13	.31261	.94988	.32911	3.03854	.495005	.977669	.517335	.482665	47	
	14	.31289	.94979	.32943	3.03556	.495388	.977628	.517761	.482239	46	
.25	15	.31316	.94970	.32975	3.03260	.495772	.977586	.518186	.481814	45	.75
	16	.31344	.94961	.33007	3.02963	.496154	.977544	.518610	.481390	44	
	17	.31372	.94952	.33040	3.02667	.496537	.977503	.519034	.480966	43	
.30	18	.31399	.94943	.33072	3.02372	.496919	.977461	.519458	.480542	42	.70
	19	.31427	.94933	.33104	3.02077	.497301	.977419	.519882	.480118	41	
	20	.31454	.94924	.33136	3.01783	.497682	.977377	.520305	.479695	40	
.35	21	.31482	.94915	.33169	3.01489	.498064	.977335	.520728	.479272	39	.65
	22	.31510	.94906	.33201	3.01196	.498444	.977293	.521151	.478849	38	
	23	.31537	.94897	.33233	3.00903	.498825	.977251	.521573	.478427	37	
.40	24	.31565	.94888	.33266	3.00611	.499204	.977209	.521995	.478005	36	.60
	25	.31593	.94878	.33298	3.00319	.499584	.977167	.522417	.477583	35	
	26	.31620	.94869	.33330	3.00028	.499963	.977125	.522838	.477162	34	
.45	27	.31648	.94860	.33363	2.99738	.500342	.977083	.523259	.476741	33	.55
	28	.31675	.94851	.33395	2.99447	.500721	.977041	.523680	.476320	32	
	29	.31703	.94842	.33427	2.99158	.501099	.976999	.524100	.475900	31	
.50	30	.31730	.94832	.33460	2.98868	.501476	.976957	.524520	.475480	30	.50
	31	.31758	.94823	.33492	2.98580	.501854	.976914	.524940	.475060	29	
	32	.31786	.94814	.33524	2.98292	.502231	.976872	.525359	.474641	28	
.55	33	.31813	.94805	.33557	2.98004	.502607	.976830	.525778	.474222	27	.45
	34	.31841	.94795	.33589	2.97717	.502984	.976787	.526197	.473803	26	
	35	.31868	.94786	.33621	2.97430	.503360	.976745	.526615	.473385	25	
.60	36	.31896	.94777	.33654	2.97144	.503735	.976702	.527033	.472967	24	.40
	37	.31923	.94768	.33686	2.96858	.504110	.976660	.527451	.472549	23	
	38	.31951	.94758	.33718	2.96573	.504485	.976617	.527868	.472132	22	
.65	39	.31979	.94749	.33751	2.96288	.504860	.976574	.528285	.471715	21	.35
	40	.32006	.94740	.33783	2.96004	.505234	.976532	.528702	.471298	20	
	41	.32034	.94730	.33816	2.95721	.505608	.976489	.529119	.470881	19	
.70	42	.32061	.94721	.33848	2.95437	.505981	.976446	.529535	.470465	18	.30
	43	.32089	.94712	.33881	2.95155	.506354	.976404	.529951	.470049	17	
	44	.32116	.94702	.33913	2.94872	.506727	.976361	.530366	.469634	16	
.75	45	.32144	.94693	.33945	2.94591	.507099	.976318	.530781	.469219	15	.25
	46	.32171	.94684	.33978	2.94309	.507471	.976275	.531196	.468804	14	
	47	.32199	.94674	.34010	2.94028	.507843	.976232	.531611	.468389	13	
.80	48	.32227	.94665	.34043	2.93748	.508214	.976189	.532025	.467975	12	.20
	49	.32254	.94656	.34075	2.93468	.508585	.976146	.532439	.467561	11	
	50	.32282	.94646	.34108	2.93189	.508956	.976103	.532853	.467147	10	
.85	51	.32309	.94637	.34140	2.92910	.509326	.976060	.533266	.466734	9	.15
	52	.32337	.94627	.34173	2.92632	.509696	.976017	.533679	.466321	8	
	53	.32364	.94618	.34205	2.92354	.510065	.975974	.534092	.465908	7	
.90	54	.32392	.94609	.34238	2.92076	.510434	.975930	.534504	.465496	6	.10
	55	.32419	.94599	.34270	2.91799	.510803	.975887	.534916	.465084	5	
	56	.32447	.94590	.34303	2.91523	.511172	.975844	.535328	.464672	4	
.95	57	.32474	.94580	.34335	2.91246	.511540	.975800	.535739	.464261	3	.05
	58	.32502	.94571	.34368	2.90971	.511907	.975757	.536150	.463850	2	
	59	.32529	.94561	.34400	2.90696	.512275	.975714	.536561	.463439	1	
1.00	60	.32557	.94552	.34433	2.90421	9.512642	9.975670	9.536972	10.463028	0	.00
Decimals	Minutes	Cos	Sin	Cot	Tan	Cos	Sin	Cot	Tan	Minutes	Decimals
		Natural Values				Common Logarithms					

71°

VALUES AND LOGARITHMS OF TRIGONOMETRIC FUNCTIONS 367

19°

Decimals	Minutes	Natural Values				Common Logarithms				Minutes	Decimals
		Sin	Cos	Tan	Cot	Sin	Cos	Tan	Cot		
.00	0	.32557	.94552	.34433	2.90421	9.512642	9.975670	9.536972	10.463028	60	1.00
	1	.32584	.94542	.34465	2.90147	.513009	.975627	.537382	.462618	59	
	2	.32612	.94533	.34498	2.89873	.513375	.975583	.537792	.462208	58	
.05	3	.32639	.94523	.34530	2.89600	.513741	.975539	.538202	.461798	57	.95
	4	.32667	.94514	.34563	2.89327	.514107	.975496	.538611	.461389	56	
	5	.32694	.94504	.34596	2.89055	.514472	.975452	.539020	.460980	55	
.10	6	.32722	.94495	.34628	2.88783	.514837	.975408	.539429	.460571	54	.90
	7	.32749	.94485	.34661	2.88511	.515202	.975365	.539837	.460163	53	
	8	.32777	.94476	.34693	2.88240	.515566	.975321	.540245	.459755	52	
.15	9	.32804	.94466	.34726	2.87970	.515930	.975277	.540653	.459347	51	.85
	10	.32832	.94457	.34758	2.87700	.516294	.975233	.541061	.458939	50	
	11	.32859	.94447	.34791	2.87430	.516657	.975189	.541468	.458532	49	
.20	12	.32887	.94438	.34824	2.87161	.517020	.975145	.541875	.458125	48	.80
	13	.32914	.94428	.34856	2.86892	.517382	.975101	.542281	.457719	47	
	14	.32942	.94418	.34889	2.86624	.517745	.975057	.542688	.457312	46	
.25	15	.32969	.94409	.34922	2.86356	.518107	.975013	.543094	.456906	45	.75
	16	.32997	.94399	.34954	2.86089	.518468	.974969	.543499	.456501	44	
	17	.33024	.94390	.34987	2.85822	.518829	.974925	.543905	.456095	43	
.30	18	.33051	.94380	.35020	2.85555	.519190	.974880	.544310	.455690	42	.70
	19	.33079	.94370	.35052	2.85289	.519551	.974836	.544715	.455285	41	
	20	.33106	.94361	.35085	2.85023	.519911	.974792	.545119	.454881	40	
.35	21	.33134	.94351	.35118	2.84758	.520271	.974748	.545524	.454476	39	.65
	22	.33161	.94342	.35150	2.84494	.520631	.974703	.545928	.454072	38	
	23	.33189	.94332	.35183	2.84229	.520990	.974659	.546331	.453669	37	
.40	24	.33216	.94322	.35216	2.83965	.521349	.974614	.546735	.453265	36	.60
	25	.33244	.94313	.35248	2.83702	.521707	.974570	.547138	.452862	35	
	26	.33271	.94303	.35281	2.83439	.522066	.974525	.547540	.452460	34	
.45	27	.33298	.94293	.35314	2.83176	.522424	.974481	.547943	.452057	33	.55
	28	.33326	.94284	.35346	2.82914	.522781	.974436	.548345	.451655	32	
	29	.33353	.94274	.35379	2.82653	.523138	.974391	.548747	.451253	31	
.50	30	.33381	.94264	.35412	2.82391	.523495	.974347	.549149	.450851	30	.50
	31	.33408	.94254	.35445	2.82130	.523852	.974302	.549550	.450450	29	
	32	.33436	.94245	.35477	2.81870	.524208	.974257	.549951	.450049	28	
.55	33	.33463	.94235	.35510	2.81610	.524564	.974212	.550352	.449648	27	.45
	34	.33490	.94225	.35543	2.81350	.524920	.974167	.550752	.449248	26	
	35	.33518	.94215	.35576	2.81091	.525275	.974122	.551153	.448847	25	
.60	36	.33545	.94206	.35608	2.80833	.525630	.974077	.551552	.448448	24	.40
	37	.33573	.94196	.35641	2.80574	.525984	.974032	.551952	.448048	23	
	38	.33600	.94186	.35674	2.80316	.526339	.973987	.552351	.447649	22	
.65	39	.33627	.94176	.35707	2.80059	.526693	.973942	.552750	.447250	21	.35
	40	.33655	.94167	.35740	2.79802	.527046	.973897	.553149	.446851	20	
	41	.33682	.94157	.35772	2.79545	.527400	.973852	.553548	.446452	19	
.70	42	.33710	.94147	.35805	2.79289	.527753	.973807	.553946	.446054	18	.30
	43	.33737	.94137	.35838	2.79033	.528105	.973761	.554344	.445656	17	
	44	.33764	.94127	.35871	2.78778	.528458	.973716	.554741	.445259	16	
.75	45	.33792	.94118	.35904	2.78523	.528810	.973671	.555139	.444861	15	.25
	46	.33819	.94108	.35937	2.78269	.529161	.973625	.555536	.444464	14	
	47	.33846	.94098	.35969	2.78014	.529513	.973580	.555933	.444067	13	
.80	48	.33874	.94088	.36002	2.77761	.529864	.973535	.556329	.443671	12	.20
	49	.33901	.94078	.36035	2.77507	.530215	.973489	.556725	.443275	11	
	50	.33929	.94068	.36068	2.77254	.530565	.973444	.557121	.442879	10	
.85	51	.33956	.94058	.36101	2.77002	.530915	.973398	.557517	.442483	9	.15
	52	.33983	.94049	.36134	2.76750	.531265	.973352	.557913	.442087	8	
	53	.34011	.94039	.36167	2.76498	.531614	.973307	.558308	.441692	7	
.90	54	.34038	.94029	.36199	2.76247	.531963	.973261	.558703	.441297	6	.10
	55	.34065	.94019	.36232	2.75996	.532312	.973215	.559097	.440903	5	
	56	.34093	.94009	.36265	2.75746	.532661	.973169	.559491	.440509	4	
.95	57	.34120	.93999	.36298	2.75496	.533009	.973124	.559885	.440115	3	.05
	58	.34147	.93989	.36331	2.75246	.533357	.973078	.560279	.439721	2	
	59	.34175	.93979	.36364	2.74997	.533704	.973032	.560673	.439327	1	
1.00	60	.34202	.93969	.36397	2.74748	9.534052	9.972986	9.561066	10.438934	0	.00
Decimals	Minutes	Cos	Sin	Cot	Tan	Cos	Sin	Cot	Tan	Minutes	Decimals
			Natural Values				Common Logarithms				

70°

20°

Decimals	Minutes	Natural Values				Common Logarithms				Minutes	Decimals
		Sin	Cos	Tan	Cot	Sin	Cos	Tan	Cot		
.00	0	.34202	.93969	.36397	2.74748	9.534052	9.972986	9.561066	10.438934	60	1.00
	1	.34229	.93959	.36430	2.74499	.534399	.972940	.561459	.438541	59	
	2	.34257	.93949	.36463	2.74251	.534745	.972894	.561851	.438149	58	
.05	3	.34284	.93939	.36496	2.74004	.535092	.972848	.562244	.437756	57	.95
	4	.34311	.93929	.36529	2.73756	.535438	.972802	.562636	.437364	56	
	5	.34339	.93919	.36562	2.73509	.535783	.972755	.563028	.436972	55	
.10	6	.34366	.93909	.36595	2.73263	.536129	.972709	.563419	.436581	54	.90
	7	.34393	.93899	.36628	2.73017	.536474	.972663	.563811	.436189	53	
	8	.34421	.93889	.36661	2.72771	.536818	.972617	.564202	.435798	52	
.15	9	.34448	.93879	.36694	2.72526	.537163	.972570	.564593	.435407	51	.85
	10	.34475	.93869	.36727	2.72281	.537507	.972524	.564983	.435017	50	
	11	.34503	.93859	.36760	2.72036	.537851	.972478	.565373	.434627	49	
.20	12	.34530	.93849	.36793	2.71792	.538194	.972431	.565763	.434237	48	.80
	13	.34557	.93839	.36826	2.71548	.538538	.972385	.566153	.433847	47	
	14	.34584	.93829	.36859	2.71305	.538880	.972338	.566542	.433458	46	
.25	15	.34612	.93819	.36892	2.71062	.539223	.972291	.566932	.433063	45	.75
	16	.34639	.93809	.36925	2.70819	.539565	.972245	.567320	.432680	44	
	17	.34666	.93799	.36958	2.70577	.539907	.972198	.567709	.432291	43	
.30	18	.34694	.93789	.36991	2.70335	.540249	.972151	.568098	.431902	42	.70
	19	.34721	.93779	.37024	2.70094	.540590	.972105	.568486	.431514	41	
	20	.34748	.93769	.37057	2.69853	.540931	.972058	.568873	.431127	40	
.35	21	.34775	.93759	.37090	2.69612	.541272	.972011	.569261	.430739	39	.65
	22	.34803	.93748	.37123	2.69371	.541613	.971964	.569648	.430352	38	
	23	.34830	.93738	.37157	2.69131	.541953	.971917	.570035	.429965	37	
.40	24	.34857	.93728	.37190	2.68892	.542293	.971870	.570422	.429578	36	.60
	25	.34884	.93718	.37223	2.68653	.542632	.971823	.570809	.429191	35	
	26	.34912	.93708	.37256	2.68414	.542971	.971776	.571195	.428805	34	
.45	27	.34939	.93698	.37289	2.68175	.543310	.971729	.571581	.428419	33	.55
	28	.34966	.93688	.37322	2.67937	.543649	.971682	.571967	.428033	32	
	29	.34993	.93677	.37355	2.67700	.543987	.971635	.572352	.427648	31	
.50	30	.35021	.93667	.37388	2.67462	.544325	.971588	.572738	.427262	30	.50
	31	.35048	.93657	.37422	2.67225	.544663	.971540	.573123	.426877	29	
	32	.35075	.93647	.37455	2.66989	.545000	.971493	.573507	.426493	28	
.55	33	.35102	.93637	.37488	2.66752	.545338	.971446	.573892	.426108	27	.45
	34	.35130	.93626	.37521	2.66516	.545674	.971398	.574276	.425724	26	
	35	.35157	.93616	.37554	2.66281	.546011	.971351	.574660	.425340	25	
.60	36	.35184	.93606	.37588	2.66046	.546347	.971303	.575044	.424956	24	.40
	37	.35211	.93596	.37621	2.65811	.546683	.971256	.575427	.424573	23	
	38	.35239	.93585	.37654	2.65576	.547019	.971208	.575810	.424190	22	
.65	39	.35266	.93575	.37687	2.65342	.547354	.971161	.576193	.423807	21	.35
	40	.35293	.93565	.37720	2.65109	.547689	.971113	.576576	.423424	20	
	41	.35320	.93555	.37754	2.64875	.548024	.971066	.576959	.423041	19	
.70	42	.35347	.93544	.37787	2.64642	.548359	.971018	.577341	.422659	18	.30
	43	.35375	.93534	.37820	2.64410	.548693	.970970	.577723	.422277	17	
	44	.35402	.93524	.37853	2.64177	.549027	.970922	.578104	.421896	16	
.75	45	.35429	.93514	.37887	2.63945	.549360	.970874	.578486	.421514	15	.25
	46	.35456	.93503	.37920	2.63714	.549693	.970827	.578867	.421133	14	
	47	.35484	.93493	.37953	2.63483	.550026	.970779	.579248	.420752	13	
.80	48	.35511	.93483	.37986	2.63252	.550359	.970731	.579629	.420371	12	.20
	49	.35538	.93472	.38020	2.63021	.550692	.970683	.580009	.419991	11	
	50	.35565	.93462	.38053	2.62791	.551024	.970635	.580389	.419611	10	
.85	51	.35592	.93452	.38086	2.62561	.551356	.970586	.580769	.419231	9	.15
	52	.35619	.93441	.38120	2.62332	.551687	.970538	.581149	.418851	8	
	53	.35647	.93431	.38153	2.62103	.552018	.970490	.581528	.418472	7	
.90	54	.35674	.93420	.38186	2.61874	.552349	.970442	.581907	.418093	6	.10
	55	.35701	.93410	.38220	2.61646	.552680	.970394	.582286	.417714	5	
	56	.35728	.93400	.38253	2.61418	.553010	.970345	.582665	.417335	4	
.95	57	.35755	.93389	.38286	2.61190	.553341	.970297	.583044	.416956	3	.05
	58	.35782	.93379	.38320	2.60963	.553670	.970249	.583422	.416578	2	
	59	.35810	.93368	.38353	2.60736	.554000	.970200	.583800	.416200	1	
1.00	60	.35837	.93358	.38386	2.60509	9.554329	9.970152	9.584177	10.415823	0	.00
Decimals	Minutes	Cos	Sin	Cot	Tan	Cos	Sin	Cot	Tan	Minutes	Decimals
		Natural Values				Common Logarithms					

69°

VALUES AND LOGARITHMS OF TRIGONOMETRIC FUNCTIONS

21°

Decimals	Minutes	Natural Values				Common Logarithms				Minutes	Decimals
		Sin	Cos	Tan	Cot	Sin	Cos	Tan	Cot		
.00	0	.35837	.93358	.38386	2.60509	9.554329	9.970152	9.584177	10.415823	60	1.00
	1	.35864	.93348	.38420	2.60283	.554658	.970103	.584555	.415445	59	
	2	.35891	.93337	.38453	2.60057	.554987	.970055	.584932	.415068	58	
.05	3	.35918	.93327	.38487	2.59831	.555315	.970006	.585309	.414691	57	.95
	4	.35945	.93316	.38520	2.59606	.555643	.969957	.585686	.414314	56	
	5	.35973	.93306	.38553	2.59381	.555971	.969909	.586062	.413938	55	
.10	6	.36000	.93295	.38587	2.59156	.556299	.969860	.586439	.413561	54	.90
	7	.36027	.93285	.38620	2.58932	.556626	.969811	.586815	.413185	53	
	8	.36054	.93274	.38654	2.58708	.556953	.969762	.587190	.412810	52	
.15	9	.36081	.93264	.38687	2.58484	.557280	.969714	.587566	.412434	51	.85
	10	.36108	.93253	.38721	2.58261	.557606	.969665	.587941	.412059	50	
	11	.36135	.93243	.38754	2.58038	.557932	.969616	.588316	.411684	49	
.20	12	.36162	.93232	.38787	2.57815	.558258	.969567	.588691	.411309	48	.80
	13	.36190	.93222	.38821	2.57593	.558583	.969518	.589066	.410934	47	
	14	.36217	.93211	.38854	2.57371	.558909	.969469	.589440	.410560	46	
.25	15	.36244	.93201	.38888	2.57150	.559234	.969420	.589814	.410186	45	.75
	16	.36271	.93190	.38921	2.56928	.559558	.969370	.590188	.409812	44	
	17	.36298	.93180	.38955	2.56707	.559883	.969321	.590562	.409438	43	
.30	18	.36325	.93169	.38988	2.56487	.560207	.969272	.590935	.409065	42	.70
	19	.36352	.93159	.39022	2.56266	.560531	.969223	.591308	.408692	41	
	20	.36379	.93148	.39055	2.56046	.560855	.969173	.591681	.408319	40	
.35	21	.36406	.93137	.39089	2.55827	.551178	.969124	.592054	.407946	39	.65
	22	.36434	.93127	.39122	2.55608	.561501	.969075	.592426	.407574	38	
	23	.36461	.93116	.39156	2.55389	.561824	.969025	.592799	.407201	37	
.40	24	.36488	.93106	.39190	2.55170	.562146	.968976	.593171	.406829	36	.60
	25	.36515	.93095	.39223	2.54952	.562468	.968926	.593542	.406458	35	
	26	.36542	.93084	.39257	2.54734	.562790	.968877	.593914	.406086	34	
.45	27	.36569	.93074	.39290	2.54516	.563112	.968827	.594285	.405715	33	.55
	28	.36596	.93063	.39324	2.54299	.563433	.968777	.594656	.405344	32	
	29	.36623	.93052	.39357	2.54082	.563755	.968728	.595027	.404973	31	
.50	30	.36650	.93042	.39391	2.53865	.564075	.968678	.595398	.404602	30	.50
	31	.36677	.93031	.39425	2.53648	.564396	.968628	.595768	.404232	29	
	32	.36704	.93020	.39458	2.53432	.564716	.968578	.596138	.403862	28	
.55	33	.36731	.93010	.39492	2.53217	.565036	.968528	.596508	.403492	27	.45
	34	.36758	.92999	.39526	2.53001	.565356	.968479	.596878	.403122	26	
	35	.36785	.92988	.39559	2.52786	.565676	.968429	.597247	.402753	25	
.60	36	.36812	.92978	.39593	2.52571	.565995	.968379	.597616	.402384	24	.40
	37	.36839	.92967	.39626	2.52357	.566314	.968329	.597985	.402015	23	
	38	.36867	.92956	.39660	2.52142	.566632	.968278	.598354	.401646	22	
.65	39	.36894	.92945	.39694	2.51929	.566951	.968228	.598722	.401278	21	.35
	40	.36921	.92935	.39727	2.51715	.567269	.968178	.599091	.400909	20	
	41	.36948	.92924	.39761	2.51502	.567587	.968128	.599459	.400541	19	
.70	42	.36975	.92913	.39795	2.51289	.567904	.968078	.599827	.400173	18	.30
	43	.37002	.92902	.39829	2.51076	.568222	.968027	.600194	.399806	17	
	44	.37029	.92892	.39862	2.50864	.568539	.967977	.600562	.399438	16	
.75	45	.37056	.92881	.39896	2.50652	.568856	.967927	.600929	.399071	15	.25
	46	.37083	.92870	.39930	2.50440	.569172	.967876	.601296	.398704	14	
	47	.37110	.92859	.39963	2.50229	.569488	.967826	.601663	.398337	13	
.80	48	.37137	.92849	.39997	2.50018	.569804	.967775	.602029	.397971	12	.20
	49	.37164	.92838	.40031	2.49807	.570120	.967725	.602395	.397605	11	
	50	.37191	.92827	.40065	2.49597	.570435	.967674	.602761	.397239	10	
.85	51	.37218	.92816	.40098	2.49386	.570751	.967624	.603127	.396873	9	.15
	52	.37245	.92805	.40132	2.49177	.571066	.967573	.603493	.396507	8	
	53	.37272	.92794	.40166	2.48967	.571380	.967522	.603858	.396142	7	
.90	54	.37299	.92784	.40200	2.48758	.571695	.967471	.604223	.395777	6	.10
	55	.37326	.92773	.40234	2.48549	.572009	.967421	.604588	.395412	5	
	56	.37353	.92762	.40267	2.48340	.572323	.967370	.604953	.395047	4	
.95	57	.37380	.92751	.40301	2.48132	.572636	.967319	.605317	.394683	3	.05
	58	.37407	.92740	.40335	2.47924	.572950	.967268	.605682	.394318	2	
	59	.37434	.92729	.40369	2.47716	.573263	.967217	.606046	.393954	1	
1.00	60	.37461	.92718	.40403	2.47509	9.573575	9.967166	9.606410	10.393590	0	.00
Decimals	Minutes	Cos	Sin	Cot	Tan	Cos	Sin	Cot	Tan	Minutes	Decimals
		Natural Values				Common Logarithms					

68°

22°

Decimals	Minutes	Natural Values				Common Logarithms				Minutes	Decimals
		Sin	Cos	Tan	Cot	Sin	Cos	Tan	Cot		
.00	0	.37461	.92718	.40403	2.47509	9.573575	9.967166	9.606410	10.393590	60	1.00
	1	.37488	.92707	.40436	2.47302	.573888	.967115	.606773	.393227	59	
	2	.37515	.92697	.40470	2.47095	.574200	.967064	.607137	.392863	58	
.05	3	.37542	.92686	.40504	2.46888	.574512	.967013	.607500	.392500	57	.95
	4	.37569	.92675	.40538	2.46682	.574824	.966961	.607863	.392137	56	
	5	.37595	.92664	.40572	2.46476	.575136	.966910	.608225	.391775	55	
.10	6	.37622	.92653	.40606	2.46270	.575447	.966859	.608588	.391412	54	.90
	7	.37649	.92642	.40640	2.46065	.575758	.966808	.608950	.391050	53	
	8	.37676	.92631	.40674	2.45860	.576069	.966756	.609312	.390688	52	
.15	9	.37703	.92620	.40707	2.45655	.576379	.966705	.609674	.390326	51	.85
	10	.37730	.92609	.40741	2.45451	.576689	.966653	.610036	.389964	50	
	11	.37757	.92598	.40775	2.45246	.576999	.966602	.610397	.389603	49	
.20	12	.37784	.92587	.40809	2.45043	.577309	.966550	.610759	.389241	48	.80
	13	.37811	.92576	.40843	2.44839	.577618	.966499	.611120	.388880	47	
	14	.37838	.92565	.40877	2.44636	.577927	.966447	.611480	.388520	46	
.25	15	.37865	.92554	.40911	2.44433	.578236	.966395	.611841	.388159	45	.75
	16	.37892	.92543	.40945	2.44230	.578545	.966344	.612201	.387799	44	
	17	.37919	.92532	.40979	2.44027	.578853	.966292	.612561	.387439	43	
.30	18	.37946	.92521	.41013	2.43825	.579162	.966240	.612921	.387079	42	.70
	19	.37973	.92510	.41047	2.43623	.579470	.966188	.613281	.386719	41	
	20	.37999	.92499	.41081	2.43422	.579777	.966136	.613641	.386359	40	
.35	21	.38026	.92488	.41115	2.43220	.580085	.966085	.614000	.386000	39	.65
	22	.38053	.92477	.41149	2.43019	.580392	.966033	.614359	.385641	38	
	23	.38080	.92466	.41183	2.42819	.580699	.965981	.614718	.385282	37	
.40	24	.38107	.92455	.41217	2.42618	.581005	.965929	.615077	.384923	36	.60
	25	.38134	.92444	.41251	2.42418	.581312	.965876	.615435	.384565	35	
	26	.38161	.92432	.41285	2.42218	.581618	.965824	.615793	.384207	34	
.45	27	.38188	.92421	.41319	2.42019	.581924	.965772	.616151	.383849	33	.55
	28	.38215	.92410	.41353	2.41819	.582229	.965720	.616509	.383491	32	
	29	.38241	.92399	.41387	2.41620	.582535	.965668	.616867	.383133	31	
.50	30	.38268	.92388	.41421	2.41421	.582840	.965615	.617224	.382776	30	.50
	31	.38295	.92377	.41455	2.41223	.583145	.965563	.617582	.382418	29	
	32	.38322	.92366	.41490	2.41025	.583449	.965511	.617939	.382061	28	
.55	33	.38349	.92355	.41524	2.40827	.583754	.965458	.618295	.381705	27	.45
	34	.38376	.92343	.41558	2.40629	.584058	.965406	.618652	.381348	26	
	35	.38403	.92332	.41592	2.40432	.584361	.965353	.619008	.380992	25	
.60	36	.38430	.92321	.41626	2.40235	.584665	.965301	.619364	.380636	24	.40
	37	.38456	.92310	.41660	2.40038	.584968	.965248	.619720	.380280	23	
	38	.38483	.92299	.41694	2.39841	.585272	.965195	.620076	.379924	22	
.65	39	.38510	.92287	.41728	2.39645	.585574	.965143	.620432	.379568	21	.35
	40	.38537	.92276	.41763	2.39449	.585877	.965090	.620787	.379213	20	
	41	.38564	.92265	.41797	2.39253	.586179	.965037	.621142	.378858	19	
.70	42	.38591	.92254	.41831	2.39058	.586482	.964984	.621497	.378503	18	.30
	43	.38617	.92243	.41865	2.38863	.586783	.964931	.621852	.378148	17	
	44	.38644	.92231	.41899	2.38668	.587085	.964879	.622207	.377793	16	
.75	45	.38671	.92220	.41933	2.38473	.587386	.964826	.622561	.377439	15	.25
	46	.38698	.92209	.41968	2.38279	.587688	.964773	.622915	.377085	14	
	47	.38725	.92198	.42002	2.38084	.587989	.964720	.623269	.376731	13	
.80	48	.38752	.92186	.42036	2.37891	.588289	.964666	.623623	.376377	12	.20
	49	.38778	.92175	.42070	2.37697	.588590	.964613	.623976	.376024	11	
	50	.38805	.92164	.42105	2.37504	.588890	.964560	.624330	.375670	10	
.85	51	.38832	.92152	.42139	2.37311	.589190	.964507	.624683	.375317	9	.15
	52	.38859	.92141	.42173	2.37118	.589489	.964454	.625036	.374964	8	
	53	.38886	.92130	.42207	2.36925	.589789	.964400	.625388	.374612	7	
.90	54	.38912	.92119	.42242	2.36733	.590088	.964347	.625741	.374259	6	.10
	55	.38939	.92107	.42276	2.36541	.590387	.964294	.626093	.373907	5	
	56	.38966	.92096	.42310	2.36349	.590686	.964240	.626445	.373555	4	
.95	57	.38993	.92085	.42345	2.36158	.590984	.964187	.626797	.373203	3	.05
	58	.39020	.92073	.42379	2.35967	.591282	.964133	.627149	.372851	2	
	59	.39046	.92062	.42413	2.35776	.591580	.964080	.627501	.372499	1	
1.00	60	.39073	.92050	.42447	2.35585	9.591878	9.964026	9.627852	10.372148	0	.00
Decimals	Minutes	Cos	Sin	Cot	Tan	Cos	Sin	Cot	Tan	Minutes	Decimals
		Natural Values				Common Logarithms					

67°

VALUES AND LOGARITHMS OF TRIGONOMETRIC FUNCTIONS

23°

Decimals	Minutes	Natural Values				Common Logarithms				Minutes	Decimals
		Sin	Cos	Tan	Cot	Sin	Cos	Tan	Cot		
.00	0	.39073	.92050	.42447	2.35585	9.591878	9.964026	9.627852	10.372148	60	1.00
	1	.39100	.92039	.42482	2.35395	.592176	.963972	.628203	.371797	59	
	2	.39127	.92028	.42516	2.35205	.592473	.963919	.628554	.371446	58	
.05	3	.39153	.92016	.42551	2.35015	.592770	.963865	.628905	.371095	57	.95
	4	.39180	.92005	.42585	2.34825	.593067	.963811	.629255	.370745	56	
	5	.39207	.91994	.42619	2.34636	.593363	.963757	.629606	.370394	55	
.10	6	.39234	.91982	.42654	2.34447	.593659	.963704	.629956	.370044	54	.90
	7	.39260	.91971	.42688	2.34258	.593955	.963650	.630306	.369694	53	
	8	.39287	.91959	.42722	2.34069	.594251	.963596	.630656	.369344	52	
.15	9	.39314	.91948	.42757	2.33881	.594547	.963542	.631005	.368995	51	.85
	10	.39341	.91936	.42791	2.33693	.594842	.963488	.631355	.368645	50	
	11	.39367	.91925	.42826	2.33505	.595137	.963434	.631704	.368296	49	
.20	12	.39394	.91914	.42860	2.33317	.595432	.963379	.632053	.367947	48	.80
	13	.39421	.91902	.42894	2.33130	.595727	.963325	.632402	.367598	47	
	14	.39448	.91891	.42929	2.32943	.596021	.963271	.632750	.367250	46	
.25	15	.39474	.91879	.42963	2.32756	.596315	.963217	.633099	.366901	45	.75
	16	.39501	.91868	.42998	2.32570	.596609	.963163	.633447	.366553	44	
	17	.39528	.91856	.43032	2.32383	.596903	.963108	.633795	.366205	43	
.30	18	.39555	.91845	.43067	2.32197	.597196	.963054	.634143	.365857	42	.70
	19	.39581	.91833	.43101	2.32012	.597490	.962999	.634490	.365510	41	
	20	.39608	.91822	.43136	2.31826	.597783	.962945	.634838	.365162	40	
.35	21	.39635	.91810	.43170	2.31641	.598075	.962890	.635185	.364815	39	.65
	22	.39661	.91799	.43205	2.31456	.598368	.962836	.635532	.364468	38	
	23	.39688	.91787	.43239	2.31271	.598660	.962781	.635879	.364121	37	
.40	24	.39715	.91775	.43274	2.31086	.598952	.962727	.636226	.363774	36	.60
	25	.39741	.91764	.43308	2.30902	.599244	.962672	.636572	.363428	35	
	26	.39768	.91752	.43343	2.30718	.599536	.962617	.636919	.363081	34	
.45	27	.39795	.91741	.43378	2.30534	.599827	.962562	.637265	.362735	33	.55
	28	.39822	.91729	.43412	2.30351	.600118	.962508	.637611	.362389	32	
	29	.39848	.91718	.43447	2.30167	.600409	.962453	.637956	.362044	31	
.50	30	.39875	.91706	.43481	2.29984	.600700	.962398	.638302	.361698	30	.50
	31	.39902	.91694	.43516	2.29801	.600990	.962343	.638647	.361353	29	
	32	.39928	.91683	.43550	2.29619	.601280	.962288	.638992	.361008	28	
.55	33	.39955	.91671	.43585	2.29437	.601570	.962233	.639337	.360663	27	.45
	34	.39982	.91660	.43620	2.29254	.601860	.962178	.639682	.360318	26	
	35	.40008	.91648	.43654	2.29073	.602150	.962123	.640027	.359973	25	
.60	36	.40035	.91636	.43689	2.28891	.602439	.962067	.640371	.359629	24	.40
	37	.40062	.91625	.43724	2.28710	.602728	.962012	.640716	.359284	23	
	38	.40088	.91613	.43758	2.28528	.603017	.961957	.641060	.358940	22	
.65	39	.40115	.91601	.43793	2.28348	.603305	.961902	.641404	.358596	21	.35
	40	.40141	.91590	.43828	2.28167	.603594	.961846	.641747	.358253	20	
	41	.40168	.91578	.43862	2.27987	.603882	.961791	.642091	.357909	19	
.70	42	.40195	.91566	.43897	2.27806	.604170	.961735	.642434	.357566	18	.30
	43	.40221	.91555	.43932	2.27626	.604457	.961680	.642777	.357223	17	
	44	.40248	.91543	.43966	2.27447	.604745	.961624	.643120	.356880	16	
.75	45	.40275	.91531	.44001	2.27267	.605032	.961569	.643463	.356537	15	.25
	46	.40301	.91519	.44036	2.27088	.605319	.961513	.643806	.356194	14	
	47	.40328	.91508	.44071	2.26909	.605606	.961458	.644148	.355852	13	
.80	48	.40355	.91496	.44105	2.26730	.605892	.961402	.644490	.355510	12	.20
	49	.40381	.91484	.44140	2.26552	.606179	.961346	.644832	.355168	11	
	50	.40408	.91472	.44175	2.26374	.606465	.961290	.645174	.354826	10	
.85	51	.40434	.91461	.44210	2.26196	.606751	.961235	.645516	.354484	9	.15
	52	.40461	.91449	.44244	2.26018	.607036	.961179	.645857	.354143	8	
	53	.40488	.91437	.44279	2.25840	.607322	.961123	.646199	.353801	7	
.90	54	.40514	.91425	.44314	2.25663	.607607	.961067	.646540	.353460	6	.10
	55	.40541	.91414	.44349	2.25486	.607892	.961011	.646881	.353119	5	
	56	.40567	.91402	.44384	2.25309	.608177	.960955	.647222	.352778	4	
.95	57	.40594	.91390	.44418	2.25132	.608461	.960899	.647562	.352438	3	.05
	58	.40621	.91378	.44453	2.24956	.608745	.960843	.647903	.352097	2	
	59	.40647	.91366	.44488	2.24780	.609029	.960786	.648243	.351757	1	
1.00	60	.40674	.91355	.44523	2.24604	9.609313	9.960730	9.643583	10.351417	0	.00
Decimals	Minutes	Cos	Sin	Cot	Tan	Cos	Sin	Cot	Tan	Minutes	Decimals
		Natural Values				Common Logarithms					

66°

372 MATHEMATICAL AND PHYSICAL TABLES

24°

Decimals	Minutes	Natural Values				Common Logarithms				Minutes	Decimals
		Sin	Cos	Tan	Cot	Sin	Cos	Tan	Cot		
.00	0	.40674	.91355	.44523	2.24604	9.609313	9.960730	9.648583	10.351417	60	1.00
	1	.40700	.91343	.44558	2.24428	.609597	.960674	.648923	.351077	59	
	2	.40727	.91331	.44593	2.24252	.609880	.960618	.649263	.350737	58	
.05	3	.40753	.91319	.44627	2.24077	.610164	.960561	.649602	.350398	57	.95
	4	.40780	.91307	.44662	2.23902	.610447	.960505	.649942	.350058	56	
	5	.40806	.91295	.44697	2.23727	.610729	.960448	.650281	.349719	55	
.10	6	.40833	.91283	.44732	2.23553	.611012	.960392	.650620	.349380	54	.90
	7	.40860	.91272	.44767	2.23378	.611294	.960335	.650959	.349041	53	
	8	.40886	.91260	.44802	2.23204	.611576	.960279	.651297	.348703	52	
.15	9	.40913	.91248	.44837	2.23030	.611858	.960222	.651636	.348364	51	.85
	10	.40939	.91236	.44872	2.22857	.612140	.960165	.651974	.348026	50	
	11	.40966	.91224	.44907	2.22683	.612421	.960109	.652312	.347688	49	
.20	12	.40992	.91212	.44942	2.22510	.612702	.960052	.652650	.347350	48	.80
	13	.41019	.91200	.44977	2.22337	.612983	.959995	.652988	.347012	47	
	14	.41045	.91188	.45012	2.22164	.613264	.959938	.653326	.346674	46	
.25	15	.41072	.91176	.45047	2.21992	.613545	.959882	.653663	.346337	45	.75
	16	.41098	.91164	.45082	2.21819	.613825	.959825	.654000	.346000	44	
	17	.41125	.91152	.45117	2.21647	.614105	.959768	.654337	.345663	43	
.30	18	.41151	.91140	.45152	2.21475	.614385	.959711	.654674	.345326	42	.70
	19	.41178	.91128	.45187	2.21304	.614665	.959654	.655011	.344989	41	
	20	.41204	.91116	.45222	2.21132	.614944	.959596	.655348	.344652	40	
.35	21	.41231	.91104	.45257	2.20961	.615223	.959539	.655684	.344316	39	.65
	22	.41257	.91092	.45292	2.20790	.615502	.959482	.656020	.343980	38	
	23	.41284	.91080	.45327	2.20619	.615781	.959425	.656356	.343644	37	
.40	24	.41310	.91068	.45362	2.20449	.616060	.959368	.656692	.343308	36	.60
	25	.41337	.91056	.45397	2.20278	.616338	.959310	.657028	.342972	35	
	26	.41363	.91044	.45432	2.20108	.616616	.959253	.657364	.342636	34	
.45	27	.41390	.91032	.45467	2.19938	.616894	.959195	.657699	.342301	33	.55
	28	.41416	.91020	.45502	2.19769	.617172	.959138	.658034	.341966	32	
	29	.41443	.91008	.45538	2.19599	.617450	.959080	.658369	.341631	31	
.50	30	.41469	.90996	.45573	2.19430	.617727	.959023	.658704	.341296	30	.50
	31	.41496	.90984	.45608	2.19261	.618004	.958965	.659039	.340961	29	
	32	.41522	.90972	.45643	2.19092	.618281	.958908	.659373	.340627	28	
.55	33	.41549	.90960	.45678	2.18923	.618558	.958850	.659708	.340292	27	.45
	34	.41575	.90948	.45713	2.18755	.618834	.958792	.660042	.339958	26	
	35	.41602	.90936	.45748	2.18587	.619110	.958734	.660376	.339624	25	
.60	36	.41628	.90924	.45784	2.18419	.619386	.958677	.660710	.339290	24	.40
	37	.41655	.90911	.45819	2.18251	.619662	.958619	.661043	.338957	23	
	38	.41681	.90899	.45854	2.18084	.619938	.958561	.661377	.338623	22	
.65	39	.41707	.90887	.45889	2.17916	.620213	.958503	.661710	.338290	21	.35
	40	.41734	.90875	.45924	2.17749	.620488	.958445	.662043	.337957	20	
	41	.41760	.90863	.45960	2.17582	.620763	.958387	.662376	.337624	19	
.70	42	.41787	.90851	.45995	2.17416	.621038	.958329	.662709	.337291	18	.30
	43	.41813	.90839	.46030	2.17249	.621313	.958271	.663042	.336958	17	
	44	.41840	.90826	.46065	2.17083	.621587	.958213	.663375	.336625	16	
.75	45	.41866	.90814	.46101	2.16917	.621861	.958154	.663707	.336293	15	.25
	46	.41892	.90802	.46136	2.16751	.622135	.958096	.664039	.335961	14	
	47	.41919	.90790	.46171	2.16585	.622409	.958038	.664371	.335629	13	
.80	48	.41945	.90778	.46206	2.16420	.622682	.957979	.664703	.335297	12	.20
	49	.41972	.90766	.46242	2.16255	.622956	.957921	.665035	.334965	11	
	50	.41998	.90753	.46277	2.16090	.623229	.957863	.665366	.334634	10	
.85	51	.42024	.90741	.46312	2.15925	.623502	.957804	.665698	.334302	9	.15
	52	.42051	.90729	.46348	2.15760	.623774	.957746	.666029	.333971	8	
	53	.42077	.90717	.46383	2.15596	.624047	.957687	.666360	.333640	7	
.90	54	.42104	.90704	.46418	2.15432	.624319	.957628	.666691	.333309	6	.10
	55	.42130	.90692	.46454	2.15268	.624591	.957570	.667021	.332979	5	
	56	.42156	.90680	.46489	2.15104	.624863	.957511	.667352	.332648	4	
.95	57	.42183	.90668	.46525	2.14940	.625135	.957452	.667682	.332318	3	.05
	58	.42209	.90655	.46560	2.14777	.625406	.957393	.668013	.331987	2	
	59	.42235	.90643	.46595	2.14614	.625677	.957335	.668343	.331657	1	
1.00	60	.42262	.90631	.46631	2.14451	9.625948	9.957276	9.668673	10.331327	0	.00
Decimals	Minutes	Cos	Sin	Cot	Tan	Cos	Sin	Cot	Tan	Minutes	Decimals
		Natural Values				Common Logarithms					

65°

VALUES AND LOGARITHMS OF TRIGONOMETRIC FUNCTIONS 373

25°

Decimals	Minutes	Natural Values				Common Logarithms				Minutes	Decimals
		Sin	Cos	Tan	Cot	Sin	Cos	Tan	Cot		
.00	0	.42262	.90631	.46631	2.14451	9.625948	9.957276	9.668673	10.331327	60	1.00
	1	.42288	.90618	.46666	2.14288	.626219	.957217	.669002	.330998	59	
	2	.42315	.90606	.46702	2.14125	.626490	.957158	.669332	.330668	58	
.05	3	.42341	.90594	.46737	2.13963	.626760	.957099	.669661	.330339	57	.95
	4	.42367	.90582	.46772	2.13801	.627030	.957040	.669991	.330009	56	
	5	.42394	.90569	.46808	2.13639	.627300	.956981	.670320	.329680	55	
.10	6	.42420	.90557	.46843	2.13477	.627570	.956921	.670649	.329351	54	.90
	7	.42446	.90545	.46879	2.13316	.627840	.956862	.670977	.329023	53	
	8	.42473	.90532	.46914	2.13154	.628109	.956803	.671306	.328694	52	
.15	9	.42499	.90520	.46950	2.12993	.628378	.956744	.671635	.328365	51	.85
	10	.42525	.90507	.46985	2.12832	.628647	.956684	.671963	.328037	50	
	11	.42552	.90495	.47021	2.12671	.628916	.956625	.672291	.327709	49	
.20	12	.42578	.90483	.47056	2.12511	.629185	.956566	.672619	.327381	48	.80
	13	.42604	.90470	.47092	2.12350	.629453	.956506	.672947	.327053	47	
	14	.42631	.90458	.47128	2.12190	.629721	.956447	.673274	.326726	46	
.25	15	.42657	.90446	.47163	2.12030	.629989	.956387	.673602	.326398	45	.75
	16	.42683	.90433	.47199	2.11871	.630257	.956327	.673929	.326071	44	
	17	.42709	.90421	.47234	2.11711	.630524	.956268	.674257	.325743	43	
.30	18	.42736	.90408	.47270	2.11552	.630792	.956208	.674584	.325416	42	.70
	19	.42762	.90396	.47305	2.11392	.631059	.956148	.674911	.325089	41	
	20	.42788	.90383	.47341	2.11233	.631326	.956089	.675237	.324763	40	
.35	21	.42815	.90371	.47377	2.11075	.631593	.956029	.675564	.324436	39	.65
	22	.42841	.90358	.47412	2.10916	.631859	.955969	.675890	.324110	38	
	23	.42867	.90346	.47448	2.10758	.632125	.955909	.676217	.323783	37	
.40	24	.42894	.90334	.47483	2.10600	.632392	.955849	.676543	.323457	36	.60
	25	.42920	.90321	.47519	2.10442	.632658	.955789	.676869	.323131	35	
	26	.42946	.90309	.47555	2.10284	.632923	.955729	.677194	.322806	34	
.45	27	.42972	.90296	.47590	2.10126	.633189	.955669	.677520	.322480	33	.55
	28	.42999	.90284	.47626	2.09969	.633454	.955609	.677846	.322154	32	
	29	.43025	.90271	.47662	2.09811	.633719	.955548	.678171	.321829	31	
.50	30	.43051	.90259	.47698	2.09654	.633984	.955488	.678496	.321504	30	.50
	31	.43077	.90246	.47733	2.09498	.634249	.955428	.678821	.321179	29	
	32	.43104	.90233	.47769	2.09341	.634514	.955368	.679146	.320854	28	
.55	33	.43130	.90221	.47805	2.09184	.634778	.955307	.679471	.320529	27	.45
	34	.43156	.90208	.47840	2.09028	.635042	.955247	.679795	.320205	26	
	35	.43182	.90196	.47876	2.08872	.635306	.955186	.680120	.319880	25	
.60	36	.43209	.90183	.47912	2.08716	.635570	.955126	.680444	.319556	24	.40
	37	.43235	.90171	.47948	2.08560	.635834	.955065	.680768	.319232	23	
	38	.43261	.90158	.47984	2.08405	.636097	.955005	.681092	.318908	22	
.65	39	.43287	.90146	.48019	2.08250	.636360	.954944	.681416	.318584	21	.35
	40	.43313	.90133	.48055	2.08094	.636623	.954883	.681740	.318260	20	
	41	.43340	.90120	.48091	2.07939	.636886	.954823	.682063	.317937	19	
.70	42	.43366	.90108	.48127	2.07785	.637148	.954762	.682387	.317613	18	.30
	43	.43392	.90095	.48163	2.07630	.637411	.954701	.682710	.317290	17	
	44	.43418	.90082	.48198	2.07476	.637673	.954640	.683033	.316967	16	
.75	45	.43445	.90070	.48234	2.07321	.637935	.954579	.683356	.316644	15	.25
	46	.43471	.90057	.48270	2.07167	.638197	.954518	.683679	.316321	14	
	47	.43497	.90045	.48306	2.07014	.638458	.954457	.684001	.315999	13	
.80	48	.43523	.90032	.48342	2.06860	.638720	.954396	.684324	.315676	12	.20
	49	.43549	.90019	.48378	2.06706	.638981	.954335	.684646	.315354	11	
	50	.43575	.90007	.48414	2.06553	.639242	.954274	.684968	.315032	10	
.85	51	.43602	.89994	.48450	2.06400	.639503	.954213	.685290	.314710	9	.15
	52	.43628	.89981	.48486	2.06247	.639764	.954152	.685612	.314388	8	
	53	.43654	.89968	.48521	2.06094	.640024	.954090	.685934	.314066	7	
.90	54	.43680	.89956	.48557	2.05942	.640284	.954029	.686255	.313745	6	.10
	55	.43706	.89943	.48593	2.05790	.640544	.953968	.686577	.313423	5	
	56	.43733	.89930	.48629	2.05637	.640804	.953906	.686898	.313102	4	
.95	57	.43759	.89918	.48665	2.05485	.641064	.953845	.687219	.312781	3	.05
	58	.43785	.89905	.48701	2.05333	.641324	.953783	.687540	.312460	2	
	59	.43811	.89892	.48737	2.05182	.641583	.953722	.687861	.312139	1	
1.00	60	.43837	.89879	.48773	2.05030	9.641842	9.953660	9.688182	10.311818	0	.00
Decimals	Minutes	Cos	Sin	Cot	Tan	Cos	Sin	Cot	Tan	Minutes	Decimals
		Natural Values				Common Logarithms					

64°

374 MATHEMATICAL AND PHYSICAL TABLES

26°

Decimals	Minutes	Natural Values				Common Logarithms				Minutes	Decimals
		Sin	Cos	Tan	Cot	Sin	Cos	Tan	Cot		
.00	0	.43837	.89879	.48773	2.05030	9.641842	9.953660	9.688182	10.311818	60	1.00
	1	.43863	.89867	.48809	2.04879	.642101	.953599	.688502	.311498	59	
	2	.43889	.89854	.48845	2.04728	.642360	.953537	.688823	.311177	58	
.05	3	.43916	.89841	.48881	2.04577	.642618	.953475	.689143	.310857	57	.95
	4	.43942	.89828	.48917	2.04426	.642877	.953413	.689463	.310537	56	
	5	.43968	.89816	.48953	2.04276	.643135	.953352	.689783	.310217	55	
.10	6	.43994	.89803	.48989	2.04125	.643393	.953290	.690103	.309897	54	.90
	7	.44020	.89790	.49026	2.03975	.643650	.953228	.690423	.309577	53	
	8	.44046	.89777	.49062	2.03825	.643908	.953166	.690742	.309258	52	
.15	9	.44072	.89764	.49098	2.03675	.644165	.953104	.691062	.308938	51	.85
	10	.44098	.89752	.49134	2.03526	.644423	.953042	.691381	.308619	50	
	11	.44124	.89739	.49170	2.03376	.644680	.952980	.691700	.308300	49	
.20	12	.44151	.89726	.49206	2.03227	.644936	.952918	.692019	.307981	48	.80
	13	.44177	.89713	.49242	2.03078	.645193	.952855	.692338	.307662	47	
	14	.44203	.89700	.49278	2.02929	.645450	.952793	.692656	.307344	46	
.25	15	.44229	.89687	.49315	2.02780	.645706	.952731	.692975	.307025	45	.75
	16	.44255	.89674	.49351	2.02631	.645962	.952669	.693293	.306707	44	
	17	.44281	.89662	.49387	2.02483	.646218	.952606	.693612	.306388	43	
.30	18	.44307	.89649	.49423	2.02335	.646474	.952544	.693930	.306070	42	.70
	19	.44333	.89636	.49459	2.02187	.646729	.952481	.694248	.305752	41	
	20	.44359	.89623	.49495	2.02039	.646984	.952419	.694566	.305434	40	
.35	21	.44385	.89610	.49532	2.01891	.647240	.952356	.694883	.305117	39	.65
	22	.44411	.89597	.49568	2.01743	.647494	.952294	.695201	.304799	38	
	23	.44437	.89584	.49604	2.01596	.647749	.952231	.695518	.304482	37	
.40	24	.44464	.89571	.49640	2.01449	.648004	.952168	.695836	.304164	36	.60
	25	.44490	.89558	.49677	2.01302	.648258	.952106	.696153	.303847	35	
	26	.44516	.89545	.49713	2.01155	.648512	.952043	.696470	.303530	34	
.45	27	.44542	.89532	.49749	2.01008	.648766	.951980	.696787	.303213	33	.55
	28	.44568	.89519	.49786	2.00862	.649020	.951917	.697103	.302897	32	
	29	.44594	.89506	.49822	2.00715	.649274	.951854	.697420	.302580	31	
.50	30	.44620	.89493	.49858	2.00569	.649527	.951791	.697736	.302264	30	.50
	31	.44646	.89480	.49894	2.00423	.649781	.951728	.698053	.301947	29	
	32	.44672	.89467	.49931	2.00277	.650034	.951665	.698369	.301631	28	
.55	33	.44698	.89454	.49967	2.00131	.650287	.951602	.698685	.301315	27	.45
	34	.44724	.89441	.50004	1.99986	.650539	.951539	.699001	.300999	26	
	35	.44750	.89428	.50040	1.99841	.650792	.951476	.699316	.300684	25	
.60	36	.44776	.89415	.50076	1.99695	.651044	.951412	.699632	.300368	24	.40
	37	.44802	.89402	.50113	1.99550	.651297	.951349	.699947	.300053	23	
	38	.44828	.89389	.50149	1.99406	.651549	.951286	.700263	.299737	22	
.65	39	.44854	.89376	.50185	1.99261	.651800	.951222	.700578	.299422	21	.35
	40	.44880	.89363	.50222	1.99116	.652052	.951159	.700893	.299107	20	
	41	.44906	.89350	.50258	1.98972	.652304	.951096	.701208	.298792	19	
.70	42	.44932	.89337	.50295	1.98828	.652555	.951032	.701523	.298477	18	.30
	43	.44958	.89324	.50331	1.98684	.652806	.950968	.701837	.298163	17	
	44	.44984	.89311	.50368	1.98540	.653057	.950905	.702152	.297848	16	
.75	45	.45010	.89298	.50404	1.98396	.653308	.950841	.702466	.297534	15	.25
	46	.45036	.89285	.50441	1.98253	.653558	.950778	.702781	.297219	14	
	47	.45062	.89272	.50477	1.98110	.653808	.950714	.703095	.296905	13	
.80	48	.45088	.89259	.50514	1.97966	.654059	.950650	.703409	.296591	12	.20
	49	.45114	.89245	.50550	1.97823	.654309	.950586	.703722	.296278	11	
	50	.45140	.89232	.50587	1.97681	.654558	.950522	.704036	.295964	10	
.85	51	.45166	.89219	.50623	1.97538	.654808	.950458	.704350	.295650	9	.15
	52	.45192	.89206	.50660	1.97395	.655058	.950394	.704663	.295337	8	
	53	.45218	.89193	.50696	1.97253	.655307	.950330	.704976	.295024	7	
.90	54	.45243	.89180	.50733	1.97111	.655556	.950266	.705290	.294710	6	.10
	55	.45269	.89167	.50769	1.96969	.655805	.950202	.705603	.294397	5	
	56	.45295	.89153	.50806	1.96827	.656054	.950138	.705916	.294084	4	
.95	57	.45321	.89140	.50843	1.96685	.656302	.950074	.706228	.293772	3	.05
	58	.45347	.89127	.50879	1.96544	.656551	.950010	.706541	.293459	2	
	59	.45373	.89114	.50916	1.96402	.656799	.949945	.706854	.293146	1	
1.00	60	.45399	.89101	.50953	1.96261	9.657047	9.949881	9.707166	10.292834	0	.00
Decimals	Minutes	Cos	Sin	Cot	Tan	Cos	Sin	Cot	Tan	Minutes	Decimals
		Natural Values				Common Logarithms					

63°

VALUES AND LOGARITHMS OF TRIGONOMETRIC FUNCTIONS 375

27°

Decimals	Minutes	Natural Values				Common Logarithms				Minutes	Decimals
		Sin	Cos	Tan	Cot	Sin	Cos	Tan	Cot		
.00	0	.45399	.89101	.50953	1.96261	9.657047	9.949881	9.707166	10.292834	60	1.00
	1	.45425	.89087	.50989	1.96120	.657295	.949816	.707478	.292522	59	
	2	.45451	.89074	.51026	1.95979	.657542	.949752	.707790	.292210	58	
.05	3	.45477	.89061	.51063	1.95838	.657790	.949688	.708102	.291898	57	.95
	4	.45503	.89048	.51099	1.95698	.658037	.949623	.708414	.291586	56	
	5	.45529	.89035	.51136	1.95557	.658284	.949558	.708726	.291274	55	
.10	6	.45554	.89021	.51173	1.95417	.658531	.949494	.709037	.290963	54	.90
	7	.45580	.89008	.51209	1.95277	.658778	.949429	.709349	.290651	53	
	8	.45606	.88995	.51246	1.95137	.659025	.949364	.709660	.290340	52	
.15	9	.45632	.88981	.51283	1.94997	.659271	.949300	.709971	.290029	51	.85
	10	.45658	.88968	.51319	1.94858	.659517	.949235	.710282	.289718	50	
	11	.45684	.88955	.51356	1.94718	.659763	.949170	.710593	.289407	49	
.20	12	.45710	.88942	.51393	1.94579	.660009	.949105	.710904	.289096	48	.80
	13	.45736	.88928	.51430	1.94440	.660255	.949040	.711215	.288785	47	
	14	.45762	.88915	.51467	1.94301	.660501	.948975	.711525	.288475	46	
.25	15	.45787	.88902	.51503	1.94162	.660746	.948910	.711836	.288164	45	.75
	16	.45813	.88888	.51540	1.94023	.660991	.948845	.712146	.287854	44	
	17	.45839	.88875	.51577	1.93885	.661236	.948780	.712456	.287544	43	
.30	18	.45865	.88862	.51614	1.93746	.661481	.948715	.712766	.287234	42	.70
	19	.45891	.88848	.51651	1.93608	.661726	.948650	.713076	.286924	41	
	20	.45917	.88835	.51688	1.93470	.661970	.948584	.713386	.286614	40	
.35	21	.45942	.88822	.51724	1.93332	.662214	.948519	.713696	.286304	39	.65
	22	.45968	.88808	.51761	1.93195	.662459	.948454	.714005	.285995	38	
	23	.45994	.88795	.51798	1.93057	.662703	.948388	.714314	.285686	37	
.40	24	.46020	.88782	.51835	1.92920	.662946	.948323	.714624	.285376	36	.60
	25	.46046	.88768	.51872	1.92782	.663190	.948257	.714933	.285067	35	
	26	.46072	.88755	.51909	1.92645	.663433	.948192	.715242	.284758	34	
.45	27	.46097	.88741	.51946	1.92508	.663677	.948126	.715551	.284449	33	.55
	28	.46123	.88728	.51983	1.92371	.663920	.948060	.715860	.284140	32	
	29	.46149	.88715	.52020	1.92235	.664163	.947995	.716168	.283832	31	
.50	30	.46175	.88701	.52057	1.92098	.664406	.947929	.716477	.283523	30	.50
	31	.46201	.88688	.52094	1.91962	.664648	.947863	.716785	.283215	29	
	32	.46226	.88674	.52131	1.91826	.664891	.947797	.717093	.282907	28	
.55	33	.46252	.88661	.52168	1.91690	.665133	.947731	.717401	.282599	27	.45
	34	.46278	.88647	.52205	1.91554	.665375	.947665	.717709	.282291	26	
	35	.46304	.88634	.52242	1.91418	.665617	.947600	.718017	.281983	25	
.60	36	.46330	.88620	.52279	1.91282	.665859	.947533	.718325	.281675	24	.40
	37	.46355	.88607	.52316	1.91147	.666100	.947467	.718633	.281367	23	
	38	.46381	.88593	.52353	1.91012	.666342	.947401	.718940	.281060	22	
.65	39	.46407	.88580	.52390	1.90876	.666583	.947335	.719248	.280752	21	.35
	40	.46433	.88566	.52427	1.90741	.666824	.947269	.719555	.280445	20	
	41	.46458	.88553	.52464	1.90607	.667065	.947203	.719862	.280138	19	
.70	42	.46484	.88539	.52501	1.90472	.667305	.947136	.720169	.279831	18	.30
	43	.46510	.88526	.52538	1.90337	.667546	.947070	.720476	.279524	17	
	44	.46536	.88512	.52575	1.90203	.667786	.947004	.720783	.279217	16	
.75	45	.46561	.88499	.52613	1.90068	.668027	.946937	.721089	.278911	15	.25
	46	.46587	.88485	.52650	1.89935	.668267	.946871	.721396	.278604	14	
	47	.46613	.88472	.52687	1.89801	.668506	.946804	.721702	.278298	13	
.80	48	.46639	.88458	.52724	1.89667	.668746	.946738	.722009	.277991	12	.20
	49	.46664	.88445	.52761	1.89533	.668986	.946671	.722315	.277685	11	
	50	.46690	.88431	.52798	1.89400	.669225	.946604	.722621	.277379	10	
.85	51	.46716	.88417	.52836	1.89266	.669464	.946538	.722927	.277073	9	.15
	52	.46742	.88404	.52873	1.89133	.669703	.946471	.723232	.276768	8	
	53	.46767	.88390	.52910	1.89000	.669942	.946404	.723538	.276462	7	
.90	54	.46793	.88377	.52947	1.88867	.670181	.946337	.723844	.276156	6	.10
	55	.46819	.88363	.52985	1.88734	.670419	.946270	.724149	.275851	5	
	56	.46844	.88349	.53022	1.88602	.670658	.946203	.724454	.275546	4	
.95	57	.46870	.88336	.53059	1.88469	.670896	.946136	.724760	.275240	3	.05
	58	.46896	.88322	.53096	1.88337	.671134	.946069	.725065	.274935	2	
	59	.46921	.88308	.53134	1.88205	.671372	.946002	.725370	.274630	1	
1.00	60	.46947	.88295	.53171	1.88073	9.671609	9.945935	9.725674	10.274326	0	.00
Decimals	Minutes	Cos	Sin	Cot	Tan	Cos	Sin	Cot	Tan	Minutes	Decimals
		Natural Values				Common Logarithms					

62°

28°

Decimals	Minutes	Natural Values				Common Logarithms				Minutes	Decimals
		Sin	Cos	Tan	Cot	Sin	Cos	Tan	Cot		
.00	0	.46947	.88295	.53171	1.88073	9.671609	9.945935	9.725674	10.274326	60	1.00
	1	.46973	.88281	.53208	1.87941	.671847	.945868	.725979	.274021	59	
	2	.46999	.88267	.53246	1.87809	.672084	.945800	.726284	.273716	58	
.05	3	.47024	.88254	.53283	1.87677	.672321	.945733	.726588	.273412	57	.95
	4	.47050	.88240	.53320	1.87546	.672558	.945666	.726892	.273108	56	
	5	.47076	.88226	.53358	1.87415	.672795	.945598	.727197	.272803	55	
.10	6	.47101	.88213	.53395	1.87283	.673032	.945531	.727501	.272499	54	.90
	7	.47127	.88199	.53432	1.87152	.673268	.945464	.727805	.272195	53	
	8	.47153	.88185	.53470	1.87021	.673505	.945396	.728109	.271891	52	
.15	9	.47178	.88172	.53507	1.86891	.673741	.945328	.728412	.271588	51	.85
	10	.47204	.88158	.53545	1.86760	.673977	.945261	.728716	.271284	50	
	11	.47229	.88144	.53582	1.86630	.674213	.945193	.729020	.270980	49	
.20	12	.47255	.88130	.53620	1.86499	.674448	.945125	.729323	.270677	48	.80
	13	.47281	.88117	.53657	1.86369	.674684	.945058	.729626	.270374	47	
	14	.47306	.88103	.53694	1.86239	.674919	.944990	.729929	.270071	46	
.25	15	.47332	.88089	.53732	1.86109	.675155	.944922	.730233	.269767	45	.75
	16	.47358	.88075	.53769	1.85979	.675390	.944854	.730535	.269465	44	
	17	.47383	.88062	.53807	1.85850	.675624	.944786	.730838	.269162	43	
.30	18	.47409	.88048	.53844	1.85720	.675859	.944718	.731141	.268859	42	.70
	19	.47434	.88034	.53882	1.85591	.676094	.944650	.731444	.268556	41	
	20	.47460	.88020	.53920	1.85462	.676328	.944582	.731746	.268254	40	
.35	21	.47486	.88006	.53957	1.85333	.676562	.944514	.732048	.267952	39	.65
	22	.47511	.87993	.53995	1.85204	.676796	.944446	.732351	.267649	38	
	23	.47537	.87979	.54032	1.85075	.677030	.944377	.732653	.267347	37	
.40	24	.47562	.87965	.54070	1.84946	.677264	.944309	.732955	.267045	36	.60
	25	.47588	.87951	.54107	1.84818	.677498	.944241	.733257	.266743	35	
	26	.47614	.87937	.54145	1.84689	.677731	.944172	.733558	.266442	34	
.45	27	.47639	.87923	.54183	1.84561	.677964	.944104	.733860	.266140	33	.55
	28	.47665	.87909	.54220	1.84433	.678197	.944036	.734162	.265838	32	
	29	.47690	.87896	.54258	1.84305	.678430	.943967	.734463	.265537	31	
.50	30	.47716	.87882	.54296	1.84177	.678663	.943899	.734764	.265236	30	.50
	31	.47741	.87868	.54333	1.84049	.678895	.943830	.735066	.264934	29	
	32	.47767	.87854	.54371	1.83922	.679128	.943761	.735367	.264633	28	
.55	33	.47793	.87840	.54409	1.83794	.679360	.943693	.735668	.264332	27	.45
	34	.47818	.87826	.54446	1.83667	.679592	.943624	.735969	.264031	26	
	35	.47844	.87812	.54484	1.83540	.679824	.943555	.736269	.263731	25	
.60	36	.47869	.87798	.54522	1.83413	.680056	.943486	.736570	.263430	24	.40
	37	.47895	.87784	.54560	1.83286	.680288	.943417	.736870	.263130	23	
	38	.47920	.87770	.54597	1.83159	.680519	.943348	.737171	.262829	22	
.65	39	.47946	.87756	.54635	1.83033	.680750	.943279	.737471	.262529	21	.35
	40	.47971	.87743	.54673	1.82906	.680982	.943210	.737771	.262229	20	
	41	.47997	.87729	.54711	1.82780	.681213	.943141	.738071	.261929	19	
.70	42	.48022	.87715	.54748	1.82654	.681443	.943072	.738371	.261629	18	.30
	43	.48048	.87701	.54786	1.82528	.681674	.943003	.738671	.261329	17	
	44	.48073	.87687	.54824	1.82402	.681905	.942934	.738971	.261029	16	
.75	45	.48099	.87673	.54862	1.82276	.682135	.942864	.739271	.260729	15	.25
	46	.48124	.87659	.54900	1.82150	.682365	.942795	.739570	.260430	14	
	47	.48150	.87645	.54938	1.82025	.682595	.942726	.739870	.260130	13	
.80	48	.48175	.87631	.54975	1.81899	.682825	.942656	.740169	.259831	12	.20
	49	.48201	.87617	.55013	1.81774	.683055	.942587	.740468	.259532	11	
	50	.48226	.87603	.55051	1.81649	.683284	.942517	.740767	.259233	10	
.85	51	.48252	.87589	.55089	1.81524	.683514	.942448	.741066	.258934	9	.15
	52	.48277	.87575	.55127	1.81399	.683743	.942378	.741365	.258635	8	
	53	.48303	.87561	.55165	1.81274	.683972	.942308	.741664	.258336	7	
.90	54	.48328	.87546	.55203	1.81150	.684201	.942239	.741962	.258038	6	.10
	55	.48354	.87532	.55241	1.81025	.684430	.942169	.742261	.257739	5	
	56	.48379	.87518	.55279	1.80901	.684658	.942099	.742559	.257441	4	
.95	57	.48405	.87504	.55317	1.80777	.684887	.942029	.742858	.257142	3	.05
	58	.48430	.87490	.55355	1.80653	.685115	.941959	.743156	.256844	2	
	59	.48456	.87476	.55393	1.80529	.685343	.941889	.743454	.256546	1	
1.00	60	.48481	.87462	.55431	1.80405	9.685571	9.941819	9.743752	10.256248	0	.00
Decimals	Minutes	Cos	Sin	Cot	Tan	Cos	Sin	Cot	Tan	Minutes	Decimals
		Natural Values				Common Logarithms					

61°

VALUES AND LOGARITHMS OF TRIGONOMETRIC FUNCTIONS

29°

Decimals	Minutes	Natural Values				Common Logarithms				Minutes	Decimals
		Sin	Cos	Tan	Cot	Sin	Cos	Tan	Cot		
.00	0	.48481	.87462	.55431	1.80405	9.635571	9.941819	9.743752	10.256248	60	1.00
	1	.48506	.87448	.55469	1.80281	.685799	.941749	.744050	.255950	59	
	2	.48532	.87434	.55507	1.80158	.686027	.941679	.744348	.255652	58	
.05	3	.48557	.87420	.55545	1.80034	.686254	.941609	.744645	.255355	57	.95
	4	.48583	.87406	.55583	1.79911	.686482	.941539	.744943	.255057	56	
	5	.48608	.87391	.55621	1.79788	.686709	.941469	.745240	.254760	55	
.10	6	.48634	.87377	.55659	1.79665	.686936	.941398	.745538	.254462	54	.90
	7	.48659	.87363	.55697	1.79542	.687163	.941328	.745835	.254165	53	
	8	.48684	.87349	.55736	1.79419	.687389	.941258	.746132	.253868	52	
.15	9	.48710	.87335	.55774	1.79296	.687616	.941187	.746429	.253571	51	.85
	10	.48735	.87321	.55812	1.79174	.687843	.941117	.746726	.253274	50	
	11	.48761	.87306	.55850	1.79051	.688069	.941046	.747023	.252977	49	
.20	12	.48786	.87292	.55888	1.78929	.688295	.940975	.747319	.252681	48	.80
	13	.48811	.87278	.55926	1.78807	.688521	.940905	.747616	.252384	47	
	14	.48837	.87264	.55964	1.78685	.688747	.940834	.747913	.252087	46	
.25	15	.48862	.87250	.56003	1.78563	.688972	.940763	.748209	.251791	45	.75
	16	.48888	.87235	.56041	1.78441	.689198	.940693	.748505	.251495	44	
	17	.48913	.87221	.56079	1.78319	.689423	.940622	.748801	.251199	43	
.30	18	.48938	.87207	.56117	1.78198	.689648	.940551	.749097	.250903	42	.70
	19	.48964	.87193	.56156	1.78077	.689873	.940480	.749393	.250607	41	
	20	.48989	.87178	.56194	1.77955	.690098	.940409	.749689	.250311	40	
.35	21	.49014	.87164	.56232	1.77834	.690323	.940338	.749985	.250015	39	.65
	22	.49040	.87150	.56270	1.77713	.690548	.940267	.750281	.249719	38	
	23	.49065	.87136	.56309	1.77592	.690772	.940196	.750576	.249424	37	
.40	24	.49090	.87121	.56347	1.77471	.690996	.940125	.750872	.249128	36	.60
	25	.49116	.87107	.56385	1.77351	.691220	.940054	.751167	.248833	35	
	26	.49141	.87093	.56424	1.77230	.691444	.939982	.751462	.248538	34	
.45	27	.49166	.87079	.56462	1.77110	.691668	.939911	.751757	.248243	33	.55
	28	.49192	.87064	.56501	1.76990	.691892	.939840	.752052	.247948	32	
	29	.49217	.87050	.56539	1.76869	.692115	.939768	.752347	.247653	31	
.50	30	.49242	.87036	.56577	1.76749	.692339	.939697	.752642	.247358	30	.50
	31	.49268	.87021	.56616	1.76629	.692562	.939625	.752937	.247063	29	
	32	.49293	.87007	.56654	1.76510	.692785	.939554	.753231	.246769	28	
.55	33	.49318	.86993	.56693	1.76390	.693008	.939482	.753526	.246474	27	.45
	34	.49344	.86978	.56731	1.76271	.693231	.939410	.753820	.246180	26	
	35	.49369	.86964	.56769	1.76151	.693453	.939339	.754115	.245885	25	
.60	36	.49394	.86949	.56808	1.76032	.693676	.939267	.754409	.245591	24	.40
	37	.49419	.86935	.56846	1.75913	.693898	.939195	.754703	.245297	23	
	38	.49445	.86921	.56885	1.75794	.694120	.939123	.754997	.245003	22	
.65	39	.49470	.86906	.56923	1.75675	.694342	.939052	.755291	.244709	21	.35
	40	.49495	.86892	.56962	1.75556	.694564	.938980	.755585	.244415	20	
	41	.49521	.86878	.57000	1.75437	.694786	.938908	.755878	.244122	19	
.70	42	.49546	.86863	.57039	1.75319	.695007	.938836	.756172	.243828	18	.30
	43	.49571	.86849	.57078	1.75200	.695229	.938763	.756465	.243535	17	
	44	.49596	.86834	.57116	1.75082	.695450	.938691	.756759	.243241	16	
.75	45	.49622	.86820	.57155	1.74964	.695671	.938619	.757052	.242948	15	.25
	46	.49647	.86805	.57193	1.74846	.695892	.938547	.757345	.242655	14	
	47	.49672	.86791	.57232	1.74728	.696113	.938475	.757638	.242362	13	
.80	48	.49697	.86777	.57271	1.74610	.696334	.938402	.757931	.242069	12	.20
	49	.49723	.86762	.57309	1.74492	.696554	.938330	.758224	.241776	11	
	50	.49748	.86748	.57348	1.74375	.696775	.938258	.758517	.241483	10	
.85	51	.49773	.86733	.57386	1.74257	.696995	.938185	.758810	.241190	9	.15
	52	.49798	.86719	.57425	1.74140	.697215	.938113	.759102	.240898	8	
	53	.49824	.86704	.57464	1.74022	.697435	.938040	.759395	.240605	7	
.90	54	.49849	.86690	.57503	1.73905	.697654	.937967	.759687	.240313	6	.10
	55	.49874	.86675	.57541	1.73788	.697874	.937895	.759979	.240021	5	
	56	.49899	.86661	.57580	1.73671	.698094	.937822	.760272	.239728	4	
.95	57	.49924	.86646	.57619	1.73555	.698313	.937749	.760564	.239436	3	.05
	58	.49950	.86632	.57657	1.73438	.698532	.937676	.760856	.239144	2	
	59	.49975	.86617	.57696	1.73321	.698751	.937604	.761148	.238852	1	
1.00	60	.50000	.86603	.57735	1.73205	9.693970	9.937531	9.761439	10.238561	0	.00
		Cos	Sin	Cot	Tan	Cos	Sin	Cot	Tan		
Decimals	Minutes	Natural Values				Common Logarithms				Minutes	Decimals

60°

378 MATHEMATICAL AND PHYSICAL TABLES

30°

Decimals	Minutes	Natural Values				Common Logarithms				Minutes	Decimals
		Sin	Cos	Tan	Cot	Sin	Cos	Tan	Cot		
.00	0	.50000	.86603	.57735	1.73205	9.698970	9.937531	9.761439	10.238561	60	1.00
	1	.50025	.86588	.57774	1.73089	.699189	.937458	.761731	.238269	59	
	2	.50050	.86573	.57813	1.72973	.699407	.937385	.762023	.237977	58	
.05	3	.50076	.86559	.57851	1.72857	.699626	.937312	.762314	.237686	57	.95
	4	.50101	.86544	.57890	1.72741	.699844	.937238	.762606	.237394	56	
	5	.50126	.86530	.57929	1.72625	.700062	.937165	.762897	.237103	55	
.10	6	.50151	.86515	.57968	1.72509	.700280	.937092	.763188	.236812	54	.90
	7	.50176	.86501	.58007	1.72393	.700498	.937019	.763479	.236521	53	
	8	.50201	.86486	.58046	1.72278	.700716	.936946	.763770	.236230	52	
.15	9	.50227	.86471	.58085	1.72163	.700933	.936872	.764061	.235939	51	.85
	10	.50252	.86457	.58124	1.72047	.701151	.936799	.764352	.235648	50	
	11	.50277	.86442	.58162	1.71932	.701368	.936725	.764643	.235357	49	
.20	12	.50302	.86427	.58201	1.71817	.701585	.936652	.764933	.235067	48	.80
	13	.50327	.86413	.58240	1.71702	.701802	.936578	.765224	.234776	47	
	14	.50352	.86398	.58279	1.71588	.702019	.936505	.765514	.234486	46	
.25	15	.50377	.86384	.58318	1.71473	.702236	.936431	.765805	.234195	45	.75
	16	.50403	.86369	.58357	1.71358	.702452	.936357	.766095	.233905	44	
	17	.50428	.86354	.58396	1.71244	.702669	.936284	.766385	.233615	43	
.30	18	.50453	.86340	.58435	1.71129	.702885	.936210	.766675	.233325	42	.70
	19	.50478	.86325	.58474	1.71015	.703101	.936136	.766965	.233035	41	
	20	.50503	.86310	.58513	1.70901	.703317	.936062	.767255	.232745	40	
.35	21	.50528	.86295	.58552	1.70787	.703533	.935988	.767545	.232455	39	.65
	22	.50553	.86281	.58591	1.70673	.703749	.935914	.767834	.232166	38	
	23	.50578	.86266	.58631	1.70560	.703964	.935840	.768124	.231876	37	
.40	24	.50603	.86251	.58670	1.70446	.704179	.935766	.768414	.231586	36	.60
	25	.50628	.86237	.58709	1.70332	.704395	.935692	.768703	.231297	35	
	26	.50654	.86222	.58748	1.70219	.704610	.935618	.768992	.231008	34	
.45	27	.50679	.86207	.58787	1.70106	.704825	.935543	.769281	.230719	33	.55
	28	.50704	.86192	.58826	1.69992	.705040	.935469	.769571	.230429	32	
	29	.50729	.86178	.58865	1.69879	.705254	.935395	.769860	.230140	31	
.50	30	.50754	.86163	.58905	1.69766	.705469	.935320	.770148	.229852	30	.50
	31	.50779	.86148	.58944	1.69653	.705683	.935246	.770437	.229563	29	
	32	.50804	.86133	.58983	1.69541	.705898	.935171	.770726	.229274	28	
.55	33	.50829	.86119	.59022	1.69428	.706112	.935097	.771015	.228985	27	.45
	34	.50854	.86104	.59061	1.69316	.706326	.935022	.771303	.228697	26	
	35	.50879	.86089	.59101	1.69203	.706539	.934948	.771592	.228408	25	
.60	36	.50904	.86074	.59140	1.69091	.706753	.934873	.771880	.228120	24	.40
	37	.50929	.86059	.59179	1.68979	.706967	.934798	.772168	.227832	23	
	38	.50954	.86045	.59218	1.68866	.707180	.934723	.772457	.227543	22	
.65	39	.50979	.86030	.59258	1.68754	.707393	.934649	.772745	.227255	21	.35
	40	.51004	.86015	.59297	1.68643	.707606	.934574	.773033	.226967	20	
	41	.51029	.86000	.59336	1.68531	.707819	.934499	.773321	.226679	19	
.70	42	.51054	.85985	.59376	1.68419	.708032	.934424	.773608	.226392	18	.30
	43	.51079	.85970	.59415	1.68308	.708245	.934349	.773896	.226104	17	
	44	.51104	.85956	.59454	1.68196	.708458	.934274	.774184	.225816	16	
.75	45	.51129	.85941	.59494	1.68085	.708670	.934199	.774471	.225529	15	.25
	46	.51154	.85926	.59533	1.67974	.708882	.934123	.774759	.225241	14	
	47	.51179	.85911	.59573	1.67863	.709094	.934048	.775046	.224954	13	
.80	48	.51204	.85896	.59612	1.67752	.709306	.933973	.775333	.224667	12	.20
	49	.51229	.85881	.59651	1.67641	.709518	.933898	.775621	.224379	11	
	50	.51254	.85866	.59691	1.67530	.709730	.933822	.775908	.224092	10	
.85	51	.51279	.85851	.59730	1.67419	.709941	.933747	.776195	.223805	9	.15
	52	.51304	.85836	.59770	1.67309	.710153	.933671	.776482	.223518	8	
	53	.51329	.85821	.59809	1.67198	.710364	.933596	.776768	.223232	7	
.90	54	.51354	.85806	.59849	1.67088	.710575	.933520	.777055	.222945	6	.10
	55	.51379	.85792	.59888	1.66978	.710786	.933445	.777342	.222658	5	
	56	.51404	.85777	.59928	1.66867	.710997	.933369	.777628	.222372	4	
.95	57	.51429	.85762	.59967	1.66757	.711208	.933293	.777915	.222085	3	.05
	58	.51454	.85747	.60007	1.66647	.711419	.933217	.778201	.221799	2	
	59	.51479	.85732	.60046	1.66538	.711629	.933141	.778488	.221512	1	
1.00	60	.51504	.85717	.60086	1.66428	9.711839	9.933066	9.778774	10.221226	0	.00
		Cos	Sin	Cot	Tan	Cos	Sin	Cot	Tan		
		Natural Values				Common Logarithms					

59°

31°

Decimals	Minutes	Natural Values				Common Logarithms				Minutes	Decimals
		Sin	Cos	Tan	Cot	Sin	Cos	Tan	Cot		
.00	0	.51504	.85717	.60086	1.66428	9.711839	9.933066	9.778774	10.221226	60	1.00
	1	.51529	.85702	.60126	1.66318	.712050	.932990	.779060	.220940	59	
	2	.51554	.85687	.60165	1.66209	.712260	.932914	.779346	.220654	58	
.05	3	.51579	.85672	.60205	1.66099	.712469	.932838	.779632	.220368	57	.95
	4	.51604	.85657	.60245	1.65990	.712679	.932762	.779918	.220082	56	
	5	.51628	.85642	.60284	1.65881	.712889	.932685	.780203	.219797	55	
.10	6	.51653	.85627	.60324	1.65772	.713098	.932609	.780489	.219511	54	.90
	7	.51678	.85612	.60364	1.65663	.713308	.932533	.780775	.219225	53	
	8	.51703	.85597	.60403	1.65554	.713517	.932457	.781060	.218940	52	
.15	9	.51728	.85582	.60443	1.65445	.713726	.932380	.731346	.218654	51	.85
	10	.51753	.85567	.60483	1.65337	.713935	.932304	.781631	.218369	50	
	11	.51778	.85551	.60522	1.65228	.714144	.932228	.781916	.218084	49	
.20	12	.51803	.85536	.60562	1.65120	.714352	.932151	.782201	.217799	48	.80
	13	.51828	.85521	.60602	1.65011	.714561	.932075	.782486	.217514	47	
	14	.51852	.85506	.60642	1.64903	.714769	.931998	.782771	.217229	46	
.25	15	.51877	.85491	.60681	1.64795	.714978	.931921	.783056	.216944	45	.75
	16	.51902	.85476	.60721	1.64687	.715186	.931845	.783341	.216659	44	
	17	.51927	.85461	.60761	1.64579	.715394	.931768	.783626	.216374	43	
.30	18	.51952	.85446	.60801	1.64471	.715602	.931691	.783910	.216090	42	.70
	19	.51977	.85431	.60841	1.64363	.715809	.931614	.784195	.215805	41	
	20	.52002	.85416	.60881	1.64256	.716017	.931537	.784479	.215521	40	
.35	21	.52026	.85401	.60921	1.64148	.716224	.931460	.784764	.215236	39	.65
	22	.52051	.85385	.60960	1.64041	.716432	.931383	.785048	.214952	38	
	23	.52076	.85370	.61000	1.63934	.716639	.931306	.785332	.214668	37	
.40	24	.52101	.85355	.61040	1.63826	.716846	.931229	.785616	.214384	36	.60
	25	.52126	.85340	.61080	1.63719	.717053	.931152	.785900	.214100	35	
	26	.52151	.85325	.61120	1.63612	.717259	.931075	.786184	.213816	34	
.45	27	.52175	.85310	.61160	1.63505	.717466	.930998	.786468	.213532	33	.55
	28	.52200	.85294	.61200	1.63398	.717673	.930921	.786752	.213248	32	
	29	.52225	.85279	.61240	1.63292	.717879	.930843	.787036	.212964	31	
.50	30	.52250	.85264	.61280	1.63185	.718085	.930766	.787319	.212681	30	.50
	31	.52275	.85249	.61320	1.63079	.718291	.930688	.787603	.212397	29	
	32	.52299	.85234	.61360	1.62972	.718497	.930611	.787886	.212114	28	
.55	33	.52324	.85218	.61400	1.62866	.718703	.930533	.788170	.211830	27	.45
	34	.52349	.85203	.61440	1.62760	.718909	.930456	.788453	.211547	26	
	35	.52374	.85188	.61480	1.62654	.719114	.930378	.788736	.211264	25	
.60	36	.52399	.85173	.61520	1.62548	.719320	.930300	.789019	.210981	24	.40
	37	.52423	.85157	.61561	1.62442	.719525	.930223	.789302	.210698	23	
	38	.52448	.85142	.61601	1.62336	.719730	.930145	.789585	.210415	22	
.65	39	.52473	.85127	.61641	1.62230	.719935	.930067	.789868	.210132	21	.35
	40	.52498	.85112	.61681	1.62125	.720140	.929989	.790151	.209849	20	
	41	.52522	.85096	.61721	1.62019	.720345	.929911	.790434	.209566	19	
.70	42	.52547	.85081	.61761	1.61914	.720549	.929833	.790716	.209284	18	.30
	43	.52572	.85066	.61801	1.61808	.720754	.929755	.790999	.209001	17	
	44	.52597	.85051	.61842	1.61703	.720958	.929677	.791281	.208719	16	
.75	45	.52621	.85035	.61882	1.61598	.721162	.929599	.791563	.208437	15	.25
	46	.52646	.85020	.61922	1.61493	.721366	.929521	.791846	.208154	14	
	47	.52671	.85005	.61962	1.61388	.721570	.929442	.792128	.207872	13	
.80	48	.52696	.84989	.62003	1.61283	.721774	.929364	.792410	.207590	12	.20
	49	.52720	.84974	.62043	1.61179	.721978	.929286	.792692	.207308	11	
	50	.52745	.84959	.62083	1.61074	.722181	.929207	.792974	.207026	10	
.85	51	.52770	.84943	.62124	1.60970	.722385	.929129	.793256	.206744	9	.15
	52	.52794	.84928	.62164	1.60865	.722588	.929050	.793538	.206462	8	
	53	.52819	.84913	.62204	1.60761	.722791	.928972	.793819	.206181	7	
.90	54	.52844	.84897	.62245	1.60657	.722994	.928893	.794101	.205899	6	.10
	55	.52869	.84882	.62285	1.60553	.723197	.928815	.794383	.205617	5	
	56	.52893	.84866	.62325	1.60449	.723400	.928736	.794664	.205336	4	
.95	57	.52918	.84851	.62366	1.60345	.723603	.928657	.794946	.205054	3	.05
	58	.52943	.84836	.62406	1.60241	.723805	.928578	.795227	.204773	2	
	59	.52967	.84820	.62446	1.60137	.724007	.928499	.795508	.204492	1	
1.00	60	.52992	.84805	.62487	1.60033	9.724210	9.928420	9.795789	10.204211	0	.00
Decimals	Minutes	Cos	Sin	Cot	Tan	Cos	Sin	Cot	Tan	Minutes	Decimals
		Natural Values				Common Logarithms					

58°

32°

Decimals	Minutes	Natural Values				Common Logarithms				Minutes	Decimals
		Sin	Cos	Tan	Cot	Sin	Cos	Tan	Cot		
.00	0	.52992	.84805	.62487	1.60033	9.724210	9.928420	9.795789	10.204211	60	1.00
	1	.53017	.84789	.62527	1.59930	.724412	.928342	.796070	.203930	59	
	2	.53041	.84774	.62568	1.59826	.724614	.928263	.796351	.203649	58	
.05	3	.53066	.84759	.62608	1.59723	.724816	.928183	.796632	.203368	57	.95
	4	.53091	.84743	.62649	1.59620	.725017	.928104	.796913	.203087	56	
	5	.53115	.84728	.62689	1.59517	.725219	.928025	.797194	.202806	55	
.10	6	.53140	.84712	.62730	1.59414	.725420	.927946	.797474	.202526	54	.90
	7	.53164	.84697	.62770	1.59311	.725622	.927867	.797755	.202245	53	
	8	.53189	.84681	.62811	1.59208	.725823	.927787	.798036	.201964	52	
.15	9	.53214	.84666	.62852	1.59105	.726024	.927708	.798316	.201684	51	.85
	10	.53238	.84650	.62892	1.59002	.726225	.927629	.798596	.201404	50	
	11	.53263	.84635	.62933	1.58900	.726426	.927549	.798877	.201123	49	
.20	12	.53288	.84619	.62973	1.58797	.726626	.927470	.799157	.200843	48	.80
	13	.53312	.84604	.63014	1.58695	.726827	.927390	.799437	.200563	47	
	14	.53337	.84588	.63055	1.58593	.727027	.927310	.799717	.200283	46	
.25	15	.53361	.84573	.63095	1.58490	.727228	.927231	.799997	.200003	45	.75
	16	.53386	.84557	.63136	1.58388	.727428	.927151	.800277	.199723	44	
	17	.53411	.84542	.63177	1.58286	.727628	.927071	.800557	.199443	43	
.30	18	.53435	.84526	.63217	1.58184	.727828	.926991	.800836	.199164	42	.70
	19	.53460	.84511	.63258	1.58083	.728027	.926911	.801116	.198884	41	
	20	.53484	.84495	.63299	1.57981	.728227	.926831	.801396	.198604	40	
.35	21	.53509	.84480	.63340	1.57879	.728427	.926751	.801675	.198325	39	.65
	22	.53534	.84464	.63380	1.57778	.728626	.926671	.801955	.198045	38	
	23	.53558	.84448	.63421	1.57676	.728825	.926591	.802234	.197766	37	
.40	24	.53583	.84433	.63462	1.57575	.729024	.926511	.802513	.197487	36	.60
	25	.53607	.84417	.63503	1.57474	.729223	.926431	.802792	.197208	35	
	26	.53632	.84402	.63544	1.57372	.729422	.926351	.803072	.196928	34	
.45	27	.53656	.84386	.63584	1.57271	.729621	.926270	.803351	.196649	33	.55
	28	.53681	.84370	.63625	1.57170	.729820	.926190	.803630	.196370	32	
	29	.53705	.84355	.63666	1.57069	.730018	.926110	.803909	.196091	31	
.50	30	.53730	.84339	.63707	1.56969	.730217	.926029	.804187	.195813	30	.50
	31	.53754	.84324	.63748	1.56868	.730415	.925949	.804466	.195534	29	
	32	.53779	.84308	.63789	1.56767	.730613	.925868	.804745	.195255	28	
.55	33	.53804	.84292	.63830	1.56667	.730811	.925788	.805023	.194977	27	.45
	34	.53828	.84277	.63871	1.56566	.731009	.925707	.805302	.194698	26	
	35	.53853	.84261	.63912	1.56466	.731206	.925626	.805580	.194420	25	
.60	36	.53877	.84245	.63953	1.56366	.731404	.925545	.805859	.194141	24	.40
	37	.53902	.84230	.63994	1.56265	.731602	.925465	.806137	.193863	23	
	38	.53926	.84214	.64035	1.56165	.731799	.925384	.806415	.193585	22	
.65	39	.53951	.84198	.64076	1.56065	.731996	.925303	.806693	.193307	21	.35
	40	.53975	.84182	.64117	1.55966	.732193	.925222	.806971	.193029	20	
	41	.54000	.84167	.64158	1.55866	.732390	.925141	.807249	.192751	19	
.70	42	.54024	.84151	.64199	1.55766	.732587	.925060	.807527	.192473	18	.30
	43	.54049	.84135	.64240	1.55666	.732784	.924979	.807805	.192195	17	
	44	.54073	.84120	.64281	1.55567	.732980	.924897	.808083	.191917	16	
.75	45	.54097	.84104	.64322	1.55467	.733177	.924816	.808361	.191639	15	.25
	46	.54122	.84088	.64363	1.55368	.733373	.924735	.808638	.191362	14	
	47	.54146	.84072	.64404	1.55269	.733569	.924654	.808916	.191084	13	
.80	48	.54171	.84057	.64446	1.55170	.733765	.924572	.809193	.190807	12	.20
	49	.54195	.84041	.64487	1.55071	.733961	.924491	.809471	.190529	11	
	50	.54220	.84025	.64528	1.54972	.734157	.924409	.809748	.190252	10	
.85	51	.54244	.84009	.64569	1.54873	.734353	.924328	.810025	.189975	9	.15
	52	.54269	.83994	.64610	1.54774	.734549	.924246	.810302	.189698	8	
	53	.54293	.83978	.64652	1.54675	.734744	.924164	.810580	.189420	7	
.90	54	.54317	.83962	.64693	1.54576	.734939	.924083	.810857	.189143	6	.10
	55	.54342	.83946	.64734	1.54478	.735135	.924001	.811134	.188866	5	
	56	.54366	.83930	.64775	1.54379	.735330	.923919	.811410	.188590	4	
.95	57	.54391	.83915	.64817	1.54281	.735525	.923837	.811687	.188313	3	.05
	58	.54415	.83899	.64858	1.54183	.735719	.923755	.811964	.188036	2	
	59	.54440	.83883	.64899	1.54085	.735914	.923673	.812241	.187759	1	
1.00	60	.54464	.83867	.64941	1.53986	9.736109	9.923591	9.812517	10.187483	0	.00
Decimals	Minutes	Cos	Sin	Cot	Tan	Cos	Sin	Cot	Tan	Minutes	Decimals
		Natural Values				Common Logarithms					

57°

VALUES AND LOGARITHMS OF TRIGONOMETRIC FUNCTIONS

33°

Decimals	Minutes	Natural Values				Common Logarithms				Minutes	Decimals
		Sin	Cos	Tan	Cot	Sin	Cos	Tan	Cot		
.00	0	.54464	.83867	.64941	1.53986	9.736109	9.923591	9.812517	10.187483	60	1.00
	1	.54488	.83851	.64982	1.53888	.736303	.923509	.812794	.187206	59	
	2	.54513	.83835	.65024	1.53791	.736498	.923427	.813070	.186930	58	
.05	3	.54537	.83819	.65065	1.53693	.736692	.923345	.813347	.186653	57	.95
	4	.54561	.83804	.65106	1.53595	.736886	.923263	.813623	.186377	56	
	5	.54586	.83788	.65148	1.53497	.737080	.923181	.813899	.186101	55	
.10	6	.54610	.83772	.65189	1.53400	.737274	.923098	.814176	.185824	54	.90
	7	.54635	.83756	.65231	1.53302	.737467	.923016	.814452	.185548	53	
	8	.54659	.83740	.65272	1.53205	.737661	.922933	.814728	.185272	52	
.15	9	.54683	.83724	.65314	1.53107	.737855	.922851	.815004	.184996	51	.85
	10	.54708	.83708	.65355	1.53010	.738048	.922768	.815280	.184720	50	
	11	.54732	.83692	.65397	1.52913	.738241	.922686	.815555	.184445	49	
.20	12	.54756	.83676	.65438	1.52816	.738434	.922603	.815831	.184169	48	.80
	13	.54781	.83660	.65480	1.52719	.738627	.922520	.816107	.183893	47	
	14	.54805	.83645	.65521	1.52622	.738820	.922438	.816382	.183618	46	
.25	15	.54829	.83629	.65563	1.52525	.739013	.922355	.816658	.183342	45	.75
	16	.54854	.83613	.65604	1.52429	.739206	.922272	.816933	.183067	44	
	17	.54878	.83597	.65646	1.52332	.739398	.922189	.817209	.182791	43	
.30	18	.54902	.83581	.65688	1.52235	.739590	.922106	.817484	.182516	42	.70
	19	.54927	.83565	.65729	1.52139	.739783	.922023	.817759	.182241	41	
	20	.54951	.83549	.65771	1.52043	.739975	.921940	.818035	.181965	40	
.35	21	.54975	.83533	.65813	1.51946	.740167	.921857	.818310	.181690	39	.65
	22	.54999	.83517	.65854	1.51850	.740359	.921774	.818585	.181415	38	
	23	.55024	.83501	.65896	1.51754	.740550	.921691	.818860	.181140	37	
.40	24	.55048	.83485	.65938	1.51658	.740742	.921607	.819135	.180865	36	.60
	25	.55072	.83469	.65980	1.51562	.740934	.921524	.819410	.180590	35	
	26	.55097	.83453	.66021	1.51466	.741125	.921441	.819684	.180316	34	
.45	27	.55121	.83437	.66063	1.51370	.741316	.921357	.819959	.180041	33	.55
	28	.55145	.83421	.66105	1.51275	.741508	.921274	.820234	.179766	32	
	29	.55169	.83405	.66147	1.51179	.741699	.921190	.820508	.179492	31	
.50	30	.55194	.83389	.66189	1.51084	.741889	.921107	.820783	.179217	30	.50
	31	.55218	.83373	.66230	1.50988	.742080	.921023	.821057	.178943	29	
	32	.55242	.83356	.66272	1.50893	.742271	.920939	.821332	.178668	28	
.55	33	.55266	.83340	.66314	1.50797	.742462	.920856	.821606	.178394	27	.45
	34	.55291	.83324	.66356	1.50702	.742652	.920772	.821880	.178120	26	
	35	.55315	.83308	.66398	1.50607	.742842	.920688	.822154	.177846	25	
.60	36	.55339	.83292	.66440	1.50512	.743033	.920604	.822429	.177571	24	.40
	37	.55363	.83276	.66482	1.50417	.743223	.920520	.822703	.177297	23	
	38	.55388	.83260	.66524	1.50322	.743413	.920436	.822977	.177023	22	
.65	39	.55412	.83244	.66566	1.50228	.743602	.920352	.823251	.176749	21	.35
	40	.55436	.83228	.66608	1.50133	.743792	.920268	.823524	.176476	20	
	41	.55460	.83212	.66650	1.50038	.743982	.920184	.823798	.176202	19	
.70	42	.55484	.83195	.66692	1.49944	.744171	.920099	.824072	.175928	18	.30
	43	.55509	.83179	.66734	1.49849	.744361	.920015	.824345	.175655	17	
	44	.55533	.83163	.66776	1.49755	.744550	.919931	.824619	.175381	16	
.75	45	.55557	.83147	.66818	1.49661	.744739	.919846	.824893	.175107	15	.25
	46	.55581	.83131	.66860	1.49566	.744928	.919762	.825166	.174834	14	
	47	.55605	.83115	.66902	1.49472	.745117	.919677	.825439	.174561	13	
.80	48	.55630	.83098	.66944	1.49378	.745306	.919593	.825713	.174287	12	.20
	49	.55654	.83082	.66986	1.49284	.745494	.919508	.825986	.174014	11	
	50	.55678	.83066	.67028	1.49190	.745683	.919424	.826259	.173741	10	
.85	51	.55702	.83050	.67071	1.49097	.745871	.919339	.826532	.173468	9	.15
	52	.55726	.83034	.67113	1.49003	.746060	.919254	.826805	.173195	8	
	53	.55750	.83017	.67155	1.48909	.746248	.919169	.827078	.172922	7	
.90	54	.55775	.83001	.67197	1.48816	.746436	.919085	.827351	.172649	6	.10
	55	.55799	.82985	.67239	1.48722	.746624	.919000	.827624	.172376	5	
	56	.55823	.82969	.67282	1.48629	.746812	.918915	.827897	.172103	4	
.95	57	.55847	.82953	.67324	1.48536	.746999	.918830	.828170	.171830	3	.05
	58	.55871	.82936	.67366	1.48442	.747187	.918745	.828442	.171558	2	
	59	.55895	.82920	.67409	1.48349	.747375	.918659	.828715	.171285	1	
1.00	60	.55919	.82904	.67451	1.48256	9.747562	9.918574	9.828987	10.171013	0	.00
Decimals	Minutes	Cos	Sin	Cot	Tan	Cos	Sin	Cot	Tan	Minutes	Decimals
		Natural Values				Common Logarithms					

56°

34°

Decimals	Minutes	Natural Values				Common Logarithms				Minutes	Decimals
		Sin	Cos	Tan	Cot	Sin	Cos	Tan	Cot		
.00	0	.55919	.82904	.67451	1.48256	9.747562	9.918574	9.828987	10.171013	60	1.00
	1	.55943	.82887	.67493	1.48163	.747749	.918489	.829260	.170740	59	
	2	.55968	.82871	.67536	1.48070	.747936	.918404	.829532	.170468	58	
.05	3	.55992	.82855	.67578	1.47977	.748123	.918318	.829805	.170195	57	.95
	4	.56016	.82839	.67620	1.47885	.748310	.918233	.830077	.169923	56	
	5	.56040	.82822	.67663	1.47792	.748497	.918147	.830349	.169651	55	
.10	6	.56064	.82806	.67705	1.47699	.748683	.918062	.830621	.169379	54	.90
	7	.56088	.82790	.67748	1.47607	.748870	.917976	.830893	.169107	53	
	8	.56112	.82773	.67790	1.47514	.749056	.917891	.831165	.168835	52	
.15	9	.56136	.82757	.67832	1.47422	.749243	.917805	.831437	.168563	51	.85
	10	.56160	.82741	.67875	1.47330	.749429	.917719	.831709	.168291	50	
	11	.56184	.82724	.67917	1.47238	.749615	.917634	.831981	.168019	49	
.20	12	.56208	.82708	.67960	1.47146	.749801	.917548	.832253	.167747	48	.80
	13	.56232	.82692	.68002	1.47053	.749987	.917462	.832525	.167475	47	
	14	.56256	.82675	.68045	1.46962	.750172	.917376	.832796	.167204	46	
.25	15	.56280	.82659	.68088	1.46870	.750358	.917290	.833068	.166932	45	.75
	16	.56305	.82643	.68130	1.46778	.750543	.917204	.833339	.166661	44	
	17	.56329	.82626	.68173	1.46686	.750729	.917118	.833611	.166389	43	
.30	18	.56353	.82610	.68215	1.46595	.750914	.917032	.833882	.166118	42	.70
	19	.56377	.82593	.68258	1.46503	.751099	.916946	.834154	.165846	41	
	20	.56401	.82577	.68301	1.46411	.751284	.916859	.834425	.165575	40	
.35	21	.56425	.82561	.68343	1.46320	.751469	.916773	.834696	.165304	39	.65
	22	.56449	.82544	.68386	1.46229	.751654	.916687	.834967	.165033	38	
	23	.56473	.82528	.68429	1.46137	.751839	.916600	.835238	.164762	37	
.40	24	.56497	.82511	.68471	1.46046	.752023	.916514	.835509	.164491	36	.60
	25	.56521	.82495	.68514	1.45955	.752208	.916427	.835780	.164220	35	
	26	.56545	.82478	.68557	1.45864	.752392	.916341	.836051	.163949	34	
.45	27	.56569	.82462	.68600	1.45773	.752576	.916254	.836322	.163678	33	.55
	28	.56593	.82446	.68642	1.45682	.752760	.916167	.836593	.163407	32	
	29	.56617	.82429	.68685	1.45592	.752944	.916081	.836864	.163136	31	
.50	30	.56641	.82413	.68728	1.45501	.753128	.915994	.837134	.162866	30	.50
	31	.56665	.82396	.68771	1.45410	.753312	.915907	.837405	.162595	29	
	32	.56689	.82380	.68814	1.45320	.753495	.915820	.837675	.162325	28	
.55	33	.56713	.82363	.68857	1.45229	.753679	.915733	.837946	.162054	27	.45
	34	.56736	.82347	.68900	1.45139	.753862	.915646	.838216	.161784	26	
	35	.56760	.82330	.68942	1.45049	.754046	.915559	.838487	.161513	25	
.60	36	.56784	.82314	.68985	1.44958	.754229	.915472	.838757	.161243	24	.40
	37	.56808	.82297	.69028	1.44868	.754412	.915385	.839027	.160973	23	
	38	.56832	.82281	.69071	1.44778	.754595	.915297	.839297	.160703	22	
.65	39	.56856	.82264	.69114	1.44688	.754778	.915210	.839568	.160432	21	.35
	40	.56880	.82248	.69157	1.44598	.754960	.915123	.839838	.160162	20	
	41	.56904	.82231	.69200	1.44508	.755143	.915035	.840108	.159892	19	
.70	42	.56928	.82214	.69243	1.44418	.755326	.914948	.840378	.159622	18	.30
	43	.56952	.82198	.69286	1.44329	.755508	.914860	.840648	.159352	17	
	44	.56976	.82181	.69329	1.44239	.755690	.914773	.840917	.159083	16	
.75	45	.57000	.82165	.69372	1.44149	.755872	.914685	.841187	.158813	15	.25
	46	.57024	.82148	.69416	1.44060	.756054	.914598	.841457	.158543	14	
	47	.57047	.82132	.69459	1.43970	.756236	.914510	.841727	.158273	13	
.80	48	.57071	.82115	.69502	1.43881	.756418	.914422	.841996	.158004	12	.20
	49	.57095	.82098	.69545	1.43792	.756600	.914334	.842266	.157734	11	
	50	.57119	.82082	.69588	1.43703	.756782	.914246	.842535	.157465	10	
.85	51	.57143	.82065	.69631	1.43614	.756963	.914158	.842805	.157195	9	.15
	52	.57167	.82048	.69675	1.43525	.757144	.914070	.843074	.156926	8	
	53	.57191	.82032	.69718	1.43436	.757326	.913982	.843343	.156657	7	
.90	54	.57215	.82015	.69761	1.43347	.757507	.913894	.843612	.156388	6	.10
	55	.57238	.81999	.69804	1.43258	.757688	.913806	.843882	.156118	5	
	56	.57262	.81982	.69847	1.43169	.757869	.913718	.844151	.155849	4	
.95	57	.57286	.81965	.69891	1.43080	.758050	.913630	.844420	.155580	3	.05
	58	.57310	.81949	.69934	1.42992	.758230	.913541	.844689	.155311	2	
	59	.57334	.81932	.69977	1.42903	.758411	.913453	.844958	.155042	1	
1.00	60	.57358	.81915	.70021	1.42815	9.758591	9.913365	9.845227	10.154773	0	.00
Decimals	Minutes	Cos	Sin	Cot	Tan	Cos	Sin	Cot	Tan	Minutes	Decimals
		Natural Values				Common Logarithms					

55°

VALUES AND LOGARITHMS OF TRIGONOMETRIC FUNCTIONS

35°

Decimals	Minutes	Natural Values				Common Logarithms				Minutes	Decimals
		Sin	Cos	Tan	Cot	Sin	Cos	Tan	Cot		
.00	0	.57358	.81915	.70021	1.42815	9.758591	9.913365	9.845227	10.154773	60	1.00
	1	.57381	.81899	.70064	1.42726	.758772	.913276	.845496	.154504	59	
	2	.57405	.81882	.70170	1.42638	.758952	.913187	.845764	.154236	58	
.05	3	.57429	.81865	.70151	1.42550	.759132	.913099	.846033	.153967	57	.95
	4	.57453	.81848	.70194	1.42462	.759312	.913010	.846302	.153698	56	
	5	.57477	.81832	.70238	1.42374	.759492	.912922	.846570	.153430	55	
.10	6	.57501	.81815	.70281	1.42286	.759672	.912833	.846839	.153161	54	.90
	7	.57524	.81798	.70325	1.42198	.759852	.912744	.847108	.152892	53	
	8	.57548	.81782	.70368	1.42110	.760031	.912655	.847376	.152624	52	
.15	9	.57572	.81765	.70412	1.42022	.760211	.912566	.847644	.152356	51	.85
	10	.57596	.81748	.70455	1.41934	.760390	.912477	.847913	.152087	50	
	11	.57619	.81731	.70499	1.41847	.760569	.912388	.848181	.151819	49	
.20	12	.57643	.81714	.70542	1.41759	.760748	.912299	.848449	.151551	48	.80
	13	.57667	.81698	.70586	1.41672	.760927	.912210	.848717	.151283	47	
	14	.57691	.81681	.70629	1.41584	.761106	.912121	.848986	.151014	46	
.25	15	.57715	.81664	.70673	1.41497	.761285	.912031	.849254	.150746	45	.75
	16	.57738	.81647	.70717	1.41409	.761464	.911942	.849522	.150478	44	
	17	.57762	.81631	.70760	1.41322	.761642	.911853	.849790	.150210	43	
.30	18	.57786	.81614	.70804	1.41235	.761821	.911763	.850057	.149943	42	.70
	19	.57810	.81597	.70848	1.41148	.761999	.911674	.850325	.149675	41	
	20	.57833	.81580	.70891	1.41061	.762177	.911584	.850593	.149407	40	
.35	21	.57857	.81563	.70935	1.40974	.762356	.911495	.850861	.149139	39	.65
	22	.57881	.81546	.70979	1.40887	.762534	.911405	.851129	.148871	38	
	23	.57904	.81530	.71023	1.40800	.762712	.911315	.851396	.148604	37	
.40	24	.57928	.81513	.71066	1.40714	.762889	.911226	.851664	.148336	36	.60
	25	.57952	.81496	.71110	1.40627	.763067	.911136	.851931	.148069	35	
	26	.57976	.81479	.71154	1.40540	.763245	.911046	.852199	.147801	34	
.45	27	.57999	.81462	.71198	1.40454	.763422	.910956	.852466	.147534	33	.55
	28	.58023	.81445	.71242	1.40367	.763600	.910866	.852733	.147267	32	
	29	.58047	.81428	.71285	1.40281	.763777	.910776	.853001	.146999	31	
.50	30	.58070	.81412	.71329	1.40195	.763954	.910686	.853268	.146732	30	.50
	31	.58094	.81395	.71373	1.40109	.764131	.910596	.853535	.146465	29	
	32	.58118	.81378	.71417	1.40022	.764308	.910506	.853802	.146198	28	
.55	33	.58141	.81361	.71461	1.39936	.764485	.910415	.854069	.145931	27	.45
	34	.58165	.81344	.71505	1.39850	.764662	.910325	.854336	.145664	26	
	35	.58189	.81327	.71549	1.39764	.764838	.910235	.854603	.145397	25	
.60	36	.58212	.81310	.71593	1.39679	.765015	.910144	.854870	.145130	24	.40
	37	.58236	.81293	.71637	1.39593	.765191	.910054	.855137	.144863	23	
	38	.58260	.81276	.71681	1.39507	.765367	.909963	.855404	.144596	22	
.65	39	.58283	.81259	.71725	1.39421	.765544	.909873	.855671	.144329	21	.35
	40	.58307	.81242	.71769	1.39336	.765720	.909782	.855938	.144062	20	
	41	.58330	.81225	.71813	1.39250	.765896	.909691	.856204	.143796	19	
.70	42	.58354	.81208	.71857	1.39165	.766072	.909601	.856471	.143529	18	.30
	43	.58378	.81191	.71901	1.39079	.766247	.909510	.856737	.143263	17	
	44	.58401	.81174	.71946	1.38994	.766423	.909419	.857004	.142996	16	
.75	45	.58425	.81157	.71990	1.38909	.766598	.909328	.857270	.142730	15	.25
	46	.58449	.81140	.72034	1.38824	.766774	.909237	.857537	.142463	14	
	47	.58472	.81123	.72078	1.38738	.766949	.909146	.857803	.142197	13	
.80	48	.58496	.81106	.72122	1.38653	.767124	.909055	.858069	.141931	12	.20
	49	.58519	.81089	.72167	1.38568	.767300	.908964	.858336	.141664	11	
	50	.58543	.81072	.72211	1.38484	.767475	.908873	.858602	.141398	10	
.85	51	.58567	.81055	.72255	1.38399	.767649	.908781	.858868	.141132	9	.15
	52	.58590	.81038	.72299	1.38314	.767824	.908690	.859134	.140866	8	
	53	.58614	.81021	.72344	1.38229	.767999	.908599	.859400	.140600	7	
.90	54	.58637	.81004	.72388	1.38145	.768173	.908507	.859666	.140334	6	.10
	55	.58661	.80987	.72432	1.38060	.768348	.908416	.859932	.140068	5	
	56	.58684	.80970	.72477	1.37976	.768522	.908324	.860198	.139802	4	
.95	57	.58708	.80953	.72521	1.37891	.768697	.908233	.860464	.139536	3	.05
	58	.58731	.80936	.72565	1.37807	.768871	.908141	.860730	.139270	2	
	59	.58755	.80919	.72610	1.37722	.769045	.908049	.860995	.139005	1	
1.00	60	.58779	.80902	.72654	1.37638	9.769219	9.907958	9.861261	10.138739	0	.00
Decimals	Minutes	Cos	Sin	Cot	Tan	Cos	Sin	Cot	Tan	Minutes	Decimals
		Natural Values				Common Logarithms					

54°

36°

Decimals	Minutes	Natural Values				Common Logarithms				Minutes	Decimals
		Sin	Cos	Tan	Cot	Sin	Cos	Tan	Cot		
.00	0	.58779	.80902	.72654	1.37638	9.769219	9.907958	9.861261	10.138739	60	1.00
	1	.58802	.80885	.72699	1.37554	.769393	.907866	.861527	.138473	59	
	2	.58826	.80867	.72743	1.37470	.769566	.907774	.861792	.138208	58	
.05	3	.58849	.80850	.72788	1.37386	.769740	.907682	.862058	.137942	57	.95
	4	.58873	.80833	.72832	1.37302	.769913	.907590	.862323	.137677	56	
	5	.58896	.80816	.72877	1.37218	.770087	.907498	.862589	.137411	55	
.10	6	.58920	.80799	.72921	1.37134	.770260	.907406	.862854	.137146	54	.90
	7	.58943	.80782	.72966	1.37050	.770433	.907314	.863119	.136881	53	
	8	.58967	.80765	.73010	1.36967	.770606	.907222	.863385	.136615	52	
.15	9	.58990	.80748	.73055	1.36883	.770779	.907129	.863650	.136350	51	.85
	10	.59014	.80730	.73100	1.36800	.770952	.907037	.863915	.136085	50	
	11	.59037	.80713	.73144	1.36716	.771125	.906945	.864180	.135820	49	
.20	12	.59061	.80696	.73189	1.36633	.771298	.906852	.864445	.135555	48	.80
	13	.59084	.80679	.73234	1.36549	.771470	.906760	.864710	.135290	47	
	14	.59108	.80662	.73278	1.36466	.771643	.906667	.864975	.135025	46	
.25	15	.59131	.80644	.73323	1.36383	.771815	.906575	.865240	.134760	45	.75
	16	.59154	.80627	.73368	1.36300	.771987	.906482	.865505	.134495	44	
	17	.59178	.80610	.73413	1.36217	.772159	.906389	.865770	.134230	43	
.30	18	.59201	.80593	.73457	1.36134	.772331	.906296	.866035	.133965	42	.70
	19	.59225	.80576	.73502	1.36051	.772503	.906204	.866300	.133700	41	
	20	.59248	.80558	.73547	1.35968	.772675	.906111	.866564	.133436	40	
.35	21	.59272	.80541	.73592	1.35885	.772847	.906018	.866829	.133171	39	.65
	22	.59295	.80524	.73637	1.35802	.773018	.905925	.867094	.132906	38	
	23	.59318	.80507	.73681	1.35719	.773190	.905832	.867358	.132642	37	
.40	24	.59342	.80489	.73726	1.35637	.773361	.905739	.867623	.132377	36	.60
	25	.59365	.80472	.73771	1.35554	.773533	.905645	.867887	.132113	35	
	26	.59389	.80455	.73816	1.35472	.773704	.905552	.868152	.131848	34	
.45	27	.59412	.80438	.73861	1.35389	.773875	.905459	.868416	.131584	33	.55
	28	.59436	.80420	.73906	1.35307	.774046	.905366	.868680	.131320	32	
	29	.59459	.80403	.73951	1.35224	.774217	.905272	.868945	.131055	31	
.50	30	.59482	.80386	.73996	1.35142	.774388	.905179	.869209	.130791	30	.50
	31	.59506	.80368	.74041	1.35060	.774558	.905085	.869473	.130527	29	
	32	.59529	.80351	.74086	1.34978	.774729	.904992	.869737	.130263	28	
.55	33	.59552	.80334	.74131	1.34896	.774899	.904898	.870001	.129999	27	.45
	34	.59576	.80316	.74176	1.34814	.775070	.904804	.870265	.129735	26	
	35	.59599	.80299	.74221	1.34732	.775240	.904711	.870529	.129471	25	
.60	36	.59622	.80282	.74267	1.34650	.775410	.904617	.870793	.129207	24	.40
	37	.59646	.80264	.74312	1.34568	.775580	.904523	.871057	.128943	23	
	38	.59669	.80247	.74357	1.34487	.775750	.904429	.871321	.128679	22	
.65	39	.59693	.80230	.74402	1.34405	.775920	.904335	.871585	.128415	21	.35
	40	.59716	.80212	.74447	1.34323	.776090	.904241	.871849	.128151	20	
	41	.59739	.80195	.74492	1.34242	.776259	.904147	.872112	.127888	19	
.70	42	.59763	.80178	.74538	1.34160	.776429	.904053	.872376	.127624	18	.30
	43	.59786	.80160	.74583	1.34079	.776598	.903959	.872640	.127360	17	
	44	.59809	.80143	.74628	1.33998	.776768	.903864	.872903	.127097	16	
.75	45	.59832	.80125	.74674	1.33916	.776937	.903770	.873167	.126833	15	.25
	46	.59856	.80108	.74719	1.33835	.777106	.903676	.873430	.126570	14	
	47	.59879	.80091	.74764	1.33754	.777275	.903581	.873694	.126306	13	
.80	48	.59902	.80073	.74810	1.33673	.777444	.903487	.873957	.126043	12	.20
	49	.59926	.80056	.74855	1.33592	.777613	.903392	.874220	.125780	11	
	50	.59949	.80038	.74900	1.33511	.777781	.903298	.874484	.125516	10	
.85	51	.59972	.80021	.74946	1.33430	.777950	.903203	.874747	.125253	9	.15
	52	.59995	.80003	.74991	1.33349	.778119	.903108	.875010	.124990	8	
	53	.60019	.79986	.75037	1.33268	.778287	.903014	.875273	.124727	7	
.90	54	.60042	.79968	.75082	1.33187	.778455	.902919	.875537	.124463	6	.10
	55	.60065	.79951	.75128	1.33107	.778624	.902824	.875800	.124200	5	
	56	.60089	.79934	.75173	1.33026	.778792	.902729	.876063	.123937	4	
.95	57	.60112	.79916	.75219	1.32946	.778960	.902634	.876326	.123674	3	.05
	58	.60135	.79899	.75264	1.32865	.779128	.902539	.876589	.123411	2	
	59	.60158	.79881	.75310	1.32785	.779295	.902444	.876852	.123148	1	
1.00	60	.60182	.79864	.75355	1.32704	9.779463	9.902349	9.877114	10.122886	0	.00
Decimals	Minutes	Cos	Sin	Cot	Tan	Cos	Sin	Cot	Tan	Minutes	Decimals
		Natural Values				Common Logarithms					

53°

VALUES AND LOGARITHMS OF TRIGONOMETRIC FUNCTIONS 385

37°

Decimals	Minutes	Natural Values				Common Logarithms				Minutes	Decimals
		Sin	Cos	Tan	Cot	Sin	Cos	Tan	Cot		
.00	0	.60182	.79864	.75355	1.32704	9.779463	9.902349	9.877114	10.122886	60	1.00
	1	.60205	.79846	.75401	1.32624	.779631	.902253	.877377	.122623	59	
	2	.60228	.79829	.75447	1.32544	.779798	.902158	.877640	.122360	58	
.05	3	.60251	.79811	.75492	1.32464	.779966	.902063	.877903	.122097	57	.95
	4	.60274	.79793	.75538	1.32384	.780133	.901967	.878165	.121835	56	
	5	.60298	.79776	.75584	1.32304	.780300	.901872	.878428	.121572	55	
.10	6	.60321	.79758	.75629	1.32224	.780467	.901776	.878691	.121309	54	.90
	7	.60344	.79741	.75675	1.32144	.780634	.901681	.878953	.121047	53	
	8	.60367	.79723	.75721	1.32064	.780801	.901585	.879216	.120784	52	
.15	9	.60390	.79706	.75767	1.31984	.780968	.901490	.879478	.120522	51	.85
	10	.60414	.79688	.75812	1.31904	.781134	.901394	.879741	.120259	50	
	11	.60437	.79671	.75858	1.31825	.781301	.901298	.880003	.119997	49	
.20	12	.60460	.79653	.75904	1.31745	.781468	.901202	.880265	.119735	48	.80
	13	.60483	.79635	.75950	1.31666	.781634	.901106	.880528	.119472	47	
	14	.60506	.79618	.75996	1.31586	.781800	.901010	.880790	.119210	46	
.25	15	.60529	.79600	.76042	1.31507	.781966	.900914	.881052	.118948	45	.75
	16	.60553	.79583	.76088	1.31427	.782132	.900818	.881314	.118686	44	
	17	.60576	.79565	.76134	1.31348	.782298	.900722	.881577	.118423	43	
.30	18	.60599	.79547	.76180	1.31269	.782464	.900626	.881839	.118161	42	.70
	19	.60622	.79530	.76226	1.31190	.782630	.900529	.882101	.117899	41	
	20	.60645	.79512	.76272	1.31110	.782796	.900433	.882363	.117637	40	
.35	21	.60668	.79494	.76318	1.31031	.782961	.900337	.882625	.117375	39	.65
	22	.60691	.79477	.76364	1.30952	.783127	.900240	.882887	.117113	38	
	23	.60714	.79459	.76410	1.30873	.783292	.900144	.883148	.116852	37	
.40	24	.60738	.79441	.76456	1.30795	.783458	.900047	.883410	.116590	36	.60
	25	.60761	.79424	.76502	1.30716	.783623	.899951	.883672	.116328	35	
	26	.60784	.79406	.76548	1.30637	.783788	.899854	.883934	.116066	34	
.45	27	.60807	.79388	.76594	1.30558	.783953	.899757	.884196	.115804	33	.55
	28	.60830	.79371	.76640	1.30480	.784118	.899660	.884457	.115543	32	
	29	.60853	.79353	.76686	1.30401	.784282	.899564	.884719	.115281	31	
.50	30	.60876	.79335	.76733	1.30323	.784447	.899467	.884980	.115020	30	.50
	31	.60899	.79318	.76779	1.30244	.784612	.899370	.885242	.114758	29	
	32	.60922	.79300	.76825	1.30166	.784776	.899273	.885504	.114496	28	
.55	33	.60945	.79282	.76871	1.30087	.784941	.899176	.885765	.114235	27	.45
	34	.60968	.79264	.76918	1.30009	.785105	.899078	.886026	.113974	26	
	35	.60991	.79247	.76964	1.29931	.785269	.898981	.886288	.113712	25	
.60	36	.61015	.79229	.77010	1.29853	.785433	.898884	.886549	.113451	24	.40
	37	.61038	.79211	.77057	1.29775	.785597	.898787	.886811	.113189	23	
	38	.61061	.79193	.77103	1.29696	.785761	.898689	.887072	.112928	22	
.65	39	.61084	.79176	.77149	1.29618	.785925	.898592	.887333	.112667	21	.35
	40	.61107	.79158	.77196	1.29541	.786089	.898494	.887594	.112406	20	
	41	.61130	.79140	.77242	1.29463	.786252	.898397	.887855	.112145	19	
.70	42	.61153	.79122	.77289	1.29385	.786416	.898299	.888116	.111884	18	.30
	43	.61176	.79105	.77335	1.29307	.786579	.898202	.888378	.111622	17	
	44	.61199	.79087	.77382	1.29229	.786742	.898104	.888639	.111361	16	
.75	45	.61222	.79069	.77428	1.29152	.786906	.898006	.888900	.111100	15	.25
	46	.61245	.79051	.77475	1.29074	.787069	.897908	.889161	.110839	14	
	47	.61268	.79033	.77521	1.28997	.787232	.897810	.889421	.110579	13	
.80	48	.61291	.79016	.77568	1.28919	.787395	.897712	.889682	.110318	12	.20
	49	.61314	.78998	.77615	1.28842	.787557	.897614	.889943	.110057	11	
	50	.61337	.78980	.77661	1.28764	.787720	.897516	.890204	.109796	10	
.85	51	.61360	.78962	.77708	1.28687	.787883	.897418	.890465	.109535	9	.15
	52	.61383	.78944	.77754	1.28610	.788045	.897320	.890725	.109275	8	
	53	.61406	.78926	.77801	1.28533	.788208	.897222	.890986	.109014	7	
.90	54	.61429	.78908	.77848	1.28456	.788370	.897123	.891247	.108753	6	.10
	55	.61451	.78891	.77895	1.28379	.788532	.897025	.891507	.108493	5	
	56	.61474	.78873	.77941	1.28302	.788694	.896926	.891768	.108232	4	
.95	57	.61497	.78855	.77988	1.28225	.788856	.896828	.892028	.107972	3	.05
	58	.61520	.78837	.78035	1.28148	.789018	.896729	.892289	.107711	2	
	59	.61543	.78819	.78082	1.28071	.789180	.896631	.892549	.107451	1	
1.00	60	.61566	.78801	.78129	1.27994	9.789342	9.896532	9.892810	10.107190	0	.00
Decimals	Minutes	Cos	Sin	Cot	Tan	Cos	Sin	Cot	Tan	Minutes	Decimals
		Natural Values				Common Logarithms					

52°

38°

Decimals	Minutes	Natural Values				Common Logarithms				Minutes	Decimals
		Sin	Cos	Tan	Cot	Sin	Cos	Tan	Cot		
.00	0	.61566	.78801	.78129	1.27994	9.789342	9.896532	9.892810	10.107190	60	1.00
	1	.61589	.78783	.78175	1.27917	.789504	.896433	.893070	.106930	59	
	2	.61612	.78765	.78222	1.27841	.789665	.896335	.893331	.106669	58	
.05	3	.61635	.78747	.78269	1.27764	.789827	.896236	.893591	.106409	57	.95
	4	.61658	.78729	.78316	1.27688	.789988	.896137	.893851	.106149	56	
	5	.61681	.78711	.78363	1.27611	.790149	.896038	.894111	.105889	55	
.10	6	.61704	.78694	.78410	1.27535	.790310	.895939	.894372	.105628	54	.90
	7	.61726	.78676	.78457	1.27458	.790471	.895840	.894632	.105368	53	
	8	.61749	.78658	.78504	1.27382	.790632	.895741	.894892	.105108	52	
.15	9	.61772	.78640	.78551	1.27306	.790793	.895641	.895152	.104848	51	.85
	10	.61795	.78622	.78598	1.27230	.790954	.895542	.895412	.104588	50	
	11	.61818	.78604	.78645	1.27153	.791115	.895443	.895672	.104328	49	
.20	12	.61841	.78586	.78692	1.27077	.791275	.895343	.895932	.104068	48	.80
	13	.61864	.78568	.78739	1.27001	.791436	.895244	.896192	.103808	47	
	14	.61887	.78550	.78786	1.26925	.791596	.895145	.896452	.103548	46	
.25	15	.61909	.78532	.78834	1.26849	.791757	.895045	.896712	.103288	45	.75
	16	.61932	.78514	.78881	1.26774	.791917	.894945	.896971	.103029	44	
	17	.61955	.78496	.78928	1.26698	.792077	.894846	.897231	.102769	43	
.30	18	.61978	.78478	.78975	1.26622	.792237	.894746	.897491	.102509	42	.70
	19	.62001	.78460	.79022	1.26546	.792397	.894646	.897751	.102249	41	
	20	.62024	.78442	.79070	1.26471	.792557	.894546	.898010	.101990	40	
.35	21	.62046	.78424	.79117	1.26395	.792716	.894446	.898270	.101730	39	.65
	22	.62069	.78405	.79164	1.26319	.792876	.894346	.898530	.101470	38	
	23	.62092	.78387	.79212	1.26244	.793035	.894246	.898789	.101211	37	
.40	24	.62115	.78369	.79259	1.26169	.793195	.894146	.899049	.100951	36	.60
	25	.62138	.78351	.79306	1.26093	.793354	.894046	.899308	.100692	35	
	26	.62160	.78333	.79354	1.26018	.793514	.893946	.899568	.100432	34	
.45	27	.62183	.78315	.79401	1.25943	.793673	.893846	.899827	.100173	33	.55
	28	.62206	.78297	.79449	1.25867	.793832	.893745	.900087	.099913	32	
	29	.62229	.78279	.79496	1.25792	.793991	.893645	.900346	.099654	31	
.50	30	.62251	.78261	.79544	1.25717	.794150	.893544	.900605	.099395	30	.50
	31	.62274	.78243	.79591	1.25642	.794308	.893444	.900864	.099136	29	
	32	.62297	.78225	.79639	1.25567	.794467	.893343	.901124	.098876	28	
.55	33	.62320	.78206	.79686	1.25492	.794626	.893243	.901383	.098617	27	.45
	34	.62342	.78188	.79734	1.25417	.794784	.893142	.901642	.098358	26	
	35	.62365	.78170	.79781	1.25343	.794942	.893041	.901901	.098099	25	
.60	36	.62388	.78152	.79829	1.25268	.795101	.892940	.902160	.097840	24	.40
	37	.62411	.78134	.79877	1.25193	.795259	.892839	.902420	.097580	23	
	38	.62433	.78116	.79924	1.25118	.795417	.892739	.902679	.097321	22	
.65	39	.62456	.78098	.79972	1.25044	.795575	.892638	.902938	.097062	21	.35
	40	.62479	.78079	.80020	1.24969	.795733	.892536	.903197	.096803	20	
	41	.62502	.78061	.80067	1.24895	.795891	.892435	.903456	.096544	19	
.70	42	.62524	.78043	.80115	1.24820	.796049	.892334	.903714	.096286	18	.30
	43	.62547	.78025	.80163	1.24746	.796206	.892233	.903973	.096027	17	
	44	.62570	.78007	.80211	1.24672	.796364	.892132	.904232	.095768	16	
.75	45	.62592	.77988	.80258	1.24597	.796521	.892030	.904491	.095509	15	.25
	46	.62615	.77970	.80306	1.24523	.796679	.891929	.904750	.095250	14	
	47	.62638	.77952	.80354	1.24449	.796836	.891827	.905008	.094992	13	
.80	48	.62660	.77934	.80402	1.24375	.796993	.891726	.905267	.094733	12	.20
	49	.62683	.77916	.80450	1.24301	.797150	.891624	.905526	.094474	11	
	50	.62706	.77897	.80498	1.24227	.797307	.891523	.905785	.094215	10	
.85	51	.62728	.77879	.80546	1.24153	.797464	.891421	.906043	.093957	9	.15
	52	.62751	.77861	.80594	1.24079	.797621	.891319	.906302	.093698	8	
	53	.62774	.77843	.80642	1.24005	.797777	.891217	.906560	.093440	7	
.90	54	.62796	.77824	.80690	1.23931	.797934	.891115	.906819	.093181	6	.10
	55	.62819	.77806	.80738	1.23858	.798091	.891013	.907077	.092923	5	
	56	.62842	.77788	.80786	1.23784	.798247	.890911	.907336	.092664	4	
.95	57	.62864	.77769	.80834	1.23710	.798403	.890809	.907594	.092406	3	.05
	58	.62887	.77751	.80882	1.23637	.798560	.890707	.907853	.092147	2	
	59	.62909	.77733	.80930	1.23563	.798716	.890605	.908111	.091889	1	
1.00	60	.62932	.77715	.80978	1.23490	9.798872	9.890503	9.908369	10.091631	0	.00
Decimals	Minutes	Cos	Sin	Cot	Tan	Cos	Sin	Cot	Tan	Minutes	Decimals
		Natural Values				Common Logarithms					

51°

39°

Decimals	Minutes	Natural Values				Common Logarithms				Minutes	Decimals
		Sin	Cos	Tan	Cot	Sin	Cos	Tan	Cot		
.00	0	.62932	.77715	.80978	1.23490	9.798872	9.890503	9.908369	10.091631	60	1.00
	1	.62955	.77696	.81027	1.23416	.799028	.890400	.908628	.091372	59	
	2	.62977	.77678	.81075	1.23343	.799184	.890298	.908886	.091114	58	
.05	3	.63000	.77660	.81123	1.23270	.799399	.890195	.909144	.090856	57	.95
	4	.63022	.77641	.81171	1.23196	.799495	.890093	.909402	.090598	56	
	5	.63045	.77623	.81220	1.23123	.799651	.889990	.909660	.090340	55	
.10	6	.63068	.77605	.81268	1.23050	.799806	.889888	.909918	.090082	54	.90
	7	.63090	.77586	.81316	1.22977	.799962	.889785	.910177	.089823	53	
	8	.63113	.77568	.81364	1.22904	.800117	.889682	.910435	.089565	52	
.15	9	.63135	.77550	.81413	1.22831	.800272	.889579	.910693	.089307	51	.85
	10	.63158	.77531	.81461	1.22758	.800427	.889477	.910951	.089049	50	
	11	.63180	.77513	.81510	1.22685	.800582	.889374	.911209	.088791	49	
.20	12	.63203	.77494	.81558	1.22612	.800737	.889271	.911467	.088533	48	.80
	13	.63225	.77476	.81606	1.22539	.800892	.889168	.911725	.088275	47	
	14	.63248	.77458	.81655	1.22467	.801047	.889064	.911982	.088018	46	
.25	15	.63271	.77439	.81703	1.22394	.801201	.888961	.912240	.087760	45	.75
	16	.63293	.77421	.81752	1.22321	.801356	.888858	.912498	.087502	44	
	17	.63316	.77402	.81800	1.22249	.801511	.888755	.912756	.087244	43	
.30	18	.63338	.77384	.81849	1.22176	.801665	.888651	.913014	.086986	42	.70
	19	.63361	.77366	.81898	1.22104	.801819	.888548	.913271	.086729	41	
	20	.63383	.77347	.81946	1.22031	.801973	.888444	.913529	.086471	40	
.35	21	.63406	.77329	.81995	1.21959	.802128	.888341	.913787	.086213	39	.65
	22	.63428	.77310	.82044	1.21886	.802282	.888237	.914044	.085956	38	
	23	.63451	.77292	.82092	1.21814	.802436	.888134	.914302	.085698	37	
.40	24	.63473	.77273	.82141	1.21742	.802589	.888030	.914560	.085440	36	.60
	25	.63496	.77255	.82190	1.21670	.802743	.887926	.914817	.085183	35	
	26	.63518	.77236	.82238	1.21598	.802897	.887822	.915075	.084925	34	
.45	27	.63540	.77218	.82287	1.21526	.803050	.887718	.915332	.084668	33	.55
	28	.63563	.77199	.82336	1.21454	.803204	.887614	.915590	.084410	32	
	29	.63585	.77181	.82385	1.21382	.803357	.887510	.915847	.084153	31	
.50	30	.63608	.77162	.82434	1.21310	.803511	.887406	.916104	.083896	30	.50
	31	.63630	.77144	.82483	1.21238	.803664	.887302	.916362	.083638	29	
	32	.63653	.77125	.82531	1.21166	.803817	.887198	.916619	.083381	28	
.55	33	.63675	.77107	.82580	1.21094	.803970	.887093	.916877	.083123	27	.45
	34	.63698	.77088	.82629	1.21023	.804123	.886989	.917134	.082866	26	
	35	.63720	.77070	.82678	1.20951	.804276	.886885	.917391	.082609	25	
.60	36	.63742	.77051	.82727	1.20879	.804428	.886780	.917648	.082352	24	.40
	37	.63765	.77033	.82776	1.20808	.804581	.886676	.917906	.082094	23	
	38	.63787	.77014	.82825	1.20736	.804734	.886571	.918163	.081837	22	
.65	39	.63810	.76996	.82874	1.20665	.804886	.886466	.918420	.081580	21	.35
	40	.63832	.76977	.82923	1.20593	.805039	.886362	.918677	.081323	20	
	41	.63854	.76959	.82972	1.20522	.805191	.886257	.918934	.081066	19	
.70	42	.63877	.76940	.83022	1.20451	.805343	.886152	.919191	.080809	18	.30
	43	.63899	.76921	.83071	1.20379	.805495	.886047	.919448	.080552	17	
	44	.63922	.76903	.83120	1.20308	.805647	.885942	.919705	.080295	16	
.75	45	.63944	.76884	.83169	1.20237	.805799	.885837	.919962	.080038	15	.25
	46	.63966	.76866	.83218	1.20166	.805951	.885732	.920219	.079781	14	
	47	.63989	.76847	.83268	1.20095	.806103	.885627	.920476	.079524	13	
.80	48	.64011	.76828	.83317	1.20024	.806254	.885522	.920733	.079267	12	.20
	49	.64033	.76810	.83366	1.19953	.806406	.885416	.920990	.079010	11	
	50	.64056	.76791	.83415	1.19882	.806557	.885311	.921247	.078753	10	
.85	51	.64078	.76772	.83465	1.19811	.806709	.885205	.921503	.078497	9	.15
	52	.64100	.76754	.83514	1.19740	.806860	.885100	.921760	.078240	8	
	53	.64123	.76735	.83564	1.19669	.807011	.884994	.922017	.077983	7	
.90	54	.64145	.76717	.83613	1.19599	.807163	.884889	.922274	.077726	6	.10
	55	.64167	.76698	.83662	1.19528	.807314	.884783	.922530	.077470	5	
	56	.64190	.76679	.83712	1.19457	.807465	.884677	.922787	.077213	4	
.95	57	.64212	.76661	.83761	1.19387	.807615	.884572	.923044	.076956	3	.05
	58	.64234	.76642	.83811	1.19316	.807766	.884466	.923300	.076700	2	
	59	.64256	.76623	.83860	1.19246	.807917	.884360	.923557	.076443	1	
1.00	60	.64279	.76604	.83910	1.19175	9.808067	9.884254	9.923814	10.076186	0	.00
Decimals	Minutes	Cos	Sin	Cot	Tan	Cos	Sin	Cot	Tan	Minutes	Decimals
		Natural Values				Common Logarithms					

50°

40°

Decimals	Minutes	Natural Values				Common Logarithms				Minutes	Decimals
		Sin	Cos	Tan	Cot	Sin	Cos	Tan	Cot		
.00	0	.64279	.76604	.83910	1.19175	9.808067	9.884254	9.923814	10.076186	60	1.00
	1	.64301	.76586	.83960	1.19105	.808218	.884148	.924070	.075930	59	
	2	.64323	.76567	.84009	1.19035	.808368	.884042	.924327	.075673	58	
.05	3	.64346	.76548	.84059	1.18964	.808519	.883936	.924583	.075417	57	.95
	4	.64368	.76530	.84108	1.18894	.808669	.883829	.924840	.075160	56	
	5	.64390	.76511	.84158	1.18824	.808819	.883723	.925096	.074904	55	
.10	6	.64412	.76492	.84208	1.18754	.808969	.883617	.925352	.074648	54	.90
	7	.64435	.76473	.84258	1.18684	.809119	.883510	.925609	.074391	53	
	8	.64457	.76455	.84307	1.18614	.809269	.883404	.925865	.074135	52	
.15	9	.64479	.76436	.84357	1.18544	.809419	.883297	.926122	.073878	51	.85
	10	.64501	.76417	.84407	1.18474	.809569	.883191	.926378	.073622	50	
	11	.64524	.76398	.84457	1.18404	.809718	.883084	.926634	.073366	49	
.20	12	.64546	.76380	.84507	1.18334	.809868	.882977	.926890	.073110	48	.80
	13	.64568	.76361	.84556	1.18264	.810017	.882871	.927147	.072853	47	
	14	.64590	.76342	.84606	1.18194	.810167	.882764	.927403	.072597	46	
.25	15	.64612	.76323	.84656	1.18125	.810316	.882657	.927659	.072341	45	.75
	16	.64635	.76304	.84706	1.18055	.810465	.882550	.927915	.072085	44	
	17	.64657	.76286	.84756	1.17986	.810614	.882443	.928171	.071829	43	
.30	18	.64679	.76267	.84806	1.17916	.810763	.882336	.928427	.071573	42	.70
	19	.64701	.76248	.84856	1.17846	.810912	.882229	.928684	.071316	41	
	20	.64723	.76229	.84906	1.17777	.811061	.882121	.928940	.071060	40	
.35	21	.64746	.76210	.84956	1.17708	.811210	.882014	.929196	.070804	39	.65
	22	.64768	.76192	.85006	1.17638	.811358	.881907	.929452	.070548	38	
	23	.64790	.76173	.85057	1.17569	.811507	.881799	.929708	.070292	37	
.40	24	.64812	.76154	.85107	1.17500	.811655	.881692	.929964	.070036	36	.60
	25	.64834	.76135	.85157	1.17430	.811804	.881584	.930220	.069780	35	
	26	.64856	.76116	.85207	1.17361	.811952	.881477	.930475	.069525	34	
.45	27	.64878	.76097	.85257	1.17292	.812100	.881369	.930731	.069269	33	.55
	28	.64901	.76078	.85308	1.17223	.812248	.881261	.930987	.069013	32	
	29	.64923	.76059	.85358	1.17154	.812396	.881153	.931243	.068757	31	
.50	30	.64945	.76041	.85408	1.17085	.812544	.881046	.931499	.068501	30	.50
	31	.64967	.76022	.85458	1.17016	.812692	.880938	.931755	.068245	29	
	32	.64989	.76003	.85509	1.16947	.812840	.880830	.932010	.067990	28	
.55	33	.65011	.75984	.85559	1.16878	.812988	.880722	.932266	.067734	27	.45
	34	.65033	.75965	.85609	1.16809	.813135	.880613	.932522	.067478	26	
	35	.65055	.75946	.85660	1.16741	.813283	.880505	.932778	.067222	25	
.60	36	.65077	.75927	.85710	1.16672	.813430	.880397	.933033	.066967	24	.40
	37	.65100	.75908	.85761	1.16603	.813578	.880289	.933289	.066711	23	
	38	.65122	.75889	.85811	1.16535	.813725	.880180	.933545	.066455	22	
.65	39	.65144	.75870	.85862	1.16466	.813872	.880072	.933800	.066200	21	.35
	40	.65166	.75851	.85912	1.16398	.814019	.879963	.934056	.065944	20	
	41	.65188	.75832	.85963	1.16329	.814166	.879855	.934311	.065689	19	
.70	42	.65210	.75813	.86014	1.16261	.814313	.879746	.934567	.065433	18	.30
	43	.65232	.75794	.86064	1.16192	.814460	.879637	.934822	.065178	17	
	44	.65254	.75775	.86115	1.16124	.814607	.879529	.935078	.064922	16	
.75	45	.65276	.75756	.86166	1.16056	.814753	.879420	.935333	.064667	15	.25
	46	.65298	.75738	.86216	1.15987	.814900	.879311	.935589	.064411	14	
	47	.65320	.75719	.86267	1.15919	.815046	.879202	.935844	.064156	13	
.80	48	.65342	.75700	.86318	1.15851	.815193	.879093	.936100	.063900	12	.20
	49	.65364	.75680	.86368	1.15783	.815339	.878984	.936355	.063645	11	
	50	.65386	.75661	.86419	1.15715	.815485	.878875	.936611	.063389	10	
.85	51	.65408	.75642	.86470	1.15647	.815632	.878766	.936866	.063134	9	.15
	52	.65430	.75623	.86521	1.15579	.815778	.878656	.937121	.062879	8	
	53	.65452	.75604	.86572	1.15511	.815924	.878547	.937377	.062623	7	
.90	54	.65474	.75585	.86623	1.15443	.816069	.878438	.937632	.062368	6	.10
	55	.65496	.75566	.86674	1.15375	.816215	.878328	.937887	.062113	5	
	56	.65518	.75547	.86725	1.15308	.816361	.878219	.938142	.061858	4	
.95	57	.65540	.75528	.86776	1.15240	.816507	.878109	.938398	.061602	3	.05
	58	.65562	.75509	.86827	1.15172	.816652	.877999	.938653	.061347	2	
	59	.65584	.75490	.86878	1.15104	.816798	.877890	.938908	.061092	1	
1.00	60	.65606	.75471	.86929	1.15037	9.816943	9.877780	9.939163	10.060837	0	.00
Decimals	Minutes	Cos	Sin	Cot	Tan	Cos	Sin	Cot	Tan	Minutes	Decimals
		Natural Values				Common Logarithms					

49°

VALUES AND LOGARITHMS OF TRIGONOMETRIC FUNCTIONS

41°

Decimals	Minutes	Natural Values				Common Logarithms				Minutes	Decimals
		Sin	Cos	Tan	Cot	Sin	Cos	Tan	Cot		
.00	0	.65606	.75471	.86929	1.15037	9.816943	9.877780	9.939163	10.060837	60	1.00
	1	.65628	.75452	.86980	1.14969	.817088	.877670	.939418	.060582	59	
	2	.65650	.75433	.87031	1.14902	.817233	.877560	.939673	.060327	58	
.05	3	.65672	.75414	.87082	1.14834	.817379	.877450	.939928	.060072	57	.95
	4	.65694	.75395	.87133	1.14767	.817524	.877340	.940183	.059817	56	
	5	.65716	.75375	.87184	1.14699	.817668	.877230	.940439	.059561	55	
.10	6	.65738	.75356	.87236	1.14632	.817813	.877120	.940694	.059306	54	.90
	7	.65759	.75337	.87287	1.14565	.817958	.877010	.940949	.059051	53	
	8	.65781	.75318	.87338	1.14498	.818103	.876899	.941204	.058796	52	
.15	9	.65803	.75299	.87389	1.14430	.818247	.876789	.941459	.058541	51	.85
	10	.65825	.75280	.87441	1.14363	.818392	.876678	.941713	.058287	50	
	11	.65847	.75261	.87492	1.14296	.818536	.876568	.941968	.058032	49	
.20	12	.65869	.75241	.87543	1.14229	.818681	.876457	.942223	.057777	48	.80
	13	.65891	.75222	.87595	1.14162	.818825	.876347	.942478	.057522	47	
	14	.65913	.75203	.87646	1.14095	.818969	.876236	.942733	.057267	46	
.25	15	.65935	.75184	.87698	1.14023	.819113	.876125	.942988	.057012	45	.75
	16	.65956	.75165	.87749	1.13961	.819257	.876014	.943243	.056757	44	
	17	.65978	.75146	.87801	1.13894	.819401	.875904	.943498	.056502	43	
.30	18	.66000	.75126	.87852	1.13828	.819545	.875793	.943752	.056248	42	.70
	19	.66022	.75107	.87904	1.13761	.819689	.875682	.944007	.055993	41	
	20	.66044	.75088	.87955	1.13694	.819832	.875571	.944262	.055738	40	
.35	21	.66066	.75069	.88007	1.13627	.819976	.875459	.944517	.055483	39	.65
	22	.66088	.75050	.88059	1.13561	.820120	.875348	.944771	.055229	38	
	23	.66109	.75030	.88110	1.13494	.820263	.875237	.945026	.054974	37	
.40	24	.66131	.75011	.88162	1.13428	.820406	.875126	.945281	.054719	36	.60
	25	.66153	.74992	.88214	1.13361	.820550	.875014	.945535	.054465	35	
	26	.66175	.74973	.88265	1.13295	.820693	.874903	.945790	.054210	34	
.45	27	.66197	.74953	.88317	1.13228	.820836	.874791	.946045	.053955	33	.55
	28	.66218	.74934	.88369	1.13162	.820979	.874680	.946299	.053701	32	
	29	.66240	.74915	.88421	1.13096	.821122	.874568	.946554	.053446	31	
.50	30	.66262	.74896	.88473	1.13029	.821265	.874456	.946808	.053192	30	.50
	31	.66284	.74876	.88524	1.12963	.821407	.874344	.947063	.052937	29	
	32	.66306	.74857	.88576	1.12897	.821550	.874232	.947318	.052682	28	
.55	33	.66327	.74838	.88628	1.12831	.821693	.874121	.947572	.052428	27	.45
	34	.66349	.74818	.88680	1.12765	.821835	.874009	.947827	.052173	26	
	35	.66371	.74799	.88732	1.12699	.821977	.873896	.948081	.051919	25	
.60	36	.66393	.74780	.88784	1.12633	.822120	.873784	.948335	.051665	24	.40
	37	.66414	.74760	.88836	1.12567	.822262	.873672	.948590	.051410	23	
	38	.66436	.74741	.88888	1.12501	.822404	.873560	.948844	.051156	22	
.65	39	.66458	.74722	.88940	1.12435	.822546	.873448	.949099	.050901	21	.35
	40	.66480	.74703	.88992	1.12369	.822688	.873335	.949353	.050647	20	
	41	.66501	.74683	.89045	1.12303	.822830	.873223	.949608	.050392	19	
.70	42	.66523	.74664	.89097	1.12238	.822972	.873110	.949862	.050138	18	.30
	43	.66545	.74644	.89149	1.12172	.823114	.872998	.950116	.049884	17	
	44	.66566	.74625	.89201	1.12106	.823255	.872885	.950371	.049629	16	
.75	45	.66588	.74606	.89253	1.12041	.823397	.872772	.950625	.049375	15	.25
	46	.66610	.74586	.89306	1.11975	.823539	.872659	.950879	.049121	14	
	47	.66632	.74567	.89358	1.11909	.823680	.872547	.951133	.048867	13	
.80	48	.66653	.74548	.89410	1.11844	.823821	.872434	.951388	.048612	12	.20
	49	.66675	.74528	.89463	1.11778	.823963	.872321	.951642	.048358	11	
	50	.66697	.74509	.89515	1.11713	.824104	.872208	.951896	.048104	10	
.85	51	.66718	.74489	.89567	1.11648	.824245	.872095	.952150	.047850	9	.15
	52	.66740	.74470	.89620	1.11582	.824386	.871981	.952405	.047595	8	
	53	.66762	.74451	.89672	1.11517	.824527	.871868	.952659	.047341	7	
.90	54	.66783	.74431	.89725	1.11452	.824668	.871755	.952913	.047087	6	.10
	55	.66805	.74412	.89777	1.11387	.824808	.871641	.953167	.046833	5	
	56	.66827	.74392	.89830	1.11321	.824949	.871528	.953421	.046579	4	
.95	57	.66848	.74373	.89883	1.11256	.825090	.871414	.953675	.046325	3	.05
	58	.66870	.74353	.89935	1.11191	.825230	.871301	.953929	.046071	2	
	59	.66891	.74334	.89988	1.11126	.825371	.871187	.954183	.045817	1	
1.00	60	.66913	.74314	.90040	1.11061	9.825511	9.871073	9.954437	10.045563	0	.00
Decimals	Minutes	Cos	Sin	Cot	Tan	Cos	Sin	Cot	Tan	Minutes	Decimals
		Natural Values				Common Logarithms					

48°

390 MATHEMATICAL AND PHYSICAL TABLES

42°

Decimals	Minutes	Natural Values				Common Logarithms				Minutes	Decimals
		Sin	Cos	Tan	Cot	Sin	Cos	Tan	Cot		
.00	0	.66913	.74314	.90040	1.11061	9.825511	9.871073	9.954437	10.045563	60	1.00
	1	.66935	.74295	.90093	1.10996	.825651	.870960	.954691	.045309	59	
	2	.66956	.74276	.90146	1.10931	.825791	.870846	.954946	.045054	58	
.05	3	.66978	.74256	.90199	1.10867	.825931	.870732	.955200	.044800	57	.95
	4	.66999	.74237	.90251	1.10802	.826071	.870618	.955454	.044546	56	
	5	.67021	.74217	.90304	1.10737	.826211	.870504	.955708	.044292	55	
.10	6	.67043	.74198	.90357	1.10672	.826351	.870390	.955961	.044039	54	.90
	7	.67064	.74178	.90410	1.10607	.826491	.870276	.956215	.043785	53	
	8	.67086	.74159	.90463	1.10543	.826631	.870161	.956469	.043531	52	
.15	9	.67107	.74139	.90516	1.10478	.826770	.870047	.956723	.043277	51	.85
	10	.67129	.74120	.90569	1.10414	.826910	.869933	.956977	.043023	50	
	11	.67151	.74100	.90621	1.10349	.827049	.869818	.957231	.042769	49	
.20	12	.67172	.74080	.90674	1.10285	.827189	.869704	.957485	.042515	48	.80
	13	.67194	.74061	.90727	1.10220	.827328	.869589	.957739	.042261	47	
	14	.67215	.74041	.90781	1.10156	.827467	.869474	.957993	.042007	46	
.25	15	.67237	.74022	.90834	1.10091	.827606	.869360	.958247	.041753	45	.75
	16	.67258	.74002	.90887	1.10027	.827745	.869245	.958500	.041500	44	
	17	.67280	.73983	.90940	1.09963	.827884	.869130	.958754	.041246	43	
.30	18	.67301	.73963	.90993	1.09899	.828023	.869015	.959008	.040992	42	.70
	19	.67323	.73944	.91046	1.09834	.828162	.868900	.959262	.040738	41	
	20	.67344	.73924	.91099	1.09770	.828301	.868785	.959516	.040484	40	
.35	21	.67366	.73904	.91153	1.09706	.828439	.868670	.959769	.040231	39	.65
	22	.67387	.73885	.91206	1.09642	.828578	.868555	.960023	.039977	38	
	23	.67409	.73865	.91259	1.09578	.828716	.868440	.960277	.039723	37	
.40	24	.67430	.73846	.91313	1.09514	.828855	.868324	.960530	.039470	36	.60
	25	.67452	.73826	.91366	1.09450	.828993	.868209	.960784	.039216	35	
	26	.67473	.73806	.91419	1.09386	.829131	.868093	.961038	.038962	34	
.45	27	.67495	.73787	.91473	1.09322	.829269	.867978	.961292	.038708	33	.55
	28	.67516	.73767	.91526	1.09258	.829407	.867862	.961545	.038455	32	
	29	.67538	.73747	.91580	1.09195	.829545	.867747	.961799	.038201	31	
.50	30	.67559	.73728	.91633	1.09131	.829683	.867631	.962052	.037948	30	.50
	31	.67580	.73708	.91687	1.09067	.829821	.867515	.962306	.037694	29	
	32	.67602	.73688	.91740	1.09003	.829959	.867399	.962560	.037440	28	
.55	33	.67623	.73669	.91794	1.03940	.830097	.867283	.962813	.037187	27	.45
	34	.67645	.73649	.91847	1.08876	.830234	.867167	.963067	.036933	26	
	35	.67666	.73629	.91901	1.08813	.830372	.867051	.963320	.036680	25	
.60	36	.67688	.73610	.91955	1.08749	.830509	.866935	.963574	.036426	24	.40
	37	.67709	.73590	.92008	1.08686	.830646	.866819	.963828	.036172	23	
	38	.67730	.73570	.92062	1.08622	.830784	.866703	.964081	.035919	22	
.65	39	.67752	.73551	.92116	1.08559	.830921	.866586	.964335	.035665	21	.35
	40	.67773	.73531	.92170	1.08496	.831058	.866470	.964588	.035412	20	
	41	.67795	.73511	.92224	1.08432	.831195	.866353	.964842	.035158	19	
.70	42	.67816	.73491	.92277	1.08369	.831332	.866237	.965095	.034905	18	.30
	43	.67837	.73472	.92331	1.08306	.831469	.866120	.965349	.034651	17	
	44	.67859	.73452	.92385	1.08243	.831606	.866004	.965602	.034398	16	
.75	45	.67880	.73432	.92439	1.08179	.831742	.865887	.965855	.034145	15	.25
	46	.67901	.73413	.92493	1.08116	.831879	.865770	.966109	.033891	14	
	47	.67923	.73393	.92547	1.08053	.832015	.865653	.966362	.033638	13	
.80	48	.67944	.73373	.92601	1.07990	.832152	.865536	.966616	.033384	12	.20
	49	.67965	.73353	.92655	1.07927	.832288	.865419	.966869	.033131	11	
	50	.67987	.73333	.92709	1.07864	.832425	.865302	.967123	.032877	10	
.85	51	.68008	.73314	.92763	1.07801	.832561	.865185	.967376	.032624	9	.15
	52	.68029	.73294	.92817	1.07738	.832697	.865068	.967629	.032371	8	
	53	.68051	.73274	.92872	1.07676	.832833	.864950	.967883	.032117	7	
.90	54	.68072	.73254	.92926	1.07613	.832969	.864833	.968136	.031864	6	.10
	55	.68093	.73234	.92980	1.07550	.833105	.864716	.968389	.031611	5	
	56	.68115	.73215	.93034	1.07487	.833241	.864598	.968643	.031357	4	
.95	57	.68136	.73195	.93088	1.07425	.833377	.864481	.968896	.031104	3	.05
	58	.68157	.73175	.93143	1.07362	.833512	.864363	.969149	.030851	2	
	59	.68179	.73155	.93197	1.07299	.833648	.864245	.969403	.030597	1	
1.00	60	.68200	.73135	.93252	1.07237	9.833783	9.864127	9.969656	10.030344	0	.00
Decimals	Minutes	Cos	Sin	Cot	Tan	Cos	Sin	Cot	Tan	Minutes	Decimals
		Natural Values				Common Logarithms					

47°

VALUES AND LOGARITHMS OF TRIGONOMETRIC FUNCTIONS

43°

Decimals	Minutes	Natural Values				Common Logarithms				Minutes	Decimals
		Sin	Cos	Tan	Cot	Sin	Cos	Tan	Cot		
.00	0	.68200	.73135	.93252	1.07237	9.833783	9.864127	9.969656	10.030344	60	1.00
	1	.68221	.73116	.93306	1.07174	.833919	.864010	.969909	.030091	59	
	2	.68242	.73096	.93360	1.07112	.834054	.863892	.970162	.029838	58	
.05	3	.68264	.73076	.93415	1.07049	.834189	.863774	.970416	.029584	57	.95
	4	.68285	.73056	.93469	1.06987	.834325	.863656	.970669	.029331	56	
	5	.68306	.73036	.93524	1.06925	.834460	.863538	.970922	.029078	55	
.10	6	.68327	.73016	.93578	1.06862	.834595	.863419	.971175	.028825	54	.90
	7	.68349	.72996	.93633	1.06800	.834730	.863301	.971429	.028571	53	
	8	.68370	.72976	.93688	1.06738	.834865	.863183	.971682	.028318	52	
.15	9	.68391	.72957	.93742	1.06676	.834999	.863064	.971935	.028065	51	.85
	10	.68412	.72937	.93797	1.06613	.835134	.862946	.972188	.027812	50	
	11	.68434	.72917	.93852	1.06551	.835269	.862827	.972441	.027559	49	
.20	12	.68455	.72897	.93906	1.06489	.835403	.862709	.972695	.027305	48	.80
	13	.68476	.72877	.93961	1.06427	.835538	.862590	.972948	.027052	47	
	14	.68497	.72857	.94016	1.06365	.835672	.862471	.973201	.026799	46	
.25	15	.68518	.72837	.94071	1.06303	.835807	.862353	.973454	.026546	45	.75
	16	.68539	.72817	.94125	1.06241	.835941	.862234	.973707	.026293	44	
	17	.68561	.72797	.94180	1.06179	.836075	.862115	.973960	.026040	43	
.30	18	.68582	.72777	.94235	1.06117	.836209	.861996	.974213	.025787	42	.70
	19	.68603	.72757	.94290	1.06056	.836343	.861877	.974466	.025534	41	
	20	.68624	.72737	.94345	1.05994	.836477	.861758	.974720	.025280	40	
.35	21	.68645	.72717	.94400	1.05932	.836611	.861638	.974973	.025027	39	.65
	22	.68666	.72697	.94455	1.05870	.836745	.861519	.975226	.024774	38	
	23	.68688	.72677	.94510	1.05809	.836878	.861400	.975479	.024521	37	
.40	24	.68709	.72657	.94565	1.05747	.837012	.861280	.975732	.024268	36	.60
	25	.68730	.72637	.94620	1.05685	.837146	.861161	.975985	.024015	35	
	26	.68751	.72617	.94676	1.05624	.837279	.861041	.976238	.023762	34	
.45	27	.68772	.72597	.94731	1.05562	.837412	.860922	.976491	.023509	33	.55
	28	.68793	.72577	.94786	1.05501	.837546	.860802	.976744	.023256	32	
	29	.68814	.72557	.94841	1.05439	.837679	.860682	.976997	.023003	31	
.50	30	.68835	.72537	.94896	1.05378	.837812	.860562	.977250	.022750	30	.50
	31	.68857	.72517	.94952	1.05317	.837945	.860442	.977503	.022497	29	
	32	.68878	.72497	.95007	1.05255	.838078	.860322	.977756	.022244	28	
.55	33	.68899	.72477	.95062	1.05194	.838211	.860202	.978009	.021991	27	.45
	34	.68920	.72457	.95118	1.05133	.838344	.860082	.978262	.021738	26	
	35	.68941	.72437	.95173	1.05072	.838477	.859962	.978515	.021485	25	
.60	36	.68962	.72417	.95229	1.05010	.838610	.859842	.978768	.021232	24	.40
	37	.68983	.72397	.95284	1.04949	.838742	.859721	.979021	.020979	23	
	38	.69004	.72377	.95340	1.04888	.838875	.859601	.979274	.020726	22	
.65	39	.69025	.72357	.95395	1.04827	.839007	.859480	.979527	.020473	21	.35
	40	.69046	.72337	.95451	1.04766	.839140	.859360	.979780	.020220	20	
	41	.69067	.72317	.95506	1.04705	.839272	.859239	.980033	.019967	19	
.70	42	.69088	.72297	.95562	1.04644	.839404	.859119	.980286	.019714	18	.30
	43	.69109	.72277	.95618	1.04583	.839536	.858998	.980538	.019462	17	
	44	.69130	.72257	.95673	1.04522	.839668	.858877	.980791	.019209	16	
.75	45	.69151	.72236	.95729	1.04461	.839800	.858756	.981044	.018956	15	.25
	46	.69172	.72216	.95785	1.04401	.839932	.858635	.981297	.018703	14	
	47	.69193	.72196	.95841	1.04340	.840064	.858514	.981550	.018450	13	
.80	48	.69214	.72176	.95897	1.04279	.840196	.858393	.981803	.018197	12	.20
	49	.69235	.72156	.95952	1.04218	.840328	.858272	.982056	.017944	11	
	50	.69256	.72136	.96008	1.04158	.840459	.858151	.982309	.017691	10	
.85	51	.69277	.72116	.96064	1.04097	.840591	.858029	.982562	.017438	9	.15
	52	.69298	.72095	.96120	1.04036	.840722	.857908	.982814	.017186	8	
	53	.69319	.72075	.96176	1.03976	.840854	.857786	.983067	.016933	7	
.90	54	.69340	.72055	.96232	1.03915	.840985	.857665	.983320	.016680	6	.10
	55	.69361	.72035	.96288	1.03855	.841116	.857543	.983573	.016427	5	
	56	.69382	.72015	.96344	1.03794	.841247	.857422	.983826	.016174	4	
.95	57	.69403	.71995	.96400	1.03734	.841378	.857300	.984079	.015921	3	.05
	58	.69424	.71974	.96457	1.03674	.841509	.857178	.984332	.015668	2	
	59	.69445	.71954	.96513	1.03613	.841640	.857056	.984584	.015416	1	
1.00	60	.69466	.71934	.96569	1.03553	9.841771	9.856934	9.984837	10.015163	0	.00
Decimals	Minutes	Cos	Sin	Cot	Tan	Cos	Sin	Cot	Tan	Minutes	Decimals
		Natural Values				Common Logarithms					

46°

44°

Decimals	Minutes	Natural Values				Common Logarithms				Minutes	Decimals
		Sin	Cos	Tan	Cot	Sin	Cos	Tan	Cot		
.00	0	.69466	.71934	.96569	1.03553	9.841771	9.856934	9.984837	10.015163	60	1.00
	1	.69487	.71914	.96625	1.03493	.841902	.856812	.985090	.014910	59	
	2	.69508	.71894	.96681	1.03433	.842033	.856690	.985343	.014657	58	
.05	3	.69529	.71873	.96738	1.03372	.842163	.856568	.985596	.014404	57	.95
	4	.69549	.71853	.96794	1.03312	.842294	.856446	.985848	.014152	56	
	5	.69570	.71833	.96850	1.03252	.842424	.856323	.986101	.013899	55	
.10	6	.69591	.71813	.96907	1.03192	.842555	.856201	.986354	.013646	54	.90
	7	.69612	.71792	.96963	1.03132	.842685	.856078	.986607	.013393	53	
	8	.69633	.71772	.97020	1.03072	.842815	.855956	.986860	.013140	52	
.15	9	.69654	.71752	.97076	1.03012	.842946	.855833	.987112	.012888	51	.85
	10	.69675	.71732	.97133	1.02952	.843076	.855711	.987365	.012635	50	
	11	.69696	.71711	.97189	1.02892	.843206	.855588	.987618	.012382	49	
.20	12	.69717	.71691	.97246	1.02832	.843336	.855465	.987871	.012129	48	.80
	13	.69737	.71671	.97302	1.02772	.843466	.855342	.988123	.011877	47	
	14	.69758	.71650	.97359	1.02713	.843595	.855219	.988376	.011624	46	
.25	15	.69779	.71630	.97416	1.02653	.843725	.855096	.988629	.011371	45	.75
	16	.69800	.71610	.97472	1.02593	.843855	.854973	.988882	.011118	44	
	17	.69821	.71590	.97529	1.02533	.843984	.854850	.989134	.010866	43	
.30	18	.69842	.71569	.97586	1.02474	.844114	.854727	.989387	.010613	42	.70
	19	.69862	.71549	.97643	1.02414	.844243	.854603	.989640	.010360	41	
	20	.69883	.71529	.97700	1.02355	.844372	.854480	.989893	.010107	40	
.35	21	.69904	.71508	.97756	1.02295	.844502	.854356	.990145	.009855	39	.65
	22	.69925	.71488	.97813	1.02236	.844631	.854233	.990398	.009602	38	
	23	.69946	.71468	.97870	1.02176	.844760	.854109	.990651	.009349	37	
.40	24	.69966	.71447	.97927	1.02117	.844889	.853986	.990903	.009097	36	.60
	25	.69987	.71427	.97984	1.02057	.845018	.853862	.991156	.008844	35	
	26	.70008	.71407	.98041	1.01998	.845147	.853738	.991409	.008591	34	
.45	27	.70029	.71386	.98098	1.01939	.845276	.853614	.991662	.008338	33	.55
	28	.70049	.71366	.98155	1.01879	.845405	.853490	.991914	.008086	32	
	29	.70070	.71345	.98213	1.01820	.845533	.853366	.992167	.007833	31	
.50	30	.70091	.71325	.98270	1.01761	.845662	.853242	.992420	.007580	30	.50
	31	.70112	.71305	.98327	1.01702	.845790	.853118	.992672	.007328	29	
	32	.70132	.71284	.98384	1.01642	.845919	.852994	.992925	.007075	28	
.55	33	.70153	.71264	.98441	1.01583	.846047	.852869	.993178	.006822	27	.45
	34	.70174	.71243	.98499	1.01524	.846175	.852745	.993431	.006569	26	
	35	.70195	.71223	.98556	1.01465	.846304	.852620	.993683	.006317	25	
.60	36	.70215	.71203	.98613	1.01406	.846432	.852496	.993936	.006064	24	.40
	37	.70236	.71182	.98671	1.01347	.846560	.852371	.994189	.005811	23	
	38	.70257	.71162	.98728	1.01288	.846688	.852247	.994441	.005559	22	
.65	39	.70277	.71141	.98786	1.01229	.846816	.852122	.994694	.005306	21	.35
	40	.70298	.71121	.98843	1.01170	.846944	.851997	.994947	.005053	20	
	41	.70319	.71100	.98901	1.01112	.847071	.851872	.995199	.004801	19	
.70	42	.70339	.71080	.98958	1.01053	.847199	.851747	.995452	.004548	18	.30
	43	.70360	.71059	.99016	1.00994	.847327	.851622	.995705	.004295	17	
	44	.70381	.71039	.99073	1.00935	.847454	.851497	.995957	.004043	16	
.75	45	.70401	.71019	.99131	1.00876	.847582	.851372	.996210	.003790	15	.25
	46	.70422	.70998	.99189	1.00818	.847709	.851246	.996463	.003537	14	
	47	.70443	.70978	.99247	1.00759	.847836	.851121	.996715	.003285	13	
.80	48	.70463	.70957	.99304	1.00701	.847964	.850996	.996968	.003032	12	.20
	49	.70484	.70937	.99362	1.00642	.848091	.850870	.997221	.002779	11	
	50	.70505	.70916	.99420	1.00583	.848218	.850745	.997473	.002527	10	
.85	51	.70525	.70896	.99478	1.00525	.848345	.850619	.997726	.002274	9	.15
	52	.70546	.70875	.99536	1.00467	.848472	.850493	.997979	.002021	8	
	53	.70567	.70855	.99594	1.00408	.848599	.850368	.998231	.001769	7	
.90	54	.70587	.70834	.99652	1.00350	.848726	.850242	.998484	.001516	6	.10
	55	.70608	.70813	.99710	1.00291	.848852	.850116	.998737	.001263	5	
	56	.70628	.70793	.99768	1.00233	.848979	.849990	.998989	.001011	4	
.95	57	.70649	.70772	.99826	1.00175	.849106	.849864	.999242	.000758	3	.05
	58	.70670	.70752	.99884	1.00116	.849232	.849738	.999495	.000505	2	
	59	.70690	.70731	.99942	1.00058	.849359	.849611	.999747	.000253	1	
1.00	60	.70711	.70711	1.00000	1.00000	9.849485	9.849485	10.000000	10.000000	0	.00
Decimals	Minutes	Cos	Sin	Cot	Tan	Cos	Sin	Cot	Tan	Minutes	Decimals
		Natural Values				Common Logarithms					

45°

Specific Gravity Conversions

$$°Bé = 145 - \frac{145}{\text{sp gr}} \text{ (heavier than } H_2O)$$

$$°Bé = \frac{140}{\text{sp gr}} - 130 \text{ (lighter than } H_2O)$$

$$°Tw = \frac{\text{sp gr } 60°/60°F - 1}{0.005}$$

$$°API = \frac{141.5}{\text{sp gr}} - 131.5$$

Sp gr 60°/60°	°Bé.	°A.P.I.	Lb/gal at 60°F, wt in air	Lb/ft³ at 60°F, wt in air	Sp gr 60°/60°	°Bé.	°A.P.I.	Lb/gal at 60°F, wt in air	Lb/ft³ at 60°F, wt in air
0.600	103.33	104.33	4.9929	37.350	0.900	25.76	25.72	7.4944	56.062
.605	101.40	102.38	5.0346	37.662	.905	24.70	24.85	7.5361	56.374
.610	99.51	100.47	5.0763	37.973	.910	23.85	23.99	7.5777	56.685
.615	97.64	98.58	5.1180	38.285	.915	23.01	23.14	7.6194	56.997
.620	95.81	96.73	5.1597	38.597	.920	22.17	22.30	7.6612	57.310
.625	94.00	94.90	5.2014	39.910	.925	21.35	21.47	7.7029	57.622
.630	92.22	93.10	5.2431	39.222	.930	20.54	20.65	7.7446	57.934
.635	90.47	91.33	5.2848	39.534	.935	19.73	19.84	7.7863	58.246
.640	88.75	89.59	5.3265	39.845	.940	18.94	19.03	7.8280	58.557
.645	87.05	87.88	5.3682	40.157	.945	18.15	18.24	7.8697	58.869
.650	85.38	86.19	5.4098	40.468	.950	17.37	17.45	7.9114	59.181
.655	83.74	84.53	5.4515	40.780	.955	16.60	16.67	7.9531	59.493
.660	82.12	82.89	5.4932	41.092	.960	15.83	15.90	7.9947	59.805
.665	80.53	81.28	5.5349	41.404	.965	15.08	15.13	8.0364	60.117
.670	78.96	79.69	5.5766	41.716	.970	14.33	14.38	8.0780	60.428
.675	77.41	78.13	5.6183	42.028	.975	13.59	13.63	8.1197	60.740
.680	75.88	76.59	5.6600	42.340	.980	12.86	12.89	8.1615	61.052
.685	74.38	75.07	5.7017	42.652	.985	12.13	12.15	8.2032	61.364
.690	72.90	73.57	5.7434	42.963	.990	11.41	11.43	8.2449	61.676
.695	71.44	72.10	5.7851	43.275	.995	10.70	10.71	8.2866	61.988
					1.000	10.00	10.00	8.3283	62.300

Sp gr 60°/60°	°Bé.	°A.P.I.	Lb/gal at 60°F, wt in air	Lb/ft³ at 60°F, wt in air	Sp gr 60°/60°	°Bé.	°Tw.	Lb/gal at 60°F, wt in air	Lb/ft³ at 60°F, wt in air
.700	70.00	70.64	5.8268	43.587					
.705	68.58	69.21	5.8685	43.899					
.710	67.18	67.80	5.9101	44.211					
.715	65.80	66.40	5.9518	44.523					
.720	64.44	65.03	5.9935	44.834					
.725	63.10	63.67	6.0352	45.146	1.005	0.72	1	8.3700	62.612
.730	61.78	62.34	6.0769	45.458	1.010	1.44	2	8.4117	62.924
.735	60.48	61.02	6.1186	45.770	1.015	2.14	3	8.4534	63.236
.740	59.19	59.72	6.1603	46.082	1.020	2.84	4	8.4950	63.547
.745	57.92	58.43	6.2020	46.394	1.025	3.54	5	8.5367	63.859
.750	56.67	57.17	6.2437	46.706	1.030	4.22	6	8.5784	64.171
.755	55.43	55.92	6.2854	47.018	1.035	4.90	7	8.6201	64.483
.760	54.21	54.68	6.3271	47.330	1.040	5.58	8	8.6618	64.795
.765	53.01	53.47	6.3688	47.642	1.045	6.24	9	8.7035	65.107
.770	51.82	52.27	6.4104	47.953	1.050	6.91	10	8.7452	65.419
.775	50.65	51.08	6.4521	47.265	1.055	7.56	11	8.7869	65.731
.780	49.49	49.91	6.4938	48.577	1.060	8.21	12	8.8286	66.042
.785	48.34	48.75	6.5355	48.889	1.065	8.85	13	8.8703	66.354
.790	47.22	47.61	6.5772	49.201	1.070	9.49	14	8.9120	66.666
.795	46.10	46.49	6.6189	49.513	1.075	10.12	15	8.9537	66.978
.800	45.00	45.38	6.6606	49.825	1.080	10.74	16	8.9954	67.290
.805	43.91	44.28	6.7023	50.137	1.085	11.36	17	9.0371	67.602
.810	42.84	43.19	6.7440	50.448	1.090	11.97	18	9.0787	67.914
.815	41.78	42.12	6.7857	50.760	1.095	12.58	19	9.1204	68.226
.820	40.73	41.06	6.8274	51.072	1.100	13.18	20	9.1621	68.537
.825	39.70	40.02	6.8691	51.384	1.105	13.78	21	9.2038	68.849
.830	38.67	38.98	6.9108	51.696	1.110	14.37	22	9.2455	69.161
.835	37.66	37.96	6.9525	52.008	1.115	14.96	23	9.2872	69.473
.840	36.67	36.95	6.9941	52.320	1.120	15.54	24	9.3289	69.785
.845	35.68	35.96	7.0358	52.632	1.125	16.11	25	9.3706	70.097
.850	34.71	34.97	7.0775	52.943	1.130	16.68	26	9.4123	70.409
.855	33.74	34.00	7.1192	53.225	1.135	17.25	27	9.4540	70.721
.860	32.79	33.03	7.1609	53.567	1.140	17.81	28	9.4957	71.032
.865	31.85	32.08	7.2026	53.879	1.145	18.36	29	9.5374	71.344
.870	30.92	31.14	7.2443	54.191	1.150	18.91	30	9.5790	71.656
.875	30.00	30.21	7.2860	54.503	1.155	19.46	31	9.6207	71.968
.880	29.09	29.30	7.3277	54.815	1.160	20.00	32	9.6624	72.280
.885	28.19	28.39	7.3694	55.127	1.165	20.54	33	9.7041	72.592
.890	27.30	27.49	7.4111	55.438	1.170	21.07	34	9.7458	72.904
.895	26.42	26.60	7.4528	55.750	1.175	21.60	35	9.7875	73.216

MATHEMATICAL AND PHYSICAL TABLES

Sp gr 60°/60°	°Bé.	°Tw.	Lb/gal at 60°F, wt in air	Lb/ft³ at 60°F, wt in air	Sp gr 60°/60°	°Bé.	°Tw.	Lb/gal at 60°F, wt in air	Lb/ft³ at 60°F, wt in air
1.180	22.12	36	9.8292	73.528	1.51	48.97	102	12.581	94.11
1.185	22.64	37	9.8709	73.840	1.52	49.61	104	12.664	94.79
1.190	23.15	38	9.9126	74.151	1.53	50.23	106	12.748	95.36
1.195	23.66	39	9.9543	74.463	1.54	50.84	108	12.831	95.98
1.200	24.17	40	9.9960	74.775	1.55	51.45	110	12.914	96.61
1.205	24.67	41	10.0377	75.087	1.56	52.05	112	12.998	97.23
1.210	25.17	42	10.0793	75.399	1.57	52.64	114	13.081	97.85
1.215	25.66	43	10.1210	75.711	1.58	53.23	116	13.165	98.48
1.220	26.15	44	10.1627	76.022	1.59	53.81	118	13.248	99.10
1.225	26.63	45	10.2044	76.334	1.60	54.38	120	13.331	99.73
1.230	27.11	46	10.2461	76.646	1.61	54.94	122	13.415	100.35
1.235	27.59	47	10.2878	76.958	1.62	55.49	124	13.498	100.97
1.240	28.06	48	10.3295	77.270	1.63	56.04	126	13.582	101.60
1.245	28.53	49	10.3712	77.582	1.64	56.59	128	13.665	102.22
1.250	29.00	50	10.4129	77.894	1.65	57.12	130	13.748	102.84
1.255	29.46	51	10.4546	78.206	1.66	57.65	132	13.832	103.47
1.260	29.92	52	10.4963	78.518	1.67	58.17	134	13.915	104.09
1.265	30.38	53	10.5380	78.830	1.68	58.69	136	13.998	104.72
1.270	30.83	54	10.5797	79.141	1.69	59.20	138	14.082	105.34
1.275	31.27	55	10.6214	79.453	1.70	59.71	140	14.165	105.96
1.280	31.72	56	10.6630	79.765	1.71	60.20	142	14.249	106.59
1.285	32.16	57	10.7047	80.077	1.72	60.70	144	14.332	107.21
1.290	32.60	58	10.7464	80.389	1.73	61.18	146	14.415	107.83
1.295	33.03	59	10.7881	80.701	1.74	61.67	148	14.499	108.46
1.300	33.46	60	10.8298	81.013	1.75	62.14	150	14.582	109.08
1.305	33.89	61	10.8715	81.325	1.76	62.61	152	14.665	109.71
1.310	34.31	62	10.9132	81.636	1.77	63.08	154	14.749	110.32
1.315	34.73	63	10.9549	81.948	1.78	63.54	156	14.832	110.95
1.320	35.15	64	10.9966	82.260	1.79	63.99	158	14.916	111.58
1.325	35.57	65	11.0383	82.572	1.80	64.44	160	14.999	112.20
1.330	35.98	66	11.0800	82.884	1.81	64.89	162	15.082	112.82
1.335	36.39	67	11.1217	83.196	1.82	65.33	164	15.166	113.45
1.340	36.79	68	11.1634	83.508	1.83	65.77	166	15.249	114.07
1.345	37.19	69	11.2051	83.820	1.84	66.20	168	15.333	114.70
1.350	37.59	70	11.2467	84.131	1.85	66.62	170	15.416	115.31
1.355	37.99	71	11.2884	84.443	1.86	67.04	172	15.499	115.94
1.360	38.38	72	11.3301	84.755	1.87	67.46	174	15.583	116.56
1.365	38.77	73	11.3718	85.067	1.88	67.87	176	15.666	117.19
1.370	39.16	74	11.4135	85.379	1.89	68.28	178	15.750	117.81
1.375	39.55	75	11.4552	85.691	1.90	68.68	180	15.832	118.43
1.380	39.93	76	11.4969	86.003	1.91	69.08	182	15.916	119.06
1.385	40.31	77	11.5386	86.315	1.92	69.48	184	16.000	119.68
1.390	40.68	78	11.5803	86.626	1.93	69.87	186	16.083	120.31
1.395	41.06	79	11.6220	86.938	1.94	70.26	188	16.166	120.93
1.400	41.43	80	11.6637	87.250	1.95	70.64	190	16.250	121.56
1.405	41.80	81	11.7054	87.562	1.96	71.02	192	16.333	122.18
1.410	42.16	82	11.7471	87.874	1.97	71.40	194	16.417	122.80
1.415	42.53	83	11.7888	88.186	1.98	71.77	196	16.500	123.43
1.420	42.89	84	11.8304	88.498	1.99	72.14	198	16.583	124.05
1.425	43.25	85	11.8721	88.810	2.00	72.50	200	16.667	124.68
1.430	43.60	86	11.9138	89.121					
1.435	43.95	87	11.9555	89.433					
1.440	44.31	88	11.9972	89.745					
1.445	44.65	89	12.0389	90.057					
1.450	45.00	90	12.0806	90.369					
1.455	45.34	91	12.1223	90.681					
1.460	45.68	92	12.1640	90.993					
1.465	46.02	93	12.2057	91.305					
1.470	46.36	94	12.2473	91.616					
1.475	46.69	95	12.2890	91.928					
1.480	47.03	96	12.3307	92.240					
1.485	47.36	97	12.3724	92.552					
1.490	47.68	98	12.4141	92.864					
1.495	48.01	99	12.4558	93.176					
1.500	48.33	100	12.4975	93.488					

Temperature Conversion

$$°F = (°C \times 9/5) + 32$$
$$°C = (°F - 32) \times 5/9$$
$$°R = °F + 459.67$$
$$°K = °C + 273.15$$

Interpolation Differences

°C	Temperature	°F	°C	Temperature	°F
0.56	1	1.8	11.11	20	36
1.11	2	3.6	16.67	30	54
1.67	3	5.4	22.22	40	72
2.22	4	7.2	27.78	50	90
2.78	5	9.0	33.33	60	108
3.33	6	10.8	38.89	70	126
3.89	7	12.6	44.44	80	144
4.44	8	14.4	50.00	90	162
5.00	9	16.2	55.56	100	180
5.56	10	18.0	555.56	1000	1800

°C	Temperature	°F	°C	Temperature	°F
−262.22	−440		−40.00	−40	−40
−256.67	−430		−34.44	−30	−22
−251.11	−420		−28.89	−20	−4
−245.56	−410		−23.35	−10	14
−240.00	−400		−17.78	0	32
−234.44	−390		−17.22	1	33.8
−228.89	−380		−16.67	2	35.6
−223.33	−370		−16.11	3	37.4
−217.78	−360		−15.56	4	39.2
−212.22	−350		−15.00	5	41.0
			−14.44	6	42.8
−206.67	−340		−13.89	7	44.6
−201.11	−330		−13.33	8	46.4
−195.56	−320		−12.78	9	48.2
−190.00	−310		−12.22	10	50.0
−184.44	−300				
			−11.67	11	51.8
−178.89	−290		−11.11	12	53.6
−173.33	−280		−10.56	13	55.4
−169.53	−**273.15**	−459.67	−10.00	14	57.2
−162.22	−260	−436	−9.44	15	59.0
−156.67	−250	−418			
			−8.89	16	60.8
−151.11	−240	−400	−8.33	17	62.6
−145.56	−230	−382	−7.78	18	64.4
−140.00	−220	−364	−7.22	19	66.2
−134.44	−210	−346	−6.67	20	68.0
−128.89	−200	−328			
			−6.11	21	69.8
−123.33	−190	−310	−5.56	22	71.6
−117.78	−180	−292	−5.00	23	73.4
−112.22	−170	−274	−4.44	24	75.2
−106.67	−160	−256	−3.89	25	77.0
−101.11	−150	−238			
			−3.33	26	78.8
−95.56	−140	−220	−2.78	27	80.6
−90.0	−130	−202	−2.22	28	82.4
−84.44	−120	−184	−1.67	29	84.2
−78.89	−110	−166	−1.11	30	86.0
−73.33	−100	−148			
			−0.56	31	87.8
−67.78	−90	−130	0	32	89.6
−62.22	−80	−121	0.56	33	91.4
−56.67	−70	−94	1.11	34	93.2
−51.11	−60	−76	1.67	35	95.0
−45.56	−50	−58			

TEMPERATURE CONVERSION

°C	Temperature	°F	°C	Temperature	°F
2.22	36	96.8	25.56	78	172.4
2.78	37	98.6	26.11	79	174.2
3.33	38	100.4	26.67	80	176.0
3.89	39	102.2			
4.44	40	104.0	27.22	81	177.8
			27.78	82	179.6
5.00	41	105.8	28.33	83	181.4
5.56	42	107.6	28.89	84	183.2
6.11	43	109.4	29.44	85	185.0
6.67	44	111.2			
7.22	45	113.0	30.00	86	186.8
7.78	46	114.8	30.56	87	188.6
8.33	47	116.6	31.11	88	190.4
8.89	48	118.4	31.67	89	192.2
9.44	49	120.2	32.22	90	194.0
10.00	50	122.0			
			32.78	91	195.8
10.56	51	123.8	33.33	92	197.6
11.11	52	125.6	33.89	93	199.4
11.67	53	127.4	34.44	94	201.2
12.22	54	129.2	35.00	95	203.0
12.78	55	131.0			
			35.56	96	204.8
13.33	56	132.8	36.11	97	206.6
13.89	57	134.6	36.67	98	208.4
14.44	58	136.4	37.22	99	210.2
15.00	59	138.2	37.78	100	212.0
15.56	60	140.0			
			43.33	110	230
16.11	61	141.8	48.89	120	248
16.67	62	143.6	54.44	130	266
17.22	63	145.4	60.00	140	284
17.78	64	147.2	65.56	150	302
18.33	65	149.0			
			71.11	160	320
18.89	66	150.8	76.67	170	338
19.44	67	152.6	82.22	180	356
20.00	68	154.4	87.78	190	374
20.56	69	156.2	93.33	200	392
21.11	70	158.0			
			100.00	212	413.6
21.67	71	159.8	104.44	220	428
22.22	72	161.6	110.00	230	446
22.78	73	163.4	115.56	240	464
23.33	74	165.2	121.11	250	482
23.89	75	167.0			
			126.67	260	500
24.44	76	168.8	132.22	270	518
25.00	77	170.6	137.78	280	536

MATHEMATICAL AND PHYSICAL TABLES

°C	Temperature	°F	°C	Temperature	°F
143.33	290	554	704.4	1300	2372
148.89	300	572	760.0	1400	2552
			815.6	1500	2732
154.44	310	590			
160.00	320	608	871.1	1600	2912
165.56	330	626	926.7	1700	3092
171.11	340	644	982.2	1800	3272
176.67	350	662	1038	1900	3452
			1093	2000	3632
182.22	360	680			
187.78	370	698	1149	2100	3812
193.33	380	716	1204	2200	3992
198.89	390	734	1260	2300	4172
204.44	400	752	1316	2400	4352
			1371	2500	4532
210.00	410	770			
215.55	420	788	1427	2600	4712
221.11	430	806	1482	2700	4892
226.66	440	824	1538	2800	5072
232.22	450	842	1593	2900	5252
			1649	3000	5432
237.77	460	860			
243.33	470	878	1704	3100	5612
248.88	480	896	1760	3200	5792
254.44	490	914	1816	3300	5972
260.00	500	932	1871	3400	6152
			1927	3500	6332
315.6	600	1112			
371.1	700	1292	1982	3600	6512
426.7	800	1472	2038	3700	6692
482.2	900	1652	2093	3800	6872
537.8	1000	1832	2149	3900	7052
			2205	4000	7232
593.3	1100	2012			
648.9	1200	2192			

CONVERSION FACTORS

Conversion Factors

Unless otherwise stated, pounds are U.S. avoirdupois, feet are U.S., and seconds are mean solar.

Decimal Prefixes

Mega-	10^6	Deka-	10	Micro-	10^{-6}
Myria-	10^4	Deci-	10^{-1}	Millimicro-	10^{-9} NANO
Kilo-	10^3	Centi-	10^{-2}	Micromicro-	10^{-12}
Hecto-	10^2	Milli-	10^{-3}		

Multiply	by	to obtain
Abamperes	10	amperes
Abamperes	2.99796×10^{10}	statamperes
Abampere-turns	12.566	gilberts
Abcoulombs	10	coulombs (abs)
Abcoulombs	2.99796×10^{10}	statcoulombs
Abcoulombs/kg	30577	statcoulombs/dyne
Abfarads	1×10^9	farads (abs)
Abfarads	8.98776×10^{20}	statfarads
Abhenries	1×10^{-9}	henries (abs)
Abhenries	1.11263×10^{-21}	stathenries
Abohms	1×10^{-9}	ohms (abs)
Abohms	1.11263×10^{-21}	statohms
Abvolts	3.33560×10^{-11}	statvolts
Abvolts	1×10^{-8}	volts (abs)
Abvolts/centimeters	2.540005×10^{-8}	volts (abs)/inch
Acre feet	43560	cubic feet
Acre feet	1233	cubic meters
Acre feet	325850	gallons (U.S.)
Acres	43560	square feet
Acres	4047	square meters
Ampere-hours (abs)	3600	coulombs (abs)
Amperes (abs)	0.1	abamperes
Amperes (abs)	1.036×10^{-5}	faradays/second
Amperes (abs)	2.9980×10^9	statamperes
Angstrom units	1×10^{-8}	centimeters
Angstrom units	3.937×10^{-9}	inches
Ares	1076	square feet
Ares	100	square meters
Assay tons	29.17	grams
Astronomical units	1.495×10^8	kilometers

MATHEMATICAL AND PHYSICAL TABLES

Multiply	by	to obtain
Astronomical units	9.290×10^7	miles
Atmospheres	1.0133	bars
Atmospheres	1.01325×10^6	dynes/square centimeter
Atmospheres	10333	kilograms/square meter
Atmospheres	14.696	pounds/square inch
Avograms	1.66036×10^{-24}	grams
Bags, cement	94	pounds of cement
Barleycorns (British)	$\frac{1}{3}$	inches
Barleycorns (British)	8.467×10^{-3}	meters
Barrels (British, dry)	5.780	cubic feet
Barrels (British, dry)	0.1637	cubic meters
Barrels (British, dry)	36	gallons (British)
Barrels, cement	170.6	kilograms
Barrels, cement	376	pounds of cement
Barrels, cranberry	3.371	cubic feet
Barrels, cranberry	0.09547	cubic meters
Barrels, oil	5.615	cubic feet
Barrels, oil	0.1590	cubic meters
Barrels, oil	42	gallons (U.S.)
Barrels (U.S. dry)	4.083	cubic feet
Barrels (U.S. dry)	0.11562	cubic meters
Barrels (U.S., liquid)	4.211	cubic feet
Barrels (U.S., liquid)	0.1192	cubic meters
Barrels (U.S., liquid)	31.5	gallons (U.S.)
Bars	10^6	dynes/square centimeter
Bars	10197	kilograms/square meter
Bars	14.50	pounds/square inch
Baryes	10^{-6}	bars
Board feet	$\frac{1}{12}$	cubic feet
Boiler horsepower	33475	Btu (mean)/hour
Boiler horsepower	34.5	pounds of water evaporated from and at 212°F (per hour)
Bolts (U.S., cloth)	120	linear feet
Bolts (U.S., cloth)	36.58	meters
Bougie decimales	1	candles (int)
Btu (mean)	251.98	calories, gram (mean)
Btu (mean)	0.55556	centigrade heat units (chu)
Btu (mean)	777.98	foot-pounds
Btu (mean)	1054.8	joules (abs)
Btu (mean)	107.565	kilogram-meters

CONVERSION FACTORS

Multiply	by	to obtain
Btu (mean)	6.876×10^{-5}	pounds of carbon to CO_2
Btu (mean)	0.29305	watt-hours
Btu (mean)/pound	0.5556	calories, gram (mean)/gram
Btu (mean)/pound/°F	1	calories, gram/gram/°C
Btu (mean)/hour (feet^2)°F	4.882	kilogram-calorie/hr (m^2)°C
Btu (mean)/hour (feet^2)°F	1.3562×10^{-4}	gram-calorie/second (cm^2)°C
Btu (mean)/hour (feet^2)°F	3.94×10^{-4}	horsepower/(ft^2)°F
Btu (mean)/hour (feet^2)°F	5.682×10^{-4}	watts/(cm^2)°C
Btu (mean)/hour (feet^2)°F	2.035×10^{-3}	watts/(in.^{2})°C
Btu (mean)/(hour)(feet^2) (°F/inch)	3.4448×10^{-4}	calories, gram (15°C)/(sec) (cm^2)(°C/cm)
Btu (mean)/(hour)(feet^2) (°F/inch)	1	chu/(hr)(ft^2)(°C/in.)
Btu (mean)/(hour)(feet^2) (°F/inch)	1.442×10^{-3}	joules (abs)/(sec)(cm^2) (°C/cm)
Btu (mean)/(hour)(feet^2) (°F/inch)	1.442×10^{-3}	watts/(cm^2)(°C/cm)
Btu (60°F)	1054.6	joules (abs)
Buckets (British, dry)	4	gallons (British)
Bushels (British)	1.03205	bushels (U.S.)
Bushels (British)	0.03637	cubic meters
Bushels (British)	1.2843	cubic feet
Bushels (U.S.)	1.2444	cubic feet
Bushels (U.S.)	0.035239	cubic meters
Bushels (U.S.)	4	pecks (U.S.)
Butts (British)	20.2285	cubic feet
Butts (British)	126	gallons (British)
Cable lengths	720	feet
Cable lengths	219.46	meters
Calories, gram (mean)	3.9685×10^{-3}	Btu (mean)
Calories, gram (mean)	0.001459	cubic feet atmospheres
Calories, gram (mean)	3.0874	foot-pounds
Calories, gram (mean)	4.186	joules (abs)
Calories, gram (mean)	0.42685	kilogram-meters
Calories, gram (mean)	0.0011628	watt-hours
Calories (thermochemical)	0.999346	calories (Int. Steam Tables)
Calories, gram (mean)/gram	1.8	Btu (mean)/pound
Candle power (spherical)	12.566	lumens
Candles (int)	0.104	carcel units

Multiply	by	to obtain
Candles (int)	1.11	Hefner units
Candles (int)	1	lumens (int)/steradian
Candles (int)/square centimeter	2919	foot-lamberts
Candles (int)/square centimeter	3.1416	lamberts
Candles (int)/square foot	3.1416	foot-lamberts
Candles (int)/square foot	3.382×10^{-3}	lamberts
Candles (int)/square inch	452.4	foot-lamberts
Candles (int)/square inch	0.4870	lamberts
Carats (metric)	3.0865	grains
Carats (metric)	0.2	grams
Centals	100	pounds
Centigrade heat units (chu)	1.8	Btu
Centigrade heat units (chu)	453.6	calories, gram (15°C)
Centigrade heat units (chu)	1897.8	joules (abs)
Centimeters	0.0328083	feet (U.S.)
Centimeters	0.3937	inches (U.S.)
Centipoises	3.60	kilograms/meter hour
Centipoises	10^{-3}	kilograms/meter second
Centipoises	2.42	pounds/foot hour
Centipoises	6.72×10^{-4}	pounds/foot second
Centipoises	2.089×10^{-5}	pound force second/square foot
Chains (engineers' or Ramden's)	100	feet
Chains (engineers' or Ramden's)	30.48	meters
Chains (surveyors' or Gunter's)	66	feet
Chains (surveyors' or Gunter's)	20.12	meters
Chaldrons (British)	32	bushels (British)
Chaldrons (U.S.)	36	bushels (U.S.)
Cheval-vapours	0.9863	horsepower
Cheval-vapours	735.5	watts (abs)
Cheval-vapours heures	2.648×10^6	joules (abs)
Chu/(hr)(ft²)(°C/in.)	1	Btu/(hr)(ft²)(°F/in.)
Circular inches	0.7854	square inches
Circular millimeters	7.854×10^{-7}	square meters
Circular mils	7.854×10^{-7}	square inches
Circumferences	360	degrees

CONVERSION FACTORS

Multiply	by	to obtain
Circumferences	400	grades
Cloves	8	pounds
Coombs (British)	4	bushels (British)
Cords	8	cord feet
Cords	128	cubic feet
Cords	3.625	cubic meters
Coulombs (abs)	0.1	abcoulombs
Coulombs (abs)	6.281×10^{18}	electronic charges
Coulombs (abs)	2.998×10^{9}	statcoulombs
Cubic centimeters	3.531445×10^{-5}	cubic feet (U.S.)
Cubic centimeters	2.6417×10^{-4}	gallons (U.S.)
Cubic centimeters	0.033814	ounces (U.S., fluid)
Cubic feet (British)	0.9999916	cubic feet (U.S.)
Cubic feet (U.S.)	28317.016	cubic centimeters
Cubic feet (U.S.)	1728	cubic inches
Cubic feet (U.S.)	7.48052	gallons (U.S.)
Cubic feet (U.S.)	28.31625	liters
Cubic foot-atmospheres	2.7203	Btu (mean)
Cubic foot-atmospheres	680.74	calories, gram (mean)
Cubic foot-atmospheres	2116	foot-pounds
Cubic foot-atmospheres	2869	joules (abs)
Cubie foot-atmospheres	292.6	kilogram-meters
Cubic foot-atmospheres	7.968×10^{-4}	kilowatt-hours
Cubic feet of common brick	120	pounds
Cubic feet of water (60°F)	62.37	pounds
Cubic feet/second	1.9834	acre-feet/day
Cubic feet/second	448.83	gallons/minute
Cubic feet/second	0.64632	million gallons/day
Cubic inches (U.S.)	16.387162	cubic centimeters
Cubic inches (U.S.)	1.0000084	cubic inches (British)
Cubic inches (U.S.)	0.55411	ounces (U.S., fluid)
Cubic meters	8.1074×10^{-4}	acre-feet
Cubic meters	8.387	barrels (U.S., liquid)
Cubic meters	35.314	cubic feet (U.S.)
Cubic meters	61023	cubic inches (U.S.)
Cubic meters	1.308	cubic yards (U.S.)
Cubic meters	264.17	gallons (U.S.)
Cubic meters	999.973	liters
Cubic yards (British)	0.76455	cubic meters
Cubic yards (British)	0.9999916	cubic yards (U.S.)
Cubic yards (U.S.)	27	cubic feet (U.S.)

Multiply	by	to obtain
Cubic yards (U.S.)	0.76456	cubic meters
Cubic yards of sand	2700	pounds
Cubits	45.720	centimeters
Cubits	1.5	feet
Days (sidereal)	86164	seconds (mean solar)
Debye units (dipole moment)	10^{18}	electrostatic units
Drachms (British, fluid)	3.5516×10^{-6}	cubic meters
Drachms (British, fluid)	0.125	ounces (British, fluid)
Drams (troy)	2.1943	drams (avoirdupois)
Drams (troy)	60	grains
Drams (troy)	3.8879351	grams
Drams (troy)	0.125	ounces (troy)
Drams (avoirdupois)	1.771845	grams
Drams (avoirdupois)	0.0625	ounces (avoirdupois)
Drams (avoirdupois)	0.00390625	pounds (avoirdupois)
Drams (U.S., fluid)	3.6967×10^{-6}	cubic meters
Drams (U.S., fluid)	0.125	ounces (fluid)
Dynes	0.00101972	grams
Dynes	2.24809×10^{-6}	pounds
Dyne-centimeters (torque)	7.3756×10^{-8}	pound-feet
Dynes/centimeter	1	ergs/square centimeter
Dynes/square centimeter	9.8692×10^{-7}	atmospheres
Dynes/square centimeter	0.01020	kilograms/square meter
Dynes/square centimeter	0.1	newtons/square meter
Dynes/square centimeter	1.450×10^{-5}	pounds/square inch
Electromagnetic cgs units, of magnetic permeability	1.1128×10^{-21}	electrostatic cgs units of magnetic permeability
Electromagnetic cgs units of mass resistance	9.9948×10^{-6}	ohms (int)-meter-gram
Electromagnetic ft lb sec units of magnetic permeability	0.0010764	electromagnetic cgs units of magnetic permeability
Electromagnetic ft lb sec units of magnetic permeability	1.03382×10^{-18}	electrostatic cgs units of magnetic permeability
Electronic charges	1.5921×10^{-19}	coulombs (abs)
Electron-volts	1.6020×10^{-12}	ergs
Electron-volts	1.0737×10^{-9}	mass units
Electron-volts	0.07386	Rydberg units of energy
Electrostatic cgs units of Hall effect	2.6962×10^{31}	electromagnetic cgs units of Hall effect
Electrostatic ft lb sec units of charge	1.1952×10^{-6}	coulombs (abs)

CONVERSION FACTORS

Multiply	by	to obtain
Electrostatic ft lb sec units of magnetic permeability	929.03	electrostatic cgs units of magnetic permeability
Ells	114.30	centimeters
Ells	45	inches
Ems, pica (printing)	0.42333	centimeters
Ems, pica (printing)	$\frac{1}{6}$	inches
Ergs	9.4805×10^{-11}	Btu (mean)
Ergs	2.3889×10^{-8}	calories, gram (mean)
Ergs	7.3756×10^{-8}	foot-pounds
Ergs	10^{-7}	joules (abs)
Ergs	1.01972×10^{-8}	kilogram-meters
Faradays	26.80	ampere-hours
Faradays	96500	coulombs (abs)
Faradays/second	96500	amperes (abs)
Farads (abs)	10^{-9}	abfarads
Farads (abs)	8.9877×10^{11}	statfarads
Fathoms	6	feet
Feet (U.S.)	30.4801	centimeters
Feet (U.S.)	1.0000028	feet (British)
Feet (U.S.)	0.30480	meters
Feet (U.S.)	1.893939×10^{-4}	miles (statute)
Feet of air (1 atmosphere, 60°F)	5.30×10^{-4}	pounds/square inch
Feet of water at 39.2°F	304.79	kilograms/square meter
Feet of water at 39.2°F	0.43352	pounds/square inch
Feet/second	1.0973	kilometers/hour
Feet/second	0.68182	miles/hour
Feet/second/second	1.0973	kilometers/hour/second
Firkins (British)	9	gallons (British)
Firkins (U.S.)	9	gallons (U.S.)
Foot-poundals	3.9951×10^{-5}	Btu (mean)
Foot-poundals	0.0421420	joules (abs)
Foot-pounds	0.0012854	Btu (mean)
Foot-pounds	0.32389	calories, gram (mean)
Foot-pounds	1.13558×10^{7}	ergs
Foot-pounds	32.174	foot-poundals
Foot-pounds	1.35582	joules (abs)
Foot-pounds	0.138255	kilogram-meters
Foot-pounds	0.013381	liter-atmospheres
Foot-pounds	3.7662×10^{-4}	watt-hours (abs)
Foot-pounds/second	4.6275	Btu (mean)/hour

Multiply	by	to obtain
Foot-pounds/second	0.0018182	horsepower
Foot-pounds/second	1.35582	watts (abs)
Furlongs	660	feet
Furlongs	201.17	meters
Furlongs	0.125	miles
Gallons (British)	4516.086	cubic centimeters
Gallons (British)	1.20094	gallons (U.S.)
Gallons (British)	10	pounds (avoirdupois) of water at 62°F
Gallons (U.S.)	3785.434	cubic centimeters
Gallons (U.S.)	0.13368	cubic feet (U.S.)
Gallons (U.S.)	231	cubic inches
Gallons (U.S.)	3.78533	liters
Gallons (U.S.)	128	ounces (U.S., fluid)
Gausses (abs)	3.3358×10^{-4}	electrostatic cgs units of magnetic flux density
Gausses (abs)	0.99966	gausses (int)
Gausses (abs)	1	maxwells (abs)/square centimeters
Gausses (abs)	6.4516	maxwells (abs)/square inch
Gausses (abs)	1	lines/square centimeter
Gilberts (abs)	0.07958	abampere turns
Gilberts (abs)	0.7958	ampere turns
Gilberts (abs)	2.998×10^{10}	electrostatic cgs units of magneto motive force
Gills (British)	5	ounces (British, fluid)
Gills (U.S.)	32	drams (fluid)
Grains	0.036571	drams (avoirdupois)
Grains	1/7000	pounds (avoirdupois)
Gram-centimeters	9.2967×10^{-3}	Btu (mean)
Gram-centimeters	2.3427×10^{-5}	calories, gram (mean)
Gram-centimeters	7.2330×10^{-5}	foot-pounds
Gram-centimeters	9.8067×10^{-5}	joules (abs)
Gram-centimeters	2.7241×10^{-8}	watt-hours
Gram-centimeters/second	9.80665×10^{-5}	watts (abs)
Grams-centimeters^2 (moment of inertia)	2.37305×10^{-6}	pounds-feet^2
Grams-centimeters^2 (moment of inertia)	3.4172×10^{-4}	pounds-inch^2
Grams/cubic centimeter	62.428	pounds/cubic foot
Grams/cubic centimeter	8.3454	pounds/gallon (U.S.)
Grams/cubic meter	0.43700	grains/cubic foot

CONVERSION FACTORS

Multiply	by	to obtain
Grams/liter	58.417	grains/gallon (U.S.)
Grams/liter	9.99973×10^{-4}	grams/cubic centimeter
Grams/liter	1000	parts per million (ppm)
Grams/liter	0.06243	pounds/cubic foot
Grams/square centimeter	0.0142234	pounds/square inch
Hands	4	inches
Hemispheres	0.5	spheres
Hemispheres	4	spherical right angles
Hemispheres	6.2832	steradians
Henries (abs)	10^9	abhenries
Henries (abs)	1.1126×10^{-12}	stathenries
Hogsheads (British)	63	gallons (British)
Hogsheads (U.S.)	8.422	cubic feet
Hogsheads (U.S.)	0.2385	cubic meters
Hogsheads (U.S.)	63	gallons (U.S.)
Horsepower	2545.08	Btu (mean)/hour
Horsepower	550	foot-pounds/second
Horsepower	0.74570	kilowatts (g = 980.665)
Horsepower, electrical	1.0004	horsepower
Horsepower (metric)	0.98632	horsepower
Horsepower-hours	2545	Btu (mean)
Horsepower-hours	1.98×10^6	foot-pounds
Horsepower-hours	2.737×10^5	kilogram-meters
Horsepower-hours	0.7457	kilowatt-hours (abs)
Hundredweights (long)	112	pounds
Inches (British)	2.540	centimeters
Inches (British)	0.9999972	inches (U.S.)
Inches (U.S.)	2.54000508	centimeters
Inches of mercury at 32°F	345.31	kilograms/square meter
Inches of mercury at 32°F	0.4912	pounds/square inch
Joules (abs)	10^7	ergs
Joules (abs)	0.73756	foot-pounds
Joules (abs)	0.101972	kilogram-meters
Joule-sec	1.5258×10^{33}	quanta
Karats (1 of gold to 24 of mixture)	41.667	milligrams/gram
Kilderkins (British)	18	gallons (British)
Kilograms	980665	dynes
Kilograms	15432.4	grains
Kilograms	35.2740	ounces (avoirdupois)
Kilograms	70.931	poundals

Multiply	by	to obtain
Kilograms	2.20462	pounds
Kilograms	9.84207×10^{-4}	tons (long)
Kilograms	0.001	tons (metric)
Kilograms	0.0011023	tons (short)
Kilogram-meters	0.0092967	Btu (mean)
Kilogram-meters	2.3427	calories, gram (mean)
Kilogram-meters	9.80665×10^{7}	ergs
Kilogram-meters	7.2330	foot-pounds
Kilogram-meters	3.6529×10^{-6}	horsepower-hours
Kilogram-meters	9.80665	joules (abs)
Kilogram-meters	2.52407×10^{-6}	kilowatt-hours (abs)
Kilogram-meters	9.579×10^{-6}	pounds water evap. at 212°F
Kilogram-meters	6.392×10^{-7}	pounds carbon to CO_2
Kilograms/square meter	9.6784×10^{-5}	atmospheres
Kilograms/square meter	98.0665	dynes/square centimeters
Kilograms/square meter	3.281×10^{-3}	feet of water at 39.2°F
Kilograms/square meter	0.1	grams/square centimeters
Kilograms/square meter	2.896×10^{-3}	inches of mercury at 32°F
Kilograms/square meter	0.07356	mm of mercury at 0°C
Kilograms/square meter	0.2048	pounds/square foot
Kilograms/square meter	0.00142234	pounds/square inch
Kilowatt-hours (abs)	3413	Btu (mean)
Kilowatt-hours (abs)	2.6552×10^{6}	foot-pounds
Kilowatt-hours (abs)	3.6709×10^{5}	kilogram-meters
Kilowatts (abs)	1.341	horsepower
Knots	1	miles (nautical)/hour
Leagues (nautical)	3	miles (nautical)
Leagues (statute)	3	miles (statute)
Light years	63274	astronomical units
Light years	9.4599×10^{12}	kilometers
Light years	5.8781×10^{12}	miles
Lignes (Paris lines)	$\frac{1}{12}$	pouces (Paris inches)
Links (Gunter's)	0.01	chains (Gunter's)
Links (Gunter's)	0.66	feet
Links (Ramden's)	1	feet
Links (Ramden's)	0.01	chains (Ramden's)
Liter-atmospheres (normal)	0.096064	Btu (mean)
Liter-atmospheres (normal)	24.206	calories, gram (mean)
Liter-atmospheres (normal)	74.735	foot-pounds
Liter-atmospheres (normal)	2.815×10^{-5}	kilowatt-hours
Liter-atmospheres (normal)	101.33	joules (abs)
Liter-atmospheres (normal)	10.33	kilogram-meters

CONVERSION FACTORS

Multiply	by	to obtain
Liters	1000.028	cubic centimeters
Liters	0.035316	cubic feet
Liters	0.26417762	gallons (U.S.)
Liters	1.0566828	quarts (U.S., liquid)
Lumens	0.07958	candle-power (spherical)
Lumens	0.00147	watts of maximum visibility radiation
Lumens/square centimeters	1	lamberts
Lumens/square centimeters/ steradian	3.1416	lamberts
Lumens/square foot	1	foot-candles
Lumens/square foot	10.764	lumens/square meter
Lumens/square feet/steradian	3.3816	millilamberts
Lumens/square meter	0.09290	foot-candles or lumens/ square foot
Lumens/square meter	10^{-4}	phots
Lux	0.09290	foot-candles
Lux	1	lumens/square meter
Lux	10^{-4}	phots
Meter-candles	1	lumens/square meter
Meters	10^{10}	angstrom units
Meters	3.280833	feet (U.S.)
Meters	39.37	inches
Meters	6.2137×10^{-4}	miles (statute)
Meters	1.09361	yards (U.S.)
Meters/second	2.23693	miles/hour
Microns	1×10^{-4}	centimeters
Miles (British)	1.6093425	kilometers
Miles (int. nautical)	1.852	kilometers
Miles (U.S. statute)	5280	feet
Miles (U.S. statute)	1609	meters
Miles (U.S. statute)	1760	yards
Miles/hour	44.7041	centimeters/second
Miles/hour	1.4667	feet/second
Miles/hour	0.86839	knots
Miles/hour/second	1.4667	feet/second/second
Milligrams/assay ton	1	ounces (troy)/ton (short)
Mils	0.001	inches
Mils	25.40	microns
Miner's inches (Colorado)	0.02604	cubic feet/second

.01 TORR = 10 MICRONS

Multiply	by	to obtain
Miner's inches (Ariz., Calif., Mont., and Ore.)	0.025	cubic feet/second
Miner's inches (Ida., Kan., Neb., Nev., N. Mex., N. Dak., S. Dak., and Utah)	0.020	cubic feet/second
Minims (British)	0.05919	cubic centimeters
Minims (U.S.)	0.06161	cubic centimeters
Months (mean calendar)	30.4202	days
Months (mean calendar)	730.1	hours
Months (mean calendar)	43805	minutes
Months (mean calendar)	2.6283×10^6	seconds
Newtons	10^5	dynes
Newtons	0.10197	kilograms
Newtons	0.22481	pounds
Noggins (British)	$\frac{1}{32}$	gallons (British)
Oersteds (abs)	1	electromagnetic cgs units of magnetizing force
Oersteds (abs)	2.9978×10^{10}	electrostatic cgs units of magnetizing force
Ohms (abs)	10^9	abohms
Ohms (abs)	1.1126×10^{-12}	statohms
Ounces (avoirdupois)	16	drams (avoirdupois)
Ounces (avoirdupois)	28.349527	grams
Ounces (avoirdupois)	0.9114583	ounces (troy)
Ounces (avoirdupois)	$\frac{1}{16}$	pounds (avoirdupois)
Ounces (troy)	31.103481	grams
Ounces (troy)	1.09714	ounces (avoirdupois)
Ounces (troy)	$\frac{1}{12}$	pounds (troy)
Ounces (U.S., fluid)	29.5737	cubic centimeters
Ounces (U.S., fluid)	1.80469	cubic inches
Ounces (U.S., fluid)	8	drams (fluid)
Ounces (U.S., fluid)	$\frac{1}{128}$	gallons (U.S.)
Paces	30	inches
Palms (British)	3	inches
Parsecs	3.084×10^{16}	meters
Parsecs	3.260	light years
Parts per million (ppm)	0.058417	grains/gallon (U.S.)
Pecks (British)	0.25	bushels (British)
Pecks (U.S.)	0.25	bushels (U.S.)
Pennyweights	24	grains
Pennyweights	1.555174	grams

CONVERSION FACTORS

Multiply	by	to obtain
Pennyweights	0.05	ounces (troy)
Perches (masonry)	24.75	cubic feet
Phots	929.0	foot-candles
Picas (printers')	$\frac{1}{6}$	inches
Pieds (French feet)	0.3249	meters
Pints (U.S., dry)	33.6003	cubic inches
Pints (U.S., liquid)	473.179	cubic centimeters
Pints (U.S., liquid)	16	ounces (U.S., fluid)
Planck's constant	6.6256×10^{-27}	erg-seconds
Pottles (British)	0.5	gallons (British)
Pouces (Paris inches)	0.02707	meters
Pouces (Paris inches)	0.08333	pieds (Paris feet)
Poundals	14.0981	grams
Poundals	0.031081	pounds
Pound-feet (torque)	1.3558×10^7	dyne-centimeters
Pounds (avoirdupois)	7000	grains
Pounds (avoirdupois)	453.5924	grams
Pounds (avoirdupois)	1.2152778	pounds (troy)
Pounds (troy)	373.2418	grams
Pounds of carbon to CO_2	14544	Btu (mean)
Pounds of water evaporated at 212°F	970.3	Btu
Pounds/square foot	4.88241	kilograms/square meter
Pounds/square inch	0.068046	atmospheres
Pounds/square inch	70.307	grams/square centimeter
Pounds/square inch	703.07	kilograms/square meter
Pounds/square inch	51.715	millimeters of mercury at 0°C
Proof (U.S.)	0.5	percent alcohol by volume
Puncheons (British)	70	gallons (British)
Quadrants	90	degrees
Quarts (U.S., liquid)	0.033420	cubic feet
Quarts (U.S., liquid)	32	ounces (U.S., fluid)
Quarts (U.S., liquid)	0.832674	quarts (British)
Quintals (metric)	100	kilograms
Quintals (long)	112	pounds
Quintals (short)	100	pounds
Quires	24	sheets
Radians	57.29578	degrees
Reams	500	sheets

Multiply	by	to obtain
Register tons (British)	100	cubic feet
Revolutions/minute	0.10472	radians/second
Revolutions/minute^2	0.0017453	radians/second/second
Reyns	6.8948×10^6	centipoises
Rods	16.5	feet
Rods	5.0292	meters
Rods	3.125×10^{-3}	miles
Roods (British)	0.25	acres
Scruples	$\frac{1}{3}$	drams (troy)
Scruples	20	grains
Sections	1	square miles
Slugs	32.174	pounds
Space, entire (solid angle)	12.566	steradians
Spans	9	inches
Square centimeters	1.07639×10^{-3}	square feet (U.S.)
Square centimeters	0.15499969	square inches (U.S.)
Square centimeter–square centimeter (moment of area)	0.024025	square inch–square inch
Square chains (Gunter's)	0.1	acres
Square chains (Gunter's)	404.7	square meters
Square chains (Ramden's)	0.22956	acres
Square chains (Ramden's)	10000	square feet
Square feet (British)	0.092903	square meters
Square feet (U.S.)	929.0341	square centimeters
Square foot–square foot (moment of area)	20736	square inch–square inch
Square inches (U.S.)	6.4516258	square centimeters
Square inches (U.S.)	7.71605×10^{-4}	square yards
Square kilometers	0.3861006	square miles (U.S.)
Square links (Gunter's)	10^{-5}	acres (U.S.)
Square links (Gunter's)	0.04047	square meters
Square meters	2.471×10^{-4}	acres (U.S.)
Square meters	10.76387	square feet (U.S.)
Square meters	1550	square inches
Square meters	3.8610×10^{-7}	square miles (statute)
Square meters	1.196	square yards (U.S.)
Square miles	640	acres
Square miles	2.78784×10^7	square feet
Square miles	2.590×10^6	square meters
Square rods	272.3	square feet

CONVERSION FACTORS

Multiply	by	to obtain
Square yards (U.S.)	1296	square inches
Square yards (U.S.)	0.8361	square meters
Statamperes	3.33560×10^{-10}	amperes (abs)
Statcoulombs	3.33560×10^{-10}	coulombs (abs)
Statcoulombs/kilogram	1.0197×10^{-6}	statcoulombs/dyne
Statfarads	1.11263×10^{-12}	farads (abs)
Stathenries	8.98776×10^{11}	henries (abs)
Statohms	8.98776×10^{11}	ohms (abs)
Statvolts	299.796	volts (abs)
Statvolts/inch	118.05	volts (abs)/centimeter
Statwebers	2.99796×10^{10}	electromagnetic cgs units of magnetic flux
Statwebers	1	electrostatic cgs units of magnetic flux
Stones (British)	6.350	kilograms
Stones (British)	14	pounds
Toises (French)	6	Paris feet (pieds)
Tons (long)	1016	kilograms
Tons (long)	2240	pounds
Tons (metric)	1000	kilograms
Tons (metric)	2204.6	pounds
Tons (short)	907.2	kilograms
Tons (short)	2000	pounds
Townships (U.S.)	23040	acres
Townships (U.S.)	36	square miles
Tuns	252	gallons
Volts (abs)	10^8	abvolts
Volts (abs)	3.336×10^{-3}	statvolts
Watts (abs)	3.41304	Btu (mean)/hour
Watts (abs)	0.01433	calories, kilogram (mean)/minute
Watts (abs)	10^7	ergs/second
Watts (abs)	0.7376	foot-pounds/second
Watts (abs)	0.0013405	horsepower (electrical)
Watts (abs)	0.10197	kilogram-meters/second
Watts/(cm^2)(°C/cm)	693.6	Btu/(hr)(ft^2)(°F/in.)
Wave length of the red line of cadmium	6.43847×10^{-7}	meters
Webers	10^3	electromagnetic cgs units
Webers	3.336×10^{-3}	electrostatic cgs units

Multiply	by	to obtain
Webers	10^8	lines
Webers	10^8	maxwells
Webers	3.336×10^{-3}	statwebers
Yards	3	feet
Yards	0.91440	meters
Years (tropical, mean solar)	365.2422	days (mean solar)
Years (tropical, mean solar)	8765.8128	hours (mean solar)
Years (tropical, mean solar)	3.155693×10^7	seconds (mean solar)
Years (tropical, mean solar)	1.00273780	years (sidereal)
Years (sidereal)	365.2564	days (mean solar)
Years (sidereal)	366.2564	days (sidereal)

INDEX

Absolute temperature, 395
  viscosity, 146
  zero, properties at, 169
Absorptivity, 200
Acceleration, angular, 112
  curvilinear motion, 114
  electron in field, 236
  force, mass relation, 115
  rectilinear motion, 112
Activity, thermodynamic, 188
Adiabatic compressibility, 169
  demagnetization, 183
  flow, 179
  process, 178
Aerodynamic heating, 163
  resistance, 161–163
Air, heat transfer, 198
Algebra, 28–46
  Boolean, 44–46
  of sets, 44
  vector, 92–94
Allowable unit stress, 119
  concrete, 133
  lumber, 132
  structural steel, 132
Alpha emission, 257
  particle, 254
  volatility ratio, 185
Alternating current, 227, 228
  machines, 231
  phase with voltage, 227

  plate resistance, 237
  sinusoidal, 227
  transient, 228
Aluminum, properties, 121
  structural sections, 143, 144
  wire table, 235
Ampere's law, 212
Amplification factor, 237
Amplifiers, 240, 241, 247
Analytic geometry, plane, 13
  solid, 24
Angle, between two lines, 13, 26
  between two planes, 26
  circular functions of, 46–50
  complex, 56–58
  degrees to radians, 346
  factor in radiation, 201
  hyperbolic, 54
  functions of, 54–56
  imaginary, 56–58
  radians to degrees, 347
  of repose, 118
  steel sections, 137, 138
  trigonometric functions of, 348–392
Angular acceleration, 112
  impulse, 117
  velocity, 112
Annuity, amount of, 278
  capital recovery, 282
  present worth of, 280
  sinking fund, 270

INDEX

Annulus, 8, 106
  flow in, 156
Anti-hyperbolic functions, 56
  -trigonometric functions, 48, 50
Arc, circular, 7, 104, 344–346
  length, 75, 346
Archimedes, principle, 148
  spiral, 9, 20
Area, of circles, 320–336
  of surface, 74
Arithmetic mean, 33
  progression, 33
Astroid, 18
Atmospheric pressure, 286
Atomic mass unit, 253
  numbers, 266
  weights, 187
Avogadro's number, 286
Axes, of inertia, 110, 111
  principal, 110
  rotated, 13, 26
Axial flow, 149–159
  stress, 119

Bars, steel, 140
Beams, 122–128
  continuous, 126
  curved, 127, 128
  diagrams, 124–126
Bearing stress, 132
Bending moment, 123, 124–126
Bernoulli's equation, 77
  measure of profitability, 272
  theorem, 152
Beta emission, 257
  particle, 254
Bias, transistor, 250–252
Binding energy, 254
Binomial theorem, 33
Black body, 200
Branch currents, 226
Briggsian logarithms, 29
Buckling, of columns, 129
  of nuclear reactors, 263
Bulk modulus, 120
Buoyancy, 148

Calculus, differential, 58–61
  integral, 62–76
  integrals, table of, 62–74

Capacitance and condensers, 214, 215
  in sinusoidal circuit, 228
  in transient circuit, 229, 230
Capacity, thermal, 169
  specific inductive, 212, 232
Capital recovery factor, 270
Cardioid, 20
Carnot cycle, 179
  refrigeration cycle, 182
Catenary, 17
Cathode-ray tube, 236
Cauchy's integral theorem, 91
Center, of gravity, 102, 104–110
  of pressure, 147
Centroid, 102
  location, table, 104–110
Channels, flow in, 156, 159
  steel, 136
Charge, electric, 212
  of a condenser, 214
Chemical activity, 188
Chemical elements, 187
Chemical equations, 186
Chemical equilibrium, 188–190
Chemical reactions, 186–190
Chords, 344, 345
Circle, 7, 15, 106
  area and circumference of, 320–336
  involute of, 20
  Mohr's, 122
  in polar coordinates, 24
Circuits, direct current, 224–226
  electrostatic, 213
  magnetic, 218
  sinusoidal, 228
  transient, 229, 230
Circular arcs, 7, 104, 344, 345
  sectors, 8, 106
  segments, 8, 106, 344, 345
Cissoid of Diocles, 22
Clapeyron equation, 174
Clausius-Clapeyron equation, 185
Coefficient of,
  performance, 182
  restitution, 117
  thermal expansion, 169
Coil, friction, 118
  heat transfer in, 198
  in magnetic field, 220–222

Columns, 127–129
Combinations, 30
Common logarithms, 29
  of hyperbolic functions, 310–319
  of numbers, 288–307
  of trigonometric functions, 348–392
Complementary function, 79
Complex angle, 56–58
  numbers, 89, 90
  substance, 146
  variables, 90–92
Composition of forces, 101, 102
Compound amount factor, 270
Compound interest, 268
  amount, 274
  present worth, 276
Compressibility, 169
  factor, 190, 191
Compressible flow, 158, 163
Compression, 119, 127, 184
  adiabatic, 184
  isothermal, 184
  refrigeration system, 182
Compressors, 184
Conchoid of Nicomedes, 23
Concurrent forces, 100
Condensers, electric, 214
Conductance, electric, 215
  thermal, 197
Conduction, electric, 209, 210
  thermal, 195, 196
Conductors, capacitance between, 215
  magnetic force between, 220
Conduits, flow in, 155–159
Cone, 11, 27, 108
Conformal mapping, 91
Conic sections, 13–15
Conservation, of energy, 116, 150
  of mass, 149
  of momentum, 149
Consistency, 145
Constants, containing e and π, 344
  defined and fundamental, 286
  electrical dimensional, 211
  gas law, 173
  Newton's second law, 2
  proportionality, 2
Constrictions, flow through, 152–155
Continuous interest, 269

Continuity equation, 149
Contraction, sudden, 153
Convection, thermal, 197–199
Conversion factors, 399–414
  of units, 3
Coordinates, systems of, 24
  transformation of, 13, 26
Copper wire table, 234
Corresponding states, 190
Cost, index numbers, 285
  minimum point, 271
Coulomb's law, 212
Couples, 102
Critical mass, 263
Cross section, nuclear, 258–260
Cube, 10, 107
Cubic equation, 32
Current, electric, 209–214
  branch, 225
  charge and discharge, 215, 217
  conduction and displacement, 214
  density, 213
  direct, 224–226
  direction of, 213, 217, 220
  eddy, 224
  magnetic field due to, 219, 220
  magnetic force on, 220
  mesh, 226
  sinusoidal, 227, 228
  stream line of, 219
  superposition of, 210
  transient, 228–230
  and voltage relations, 209, 210
Curtate cycloid, 18
Curves, 13–24, 104
  conic sections, 13–15
  exponential, 17
  higher plane, 15–24
  logarithmic, 17
  radius of curvature, 16
Cusp, 16
Cycle, Brayton, 181
  Carnot, 179, 182
  current, 227
  Diesel, 181
  efficiency, 179
  neutron, 261–263
  Otto, 180
  Rankine, 180

Cycle, refrigeration, 182, 183
Cycloid, 9, 18
Cylinders, 11, 108
  flow resistance of, 162
  heat transfer to, 198
  under pressure, 129

Damped wave, 17
Darcy's law, 155
  friction factor, 157
Decay, radioactive, 257
Decimal equivalents, 320
  prefixes, 399
Defined constants, 286
Definite integrals, 73, 74
Deflection, of beams, 124–126
  electrostatic, 236
  magnetic, 221, 236
Deformation, mechanical, 119
Degree, API, 393
  Baumé, 393
  decimals of, 347
  in radians, 346
  temperature scales, 395
  Twaddell, 393
Density, conversion table, 393
  current, 213, 265
  dielectric flux, 213
  magnetic flux, 217
Derivatives, of functions, 60
  partial, 60, 61
Determinants, 39, 43
  transistor relations, 244
Dielectric constant, 211, 218
Dielectric flux 213, 214
Diesel cycle, 181
Differential calculus, 58–61
  equations, 76–82
  vector operators, 95
Diffusion length, 263
Dilatant substance, 146
Dimensional constants, 2
  electric, 211
  gas law, 173
  Newton's second law, 2
Dimensionless numbers, 2, 194
Dimensions and units, 1–5
Direct-current, motors and generators, 230
  networks, 224–226

Direction, of current, 213, 217, 220
  of induced emf, 216
  of magnetizing force, 219, 220
Discharge, coefficients of, 152
  of a condenser, 215
Discounted cash flow, 272
Disks, flow resistance, 161, 162
  stress in, 130
Displacement current, 214
  of a particle, 111
Dose rate, radiation, 264
Drag coefficients, 161, 162
Ducts, flow through, 155–159
Durand's rule, 12
Dynamics, electron, 236
  fluid, 148–164
  mechanics, 114–119

Economic choice, 271–273
Eddy currents, 224
Effective current, 210
Einstein's law, 254
Elastic curve, 123
Elasticity, 119
  modulus of, 120
Electromagnetic induction, 222–224
  spectrum, 210
  units, 211
Electromagnetism, 216–224
Electromotive force, 210
  direction of, 224
  induced, 222
Electron dynamics, 236, 237
  emission, 257
  properties of, 254
  tubes, 237–241
  volt, 253
Electronics, 236–252
Electrostatics, 212–216
  units, 211
Ellipse, 8, 14, 15, 106
Ellipsoids, 12, 27, 110
Emission, radioactive, 257
Emissivity, 200
Energy, balance, 116
  conservation of, 116
  conversion of mass and, 254–256
  electric, 212
  electrostatic, 216
  fission, 256

Energy, internal, 167
  kinetic, 116
  losses in flow, 151–159
    valves and fittings, 159
  magnetic, 223
  mechanical, 116
  neutron thermal, 260
  potential, 116
  radiant, 200, 256, 263
Engines, heat, 179–183
Enthalpy diagram, 171
Entropy diagram, 171
Epicycloid, 9, 19
Epitrochoid, 19
Equiangular spiral, 21
Equilibrium constant, 188–190
  forces in, 100–102
  hydrostatic, 146
  mixture, 189
  phase, 185
Equipotential surface, 213
Equivalent diameter, 156
  interest, 269, 283
  length, 158
  sinusoidal current, 210, 227
Error, probable, 343
Escape velocity, 287
Euler's column formula, 129
  homogeneous equation, 80
  rigid body equations, 115
Expansion, sudden, 153
Exponential curve, 17
  values and logarithms, 310–319

Factorials, 30, 336
Faraday's induction law, 212, 217
Field, electric, 212
  current in, 213
  energy, 216
  forces in, 216
  intensity, 213
  strength, 213
  magnetic, 216
  energy, 223
  force in, 220, 221
  intensity, 218–220
Fission energy, 256
  reaction, 255
Flow, adiabatic, 179
  in closed conduits, 155–159

  compressible, 158, 163
  friction factors, 156, 157
  around immersed bodies, 161–163
  isothermal, 159
  laminar, 156, 162
  in open channels, 159–161
  processes, 179
  steady, 146, 179
  through constrictions, 152–155
  through porous media, 155
  uniform, 146
Fluid, 145, 146
  consistency, 145
  dynamics, 148–164
  statics, 146–148
  viscosity, 146
Flux, dielectric, 213, 214
  magnetic, 216–218
  neutron, 258
Force and mass, 2, 100, 115
  couple, 102
  in electric field, 216
  in magnetic field, 220, 221
  moment of, 102
  parallelepiped and polygon, 101
Form factor, 227
Fourier number, 194
  series, 36
  thermal equation, 195
Fractional powers of numbers, 340–342
Friction, 117–119
  belt or coil, 118, 119
  coefficients of, 118
  factors, 158
  hydrodynamic, 118
Fugacity, 191
Functions of a complex variable, 89–92

Gamma emission, 257
Gas, emissivity, 203
  law, 173
Gauss' laws, 212
  theorem, 96
Gaussian surface, 217
Generalized displacements, 166
  forces, 166
  thermodynamic functions, 175, 176
Geometric buckling, 263

Geometric mean, 33
  progression, 33
Geometry, plane analytic, 13–24
  mensuration, 6–12
  solid analytic, 24–29
Gordon's column formula, 129
Gravitational constant, 286
Gravity, acceleration of, 286
  center of, 102
  of lines and shapes, 104–109
Gray surface, 200
Greek alphabet, inside cover
Green's theorem, 96
Gyration, radius of, 103
  structural sections, 134–144
  surfaces and bodies, 104–110

H beams, aluminum, 144
  steel, 134
Hagen-Poiseuille equation, 156
Half-life, 257
Harmonic mean, 33
  motion, 113
  progression, 33
  series, 33
Heat, absorptivity, 200
  capacity, 173
  conduction, 195, 196
  conductivity, 195
  convection, 197–200
  emissivity, 200
  engines, 179–183
  internal generation of, 196
  radiation, 200–207
  reflectivity, 200
  transmissivity, 200
Heat transfer, coefficient, 197
  coefficients for deposits, 200
  in fluidized solids, 199
  by forced convection, 197–199
  inside tubes, 198
  from luminous flames, 205–207
  by natural convection, 197
  from nonluminous gases, 203–205
  outside tubes, 198
  overall coefficient, 199
  in packed beds, 199
  by surface radiation, 201, 202
Helix, 9

Hemisphere, 11, 109
Henry's law, 185
Hermitian forms, 43
  matrix, 43
Homogeneous equations, 1, 76
Hooke's law, 120, 121
Hybrid transistor parameters, 242
Hyperbola, 8, 14, 15
Hyperbolic angles, 54
  functions, 54
  values and logarithms, 310–319
  trigonometry, 54–58
Hyperboloid, 12, 27
Hypocycloid, 9, 18
Hypotrochoid, 18
Hysteresis loss, 224

I beams, aluminum, 143
  steel, 135
Ideal, compressor, 184
  fluid, 146
  gas, 173
Impact, 117
Impulse, 117
Inches to decimal feet, 336
Incompressible flow, 151–158
Indefinite integral, 62
Inductance, 222
  mutual, 223
Induction data, 233
  electromagnetic, 222–224
  Faraday's law of, 212
  motors, 231
Inductive capacitance, 214
Infinite series, 34–38
Insulators, electric, 232
Integral calculus, 62–76
Integrals, table of, 62–74
Intensity, electric field, 213
  magnetic field, 218–220
  radiation, 265
Interest, formulas, 268–270
  rate of return, 272
  tables, 274–283
Internal energy, 167
Involute of a circle, 20
Irreversible processes, 166, 178
Isentropic flow, 179
  process, 178
Isothermal compressibility, 169

Isotopes, 266, 267

Joule-Thomson coefficient, 169
Joule's law, 210

Kinematic viscosity, 146
Kinematics, 111–114
 table of equations, 112
Kinetic energy, 116
 equations, 116
 friction, 117
Kirchhoff's laws, electric network, 210
 thermal radiation, 201

Lambert's law, 201
Laminar flow, 156, 162
 heat transfer with, 198
Laplace transformation, 82–89
 table of transforms, 87–89
 theorems, 86
Laurent series, 91
Leakage current, 215
Left-hand rule, 220
Lemniscate, 21
Length, diffusion, 263
 equivalent, 158
Lenz's law, 217
Life expectancy, 284
Lift, 161
Light speed, 286
Limacon of Pascal, 19
Linear dependence, 41
 transformations, 42
Lines, equations of, 13
Linkages, flux, 217
Logarithmic curve, 17
 mean, 197
 spiral, 21
Logarithms, 28, 29
 of exponentials, 310–319
 of hyperbolic functions, 310–319
 natural, 308, 309
 of numbers, common, 288–307
 of trigonometric functions, 348–392
Lumber, properties, 142
 allowable unit stress, 132
Luminous flames, 205

Machines, electrical, 230, 231
 thermodynamic, 179, 184

Maclaurin's series, 35
Magnetic attraction, 221
 circuit, 218
 energy, 223
 field intensity, 218
 flux, 216–218
 forces, 220, 221
 induction, 222–224
 intensity, 218–220
 linkages, 219
 permeability, 218
 of free space, 211
 permeance, 218
 properties of materials, 233
 reluctance, 218
Magnetomotive force, 218
Manometer, 147
Mass, atomic, 253
 acceleration, and force, 2, 100, 115
 center of, 102, 104–110
 defect, 254
 of electron, 254
 of moving body, 254
 of planets, 287
 velocity, 145
Matrices and determinants, 39–44
Matrix operations, 40
 transistor relations, 243–246
Maximum discharge rate, 154
Maxwell's relations, electromagnetic, 212
 thermodynamic, 168
Mean free path, 260
Mechanical energy balance, 151
Mechanics, of fluids, 145–163
 of materials, 119–144
 of rigid bodies, 92–119
Mensuration, 6–12
Mesh currents, 226
Metric units, 4, 5
Minutes, in decimals of a degree, 347
 in radians, 346
MKSC units, 4, 5
Modulus, bulk, 120
 of elasticity, 120
 section, 122
 shear, 120
 Young's, 120
Mohr's circle, 122
Mollweide's check formulas, 52

Moment, 75, 103
  bending, 123
  of a couple, 102
  diagram of beams, 124–126
  of force, 102
  of inertia, 75, 103, 110
    axes, 110, 111
    of structural sections, 134–144
    of surfaces and bodies, 104–110
  in shafts, 130
  statical, 75, 103
Momentum, 117
  conservation of, 149
Money, time value of, 268–270
Mortality table, 284
Motion, curvilinear, 114, 115
  of electron, 236, 237
  Newton's laws of, 114
  of particle, 111
  of projectile, 114
  rectilinear and rotational, 112
  simple harmonic, 113
Mutual conductance, 237
  inductance, 223

Naperian logarithms, 29
  of numbers, 308, 309
Natural convection, 197
Neil's parabola, 17
Networks, D-C, 224–226
  transistor, 241–242
Neutron cycle, 261–263
  diffusion length, 263
  emission, 257
  flux, 258
  moderators, 262
  properties, 254
  thermal, 260
Newton's drag law, 162
  hydrodynamic friction, 118
  laws of motion, 114
  viscosity coefficient, 146
Nonblack surfaces, 202
Noncoplanar forces, 101
Non-flow processes, 177, 178
Nonluminous gases, 203
Normal emissivities, 207
Normal stress, 119
Normal to curve, 16
Nozzles, 153

Nuclear constants, 254
  cross sections, 258–260
  flux, 258
  fuels, 262
  reactions, 255–260
  reactors, 260–263
Nusselt number, 194

Ohm's law, 209, 210
Open channels, flow in, 159
Orbital electron capture, 257
Orifice, 153
Otto cycle, 180
Overall coefficients, 199

Packed beds, flow in, 154
  heat transfer in, 199
Parabola, 8, 14, 15
  Neil's semicubic, 17
Parabolic segment, 8, 106
Paraboloid, 12, 28, 110
Parallel circuits, 224
  lines, 13
  planes, 26
  plates, capacitance, 214
Parallelepiped, 10, 107
  of forces, 101
Parallelogram, 6
  law, 101
Partial derivatives, 60, 61
  differential equations, 80–82
Particle, motions of, 111
Particles, elementary, 111–113
Peclet number, 194
Peltier effect, 183
Pendulum, 114
Periodic interest, 268
Permeability, 218
  of free space, 211
Permeance, 218
Permutations, 30
Perpendicular lines, 13
Phase angle, 227
  difference, 227
  equilibria, 185
Physical quantities, 1
  dimensions and units, 4, 5
Pi, constants, 344
Pipe, properties of, 141
Planck's constant, 286

INDEX 423

Plane curves, 13-24
  curvilinear figures, 7-9
  rectilinear figures, 6
  surfaces, 6-8, 104-106
Planets, properties of, 287
Plates, drag coefficient of, 162
  heat conduction in, 197
  heat transfer to, 197, 198
  magnetic force between, 220
  magnetic intensity near, 220
  radiation between, 202
  stresses in, 130
Point, distance from line, 13
  distance from plane, 26
  of minimum cost, 271
Poisson's ratio, 120
Polar coordinates, 24
Polygon, 7, 105
  of force, 101
  spherical, 12
Polyhedrons, 10
Polytropic process, 178
Potential, electric, 212
  difference, 213
  energy, 116
  magnetic, 223
Power, electric, 212
  of electrical machines, 231
  factor, 212
  series, 33
  with sinusoidal current, 227
  work, and energy, 115-117
Powers of numbers, 320-336
  fractional, 340-342
  higher, 337-339
Prandtl number, 194
Present worth, 268
  of annuity, 280
  measure of profitability, 273
Pressure, atmospheric, 286
  center of, 147
  stresses, 129, 130
Prism, 10
Prismatoid, 10
Probability, 30
Probable error, 343
Product of inertia, 103
Profitability, measures of, 271-273
Progressions, 33
Projectile, motion of, 114

Prolate cycloid, 18
Proportions, 29
Proportional limit, 121
Proportionality constants, 2
Proton, 254
Pyramid, 10, 107

Quadratic equation, 31
  forms, 42
Quadric surfaces, 27
Quadrilateral, 7, 105
Quantities, physical, 1

Radians in degrees, 347
Radiation, 200-207
  angle factor, 201, 202
  biological effectiveness, 264
  black body, 201
  electromagnetic spectrum of, 210
  emissivities, representative, 207
  exposure, units of, 263, 264
  geometry, 202
  Kirchhoff's law, 201
  Lambert's cosine law, 201
  from luminous flames, 205-207
  shielding, 264, 265
  Stefan-Boltzman law, 201
  between surfaces, 201, 202
Radioactive decay, 257, 258
  fission reaction, 255
Radius of curvature, 16
  of gyration, 103
  structural sections, 134-144
  surfaces and bodies, 104-110
Rankine column formula, 129
  cycle, 180
  temperature scale, 395
Raoult's law, 185
Rationalized units, 3
Reactive power, 227
Reciprocals of numbers, 320-326
Rectangle, 6, 105
Rectilinear motion, 112
  plane figures, 6
Redlich-Kwong equation, 190
Reduced equation of state, 190
Reflectivity, 200
Refractory surfaces, 201
Refrigeration systems, 182, 183
Reinforced concrete, 133

Relative volatility, 185
Reluctance, magnetic, 218
Resistance, electric, 210
  circuits, 224, 228–230
  dynamic plate, 237
  Ohm's law, 209, 210
  in parallel, 224
  in series, 224
  thermal, 196
  to flow, form, 162
Resolution of forces, 100–102
Restitution, coefficient of, 117
Resultant of forces, 100
Reversible processes, 166
Revolution, bodies of, 107
Reynolds analogy, 155
  number, 155, 194
Rhomboid, 6
Rhombus, 6
Riccati's equation, 77
Right-hand screw law, 217
Rods, properties, 107
  dimensions, 140
Roentgen, 263
Rollers, stresses in, 130
Root mean square, current and voltage, 210, 227
Roots, of equations, 32
  of numbers, table, 320–336
Roses, four-leaved, 22
  three-leaved, 21
Rotation, motion, 112
Rotometer, 154
Roughness, to flow, 157, 161
  Manning factor, 161

Saturated steam, table, 179
Secant column formula, 129
Seconds, decimals of degree, 347
Section modulus, 122
  conic, 13–15
  structural sections, 134–144
Sector, circular, 8, 106
  spherical, 11, 109
Seebeck effect, 182
Segments, circular, 8, 106, 344, 345
  of a curve, 75
  hyperbolic, 8
  parabolic, 8, 106
  spherical, 12, 109

Self-inductance, 222
Semicircle, 8, 106
Semicubic parabola, 17
Series, 33–38
  circuits, 228–230
  functions expanded in, 37, 38
  infinite, 34–38
  tests for convergence, 34
  uniform annual, 270
Shafts, 130
Shear, fluid, 146
  mechanical, 119
  modulus, 120
  vertical, 123
Shielding, radiation, 264, 265
Simpson's rule, 12
Singular point, 16
Sinking fund deposit, 270
Sinusoidal currents and voltages, 227, 228
Slenderness ratio, 127
Slide rule, 29
Solar system, 287
Solenoid, 220
Solids, mensuration, 10–12
Space, free, dielectric constant of, 211
  magnetic permeability of, 211
Specific gravity, conversion table, 393, 394
Specific heats, 169
  of gases, 170
Speed, of light, 286
  synchronous, 231
Sphere, 11, 27, 109
  drag coefficients, 161, 162
  thermal conduction, 195, 196
Spherical coordinates, 24, 97
  polygon, 12
  sector, 11, 109
  segment, 12, 109
  triangle, 12
  trigonometry, 52–54
  zone, 12
Spirals, 9, 20, 21
Spontaneous processes, 165
Square, 6, 105
Stability of submerged bodies, 148
Statical moment, 103
Statics, 100
  fluid, 146–148

Steady flow, 146
Steam table, 172
Steel, bars, 140
  structural sections, 134–144
  unit stresses, allowable, 132
Stefan-Boltzmann law, 201
Stoichiometry, 186
Stokes' law, 162
  theorem, 96
Strain, 119
Stress, 119
  allowable unit, 132, 133
  in fluid flow, 146
  from temperature change, 129
Strophoid, 22
Structural materials, 121
Sudden contraction, 153
  expansion, 153
Sun, properties of, 287
Surface, area, 74
  electric equipotential, 213
  moment of inertia, 104
  neutral, 122
Surfaces, 6–8, 104–106
  in contact, 130
  quadric, 27
  of revolution, 107
Symbolic logic, 45, 46
Systems of units, 4, 5

Taylor's series, 35, 36
Tees, steel, 139
Temperature conversion, 395–398
Thermal capacity, 169
  conduction, 195, 196
  conductivity, 195
  efficiency, 179
  radiation, 200–207
Thermodynamic, engines, 179–184
  equation of state of, 174
  Maxwell relations in, 168
  potentials for work of, 167
  processes in, 177–179
  properties of, 168
    at absolute zero, 169
    generalized, 190–192
    of ideal gas, 173
    physically measurable, 169
  simple systems, 169

Thermodynamics, of chemical reactions, 186–190
  general equations of, 167
  laws of, 166
  of phase equilibria, 185
  of simple systems, 169–179
Thermoelectric refrigeration, 183
Thixotropic substance, 146
Timber, allowable unit stress, 132
  properties, 142
Time value of money, 268–270
Torque, 102
  on coil in magnetic field, 221
  on a shaft, 130
  work and power of a, 115, 116
Torsional shear stress, 130
Torus, 12, 110
  magnetic force within, 220
Tractrix, 23
Transformation, of coordinates, 13, 26
  linear, of matrices, 42
  method, Laplace, 82–89
Transient circuits, 228–230
  current and voltage, 229
Transistor bias, 250–252
  electron tube equivalents, 238
  impedance, 241
  matrix interrelations, 243–246
  modes of connection, 245
  parameters, 242
  power gain, 242
  terminated networks, 247–249
  two-port networks, 242
Translation, motion, 112
Transmission line, capacitance, 215
  inductance, 222
Transmissivity, 200
Transverse loads, 129
Trapezium, 7
Trapezoid, 6, 105
Triangle, 6, 105
  law, 101
  solution of, 51–52
  spherical, 12, 54
Trigonometric functions, 46–50
  values and logarithms, 348–392
Trigonometry, hyperbolic, 54–58
  plane, 46–52
  spherical, 52–54

Trochoid, 17
True stress and strain, 119, 120
Truncated prism, 10
Truth table, 46
Tubes, electron, 237–241
    amplifier coupling, 241
    characteristics, 239
    equivalent circuits, 240
    drag coefficients, 162
    flow through, 155
        across banks of, 155
    tension in, 130
Twist, angle of, 131

Ultimate strength, 121
Unit annuity, 282
    stress, 119
    systems, 4, 5
Units, atomic, 253
    conversion of, 3
    conversion factors, 399–414
    dimensions, 4, 5
    of radiation exposure, 263, 264
    of viscosity, 163
Unsteady conduction, 196

Vacuum tubes, *see* Tubes, electron
Value, present, *see* Present worth
Van der Wall's equation, 190
Vapor-liquid equilibria, 185
Vector analysis, 92–97
    diagrams, 101
    theorems, 96
Vectors, alegbra of, 92–94
    differentiation of, 94, 95
    integration of, 95
Velocity, angular, 112
    curvilinear motion, 114, 115
    effect on mass, 254, 255
    of light, 286
    mass, 145

    rectilinear motion, 112
    terminal settling, 163
Venturi meter, 153
Vertical deflection, 126
    shear, 123
Viscosity, 146
    units, 163
Viscous flow, 156, 162
    heat transfer with, 198
Volatility ratio, 185
Voltage, sinusoidal, 227, 228
    transient, 228–230
Volume, of ideal gas, 286
    by integration, 75

Water hammer, 159
Wave, damped, 17
    frequency modulated, 24
Wavelength of radiation, 210
Wedge, 10, 11
Weight of structural sections, 134–144
Weirs, 160
Wide flange beams, 134, 144
Wire, aluminum, 235
    copper, 234
Witch of Agnesi, 23
Wood, allowable stresses, 132
Work, of a force, 115
    of a varying magnetic flux, 223
    power, and energy, 115–117
    of a torque, 115

Yield point, 121
Yield strength, 121
Young's modulus, *see* Modulus of elasticity

Zero, absolute, 396
    properties at, 169, 170
Zeroth law, 166
Zone, spherical, 12

PRINCIPAL SYMBOLS

S section modulus (2)
surface area (3, 6)
entropy (4)
slip (6)
future amount (8)

s distance (2, 6)

T temperature
torque (2)

t time
thickness (2)

U internal energy (2, 3)
overall coefficient of heat transfer (5)

u speed, velocity (3, 4)
axial velocity (3, 4)

V volume (2, 4)
shearing force (2)
average velocity (3, 7)
voltage, potential (6)

v speed, velocity (2, 6)
radial velocity (3)
instantaneous voltage (6)

W work

w weight
mass flow rate (3)

X reactance (6)

x rectangular coordinate
axial coordinate (3)
mole fraction (4)
distance, thickness (6, 7)

Y admittance (6)

y rectangular coordinate
mole fraction (4)
admittance parameter (6)

Z elevation (3, 4)
impedance (6)
atomic number (7)

z rectangular coordinate
compressibility factor (4)
impedance parameter (6)
number of protons (7)

α angle
angular acceleration (2)
thermal coefficient of linear expansion (2)
volatility ratio (4)
absorptivity (5)
alpha particle (7)

β thermal coefficient of volume expansion (2, 4, 5)
beta particle (7)

γ unit shear strain (2)
ratio of specific heats (3, 4)
electric conductivity (6)
gamma radiation (7)